Special Volume:
Computational Models for the Human Body
Guest Editor: N. Ayache

Handbook of Numerical Analysis

General Editor:

P.G. Ciarlet

Laboratoire Jacques-Louis Lions
Université Pierre et Marie Curie
4 Place Jussieu
75005 PARIS, France

and

Department of Mathematics
City University of Hong Kong
Tat Chee Avenue
KOWLOON, Hong Kong

ELSEVIER
NORTH
HOLLAND

Amsterdam • Boston • Heidelberg • London • New York • Oxford • Paris
San Diego • San Francisco • Singapore • Sydney • Tokyo

Volume XII

Special Volume: Computational Models for the Human Body

Guest Editor:

N. Ayache

INRIA
2004 Route des Lucioles
06902 Sophia–Antipolis
France

2004

ELSEVIER
NORTH
HOLLAND

Amsterdam • Boston • Heidelberg • London • New York • Oxford • Paris
San Diego • San Francisco • Singapore • Sydney • Tokyo

ELSEVIER B.V.
Sara Burgerhartstraat 25
P.O. Box 211, 1000 AE Amsterdam
The Netherlands

ELSEVIER Inc.
525 B Street, Suite 1900
San Diego, CA 92101-4495
USA

ELSEVIER Ltd
The Boulevard, Langford Lane
Kidlington, Oxford OX5 1GB
UK

ELSEVIER Ltd
84 Theobalds Road
London WC1X 8RR
UK

First edition 2004

Library of Congress Cataloging in Publication Data
A catalog record is available from the Library of Congress.

British Library Cataloguing in Publication Data
A catalogue record is available from the British Library.

ISBN: 0-444-51566-6
ISSN (Series): 1570-8659

⊗ The paper used in this publication meets the requirements of ANSI/NISO Z39.48-1992 (Permanence of Paper).
Printed in the Netherlands

General Preface

In the early eighties, when Jacques-Louis Lions and I considered the idea of a *Handbook of Numerical Analysis*, we carefully laid out specific objectives, outlined in the following excerpts from the "General Preface" which has appeared at the beginning of each of the volumes published so far:

> During the past decades, giant needs for ever more sophisticated mathematical models and increasingly complex and extensive computer simulations have arisen. In this fashion, two indissociable activities, *mathematical modeling* and *computer simulation*, have gained a major status in all aspects of science, technology and industry.
>
> In order that these two sciences be established on the safest possible grounds, mathematical rigor is indispensable. For this reason, two companion sciences, *Numerical Analysis* and *Scientific Software*, have emerged as essential steps for validating the mathematical models and the computer simulations that are based on them.
>
> *Numerical Analysis* is here understood as the part of *Mathematics* that describes and analyzes all the numerical schemes that are used on computers; its objective consists in obtaining a clear, precise, and faithful, representation of all the "information" contained in a mathematical model; as such, it is the natural extension of more classical tools, such as analytic solutions, special transforms, functional analysis, as well as stability and asymptotic analysis.
>
> The various volumes comprising the *Handbook of Numerical Analysis* will thoroughly cover all the major aspects of Numerical Analysis, by presenting accessible and in-depth surveys, which include the most recent trends.
>
> More precisely, the Handbook will cover the *basic methods of Numerical Analysis*, gathered under the following general headings:
>
> – Solution of Equations in $\mathbb{R}^n$,
> – Finite Difference Methods,
> – Finite Element Methods,
> – Techniques of Scientific Computing.

It will also cover the *numerical solution of actual problems of contemporary interest in Applied Mathematics*, gathered under the following general headings:

- Numerical Methods for Fluids,
- Numerical Methods for Solids.

In retrospect, it can be safely asserted that Volumes I to IX, which were edited by both of us, fulfilled most of these objectives, thanks to the eminence of the authors and the quality of their contributions.

After Jacques-Louis Lions' tragic loss in 2001, it became clear that Volume IX would be the last one of the type published so far, i.e., edited by both of us and devoted to some of the general headings defined above. It was then decided, in consultation with the publisher, that each future volume will instead be devoted to a single "*specific application*" and called for this reason a "*Special Volume*". "*Specific applications*" will include Mathematical Finance, Meteorology, Celestial Mechanics, Computational Chemistry, Living Systems, Electromagnetism, Computational Mathematics etc. It is worth noting that the inclusion of such "specific applications" in the *Handbook of Numerical Analysis* was part of our initial project.

To ensure the continuity of this enterprise, I will continue to act as Editor of each Special Volume, whose conception will be jointly coordinated and supervised by a Guest Editor.

P.G. CIARLET
July 2002

Foreword

Computational Models for the Human Body constitute an emerging and rapidly progressing area of research whose primary objective is to provide a better understanding of the physiological and mechanical behavior of the human body and to design tools for their realistic numerical simulations. This volume describes concrete examples of such computational models. Although far from being exhaustive, it covers a large range of methods and an illustrative set of applications, and proposes a number of well-defined mathematical and numerical modeling of physical problems (including the analysis of existence and uniqueness of solutions for instance), followed by various numerical simulations.

Medical applications are addressed first, because physiological and biomechanical models of the human body already play a prominent role in the prevention, diagnosis and therapy of many diseases. The generalized introduction of such models in medicine will in fact strongly contribute to the development of a more *individualized* and *preventive* medicine. In effect, through the continuous progress of medical imaging during the past decades, it is currently possible to extract an increasing flow of anatomical or functional information on any individual, with an increasingly accurate resolution in space and time. The overwhelming quantity of available signals and images makes a direct analysis of the data more and more difficult, when not impossible. New computational models are necessary to capture those parameters that are pertinent to analyze the human system under study or to simulate it. There is also a number of important non-medical applications of these computational models which cover numerous human activities, like driving (safer design of vehicles), working (better ergonomy of workplaces), exercising (more efficient training of athletes), entertaining (simulation for movies), etc.

There are basically three levels of design for human models. The first level is mainly *geometrical* and addresses the construction of a digital description of the anatomy, often acquired from medical imagery. The second level is *physical*, involving mainly the biomechanical modeling of various tissues, organs, vessels, muscles or bone structures. The third level is *physiological*, involving a modeling of the functions of the major biological systems (e.g., cardiovascular, respiratory, digestive, hormonal, muscular, central or peripheral nervous system, etc.) or some pathological metabolism (e.g., evolution of cancerous or inflammatory lesions, formation of vessel stenoses, etc.). A fourth level (not described in this volume) would be cognitive, modeling the higher functions of the human brain. These different levels of modeling are closely related to each other, and several physiological systems may interact together (e.g., the cardiopulmonary interaction). The choice of the resolution at which each level is described is important, and may vary from microscopic to macroscopic, ideally through multiscale descriptions.

The first three chapters of this volume study three important physiological models (vascular, cardiac, and tumoral) from a mathematical and numerical perspective. The chapter by Alfio Quarteroni and Luca Formaggia addresses the problem of developing models for the numerical simulation of the human circulatory system, focussing on the analysis of haemodynamics in arteries. Applications include the prediction (and therefore the possible prevention) of stenoses (a local reduction of the lumen of the artery), a leading cause of cardiovascular accidents. The chapter by Mary Belik, Taras Usyk and Andrew McCulloch describes computational methods for modeling and simulating the cardiac electromechanical function. These methods provide tools to predict physiological function from quantitative measurements of tissue, cellular or molecular structures. Applications include a better understanding of cardiac pathologies, and a quantitative modeling of their evolution from various sources of measurements, including medical imagery. The chapter by Jesús Ildefonso Díaz and José Ignacio Tello studies the mathematical properties of a simple model of tumor growth. Proofs are given for the existence and uniqueness of solutions and numerical simulations of the model are presented.

The next two chapters are dedicated to the simulation of deformations inside the human body in two different contexts. The chapter by Eberhard Haug, Hyung-Yun Choi, Stéphane Robin and Muriel Beaugonin describes computational models for crash and impact simulation. It presents the latest generation of virtual human models used to study the consequences of car accidents on organs and important anatomical structures. These models allow the interactive design of safer vehicles with an unrivaled flexibility. The chapter by Hervé Delingette and Nicholas Ayache describes computational models of soft tissue useful for surgery simulation. The real-time constraint imposed by the necessary realism of a training system leads to specific models which are applied to the simulation of minimally invasive digestive surgery, including liver surgery.

The last two chapters describe computational models dedicated to image-guided intervention and diagnosis. The chapter by Xenophon Papademetris, Oskar Skrinjar and James Duncan describes computational models of organs used to predict and track deformations of tissues from sparse information acquired through medical imaging. These models rest on a successful combination of biomechanical modeling with medical image analysis, with an application to image-guided neurosurgery and an application to the image-based quantitative analysis of cardiac diseases. The chapter by Fred Azar, Dimitris Metaxas and Mitchell Schnall presents a computational model of the breast used to predict deformations during interventions. The main applications are for image-guided clinical biopsies and for image-guided therapy.

Before concluding this introduction, I wish to wholeheartedly thank all the authors for their essential contributions, their patience and confidence during all the genesis process of this book. Special thanks are due to my colleague Hervé Delingette, whose advice was extremely helpful from the very beginning. I wish to thank several colleagues for their important help and the many improvements they suggested: Michel Audette, Chris Berenbruch, Mark Chaplain, Olivier Clatz, Stéphane Lanteri, Denis Laurendeau, Philippe Meseure, Serge Piperno, Jean-Marc Schwartz, Brian Sleeman, Michel

Sorine, Matthias Teschner, Marc Thiriet, Marina Vidrascu. I also wish to thank Gilles Kahn, Scientific Director of INRIA, who has been extremely supportive of this project originating from our institute.

Finally, I wish to honor the memory of Jacques-Louis Lions, who contacted me for the first time at the end of November 1999 with the proposition to work on this project. The original title changed several times, before finally converging towards its final title after recent discussions with Philippe Ciarlet, to whom will go my final thanks, for his great encouragements and confidence.

NICHOLAS AYACHE
Sophia–Antipolis, France
1st November 2003

Contents of Volume XII

Contents of the Handbook

VOLUME VII

SOLUTION OF EQUATIONS IN $\mathbb{R}^n$ (PART 3)

TECHNIQUES OF SCIENTIFIC COMPUTING (PART 3)

VOLUME VIII

SOLUTION OF EQUATIONS IN $\mathbb{R}^n$ (PART 4)

TECHNIQUES OF SCIENTIFIC COMPUTING (PART 4)

NUMERICAL METHODS FOR FLUIDS (PART 2)

VOLUME IX

NUMERICAL METHODS FOR FLUIDS (PART 3)

VOLUME X

SPECIAL VOLUME: COMPUTATIONAL CHEMISTRY

VOLUME XI

SPECIAL VOLUME: FOUNDATIONS OF COMPUTATIONAL MATHEMATICS

VOLUME XII

SPECIAL VOLUME: COMPUTATIONAL MODELS FOR THE HUMAN BODY

Special Volume:
Computational Models
for the Human Body

Mathematical Modelling and Numerical Simulation of the Cardiovascular System

Alfio Quarteroni [a,b], Luca Formaggia [b]

[a]*Institute of Mathematics, EPFL, Lausanne, Switzerland*

[b]*MOX, Department of Mathematics, Politecnico di Milano, Milano, Italy*

E-mail addresses: alfio.quarteroni@epfl.ch (A. Quarteroni), luca.formaggia@mate.polimi.it (L. Formaggia)

Computational Models for the Human Body
Special Volume (N. Ayache, Guest Editor) of
HANDBOOK OF NUMERICAL ANALYSIS, VOL. XII
P.G. Ciarlet (Editor)

ISSN 1570-8659
DOI 10.1016/S1570-8659(03)12001-7

Contents

CHAPTER I

1. Introduction

In these notes we will address the problem of developing models for the numerical simulation of the human circulatory system. In particular, we will focus our attention on the problem of haemodynamics in large human arteries.

Indeed, the mathematical investigation of blood flow in the human circulatory system is certainly one of the major challenges of the next years. The social and economical relevance of these studies is highlighted by the unfortunate fact that cardiovascular diseases represent the major cause of death in developed countries.

Altered flow conditions, such as separation, flow reversal, low and oscillatory shear stress areas, are now recognised by the medical research community as important factors in the development of arterial diseases. An understanding of the local haemodynamics can then have useful applications for the medical research and, in a longer term perspective, to surgical planning and therapy. The development of effective and accurate numerical simulation tools could play a crucial role in this process.

Besides their possible role in medical research, another possible use of numerical models of vascular flow is to form the basis for simulators to be used as training systems. For instance, a technique now currently used to cure a stenosis (a pathological restriction of an artery, usually due to fat deposition) is angioplasty. It consists of inflating a balloon positioned in the stenotic region by the help of a catheter. The balloon should squash the stenosis and approximately restore the original lumen area. The success of the procedure depends, among other things, on the sensitivity of the surgeon and his ability of placing the catheter in the right position. A training system which couples virtual reality techniques with the simulation of the flow field around the catheter, the balloon and the vessel walls, employing geometries extracted from real patients, could well serve as training bed for new vascular surgeons. A similar perspective could provide specific design indications concerning the realisations of surgical operations. For instance, numerical simulations could help the surgeon in understanding how the different surgical solutions may affect blood circulation and guide the selection of the most appropriate procedure for a specific patient.

In such "virtual surgery" environments, the outcome of alternative treatment plans for the individual patient can be foreseen by simulations. This numerical approach is one of the aspects of a new paradigm of the clinical practice, which is referred to as "predictive medicine" (see TAYLOR, DRANEY, KU, PARKER, STEELE, WANG and ZARINS [1999]).

Since blood flow interacts mechanically with the vessel walls, it gives rise to a rather complex fluid–structure interaction problem which requires algorithms able to correctly describe the energy transfer between the fluid (typically modelled by the Navier–Stokes equations) and the structure. This is indeed one of the main subjects of these notes, which will adopt the following steps:

(1) *Analysis of the physical problem.* We illustrate problems related to haemodynamics, focusing on those aspects which are more relevant to human physiology. This will allow us to identify the major mathematical variables useful for our investigation. This part will be covered in Section 2.

(2) *Mathematical modelling.* Starting from some basic physical principles, we will derive the partial differential equations which link the variables relevant to the problem. We will address some difficulties associated to the specific characteristics of these equations. Problems such as existence, uniqueness and data dependence of the solution will be briefly analysed. In particular, in Section 5 we will deal with models for the fluid flow and recall the derivation of the incompressible Navier–Stokes equations starting from the basic principles of conservation of mass and momentum. In Section 15 the attention will be instead focused on the dynamics of the vessel wall structure. Some simple, yet effective, mathematical models for the vessel wall displacement will be derived and discussed.

(3) *Numerical modelling.* We present different schemes which can be employed to solve the equations that have been derived and discuss their properties. In particular, Section 10 deals with some relevant mathematical aspects related to the numerical solution of the equations governing the flow field, while Section 18 is dedicated to the coupled fluid–structure problem.

    Reduced models which make use of a one dimensional description of blood flow in arteries are often used to study the propagation of average pressure and mass flow on segments of the arterial tree. In Section 20 we present the derivation of a model of this type, together with a brief analysis of its main mathematical characteristics.

(4) *Numerical simulation.* A final section is dedicated to numerical results obtained on relevant test cases.

2. A brief description of the human vascular system

The major components of the cardiovascular system are the heart, the arteries and the veins. It is usually subdivided into two main parts: the *large circulation* system and the *small circulation* system, as shown in Fig. 2.1. The former brings oxygenated blood from the heart left ventricle to the various organs (arterial system) and then brings it back to right atrium (venous system). The latter pumps the venous blood into the pulmonary artery, where it enters the pulmonary system, get oxygenated and is finally received by the heart left atrium, ready to be sent to the large circulation system.

Fig. 2.2 shows a picture of the human heart. Its functioning is very complex and various research teams are currently trying to develop satisfactory mathematical models of its mechanics, which involves, among other things, the study of the electro-chemical activation of the muscle cells. We will not cover this aspect in these notes, where we rather concentrate on vascular flow and, in particular, flow in arteries.

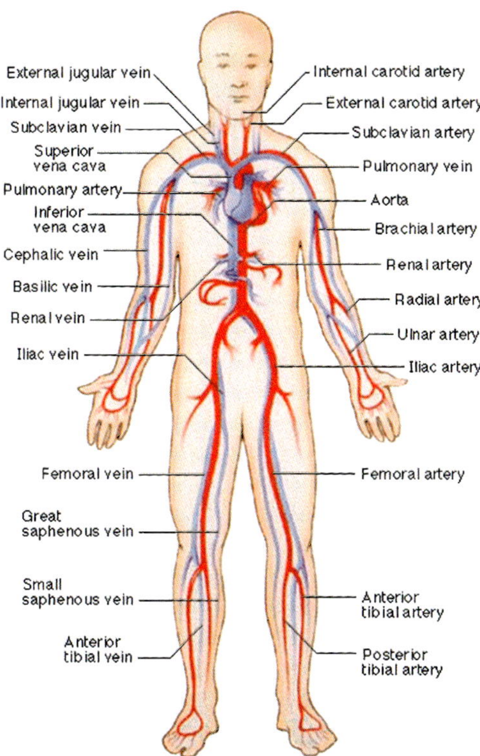

FIG. 2.1. The human circulatory system. The human cardiovascular system has the task of supplying the human organs with blood. Its correct working is obviously crucial and depends on many parameters: external temperature, muscular activity, state of health, just to mention a few. The blood pressure and flow rate then change according to the body needs.

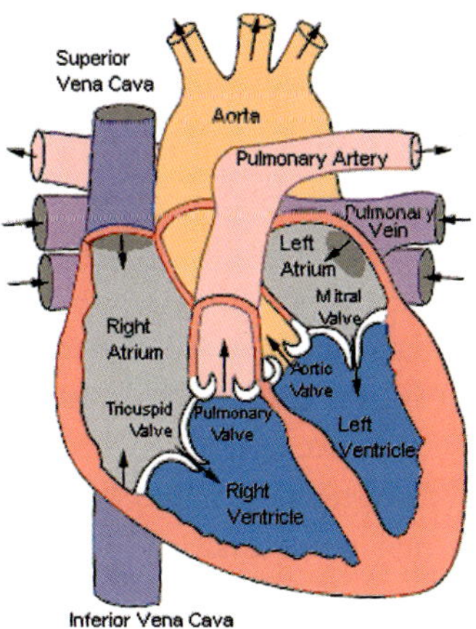

FIG. 2.2. The human heart. Courtesy of the Texas Heart®Institute.

Arteries can be regarded as hollow tubes with strongly variable diameters and can be subdivided into *large arteries*, *medium arteries* and *arterioles and capillaries*. The main role of large arteries (1–3 cm of diameter) is to carry a substantial blood flow rate from the heart to the periphery and to act as a "compliant system". They deform under blood pressure and by doing so they are capable of storing elastic energy during the systolic phase and return it during the diastolic phase. As a result the blood flow is more regular than it would be if the large arteries were rigid. We then have a *fluid–structure* interaction problem. The blood may be considered a homogeneous fluid, with "standard" behaviour (Newtonian fluid), the wall may be considered elastic (or mildly visco-elastic).

The smaller arteries (0.2 mm–1 cm of diameter) are characterised by a strong branching. The vessel may in general be considered rigid (apart in the heart, where the vessel movement is mainly determined by the heart motion). Yet, the blood begins to show "non-standard" behaviour typical of a shear-thinning (non-Newtonian) fluid.

The arterioles have an important muscular activity, which is aimed at regulating blood flow to the periphery. Consequently, the vessel wall mechanical characteristics may change depending on parameters such as blood pressure and others. At the smallest levels (capillaries), blood cannot be modelled anymore as a homogeneous fluid, as the dimension of the particles are now of the same order of that of the vessel. Furthermore, the effect of wall permeability on the blood flow becomes important.

The previous subdivision is not a mere taxonomy: the morphology of the vessel walls and the physical characteristics of blood change in dependence of the type of vessel.

Indeed, the blood is not a fluid but a suspension of particles in a fluid called *plasma*. Blood particles must be taken into account in the rheological model in smaller arterioles and capillaries since their size becomes comparable to that of the vessel. The most important blood particles are:

- red cells (erythrocytes), responsible for the exchange of oxygen and carbon-dioxide with the cells;
- white cells (leukocytes), which play a major role in the human immune system;
- platelets (thrombocytes), main responsible for blood coagulation.

Here, we will limit to flow in *large/medium sized vessels*. We have mentioned that the vascular system is highly complex and able to regulate itself: an excessive decrease in blood pressure will cause the smaller arteries (arterioles) to contract and the heart rate increase. On the contrary, an excessive blood pressure is counter-reacted by a relaxation of the arterioles wall (which causes a reduction of the periphery resistance to the flow) and decreasing the heart beat. Yet, it may happen that some pathological conditions develop, for example, the arterial wall may become more rigid, due to illness or excessive smoking habits, fat may accumulate in some areas causing a stenosis, that is a reduction of the vessel section as illustrated in Fig. 2.3, aneurysms may develop. The consequence of these pathologies on the blood field as well as the possible outcome of a surgical intervention may be studied by numerical tools.

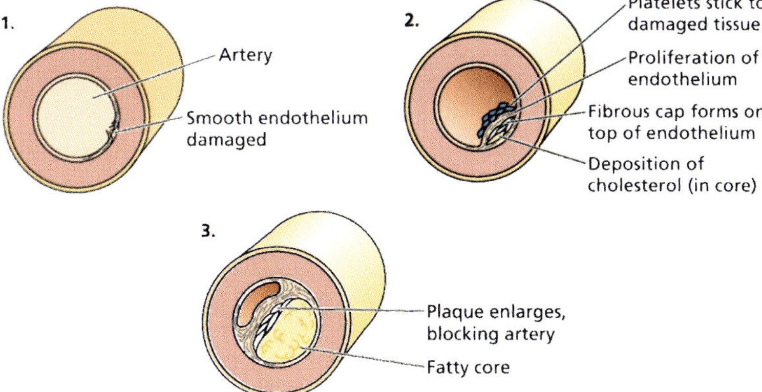

FIG. 2.3. The deposition of lipids and cholesterol in the inner wall of an artery (frequently a coronary) can cause a stenosis and eventually a dramatic reduction (or even the interruption) of blood flow. Images taken from "Life: the Science of Biology" by W.K. Purves et al., fourth edition, published by Sinauer Associates Inc. and W.H. Freeman and Company.

3. The main variables for the mathematical description of blood flow

The principal quantities which describe blood flow are the *velocity* $\mathbf{u}$ and *pressure* P. Knowing these fields allows the computation of the *stresses* to which an arterial wall is subjected due to the blood movement. Since we will treat fluid–structure interaction problems, the *displacement* of the vessel wall due to the action of the flow field is another quantity of relevance. Pressure, velocity and vessel wall displacement will be functions of time and the spatial position.

The knowledge of the *temperature* field may also be relevant in some particular context, such as the hyperthermia treatment, where some drugs are activated through an artificial localised increase in temperature. Temperature may also have a notable influence on blood properties, in particular on blood viscosity. Yet, this aspect is relevant only in the flow through very small arterioles/veins and in the capillaries, a subject which is not covered in these notes.

Another aspect of blood flow which we will not cover in these notes, is the *chemical interaction* with the vessel wall, which is relevant both for the physiology of the blood vessels and for the development of certain vascular diseases. Not mentioning the potential relevance of such investigation for the study of the propagation/absorption of pharmaceutical chemicals. Some numerical models and numerical studies for the chemical transport/diffusion process in blood and through arterial wall may be found in RAPPITSCH and PERKTOLD [1996], QUARTERONI, VENEZIANI and ZUNINO [2002].

4. Some relevant issues

Among the difficulties in the modelling of blood flow in large vessels, we mention the following ones:

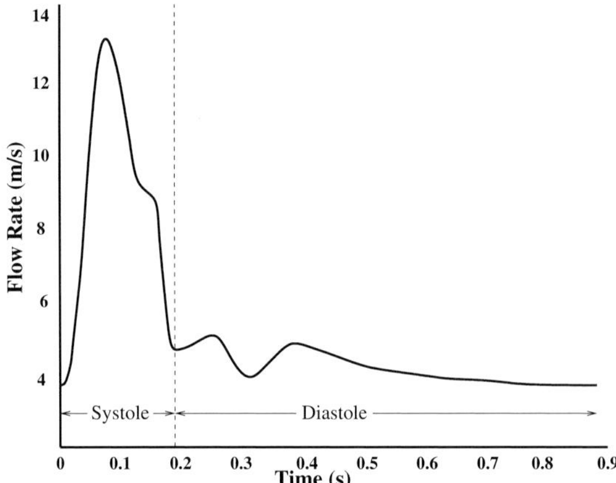

FIG. 4.1. A typical flow rate in an artery during the cardiac cycle.

- *The flow is transient.* Blood flow is obviously pulsatile. This means that one cannot neglect the time by considering a "steady state" solution, function only of the spatial position, as it is often done in many other situations (for example, the study of the flow field around an aeroplane or a car). With some approximation one may think the blood flow to be periodic in time. Yet, this is usually true only for relatively short periods, since the various human activities require to change the amount of blood sent to the various organs.

  The cardiac cycle can be subdivided into two phases. The *systole* corresponds to the instant in which the heart is pumping the blood into the arterial system. The systolic period is then characterised by the highest flow rate. The *diastole*, instead, corresponds to the instant in which the heart is filling up with the blood coming from the venous system and the aortic valve is closed. The blood flow is then at its minimum. Fig. 4.1 illustrates a typical flow rate curve on a large artery during the cardiac cycle.

  Unsteady flow is usually much more complex than its steady counterpart. For instance, if we consider a steady flow of a fluid like water inside an "infinitely long" cylindrical tube, it is possible to derive the analytical steady state solution (also called the *Poiseuille flow* solution), characterised by a parabolic velocity profile. Transient flow in the same geometrical configuration becomes much more complex. The solution may still be obtained analytically if we assume time periodicity, giving rise to the so-called *Womersley flow* (WOMERSLEY [1955]), whose expression may be found, for instance, in QUARTERONI, TUVERI and VENEZIANI [2000]. Just as an example, in Fig. 4.2 we show the velocity profile in a tube for a Poiseuille and for a Womersley flow (the latter, obviously, at a given instant) (from VENEZIANI [1998]).
- *The wall interacts mechanically with the flow field.* This aspect is relevant for relatively large vessels. In the aorta, for example, the radius may vary in a range of 5 to 10% between diastole and systole. This is quite a large displacement, which affects the flow field. The fluid–structure interaction problem is the responsible for

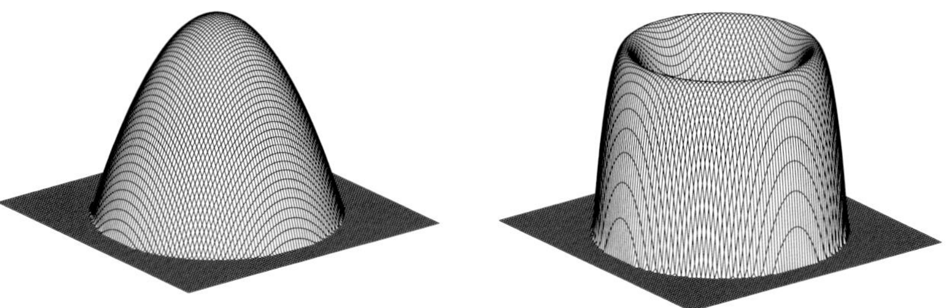

FIG. 4.2. Three-dimensional velocity profiles for a Poiseuille flow (left) and Womersley unsteady flow at a given instant (right).

the propagation of pulse pressure waves. Indeed, no propagative phenomena would otherwise occur in an incompressible fluid like blood. The interaction problem is a rather complex one, since the time scales associated to the interaction phenomena are two orders of magnitude greater than those associated to the bulk flow field.

In arterioles and capillaries the movement of the wall may be considered negligible.

- *Lack of boundary data.* We are normally interested in modelling only a section of the cardiovascular system by means of partial differential equations. A proper setting of a differential problem requires to provide appropriate conditions at the domain boundary, i.e., on the sections at the ends of the region of interest. For instance, let us consider Fig. 4.3. "Standard" conditions for the inlet section Γ^{in} and the outlet sections Γ^{out}, may be derived from the analysis of the differential equations governing the fluid flow. A possible choice is to prescribe all components of the velocity on Γ^{in} and the velocity derivative along the normal direction

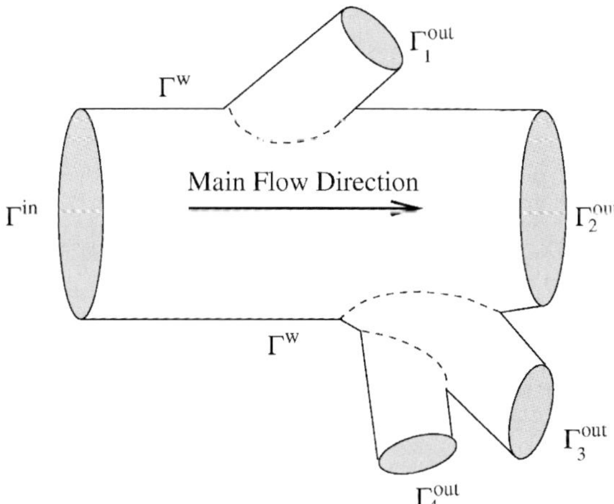

FIG. 4.3. An example of a computational domain made of a section of vascular system. We need to provide proper boundary conditions at Γ^{in}, Γ^{out} and Γ^{w}.

(or the normal stress components) on Γ^{out}. Unfortunately, in practise one never has enough data for prescribing all these conditions. Normally, only "averaged" data are available (mean velocity and mean pressure), which are not sufficient for a "standard" treatment of the mathematical problem. One has thus to devise alternative formulations for the boundary conditions which, on one hand reflect the physics and exploit the available data, on the other hand, permit to formulate a mathematically well posed problem. In these notes we will not investigate this particular aspect. A possible formulation for the flow boundary conditions which is particularly suited for vascular flow problems is illustrated and analysed in FORMAGGIA, GERBEAU, NOBILE and QUARTERONI [2002].

We have not used the terms "inflow" and "outflow" to indicate boundary conditions at Γ^{in} and Γ^{out} since they would be incorrect. Indeed, outflow would indicate the normal component of the velocity is everywhere positive (while it is negative at an inflow section). However, in vascular problems, this assumption is seldom true because the pulsating nature of blood flow might (and typically does) induce a flow reversal on portions of an artery during the cardiac beat. Indeed, the Womersley solution (WOMERSLEY [1955]) of a pulsatile flow in circular cylinders, which provides a reasonable approximation of the general flow pattern encountered in arteries, shows a periodic flow reversal.

In the medical literature, one encounters the terms "proximal" to indicate the section which is reached first by the flow exiting from the heart, while "distal" is the term associated to the sections which are farther from the heart. Here we have preferred instead the terms "inlet" and "outlet" which refer to the behaviour of the mean flow rate across the section. At an inlet (outlet) section the mean flow is entering (exiting) the vascular element under consideration.

Some of the problems which the simulation of blood flow in large arteries may help in answering are summarised below.

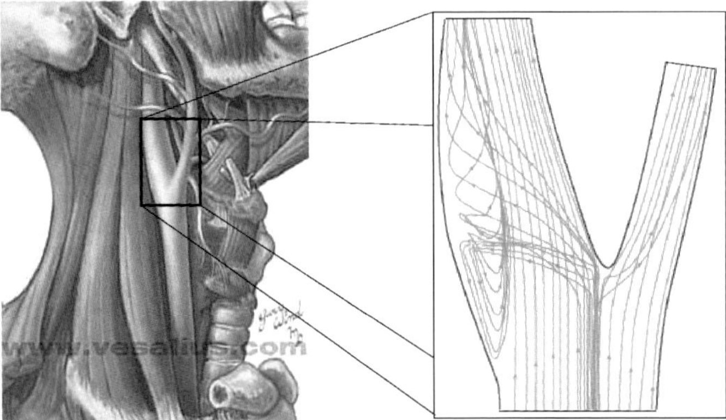

FIG. 4.4. Recirculation in the carotid bifurcation. On the left we illustrate the location of the carotid bifurcation. The image on the right shows the particle path during the diastolic period in a model of the carotid bifurcation. A strong recirculation occurs inside the carotid sinus. The image on the left is courtesy of vesalius.com.

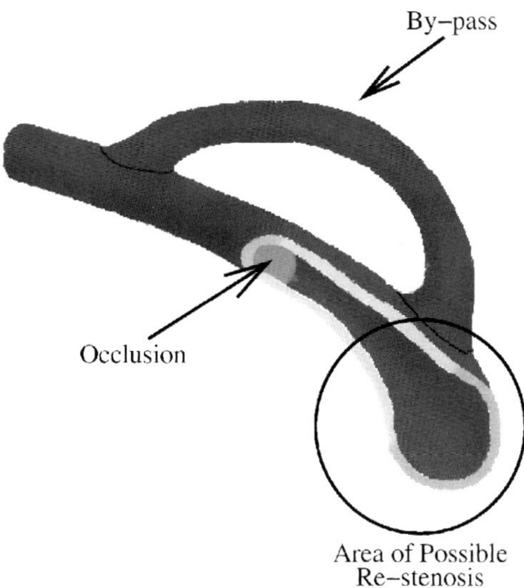

FIG. 4.5. A schematic example of a coronary by-pass. The alteration of the flow field due to the by-pass may cause the formation of a new stenosis, typically immediately downstream the by-pass.

- *Study of the physiological behaviour of vessel walls.* For example, are there any characteristics of the flow field which may be related to the formation of stenoses? In particular, in some sites like the carotid bifurcation (see Fig. 4.4) it is quite usual to have a reversal of the flow during the cardiac cycle which generates a *recirculation zone*. These recirculation zones have been found to be possible sites for fat accumulation and, consequently, the appearance of stenosis. There is some evidence that one of the factors which prompt fat accumulation is linked to the oscillatory nature of the vessel wall stresses induced by the fluid in the flow reversal zone. Wall stresses are quantities very difficult to measure "in vivo" while are easily computed once the flow field is known. Numerical simulations may then help in assessing the effectiveness of such theory.
- *Study of post-surgical situations.* Is it possible to predict the flow behaviour after the geometry has been modified by a surgical operation like a by-pass (see Fig. 4.5)? It has been found that the flow pattern in the by-pass region may affect the insurgence of post-surgery pathologies. Again, a zone with recirculating or stagnant fluid has negative consequences. Numerical simulations may allow to predict the post-surgery flow pattern and determine, say, the best by-pass configuration.

CHAPTER II

5. The derivation of the equations for the flow field

The flow field is governed by a set of partial differential equations in a region whose boundary changes in time. Their derivation, moving from the basic physical principles of conservation of mass and momentum, is the scope of this chapter.

6. Some nomenclature

The space $\mathbb{R}^3$ is equipped with a Cartesian coordinate system defined by the orthonormal basis $(\mathbf{e}_1, \mathbf{e}_2, \mathbf{e}_3)$, where

$$\mathbf{e}_1 = \begin{bmatrix} 1 \\ 0 \\ 0 \end{bmatrix}, \qquad \mathbf{e}_2 = \begin{bmatrix} 0 \\ 1 \\ 0 \end{bmatrix}, \qquad \mathbf{e}_3 = \begin{bmatrix} 0 \\ 0 \\ 1 \end{bmatrix}.$$

Vectors are understood as column vectors. A vector $\mathbf{f} \in \mathbb{R}^3$ may then be written as

$$\mathbf{f} = \sum_{i=1}^{3} f_i \mathbf{e}_i,$$

where f_i is the ith component of $\mathbf{f}$ with respect to the chosen basis. Vectors will be always indicated using bold letters while their components will be generally denoted by the same letter in normal typeface. Sometimes, when necessary for clarity, we will indicate the ith component of a vector $\mathbf{f}$ by $(\mathbf{f})_i$ or simply f_i. These definitions apply to vectors in $\mathbb{R}^2$ as well. With the term *domain* we will indicate an open, bounded, connected subset of $\mathbb{R}^N$, $N = 2, 3$, with orientable boundary. We will indicate with $\mathbf{n}$ the outwardly oriented unit vector normal to the boundary. We will also assume that the domain boundary be Lipschitz continuous (for instance, a piece-wise polynomial, or a C^1 curve). In Fig. 6.1 some admissible domains are shown. If a quantity f (like temperature or pressure) takes a scalar value on a domain Ω, we say that the quantity defines a *scalar field* on Ω, which we will indicate with $f : \Omega \to \mathbb{R}$. If instead a quantity $\mathbf{f}$ associates to each point in Ω a vector (as in the case of the velocity), we say that it defines a *vector field* on Ω, and we will indicate it with $\mathbf{f} : \Omega \to \mathbb{R}^3$. Finally, if a quantity $\mathbf{T}$ associates to each point in Ω a $\mathbb{R}^{N \times N}$ matrix, we will say that it defines a (second order) *tensor* field on Ω if it obeys the ordinary transformation rules for tensors (ARIS [1962]). Its components will be indicated by either $(\mathbf{T})_{ij}$, or simply T_{ij}, with $i, j = 1, \dots, 3$.

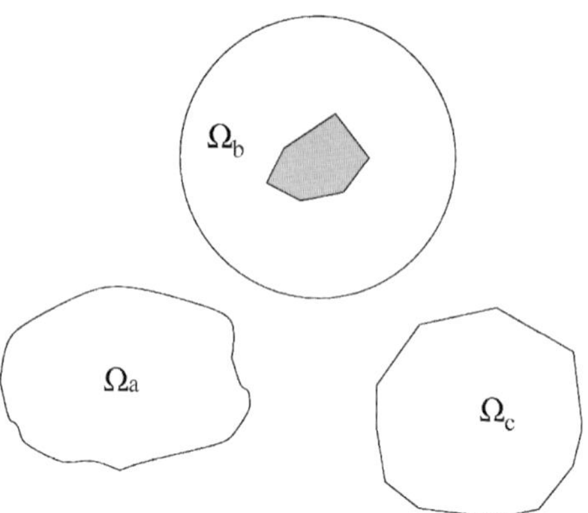

FIG. 6.1. Example of admissible domains. Ω_a has a boundary formed by piece-wise C^1 curves. Ω_b is a multi-connected domain, with a polygonal internal and a C^∞ external boundary. Finally, Ω_c has a polygonal boundary.

Given a function $f : \Omega \to \mathbb{R}$, $\mathbf{x} \to f(\mathbf{x})$, and a domain $V \subset \Omega$, we will use the shorthand notation

$$\int_V f$$

to indicate the integral

$$\int_V f(\mathbf{x})\,\mathrm{d}\mathbf{x},$$

and

$$\int_{\partial V} f$$

to indicate the surface (or line) integral

$$\int_{\partial V} f\,\mathrm{d}\sigma,$$

unless the context requires otherwise.

When referring to a physical quantity f, we will indicate with $[f]$ its measure units (in the international system). For instance, if $\mathbf{v}$ indicates a velocity, $[\mathbf{v}] = \mathrm{m/s}$, where m stands for meters and s for seconds.

7. The motion of continuous media

In order to derive the differential equations which govern the fluid motion, we need to introduce some kinematic concepts and quantities. The kinematics of a continuous medium studies the property of the motion of a medium which may be thought as continuously occupying, at each time, a portion of space. This allows the use of standard methods of analysis. We will set the derivation in $\mathbb{R}^3$, since this is the natural spatial

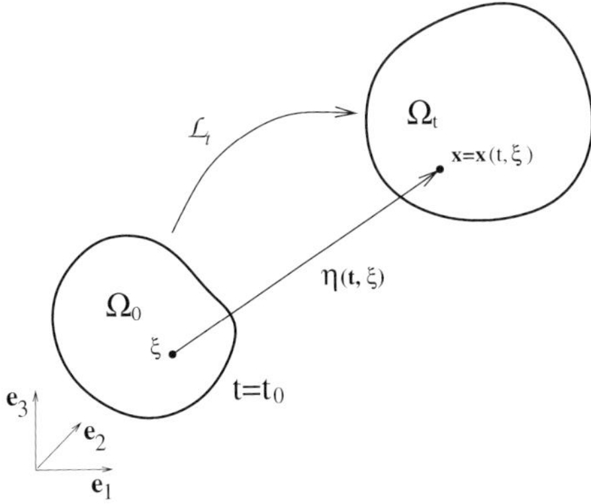

FIG. 7.1. The Lagrangian mapping.

dimension. However, the definitions and final differential equations are valid also in $\mathbb{R}^2$. Furthermore, we will assume that the motion will take place during a time interval $I = (t_0, t_1)$.

The motion itself is described by a family of mappings $\mathcal{L}_t$ which associate the position $\mathbf{x}$ of a fluid particle at time $t \in I$ to a point $\boldsymbol{\xi} \in \Omega_0$, Ω_0 being the domain occupied by the fluid at the reference initial time t_0. More precisely, we denote with Ω_t the portion of space occupied by the fluid at time t and we indicate with $\mathcal{L}_t$ the mapping

$$\mathcal{L}_t : \Omega_0 \to \Omega_t, \qquad \boldsymbol{\xi} \to \mathbf{x} = \mathbf{x}(t, \boldsymbol{\xi}) = \mathcal{L}_t(\boldsymbol{\xi}),$$

which will be denoted *Lagrangian* mapping at time t (see Fig. 7.1). We assume that $\mathcal{L}_t$ is continuous and invertible in $\overline{\Omega_0}$, with continuous inverse.

We call Ω_0 the *reference* configuration, while Ω_t is called *current* (or spatial) configuration. The position of the material particle located at the point $\mathbf{x}$ in the current configuration Ω_t is a function of time and of the position of the same material particle at the reference time.

We may thus relate the variables $(t, \mathbf{x})$ to $(t, \boldsymbol{\xi})$. The former couple is referred to as the *Eulerian* variables while the latter are called the *Lagrangian* variables.

It is worthwhile to point out that when using the Eulerian variables as independent variables, we are concentrating our attention on a position in space $\mathbf{x} \in \Omega_t$ and on the fluid particle which, at that particular time, is located at $\mathbf{x}$. When using the Lagrangian variables as independent variables (Lagrangian frame), we are instead targeting the fluid particle "labelled" $\boldsymbol{\xi}$ (that is the fluid particle which was located at position $\boldsymbol{\xi}$ at the reference time). That is, we are following the *trajectory* $T_{\boldsymbol{\xi}}$ of fluid particle $\boldsymbol{\xi} \in \Omega_0$, defined as

$$T_{\boldsymbol{\xi}} = \left\{ \left(t, \mathbf{x}(t, \boldsymbol{\xi})\right) \colon t \in I \right\}. \tag{7.1}$$

The basic principles of mechanics are more easily formulated with reference to the moving particles, thus in the Lagrangian frame. Yet, in practice it is more convenient to work with the Eulerian variables. Therefore, we need to rewrite the equations stemming from those basic principles into the Eulerian frame. We will see later on that for the numerical approximation of the problem at hand it will be necessary to introduce yet another, intermediate, frame of reference, called *Arbitrary Lagrangian Eulerian*.

Being the mapping surjective, a quantity associated with the fluid may be described as function of either the Lagrangian or the Eulerian variables, depending on convenience. We will in general use the same symbol for the functions which describe the evolution of the same quantity in the Lagrangian and in the Eulerian frame, unless the context needs otherwise. In the latter case, we will mark with the hat symbol "^" a quantity expressed as function of the Lagrangian variables, that is, if $f : I \times \Omega_t \to \mathbb{R}$ we have the equality

$$\hat{f}(t, \boldsymbol{\xi}) = f(t, \mathbf{x}), \quad \text{with } \mathbf{x} = \mathcal{L}_t(\boldsymbol{\xi}).$$

We will often use the following alternative notation:

$$\hat{f} = f \circ \mathcal{L}_t \quad \text{or} \quad f = \hat{f} \circ \mathcal{L}_t^{-1}$$

with the understanding that the composition operator applies only to the spatial variables.

The symbol ∇ is used exclusively to indicate the gradient *with respect to the Eulerian variable* $\mathbf{x}$. When we need to indicate the gradient with respect to the Lagrangian variable $\boldsymbol{\xi}$, we will use the symbol $\nabla_{\boldsymbol{\xi}}$, that is

$$\nabla_{\boldsymbol{\xi}} \hat{f} = \sum_{i=1}^{3} \frac{\partial \hat{f}}{\partial \xi_i} \mathbf{e}_i .$$

The same convention applies to other spatial differential operators (divergence, Laplacian, etc.) as well.

In the following we will put $I \times \Omega_t = \{(t, \mathbf{x}) \colon t \in I,\ x \in \Omega_t\}$ (note the little abuse of notation since technically it is not a Cartesian product).

7.1. *The velocity*

The *fluid velocity* is the major kinematic quantity of our problem. In the Lagrangian frame it is expressed by means of a vector field $\hat{\mathbf{u}} = \hat{\mathbf{u}}(t, \boldsymbol{\xi})$ defined as

$$\hat{\mathbf{u}} = \frac{\partial \mathbf{x}}{\partial t}, \quad \text{i.e.,} \quad \hat{\mathbf{u}}(t, \boldsymbol{\xi}) = \frac{\partial}{\partial t} \mathbf{x}(t, \boldsymbol{\xi}). \tag{7.2}$$

$\hat{\mathbf{u}}$ is called the *Lagrangian* velocity field (or velocity field in the Lagrangian frame), and it denotes the time derivative along the trajectory $T_{\boldsymbol{\xi}}$ of the fluid particle $\boldsymbol{\xi}$. The velocity $\mathbf{u}$ on the Eulerian frame is defined for $(t, \mathbf{x}) \in I \times \Omega_t$ as

$$\mathbf{u} = \hat{\mathbf{u}} \circ \mathcal{L}_t^{-1}, \quad \text{i.e.,} \quad \mathbf{u}(t, \mathbf{x}) = \hat{\mathbf{u}}\big(t, \mathcal{L}_t^{-1}(\mathbf{x})\big).$$

EXAMPLE 7.1. Let us consider a 2D case and the following movement law, for $t \geqslant 0$:

$$x_1 = \xi_1 e^t, \quad \xi_1 \in (-1, 1), \qquad x_2 = \xi_2, \quad \xi_2 \in (-1, 1).$$

The domain at time $t > 0$ occupies the rectangle $(-e^t, e^t) \times (-1, 1)$. The mapping is clearly invertible for all $t \geqslant 0$.

We have

$$\hat{u}_1 = \partial x_1/\partial t = \xi_1 e^t, \qquad \hat{u}_2 = \partial x_2/\partial t = 0.$$

We can immediately compute the velocity field as function of the Eulerian variable as

$$u_1 = x_1, \qquad u_2 = 0.$$

Once the velocity field and the reference configuration is known, the motion may be derived by solving the following Cauchy problem:

For any $\boldsymbol{\xi} \in \Omega_0$, find the function $\mathbf{x} = \mathbf{x}(t, \boldsymbol{\xi})$ which satisfies

$$\begin{cases} \dfrac{\partial \mathbf{x}}{\partial t}(t, \boldsymbol{\xi}) = \hat{\mathbf{u}}(t, \boldsymbol{\xi}), & \forall t \in I, \\ \mathbf{x}(t_0, \boldsymbol{\xi}) = \boldsymbol{\xi}. \end{cases}$$

7.2. *The material derivative*

We can relate time derivatives computed with respect to the different frames. The *material* (or *Lagrangian*) time derivative of a function f, which we will denote Df/Dt, is defined as the time derivative in the Lagrangian frame, yet expressed as function of the Eulerian variables.

That is, if $f : I \times \Omega_t \to \mathbb{R}$ and $\hat{f} = f \circ \mathcal{L}_t$,

$$\frac{Df}{Dt} : I \times \Omega_t \to \mathbb{R}, \qquad \frac{Df}{Dt}(t, \mathbf{x}) = \frac{\partial \hat{f}}{\partial t}(t, \boldsymbol{\xi}), \quad \boldsymbol{\xi} = \mathcal{L}_t^{-1}(\mathbf{x}). \tag{7.3}$$

Therefore, for any fixed $\boldsymbol{\xi} \in \Omega_0$ we may also write

$$\frac{Df}{Dt}(t, \mathbf{x}) = \frac{\mathrm{d}}{\mathrm{d}t} f\big(t, \mathbf{x}(t, \boldsymbol{\xi})\big),$$

by which we can observe that the material derivative represents the rate of variation of f along the trajectory $T_{\boldsymbol{\xi}}$.

By applying the chain-rule of derivation of composed functions, we have

$$\frac{Df}{Dt} = \frac{\partial f}{\partial t} + \mathbf{u} \cdot \nabla f. \tag{7.4}$$

Indeed,

$$\frac{Df}{Dt} = \left[\frac{\partial}{\partial t}(f \circ \mathcal{L}_t)\right] \circ \mathcal{L}_t^{-1} = \frac{\partial f}{\partial t} + \nabla f \cdot \left(\frac{\partial \mathbf{x}}{\partial t} \circ \mathcal{L}_t^{-1}\right) = \frac{\partial f}{\partial t} + \mathbf{u} \cdot \nabla f.$$

A quantity which satisfies

$$\frac{\partial f}{\partial t} = 0$$

is called *stationary*, and a motion for which

$$\frac{\partial \mathbf{u}}{\partial t} = 0$$

is said a *stationary motion*.

EXAMPLE 7.2. Let us consider again the motion of Example 7.1 and consider the function $f(x_1, x_2) = 3x_1 + x_2$ (which is independent of t). The application of relation (7.4) gives

$$\frac{Df}{Dt} = 0 + \begin{bmatrix} x_1 \\ 0 \end{bmatrix} \cdot \begin{bmatrix} 3 \\ 1 \end{bmatrix} = 3x_1.$$

On the other hand,

$$\hat{f} = 3\xi_1 e^t + \xi_2 \quad \text{and} \quad \frac{\partial \hat{f}}{\partial t} = 3\xi_1 e^t,$$

by which we deduce that,

$$\frac{\partial \hat{f}}{\partial t} \circ \mathcal{L}_t^{-1} = 3x_1.$$

This example, besides verifying relation (7.4), shows that a function $f = f(t, \mathbf{x})$ with $\partial f / \partial t = 0$ in general has $Df/Dt \neq 0$.

7.3. *The acceleration*

In the Lagrangian frame the acceleration is a vector field $\hat{\mathbf{a}} : I \times \Omega_0 \to \mathbb{R}^3$ defined as

$$\hat{\mathbf{a}} = \frac{\partial \hat{\mathbf{u}}}{\partial t} = \frac{\partial^2 \mathbf{x}}{\partial t^2}.$$

By recalling the definition of material derivative, we may write the acceleration in Eulerian frame as

$$\mathbf{a} = \frac{D\mathbf{u}}{Dt} = \frac{\partial \mathbf{u}}{\partial t} + (\boldsymbol{u} \cdot \nabla)\mathbf{u}. \tag{7.5}$$

Component-wise,

$$a_i = \frac{\partial u_i}{\partial t} + \sum_{j=1}^{3} u_j \frac{\partial u_i}{\partial x_j}. \tag{7.6}$$

7.4. *The deformation gradient*

Another kinematic quantity necessary for the derivation of the mathematical model is the *deformation gradient* $\widehat{\mathbf{F}}_t$, which is defined, for each $t \in I$, as

$$\widehat{\mathbf{F}}_t : \Omega_0 \to \mathbb{R}^{N \times N}, \quad \widehat{\mathbf{F}}_t = \nabla_m \mathcal{L}_t = \frac{\partial \mathbf{x}}{\partial \boldsymbol{\xi}}. \tag{7.7}$$

Component-wise,

$$(\widehat{\mathbf{F}}_t)_{ij} = \frac{\partial x_i}{\partial \boldsymbol{\xi}_j}.$$

In particular, its determinant,

$$\widehat{J}_t = \det \widehat{\mathbf{F}}_t, \tag{7.8}$$

is called the Jacobian of the mapping $\mathcal{L}_t$. As usual, its counterpart in the Eulerian frame is indicated J_t.

It is possible to show that the time continuity and the invertibility of the Lagrangian mapping is sufficient to have, for all $t \in I$,

$$\widehat{J}_t(\boldsymbol{\xi}) > 0 \quad \forall \boldsymbol{\xi} \in \Omega_0. \tag{7.9}$$

The importance of J_t is clearly linked to the rule which transforms integrals from the current to the reference configuration. We recall the following theorem of elementary calculus (without providing its proof).

THEOREM 7.1. *Let $V_t \subset \Omega_t$ be a subdomain of Ω_t and let us consider the function $f : I \times V_t \to \mathbb{R}$. Then f is integrable on V_t if and only if $(f \circ \mathcal{L}_t) J_t$ is integrable on $V_0 = \mathcal{L}_t^{-1}(V_t)$, and*

$$\int_{V_t} f(t, \mathbf{x}) \, \mathrm{d}\mathbf{x} = \int_{V_0} \hat{f}(t, \boldsymbol{\xi}) \, \widehat{J}_t(\boldsymbol{\xi}) \, \mathrm{d}\boldsymbol{\xi},$$

where $\hat{f}(t, \boldsymbol{\xi}) = f(t, \mathcal{L}_t(\boldsymbol{\xi}))$. In short,

$$\int_{V_t} f = \int_{V_0} \hat{f} \widehat{J}_t.$$

7.5. *The Reynolds transport theorem*

An interesting property of the Jacobian is that its time derivative is linked to the divergence of the velocity field.

LEMMA 7.1. *Let J_t denote the Jacobian* (7.8) *in the Eulerian frame. Then*

$$\frac{D}{Dt} J_t = J_t \operatorname{div} \mathbf{u}. \tag{7.10}$$

This relation is sometimes called Euler expansion formula.

PROOF. We have, by direct application of the chain-rule,

$$\nabla_{\boldsymbol{\xi}} \hat{\mathbf{u}} = \nabla_{\boldsymbol{\xi}} (\mathbf{u} \circ \mathcal{L}_t) = \widehat{\nabla \mathbf{u}} \, \nabla_{\boldsymbol{\xi}} \mathcal{L}_t = \widehat{\nabla \mathbf{u}} \, \widehat{\mathbf{F}}_t.$$

On the other hand, by recalling the definition of the velocity (7.2),

$$\nabla_{\boldsymbol{\xi}} \hat{\mathbf{u}} = \nabla_{\boldsymbol{\xi}} \left(\frac{\partial \mathbf{x}}{\partial t} \right) = \frac{\partial}{\partial t} \nabla_{\boldsymbol{\xi}} \mathbf{x} = \frac{\partial \widehat{\mathbf{F}}_t}{\partial t}.$$

Thus, we may write

$$\widehat{\mathbf{F}}_{t+\varepsilon} = \widehat{\mathbf{F}}_t + \varepsilon \frac{\partial \widehat{\mathbf{F}}_t}{\partial t} + \mathbf{o}(\varepsilon) = \widehat{\mathbf{F}}_t + \varepsilon \widehat{\nabla \mathbf{u}} \, \widehat{\mathbf{F}}_t + \mathbf{o}(\varepsilon) = (\mathbf{I} + \varepsilon \widehat{\nabla \mathbf{u}}) \, \widehat{\mathbf{F}}_t + \mathbf{o}(\varepsilon).$$

We now exploit the well-known result that for any matrix $\mathbf{A}$,

$$\det(\mathbf{I}+\varepsilon\mathbf{A}) = 1+\varepsilon\,\mathrm{tr}\,\mathbf{A}+o(\varepsilon),$$

where $\mathrm{tr}\,\mathbf{A}=\sum_i A_{ii}$ denotes the trace of the matrix A, to write

$$\widehat{J}_{t+\varepsilon} = \det(\widehat{\mathbf{F}}_{t+\varepsilon}) = (1+\varepsilon\,\mathrm{tr}\,\widehat{\nabla}\mathbf{u})\widehat{J}_t + o(\varepsilon) = (1+\varepsilon\,\mathrm{div}\,\hat{\mathbf{u}})\widehat{J}_t + o(\varepsilon).$$

We have used the identity $\mathrm{tr}\,\widehat{\nabla\mathbf{u}} = \mathrm{div}\,\hat{\mathbf{u}}$. Then, by applying the definition of material derivative and exploiting the continuity of the Lagrangian mapping, we may write

$$\frac{DJ_t}{Dt} = \left(\lim_{\varepsilon\to 0}\frac{\widehat{J}_{t+\varepsilon}-\widehat{J}_t}{\varepsilon}\right)\circ\mathcal{L}_t^{-1} = (\mathrm{div}\,\hat{\mathbf{u}}\,\widehat{J}_t)\circ\mathcal{L}_t^{-1} = \mathrm{div}\,\mathbf{u}\,J_t. \qquad \square$$

EXAMPLE 7.3. For the movement law given by Example 7.1, we have

$$\widehat{J}_t = \det\begin{bmatrix} e^t & 0\\ 0 & 1\end{bmatrix} = e^t$$

and $J_t = e^t$ as well. We may verify directly relation (7.10) since

$$J_t\,\mathrm{div}\,\mathbf{u} = e^t(1+0) = e^t = \frac{\mathrm{d}}{\mathrm{d}t}\widehat{J}_t = (\text{by relation (7.3)}) = \frac{D}{Dt}J_t.$$

We have now the following fundamental result.

THEOREM 7.2 (Reynolds transport theorem). *Let $V_0\subset\Omega_0$, and $V_t\subset\Omega_t$ be its image under the mapping $\mathcal{L}_t$. Let $f: I\times\Omega_t\to\mathbb{R}$ be a continuously differentiable function with respect to both variables $\mathbf{x}$ and t. Then,*

$$\frac{\mathrm{d}}{\mathrm{d}t}\int_{V_t} f = \int_{V_t}\left(\frac{Df}{Dt}+f\,\mathrm{div}\,\mathbf{u}\right) = \int_{V_t}\left(\frac{\partial f}{\partial t}+\mathrm{div}(f\mathbf{u})\right). \tag{7.11}$$

PROOF. Thanks to Theorem 7.1 and relations (7.10) and (7.3), we have

$$\begin{aligned}\frac{\mathrm{d}}{\mathrm{d}t}\int_{V_t} f(t,\mathbf{x})\,\mathrm{d}\mathbf{x} &= \frac{\mathrm{d}}{\mathrm{d}t}\int_{V_0}\hat{f}(t,\boldsymbol{\xi})\widehat{J}_t(\boldsymbol{\xi})\,\mathrm{d}\boldsymbol{\xi} = \int_{V_0}\frac{\partial}{\partial t}\big[\hat{f}(t,\boldsymbol{\xi})\widehat{J}_t(\boldsymbol{\xi})\big]\,\mathrm{d}\boldsymbol{\xi}\\ &= \int_{V_0}\left[\frac{\partial\hat{f}}{\partial t}(t,\boldsymbol{\xi})\widehat{J}_t(\boldsymbol{\xi}) + \hat{f}(t,\boldsymbol{\xi})\frac{\partial}{\partial t}\widehat{J}_t(\boldsymbol{\xi})\right]\mathrm{d}\boldsymbol{\xi}.\end{aligned}$$

We now use Theorem 7.1 and the definition of material derivative (7.3) to write

$$\int_{V_0}\frac{\partial\hat{f}}{\partial t}(t,\boldsymbol{\xi})\widehat{J}_t(\boldsymbol{\xi})\,\mathrm{d}\boldsymbol{\xi} = \int_{V_t}\frac{Df}{Dt}(t,\mathbf{x})\,\mathrm{d}\mathbf{x}.$$

Furthermore, we exploit again the definition of material derivative (7.3) in order to rewrite relation (7.10) in the following equivalent form:

$$\frac{\partial}{\partial t}\widehat{J}_t(\boldsymbol{\xi}) = \widehat{J}_t(\boldsymbol{\xi})\,\mathrm{div}\,\mathbf{u}\big(t,\mathbf{x}(t,\boldsymbol{\xi})\big).$$

Consequently,

$$\begin{aligned}\frac{\mathrm{d}}{\mathrm{d}t}\int_{V_t} f(t,\mathbf{x})\,\mathrm{d}\mathbf{x} &= \int_{V_t} \frac{Df}{Dt}(t,\mathbf{x})\,\mathrm{d}\mathbf{x} + \int_{V_0} \hat{f}(t,\boldsymbol{\xi})J_t\big(\mathbf{x}(t,\boldsymbol{\xi})\big)\operatorname{div}\mathbf{u}\big(t,\mathbf{x}(t,\boldsymbol{\xi})\big)\,\mathrm{d}\boldsymbol{\xi}\\ &= \int_{V_t} \frac{Df}{Dt}(t,\mathbf{x})\,\mathrm{d}\mathbf{x} + \int_{V_t} f(t,\mathbf{x})\operatorname{div}\mathbf{u}(t,\mathbf{x})\,\mathrm{d}\mathbf{x}\\ &= \int_{V_t} \frac{Df}{Dt}(t,\mathbf{x}) + f(t,\mathbf{x})\operatorname{div}\mathbf{u}(t,\mathbf{x})\,\mathrm{d}\mathbf{x}.\end{aligned}$$

The second equality in (7.11) is a consequence of (7.4). □

Relation (7.11) is given the name of *Reynolds transport formula*, or simply *transport formula* (sometimes the name *convection formula* is used as well).

By the application of the divergence theorem the previous expression becomes

$$\frac{\mathrm{d}}{\mathrm{d}t}\int_{V_t} f = \int_{V_t}\frac{\partial f}{\partial t} + \int_{\partial V_t} f\mathbf{u}\cdot\mathbf{n}.$$

8. The derivation of the basic equations of fluid mechanics

In the sequel, the symbol V_t will always be used to indicate a *material volume* at time t, i.e., V_t is the image under the Lagrangian mapping of a subdomain $V_0 \subset \Omega_0$, i.e., $V_t = \mathcal{L}_t(V_0)$ (as already done in Theorem 7.2).

8.1. Continuity equation or mass conservation

We assume that there exists a strictly positive, measurable function $\rho : I \times \Omega_t \to \mathbb{R}$, called *density* such that on each $V_t \subset \Omega_t$,

$$\int_{V_t} \rho = m(V_t),$$

where $m(V_t)$ is the mass of the material contained in V_t. The density ρ has dimensions $[\rho] = \mathrm{kg/m^3}$.

A fundamental principle of classical mechanics, called principle of *mass conservation*, states that mass is neither created nor destroyed during the motion. This principle translates into the following mathematical statement.

Given any material volume $V_t \subset \Omega_t$ the following equality holds:

$$\frac{\mathrm{d}}{\mathrm{d}t}\int_{V_t} \rho = 0.$$

We can apply the transport theorem, obtaining

$$\int_{V_t}\left(\frac{D\rho}{Dt} + \rho\operatorname{div}\mathbf{u}\right) = 0. \tag{8.1}$$

By assuming that the terms under the integral are continuous, the arbitrariness of V_t allows us to write the *continuity equation* in differential form

$$\frac{\partial \rho}{\partial t} + \operatorname{div} \rho \mathbf{u} = 0.$$

In these cases for which we can make the assumption that ρ is constant (like for blood flow), we obtain

$$\operatorname{div} \mathbf{u} = 0. \tag{8.2}$$

Relation (8.2), which has been derived from the continuity equation in the case of a *constant density* fluid (sometimes also called incompressible fluid), is indeed a kinematic constraint. Thanks to (7.10), relation (8.2) is equivalent to

$$\frac{D}{Dt} J_t = 0, \tag{8.3}$$

which is the *incompressibility constraint*. A flow which satisfies the incompressibility constraint is called *incompressible*. By the continuity equation, we derive the following implication:

$$\text{constant density fluid} \quad \Rightarrow \quad \text{incompressible flow},$$

whereas the converse is not true in general.

By employing the transport formula (7.11) with $f = 1$, we may note that the incompressibility constraint is equivalent to

$$\frac{\mathrm{d}}{\mathrm{d}t} \int_{V_t} \mathrm{d}\mathbf{x} = 0 \quad \forall V_t \subset \Omega_t,$$

which means that the only possible motions of an incompressible flow are those which preserve the fluid volume.

8.2. *The momentum equation*

Another important principle allows the derivation of an additional set of differential equations, that is the principle of *conservation of momentum*. It is an extension of the famous Newton law, “force = mass × acceleration”, to a continuous medium.

REMARK 8.1. In the dimension unit specifications we will use the symbol Ne to indicate the *Newtons* (the dimension units of a force), Ne $= \mathrm{kg\,m/s^2}$, instead of the more standard symbol N, since we have used the latter to indicate the number of space dimensions.

Three different types of forces may be acting on the material inside Ω_t:

- *Body forces*. These forces are proportional to the mass. They are normally represented by introducing a vector field $\mathbf{f}^b : I \times \Omega_t \to \mathbb{R}^3$, called *specific body force*, whose dimension unit, $[\mathbf{f}^b] = \mathrm{Ne}/\mathrm{kg} = \mathrm{m/s^2}$, is that of an acceleration. The body force acting on a volume V_t is given by

$$\int_{V_t} \rho \mathbf{f}^b,$$

whose dimension unit is clearly Ne. An example is the gravity force, given by $\mathbf{f}^b = -g\mathbf{e}_3$, where $\mathbf{e}_3$ represents the vertical direction and g the gravitational acceleration.

- *Applied surface forces.* They represent that part of the forces which are imposed on the medium through its surface. We will assume that they may be represented through a vector field $\mathbf{t}^e : I \times \Gamma_t^n \to \mathbb{R}^3$, called *applied stresses*, defined on a measurable subset of the domain boundary $\Gamma_t^n \subset \partial\Omega_t$ and with dimension unit $[\mathbf{t}^e] = \mathrm{Ne}/\mathrm{m}^2$. The resultant force acting through the surface is then given by

$$\int_{\Gamma_t^n} \mathbf{t}^e.$$

An example of a surface stress is that caused by the friction of the air flowing over the surface of a lake.

- *Internal "continuity" forces.* These are the forces that the continuum media particles exert on each other and are responsible for maintaining material continuity during the movement. To model these forces let us recall the following principle, due to Cauchy.

8.2.1. The Cauchy principle

There exists a vector field $\mathbf{t}$, called *Cauchy stress*,

$$\mathbf{t}: I \times \Omega_t \times \mathbf{S}_1 \to \mathbb{R}^3$$

with

$$\mathbf{S}_1 = \{\mathbf{n} \in \mathbb{R}^3 : |\mathbf{n}| = 1\}$$

such that its integral on the surface of any material domain $V_t \subset \Omega_t$, given by

$$\int_{\partial V_t} \mathbf{t}(t, \mathbf{x}, \mathbf{n})\,\mathrm{d}\sigma \tag{8.4}$$

is *equivalent* to the resultant of the material continuity forces acting on V_t. In (8.4), $\mathbf{n}$ indicates the outward normal of ∂V_t.

Furthermore, we have that

$$\mathbf{t} = \mathbf{t}^e \quad \text{on } \partial V_t \cap \Gamma_t^n.$$

This principle is of fundamental importance because it states that the only dependence of the internal forces on the geometry of ∂V_t is through $\mathbf{n}$.

We may now state the following *principle of conservation of linear momentum.*

For any $t \in I$, on any sub-domain $V_t \subset \Omega_t$ *completely contained in* Ω_t, the following relation holds:

$$\frac{\mathrm{d}}{\mathrm{d}t}\int_{V_t} \rho(t, \mathbf{x})\mathbf{u}(t, \mathbf{x})\,\mathrm{d}\mathbf{x} = \int_{V_t} \rho(t, \mathbf{x})\mathbf{f}^b(t, \mathbf{x})\,\mathrm{d}\mathbf{x} + \int_{\partial V_t} \mathbf{t}(t, \mathbf{x}, \mathbf{n})\,\mathrm{d}\sigma, \tag{8.5}$$

where all terms dimension unit is Ne. Relation (8.5) expresses the property that the variation of the *linear momentum* of V_t (represented by the integral at the left-hand side) is balanced by the resultant of the internal and body forces.

With some further assumptions on the regularity of the Cauchy stresses, we are now able to relate the internal continuity forces to a *tensor field*, as follows.

THEOREM 8.1 (Cauchy stress tensor theorem). *Let us assume that* $\forall t \in I$, *the body forces* $\mathbf{f}^b$, *the density* ρ *and* $(D/Dt)\mathbf{u}$ *are all bounded functions on* Ω_t *and that the Cauchy stress vector field* $\mathbf{t}$ *is continuously differentiable with respect to the variable* $\mathbf{x}$ *for each* $\mathbf{n} \in \mathbf{S}_1$, *and continuous with respect to* $\mathbf{n}$. *Then, there exists a continuously differentiable symmetric*[1] *tensor field, called* Cauchy stress tensor,

$$\mathbf{T} : I \times \overline{\Omega_t} \to \mathbb{R}^{3\times 3}, \quad [\mathbf{T}] = \text{Ne}/\text{m}^2,$$

such that

$$\mathbf{t}(t, \mathbf{x}, \mathbf{n}) = \mathbf{T}(t, \mathbf{x}) \cdot \mathbf{n}, \quad \forall t \in I,\ \forall \mathbf{x} \in \Omega_t,\ \forall \mathbf{n} \in \mathbf{S}_1.$$

The proof is omitted. The interested reader may refer to ARIS [1962], SERRIN [1959]. Therefore, under the hypotheses of the Cauchy theorem, we have

$$\mathbf{T} \cdot \mathbf{n} = \mathbf{t}^e, \quad \text{on } \partial V_t \cap \Gamma_t^n, \tag{8.6}$$

and that the resultant of the internal forces on V_t is expressed by

$$\int_{\partial V_t} \mathbf{T} \cdot \mathbf{n}, \tag{8.7}$$

and we may rewrite the principle of conservation of linear momentum (8.5) as follows.

For all $t \in I$, on any sub-domain $V_t \subset \Omega_t$ *completely contained in* Ω_t, the following relation holds:

$$\frac{\mathrm{d}}{\mathrm{d}t} \int_{V_t} \rho \mathbf{u} = \int_{V_t} \rho \mathbf{f}^b + \int_{\partial V_t} \mathbf{T} \cdot \mathbf{n}. \tag{8.8}$$

Since ρ is constant and $\operatorname{div} \mathbf{u} = 0$, by invoking the transport formula (7.11), we obtain

$$\frac{\mathrm{d}}{\mathrm{d}t} \int_{V_t} \rho \mathbf{u} = \int_{V_t} \left(\frac{D}{Dt}(\rho \mathbf{u}) + \rho \mathbf{u} \operatorname{div} \mathbf{u} \right) = \int_{V_t} \rho \frac{D\mathbf{u}}{Dt}.$$

By using the divergence theorem and assuming that $\operatorname{div} \mathbf{T}$ is integrable, relation (8.8) becomes

$$\int_{V_t} \left[\rho \frac{D\mathbf{u}}{Dt} - \mathbf{div}\, \mathbf{T} - \rho \mathbf{f}^b \right] = 0.$$

Thanks to the arbitrariness of V_t and under the hypothesis that the terms under the integrals are continuous in space, we derive the following differential equation:

$$\rho \frac{D\mathbf{u}}{Dt} - \mathbf{div}\, \mathbf{T} = \rho \mathbf{f}^b \quad \text{in } \Omega_t. \tag{8.9}$$

[1] The symmetry of the Cauchy tensor may indeed be derived from the conservation of angular momentum.

REMARK 8.2. In deriving (8.9), we have assumed that V_t is completely contained into Ω_t. We may however extend the derivation to the case where V_t has a part of boundary in common with Γ_t^n. In that case, we should use in place of (8.8) the following:

$$\frac{\mathrm{d}}{\mathrm{d}t}\int_{V_t}\rho\mathbf{u} = \int_{\partial V_t\setminus\Gamma_t^n}\mathbf{T}\cdot\mathbf{n} + \int_{\partial V_t\cap\Gamma_t^n}\mathbf{t}^e + \int_{V_t}\rho\mathbf{f}^b. \tag{8.10}$$

Even now we would re-obtain (8.9) in view of property (8.6) of the Cauchy stress tensor, which should now be regarded as *boundary condition*.

We may note that $D\mathbf{u}/Dt$ is indeed the fluid acceleration. Referring to relation (7.5), it may be written as

$$\frac{D\mathbf{u}}{Dt} = \frac{\partial\mathbf{u}}{\partial t} + (\mathbf{u}\cdot\nabla)\mathbf{u},$$

where $(\mathbf{u}\cdot\nabla)\mathbf{u}$ is a vector whose components are

$$\big((\mathbf{u}\cdot\nabla)\mathbf{u}\big)_i = \sum_{j=1}^{3} u_j\frac{\partial u_i}{\partial x_j}, \quad i = 1,\ldots,3.$$

For ease of notation, from now on we will omit the subscript b to indicate the body force density applied to the fluid, which will be indicated just as $\mathbf{f}$.

Relation (8.9) may finally be written as

$$\rho\frac{\partial\mathbf{u}}{\partial t} + \rho(\mathbf{u}\cdot\nabla)\mathbf{u} - \mathbf{div}\,\mathbf{T} = \rho\mathbf{f}. \tag{8.11}$$

Component-wise,

$$\rho\frac{\partial u_i}{\partial t} + \rho\sum_{j=1}^{3} u_j\frac{\partial u_i}{\partial x_j} - \sum_{j=1}^{3}\frac{\partial T_{ij}}{\partial x_j} = \rho f_i^b, \quad i = 1,\ldots,3.$$

The non-linear term $\rho(\mathbf{u}\cdot\nabla)\mathbf{u}$ is called the *convective term*.

REMARK 8.3. We note the convective term may be written in the so-called *divergence form* $\mathbf{div}(\mathbf{u}\otimes\mathbf{u})$, where

$$(\mathbf{div}\,\mathbf{u}\otimes\mathbf{u})_i = \sum_{j=1}^{3}\frac{\partial}{\partial x_j}(u_i u_j), \quad i = 1,\ldots,3.$$

Indeed, thanks to the incompressibility of the fluid,

$$(\mathbf{u}\cdot\nabla)\mathbf{u} = (\mathbf{u}\cdot\nabla)\mathbf{u} + \mathbf{u}\,\mathrm{div}\,\mathbf{u} = \mathbf{div}(\mathbf{u}\otimes\mathbf{u}).$$

The momentum equation in divergence form is then

$$\rho\frac{\partial\mathbf{u}}{\partial t} + \rho\,\mathbf{div}(\mathbf{u}\otimes\mathbf{u} - \mathbf{T}) = \rho\mathbf{f}. \tag{8.12}$$

8.3. The constitutive law

In order to close the system of Eqs. (8.2) and (8.11) just derived, we need to link the Cauchy stress tensor to the kinematic quantities, and in particular, the velocity field. Such a relation, called *constitutive law*, provides a characterization of the mechanical behavior of the particular fluid under consideration.

The branch of science which studies the behavior of a moving fluid and in particular the relation between stresses and kinematic quantities is called *rheology*. We have already anticipated in the introduction that blood rheology could be complex, particularly in vessels with small size.

Here, we will assume for the fluid a *Newtonian behavior* (an approximation valid for many fluids and also for blood flow in large vessels, which is the case in our presentation). *In a Newtonian incompressible fluid*, the Cauchy stress tensor may be written as a linear function of the velocity derivatives (SERRIN [1959]), according to

$$\mathbf{T} = -P\mathbf{I} + \mu\left(\nabla\mathbf{u} + \nabla\mathbf{u}^{\mathrm{T}}\right), \tag{8.13}$$

where P is a scalar function called *pressure*, $\mathbf{I}$ is the identity matrix, μ is the *dynamic viscosity* of the fluid and is a positive quantity. The tensor

$$\mathbf{D}(\mathbf{u}) = \frac{(\nabla\mathbf{u} + \nabla\mathbf{u}^{\mathrm{T}})}{2}, \qquad D_{ij} = \frac{1}{2}\left(\frac{\partial u_i}{\partial x_j} + \frac{\partial u_j}{\partial x_i}\right), \; i = 1, \ldots, 3, \; j = 1, \ldots, 3,$$

is called the *strain rate* tensor. Then,

$$\mathbf{T} = -P\mathbf{I} + 2\mu\mathbf{D}(\mathbf{u}).$$

The term $2\mu\mathbf{D}(\mathbf{u})$ in the definition of the Cauchy stress tensor is often referred to as *viscous stress* component of the stress tensor. We have that $[P] = \mathrm{Ne}/\mathrm{m}^2$ and $[\mu] = \mathrm{kg/m\,s}$. The viscosity may vary with respect to time and space. For example, it may depend on the fluid temperature. The assumption of Newtonian fluid, however, implies that μ is *independent from kinematic quantities*. Simple models for non-Newtonian fluids, often used for blood flow simulations, express the viscosity as function of the strain rate, that is $\mu = \mu(\mathbf{D}(\mathbf{u}))$. The treatment of such cases is rather complex and will not be considered here, the interested reader may consult, for instance, RAJAGOPAL [1993], COKELET [1987].

We now recall that, if P is a scalar and $\boldsymbol{\Sigma}$ a vector field, then

$$\mathbf{div}(P\boldsymbol{\Sigma}) = \nabla P\boldsymbol{\Sigma} + P\,\mathbf{div}\,\boldsymbol{\Sigma},$$

and, therefore,

$$\mathbf{div}(P\mathbf{I}) = \nabla P\mathbf{I} + P\,\mathbf{div}\,\mathbf{I} = \nabla P.$$

The momentum equation may then be written as

$$\rho\frac{\partial\mathbf{u}}{\partial t} + \rho(\mathbf{u}\cdot\nabla)\mathbf{u} + \nabla P - 2\,\mathbf{div}\big(\mu\mathbf{D}(\mathbf{u})\big) = \rho\mathbf{f}.$$

Since ρ is constant, it is sometimes convenient to introduce the kinematic viscosity $\nu = \mu/\rho$, with $[\nu] = \mathrm{m}^2/\mathrm{s}$, and to write

$$\frac{\partial \mathbf{u}}{\partial t} + (\mathbf{u}\cdot\nabla)\mathbf{u} + \nabla p - 2\,\mathbf{div}\big(\nu\mathbf{D}(\mathbf{u})\big) = \mathbf{f}, \tag{8.14}$$

where $p = P/\rho$ is a scaled pressure (with $[p] = \mathrm{m}^2/\mathrm{s}^2$).

REMARK 8.4. Under the additional hypothesis that ν is constant, the momentum equation may be further elaborated by considering that

$$\mathbf{div}\,\nabla\mathbf{u} = \Delta\mathbf{u},$$
$$\mathbf{div}\,\nabla\mathbf{u}^{\mathrm{T}} = \nabla(\operatorname{div}\mathbf{u}) = (\text{by relation (8.2)}) = \mathbf{0}.$$

Consequently, the momentum equation for an incompressible Newtonian fluid with constant viscosity may be written in the alternative form

$$\frac{\partial \mathbf{u}}{\partial t} + (\mathbf{u}\cdot\nabla)\mathbf{u} + \nabla p - \nu\Delta\mathbf{u} = \mathbf{f}. \tag{8.15}$$

However, for reasons that will appear clear later on (and that have to see with the different natural boundary conditions associated with the two formulations), we prefer to use the Navier–Stokes equations in the form (8.14), even when considering a constant viscosity.

9. The Navier–Stokes equations

The set of differential equations formed by the continuity equation and the momentum equations in the form derived in the previous section provides the *Navier–Stokes equations* for incompressible fluids.

They are, in particular, valid on any fixed spatial domain Ω which is for all times of interest inside the portion of space filled by the fluid, i.e., $\Omega \subset \Omega_t$. Indeed, in most cases, as with the flow around a car or an aeroplane, the flow motion is studied in a fixed domain Ω (usually called computational domain) embodying the region of interest. We will see in Section 18 that this is not possible anymore when considering the fluid–structure interaction problem arising when blood is flowing in a large artery.

Yet, before addressing this more complex situation, we will analyze the Navier–Stokes equations in a fixed domain, that is, we will consider, for any $t \in I$, the system of equations

$$\begin{aligned}
&\frac{\partial \mathbf{u}}{\partial t} + (\mathbf{u}\cdot\nabla)\mathbf{u} + \nabla p - 2\,\mathbf{div}\big(\nu\mathbf{D}(\mathbf{u})\big) = \mathbf{f}, \quad \text{in } \Omega,\\
&\operatorname{div}\mathbf{u} = 0, \quad \text{in } \Omega.
\end{aligned} \tag{9.1}$$

Furthermore, we need to prescribe the initial status of the fluid velocity, for instance

$$\mathbf{u}(t = t_0, \mathbf{x}) = \mathbf{u}_0(\mathbf{x}), \quad \mathbf{x} \in \Omega. \tag{9.2}$$

The principal unknowns are the velocity $\mathbf{u}$ and the "scaled" pressure $p = P/\rho$.

Let us take a practical case-study, namely the blood flow in an artery, for example the carotid (ref. Fig. 2.3), which we will here consider rigid. We proceed by identifying the area of interest, which may be the carotid sinus, and a domain Ω which will contain that area and which extends into the vessels up to a certain distance. For obvious practical reasons we will need to "truncate" the domain at certain sections. Inside such domain, the Navier–Stokes equations are valid, yet in order to solve them we need to provide appropriate boundary conditions.

9.1. Boundary conditions for the Navier–Stokes equations

The Navier–Stokes equations must be supplemented by proper boundary conditions that allow the determination of the velocity field up to the boundary of the computational domain Ω. The more classical boundary conditions which are mathematically compatible with the Navier–Stokes equations are:

(1) *Applied stresses* (or *Neumann* boundary condition). We have already faced this condition when discussing the Cauchy principle. With the current definition for the Cauchy stresses it becomes

$$\mathbf{T}\cdot\mathbf{n} = -P\mathbf{n} + 2\mu\mathbf{D}(\mathbf{u})\cdot\mathbf{n} = \mathbf{t}^e \quad \text{on } \Gamma^n \subset \partial\Omega, \tag{9.3}$$

where Γ^n is a measurable subset (possibly empty) of the whole boundary $\partial\Omega$.

(2) *Prescribed velocity* (or *Dirichlet* boundary condition). A given velocity field is imposed on Γ^d, a measurable subset of $\partial\Omega$ (which may be empty). This means that a vector field

$$\mathbf{g}: I \times \Gamma^d \to \mathbb{R}^3$$

is prescribed and we impose that

$$\mathbf{u} = \mathbf{g} \quad \text{on } \Gamma^d.$$

Since $\operatorname{div}\mathbf{u} = 0$ in Ω, it must be noted that if $\Gamma^d = \partial\Omega$ then at any time $\mathbf{g}$ must satisfy the following compatibility condition:

$$\int_{\partial\Omega} \mathbf{g}\cdot\mathbf{n} = 0. \tag{9.4}$$

Clearly, for a proper boundary conditions specification we must have $\Gamma^n \cup \Gamma^d = \partial\Omega$.

The conditions to apply are normally driven by physical considerations. For instance, for a viscous fluid ($\mu > 0$) like the one we are considering here, physical considerations lead to impose the homogeneous Dirichlet condition $\mathbf{u} = \mathbf{0}$ at a solid fixed boundary. When dealing with an "artificial boundary", that is a boundary which truncates the space occupied by the fluid (for computational reasons) the choice of appropriate conditions is often more delicate and should in any case guarantee the well-posedness of the resulting differential problem.

For example, for the flow field inside a 2D model for the carotid artery such as the one shown in Fig. 9.1, we could impose a Dirichlet boundary condition on Γ^{in}, by prescribing a velocity field $\mathbf{g}$.

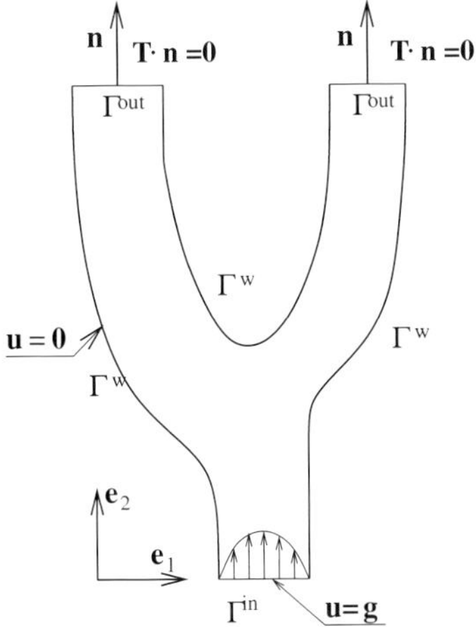

FIG. 9.1. A possible boundary subdivision for the flow in a carotid bifurcation.

On the "wall" boundary Γ^w, which is in this case assumed to be fixed, we will impose homogeneous Dirichlet conditions, that is $\mathbf{u} = \mathbf{0}$ on Γ^w. When we will consider the coupled problem between fluid and vessel wall, Γ^w will be moving, hence the homogeneous Dirichlet condition will be replaced by $\mathbf{u} = \mathbf{w}$, where $\mathbf{w}$ is the wall velocity.

At the exit Γ^{out}, we could, for instance, impose homogeneous Neumann conditions, i.e., relation (9.3) with $\mathbf{t}^e = \mathbf{0}$. For the case illustrated in Fig. 9.1 and with that choice of coordinate basis, it becomes (derivation left as exercise)

$$\mu\left(\frac{\partial u_1}{\partial x_2} + \frac{\partial u_2}{\partial x_1}\right) = 0, \qquad -P + 2\mu\frac{\partial u_2}{\partial x_2} = 0.$$

REMARK 9.1. We anticipate the fact (without providing the proof) that this choice of boundary conditions, with the hypothesis that at Γ^{out} the velocity satisfies everywhere the condition $\mathbf{u} \cdot \mathbf{n} > 0$, is sufficient to guarantee that the solution of the Navier–Stokes problem exists and is continuously dependent from the data (initial solution, boundary conditions, forcing terms), provided that the initial data and forcing term are sufficiently small.

Unfortunately, the homogeneous Neumann condition, which indeed would simulate a discharge into the open air, is rather unphysical for the case of a human vessel. As a matter of fact, it neglects completely the presence of the remaining part of the circulatory system. The difficulty in devising proper boundary conditions for this specific problem was already mentioned in Section 2 of these notes. The matter is still open and is the subject of active research. A possibility is provided by coupling the Navier–Stokes equations on the section of the arterial tree of interest with reduced models, like the one that will be presented in Section 20, which are able to represent, though in a simplified way, the presence of the remaining part of the circulatory system. Techniques of this type has been used and analysed in FORMAGGIA, NOBILE, QUARTERONI and VENEZIANI [1999], FORMAGGIA, GERBEAU, NOBILE and QUARTERONI [2001].

CHAPTER III

10. The incompressible Navier–Stokes equations and their approximation

In this section we introduce the weak formulation of the Navier–Stokes equations for constant density (incompressible) fluids. Then, we address basic issues concerning the approximation of these equations in the context of the finite element method.

10.1. Some functional spaces

For the following discussion we need to introduce some Sobolev spaces for vector functions. We assume that the reader is already acquainted with the main definitions and results on Sobolev spaces in one dimension. A simple introduction is provided in REDDY [1998]. For a deeper insight see, for instance, BREZIS [1983].

We will indicate with $\mathbf{L}^p(\Omega)$ $(1 \leqslant p \leqslant \infty)$ the space of vector functions $\mathbf{f}: \Omega \to \mathbb{R}^N$ (with $N = 2$ or 3) whose components belong to $L^p(\Omega)$. Its norm is

$$\|\mathbf{f}\|_{\mathbf{L}^p(\Omega)} = \left(\sum_{i=1}^{N} \|f_i\|^p_{L^p(\Omega)}\right)^{1/p}, \quad 1 \leqslant p < \infty,$$

and

$$\|\mathbf{f}\|_{\mathbf{L}^\infty(\Omega)} = \inf\bigl\{C \in \mathbb{R}\colon |f_i| \leqslant C,\ i = 1, \ldots, N, \text{ a.e. in } \Omega\bigr\},$$

where "a.e." stands for "almost everywhere". We will use the same notation for tensor fields, i.e., we will also indicate with $\mathbf{L}^p(\Omega)$ the space of tensor fields $\mathbf{T}: \Omega \to \mathbb{R}^{N\times N}$ whose components belongs to $L^p(\Omega)$. In this case

$$\|\mathbf{T}\|_{\mathbf{L}^p(\Omega)} = \left(\sum_{i=1}^{N}\sum_{j=1}^{N} \|T_{ij}\|^p_{L^p(\Omega)}\right)^{1/p}, \quad 1 \leqslant p < \infty.$$

Analogously, a vector (or a tensor) function $\mathbf{f}$ belongs to $\mathbf{H}^m(\Omega)$ if all its components belong to $H^m(\Omega)$, and we have

$$\|\mathbf{f}\|_{\mathbf{H}^m(\Omega)} = \left(\sum_{i=1}^{N} \|f_i\|^2_{H^m(\Omega)}\right)^{1/2},$$

while its semi-norm is

$$|\mathbf{f}|_{\mathbf{H}^m(\Omega)} = \left(\sum_{i=1}^{N} |f_i|^2_{H^m(\Omega)}\right)^{1/2}.$$

It is understood that, when $m = 0$,

$$\mathbf{H}^0(\Omega) \equiv \mathbf{L}^2(\Omega).$$

When equipped with the following scalar product:

$$(\mathbf{f}, \mathbf{g})_{\mathbf{H}^m(\Omega)} = \sum_{i=1}^{N} (f_i, g_i)_{H^m(\Omega)}, \quad \mathbf{f}, \mathbf{g} \in \mathbf{H}^m(\Omega),$$

the space $\mathbf{H}^m(\Omega)$ is a Hilbert space.

To ease notation, we will often use the following short-hand notation for the L^2 scalar products:

$$(\mathbf{v}, \mathbf{w}) \equiv (\mathbf{v}, \mathbf{w})_{\mathbf{L}^2(\Omega)}, \qquad (p, q) \equiv (p, q)_{L^2(\Omega)}.$$

We note that the L^2 scalar product of two tensor fields $\mathbf{T}$ and $\mathbf{G}$ belonging to $\mathbf{L}^2(\Omega)$ is defined as

$$(\mathbf{T}, \mathbf{G}) \equiv (\mathbf{T}, \mathbf{G})_{\mathbf{L}^2(\Omega)} = \int_\Omega \mathbf{T} : \mathbf{G} = \sum_{i=1}^{N} \sum_{j=1}^{N} \int_\Omega T_{ij} G_{ij}.$$

For our purposes we will usually have $m = 1$. In that case we have the equality

$$\|\mathbf{f}\|^2_{\mathbf{H}^1(\Omega)} = \|\mathbf{f}\|^2_{\mathbf{L}^2(\Omega)} + \|\nabla \mathbf{f}\|^2_{\mathbf{L}^2(\Omega)}.$$

We often utilise the space $\mathbf{H}^1_0(\Omega)$ defined as

$$\mathbf{H}^1_0(\Omega) = \{\mathbf{v} \in \mathbf{H}^1(\Omega)\colon \ \mathbf{v}|_{\partial\Omega} = 0\}.$$

We will consider *bounded* domains Ω with regular (i.e., *Lipschitz continuous*) boundary $\partial\Omega$, so that both the Sobolev embedding theorems in $\mathbb{R}^N$ and the *Green* integration formula hold. Some important results are here recalled, without providing the demonstration, which may be found in LIONS and MAGENES [1968] or BREZIS [1983].

THEOREM 10.1 (Sobolev embeddings (simplified form)). *Let Ω be a bounded domain of $\mathbb{R}^N$ with Lipschitz continuous boundary. The following properties hold*:

$$\begin{cases} \textit{If } 0 \leqslant s < \dfrac{N}{2}, & \mathbf{H}^s(\Omega) \hookrightarrow \mathbf{L}^p(\Omega), \quad p = \dfrac{2N}{N-2s}, \\ \textit{If } s = \dfrac{N}{2}, & \mathbf{H}^s(\Omega) \hookrightarrow \mathbf{L}^q(\Omega), \quad 2 \leqslant q < \infty, \\ \textit{If } s > \dfrac{N}{2}, & \mathbf{H}^s(\Omega) \hookrightarrow [C^0(\overline{\Omega})]^N, \end{cases}$$

where $A \hookrightarrow B$ means that A is included in B with continuous embedding.

THEOREM 10.2 (Green integration formula). *Let Ω be a bounded domain of $\mathbb{R}^N$ with Lipschitz continuous boundary and let* $\mathbf{n}$ *denote the unit outer normal along $\partial\Omega$. Let*

$u, v \in H^1(\Omega)$, then the integral

$$\int_{\partial\Omega} u v n_i$$

exists and is finite for each component n_i of **n**. *In addition we have*

$$\int_\Omega \frac{\partial u}{\partial x_i} v = -\int_\Omega u \frac{\partial v}{\partial x_i} + \int_{\partial\Omega} u v n_i, \quad i = 1, \ldots, N.$$

LEMMA 10.1 (Poincaré inequality – multidimensional case). *Let $f : \mathbb{R}^N \to \mathbb{R}$ be a function of $\mathbf{H}^1(\Omega)$, with $f = 0$ on $\Gamma \subset \partial\Omega$ of strictly positive measure. Then there exists a positive constant C_P (depending only on the domain Ω and on Γ), such that*

$$\|f\|_{L_2(\Omega)} \leqslant C_P \|\nabla f\|_{\mathbf{L}_2(\Omega)}. \tag{10.1}$$

LEMMA 10.2. *Let Ω be a bounded and connected subset of $\mathbb{R}^N$, where $N = 2$ or 3. Furthermore, let us assume that the velocity field $\mathbf{u} \in \mathbf{H}^1(\Omega)$ vanishes on $\Gamma \subset \partial\Omega$ of strictly positive measure. Then, there exists a constant $C_K > 0$ so that the following inequality holds*:

$$\int_\Omega \mathbf{D}(\mathbf{u}) : \mathbf{D}(\mathbf{u}) \geqslant C_K \|\nabla \mathbf{u}\|^2_{\mathbf{L}^2(\Omega)}. \tag{10.2}$$

This theorem is a consequence of the *Korn inequality*, whose precise statement may be found, for instance, in CIARLET [1988], DUVAUT and LIONS [1976].

LEMMA 10.3 (Gronwall lemma). *Let f be a non-negative function which is integrable in $I = (t_0, t_1)$ and g and ϕ be two continuous functions in I, with g non-decreasing. If*

$$\phi(t) \leqslant g(t) + \int_{t_0}^t f(\tau)\phi(\tau)\,\mathrm{d}\tau \quad \forall t \in I, \tag{10.3}$$

then

$$\phi(t) \leqslant g(t) \exp \int_{t_0}^t f(\tau)\,\mathrm{d}\tau \quad \forall t \in I. \tag{10.4}$$

11. Weak form of Navier–Stokes equations

The incompressible Navier–Stokes equations read

$$\frac{\partial \mathbf{u}}{\partial t} + (\mathbf{u}\cdot\nabla)\mathbf{u} + \nabla p - 2\,\mathbf{div}\big(\nu \mathbf{D}(\mathbf{u})\big) = \mathbf{f}, \quad \text{in } \Omega,\ t \in I, \tag{11.1a}$$

$$\operatorname{div} \mathbf{u} = 0, \quad \text{in } \Omega,\ t \in I, \tag{11.1b}$$

$$\mathbf{u} = \mathbf{u}_0, \quad \text{in } \Omega,\ t = t_0. \tag{11.1c}$$

We assume that ν is a bounded strictly positive function, precisely we assume that there exist two constants $\nu_0 > 0$ and $\nu_1 > 0$ such that $\forall t \in I$,

$$\nu_0 \leqslant \nu \leqslant \nu_1 \quad \text{almost everywhere in } \Omega.$$

We consider the case in which the system of differential equations (11.1) is equipped with the following boundary conditions:

$$\mathbf{u}=\mathbf{g} \quad \text{on } \Gamma^d,\ t\in I, \tag{11.2a}$$

$$-p\mathbf{n}+2\nu\mathbf{D}(\mathbf{u})\cdot\mathbf{n}=\mathbf{h} \quad \text{on } \Gamma^n,\ t\in I, \tag{11.2b}$$

We have indicated with Γ^d and Γ^n the portions of $\partial\Omega$ where Dirichlet and Neumann boundary conditions are applied, respectively. We must have $\Gamma^d\cup\Gamma^n=\partial\Omega$.

REMARK 11.1. If $\Gamma^d=\partial\Omega$ we call the problem formed by (11.1) and (11.2) a *Dirichlet problem*. We will instead use the term *Neumann problem* when $\Gamma^n=\partial\Omega$. The conditions $\mathbf{g}=\mathbf{0}$ and $\mathbf{h}=\mathbf{0}$ are called *homogeneous boundary conditions*.

In the case of a Dirichlet problem, the boundary datum has to satisfy the following compatibility relation for all $t\in I$:

$$\int_{\partial\Omega}\mathbf{g}\cdot\mathbf{n}=0.$$

REMARK 11.2. For the problem at hand, we normally have $\mathbf{f}=\mathbf{0}$, since the only external force which one may eventually consider in blood flow is the gravity force. Even in this case, we may still adopt the Navier–Stokes equations with $\mathbf{f}=\mathbf{0}$ by replacing p with $p^*(t,\mathbf{x})=p(t,\mathbf{x})+gz(\mathbf{x})\mathbf{e}_z$, where g is the gravity acceleration, $\mathbf{e}_z$ the unit vector defining the vertical direction (upwardly oriented) and $z(\mathbf{x})$ the (known) quota of point $\mathbf{x}$ with respect to a reference horizontal plane. Yet, for the sake of completeness, many of the derivations of this as well as the following sections refer to the general case $\mathbf{f}\neq\mathbf{0}$.

The *weak form* of the Navier–Stokes equations is (formally) obtained by taking the scalar product of the momentum equations with a vector function $\mathbf{v}$ belonging to a functional space $\mathbf{V}$ (called *test function space*), which will be better specified later on, integrating over Ω and applying the Green integration formula. We operate similarly on the continuity equation, by multiplying it by a function $q\in Q$ and integrating. Also the space Q will be specified at a later stage.

We formally obtain

$$\begin{aligned}&\left(\frac{\partial\mathbf{u}}{\partial t},\mathbf{v}\right)+\big((\mathbf{u}\cdot\nabla)\mathbf{u},\mathbf{v}\big)+2\int_\Omega \nu\mathbf{D}(\mathbf{u}):\mathbf{D}(\mathbf{v})-(p,\operatorname{div}\mathbf{v})\\ &\quad=(\mathbf{f},\mathbf{v})+\int_{\partial\Omega}\mathbf{v}\cdot\big(2\nu\mathbf{D}(\mathbf{u})\cdot\mathbf{n}-p\mathbf{n}\big),\\ &(\operatorname{div}\mathbf{u},q)=0.\end{aligned}$$

We have exploited the identity

$$\int_\Omega \nu\mathbf{D}(\mathbf{u}):\nabla\mathbf{v}=\int_\Omega \nu\mathbf{D}(\mathbf{u}):\mathbf{D}(\mathbf{v}),$$

which derives from the symmetry of the tensor $\mathbf{D}(\mathbf{u})$.

The boundary term may now be split into two parts,

$$\int_{\partial\Omega} \mathbf{v}\cdot\big(2\nu\mathbf{D}(\mathbf{u})\cdot\mathbf{n} - p\mathbf{n}\big) = \int_{\Gamma^d} \mathbf{v}\cdot\big(2\nu\mathbf{D}(\mathbf{u})\cdot\mathbf{n} - p\mathbf{n}\big) + \int_{\Gamma^n} \mathbf{v}\cdot\mathbf{h}.$$

We note that the contribution from the Neumann boundary is now a *given data*, while contribution from the Dirichlet boundary can be eliminated by appropriately choosing the test space $\mathbf{V}$.

By inspection, we may recognise that all terms make sense if we choose as test function spaces

$$\mathbf{V} = \big\{\mathbf{v}\in\mathbf{H}^1(\Omega)\colon\ \mathbf{v}|_{\Gamma^d} = \mathbf{0}\big\},$$

$$Q = \left\{q\in L^2(\Omega)\colon\ \text{with} \int_\Omega q = 0 \text{ if } \Gamma^d = \partial\Omega\right\},$$

and if we seek, at each time t, the velocity in

$$\mathbf{V}_\mathbf{g} = \big\{\mathbf{u}\in\mathbf{H}^1(\Omega)\colon\ \mathbf{u}|_{\Gamma^d} = \mathbf{g}\big\}$$

and the pressure in Q.

REMARK 11.3. The request that Q is formed by functions with zero mean on Ω when we treat a Dirichlet problem derives from the fact that in such a case the pressure is determined only up to a constant, as it appears in the equations only through its gradient. To compute a unique value for the pressure it is then necessary to fix the constant. This is obtained by the zero-mean constraint.

Finally, the *weak form* of the Navier–Stokes problem (11.1) and (11.2), reads:

Find, $\forall t\in I$, $\mathbf{u}(t)\in\mathbf{V}_\mathbf{g}$ *and* $p(t)\in Q$ *such that*

$$\begin{cases} \left(\dfrac{\partial\mathbf{u}}{\partial t}, \mathbf{v}\right) + a(\mathbf{u},\mathbf{v}) + c(\mathbf{u},\mathbf{u},\mathbf{v}) + b(\mathbf{v},p) = (\mathbf{f},\mathbf{v}) + \int_{\Gamma^n}\mathbf{v}\cdot\mathbf{h}, & \forall\mathbf{v}\in\mathbf{V}, \\ b(\mathbf{u},q) = 0, \quad \forall q\in Q, \end{cases} \tag{11.3}$$

where

$$a(\mathbf{u},\mathbf{v}) = 2\int_\Omega \nu\mathbf{D}(\mathbf{u}):\mathbf{D}(\mathbf{v}), \tag{11.4}$$

$$c(\mathbf{w},\mathbf{u},\mathbf{v}) = \int_\Omega (\mathbf{w}\cdot\nabla)\mathbf{u}\cdot\mathbf{v}, \tag{11.5}$$

$$b(\mathbf{v},p) = -\int_\Omega p\,\operatorname{div}\mathbf{v}. \tag{11.6}$$

11.1. The homogeneous Dirichlet problem

In this section we will focus on the homogeneous Dirichlet problem, that is the case when $\Gamma^d = \partial\Omega$ and $\mathbf{g} = \mathbf{0}$ in (11.2a). Therefore,

$$\mathbf{V} = \mathbf{H}_0^1(\Omega), \qquad Q = L_0^2(\Omega) = \left\{q\in L^2(\Omega),\ \int_\Omega q = 0\right\} \tag{11.7}$$

and the weak form reads:

Find, $\forall t \in I$, $\mathbf{u}(t) \in \mathbf{V}$ and $p(t) \in Q$ such that

$$\begin{cases} \left(\dfrac{\partial \mathbf{u}}{\partial t}, \mathbf{v}\right) + a(\mathbf{u}, \mathbf{v}) + c(\mathbf{u}, \mathbf{u}, \mathbf{v}) + b(\mathbf{v}, p) = (\mathbf{f}, \mathbf{v}), & \forall \mathbf{v} \in \mathbf{V}, \\ b(\mathbf{u}, q) = 0, \quad \forall q \in Q. \end{cases} \tag{11.8}$$

LEMMA 11.1. *The forms $a : \mathbf{V} \times \mathbf{V} \to \mathbb{R}$, $c : \mathbf{V} \times \mathbf{V} \times \mathbf{V} \to \mathbb{R}$ and $b : \mathbf{V} \times Q \to \mathbb{R}$ are continuous with respect to their arguments. In addition, $a(\cdot, \cdot)$ is coercive, i.e., $\exists \alpha > 0$ such that*

$$a(\mathbf{v}, \mathbf{v}) \geqslant \alpha \|\mathbf{v}\|^2_{\mathbf{H}^1(\Omega)}, \quad \forall \mathbf{v} \in \mathbf{V}.$$

PROOF. The continuity of the bilinear forms a and b is an immediate consequence of the Cauchy–Schwarz inequality. Indeed, $\forall \mathbf{u}, \mathbf{v} \in V$ and $\forall q \in Q$,

$$\begin{aligned} |a(\mathbf{u}, \mathbf{v})| &\leqslant \nu_1 |\mathbf{u}|_{\mathbf{H}^1(\Omega)} |\mathbf{v}|_{\mathbf{H}^1(\Omega)} \leqslant \nu_1 \|\mathbf{u}\|_{\mathbf{H}^1(\Omega)} \|\mathbf{v}\|_{\mathbf{H}^1(\Omega)}, \\ |b(\mathbf{u}, p)| &\leqslant \|\operatorname{div} \mathbf{u}\|_{L^2(\Omega)} \|p\|_{L^2(\Omega)} \leqslant \|\mathbf{u}\|_{\mathbf{H}^1(\Omega)} \|p\|_{L^2(\Omega)}. \end{aligned}$$

For the tri-linear form c we first have to note that thanks to the Sobolev embedding theorem $\mathbf{H}^1(\Omega) \hookrightarrow \mathbf{L}^6(\Omega)$ (as $N = 2, 3$) and consequently $\mathbf{H}^1(\Omega) \hookrightarrow \mathbf{L}^4(\Omega)$. Then, $\mathbf{wu} \in \mathbf{L}^2(\Omega)$, and considering the expression of $c(\cdot, \cdot, \cdot)$ component-wise, we have

$$\int_\Omega w_i \frac{\partial u_k}{\partial x_i} v_k \leqslant \|w_i v_k\|_{L^2(\Omega)} \left\| \frac{\partial u_k}{\partial x_i} \right\|_{L^2(\Omega)} \leqslant \|w_i\|_{L^4(\Omega)} \|v_k\|_{L^4(\Omega)} \left\| \frac{\partial u_k}{\partial x_i} \right\|_{L^2(\Omega)}.$$

Then

$$\begin{aligned} \int_\Omega w_i \frac{\partial u_k}{\partial x_i} v_k &\leqslant C \|w_i\|_{H^1(\Omega)} \left\| \frac{\partial u_k}{\partial x_i} \right\|_{L^2(\Omega)} \|v_k\|_{H^1(\Omega)} \\ &\leqslant C \|w_i\|_{H^1(\Omega)} |u_k|_{H^1(\Omega)} \|v_k\|_{H^1(\Omega)} \\ &\leqslant C \|w_i\|_{H^1(\Omega)} \|u_k\|_{H^1(\Omega)} \|v_k\|_{H^1(\Omega)}, \end{aligned} \tag{11.9}$$

where C is a positive constant.

It follows that, $\forall \mathbf{u}, \mathbf{v}, \mathbf{v} \in \mathbf{V}$,

$$c(\mathbf{w}, \mathbf{u}, \mathbf{v}) \leqslant C_1 \|\mathbf{w}\|_{\mathbf{H}^1(\Omega)} \|\mathbf{u}\|_{\mathbf{H}^1(\Omega)} \|\mathbf{v}\|_{\mathbf{H}^1(\Omega)},$$

by which the continuity of the tri-linear form is proved (C_1 is a positive constant).

The coercivity of the linear form a derives from inequalities (10.1) and (10.2), since

$$a(\mathbf{v}, \mathbf{v}) \geqslant 2\nu_0 \int_\Omega \mathbf{D}(\mathbf{v}) : \mathbf{D}(\mathbf{v}) \geqslant 2\nu_0 C_K |\mathbf{v}|^2_{\mathbf{H}^1(\Omega)} \geqslant \alpha \|\mathbf{v}\|^2_{\mathbf{H}^1(\Omega)}, \quad \forall \mathbf{v} \in \mathbf{V}, \tag{11.10}$$

with $\alpha = (2\nu_0 C_K)/(C_P^2 + 1)$, being C_P and C_K the constants in (10.1) and (10.2), respectively. □

We now introduce the space

$$\mathbf{V}_{\text{div}} = \{\mathbf{v} \in V : \operatorname{div} \mathbf{v} = 0 \text{ a.e. in } \Omega\}.$$

THEOREM 11.1. *If* $\mathbf{u}$ *is a solution of the weak formulation* (11.8), *then* $\mathbf{u}(t) \in \mathbf{V}_{\text{div}}$ *for all* $t \in I$ *and it satisfies*

$$\left(\frac{\partial \mathbf{u}}{\partial t}, \mathbf{v}\right) + a(\mathbf{u}, \mathbf{v}) + c(\mathbf{u}, \mathbf{u}, \mathbf{v}) = (\mathbf{f}, \mathbf{v}), \quad \forall \mathbf{v} \in \mathbf{V}_{\text{div}},\ t \in I. \tag{11.11}$$

Conversely, if, $\forall t \in I$, $\mathbf{u}(t) \in \mathbf{V}_{\text{div}}$ *is a solution of* (11.11) *and* $\partial \mathbf{u}/\partial t \in \mathbf{L}^2(\Omega)$, *then there exists a unique* $p \in Q$ *such that* $(\mathbf{u}, p)$ *satisfies* (11.8).

PROOF. The first part of the proof is trivial. If $\mathbf{u}$ satisfies (11.8) then it belongs to $\mathbf{V}_{\text{div}}$ and it satisfies (11.11), since $\mathbf{V}_{\text{div}} \subset \mathbf{V}$.

The demonstration of the inverse implication requires first to state the following result.

LEMMA 11.2. *Let* Ω *be a domain of* $\mathbb{R}^N$ *and let* $L \in \mathbf{V}'$. *Then* $L(\mathbf{v}) = 0$, $\forall \mathbf{v} \in \mathbf{V}_{\text{div}}$ *if and only if there exists a function* $p \in L^2(\Omega)$ *such that*

$$L(\mathbf{v}) = (p, \operatorname{div} \mathbf{v}), \quad \forall \mathbf{v} \in \mathbf{V}.$$

For the proof see Lemma 2.1 of GIRAULT and RAVIART [1986].

The application L defined as

$$L(\mathbf{v}) = \left(\frac{\partial \mathbf{u}}{\partial t}, \mathbf{v}\right) + a(\mathbf{u}, \mathbf{v}) + c(\mathbf{u}, \mathbf{u}, \mathbf{v}) - (\mathbf{f}, \mathbf{v}), \quad \forall \mathbf{v} \in \mathbf{V},$$

belongs to $\mathbf{V}'$, being a linear continuous functional on $\mathbf{V}$. We can therefore apply Lemma 11.2 and obtain the desired result. □

12. An energy inequality for the Navier–Stokes equations

We now prove an energy inequality for problem (11.8), by which we may assess a continuous dependence of the solution from the given data.

THEOREM 12.1 (Energy inequalities). *Let* $\mathbf{u}(t) \in \mathbf{V}_{\text{div}}$ *be a solution of* (11.8), $\forall t \in I$. *Then the following inequalities hold*:

$$\|\mathbf{u}(t)\|^2_{\mathbf{L}^2(\Omega)} + C_1 \int_0^t \|\nabla \mathbf{u}(\tau)\|^2_{\mathbf{L}^2(\Omega)} \,\mathrm{d}\tau \leqslant \left(\|\mathbf{u}_0\|^2_{\mathbf{L}^2(\Omega)} + \int_0^t \|\mathbf{f}(\tau)\|^2_{\mathbf{L}^2(\Omega)} \,\mathrm{d}\tau\right) e^t,$$

where $C_1 = 4\nu_0 C_K$, *and*

$$\|\mathbf{u}(t)\|^2_{\mathbf{L}^2(\Omega)} + C_2 \int_0^t \|\nabla \mathbf{u}(\tau)\|^2_{\mathbf{L}^2(\Omega)} \,\mathrm{d}\tau \leqslant \|\mathbf{u}_0\|^2_{\mathbf{L}^2(\Omega)} + \frac{C_P}{C_2} \int_0^t \|\mathbf{f}(\tau)\|^2_{\mathbf{L}^2(\Omega)} \,\mathrm{d}\tau,$$

where $C_2 = 2\nu_0 C_K$. *Here,* C_K *and* C_P *are the constants in the Poincaré inequality* (10.1) *and in* (10.2), *respectively.*

We first prove the following result.

LEMMA 12.1. *If* $\mathbf{u}$ *is a solution of* (11.8) *then* $c(\mathbf{u},\mathbf{u},\mathbf{u})=0$.

PROOF. It follows from the Green formula and the fact that $\mathbf{u}|_{\partial\Omega}=\mathbf{0}$. Indeed,

$$c(\mathbf{u},\mathbf{u},\mathbf{u})=\int_\Omega(\mathbf{u}\cdot\nabla)\mathbf{u}\cdot\mathbf{u}=\int_\Omega\frac{1}{2}\nabla\big(|\mathbf{u}|^2\big)\cdot\mathbf{u}$$
$$=-\frac{1}{2}\int_\Omega|\mathbf{u}|^2\operatorname{div}\mathbf{u}+\frac{1}{2}\int_{\partial\Omega}|\mathbf{u}|^2\mathbf{u}\cdot\mathbf{n}.$$

Now, the last integral is zero since $\mathbf{u}=0$ on $\partial\Omega$. Moreover, for the same reason

$$\int_\Omega\operatorname{div}\mathbf{u}=\int_{\partial\Omega}\mathbf{u}\cdot\mathbf{n}=0.$$

Then, if we set

$$c=\int_\Omega|\mathbf{u}|^2,$$

we have

$$\int_\Omega|\mathbf{u}|^2\operatorname{div}\mathbf{u}=\int_\Omega|\mathbf{u}|^2\operatorname{div}\mathbf{u}-c\int_\Omega\operatorname{div}\mathbf{u}=\int_\Omega\big(|\mathbf{u}|^2-c\big)\operatorname{div}\mathbf{u}$$
$$=b\big(\mathbf{u},\big(|\mathbf{u}|^2-c\big)\big)=0,$$

where the last equality is obtained since $(|\mathbf{u}|^2-c)\in Q$ and $b(\mathbf{u},q)=0,\ \forall q\in Q$. □

We now give the demonstration of Theorem 12.1.

PROOF. For all fixed t, take $\mathbf{v}=\mathbf{u}(t)$ in the momentum equation of (11.11). We have

$$\frac{1}{2}\frac{\mathrm{d}}{\mathrm{d}t}\|\mathbf{u}\|^2_{\mathbf{L}^2(\Omega)}+c(\mathbf{u},\mathbf{u},\mathbf{u})+b(\mathbf{u},p)+a(\mathbf{u},\mathbf{u})=(\mathbf{f},\mathbf{u}).\tag{12.1}$$

Then,

$$\frac{1}{2}\frac{\mathrm{d}}{\mathrm{d}t}\|\mathbf{u}\|^2_{\mathbf{L}^2(\Omega)}+a(\mathbf{u},\mathbf{u})=(\mathbf{f},\mathbf{u}).$$

Now, thanks to (10.2),

$$a(\mathbf{u},\mathbf{u})=2\int_\Omega\nu\mathbf{D}(\mathbf{u}):\mathbf{D}(\mathbf{u})\geqslant 2\nu_0C_K\|\nabla\mathbf{u}\|^2_{\mathbf{L}^2(\Omega)},$$

then

$$\frac{\mathrm{d}}{\mathrm{d}t}\|\mathbf{u}\|^2_{\mathbf{L}^2(\Omega)}+4\nu_0C_K\|\nabla\mathbf{u}\|^2_{\mathbf{L}^2(\Omega)}\leqslant 2(\mathbf{f},\mathbf{u})\leqslant\frac{1}{2\varepsilon}\|\mathbf{f}\|^2_{\mathbf{L}^2(\Omega)}+2\varepsilon\|\mathbf{u}\|^2_{\mathbf{L}^2(\Omega)},\tag{12.2}$$

for any $\varepsilon>0$. By choosing $\varepsilon=1/2$ and integrating between t_0 and t, we have

$$\big\|\mathbf{u}(t)\big\|^2_{\mathbf{L}^2(\Omega)}+4\nu_0C_K\int_{t_0}^t\big\|\nabla\mathbf{u}(\tau)\big\|^2_{\mathbf{L}^2(\Omega)}\,\mathrm{d}\tau$$
$$\leqslant\int_{t_0}^t\big\|\mathbf{f}(\tau)\big\|^2_{\mathbf{L}^2(\Omega)}\,\mathrm{d}\tau+\int_{t_0}^t\big\|\mathbf{u}(\tau)\big\|^2_{\mathbf{L}^2(\Omega)}\,\mathrm{d}\tau+\|\mathbf{u}_0\|^2_{\mathbf{L}^2(\Omega)}.$$

We apply Gronwall lemma (Lemma 10.3) by identifying

$$\left\|\mathbf{u}(t)\right\|^2_{\mathbf{L}^2(\Omega)} + 4\nu_0 C_K \int_{t_0}^{t} \left\|\nabla\mathbf{u}(\tau)\right\|^2_{\mathbf{L}^2(\Omega)} \,\mathrm{d}\tau$$

with $\phi(t)$, obtaining the first inequality.

By using instead the Poincaré inequality on the last term of (12.2), and by taking $\varepsilon = (\nu_0 C_K)/C_P^2$, we obtain

$$\frac{\mathrm{d}}{\mathrm{d}t}\|\mathbf{u}\|^2_{\mathbf{L}^2(\Omega)} + 2\nu_0 C_K \|\nabla\mathbf{u}\|^2_{\mathbf{L}^2(\Omega)} \leqslant \frac{C_P^2}{2\nu_0 C_K}\|\mathbf{f}\|^2_{\mathbf{L}^2(\Omega)}.$$

By integrating between t_0 and t, we obtain the second inequality of the theorem. □

REMARK 12.1. In the case where $\mathbf{f} = \mathbf{0}$ we may derive the simpler estimate

$$\left\|\mathbf{u}(t)\right\|^2_{\mathbf{L}^2(\Omega)} + 4\nu_0 C_K \int_{0}^{t} \left\|\nabla\mathbf{u}(\tau)\right\|^2_{\mathbf{L}^2(\Omega)} \,\mathrm{d}\tau \leqslant \|\mathbf{u}_0\|^2_{\mathbf{L}^2(\Omega)}, \qquad \forall t \geqslant t_0.$$

13. The Stokes equations

The space discretisation of the Navier–Stokes equations give rise to a *non-linear* set of ordinary differential equations because of the presence of the convective term. This makes both the analysis and the numerical solution more difficult. In some cases, when the fluid is highly viscous, the contribution of the non-linear convective term may be neglected. The key parameter which allow us to make that decision is the *Reynolds number Re*, which is an a-dimensional number defined as

$$Re = \frac{|\mathbf{u}|L}{\nu},$$

where L represents a length-scale for the problem at hand and $|\mathbf{u}|$ the Euclidean norm of the velocity. For the flow in a tube L is the tube diameter.

Contrary to other fluid dynamic situations, the high variation in time and space of the velocity in the vascular system does not allow to select a single representative value of the Reynolds number,[2] nevertheless in the situations where $\mathrm{Re} \ll 1$ (for instance, flow in smaller arteries or capillaries) we may say that the convective term is negligible compared to the viscous contribution and may be discarded. We have then the *Stokes* equations, which read (in the case of homogeneous Dirichlet conditions):

$$\frac{\partial\mathbf{u}}{\partial t} + \nabla p - 2\,\mathbf{div}\big(\nu\mathbf{D}(\mathbf{u})\big) = \mathbf{f}, \quad \text{in } \Omega,\ t \in I, \tag{13.1a}$$

$$\operatorname{div}\mathbf{u} = 0, \quad \text{in } \Omega,\ t \in I, \tag{13.1b}$$

$$\mathbf{u} = \mathbf{0}, \quad \text{on } \partial\Omega,\ t \in I, \tag{13.1c}$$

$$\mathbf{u} = \mathbf{u}_0, \quad \text{in } \Omega,\ t = t_0. \tag{13.1d}$$

The corresponding weak form reads:

[2] Another a-dimensional number which measures the relative importance of inertia versus viscous in oscillatory flow is the Womersley number (FUNG [1984]).

Find, $\forall t \in I$, $\mathbf{u}(t) \in \mathbf{V}$, $p(t) \in Q$, *such that*

$$\left(\frac{\partial \mathbf{u}}{\partial t}, \mathbf{v}\right) + a(\mathbf{u}, \mathbf{v}) + b(\mathbf{v}, p) = (\mathbf{f}, \mathbf{v}), \quad \forall \mathbf{v} \in \mathbf{V},$$
$$b(\mathbf{u}, q) = 0, \quad \forall q \in Q. \tag{13.2}$$

In the case of a *steady problem*, that is when we consider $\partial\mathbf{u}/\partial t = \mathbf{0}$, the solution $(\mathbf{u}, p)$ of the Stokes problem (13.2) is a saddle point for the functional

$$\mathcal{S}(\mathbf{v}, q) = \frac{1}{2}a(\mathbf{v}, \mathbf{v}) + b(\mathbf{v}, q) - (\mathbf{f}, \mathbf{v}), \quad \mathbf{v} \in \mathbf{V},\ q \in Q.$$

This means

$$\mathcal{S}(\mathbf{u}, p) = \min_{\mathbf{v}\in\mathbf{V}} \max_{q\in Q} \mathcal{S}(\mathbf{v}, q).$$

In this respect, the pressure p may be considered as a *Lagrange multiplier* associated to the incompressibility constraint.

REMARK 13.1. In those cases where $Re \gg 1$ (*high Reynolds number flows*) the flow becomes *unstable*. High frequency fluctuations in the velocity and pressure field appear, which might give rise to *turbulence*. This phenomenon is particularly complex and its numerical simulation may be extremely difficult. To make the problem amenable to numerical solution it is often necessary to adopt a *turbulence model*, which allows to give a more or less accurate description of the effect of turbulence on the main flow variables.

In normal physiological situations, the typical values of the Reynolds number reached in the cardiovascular system do not allow the formation of full scale turbulence. Some flow instabilities may occur only at the exit of the aortic valve and limited to the systolic phase. Indeed, in this region the Reynolds number may reach the value of few thousands only for the portion of the cardiac cycle corresponding the peak systolic velocity. Therefore, there is no sufficient time for a full turbulent flow to develop.

The situation is different in some pathological circumstances, e.g., in the presence of a stenotic artery. The increase of the velocity at the location of the vessel restriction may induce turbulence to develop. This fact could explain the high increase in the noise caused by the blood stream in this situation.

14. Numerical approximation of Navier–Stokes equations

In this section we give a very short account on possible numerical methods for the solution of the Navier–Stokes equations. This subject is far from being simple, and we will not make any attempt to be exhaustive. The interested reader can consult, for instance, QUARTERONI and VALLI [1994], Chapters 9, 10 and 13, and the classic books on the subject by GIRAULT and RAVIART [1986] and TEMAM [1984].

Here, we will simply mention a few methods to advance the Navier–Stokes equations from a given time-level to a new one and we will point out some of the mathematical problems that have to be faced. For the sake of simplicity we will confine ourselves to the homogeneous Dirichlet problem (11.8).

14.1. Time advancing by finite differences

The Navier–Stokes problem (9.1) (equivalently, its weak form (11.8)) can be advanced in time by suitable finite difference schemes.

The simulation will cover the interval $I = (0, T)$ which we subdivide into sub-intervals (time-steps) $I^k = (t^k, t^{k+1})$ with $k = 0, \ldots, N$ and where $t^{k+1} - t^k = \Delta t$ is constant. We have thus partitioned the space-time domain $I \times \Omega$ into several *time-slabs* $I^k \times \Omega$. We assume that on each slab we know the solution at $t = t^k$ and that we wish to find the solution at $t = t^{k+1}$. Clearly, for the first time slab the assumption is true since at $t = 0$ the approximate solution is obtained from the initial data. If we treat the time slabs in their natural order, as soon as the solution on the kth time slab has been found, it is made available as initial condition for the computation on the next time slab. This is a *time-advancing* procedure.

We will indicate by $(\mathbf{u}^k, p^k)$ the approximate solution at time t^k, that is

$$(\mathbf{u}^k, p^k) \approx (\mathbf{u}(t^k), p(t^k)).$$

A family of simple time-advancing schemes is obtained by using the Taylor expansion formula to write

$$\frac{\partial \mathbf{u}}{\partial t}(t^{k+1}) = \frac{\mathbf{u}(t^{k+1}) - \mathbf{u}(t^k)}{\Delta t} + \mathrm{O}(\Delta t).$$

Then, by making the first order approximation

$$\frac{\partial \mathbf{u}}{\partial t}(t^{k+1}) \approx \frac{\mathbf{u}^{k+1} - \mathbf{u}^k}{\Delta t},$$

into (9.1), we may write the following time-stepping scheme to calculate $\mathbf{u}^{k+1}$ and p^{k+1}:

$$\frac{\mathbf{u}^{k+1} - \mathbf{u}^k}{\Delta t} - 2\,\mathbf{div}\,\nu\mathbf{D}(\mathbf{u}^{k+1}) + (\mathbf{u}^* \cdot \nabla)\mathbf{u}^{**} + \nabla p^{k+1} = \mathbf{f}^{k+1}, \quad \text{in } \Omega, \tag{14.1a}$$

$$\operatorname{div} \mathbf{u}^{k+1} = 0, \quad \text{in } \Omega, \tag{14.1b}$$

$$\mathbf{u}^{k+1} = \mathbf{0}, \quad \text{on } \partial\Omega. \tag{14.1c}$$

Here, $\mathbf{f}^{k+1}$ stands for $\mathbf{f}(t^{k+1})$.

The value of $\mathbf{u}^*$ and $\mathbf{u}^{**}$ in the non-linear convective term may be taken, for instance, as follows:

$$(\mathbf{u}^* \cdot \nabla)\mathbf{u}^{**} = \begin{cases} (\mathbf{u}^k \cdot \nabla)\mathbf{u}^k, & \text{fully explicit treatment,} \\ (\mathbf{u}^k \cdot \nabla)\mathbf{u}^{k+1}, & \text{semi-implicit treatment,} \\ (\mathbf{u}^{k+1} \cdot \nabla)\mathbf{u}^{k+1}, & \text{fully implicit treatment.} \end{cases}$$

In the case of the fully implicit treatment, Eqs. (14.1) give rise to a non-linear system. The semi-implicit and fully explicit treatments, instead, perform a *linearisation* of the convective term, thus eliminating the non-linearity.

Let us consider the scheme resulting from the *fully explicit treatment* of the convective term. Problem (14.1) is then rewritten as

$$\frac{1}{\Delta t}\mathbf{u}^{k+1} - 2\,\mathbf{div}\big(\nu\mathbf{D}(\mathbf{u}^{k+1})\big) + \nabla p^{k+1} = \mathbf{f}^{k+1} + \frac{1}{\Delta t}\mathbf{u}^k - \big(\mathbf{u}^k\cdot\nabla\big)\mathbf{u}^k \quad \text{in } \Omega, \tag{14.2a}$$

$$\operatorname{div}\mathbf{u}^{k+1} = 0, \quad \text{in } \Omega, \tag{14.2b}$$

$$\mathbf{u}^{k+1} = \mathbf{0}, \quad \text{on } \partial\Omega. \tag{14.2c}$$

We will now denote $\mathbf{u}^{k+1}$ and p^{k+1} by $\mathbf{w}$ and π, respectively, and by $\mathbf{q}$ and a_0 the quantities

$$\mathbf{q} = \mathbf{f}^{k+1} + \frac{1}{\Delta t}\mathbf{u}^k - \big(\mathbf{u}^k\cdot\nabla\big)\mathbf{u}^k, \qquad a_0 = \frac{1}{\Delta t}. \tag{14.3}$$

Problem (14.2) may be written in the form

$$a_0\mathbf{w} - 2\,\mathbf{div}\big(\nu\mathbf{D}(\mathbf{w})\big) + \nabla\pi = \mathbf{q}, \quad \text{in } \Omega, \tag{14.4a}$$

$$\operatorname{div}\mathbf{w} = 0, \quad \text{in } \Omega, \tag{14.4b}$$

$$\mathbf{w} = \mathbf{0}, \quad \text{on } \partial\Omega, \tag{14.4c}$$

which is called the *generalised Stokes problem*.

A characteristic treatment of the time derivative would also lead at each time step to a generalised Stokes problem (see Section 14.3).

For its approximation, a Galerkin finite element procedure can be set up by considering two finite element spaces $\mathbf{V}_h$ for the velocity and Q_h for the pressure, and seeking $\mathbf{w}_h \in \mathbf{V}_h$ and $\pi_h \in Q_h$ such that

$$\begin{cases} \tilde{a}(\mathbf{w}_h, \mathbf{v}_h) + b(\mathbf{w}_h, \pi_h) = (\mathbf{q}, \mathbf{v}_h), & \forall \mathbf{v}_h \in \mathbf{V}_h, \\ b(\mathbf{w}_h, q_h) = 0, & \forall q_h \in Q_h, \end{cases} \tag{14.5}$$

where $\tilde{a}(\mathbf{w}, \mathbf{v}) = a_0(\mathbf{w}, \mathbf{v}) + a(\mathbf{w}, \mathbf{v})$.

The algebraic form of problem (14.5) is derived by denoting with

$$\{\boldsymbol{\varphi}_i,\ i = 1, \ldots, N_{\mathbf{V}_h}\}, \qquad \{\psi_i,\ i = 1, \ldots, N_{Q_h}\}$$

the bases of $\mathbf{V}_h$ and Q_h, respectively. Here $N_{\mathbf{V}_h} = \dim(\mathbf{V}_h)$ and $N_{Q_h} = \dim(Q_h)$. Then, by setting

$$\mathbf{w}_h(\mathbf{x}) = \sum_{i=1}^{N_{\mathbf{V}_h}} w_i\boldsymbol{\varphi}_i(\mathbf{x}), \qquad p_h(\mathbf{x}) = \sum_{i=1}^{N_{Q_h}} \pi_i\psi_i(\mathbf{x}), \tag{14.6}$$

we obtain the following system from (14.5):

$$\begin{pmatrix} C & D^{\mathrm{T}} \\ D & 0 \end{pmatrix}\begin{pmatrix} \mathbf{W} \\ \boldsymbol{\Pi} \end{pmatrix} = \begin{pmatrix} \mathbf{F}_s \\ \mathbf{0} \end{pmatrix}, \tag{14.7}$$

where $\mathbf{W}$, $\boldsymbol{\Pi}$ and $\mathbf{F}_s$ denote three vectors defined respectively as

$$(\mathbf{W})_i = w_i, \qquad (\boldsymbol{\Pi})_i = \pi_i, \qquad (\mathbf{F}_s)_i = (\mathbf{q}, \boldsymbol{\varphi}_i),$$

while C, K and D are matrices whose components are defined as

$$(C)_{ij} = \tilde{a}(\boldsymbol{\varphi}_j, \boldsymbol{\varphi}_i), \qquad (D)_{ij} = b(\boldsymbol{\varphi}_j, \psi_i).$$

The global matrix

$$A = \begin{pmatrix} C & D^{\mathrm{T}} \\ D & 0 \end{pmatrix} \tag{14.8}$$

is a square matrix with dimension $(N_{\mathbf{V}_h} + N_{Q_h}) \times (N_{\mathbf{V}_h} + N_{Q_h})$.

In the case of a finite element approximation, p_i represents the pressure at the ith mesh node. The interpretation of w_i is made more complex by the fact that the velocity is a vector function, while w_i is a scalar. Let us assume that we are considering a three-dimensional problem and let the basis for $\mathbf{V}_h$ be chosen by grouping the vector functions $\boldsymbol{\varphi}_i$ into 3 families, as follows:

$$\boldsymbol{\varphi}_i^1 = \begin{bmatrix} \varphi_i \\ 0 \\ 0 \end{bmatrix}, \qquad \boldsymbol{\varphi}_i^2 = \begin{bmatrix} 0 \\ \varphi_i \\ 0 \end{bmatrix}, \qquad \boldsymbol{\varphi}_i^3 = \begin{bmatrix} 0 \\ 0 \\ \varphi_i \end{bmatrix}.$$

Finally, let $M_{\mathbf{V}_h} = N_{\mathbf{V}_h}/3$. Then, we may rewrite the first expansion in (14.6) as

$$\mathbf{w}_h(\mathbf{x}) = \sum_{i=1}^{M_{\mathbf{V}_h}} \sum_{j=1}^{3} w_i^j \boldsymbol{\varphi}_i^j(\mathbf{x}),$$

where w_i^j here represents the jth component of $\mathbf{w}$ at the ith mesh node.

LEMMA 14.1. *If* $\ker D^{\mathrm{T}} = \mathbf{0}$, *then matrix A is non-singular.*

PROOF. We first prove the non-singularity of C. For any $\mathbf{W} \in \mathbb{R}^{N_{\mathbf{V}_h}}$, $\mathbf{W} \neq \mathbf{0}$,

$$\mathbf{W}^{\mathrm{T}} C \mathbf{W} = \sum_{i=1}^{N_{\mathbf{V}_h}} \sum_{j=1}^{N_{\mathbf{V}_h}} w_i w_j C_{ij} = \tilde{a}(\mathbf{w}, \mathbf{w}) > 0,$$

where $\mathbf{w} = \sum_{i=1}^{N_{\mathbf{V}_h}} w_i \boldsymbol{\varphi}_i$. Consequently, C is positive-definite, and thus non-singular. From (14.7) we have

$$\mathbf{W} - C^{-1}(\mathbf{F}_s - D^{\mathrm{T}} \boldsymbol{\Pi}), \qquad D\mathbf{W} = \mathbf{0}.$$

Then we may formally compute the discrete pressure terms by

$$-(DC^{-1}D^{\mathrm{T}})\boldsymbol{\Pi} = -DC^{-1}\mathbf{F}_s.$$

Proving that A is non-singular thus reduces to show that the matrix

$$S = DC^{-1}D^{\mathrm{T}}$$

is non-singular. If we take any $\mathbf{q} \in \mathbb{R}^{N_{Q_h}}$ with $|\mathbf{q}| \neq 0$ we have by hypothesis that $D^{\mathrm{T}}\mathbf{q} \neq 0$. Then

$$\mathbf{q}^{\mathrm{T}} S \mathbf{q} = (D^{\mathrm{T}}\mathbf{q})^{\mathrm{T}} C^{-1} D^{\mathrm{T}} \mathbf{q} \neq 0,$$

since C^{-1} is symmetric positive definite. Thus, matrix S (which is clearly symmetric) has all eigenvalues different from zero and, consequently, is non-singular. This concludes the proof. □

The scheme we have presented, with an explicit treatment of just the convective term, is only one of the many possible ways of producing a time discretisation of the Navier–Stokes equations. Another choice is to resort to a fully implicit scheme.

14.2. *Fully implicit schemes*

By employing in (14.1) a full implicit treatment of the convective part, we would obtain a non-linear system of the following type:

$$\begin{pmatrix} E(\mathbf{W}) & D^{\mathrm{T}} \\ D & 0 \end{pmatrix}\begin{pmatrix} \mathbf{W} \\ \boldsymbol{\Pi} \end{pmatrix} = \begin{pmatrix} \mathbf{F}_s \\ \mathbf{0} \end{pmatrix}, \tag{14.9}$$

where now the matrix E is a function of the unknown velocity,

$$\big(E(\mathbf{W})\big)_{ij} = \tilde{a}(\boldsymbol{\varphi}_i, \boldsymbol{\varphi}_j) + c\big(\mathbf{u}^{k+1}, \boldsymbol{\varphi}_j, \boldsymbol{\varphi}_i\big) = C_{ij} + \sum_{m=1}^{N_{\mathbf{V}_h}} c(\boldsymbol{\varphi}_m, \boldsymbol{\varphi}_j, \boldsymbol{\varphi}_i) W_m.$$

A possible way to solve it is to resort to *Newton's method*:

Given $\binom{\mathbf{W}^0}{\boldsymbol{\Pi}^0}$, solve for $l = 0, \ldots,$

$$\begin{pmatrix} \dfrac{\partial E}{\partial \mathbf{W}}(\mathbf{W}^l) \cdot \mathbf{W}^l + E(\mathbf{W}^l) & D^{\mathrm{T}} \\ D & 0 \end{pmatrix}\begin{pmatrix} \mathbf{W}^{l+1} - \mathbf{W}^l \\ \boldsymbol{\Pi}^{l+1} - \boldsymbol{\Pi}^l \end{pmatrix}$$
$$= \begin{pmatrix} \mathbf{F}_s \\ \mathbf{0} \end{pmatrix} - \begin{pmatrix} E(\mathbf{W}^l) & D^{\mathrm{T}} \\ D & 0 \end{pmatrix}\begin{pmatrix} \mathbf{W}^l \\ \boldsymbol{\Pi}^l \end{pmatrix}, \tag{14.10}$$

until a suitable convergence criterion is met.

The solution of a non-linear system is now reduced to a series of solutions of linear systems. Going back to the Navier–Stokes equations, we may note that a full implicit scheme would require to solve *at each time step* a series of linear systems of form (14.10), that resembles the Stokes problem. The resulting numerical scheme is thus in general very *computationally intensive*.

14.3. *Semi-Lagrangian schemes*

An alternative way to treat the non-linear term in the Navier–Stokes equations is obtained by performing an operator splitting that separates the effect of the convective term. The technique exploits the fact that the convective term is indeed the material derivative of $\mathbf{u}$,

$$\frac{\partial \mathbf{u}}{\partial t} + \mathbf{u} \cdot \nabla \mathbf{u} = \frac{D\mathbf{u}}{Dt},$$

that is the derivative along the particle trajectories T_ξ (also called characteristic lines) defined in Section 7.

On each time-slab I^k we then have that

$$\int_{t^k}^{t^{k+1}} \frac{D\mathbf{u}(\tau,\mathbf{x})}{D\tau}\,\mathrm{d}\tau = \mathbf{u}(t^{k+1},\mathbf{x}) - \mathbf{u}(t^{k+1},\mathbf{x}^*) \approx \mathbf{u}^{k+1}(\mathbf{x}) - \mathbf{u}^k(\mathbf{x}^*), \tag{14.11}$$

where $\mathbf{x}^*$ is position at $t = t^k$ of the fluid particle located in $\mathbf{x}$ at $t = t^{k+1}$, i.e., $\mathbf{x}^* = \mathbf{y}_\mathbf{x}(1)$ where $\mathbf{y}_\mathbf{x}(s)$ is the solution of

$$\begin{cases} \dfrac{\mathrm{d}\mathbf{y}_\mathbf{x}(s)}{\mathrm{d}s} = -\mathbf{u}\big(t^{k+1} - s\Delta t, \mathbf{y}_\mathbf{x}(s)\big), \\ \mathbf{y}_\mathbf{x}(0) = \mathbf{x}. \end{cases} \tag{14.12}$$

The point $\mathbf{x}^*$ is often denoted as the "foot" of the characteristic line $\mathbf{y}_\mathbf{x}$.

This interpretation leads to the *semi-Lagrangian* schemes, so called because we treat the convective operator in the Lagrangian frame. For instance, a backward Euler semi-Lagrangian scheme will lead at each time step I^k a generalised Stokes problem like (14.4), where now

$$\mathbf{q}(\mathbf{x}) = \mathbf{f}^{k+1}(\mathbf{x}) + \frac{1}{\Delta t}\mathbf{u}^k(\mathbf{x}^*),$$

that may then be treated by a Galerkin finite element procedure as described in Section 14.1.

Clearly, system (14.12) has to be approximated as well. A first-order approximation leads to

$$\mathbf{x}^* = \mathbf{x} - \mathbf{u}^k(\mathbf{x})\Delta t.$$

This explicit treatment will eventually entail a stability condition which depends on the fluid velocity. Higher-order schemes may be devised as well, see, for instance, BOUKIR, MADAY, MÉTIVET and RAZAFINDRAKOTO [1997].

The major drawback of semi-Lagrangian schemes is the computation of the approximation of $\mathbf{u}^k(\mathbf{x}^*)$. In a finite element context it requires to locate the mesh element where the foot of the characteristic passing through each mesh point lies (or each quadrature point if a quadrature rule is used to compute the space integrals). An efficient implementation calls for the use of special data structures. Furthermore, a proper treatment is needed when $\mathbf{x}^*$ falls outside the computational domain. In that case the boundary conditions have to be properly taken into account.

14.4. Projection methods

We now follow another route for the solution of the incompressible Navier–Stokes equations which does not lead to a Stokes problem but to a series of simpler systems of partial differential equations. We start from the Navier–Stokes equations already discretised in time and we will consider again a single time step, that is

$$\frac{\mathbf{u}^{k+1} - \mathbf{u}^k}{\Delta t} + (\mathbf{u}^k\cdot\nabla)\mathbf{u}^{k+1} - 2\,\mathbf{div}\big(\nu\mathbf{D}(\mathbf{u}^{k+1})\big) + \nabla p^{k+1} = \mathbf{f}^{k+1}, \quad \text{in } \Omega, \tag{14.13}$$

plus (14.1b) and (14.1c). Here, for the sake of simplicity (and without any loss of generality) we have chosen a semi-implicit treatment of the convective term. We wish now to split the system in order to consider the effects of the velocity and the pressure terms separately. We define an intermediate velocity $\tilde{\mathbf{u}}$, obtained by solving the momentum equation where the pressure contribution has been dropped, precisely

$$\frac{\tilde{\mathbf{u}}-\mathbf{u}^k}{\Delta t}+\left(\mathbf{u}^k\cdot\nabla\right)\tilde{\mathbf{u}}-2\,\mathbf{div}\big(\nu\mathbf{D}(\tilde{\mathbf{u}})\big)=\mathbf{f}^{k+1},\quad \text{in } \Omega, \tag{14.14a}$$

$$\tilde{\mathbf{u}}=\mathbf{0},\quad \text{on } \partial\Omega. \tag{14.14b}$$

We may recognise that (14.14a) is now a problem on the velocity only, which could be re-interpreted as the time discretisation of a parabolic differential equation of the following type:

$$\frac{\partial\tilde{\mathbf{u}}}{\partial t}+(\mathbf{w}\cdot\nabla)\tilde{\mathbf{u}}-2\,\mathbf{div}\big(\nu\mathbf{D}(\tilde{\mathbf{u}})\big)=\mathbf{f},$$

with $\mathbf{w}$ a given vector field. At this stage, we cannot impose the incompressibility condition because we would obtain an over-constrained system.

We then consider the contribution given by the pressure term and the incompressibility constraint, that is

$$\frac{\mathbf{u}^{k+1}-\tilde{\mathbf{u}}}{\Delta t}+\nabla p^{k+1}=\mathbf{0},\quad \text{in } \Omega, \tag{14.15a}$$

$$\operatorname{div}\mathbf{u}^{k+1}=0,\quad \text{in } \Omega. \tag{14.15b}$$

System (14.15) depends on both the velocity and pressure, yet we may derive an *equation only for the pressure* by taking (formally) the divergence of (14.15a) and exploiting the incompressibility constraint (14.15b). That is,

$$\begin{aligned} 0=\operatorname{div}\left(\frac{\mathbf{u}^{k+1}-\tilde{\mathbf{u}}}{\Delta t}+\nabla p^{k+1}\right)&=-\frac{1}{\Delta t}\operatorname{div}\tilde{\mathbf{u}}+\operatorname{div}\nabla p^{k+1}\\ &=-\frac{1}{\Delta t}\operatorname{div}\tilde{\mathbf{u}}+\Delta p^{k+1}, \end{aligned}$$

by which we obtain a *Poisson equation* for the pressure in the form

$$\Delta p^{k+1}=\frac{1}{\Delta t}\operatorname{div}\tilde{\mathbf{u}},\quad \text{in } \Omega. \tag{14.16}$$

Eq. (14.16) must be supplemented by boundary conditions, which are *not directly available from the original problem* (14.13). For that, we need to resort to the following theorem, also known as *Ladhyzhenskaja theorem*.

THEOREM 14.1 (Helmholtz decomposition principle). *Let Ω be a domain of $\mathbb{R}^N$ with smooth boundary. Any vector function $\mathbf{v}\in\mathbf{L}^2(\Omega)$ (with $N=2,3$) can be uniquely represented as $\mathbf{v}=\mathbf{w}+\nabla\psi$ with $\mathbf{w}\in\mathbf{H}_{\mathrm{div}}(\Omega)$, where*

$$\mathbf{H}_{\mathrm{div}}(\Omega)=\left\{\mathbf{w}\colon\ \mathbf{w}\in\mathbf{L}^2(\Omega),\ \operatorname{div}\mathbf{w}=0,\ \textit{a.e. } \mathbf{w}\cdot\mathbf{n}=0 \textit{ on } \partial\Omega\right\},$$

and $\psi\in H^1(\Omega)$.

The proof is rather technical and is here omitted. An outline, valid for the case $\mathbf{v} \in \mathbf{H}^1(\Omega)$, is given in CHORIN and MARSDEN [1990]. A more general demonstration is found in TEMAM [1984], Theorems 1 and 5.

If we now consider the expression

$$\tilde{\mathbf{u}} = \mathbf{u}^{k+1} + \nabla\left(\Delta t p^{k+1}\right), \tag{14.17}$$

derived from (14.15a), we may identify $\tilde{\mathbf{u}}$ with $\mathbf{v}$ and $(\Delta t p^{k+1})$ with ψ in the Helmholtz decomposition principle. Then, the natural space for $\mathbf{u}^{k+1}$ is $\mathbf{H}_{\mathrm{div}}(\Omega)$, by which we should impose

$$\mathbf{u}^{k+1} \cdot \mathbf{n} = 0, \quad \text{on } \partial\Omega. \tag{14.18}$$

Unfortunately, (14.18) is still a condition on the velocity, while we are looking for a boundary condition for the pressure. The latter is found by considering the normal component of (14.17) on the boundary,

$$\tilde{\mathbf{u}} \cdot \mathbf{n} = \mathbf{u}^{k+1} \cdot \mathbf{n} + \Delta t \nabla p^{k+1} \cdot \mathbf{n}, \quad \text{on } \partial\Omega,$$

and noting that on $\partial\Omega$ we have $\tilde{\mathbf{u}} \cdot \mathbf{n} = 0$, because of (14.14b), and $\mathbf{u}^{k+1} \cdot \mathbf{n} = 0$. Then,

$$\nabla p^{k+1} \cdot \mathbf{n} = \frac{\partial p^{k+1}}{\partial n} = 0, \quad \text{on } \partial\Omega,$$

which is a *homogeneous Neumann boundary condition* for the Poisson problem (14.16).

The *projection method* here presented for the solution of the Navier–Stokes equations consists then in solving at each time-step a sequence of simpler problems, listed in the following:

(1) *Advection–diffusion problem for the velocity* $\tilde{\mathbf{u}}$. Solve problem (14.14a)–(14.14b).

(2) *Poisson problem for the pressure*

$$\Delta p^{k+1} = \frac{1}{\Delta t} \operatorname{div} \tilde{\mathbf{u}}, \quad \text{in } \Omega, \tag{14.19a}$$

$$\frac{\partial}{\partial n} p^{k+1} = 0, \quad \text{on } \partial\Omega. \tag{14.19b}$$

(3) *Computation of* $\mathbf{u}^{k+1}$ (this is an explicit step)

$$\mathbf{u}^{k+1} = \tilde{\mathbf{u}} - \Delta t \nabla p^{k+1}. \tag{14.20}$$

For an analysis of projection methods as well as the set-up of higher order schemes the reader may consult PROHL [1997] and GUERMOND [1999]. We point out that projection schemes may also be used in conjunction with the semi-Lagrangian treatment of the convective term (ACHDOU and GUERMOND [2000]).

14.5. *Algebraic factorisation methods*

An alternative way of reducing the computational cost of the solution of the full Navier–Stokes problem is to operate at algebraic level. We will consider the generalised Stokes problem in its algebraic form (14.7). This is the typical system that arises at each time step of a time advancing scheme for the solution of the Navier–Stokes by a finite element method, when the convective term is treated explicitly. In this case, the matrix C has the form

$$C = \frac{M}{\Delta t} + K + B,$$

where M is the mass matrix, K the stiffness matrix and B the matrix arising from the explicit treatment of the convective term.

The matrix D derives from the discretisation of the divergence term, while D^{T} represents a discrete gradient operator. We may formally solve for $\mathbf{W}$,

$$\mathbf{W} = C^{-1}\big(\mathbf{F}_s - D^{\mathrm{T}}\boldsymbol{\Pi}\big), \tag{14.21}$$

and by substituting into (14.7), we have

$$DC^{-1}D^{\mathrm{T}}\boldsymbol{\Pi} = DC^{-1}\mathbf{F}_s. \tag{14.22}$$

The matrix $DC^{-1}D^{\mathrm{T}}$ is called *Stokes pressure matrix* and is somehow akin to a discrete Laplace operator. Having obtained $\boldsymbol{\Pi}$ from (14.22), we can then compute the velocity by solving (14.21).

However, the inversion of C is in general prohibitive in terms of memory and computational cost (indeed, C is sparse, but C^{-1} is not).

A way to simplify the computation can be found by recognising that steps (14.22) and (14.21) may be derived from the following LU factorisation of the global matrix A:

$$A = \begin{pmatrix} C & D^{\mathrm{T}} \\ D & 0 \end{pmatrix} = \begin{pmatrix} C & 0 \\ D & -DC^{-1}D^{\mathrm{T}} \end{pmatrix} \begin{pmatrix} I_{\mathbf{W}} & C^{-1}D^{\mathrm{T}} \\ D & I_{\Pi} \end{pmatrix} = LU, \tag{14.23}$$

where $I_{\mathbf{W}}$ and I_{Π} indicate the identity matrices of dimension equal to the number of velocity and pressure degrees of freedom, respectively. We then consider the LU solution

$$\begin{cases} C\widetilde{\mathbf{W}} = \mathbf{F}_s, \\ D\widetilde{\mathbf{W}} - DC^{-1}D^{\mathrm{T}}\widetilde{\boldsymbol{\Pi}} = 0, \end{cases} \qquad \begin{cases} \mathbf{W} + C^{-1}D^{\mathrm{T}}\boldsymbol{\Pi} = \widetilde{\mathbf{W}}, \\ \boldsymbol{\Pi} = \widetilde{\boldsymbol{\Pi}}, \end{cases}$$

where $\widetilde{\mathbf{W}}$ and $\widetilde{\boldsymbol{\Pi}}$ are intermediate velocities and pressures.

The scheme may be written in the following alternative form:

Intermediate velocity $\quad C\widetilde{\mathbf{W}} = \mathbf{F}_s, \qquad$ (14.24a)

Pressure computation $\quad -DC^{-1}D^{\mathrm{T}}\boldsymbol{\Pi} = -D\widetilde{\mathbf{W}}, \qquad$ (14.24b)

Velocity update $\quad \mathbf{W} = \widetilde{\mathbf{W}} - C^{-1}D^{\mathrm{T}}\boldsymbol{\Pi}. \qquad$ (14.24c)

The key to reduce complexity is to replace C by a matrix simpler to invert, which, however, is "similar" to C, in a sense that we will make precise. This technique is called *inexact factorisation*. In practise, we replace A in (14.23) by an approximation A^* obtained by replacing in the LU factorisation the matrix C^{-1} by convenient approximations, which we indicate by H_1 and H_2, that is

$$\begin{aligned} A^* = L^*U^* &= \begin{pmatrix} C & 0 \\ D & -DH_1D^{\mathrm{T}} \end{pmatrix} \begin{pmatrix} I_{\mathbf{W}} & H_2D^{\mathrm{T}} \\ D & I_{\Pi} \end{pmatrix} \\ &= \begin{pmatrix} C & CH_2D^{\mathrm{T}} \\ D & D(H_2 - H_1)D^{\mathrm{T}} \end{pmatrix}. \end{aligned} \tag{14.25}$$

If we choose $H_2 = H_1$, the discrete continuity equation is unaltered, that means that the approximated system still guarantees *mass conservation* at discrete level. If $H_2 = C^{-1}$, the discrete momentum equations are unaltered, and the resulting scheme satisfies the discrete conservation of momentum. In particular, we can consider the two special cases

$$H_1 = H_2 = H \quad \Rightarrow \quad A^* = \begin{pmatrix} C & CHD^{\mathrm{T}} \\ D & 0 \end{pmatrix},$$

$$H_1 = C^{-1} \neq H_2 \quad \Rightarrow \quad A^* = \begin{pmatrix} C & D^{\mathrm{T}} \\ D & Q \end{pmatrix}, \quad Q = D\big(H_1 - C^{-1}\big)D^{\mathrm{T}}.$$

14.5.1. The algebraic Chorin–Temam scheme

We note that

$$C = \frac{M}{\Delta t} + K + B = \frac{1}{\Delta t}\big(M + \Delta t(K + B)\big) = \frac{1}{\Delta t} M\big(I_{\mathbf{W}} + \Delta t M^{-1}(K + B)\big).$$

We recall the well-known Neumann expansion formula (MEYER [2000])

$$(I + \varepsilon A)^{-1} = \sum_{j=0}^{\infty} (-1)^j (\varepsilon A)^j,$$

which converges for any matrix A and any positive number ε small enough to guarantee that the spectral radius of εA is strictly less than one. We can apply this formula to C^{-1} to get

$$\begin{aligned} C^{-1} &= \Delta t\big(I_{\mathbf{W}} + \Delta t M^{-1}(K + B)\big)^{-1} M^{-1} \\ &= \Delta t \sum_{j=0}^{\infty} (-1)^j \big[\Delta t M^{-1}(K + B)\big]^j M^{-1} \\ &= \Delta t\big(I_{\mathbf{W}} - \Delta t M^{-1} S + \cdots\big) M^{-1}, \end{aligned} \tag{14.26}$$

where we have put $S = K + B$.

A way to find a suitable approximation is to replace C^{-1} with a few terms of the series. The simplest choice considers just a first order approximation, which corresponds to put into (14.25)

$$H_1 = H_2 = H = \Delta t M^{-1}. \tag{14.27}$$

Consequently,

$$A^* = A_{CT} = \begin{pmatrix} C & \Delta t C M^{-1} D^{\mathrm{T}} \\ D & 0 \end{pmatrix} = \begin{pmatrix} C & D^{\mathrm{T}} + \Delta t S M^{-1} D^{\mathrm{T}} \\ D & 0 \end{pmatrix}. \tag{14.28}$$

The scheme obtained by applying the corresponding LU decomposition reads:

Intermediate velocity $\quad C\widetilde{\mathbf{W}} = \mathbf{F}_s,$ (14.29a)

Pressure computation $\quad -\Delta t D M^{-1} D^{\mathrm{T}} \boldsymbol{\Pi} = -D\widetilde{\mathbf{W}},$ (14.29b)

Velocity update $\quad \mathbf{W} = \widetilde{\mathbf{W}} - \Delta t M^{-1} D^{\mathrm{T}} \boldsymbol{\Pi}.$ (14.29c)

This algorithm is known as *algebraic Chorin–Temam scheme*. Comparing with the standard projection method, we may note that the algebraic scheme replaces in the pressure computation step (14.29b) the Laplace operator of the Poisson problem (14.19) with a "discrete Laplacian" $DM^{-1}D^{\mathrm{T}}$, which incorporates the boundary condition of the original problem. No additional boundary condition is required for the pressure, contrary to the standard (differential type) scheme.

REMARK 14.1. The finite element mass matrix M is sparse and with the same graph structure as C. Therefore, it may seem that there is little gain in the computational efficiency with respect to the original factorisation (14.24). However, the matrix M may be approximated by a diagonal matrix called *lumped mass matrix* (QUARTERONI and VALLI [1994]), whose inversion is now trivial.

REMARK 14.2. It is possible to write the algebraic Chorin–Temam scheme in incremental form, as it has been done for its differential counterpart.

14.5.2. The Yosida scheme

If we make the special choice

$$H_1 = \Delta t M^{-1}, \qquad H_2 = C^{-1}, \tag{14.30}$$

we have

$$A^* = A_Y = \begin{pmatrix} C & D^{\mathrm{T}} \\ D & Q \end{pmatrix} \quad \text{with } Q = -D\big(\Delta t M^{-1} - C^{-1}\big)D^{\mathrm{T}}. \tag{14.31}$$

The corresponding scheme reads

$$\textit{Intermediate velocity} \qquad C\widetilde{\mathbf{W}} = \mathbf{F}_s, \tag{14.32a}$$

$$\textit{Pressure computation} \qquad -\Delta t D M^{-1} D^{\mathrm{T}} \boldsymbol{\Pi} = -D\widetilde{\mathbf{W}}, \tag{14.32b}$$

$$\textit{Velocity update} \qquad \mathbf{W} = \widetilde{\mathbf{W}} - \Delta t C^{-1} D^{\mathrm{T}} \boldsymbol{\Pi}. \tag{14.32c}$$

The last step (14.32c) is more expensive than its counterpart (14.29c) in the Chorin–Temam scheme, since now we need to invert the full matrix C. An analysis of this method is found in QUARTERONI, SALERI and VENEZIANI [1999].

REMARK 14.3. If we consider the Stokes problem, we have $C = (\Delta t)^{-1}M + K$ and consequently the matrix $Q = -D(\Delta t M^{-1} - C^{-1})D^{\mathrm{T}}$ in (14.31) may be written as

$$Q = -\Delta t D\big[I_{\mathbf{W}} - (I_{\mathbf{W}} + \Delta t K)^{-1}\big]D^{\mathrm{T}} = -(\Delta t)^2 D Y D^{\mathrm{T}},$$

where

$$Y = \frac{1}{\Delta t}\big[I_{\mathbf{W}} - (I_{\mathbf{W}} + \Delta t K)^{-1}\big],$$

may be regarded as the *Yosida* regularisation of K, which is the discretisation of the Laplace operator. That is Q may be interpreted as the discretisation of the differential operator

$$(\Delta t)^2 \operatorname{div}(\mathcal{Y}_{\Delta t} \nabla),$$

where $\mathcal{Y}_{\Delta t}$ is the Yosida operator (BREZIS [1983]).

REMARK 14.4. An incremental form may be found as follows. If $\boldsymbol{\Pi}^n$ represents the known value of the pressure degrees of freedom from the previous time step, we have

$$\begin{aligned}
&\textit{Intermediate velocity} && C\widetilde{\mathbf{W}} = \mathbf{F}_s - D^{\mathrm{T}}\boldsymbol{\Pi}^n,\\
&\textit{Pressure increment} && -\Delta t D M^{-1} D^{\mathrm{T}}(\boldsymbol{\Pi} - \boldsymbol{\Pi}^n) = -D\widetilde{\mathbf{W}},\\
&\textit{Velocity update} && \mathbf{W} = \widetilde{\mathbf{W}} - \Delta t C^{-1} D^{\mathrm{T}}(\boldsymbol{\Pi} - \boldsymbol{\Pi}^n).
\end{aligned}$$

More details on algebraic fractional step methods may be found in PEROT [1993] and QUARTERONI, SALERI and VENEZIANI [2000].

A major advantage of the algebraic factorisation schemes with respect to projection methods is that they do not require to devise special boundary conditions for the pressure problem, a task which is not always trivial.

All the techniques here presented may be extended to moving domains using the procedure that will be illustrated in Section 18. In a moving domain context the various matrices of the final algebraic system have to be recomputed at each times step to reflect the change of domain geometry. As a consequence, a fully implicit approach is even less computationally attractive, and factorisation schemes (at differential or algebraic level), possibly with a semi-Lagrangian treatment of the convective term, are normally preferred.

In the context of haemodynamics, algebraic factorisation schemes are particularly attractive because of their flexibility with respect to the application of boundary conditions. In particular, they can easily accommodate defective boundary conditions (FORMAGGIA, GERBEAU, NOBILE and QUARTERONI [2002]).

CHAPTER IV

15. Mathematical modelling of the vessel wall

The vascular wall has a very complex nature and devising an accurate model for its mechanical behaviour is rather difficult. Its structure is indeed formed by many layers with different mechanical characteristics (FUNG [1993], HOLZAPFEL, GASSER and OGDEN [2000]) (see Fig. 15.1). Moreover, experimental results obtained by specimens are only partially significant. Indeed, the vascular wall is a living tissue with the presence of muscular cells which contribute to its mechanical behaviour. It may then be expected that the dead tissue used in the laboratory will have different mechanical characteristics than the living one. Moreover, the arterial mechanics depend also on the type of the surrounding tissues, an aspect almost impossible to reproduce in a laboratory. We are then facing a problem whose complexity is enormous. It is the role of mathematical modelling to find reasonable simplifying assumptions by which major physical characteristics remain present, yet the problem becomes amenable to numerical analysis and computational solution.

The set up of a general mathematical model of the mechanics of a solid continuum may follow the same general route that we have indicated for fluid mechanics. In particular, it is possible to identify again a Cauchy stress tensor $\mathbf{T}$. The major difference between solids and fluids is in the constitutive relation which links $\mathbf{T}$ to kinematics

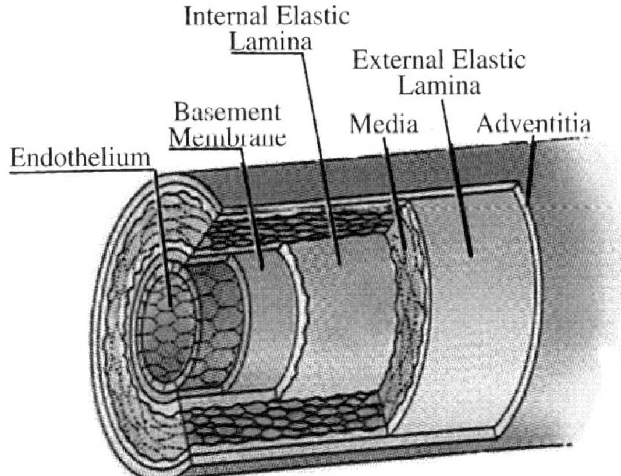

FIG. 15.1. The vessel wall is formed by many layers made of tissues with different mechanical characteristics. Image taken from "Life: the Science of Biology" by W.K. Purves et al., fourth edition, published by Sinauer Associates Inc. and W.H. Freeman and Company.

quantities. We have seen in Section 5 that for a fluid such a kinematic quantity is the velocity gradient or, more precisely, the strain rate $\mathbf{D}$. For a solid, the Cauchy stress tensor is instead a function of the *deformation gradient*, which we have already defined in (7.7). That is, the *constitutive law* for a solid may be written as

$$\mathbf{T} = \mathbf{T}(\mathbf{F}_t).$$

If we assume that both the deformation gradient and the displacements are small, under the hypothesis of *linear elasticity* and homogeneous material it is possible to derive relatively simple relations for $\mathbf{T}$. For sake of space, we will not pursue that matter here. The interested reader may consult, for instance, the book by L.A. Segel (SEGEL [1987], Chapter 4), or, for a more extensive treatment, the book by P.G. Ciarlet (CIARLET [1988]).

Another possible situation is the one that involves a constitutive law of the form

$$\mathbf{T} = \mathbf{T}(\mathbf{D}, \mathbf{F}_t), \tag{15.1}$$

which describes the mechanical behaviour of a material with characteristics intermediate to those of a liquid and a solid. In such case, the continuum is said to be *viscoelastic*. An example of such behaviour is given by certain plastics or by liquid suspensions. In particular, also blood exhibits a viscoelastic nature, particularly when flowing in small vessels, e.g. in arterioles and capillaries. Indeed, in that case the presence of suspended particles and their interaction during the motion strongly affect the blood mechanical behaviour. Again, we will not cover this topic here. The book by Y.C. Fung (FUNG [1993]) may be used by the reader interested on the peculiar aspects of the mechanics of living tissues.

The geometry of a section of an artery where no branching is present may be described by using a curvilinear cylindrical coordinate system (r, θ, z) with the corresponding base unit vectors $\mathbf{e}_r$, $\mathbf{e}_\theta$ and $\mathbf{e}_z$, where $\mathbf{e}_z$ is aligned with the axis of the artery, as shown in Fig. 15.2.

Clearly, the vessel structure may be studied using full three-dimensional models, which may also account for its multilayer nature. However, it is common practice to

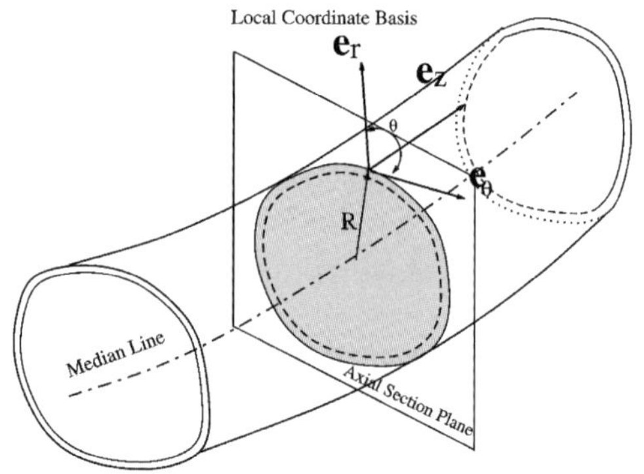

FIG. 15.2. A model of a "realistic" section of an artery with the principal geometrical parameters.

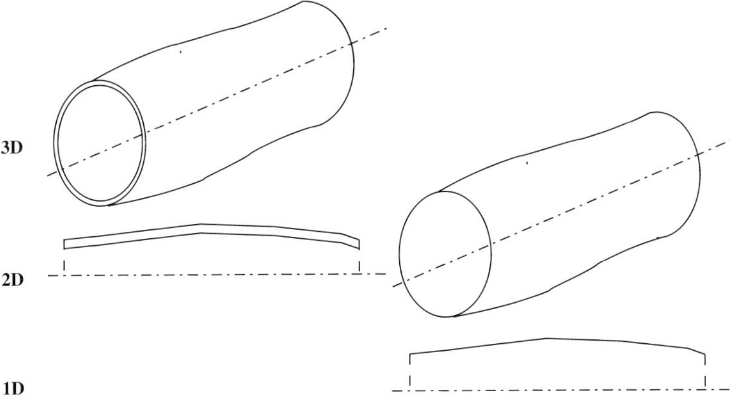

FIG. 15.3. Different models for arterial wall mechanics.

resort to simplified 2D or even 1D mechanical models in order to reduce the overall computational complexity when the final aim is to study the coupled fluid–structure problem. In Fig. 15.3 we sketch some of the approximations normally made. A 2D model may be obtained by either resorting to a shell-type description or considering longitudinal sections ($\theta =$ const.) of the vessels. In the first case we exploit the fact that the effective wall thickness is relatively small to reduce the whole structure to a surface. A rigorous mathematical derivation (for the linear case) may be found in CIARLET [1998]. In the second case we neglect the variations of the stresses in the circumferential direction. In this way we are able to eliminate all terms containing derivatives with respect to θ in the equations and we may consider each plane $\theta =$ const. independently. The resulting displacement field will depend only parametrically on θ. If, in addition, we assume that the problem has an axial symmetry (which implies the further assumption of a straight axis) the dependence on θ is completely neglected. In this case, also the fluid would be described by a 2D axi-symmetric model.

The simplest models, called 1D models, are derived by making the same assumption on the wall thickness made for the shell model, yet starting from a 2D model. The structure will then be represented by a line on a generic longitudinal section, as shown in the last picture of Fig. 15.3.

Even with all these simplifying assumptions an accurate model of the vessel wall mechanics is rather complex. Therefore, in these notes we will only present the simplest models, whose derivation is now detailed.

16. Derivation of 1D models of vessel wall mechanics

We are going to introduce a hierarchy of 1D models for the vessel structure of variable complexity. We first present the assumptions common to all models.

The relatively small thickness of the vessel wall allows us to use as basis model a shell model, where the vessel wall geometry is fully described by its median surface, see Fig. 16.1.

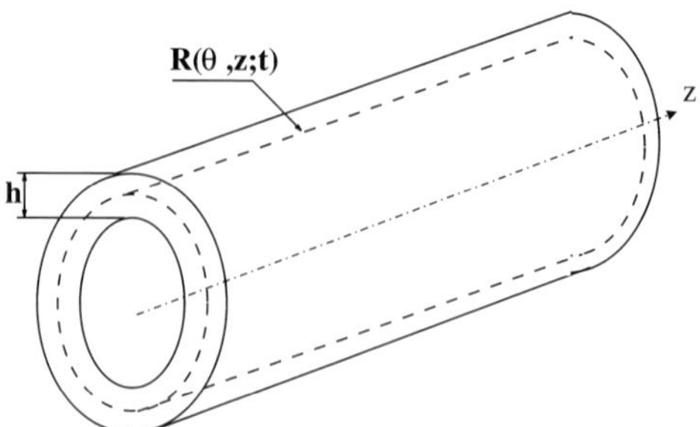

FIG. 16.1. A cylindrical model of the vessel geometry. The latter is approximated, at any time t, by a surface $r = R(\theta, z; t)$, which is outlined with dashed lines in figure.

We take as *reference configuration* Γ_0^w the one assumed by the vessel at rest when filled with fluid with zero velocity and whose pressure is equal to the pressure P_{ext} exerted by the tissues external to the vessel. Although in principle P_{ext} can change along the vessel (for instance, because of the effect of gravity), for the sake of simplicity (and without any loss of generality) we will consider only the case where P_{ext} is constant.

The cylindrical-like aspect of sections of the arterial system allows us to derive simplified mathematical models for the movement of the arterial wall assuming a straight cylindrical geometry. We thus assume that the reference configuration Γ_0^w be a cylindrical surface with radius R_0 (a regular strictly positive function of z), i.e.,

$$\Gamma_0^w = \{(r, \theta, z)\colon r = R_0(z),\ \theta \in [0, 2\pi),\ z \in [0, L]\},$$

where L indicates the length of the arterial element under consideration. In our cylindrical coordinate system (r, θ, z), the z coordinate is aligned along the vessel axes and a plane $z = \bar{z}$ (= constant) defines an *axial section*.

We assume that the displacement vector $\boldsymbol{\eta}$ has only a radial component, that is

$$\boldsymbol{\eta} = \eta \mathbf{e}_r = (R - R_0)\mathbf{e}_r, \tag{16.1}$$

where $R = R(\theta, z; t)$ is the function that provides, at each t, the radial coordinate $r = R(\theta, z; t)$ of the wall surface. The *current configuration* Γ_t^w at time t of the vessel surface is then given by

$$\Gamma_t^w = \{(r, \theta, z)\colon r = R(\theta, z; t),\ \theta \in [0, 2\pi),\ z \in [0, L]\}.$$

As a consequence, the length of the vessel does not change with time. We will indicate with $\mathbf{n}$ the outwardly oriented unit normal to the surface Γ_t^w at a given point. In Fig. 16.2 we sketch the reference and current configuration for the model of the section of an artery.

Another important assumption is that of *plain stresses*. We neglect the stress components along the normal direction $\mathbf{n}$, i.e., we assume that the stresses lie on the vessel surface.

We itemise here the main assumptions:

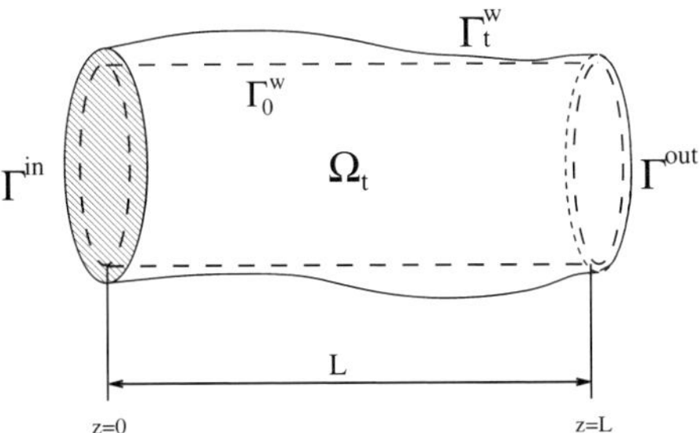

FIG. 16.2. The reference configuration Γ_0^w used for the derivation of our models is that of a circular cylinder. Γ_t^w indicates the current configuration at a given time t, while Ω_t is the domain occupied by the fluid.

(A1) *Small thickness and plain stresses.* The vessel wall thickness h is sufficiently small to allow a shell-type representation of the vessel geometry. In addition, we will also suppose that it is constant in the reference configuration. The vessel structure is subjected to plain stresses.

(A2) *Cylindrical reference geometry and radial displacements.* The reference vessel configuration is described by a circular cylindrical surface with straight axes.[3] The displacements are only in the radial direction.

(A3) *Small deformation gradients.* We assume that the deformation gradients are small, so that the structure basically behaves like a linear elastic solid and $\partial R/\partial\theta$ and $\partial R/\partial z$ remain uniformly bounded during motion.

(A4) *Incompressibility.* The vessel wall tissue is incompressible, i.e., it maintains its volume during the motion. This is a reasonable assumption since biological tissues are indeed nearly incompressible.

The models that we are going to illustrate could be derived from the general laws of solid mechanics. Yet, this is not the route we will follow, preferring to describe them in a more direct way, while trying to give some insight on the physical meaning of the various terms that we are about to introduce.

16.1. *Forces acting on the vessel wall*

Let us consider the vessel configuration at a given time t and a generic point on the vessel surface of coordinates $\theta = \bar{\theta}$, $z = \bar{z}$ and $r = R(\bar{\theta}, \bar{z}; t)$, with $\bar{z} \in (0, L)$ and $\theta \in (0, 2\pi)$. In the following derivation, if not otherwise indicated, all quantities are computed at location $(R(\bar{\theta}, \bar{z}; t), \bar{\theta}, \bar{z})$ and at time t.

We will indicate with $\mathrm{d}\sigma$ the measure of the following elemental surface:

$$\mathrm{d}S = \left\{(r, \theta, z)\colon\ r = R(\theta, z; t),\ \theta \in \left[\bar{\theta} - \frac{\mathrm{d}\theta}{2}, \bar{\theta} + \frac{\mathrm{d}\theta}{2}\right],\ z \in \left[\bar{z} - \frac{\mathrm{d}z}{2}, \bar{z} + \frac{\mathrm{d}z}{2}\right]\right\}.$$

[3]This assumption may be partially dispensed with, by assuming that the reference configuration is "close" to that of a circular cylinder. The model here derived may be supposed valid also in that situation.

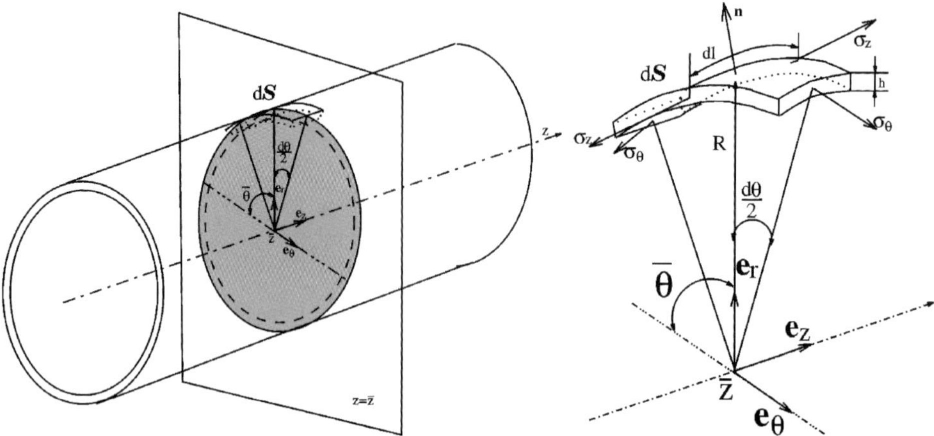

FIG. 16.3. A cylindrical model of the vessel geometry (left) and the infinitesimal portion of vessel wall used for the derivation of the equations (right).

In Fig. 16.3 we have also indicated the two main stresses, the circumferential stress and the longitudinal stress $\boldsymbol{\sigma}_\theta$ and $\boldsymbol{\sigma}_z$, which represent the internal forces acting on the portion under consideration.

We may derive the following expression for $\mathbf{n}$ and $\mathrm{d}\sigma$:

$$\mathbf{n} = (R_0 g)^{-1}\left(R\mathbf{e}_r - \frac{\partial R}{\partial \theta}\mathbf{e}_\theta - R\frac{\partial R}{\partial z}\mathbf{e}_z\right), \tag{16.2}$$

$$\mathrm{d}\sigma = g R_0\, \mathrm{d}\theta\, \mathrm{d}z = g\, \mathrm{d}\sigma_0, \tag{16.3}$$

where

$$g = \frac{R}{R_0}\sqrt{1 + \left(\frac{1}{R}\frac{\partial R}{\partial \theta}\right)^2 + \left(\frac{\partial R}{\partial z}\right)^2},$$

and $\sigma_0 = R_0\, \mathrm{d}\theta\, \mathrm{d}z$ is the measure of the image of $\mathrm{d}\mathcal{S}$ in the reference configuration Γ_0^w. In particular, we have

$$\mathbf{n} \cdot \mathbf{e}_r = \frac{R}{R_0} g^{-1} \tag{16.4}$$

and

$$\mathbf{n} \cdot \mathbf{e}_r\, \mathrm{d}\sigma = R\, \mathrm{d}\theta\, \mathrm{d}z. \tag{16.5}$$

The linear dimension of the elemental surface $\mathrm{d}\mathcal{S}$ along the longitudinal direction has been indicated with $\mathrm{d}l$. It can be easily verified that

$$\mathrm{d}l = \sqrt{1 + \left(\frac{\partial R}{\partial z}\right)^2}\, \mathrm{d}z. \tag{16.6}$$

Let us now consider the external forces acting through the elemental surface $\mathrm{d}\mathcal{S}$.

- *Forces from the surrounding tissues.* As the tissue surrounding the vessel interacts with the vessel wall structure by exerting a constant pressure P_{ext}, the resulting force acting on $\mathrm{d}S$ is simply given by

$$\mathbf{f}_{\text{tissue}} = -P_{\text{ext}}\mathbf{n}\,\mathrm{d}\sigma + \mathrm{o}(\mathrm{d}\sigma). \tag{16.7}$$

- *Forces from the fluid.* The forces the fluid exerts on the vessel wall are represented by the Cauchy stresses on the wall. Then, if we indicate with $\mathbf{T}_f$ the Cauchy stress tensor for the fluid, we have

$$\mathbf{f}_{\text{fluid}} = -\mathbf{T}_f \cdot \mathbf{n}\,\mathrm{d}\sigma + \mathrm{o}(\mathrm{d}\sigma) = P\mathbf{n}\,\mathrm{d}\sigma - 2\mu\mathbf{D}(\mathbf{u}) \cdot \mathbf{n}\,\mathrm{d}\sigma + \mathrm{o}(\mathrm{d}\sigma). \tag{16.8}$$

16.2. The independent ring model

The independent ring model is expressed by a differential equation for the time evolution of η, for each z and θ. For the derivation of this model, we will make some additional assumptions:

(IR-1) *Dominance of circumferential stresses* $\boldsymbol{\sigma}_\theta$. The stresses $\boldsymbol{\sigma}_z$ acting along longitudinal direction are negligible with respect to $\boldsymbol{\sigma}_\theta$ and are thus neglected when writing the momentum equation.

(IR-2) *Cylindrical configuration.* The vessel remains a circular cylinder during motion, i.e., $\partial R/\partial\theta = 0$. This hypothesis may be partially dispensed with, by allowing small circumferential variations of the radius, yet we will neglect $\partial R/\partial\theta$ in our model.

(IR-3) *Linear elastic behaviour.* Together with hypotheses (IR-1) and (IR-2) it allows us to write that the circumferential stress is proportional to the relative circumferential elongation, i.e.,

$$\sigma_\theta = \frac{E}{1-\xi^2}\frac{\eta}{R_0}, \tag{16.9}$$

where ξ is the *Poisson ratio* (which may be taken equal to 0.5 thanks to the hypothesis (A4)) and E is the Young modulus.[4]

We will write the balance of momentum along the radial direction by analysing the system of forces acting on $\mathrm{d}S$. We have already examined the external forces, we need now to look in more details at the effect of the internal forces, which, by assumption, are only due to the circumferential stress $\boldsymbol{\sigma}_\theta$.

We may note in Fig. 16.4 that the two vectors

$$\mathbf{e}_\theta\left(\bar\theta + \frac{\mathrm{d}\theta}{2}\right) \quad \text{and} \quad \mathbf{e}_\theta\left(\bar\theta - \frac{\mathrm{d}\theta}{2}\right)$$

form with $\mathbf{e}_r$ an angle of $\pi/2 + \mathrm{d}\theta/2$ and $-(\pi/2 + \mathrm{d}\theta/2)$, respectively. The component of the resultant of the internal forces on the radial direction is then

$$f_{\text{int}} = \left(\sigma_\theta \mathbf{e}_\theta\left(\bar\theta + \frac{\mathrm{d}\theta}{2}\right) + \sigma_\theta \mathbf{e}_\theta\left(\bar\theta - \frac{\mathrm{d}\theta}{2}\right)\right) \cdot \mathbf{e}_r h\,\mathrm{d}l$$

[4] The presence of the term $1-\xi^2$ is due to the assumption of planar stresses. Some authors (like FUNG [1984]) consider that the hypothesis of mono-axial stresses is more realistic for the problem at hand. In that case one has to omit the term $1-\xi^2$ from the stress–strain relation and write simply $\sigma_\theta = E\eta R_0^{-1}$.

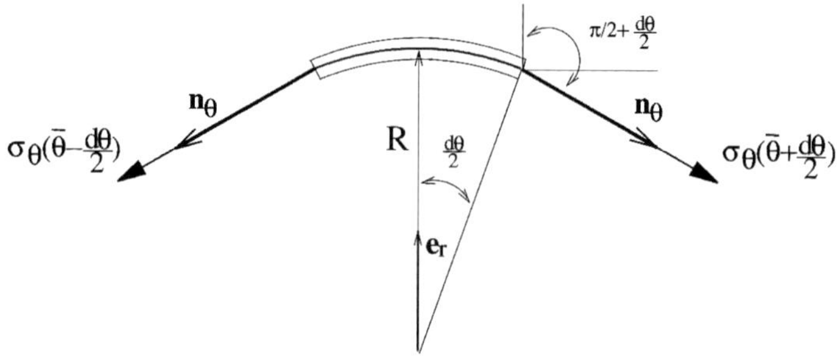

FIG. 16.4. Computation of the angle between $\boldsymbol{\sigma}_\theta$ and the radial direction $\mathbf{e}_r$.

$$= -2\sigma_\theta \sin\frac{\mathrm{d}\theta}{2} h \,\mathrm{d}l = -\sigma_\theta h \,\mathrm{d}\theta \,\mathrm{d}l + \mathrm{o}(\mathrm{d}\theta \,\mathrm{d}l). \tag{16.10}$$

Owing to the incompressibility assumption (A4), the volume in the current configuration is unchanged with respect to that in the reference configuration, i.e.,

$$hR \,\mathrm{d}\theta \,\mathrm{d}l = h_0 R_0 \,\mathrm{d}\theta \,\mathrm{d}z.$$

Then, being $\mathrm{o}(\mathrm{d}l) = \mathrm{o}(\mathrm{d}z)$, we may write (16.10) as

$$f_{\mathrm{int}} = -\frac{\sigma_\theta}{R} h_0 R_0 + \mathrm{o}(\mathrm{d}\theta \,\mathrm{d}z) = \frac{Eh_0}{1-\xi^2}\frac{\eta}{R} \,\mathrm{d}\theta \,\mathrm{d}z + \mathrm{o}(\mathrm{d}\theta \,\mathrm{d}z).$$

Finally, the mass of the portion of vessel wall under consideration is

$$\mathrm{mass} = \rho_w h R \,\mathrm{d}\theta \,\mathrm{d}l = \rho_w h_0 R_0 \,\mathrm{d}\theta \,\mathrm{d}z,$$

where ρ_w is the density of the vessel tissue, whereas the acceleration along the radial direction is given by

$$\frac{\partial^2 R}{\partial t^2} = \frac{\partial^2 \eta}{\partial t^2}.$$

By balancing the resultant of the internal and external forces, provided in (16.7) and (16.8), with the inertia term, we have

$$\rho_w h_0 R_0 \frac{\partial^2 \eta}{\partial t^2} \,\mathrm{d}\theta \,\mathrm{d}z + \frac{Eh_0}{1-\xi^2}\frac{\eta}{R} \,\mathrm{d}\theta \,\mathrm{d}z$$
$$= -\big(2\mu \mathbf{D}(\mathbf{u}) \cdot \mathbf{n}\big) \cdot \mathbf{e}_r \,\mathrm{d}\sigma + (P - P_{\mathrm{ext}})\mathbf{n} \cdot \mathbf{e}_r \,\mathrm{d}\sigma + \mathrm{o}(\mathrm{d}\theta \,\mathrm{d}z). \tag{16.11}$$

By dividing either side by $\mathrm{d}\theta \,\mathrm{d}z$ and passing to the limit for $\mathrm{d}\theta \to 0$ and $\mathrm{d}z \to 0$, and recalling that $\mathrm{d}\sigma = gR_0 \,\mathrm{d}\theta \,\mathrm{d}z = R(\mathbf{n}\cdot\mathbf{e}_r)^{-1} \,\mathrm{d}\theta \,\mathrm{d}z$, thanks to (16.2) and (16.4), we obtain

$$\rho_w h_0 R_0 \frac{\partial^2 \eta}{\partial t^2} + \frac{Eh_0}{1-\xi^2}\frac{\eta}{R} = -\big(2\mu \mathbf{D}(\mathbf{u}) \cdot \mathbf{n}\big) \cdot \mathbf{e}_r g R_0 + (P - P_{\mathrm{ext}})R.$$

Since the derivation has been made by considering an arbitrary plane $\theta = \bar{\theta}$ and time t, we may finally obtain the *independent ring model*

$$\frac{\partial^2 \eta}{\partial t^2} + b\eta = H, \quad \text{in } \Gamma_0^w,\ t \in I, \tag{16.12}$$

where

$$b = \frac{E}{\rho_w(1-\xi^2)R_0^2}, \tag{16.13}$$

is a positive coefficient linked to the wall mechanical properties, while

$$\begin{aligned} H &= \frac{1}{\rho_w h_0}\left[\frac{R}{R_0}(P - P_{\text{ext}}) - 2g\mu\big(\mathbf{D}(\mathbf{u})\cdot\mathbf{n}\big)\cdot\mathbf{e}_r\right] \\ &= \frac{\rho}{\rho_w h_0}\left[\frac{R}{R_0}(p - p_{\text{ext}}) - 2g\nu\big(\mathbf{D}(\mathbf{u})\cdot\mathbf{n}\big)\cdot\mathbf{e}_r\right], \end{aligned} \tag{16.14}$$

is the *forcing term* which accounts for the action of external forces.

REMARK 16.1. Often, the term R/R_0 in the right-hand side of (16.14) is neglected as well as the contribution to the forcing term due to the fluid viscous stresses. In this case, we have just

$$H = \frac{P - P_{\text{ext}}}{\rho_w h_0} \tag{16.15}$$

and the forcing term does not depend anymore on the current geometrical configuration.

By neglecting the acceleration term in (16.12), we obtain the following *algebraic model*, which is often found in the medical and bioengineering literature:

$$b\eta = H, \quad \text{in } \Gamma_0^w,\ t \in I, \tag{16.16}$$

according to which the wall displacement is proportional to the normal component of the applied external stresses.

REMARK 16.2. One may account for the *viscoelastic* nature of the vessel wall structure even in this simple model by adding to the constitutive relation (16.9) a term proportional to the displacement velocity, as in a simple Voigt–Kelvin model (FUNG [1993]), that is by writing

$$\sigma_\theta = \frac{E}{1-\xi^2}\frac{\eta}{R_0} + \frac{\gamma}{R_0}\frac{\partial \eta}{\partial t},$$

where γ (whose unit is $[\gamma] = \text{kg/m s}$) is a positive constant damping parameter.

Then, the resulting differential equation would read:

$$\frac{\partial^2\eta}{\partial t^2} + \frac{\gamma}{R_0^2\rho_w h_0}\frac{\partial\eta}{\partial t} + b\eta = H, \quad \text{in } \Gamma_0^w,\ t \in I. \tag{16.17}$$

We may note that the term $\frac{1}{R_0}\frac{\partial\eta}{\partial t}$ plays the role of the strain rate $\mathbf{D}$ into the general relation for viscoelastic materials (15.1).

Models (16.12), (16.16) and (16.17) are all apt to provide a solution η for every possible value of θ. In principle, since no differentiation with respect to θ is present in the model, nothing would prevent us to get significant variations of η with θ (or even a discontinuity), which would contradict assumption (IR-2). This potential drawback could be eliminated by enriching the models with further terms involving derivatives along θ, as in the case of models derived from shell theory (CIARLET [2000]). On the other hand, a more heuristic and less rigorous argument can be put forward moving from (16.16). Since b is relatively large, smooth variations of the forcing term H with respect to θ are damped to tiny one on η. This observation may be extended also to models (16.12) and (16.17) in view of the fact that for the problems at hand the term $b\eta$ dominates the other terms on the left-hand side. Similar considerations apply to the model that we will introduce in the next subsection.

16.3. The generalised string model

A more complete model (QUARTERONI, TUVERI and VENEZIANI [2000]) considers also the effects of the longitudinal stresses $\boldsymbol{\sigma}_z$. Experimental and physiological analysis (FUNG [1993]) show that vessel walls are in a "pre-stressed" state. In particular, when an artery is extracted from a body tends to "shrink", i.e., to reduce its length. This fact implies that arteries in the human body are normally subjected to a longitudinal tension.

At the base of the generalised string model is the assumption that this longitudinal tension is indeed the dominant component of the longitudinal stresses.

More precisely, let us refer to Fig. 16.5; we replace assumption (IR-1) by the following:

(GS-1) The longitudinal stress $\boldsymbol{\sigma}_z$ is not negligible and, in particular,

$$\boldsymbol{\sigma}_z = \pm\sigma_z \boldsymbol{\tau}, \tag{16.18}$$

where $\boldsymbol{\tau}$ is the unitary vector tangent to the curve

$$r = R(\bar{\theta}, z; t), \tag{16.19}$$

and its modulus σ_z is constant. Moreover, we assume that it is a *traction stress* (that is with a versus equal to that of the normal to the surface on which it applies).

We also maintain assumption (IR-2) of the independent ring model. When considering the forces acting on $\mathrm{d}\mathcal{S}$, we have now a further term, namely (referring again to Fig. 16.5)

$$\begin{aligned}\mathbf{f}_z &= \big[\boldsymbol{\sigma}_z(\bar{z} + \mathrm{d}z/2) + \boldsymbol{\sigma}_z(\bar{z} - \mathrm{d}z/2)\big] h R \,\mathrm{d}\theta \\ &= \sigma_z \frac{\boldsymbol{\tau}(\bar{z} + \mathrm{d}z/2) - \boldsymbol{\tau}(\bar{z} - \mathrm{d}z/2)}{\mathrm{d}l} \mathrm{d}l h R \,\mathrm{d}\theta = \sigma_z \frac{\mathrm{d}\boldsymbol{\tau}}{\mathrm{d}l} R_0 h_0 \,\mathrm{d}l \,\mathrm{d}\theta + \mathbf{o}(\mathrm{d}z \,\mathrm{d}\theta).\end{aligned}$$

We now exploit the Frenet–Serret formulae to write

$$\frac{\mathrm{d}\boldsymbol{\tau}}{\mathrm{d}l} = \kappa \mathbf{n},$$

where κ is the curvature of the line $r = R(\bar{\theta}, z; t)$, whose expression is

$$\kappa = \frac{\partial^2 R}{\partial z^2}\left[1 + \left(\frac{\partial R}{\partial z}\right)^2\right]^{-3/2}. \tag{16.20}$$

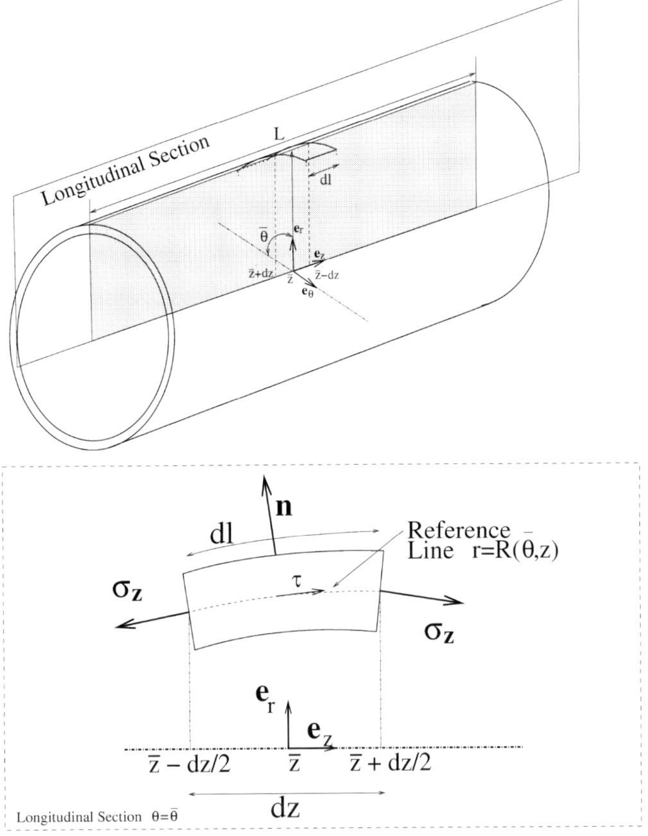

FIG. 16.5. A cylindrical model of the vessel geometry (top) and quantities on a longitudinal section (bottom).

By recalling (16.6) and (16.4), we obtain

$$\mathbf{f}_z \cdot \mathbf{e}_r = \sigma_z \frac{\partial^2 R}{\partial z^2}\left[1 + \left(\frac{\partial R}{\partial z}\right)^2\right]^{-3/2} R_0 h_0 \,\mathrm{d}z\,\mathrm{d}\theta + o(\mathrm{d}z\,\mathrm{d}\theta).$$

We eliminate the geometric non-linearity in the model by neglecting the term $(\partial R/\partial z)^2$. Furthermore, we replace $\partial^2 R/\partial z^2$ by $\partial^2 \eta/\partial z^2$.[5]

By proceeding like in the previous section, we may modify the independent ring model into the following differential equation:

$$\frac{\partial^2 \eta}{\partial t^2} - a\frac{\partial^2 \eta}{\partial z^2} + b\eta = H, \quad \text{in } \Gamma_0^w,\ t \in I, \tag{16.21}$$

where

$$a = \frac{\sigma_z}{\rho_w h_0}.$$

[5]This last equality is clearly true whenever R_0 is varying linearly with z.

The final *generalised string* model is obtained by adding to the expression for $\boldsymbol{\sigma}_z$ in (16.18) a term

$$c\frac{\partial}{\partial t}\frac{\partial \eta}{\partial z}, \qquad c>0,$$

which is a viscoelastic term linking the longitudinal stress to the rate of rotation of the structure. For small displacements, $\partial\eta/\partial z$ is indeed proportional to the angle of rotation around the circumferential direction of the structure, with respect to the reference configuration.

The result is

$$\frac{\partial^2\eta}{\partial t^2}-a\frac{\partial^2\eta}{\partial z^2}+b\eta-c\frac{\partial^3\eta}{\partial t\partial z^2}=H, \quad \text{in } \Gamma_0^w,\ t\in I. \tag{16.22}$$

17. Analysis of vessel wall models

In the following we will provide some a-priori estimates for the differential models just proposed.

We recall Poincaré inequality for the one-dimensional case.

LEMMA 17.1 (Poincaré inequality – one-dimensional case). *Let* $f\in H^1(a,b)$ *with* $f(a)=0$. *Then there exists a positive constant* C_p *such that*

$$\|f\|_{L^2(0,L)}\leqslant C_p\left\|\frac{\mathrm{d}f}{\mathrm{d}x}\right\|_{L^2(0,L)}. \tag{17.1}$$

PROOF. For all $x\in[a,b]$ we have,

$$f(x)=f(a)+\int_a^x\frac{\mathrm{d}f}{\mathrm{d}x}(\tau)\,\mathrm{d}\tau=\int_a^x\frac{\mathrm{d}f}{\mathrm{d}x}(\tau)\,\mathrm{d}\tau.$$

Then,

$$\begin{aligned}\int_a^b f^2(s)\,\mathrm{d}s&=\int_a^b\left(\int_a^s\frac{\mathrm{d}f}{\mathrm{d}x}(\tau)\,\mathrm{d}\tau\right)^2\mathrm{d}s\\&\leqslant\int_a^b\left(\left(\int_a^s 1^2\,\mathrm{d}\tau\right)^{1/2}\left\{\int_a^s\left[\frac{\mathrm{d}f}{\mathrm{d}x}(\tau)\right]^2\mathrm{d}\tau\right\}^{1/2}\right)^2\mathrm{d}s\\&\qquad\text{(by Cauchy–Schwarz inequality)}\\&\leqslant\int_a^b(b-a)\left\|\frac{\mathrm{d}f}{\mathrm{d}x}\right\|^2_{L^2(a,b)}\mathrm{d}s=(b-a)^2\left\|\frac{\mathrm{d}f}{\mathrm{d}x}\right\|^2_{L^2(a,b)},\end{aligned}$$

by which inequality (17.1) is proved by taking $C_p=(b-a)$. The same inequality holds if $f(b)=0$. □

Thanks to the fact that no derivatives with respect to the variable θ are present in the equations, we may carry out some further analysis of the structure models illustrated so far by considering the equations for a fixed value of θ and z.

We will consider Eq. (16.12) and address then the following problem:

$$\frac{\partial^2 \eta}{\partial t^2} + b\eta = H, \quad \text{in } \Gamma_0^w,\ t \in I, \tag{17.2}$$

with the following initial values for the displacement and its time rate:

$$\eta = \eta_0, \qquad \frac{\partial \eta}{\partial t} = \eta_1, \quad \text{in } \Gamma_0^w,\ t = t_0. \tag{17.3}$$

We also introduce the space $L^2(I; L^2(\Gamma_0^w))$ of functions $f : \Gamma_0^w \times I \to \mathbb{R}$ that are square integrable in Γ_0^w for almost every (a.e.) $t \in I$ and such that

$$\int_{t_0}^{t_1} \|f(\tau)\|^2_{L^2(\Gamma_0^w)}\, \mathrm{d}\tau < \infty.$$

LEMMA 17.2. *If $H \in L^2(I; L^2(\Gamma_0^w))$, the following inequality holds for a.e. $t \in I$:*

$$\left\| \frac{\partial \eta}{\partial t}(t) \right\|^2_{L^2(\Gamma_0^w)} + b\|\eta(t)\|^2_{L^2(\Gamma_0^w)} \leqslant \left(\|\eta_1\|^2_{L^2(\Gamma_0^w)} + b\|\eta_0\|^2_{L^2(\Gamma_0^w)} + \int_{t_0}^{t} \|H(\tau)\|^2_{L^2(\Gamma_0^w)}\, \mathrm{d}\tau \right) e^{(t-t_0)}. \tag{17.4}$$

PROOF. It can be obtained by multiplying (17.2) by $\partial\eta/\partial t$ and applying Gronwall lemma (Lemma 10.3). □

Relation (17.4) asserts that the sum of the total kinetic and elastic potential energy associated to Eq. (17.2) is bounded, at each time t, by a quantity which depends only on the initial condition and the forcing term.

Let us consider the generalised string model (16.22) with the following initial and boundary conditions:

$$\eta = \eta_0, \qquad \frac{\partial \eta}{\partial t} = \eta_1 \quad \text{in } \Gamma_0^w,\ t = t_0, \tag{17.5a}$$

$$\eta|_{z=0} = \alpha, \qquad \eta|_{z=L} = \beta, \quad t \in I. \tag{17.5b}$$

Let us define the following energy function:

$$e_s(t) = \frac{1}{2}\left(\left\| \frac{\partial \eta}{\partial t}(t) \right\|^2_{L^2(\Gamma_0^w)} + a\left\| \frac{\partial \eta}{\partial z}(t) \right\|^2_{L^2(\Gamma_0^w)} + b\|\eta(t)\|^2_{L^2(\Gamma_0^w)} \right). \tag{17.6}$$

LEMMA 17.3. *If $H \in L^2(I; L^2(\Gamma_0^w))$ and $\alpha = \beta = 0$, the following inequality holds for a.e. $t \in I$:*

$$e_s(t) + \frac{c}{2}\int_{t_0}^{t} \left\| \frac{\partial^2 \eta}{\partial t\, \partial z}(\tau) \right\|^2_{L^2(\Gamma_0^w)} \mathrm{d}\tau \leqslant e_s(0) + k\int_{t_0}^{t_1} \|H(\tau)\|^2_{L^2(\Gamma_0^w)}\, \mathrm{d}\tau, \tag{17.7}$$

where $k = C_p^2/(2c)$ and C_p is the Poincaré constant.

PROOF. We use the short-hand notations $\dot{\eta}$ and $\ddot{\eta}$ for the time derivatives of η. We first multiply the generalised string equation (16.22) by $\dot{\eta}$ and integrate w.r.t. to z:

$$\begin{aligned}\int_0^L \dot{\eta}\ddot{\eta} - a\int_0^L \dot{\eta}\frac{\partial^2\eta}{\partial z^2} - c\int_0^L \dot{\eta}\frac{\partial^3\eta}{\partial t\partial z^2} + b\int_0^L \dot{\eta}\eta \\ = \frac{1}{2}\frac{\mathrm{d}}{\mathrm{d}t}\int_0^L \dot{\eta}^2 + a\int_0^L \frac{\partial^2\eta}{\partial t\partial z}\frac{\partial\eta}{\partial z} - a\left[\dot{\eta}\frac{\partial\eta}{\partial z}\right]_0^L + c\int_0^L\left(\frac{\partial^2\eta}{\partial t\partial z}\right)^2 \\ - c\left[\frac{\partial^2\eta}{\partial t\partial z}\dot{\eta}\right]_0^L + \frac{b}{2}\frac{\mathrm{d}}{\mathrm{d}t}\int_0^L \eta^2 = \int_0^L \dot{\eta}H.\end{aligned} \tag{17.8}$$

By exploiting the homogeneous boundary conditions and the fact that

$$\frac{\partial^2\eta}{\partial t\partial z}\frac{\partial\eta}{\partial z} = \frac{1}{2}\frac{\partial}{\partial t}\left(\frac{\partial\eta}{\partial z}\right)^2,$$

we have

$$\frac{1}{2}\frac{\mathrm{d}}{\mathrm{d}t}\int_0^L \dot{\eta}^2 + \frac{a}{2}\frac{\mathrm{d}}{\mathrm{d}t}\int_0^L \frac{\partial\eta}{\partial z}^2 + c\int_0^L\left(\frac{\partial^2\eta}{\partial t\partial z}\right)^2 + \frac{b}{2}\frac{\mathrm{d}}{\mathrm{d}t}\int_0^L \eta^2 = \int_0^L \dot{\eta}H.$$

Thanks to the hypothesis of axial symmetry, we have

$$\frac{\mathrm{d}e_s}{\mathrm{d}t} + c\left\|\frac{\partial^2\eta}{\partial t\partial z}\right\|^2_{L^2(\Gamma_0^w)} = \int_{\Gamma_0^w} \dot{\eta}H. \tag{17.9}$$

The application the Cauchy–Schwarz, Young and Poincaré inequalities to the right-hand side gives

$$\begin{aligned}\frac{\mathrm{d}e_s}{\mathrm{d}t} + c\left\|\frac{\partial^2\eta}{\partial t\partial z}\right\|^2_{L^2(\Gamma_0^w)} &\leqslant \frac{1}{4\varepsilon}\|H\|^2_{L^2(\Gamma_0^w)} + \varepsilon\|\dot{\eta}\|^2_{L^2(\Gamma_0^w)} \\ &\leqslant \frac{1}{4\varepsilon}\|H\|^2_{L^2(\Gamma_0^w)} + C_p^2\varepsilon\left\|\frac{\partial^2\eta}{\partial t\partial z}\right\|^2_{L^2(\Gamma_0^w)}\end{aligned}$$

for any positive ε. If we choose ε such that $C_p^2\varepsilon = c/2$ and integrate in time between t_0 and t, we finally obtain the desired result. □

CHAPTER V

18. The coupled fluid structure problem

In this part we will treat the situation arising when the flow in a vessel interacts mechanically with the wall structure. This aspect is particularly relevant for blood flow in large arteries, where the vessel wall radius may vary up to 10% because of the forces exerted by the flowing blood stream.

We will first illustrate a framework for the Navier–Stokes equations in a moving domain which is particularly convenient for the analysis and for the set up of numerical solution methods.

18.1. The Arbitrary Lagrangian Eulerian (ALE) formulation of the Navier–Stokes equation

In Section 9 we have introduced the Navier–Stokes equations in a fixed domain Ω, according to the *Eulerian* approach where the independent spatial variables are the coordinates of a fixed Eulerian system. We now consider the case where the domain is moving. In practical situations, such as the flow inside a portion of a compliant artery, we have to compute the flow solution in a *computational domain* Ω_t varying with time.

The boundary of Ω_t may in general be subdivided into two parts. The first part coincides with the physical fluid boundary, i.e., the vessel wall. In the example of Fig. 18.1, this part is represented by Γ_t^w, which is moving under the effect of the flow field. The other part of $\partial\Omega_t$ corresponds to "fictitious boundaries" (also called artificial boundaries) which delimit the region of interest. They are necessary because solving the fluid equation on the whole portion of space occupied by the fluid under study is in general

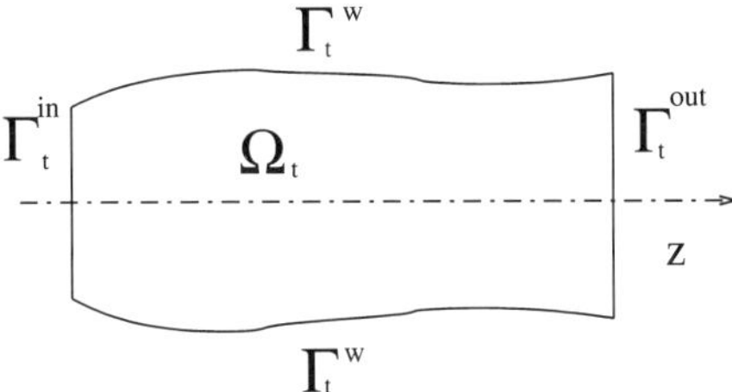

FIG. 18.1. The longitudinal section of a model of an artery. The vessel wall Γ_t^w is moving. The location along the z axis of Γ_t^{in} and Γ_t^{out} are fixed.

impractical, if not impossible. In our case, that would mean solving the whole circulatory system!

In the example of Fig. 18.1, the artificial boundaries are the inlet and outlet boundaries, there indicated by Γ_t^{in} and Γ_t^{out}, respectively. The location of these boundaries is fixed a priori. More precisely, Γ_t^{in} and Γ_t^{out} may change with time because of the displacement of Γ_t^w, however they remain planar and their position along the vessel axis is fixed.

Clearly in this case the Eulerian approach becomes impractical.

A possible alternative would be to use the *Lagrangian approach*. Here, we identify the computational domain on a reference configuration Ω_0 and the corresponding domain in the current configuration, which we indicate with $\Omega_{\mathcal{L}_t}$, will be provided by the Lagrangian mapping (which has been introduced in Section 7), i.e.,

$$\Omega_{\mathcal{L}_t} = \mathcal{L}_t(\Omega_0), \quad t \in I. \tag{18.1}$$

Fig. 18.2 illustrates the situation for the flow inside an artery whose wall is moving. Since the fluid velocity at the wall is equal to the wall velocity, the Lagrangian mapping effectively maps Γ_0^w to the correct wall position Γ_t^w at each time t. However, the "fictitious" boundaries Γ_0^{in} and Γ_0^{out} in the reference configuration will now be transported along the fluid trajectories, into $\Gamma_{\mathcal{L}_t}^{\text{in}}$ and $\Gamma_{\mathcal{L}_t}^{\text{out}}$. This is clearly not acceptable, particularly if one wants to study the problem for a relatively large time interval. Indeed, the domain rapidly becomes highly distorted.

The ideal situation would then be that indicated in Fig. 18.2(b). Even if the wall is moving, one would like to keep the inlet and outlet boundaries at the same spatial location along the vessel axis.

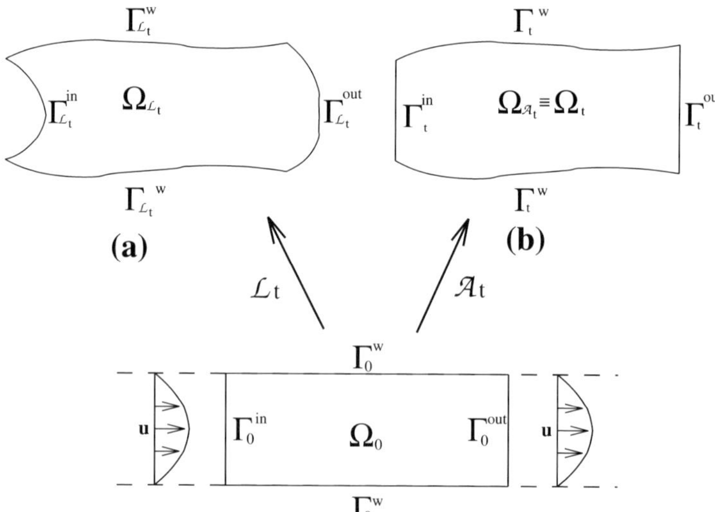

FIG. 18.2. Comparison between the Lagrangian and the ALE approach. The reference computational domain Ω_0 is mapped by (a) the Lagrangian mapping $\mathcal{L}_t$ and by (b) the Arbitrary Lagrangian Eulerian mapping.

With that purpose, we introduce the *Arbitrary Lagrangian Eulerian* (ALE) mapping

$$\mathcal{A}_t : \Omega_0 \to \Omega_{\mathcal{A}_t}, \qquad \mathbf{Y} \to \mathbf{y}(t,\mathbf{Y}) = \mathcal{A}_t(\mathbf{Y}), \tag{18.2}$$

which provides the spatial coordinates $(t, \mathbf{y})$ in terms of the so-called *ALE coordinates* $(t, \mathbf{Y})$, with the basic requirement that $\mathcal{A}_t$ retrieves, at each time $t \in I$, the desired computational domain, i.e.,

$$\Omega_{\mathcal{A}_t} \equiv \mathcal{A}_t(\Omega_0) = \Omega_t, \quad \forall t \in I.$$

The ALE mapping should be continuous and bijective in $\overline{\Omega_0}$. Once given, we may define the *domain* velocity field as

$$\widetilde{\mathbf{w}}(t,\mathbf{Y}) = \frac{\partial}{\partial t}\mathbf{y}(t,\mathbf{Y}), \tag{18.3}$$

which in the spatial coordinates is expressed as

$$\mathbf{w} = \widetilde{\mathbf{w}} \circ \mathcal{A}_t^{-1}, \quad \text{i.e.,} \quad \mathbf{w}(t,\mathbf{y}) = \widetilde{\mathbf{w}}\big(t, \mathcal{A}_t^{-1}(\mathbf{y})\big). \tag{18.4}$$

Similarly to what has been done for the Lagrangian mapping in Section 7 we use the convention of indicating by $\tilde{f}$ the composition of a function f with the ALE mapping, i.e., $\tilde{f} = f \circ \mathcal{A}_t$.

We define the ALE trajectory $T_{\mathbf{Y}}$ for every $\mathbf{Y} \in \Omega_0$ as

$$T_{\mathbf{Y}} = \big\{\big(t, \mathbf{y}(t,\mathbf{Y})\big) \colon t \in I\big\} \tag{18.5}$$

and the ALE derivative of a function f, which we denote by $(D^{\mathcal{A}}/Dt\, f)$, as the time derivative along a trajectory $T_{\mathbf{Y}}$, that is if

$$f : I \times \Omega_t \to \mathbb{R},$$

then

$$\frac{D^{\mathcal{A}}}{Dt} f : I \times \Omega_t \to \mathbb{R}, \qquad \frac{D^{\mathcal{A}}}{Dt} f(t,\mathbf{y}) = \frac{\partial \tilde{f}}{\partial t}(t,\mathbf{Y}), \qquad \mathbf{Y} = \mathcal{A}_t^{-1}(\mathbf{y}). \tag{18.6}$$

Similarly to what already obtained for the Lagrangian mapping (relation (7.4)), we have

$$\frac{D^{\mathcal{A}}}{Dt} f = \frac{\partial f}{\partial t} + \mathbf{w} \cdot \nabla f, \tag{18.7}$$

where now the gradient is made with respect to the $\mathbf{y}$-coordinates.

The Jacobian of the ALE mapping $J^{\mathcal{A}_t}$, defined as

$$J^{\mathcal{A}_t} = \det\left(\frac{\partial \mathbf{y}}{\partial \mathbf{Y}}\right), \tag{18.8}$$

is, for all $t \in I$, a positive quantity because the ALE mapping is surjective and at time t_0 is equal to the identity mapping. It satisfies the following relation:

$$\frac{D^{\mathcal{A}}}{Dt} J^{\mathcal{A}_t} = J^{\mathcal{A}_t} \operatorname{div} \mathbf{w}. \tag{18.9}$$

Again in a way all analogous to what seen for the Lagrangian mapping, we may derive the following result.

THEOREM 18.1 (ALE transport theorem). *Let $V_0 \subset \Omega_0$, and let $V^{\mathcal{A}_t} \subset \Omega_t$ be its image under the mapping $\mathcal{A}_t$. Furthermore, let $f : I \times \Omega_t \to \mathbb{R}$ be continuously differentiable with respect to both variables. Then*

$$\begin{aligned}\frac{\mathrm{d}}{\mathrm{d}t}\int_{V^{\mathcal{A}_t}} f &= \int_{V^{\mathcal{A}_t}} \left(\frac{D^{\mathcal{A}}}{Dt} f + f \operatorname{div} \mathbf{w}\right) = \int_{V^{\mathcal{A}_t}} \left(\frac{\partial f}{\partial t} + \operatorname{div}(f\mathbf{w})\right) \\ &= \int_{V^{\mathcal{A}_t}} \frac{\partial f}{\partial t} + \int_{\partial V^{\mathcal{A}_t}} f\mathbf{w}\cdot\mathbf{n}. \qquad (18.10)\end{aligned}$$

The proof is similar to that of Theorem 7.2 and is omitted.

The Navier–Stokes equations (9.1) are clearly valid on Ω_t, yet it may be convenient to recast them in order to put into evidence the ALE time derivative. We obtain, by a straightforward application of (18.7) to (9.1),

$$\begin{aligned}&\frac{D^{\mathcal{A}}}{Dt}\mathbf{u} + \left[(\mathbf{u}-\mathbf{w})\cdot\nabla\right]\mathbf{u} + \nabla p - 2\,\mathbf{div}\big(\nu\mathbf{D}(\mathbf{u})\big) = \mathbf{f}, \\ &\operatorname{div}\mathbf{u} = 0, \qquad (18.11)\end{aligned}$$

in Ω_t and for all $t \in I$.

18.2. Coupling with the structure model

We now study the properties of the coupled fluid–structure problem, using for the structure the generalised string model (16.22). Referring to Fig. 16.2, we recall that Γ_t^w is the current configuration of the vessel structure, while Γ_0^w is the reference configuration in which the structure equation is written. We also recall that we take $\mathbf{n}$ always to be the outwardly vector normal to the fluid domain boundary.

We will then address the following problem:

For all $t \in I$, find $\mathbf{u}$, p, η such that

$$\begin{aligned}&\frac{D^{\mathcal{A}}}{Dt}\mathbf{u} + \left[(\mathbf{u}-\mathbf{w})\cdot\nabla\right]\mathbf{u} + \nabla p - 2\,\mathbf{div}\big(\nu\mathbf{D}(\mathbf{u})\big) = \mathbf{f}, \\ &\operatorname{div}\mathbf{u} = 0, \quad \text{in } \Omega_t, \qquad (18.12)\end{aligned}$$

and

$$\frac{\partial^2\eta}{\partial t^2} - a\frac{\partial^2\eta}{\partial z^2} + b\eta - c\frac{\partial^3\eta}{\partial t\partial^2 z} = H, \quad \text{in } \Gamma_0^w \qquad (18.13)$$

with the following initial conditions for $t = t_0$:

$$\mathbf{u} = \mathbf{u}_0, \qquad \mathbf{x} \in \Omega_0, \qquad (18.14a)$$

$$\eta = \eta_0, \quad \dot{\eta} = \eta_1, \quad \text{in } \Gamma_0^w, \qquad (18.14b)$$

boundary conditions for $t \in I$,

$$\big[2\nu\mathbf{D}(\mathbf{u}) - (p - P_{\text{ext}})\mathbf{I}\big]\cdot\mathbf{n} = \mathbf{0}, \quad \text{on } \Gamma_t^{\text{out}}, \qquad (18.15a)$$

$$\mathbf{u} = \mathbf{g}, \quad \text{on } \Gamma_t^{\text{in}}, \tag{18.15b}$$

$$\eta|_{z=0} = \alpha, \qquad \eta|_{z=L} = \beta, \tag{18.15c}$$

and the interface condition

$$\tilde{\mathbf{u}} = \mathbf{u} \circ \mathcal{A}_t = \frac{\partial \eta}{\partial t}\mathbf{e}_r, \quad \text{on } \Gamma_0^w, \ t \in I. \tag{18.16}$$

Another interface condition is implicitly provided by the fact that the forcing term H is function of the fluid variables (see (16.14)).

Here, $\mathbf{u}_0$, $\mathbf{g}$, α and β are given functions, H is the forcing term (16.14) and $\mathcal{A}_t$ is an ALE mapping such that $\mathcal{A}_t^{-1}(\partial\Omega_t) = \Gamma^{\text{in}} \cup \Gamma^{\text{out}} \cup \Gamma_0^w$. We have used the ALE form for the Navier–Stokes equations since it is best suited in view of the numerical solution, as it will be detailed in the next section.

We may then recognise the sources of the coupling between the fluid and the structure models, which are twofold (in view of a possible iterative solution strategy):

- fluid $\to$ structure. The fluid solution provides the value of H, which is function of the fluid stresses at the wall.
- structure $\to$ fluid. The movement of the vessel wall changes the geometry on which the fluid equations must be solved. In addition, the proper boundary conditions for the fluid velocity in correspondence to vessel wall are not anymore homogeneous Dirichlet conditions, but they impose the equality between the fluid and the structure velocity. They express the fact that the fluid particle in correspondence of the vessel wall should move at the same velocity as the wall.

Note that we have made some changes with respect to the nomenclature used in (9.1) to indicate that the domain is now moving. We rewrite the expression of the forcing term H, given in (16.14), by noting that while the fluid velocity and pressure are written in the current configuration, H lives in the reference configuration for the vessel wall Γ_0^w. Therefore, following the nomenclature introduced in the previous subsection, we write

$$H = \frac{\rho}{\rho_w h_0}\left[(\tilde{p} - p_0)\frac{R}{R_0} - 2g\tilde{\nu}\big(\widetilde{\mathbf{D}(\mathbf{u}) \cdot \mathbf{n}}\big) \cdot \mathbf{e}_r\right]. \tag{18.17}$$

18.3. An energy inequality for the coupled problem

In this section we will obtain an a-priori inequality for the coupled fluid–structure problem just presented. We will consider only the case of homogeneous boundary conditions, that is

$$\mathbf{g} = \mathbf{0}, \qquad \alpha = \beta = 0,$$

for the coupled problem (18.12)–(18.16).

LEMMA 18.1. *The coupled problem* (18.12)–(18.16) *with* $\mathbf{g} = \mathbf{0}$ *and* $\alpha = \beta = 0$ *satisfies the following energy equality for all* $t \in I$:

$$\frac{\mathrm{d}}{\mathrm{d}t}\left[\frac{\omega}{2}\|\mathbf{u}(t)\|_{\mathbf{L}^2(\Omega_t)} + e_s(t)\right] + 2\omega\int_{\Omega_t} \nu\mathbf{D}(\mathbf{u}) : \mathbf{D}(\mathbf{u}) + c\left\|\frac{\partial^2\eta}{\partial z\partial t}\right\|^2_{L^2(\Gamma_0^w)}$$

$$+\frac{\omega}{2}\int_{\Gamma_t^{\text{out}}}|\mathbf{u}|^2\mathbf{u}\cdot\mathbf{n}=\omega\int_{\Omega_t}\mathbf{f}\cdot\mathbf{u}, \tag{18.18}$$

where e_s was defined in (17.6) *and*

$$\omega=\frac{\rho}{\rho_w h_0}. \tag{18.19}$$

Moreover, if we assume that the net kinetic energy flux is non-negative on the outlet section, i.e.,

$$\int_{\Gamma_t^{\text{out}}}|\mathbf{u}|^2\mathbf{u}\cdot\mathbf{n}\geqslant 0\quad\forall t\in I, \tag{18.20}$$

we obtain the a-priori *energy estimate*

$$\begin{aligned}&\frac{\omega}{2}\left\|\mathbf{u}(t)\right\|_{\mathbf{L}^2(\Omega_t)}+e_s(t)+C_K\omega\nu_0\int_{t_0}^{t}\left\|\nabla\mathbf{u}(\tau)\right\|^2_{\mathbf{L}^2(\Omega_\tau)}\,\mathrm{d}\tau\\&\quad+c\int_{t_0}^{t}\left\|\frac{\partial^2\eta}{\partial z\partial t}(\tau)\right\|^2_{L^2(\Gamma_w^0)}\mathrm{d}\tau\\&\leqslant\frac{\omega}{2}\|\mathbf{u}_0\|_{\mathbf{L}^2(\Omega_t)}+e_s(t_0)+\frac{\omega C_P^2}{4\nu C_K}\int_{t_0}^{t}\|\mathbf{f}(\tau)\|^2_{\mathbf{L}^2(\Omega_\tau)}\,\mathrm{d}\tau,\qquad t\in I.\end{aligned} \tag{18.21}$$

PROOF. We recall expression (17.9) and we recast the right-hand side on the current configuration Γ_t^w. By exploiting (16.3) and (16.4), we have

$$\begin{aligned}\int_{\Gamma_0^w}H\frac{\partial\eta}{\partial t}\,\mathrm{d}\sigma_0&=\frac{\rho}{\rho_w h_0}\int_{\Gamma_0^w}\left[\frac{R}{R_0}(\tilde{p}-p_{\text{ext}})-2g\nu\big(\widetilde{\mathbf{D}(\mathbf{u})\cdot\mathbf{n}}\big)\cdot\mathbf{e}_r\right]\frac{\partial\eta}{\partial t}\,\mathrm{d}\sigma_0\\&=\omega\int_{\Gamma_0^w}\left[(\tilde{p}-p_{\text{ext}})\mathbf{n}\cdot\mathbf{e}_r-2\nu\big(\widetilde{\mathbf{D}(\mathbf{u})\cdot\mathbf{n}}\big)\cdot\mathbf{e}_r\right]\frac{\partial\eta}{\partial t}g\,\mathrm{d}\sigma_0\\&=\omega\int_{\Gamma_0^w}\left[(\tilde{p}-p_{\text{ext}})\mathbf{n}-2\nu\big(\widetilde{\mathbf{D}(\mathbf{u})\cdot\mathbf{n}}\big)\right]\cdot\tilde{\mathbf{u}}g\,\mathrm{d}\sigma_0\\&=\omega\int_{\Gamma_t^w}\left[(p-p_{\text{ext}})\mathbf{n}-2\nu\big(\mathbf{D}(\mathbf{u})\cdot\mathbf{n}\big)\right]\cdot\mathbf{u}\,\mathrm{d}\sigma,\end{aligned}$$

where we have used the interface conditions (18.16). Then,

$$\frac{1}{2}\frac{\mathrm{d}e_s}{\mathrm{d}t}+c\left\|\frac{\partial^2\eta}{\partial z\partial t}\right\|^2_{L^2(\Gamma_0^w)}=\omega\int_{\Gamma_t^w}\left[(p-p_{\text{ext}})\mathbf{n}-2\nu\big(\mathbf{D}(\mathbf{u})\cdot\mathbf{n}\big)\right]\cdot\mathbf{u}\,\mathrm{d}\sigma. \tag{18.22}$$

As for the fluid equations, we follow the same route of Theorem 12.1. In particular, we begin by multiplying (18.12) by $\mathbf{u}$ and integrating over Ω_t, obtaining

$$\int_{\Omega_t}\mathbf{u}\cdot\frac{D^{\mathcal{A}}}{Dt}\mathbf{u}+\int_{\Omega_t}\mathbf{u}\cdot\big[(\mathbf{u}-\mathbf{w})\cdot\nabla\big]\mathbf{u}+\int_{\Omega_t}\mathbf{u}\cdot\big(\nabla p-2\nu\,\mathbf{div}\,\mathbf{D}(\mathbf{u})\big)=(\mathbf{f},\mathbf{u}). \tag{18.23}$$

We now analyse each term in turn. By exploiting the ALE transport theorem (18.10), we may derive that

$$\int_{\Omega_t} \mathbf{u} \cdot \frac{D^{\mathcal{A}}}{Dt}\mathbf{u} = \int_{\Omega_0} J_t \tilde{\mathbf{u}} \cdot \frac{\partial \tilde{\mathbf{u}}}{\partial t} = \frac{1}{2}\int_{\Omega_0} J_t \frac{\partial |\tilde{\mathbf{u}}|^2}{\partial t}$$
$$= \frac{1}{2}\int_{\Omega_t} \frac{D^{\mathcal{A}}}{Dt}|\mathbf{u}|^2 = \frac{1}{2}\frac{\mathrm{d}}{\mathrm{d}t}\int_{\Omega_t} |\mathbf{u}|^2 - \frac{1}{2}\int_{\Omega_t} |\mathbf{u}|^2 \operatorname{div} \mathbf{w}. \tag{18.24}$$

The convective term gives

$$\int_{\Omega_t} \mathbf{u} \cdot \big[(\mathbf{u} - \mathbf{w}) \cdot \nabla\big]\mathbf{u}$$
$$= -\frac{1}{2}\int_{\Omega_t} |\mathbf{u}|^2 \operatorname{div} \mathbf{u} + \frac{1}{2}\int_{\Omega_t} |\mathbf{u}|^2 \operatorname{div} \mathbf{w} + \frac{1}{2}\int_{\partial\Omega_t} |\mathbf{u}|^2 (\mathbf{u} - \mathbf{w}) \cdot \mathbf{n}$$
$$= \frac{1}{2}\int_{\Omega_t} |\mathbf{u}|^2 \operatorname{div} \mathbf{w} + \frac{1}{2}\int_{\Gamma_t^{\mathrm{out}}} |\mathbf{u}|^2 \mathbf{u} \cdot \mathbf{n}, \tag{18.25}$$

since $\operatorname{div} \mathbf{u} = 0$ in Ω_t while $\mathbf{w} = \mathbf{u}$ on Γ_t^w and $\mathbf{w} = \mathbf{0}$ on $\partial\Omega_t \setminus \Gamma_t^w$.

The other terms provide

$$\int_{\Omega_t} \mathbf{u} \cdot \nabla p = (\text{since } p_{\mathrm{ext}} = \text{const.}) \int_{\Omega_t} \mathbf{u} \cdot \nabla (p - p_{\mathrm{ext}})$$
$$= -\int_{\Omega_t} (p - p_{\mathrm{ext}}) \operatorname{div} \mathbf{u} + \int_{\partial\Omega_t} (p - p_{\mathrm{ext}}) \mathbf{u} \cdot \mathbf{n}$$
$$= \int_{\Gamma_t^{\mathrm{out}}} (p - p_{\mathrm{ext}}) \mathbf{u} \cdot \mathbf{n} + \int_{\Gamma_t^w} (p - p_{\mathrm{ext}}) \mathbf{u} \cdot \mathbf{n} \tag{18.26}$$

and

$$\int_{\Omega_t} \nu \mathbf{u} \cdot \mathbf{div}\, \mathbf{D}(\mathbf{u}) = -\int_{\Omega_t} \nu \nabla \mathbf{u} : \mathbf{D}(\mathbf{u}) + \int_{\partial\Omega_t} \nu \mathbf{u} \cdot \mathbf{D}(\mathbf{u}) \cdot \mathbf{n}$$
$$= -\int_{\Omega_t} \nu \mathbf{D}(\mathbf{u}) : \mathbf{D}(\mathbf{u}) + \int_{\partial\Omega_t} \nu \mathbf{u} \cdot \mathbf{D}(\mathbf{u}) \cdot \mathbf{n}$$
$$= -\int_{\Omega_t} \nu \mathbf{D}(\mathbf{u}) : \mathbf{D}(\mathbf{u}) + \int_{\Gamma_t^{\mathrm{out}}} \nu \big(\mathbf{D}(\mathbf{u}) \cdot \mathbf{n}\big) \cdot \mathbf{u}$$
$$+ \int_{\Gamma_t^w} \nu \big(\mathbf{D}(\mathbf{u}) \cdot \mathbf{n}\big) \cdot \mathbf{u}, \tag{18.27}$$

where we have exploited again the symmetry of $\mathbf{D}(\mathbf{u})$.

Using the results obtained in (18.24)–(18.27) into (18.23), rearranging the terms and recalling the boundary condition (18.15a), we can write

$$\frac{1}{2}\frac{\mathrm{d}}{\mathrm{d}t}\|\mathbf{u}\|^2_{\mathbf{L}^2(\Omega_t)} + 2\int_{\Omega_t} \nu \mathbf{D}(\mathbf{u}) : \mathbf{D}(\mathbf{u}) + \frac{1}{2}\int_{\Gamma_t^{\mathrm{out}}} |\mathbf{u}|^2 \mathbf{u} \cdot \mathbf{n}$$
$$+ \int_{\Gamma_t^w} \big[(p - p_{\mathrm{ext}})\mathbf{n} - 2\nu \mathbf{D}(\mathbf{u}) \cdot \mathbf{n}\big] \cdot \mathbf{u} = \int_{\Omega_t} \mathbf{f} \cdot \mathbf{u}.$$

We now recall expression (18.22) and recognise the equivalence of the integrals over Γ_t^w, which express the exchange of power (rate of energy) between fluid and structure. We multiply then the last equality by ω and add it to (18.22), obtaining (18.18).

Using (18.20), (10.2) and the fact that $\nu \geqslant \nu_0 > 0$,

$$\frac{\mathrm{d}}{\mathrm{d}t}\left[\frac{\omega}{2}\|\mathbf{u}(t)\|_{\mathbf{L}^2(\Omega_t)} + e_s(t)\right] + 2C_K\omega\nu_0\|\nabla\mathbf{u}\|^2_{\mathbf{L}^2(\Omega_t)} + c\left\|\frac{\partial^2\eta}{\partial z\partial t}\right\|^2_{L^2(\Gamma_0^w)}$$

$$\leqslant \frac{\mathrm{d}}{\mathrm{d}t}\left[\frac{\omega}{2}\|\mathbf{u}(t)\|_{\mathbf{L}^2(\Omega_t)} + e_s(t)\right] + 2\omega\int_{\Omega_t}\nu\mathbf{D}(\mathbf{u}):\mathbf{D}(\mathbf{u}) + c\left\|\frac{\partial^2\eta}{\partial z\partial t}\right\|^2_{L^2(\Gamma_0^w)}$$

$$\leqslant \omega\int_{\Omega_t}\mathbf{f}\cdot\mathbf{u} \leqslant \frac{\omega}{4\varepsilon}\|\mathbf{f}\|^2_{\mathbf{L}^2(\Omega_t)} + \omega\varepsilon\|\mathbf{u}\|^2_{\mathbf{L}^2(\Omega_t)} \leqslant \frac{\omega}{4\varepsilon}\|\mathbf{f}\|^2_{\mathbf{L}^2(\Omega_t)} + C_P^2\omega\varepsilon\|\nabla\mathbf{u}\|^2_{\mathbf{L}^2(\Omega_t)},$$

for any positive ε. To derive the last inequality we have applied the Poincaré inequality (10.1).

The desired result is then obtained by taking $\varepsilon = (\nu_0 C_K)/C_P^2$ and integrating in time between t_0 and t. □

This last result shows that the energy associated to the coupled problem is bounded, at any time, by quantities which depend only on the initial condition and the applied volume forces. Moreover, since in blood flow simulation we neglect the volume force term $\mathbf{f}$ in the Navier–Stokes equations, estimate (18.21) simplifies into

$$\frac{\omega}{2}\|\mathbf{u}(t)\|_{\mathbf{L}^2(\Omega_t)} + e_s(t) + 2C_K\omega\nu_0\int_{t_0}^t\|\nabla\mathbf{u}(\tau)\|^2_{\mathbf{L}^2(\Omega_\tau)}\,\mathrm{d}\tau$$

$$+ c\int_{t_0}^t\left\|\frac{\partial^2\eta}{\partial z\partial t}(\tau)\right\|^2_{L^2(\Gamma_w^0)}\mathrm{d}\tau$$

$$\leqslant \frac{\omega}{2}\|\mathbf{u}_0\|_{\mathbf{L}^2(\Omega_t)} + e_s(t_0), \quad \forall t \in I.$$

REMARK 18.1. We may note that the non-linear convective term in the Navier–Stokes equations is crucial to obtain the stability result, because it generates a boundary term which compensates that coming from the treatment of the velocity time derivative. These two contributions are indeed only present in the case of a moving boundary.

REMARK 18.2. Should we replace the boundary condition (18.15a) by

$$2\nu\mathbf{D}(\mathbf{u})\cdot\mathbf{n} - \left(p - p_{\mathrm{ext}} + \frac{1}{2}|\mathbf{u}|^2\right)\mathbf{n} = \mathbf{0} \quad \text{on } \Gamma_t^{\mathrm{out}},\ t \in I, \tag{18.28}$$

we would obtain the stability results without the restrictions on the outlet velocity (18.20).

Let us note that the above boundary condition amounts to imposing a zero value for the *total stress* at the outflow surface.

REMARK 18.3. Under slightly different assumptions, that is periodic boundary conditions in space and the presence of a further dissipative term proportional to $\partial^4\eta/\partial z^4$ in the generalised string model, BEIRÃO DA VEIGA [2004] has recently proven an existence result of strong solutions to the coupled fluid–structure problem. The well posedness of fluid–structure interaction solutions in more general settings is still a largely open problem. A review of recent theoretical results may be found in GRANDMONT and MADAY [2000].

The hypothesis (18.20) is obviously satisfied if Γ_t^{out} is indeed an *outflow section*, i.e., $\mathbf{u}\cdot\mathbf{n}\geqslant 0$ for all $\mathbf{x}\in\Gamma_t^{\text{out}}$. As already pointed out, this is seldom true for vascular flow, particularly in large arteries.

We may observe that the "viscoelastic term" $-c(\partial^3\eta/(\partial t\partial^2 z))$ in (16.22) allows to obtain the appropriate regularity of the velocity field $\mathbf{u}$ on the boundary (see NOBILE [2001]).

In the derivation of the energy inequality (18.21), we have considered homogeneous boundary conditions both for the fluid and the structure. However, the conditions $\eta=0$ at $z=0$ and $z=L$, which correspond to hold the wall fixed at the two ends, are not realistic in the context of blood flow. Since the model (16.22) for the structure is of propagative type, the first order absorbing boundary conditions

$$\frac{\partial\eta}{\partial t}-\sqrt{a}\frac{\partial\eta}{\partial z}=0 \quad \text{at } z=0, \tag{18.29}$$

$$\frac{\partial\eta}{\partial t}+\sqrt{a}\frac{\partial\eta}{\partial z}=0 \quad \text{at } z=L \tag{18.30}$$

look more suited to the problem at hand. An inequality of the type (18.21) could still be proven. Indeed, the boundary term which appears in (17.8) would now read

$$-\left[a\frac{\partial\eta}{\partial z}\frac{\partial\eta}{\partial t}+c\frac{\partial^2\eta}{\partial z\partial t}\frac{\partial\eta}{\partial t}\right]_{z=0}^{z=L}=\sqrt{a}\left[\left(\frac{\partial\eta}{\partial t}\bigg|_{z=0}\right)^2+\left(\frac{\partial\eta}{\partial t}\bigg|_{z=L}\right)^2\right]$$
$$+\frac{c}{2}\sqrt{a}\frac{\mathrm{d}}{\mathrm{d}t}\left[\left(\frac{\partial\eta}{\partial t}\bigg|_{z=0}\right)^2+\left(\frac{\partial\eta}{\partial t}\bigg|_{z=L}\right)^2\right].$$

This term, integrated in time, would eventually appear on the left-hand side of inequality (18.21). We may note, however, that we obtain both for $z=0$ and $z=L$ the following expression:

$$\sqrt{a}\int_{t_0}^{t}\left(\frac{\partial\eta}{\partial t}(\tau)\right)^2\mathrm{d}\tau+\frac{c}{2}\sqrt{\frac{1}{a}}\left(\frac{\partial\eta}{\partial t}(t)\right)^2=\frac{c}{2}\sqrt{\frac{1}{a}}\left(\frac{\partial\eta}{\partial t}(t_0)\right)^2. \tag{18.31}$$

This additional term is positive and depends only on initial conditions.

Yet, conditions (18.29) and (18.30) are not compatible with the homogeneous Dirichlet boundary conditions for the fluid; indeed, if $\eta|_{z=0}\neq 0$ and $\mathbf{u}=\mathbf{0}$ on Γ_t^{in}, the trace of $\mathbf{u}$ on the boundary is discontinuous and thus not compatible with the regularity required on the solution of (18.12) (see, e.g., QUARTERONI and VALLI [1994]).

A possible remedy consists of changing the condition $\mathbf{u}=\mathbf{0}$ on Γ_t^{in} into

$$\mathbf{u}\cdot\mathbf{e}_z=g_z\circ\mathcal{A}_t^{-1}, \qquad (\mathbf{T}\cdot\mathbf{n})\times\mathbf{e}_z=\mathbf{0}$$

on Γ_t^{in}, where g_z is a given function defined on Γ_0^{in}, with $g_z = 0$ on $\partial\Gamma_0^{\text{in}}$. Here $\mathbf{T}$ is the stress tensor defined in (8.13). An energy inequality for the coupled problem can be derived also in this case with standard calculations, taking a suitable harmonic extension $\tilde{g}_z$ of the non-homogeneous data g_z. The calculations are here omitted for the sake of brevity.

19. An iterative algorithm to solve the coupled fluid–structure problem

In this section we outline an algorithm that at each time-level allows the decoupling of the sub-problem related to the fluid from that related to the vessel wall. As usual, t^k, $k = 0, 1, \ldots$ denotes the kth discrete time level; $\Delta t > 0$ is the time-step, while v^k is the approximation of the function (scalar or vector) v at time t^k.

The numerical solution of the fluid–structure interaction problem (18.12), (18.13) will be carried out by constructing a suitable finite element approximation of each sub-problem. In particular, for the fluid we need to devise a finite element formulation suitable for moving domains (or, more precisely, moving grids). In this respect, the ALE formulation will provide an appropriate framework.

To better illustrate the situation, we refer to Fig. 19.1 where we have drawn a 2D fluid–structure interaction problem. The fluid domain is Ω_t and the movement of its upper boundary Γ_t^w is governed by a generalised string model. This geometry could be derived from an axisymmetric model of the flow inside a cylindrical vessel. However, in this case we should employ the Navier–Stokes equations in axisymmetric coordinates. Since this example is only for the purpose of illustrating a possible set-up for a coupled fluid–structure algorithm, for the sake of simplicity we consider here a two-dimensional fluid–structure problem governed by Eqs. (18.12), (18.13), with interface conditions (18.16), initial and boundary conditions (18.14) and the additional condition

$$\mathbf{u}|_{\Gamma^0} = \mathbf{0}, \quad t \in I.$$

The algorithm here presented may be readily extended to three-dimensional problems.

The structure on Γ_0^w will be discretised by means of a finite element triangulation $\mathcal{T}_h^s$, like the one we illustrate in Fig. 19.2. We have considered the space S_h of piece-wise linear continuous (P1) finite elements functions to represent the approximate vessel wall displacement η_h. In the same figure we show the position at time t of the discretised vessel wall boundary $\Gamma_{t,h}^w$, corresponding to a given value of the discrete displacement field $\eta_h \in S_h$. Consequently, the fluid domain will be represented at every time by a polygon, which we indicate by $\Omega_{t,h}$. Its triangulation $\mathcal{T}_{t,h}^f$ will be

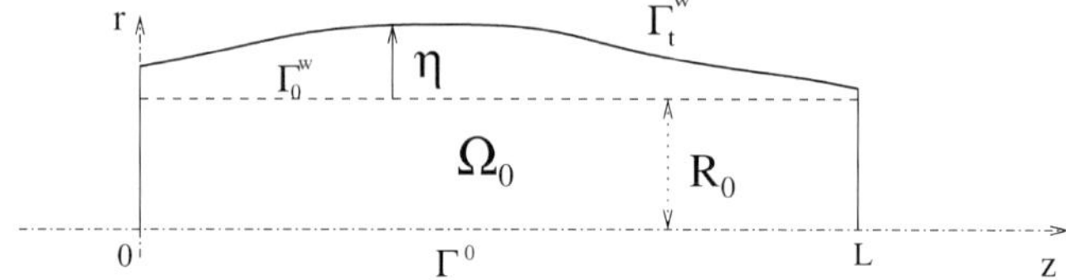

FIG. 19.1. A simple fluid–structure interaction problem.

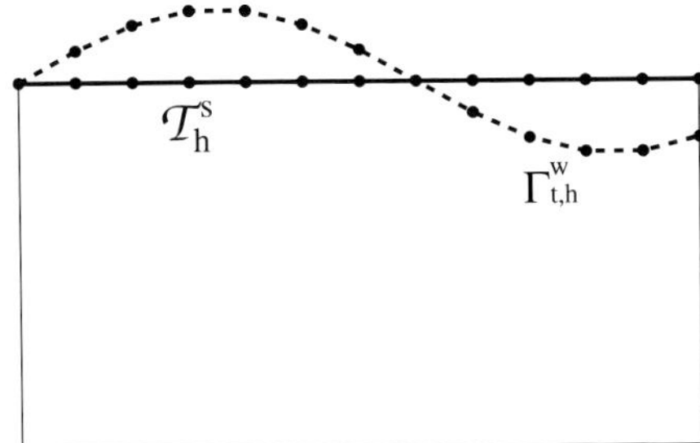

FIG. 19.2. Position of the discretised vessel wall corresponding to a possible value of η_h.

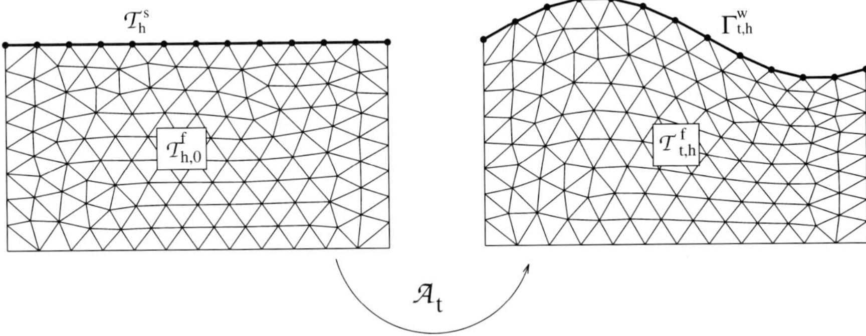

FIG. 19.3. The triangulation used for the fluid problem at each time t is the image through a map $\mathcal{A}_t$ of a mesh constructed on Ω_0.

constructed as the image by an appropriate ALE mapping $\mathcal{A}_t$ of a triangulation $\mathcal{T}_{0,h}^f$ of Ω_0, as shown in Fig. 19.3. Correspondingly, $\Omega_{t,h} = \mathcal{A}_t\Omega_{0,h}$, where $\Omega_{0,h}$ is the approximation of Ω_0 induced by the triangulation $\mathcal{T}_{t,h}^f$ (clearly, if Ω_0 has a polygonal boundary we have $\Omega_{0,h} = \Omega_0$.) The trace of $\mathcal{T}_{0,h}^f$ on Γ_0^w will coincide with the "triangulation" $\mathcal{T}_h^s$ of the vessel wall, thus we consider *geometrically conforming* finite elements between the fluid and the structure. The possibility of using a geometrically non-conforming finite element representation has been investigated in GRANDMONT and MADAY [1998].

We then have to face the following problem. Suppose that we know at $t = t^{k+1}$ a discrete displacement field η_h^{k+1} and thus the corresponding position of the domain boundary $\partial\Omega_{t^{k+1},h}$. How to build a map $\mathcal{A}_{t^{k+1}}$ such that $\mathcal{A}_{t^{k+1}}(\mathcal{T}_{0,h}^f)$ is an acceptable finite element mesh for the fluid domain? This task is in general not simple. However, if we can assume that $\Omega_{t,h}$ is convex for all t and that the displacements are relatively small, the technique known as *harmonic extension* may well serve the purpose. Let $\mathbf{X}_h$ be the P1 finite element vector space associated to $\mathcal{T}_{0,h}^f$, while

$$\mathbf{X}_h^0 = \{\mathbf{w}_h \in \mathbf{X}_h\colon\ \mathbf{w}_h|_{\partial\Omega_{0,h}} = \mathbf{0}\}$$

and let $\mathbf{g}_h : \partial\Omega_{0,h} \to \partial\Omega_{t^{k+1},h}$ be the function describing the fluid domain boundary. We build the map by seeking $\mathbf{y}_h \in \mathbf{X}_h$ such that

$$\int_{\Omega_0} \nabla\mathbf{y}_h : \nabla\mathbf{z}_h = 0 \quad \forall\mathbf{z}_h \in \mathbf{X}_h^0, \qquad \mathbf{y}_h = \mathbf{g}_h, \quad \text{on } \partial\Omega_{0,h}, \tag{19.1}$$

and then setting $\mathcal{A}_{t^{k+1}}(\mathbf{Y}) = \mathbf{y}_h(\mathbf{Y})$, $\forall\mathbf{Y} \in \Omega_{0,h}$. This technique has indeed been adopted for the mesh in Fig. 19.3. From a practical point of view, the value of $\mathbf{y}_h$ in correspondence to the nodes of $\mathcal{T}_{0,h}^f$ gives the position of the corresponding node in $\mathcal{T}_{t,h}^f$ at time t^{k+1}. A more general discussion on the construction of the ALE mapping may be found in FORMAGGIA and NOBILE [1999], NOBILE [2001] as well as in GASTALDI [2001].

REMARK 19.1. Adopting P1 elements for the construction of the ALE map ensures that the triangles of $\mathcal{T}_{h,0}^f$ are mapped into triangles, thus $\mathcal{T}_{h,t}^f$ is a valid triangulation, under the requirement of invertibility of the map (which is assured if the domain is convex and the wall displacements are small).

As for the time evolution, we may adopt a linear time variation within each time slab $[t^k, t^{k+1}]$ by setting

$$\mathcal{A}_t = \frac{t - t^k}{\Delta t}\mathcal{A}_{t^{k+1}} - \frac{t - t^{k+1}}{\Delta t}\mathcal{A}_{t^k}, \quad t \in \left[t^k, t^{k+1}\right].$$

Then, the corresponding domain velocity $\mathbf{w}_h$ will be constant on each time slab.

We are now in the position of describing a possible finite element scheme for both the structure and the fluid problem, to be adopted in the sub-structuring algorithm. We first give more details on the adopted finite element discretisation.

19.1. The discretisation of the structure

For the structure we consider a *mid-point* scheme. We introduce the additional variable $\dot{\eta}^k$ which is the approximation of the displacement velocity at time t^k.

The time advancing scheme reads:

$\forall k \geqslant 0$ *find* η^{k+1} *and* $\dot{\eta}^{k+1}$ *that satisfy the following system*:

$$\frac{\eta^{k+1} - \eta^k}{\Delta t} = \frac{\dot{\eta}^k + \dot{\eta}^{k+1}}{2}, \tag{19.2a}$$

$$\frac{\dot{\eta}^{k+1} - \dot{\eta}^k}{\Delta t} - a\frac{\partial^2}{\partial z^2}\frac{\eta^k + \eta^{k+1}}{2} + b\frac{\eta^{k+1} + \eta^k}{2} - c\frac{\partial^2}{\partial z^2}\frac{\dot{\eta}^{k+1} + \dot{\eta}^k}{2} = H^{k+1/2}, \tag{19.2b}$$

with

$$\eta^{k+1}|_{z=0} = \alpha\left(t^{k+1}\right), \qquad \eta^{k+1}|_{z=L} = \beta\left(t^{k+1}\right), \tag{19.3a}$$

and

$$\dot{\eta}^{k+1}|_{z=0} = \frac{\partial}{\partial t}\alpha\big(t^{k+1}\big), \quad \dot{\eta}^{k+1}|_{z=L} = \frac{\partial}{\partial t}\beta\big(t^{k+1}\big), \tag{19.3b}$$

while the value of η^0 and $\dot{\eta}^0$ are given by the initial conditions.

Here, $H^{k+1/2}$ is a suitable approximation of H at time $t^k + \frac{1}{2}\Delta t$ which in the context of a sub-structuring iteration for the coupled problem is a known quantity and whose calculation from the Navier–Stokes data will be made precise later.

System (19.2) is then discretised in space by taking $\eta_h^k \in S_h$ and $\dot{\eta}_h^k \in S_h$. We set $S_h^0 = \{s_h \in S_h\colon\ s_h(0) = 0,\ s_h(L) = 0\}$ and the finite element problem reads:

For all $k \geqslant 0$ find $\eta_h^{k+1} \in S_h$ and $\dot{\eta}_h^{k+1} \in S_h$ that satisfy the following system:

$$\big(2\eta_h^{k+1} - \Delta t\dot{\eta}_h^{k+1}, s_h\big) = \big(2\eta_h^k + \Delta t\dot{\eta}_h^k, s_h\big), \tag{19.4a}$$

$$\begin{aligned}
&\left(\frac{1}{\Delta t}\dot{\eta}_h^{k+1} + \frac{b}{2}\eta_h^{k+1}, s_h\right) + \frac{1}{2}\left(a\frac{\partial \eta_h^{k+1}}{\partial z} + c\frac{\partial \dot{\eta}_h^{k+1}}{\partial z}, \frac{\partial s_h}{\partial z}\right)\\
&\quad = \big(H^{k+1/2}, s_h\big) + \left(\frac{1}{\Delta t}\dot{\eta}_h^k + \frac{b}{2}\eta_h^k, s_h\right) - \frac{1}{2}\left(a\frac{\partial \eta_h^k}{\partial z} + c\frac{\partial \dot{\eta}_h^k}{\partial z}, \frac{\partial s_h}{\partial z}\right),
\end{aligned} \tag{19.4b}$$

$\forall s_h \in S_h^0$, together with the boundary conditions

$$\eta_h^{k+1}|_{z=0} = \alpha\big(t^{k+1}\big), \quad \eta_h^{k+1}|_{z=L} = \beta\big(t^{k+1}\big), \tag{19.5a}$$

$$\dot{\eta}_h^{k+1}|_{z=0} = \frac{\partial}{\partial t}\alpha\big(t^{k+1}\big), \quad \dot{\eta}_h^{k+1}|_{z=L} = \frac{\partial}{\partial t}\beta\big(t^{k+1}\big), \tag{19.5b}$$

and the initial conditions

$$\eta_h^0 = \pi_{S_h}\eta^0, \qquad \eta_h^0 = \pi_{S_h}\eta^0,$$

being π^{S_h} the standard interpolation operator upon S_h.

19.2. The discretisation of the fluid problem

In the frame of our splitting scheme the velocity field at Γ_t^w as well as the current domain configuration are provided by the calculation of η_h; they can thus be considered as given data. We consider the following finite element spaces. $\widetilde{Q}_h$ is the space of continuous piece-wise linear finite elements, while $\widetilde{\mathbf{V}}_h$ is that of vector functions whose components are in the space $\widetilde{V}_h$ of continuous piece-wise quadratic (or P1-isoP2) finite elements. Both refer to the triangulation $\mathcal{T}_{0,h}^f$ of Ω_0. For a precise definition of these finite element spaces the reader may refer to QUARTERONI and VALLI [1994] or BREZZI and FORTIN [1991].

We will also need to define

$$\widetilde{\mathbf{V}}_h^0 = \{\tilde{\mathbf{v}}_h \in \widetilde{\mathbf{V}}_h\colon\ \tilde{\mathbf{v}}_h|_{\partial\Omega_0\setminus\Gamma_0^{\text{in}}} = \mathbf{0}\}$$

and the space $\widetilde{V}_h^{\Gamma_0^w}$ formed by function in Γ_0^w which are the trace of a function in $\widetilde{V}_h$.

The corresponding spaces on the current configuration will be given by

$$Q_{h,t} = \{q_h\colon\ q_h \circ \mathcal{A}_t \in \widetilde{Q}_h\}, \qquad \mathbf{V}_{h,t} = \{\mathbf{v}_h\colon\ \mathbf{v}_h \circ \mathcal{A}_t \in \widetilde{\mathbf{V}}_h\},$$

and analogously for $\mathbf{V}^0_{h,t}$.

REMARK 19.2. The functions belonging to $Q_{h,t}$ and $\mathbf{V}_{h,t}$ depend also on time through the ALE mapping. A thorough presentation of finite element spaces in an ALE framework is contained in FORMAGGIA and NOBILE [1999] and NOBILE [2001].

We will employ an implicit Euler time advancing scheme with a semi-explicit treatment of the convective term. Let us assume that the solution $(\mathbf{u}_h^k, p_h^k)$ at time step t^k is known, as well as the domain configuration $\Omega_{t^{k+1},h}$ at time t^{k+1} (and thus the corresponding ALE map).

The numerical solution at t^{k+1} can be computed as follows:

Find $\mathbf{u}_h^{k+1} \in \mathbf{V}_{h,t^{k+1}}$ *and* $p_h^{k+1} \in Q_{h,t^{k+1}}$ *such that*

$$\begin{aligned}
&\frac{1}{\Delta t}\big(\mathbf{u}^{k+1}, \tilde{\mathbf{v}}_h\big)_{k+1} - c_{k+1/2}\big(\mathbf{w}^{k+1/2}, \mathbf{u}^{k+1}, \tilde{\mathbf{v}}_h\big) + c_{k+1}\big(\mathbf{u}^k, \mathbf{u}^{k+1}, \tilde{\mathbf{v}}_h\big)\\
&\qquad + d_{k+1/2}\big(\mathbf{w}^{k+1/2}, \mathbf{u}^{k+1}, \tilde{\mathbf{v}}_h\big) + b_{k+1}\big(\tilde{\mathbf{v}}_h, p^{k+1}\big) + a_{k+1}\big(\mathbf{u}^{k+1}, \tilde{\mathbf{v}}_h\big)\\
&\quad = \big(\mathbf{f}^{k+1}, \tilde{\mathbf{v}}_h\big)_{k+1} + \frac{1}{\Delta t}\big(\mathbf{u}^k, \tilde{\mathbf{v}}_h\big)_k, \quad \forall \tilde{\mathbf{v}}_h \in \widetilde{\mathbf{V}}^0_h
\end{aligned} \tag{19.6a}$$

$$b_{k+1}\big(\mathbf{u}^{k+1}, \tilde{q}_h\big) = 0, \quad \forall \tilde{q}_h \in \widetilde{Q}_h, \tag{19.6b}$$

and

$$\mathbf{u}_h^{k+1} = \mathbf{g}_h^{k+1}, \quad \text{on } \Gamma^{\text{in}}_{t^{k+1}}, \tag{19.7a}$$

$$\mathbf{u}_h^{k+1} = \big(\Pi_h^{\Gamma_0^w} \dot{\eta}_h^{k+1}\big) \circ \mathcal{A}^{-1}_{t^{k+1}} \mathbf{e}_r, \quad \text{on } \Gamma^w_{t^{k+1}}. \tag{19.7b}$$

We have defined

$$(\mathbf{w}, \tilde{\mathbf{v}})_k = \int_{\Omega_{t^k}} \mathbf{w} \cdot \big(\tilde{\mathbf{v}} \circ \mathcal{A}^{-1}_{t^k}\big),$$

$$c_k(\mathbf{w}, \mathbf{z}, \tilde{\mathbf{v}}) = \int_{\Omega_{t^k}} \big((\mathbf{w}\cdot\nabla)\mathbf{z}\big) \cdot \big(\tilde{\mathbf{v}} \circ \mathcal{A}^{-1}_{t^k}\big),$$

$$d_k(\mathbf{w}, \mathbf{z}, \tilde{\mathbf{v}}) = \int_{\Omega_{t^k}} (\operatorname{div} \mathbf{w})\mathbf{z} \cdot \big(\tilde{\mathbf{v}} \circ \mathcal{A}^{-1}_{t^k}\big),$$

$$b_k(\mathbf{w}, \tilde{q}) = \int_{\Omega_{t^k}} \operatorname{div} \mathbf{w}\, \big(\tilde{q} \circ \mathcal{A}^{-1}_{t^k}\big), \qquad b_k(\widetilde{\mathbf{w}}, q) = \int_{\Omega_{t^k}} \operatorname{div}\big(\widetilde{\mathbf{w}} \circ \mathcal{A}^{-1}_{t^k}\big) q,$$

$$a_k(\mathbf{w}, \tilde{\mathbf{v}}) = \int_{\Omega_{t^k}} 2\nu \mathbf{D}(\mathbf{w}) : \mathbf{D}\big(\tilde{\mathbf{v}} \circ \mathcal{A}^{-1}_{t^k}\big).$$

The function $\mathbf{g}_h^{k+1}$ is the finite element interpolant of the boundary data $\mathbf{g}(t^{k+1})$ on the space of restrictions of $\mathbf{V}_{h,t^{k+1}}$ on $\Gamma^{\text{in}}_{t^{k+1}}$. Moreover, $\Pi_h^{\Gamma_0^w} : S_h \to \widetilde{V}_h^{\Gamma_0^w}$ is the interpolation operator required to project the discrete vessel velocity computed by the structure solver on the trace space of discrete fluid velocity on the vessel wall. Since we are using geometrically conforming finite elements, this operator is quite simple to build up.

It is understood that when the approximation of $\mathbf{u}$ and $\mathbf{w}$ in (19.6) are not evaluated at the same time as the integral, they need to be mapped on the correct domain by means of the ALE transformation.

REMARK 19.3. The term involving the domain velocity $\mathbf{w}$ has been computed on the intermediate geometry $\Omega_{t^{k+1/2}}$ in order to satisfy the so-called Geometry Conservation Law (GCL) (GUILLARD and FARHAT [2000]). A discussion on the significance of the GCL for the problem at hand may be found in NOBILE [2001].

19.3. *Recovering the forcing term for the vessel wall*

We need now to compute the forcing term $H^{k+1/2}$ in (19.4) as the residual of the discrete momentum equation (19.6b) for time step t^{k+1}. Let us define

$$\begin{aligned} R_h^{k+1}(\tilde{\mathbf{v}}_h) &= \left(\mathbf{f}^{k+1}, \tilde{\mathbf{v}}_h\right)_{k+1} + \frac{1}{\Delta t}\left(\mathbf{u}^k, \tilde{\mathbf{v}}_h\right)_k - \frac{1}{\Delta t}\left(\mathbf{u}^{k+1}, \tilde{\mathbf{v}}_h\right)_{k+1} \\ &\quad + c_{k+1/2}\left(\mathbf{w}^{k+1/2}, \mathbf{u}^{k+1}, \tilde{\mathbf{v}}_h\right) - c_{k+1}\left(\mathbf{u}^k, \mathbf{u}^{k+1}, \tilde{\mathbf{v}}_h\right) \\ &\quad - d_{k+1/2}\left(\mathbf{w}^{k+1/2}, \mathbf{u}^{k+1}, \tilde{\mathbf{v}}_h\right) - b_{k+1}\left(\tilde{\mathbf{v}}_h, p^{k+1}\right) - a_{k+1}\left(\mathbf{u}^{k+1}, \tilde{\mathbf{v}}_h\right), \quad \forall \tilde{\mathbf{v}}_h \in \widetilde{\mathbf{V}}_h. \end{aligned}$$

Note that $R_h^{k+1}(\tilde{\mathbf{v}}_h) = 0$, for all $\tilde{\mathbf{v}}_h \in \widetilde{\mathbf{V}}_h^0$. We define the following operator:

$$\boldsymbol{\mathcal{S}}_h : S_h \to \widetilde{\mathbf{V}}_h, \qquad \boldsymbol{\mathcal{S}}_h s_h = \left[\mathcal{R}_h\left(\Pi_h^{\Gamma_0^w} s_h\right)\right]\mathbf{e}_r,$$

where $\mathcal{R}_h : \widetilde{V}_h^{\Gamma_0^w} \to \widetilde{V}_h$ is a finite element extension operator such that

$$(\mathcal{R}_h v_h)|_{\Gamma_0^w} = v_h, \quad \forall v_h \in \widetilde{V}_h^{\Gamma_0^w},$$

for instance the one obtained by extending by zero at all internal nodes (see QUARTERONI and VALLI [1999]). We then take

$$\left(H^{k+1/2}, s_h\right) = \frac{\omega}{2}\left[R_h^{k+1}(\boldsymbol{\mathcal{S}}_h s_h) + R_h^k(\boldsymbol{\mathcal{S}}_h s_h)\right]. \tag{19.8}$$

19.4. *The algorithm*

We are now in the position of describing an iterative algorithm for the solution of the coupled problem. As usual, we assume to have all quantities available at $t = t^k$, $k \geqslant 0$, provided either by previous calculations or by the initial data and we wish to advance to the new time step t^{k+1}. For ease of notation we here omit the subscript h, with the understanding that we are referring exclusively to finite element quantities.

The algorithm requires to choose a *tolerance* $\tau > 0$, which is used to test the convergence of the procedure, and a *relaxation parameter* $0 < \theta \leqslant 1$. In the following, the subscript $j \geqslant 0$ denotes the sub-iteration counter.

The algorithm reads:

(A1) Extrapolate the vessel wall structure displacement and velocity:

$$\eta_{(0)}^{k+1} = \eta^k + \Delta t \dot{\eta}^k, \qquad \dot{\eta}_{(0)}^{k+1} = \dot{\eta}^k.$$

(A2) Set $j = 0$.

(A2.1) By using $\eta_{(j)}^{k+1}$, compute the new grid for the fluid domain Ω_t and the ALE map by solving (19.1).

(A2.2) Solve the Navier–Stokes problem (19.6) to compute $\mathbf{u}_{(j+1)}^{k+1}$ and $p_{(j+1)}^{k+1}$, using as velocity on the wall boundary the one calculated from $\dot{\eta}_{(j)}^{k+1}$.

(A2.3) Solve (19.4) to compute η_*^{k+1} and $\dot{\eta}_*^{k+1}$ using as forcing term the one recovered from $\mathbf{u}_{(j+1)}^{k+1}$ and $p_{(j+1)}^{k+1}$ using (19.8).

(A2.4) Unless $\|\eta_*^{k+1} - \eta_{(j)}^{k+1}\|_{L^2(\Gamma_0^w)} + \|\dot{\eta}_*^{k+1} - \dot{\eta}_{(j)}^{k+1}\|_{L^2(\Gamma_0^w)} \leqslant \tau$, set

$$\eta_{(j+1)}^{k+1} = \theta \eta_{(j)}^{k+1} + (1-\theta)\eta_*^{k+1},$$
$$\dot{\eta}_{(j+1)}^{k+1} = \theta \dot{\eta}_{(j)}^{k+1} + (1-\theta)\dot{\eta}_*^{k+1},$$

and $j \leftarrow j+1$. Then return to step (A2.1).

(A3) Set

$$\eta^{k+1} = \eta_*^{k+1}, \qquad \dot{\eta}^{k+1} = \dot{\eta}_*^{k+1}.$$
$$\mathbf{u}^{k+1} = \mathbf{u}_{(j+1)}^{k+1}, \qquad p^{k+1} = p_{(j+1)}^{k+1}.$$

If the algorithm converges, $\lim_{j\to\infty} \mathbf{u}_{(j)}^{k+1} = \mathbf{u}^{k+1}$ and $\lim_{j\to\infty} \eta_{(j)}^{k+1} = \eta^{k+1}$, where $\mathbf{u}^{k+1}$ and η^{k+1} are the solution at time step t^{k+1} of the coupled problem.

The algorithm entails, at each sub-iteration, the computation of the generalised string equation (19.4)–(19.5), the Navier–Stokes equations and the solution of two Laplace equations (19.1), one for every displacement component.

Improvements on the computational efficiency of the coupled procedure just described may be obtained either by employing standard acceleration techniques like Aitken extrapolation, or by using an altogether different approach to the non-linear problem (like Newton–Krylov techniques or multilevel schemes). The matter is still the subject of current active research investigations.

More explicit schemes for the fluid–structure interaction problem, known as "serial staggered" procedures, have been successfully applied to aeroelastic 2D and 3D problems (FARHAT, LESOINNE and MAMAN [1995], FARHAT and LESOINNE [2000], PIPERNO and FARHAT [2001]). However, it has been found that in the case of an incompressible fluid they become unstable when the density of the structure mass is comparable to that of the fluid (LE TALLEC and MOURO [2001]), which is unfortunately our situation. An analysis of decoupling technique for unsteady fluid structure interaction, carried out on a simplified, yet representative, one-dimensional model may be found in GRANDMONT, GUIMET and MADAY [2001].

CHAPTER VI

20. One-dimensional models of blood flow in arteries

In this section we introduce a simple 1D model to describe the flow motion in arteries and its interaction with the wall displacement. In the absence of branching, a short section of an artery may be considered as a cylindrical compliant tube. As before we denote by $I = (t_0, t_1)$ the time interval of interest and by Ω_t the spatial domain which is supposed to be a circular cylinder filled with blood. The reason why one-dimensional models for blood flow may be attractive is that full 3D investigations are quite computationally expensive. Yet, in many situations we might desire to have just information of the evolution of averaged quantities along the arterial tree, such as mass flux and average pressure. In this context simplified models are able to provide an reasonable answer in short times.

As already done in Section 15, we will employ cylindrical coordinates and indicate with $\mathbf{e}_r$, $\mathbf{e}_\theta$ and $\mathbf{e}_z$ the radial, circumferential and axial unit vectors, respectively, and with (r, θ, z) the corresponding coordinates system. The vessel extends from $z = 0$ to $z = L$ and the vessel length L is constant with time.

The basic model is deduced by making the following assumptions, some of which are analogous to the ones made in Section 16:

(A1) *Axial symmetry*. All quantities are independent from the angular coordinate θ. As a consequence, every axial section $z =$ const. remains circular during the wall motion. The tube radius R is a function of z and t.

(A2) *Radial displacements*. The wall displaces along the radial direction solely, thus at each point on the tube surface we may write $\boldsymbol{\eta} = \eta \mathbf{e}_r$, where $\eta = R - R_0$ is the displacement with respect to the reference radius R_0.

(A3) *Constant pressure*. We assume that the pressure P is constant on each section, so that it depends only on z and t.

(A4) *No body forces*. We neglect body forces (the inclusion of the gravity force, if needed, is straightforward); thus we put $\mathbf{f} = \mathbf{0}$ in the momentum equation (11.1a).

(A5) *Dominance of axial velocity*. The velocity components orthogonal to the z axis are negligible compared to the component along z. The latter is indicated by u_z and its expression in cylindrical coordinates reads

$$u_z(t, r, z) = \bar{u}(t, z) s\left(\frac{r}{R(t, z)}\right), \tag{20.1}$$

where $\bar{u}$ is the *mean velocity* on each axial section and $s:\mathbb{R}\to\mathbb{R}$ is a *velocity profile*.[6]

A generic axial section will be indicated by $\mathcal{S}=\mathcal{S}(t,z)$. Its measure A is given by

$$A(t,z)=\text{meas}\big(\mathcal{S}(t,z)\big)=\pi R^2(t,z)=\pi\big(R_0(z)+\eta(t,z)\big)^2. \tag{20.2}$$

The mean velocity $\bar{u}$ is then given by

$$\bar{u}=A^{-1}\int_{\mathcal{S}} u_z\,\mathrm{d}\sigma,$$

and from (20.1) and the definition of $\bar{u}$ it follows that

$$\int_0^1 s(y)y\,\mathrm{d}y=\frac{1}{2}.$$

We will indicate with α the *momentum-flux correction coefficient* (sometimes called Coriolis coefficient), defined as

$$\alpha=\frac{\int_{\mathcal{S}} u_z^2\,\mathrm{d}\sigma}{A\bar{u}^2}=\frac{\int_{\mathcal{S}} s^2\,\mathrm{d}\sigma}{A}, \tag{20.3}$$

where the dependence of the various quantities on the spatial and time coordinates is understood. It is immediate to verify that $\alpha\geqslant 1$. In general, this coefficient will vary in time and space, yet in our model it is taken constant as a consequence of (20.1).

One possible choice for the profile law is the parabolic profile $s(y)=2(1-y^2)$, which corresponds to the Poiseuille solution characteristic of steady flows in circular tubes. In this case we have $\alpha=4/3$. However, for blood flow in arteries it has been found that the velocity profile is, on average, rather flat. Indeed, a profile law often used for blood flow in arteries (see, for instance, SMITH, PULLAN and HUNTER [2003]) is a power law of the type $s(y)=\gamma^{-1}(\gamma+2)(1-y^\gamma)$, with typically $\gamma=9$ (the value $\gamma=2$ gives again the parabolic profile). Correspondingly, we have $\alpha=1.1$. Furthermore, we will see that the choice $\alpha=1$, which indicates a completely flat velocity profile, would lead to a certain simplification in our analysis.

The mean flux Q, defined as

$$Q=\int_{\mathcal{S}} u_z\,\mathrm{d}\sigma=A\bar{u},$$

is one of the main variables of our problem, together with A and the pressure P.

20.1. *The derivation of the model*

There are (at least) three ways of deriving our model. The first one moves from the incompressible Navier–Stokes equations with constant viscosity and performs an asymptotic analysis by assuming that the ratio R_0/L is small, thus discarding the higher order terms with respect to R_0/L (see BARNARD, HUNT, TIMLAKE and VARLEY [1966]).

[6]The fact that the velocity profile does not vary is in contrast with experimental observations and numerical results carried out with full scale models. However, it is a necessary assumption for the derivation of the reduced model. One may then think s as being a profile representative of an average flow configuration.

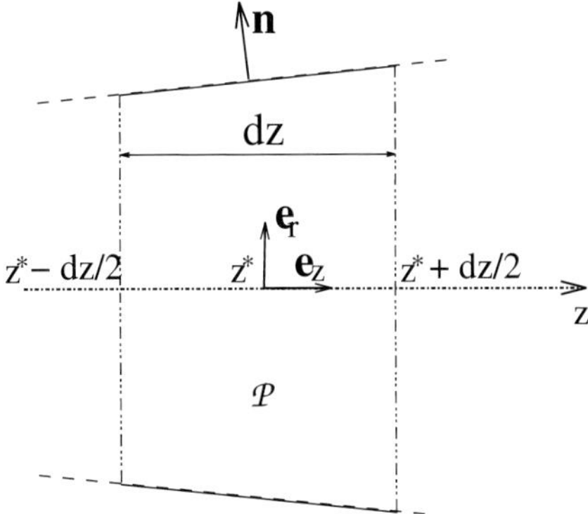

FIG. 20.1. A longitudinal section ($\theta =$ const.) of the tube and the portion between $z = z^* - \mathrm{d}z/2$ and $z = z^* + \mathrm{d}z/2$ used for the derivation of the 1D reduced model.

The second approach derives the model directly from the basic conservation laws written in integral form. The third approach consists of integrating the Navier–Stokes equations on a generic section $\mathcal{S}$.

We will indicate with Γ_t^w the wall boundary of Ω_t, which now reads

$$\Gamma_t^w = \{(r, \theta, z)\colon r = R(z, t),\ \theta \in [0, 2\pi),\ z \in (0, L)\}$$

while $\mathbf{n}$ is the outwardly oriented normal to $\partial\Omega_t$. Under the previous assumption, the momentum and continuity equations along z are:

$$\frac{\partial u_z}{\partial t} + \operatorname{div}(u_z \mathbf{u}) + \frac{1}{\rho}\frac{\partial P}{\partial z} - \nu \Delta u_z = 0, \quad z \in (0, L),\ t \in I, \tag{20.4a}$$

$$\operatorname{div} \mathbf{u} = 0, \quad z \in (0, L),\ t \in I, \tag{20.4b}$$

and on the tube wall we have

$$\mathbf{u} = \dot{\boldsymbol{\eta}}, \quad \text{on } \Gamma_t^w,\ t \in I.$$

We have written the convective term in divergence form, like in (8.12), because it simplifies the further derivation.

To ease notation, in this section we will omit to explicitly indicate the time dependence, with the understanding that all variables are considered at time t. Let us consider the portion $\mathcal{P}$ of Ω_t, sketched in Fig. 20.1, comprised between $z = z^* - \mathrm{d}z/2$ and $z = z^* + \mathrm{d}z/2$, with $z^* \in (0, L)$ and $\mathrm{d}z > 0$ small enough so that $z^* + \mathrm{d}z/2 < L$ and $z^* - \mathrm{d}z/2 > 0$. The part of $\partial\mathcal{P}$ laying on the tube wall is indicated by $\Gamma_{\mathcal{P}}^w$. The reduced model is derived by integrating (20.4b) and (20.4a) on $\mathcal{P}$ and passing to the limit as $\mathrm{d}z \to 0$, assuming that all quantities are smooth enough.

We will first illustrate a result derived from the application of the ALE transport theorem (Theorem 18.1) to $\mathcal{P}$.

LEMMA 20.1. *Let $f : \Omega_t \times I \to \mathbb{R}$ be an axisymmetric function, i.e., $\partial f/\partial\theta = 0$. Let us indicate by f_w the value of f on the wall boundary and by $\bar{f}$ its mean value on each* *axial section, defined by*

$$\bar{f} = A^{-1}\int_{\mathcal{S}} f \, \mathrm{d}\sigma .$$

We have the following relation:

$$\frac{\partial}{\partial t}(A\bar{f}) = A\overline{\frac{\partial f}{\partial t}} + 2\pi R\dot{\eta} f_w . \tag{20.5}$$

In particular, taking $f = 1$ *yields*

$$\frac{\partial A}{\partial t} = 2\pi R\dot{\eta}. \tag{20.6}$$

PROOF. The application of (18.10) to $\mathcal{P}$ gives

$$\frac{\mathrm{d}}{\mathrm{d}t}\int_{\mathcal{P}} f = \int_{\mathcal{P}} \frac{\partial f}{\partial t} + \int_{\partial\mathcal{P}} f\mathbf{g}\cdot\mathbf{n}, \tag{20.7}$$

where $\mathbf{g}$ denotes the velocity of the boundary of $\mathcal{P}$, i.e.,

$$\mathbf{g} = \begin{cases} \dot{\boldsymbol{\eta}} & \text{on } \Gamma_{\mathcal{P}}^{w}, \\ \mathbf{0} & \text{on } \partial\mathcal{P}\setminus\Gamma_{\mathcal{P}}^{w}. \end{cases} \tag{20.8}$$

Then, by applying the mean-value theorem to both sides of (20.7), we have

$$\frac{\mathrm{d}}{\mathrm{d}t}\big[A(z^*)\bar{f}(z^*)\,\mathrm{d}z + \mathrm{o}(\mathrm{d}z)\big] = A\overline{\frac{\partial f}{\partial t}} + \mathrm{o}(\mathrm{d}z) + \int_{\Gamma_{\mathcal{P}}^{w}} f\dot{\eta}\mathbf{e}_r\cdot\mathbf{n}.$$

We recall relation (16.5), already used in the derivation of the models for the wall structure dynamics, to write

$$\int_{\Gamma_{\mathcal{P}}^{w}} f\dot{\eta}\mathbf{e}_r\cdot\mathbf{n}$$
$$= \int_0^{2\pi}\int_{z^*-\mathrm{d}z/2}^{z^*+\mathrm{d}z/2} f\dot{\eta}R\,\mathrm{d}z\,\mathrm{d}\theta = \big[2\pi\dot{\eta}(z^*)R(z^*)f_w(z^*)\,\mathrm{d}z + \mathrm{o}(\mathrm{d}z)\big]. \tag{20.9}$$

By substituting into (20.7), dividing by $\mathrm{d}z$ and passing to the limit as $\mathrm{d}z \to 0$, we obtain the desired result. □

We are now ready to derive our reduced model. We start first from the continuity equation. Using the divergence theorem, we obtain

$$\begin{aligned} 0 = \int_{\mathcal{P}} \operatorname{div}\mathbf{u} &= -\int_{\mathcal{S}^-} u_z + \int_{\mathcal{S}^+} u_z + \int_{\Gamma_{\mathcal{P}}^{w}} \mathbf{u}\cdot\mathbf{n} \\ &= -\int_{\mathcal{S}^-} u_z + \int_{\mathcal{S}^+} u_z + \int_{\Gamma_{\mathcal{P}}^{w}} \dot{\boldsymbol{\eta}}\cdot\mathbf{n}. \end{aligned} \tag{20.10}$$

We have exploited (20.8) and the fact that $\mathbf{n} = -\mathbf{e}_z$ on $\mathcal{S}^-$ while $\mathbf{n} = \mathbf{e}_z$ on $\mathcal{S}^+$. Now, since $\dot{\boldsymbol{\eta}} = \dot{\eta}\mathbf{e}_r$, we deduce

$$\int_{\Gamma_{\mathcal{P}}^{w}} \dot{\boldsymbol{\eta}}\cdot\mathbf{n} = \big[2\dot{\eta}\pi R(z^*)\,\mathrm{d}z + \mathrm{o}(\mathrm{d}z)\big] = (\text{by (20.6)}) = \frac{\partial}{\partial t}A(z^*)\,\mathrm{d}z + \mathrm{o}(\mathrm{d}z).$$

By substituting into (20.10), using the definition of Q, and passing to the limit as $\mathrm{d}z \to 0$, we finally obtain

$$\frac{\partial A}{\partial t} + \frac{\partial Q}{\partial z} = 0,$$

which is the reduced form of the continuity equation.

We will now consider all terms in the momentum equation in turn. Again, we will integrate them over $\mathcal{P}$ and consider the limit as $\mathrm{d}z$ tends to zero,

$$\int_{\mathcal{P}} \frac{\partial u_z}{\partial t} = \frac{\mathrm{d}}{\mathrm{d}t}\int_{\mathcal{P}} u_z - \int_{\partial\mathcal{P}} u_z \mathbf{g}\cdot\mathbf{n} = \frac{\mathrm{d}}{\mathrm{d}t}\int_{\mathcal{P}} u_z.$$

In order to eliminate the boundary integral, we have exploited the fact that $u_z = 0$ on $\Gamma_{\mathcal{P}}^w$ and $\mathbf{g} = \mathbf{0}$ on $\mathcal{S}^-$ and $\mathcal{S}^+$. We may then write

$$\int_{\mathcal{P}} \frac{\partial u_z}{\partial t} = \frac{\partial}{\partial t}\big[A(z^*)\bar{u}(z^*)\,\mathrm{d}z + \mathrm{o}(\mathrm{d}z)\big] = \frac{\partial Q}{\partial t}(z^*)\,\mathrm{d}z + \mathrm{o}(\mathrm{d}z).$$

Moreover, we have

$$\begin{aligned}\int_{\mathcal{P}} \mathrm{div}(u_z\mathbf{u}) &= \int_{\partial\mathcal{P}} u_z\mathbf{u}\cdot\mathbf{n} = -\int_{\mathcal{S}^-} u_z^2 + \int_{\mathcal{S}^+} u_z^2 + \int_{\Gamma_{\mathcal{P}}^w} u_z\mathbf{g}\cdot\mathbf{n}\\ &= \alpha\left[A\left(z^* + \frac{\mathrm{d}z}{2}\right)\bar{u}^2\left(z^* + \frac{\mathrm{d}z}{2}\right) - A\left(z^* - \frac{\mathrm{d}z}{2}\right)\bar{u}^2\left(z^* - \frac{\mathrm{d}z}{2}\right)\right]\\ &= \frac{\partial\alpha A\bar{u}^2}{\partial z}(z^*)\,\mathrm{d}z + \mathrm{o}(\mathrm{d}z).\end{aligned}$$

Again, we have exploited the condition $u_z = 0$ on $\Gamma_{\mathcal{P}}^w$.

Since the pressure is assumed to be constant on each section, we obtain

$$\begin{aligned}\int_{\mathcal{P}} \frac{\partial P}{\partial z} &= -\int_{\mathcal{S}^-} P + \int_{\mathcal{S}^+} P + \int_{\Gamma_{\mathcal{P}}^w} Pn_z\\ &= A\left(z^* + \frac{\mathrm{d}z}{2}\right)P\left(z^* + \frac{\mathrm{d}z}{2}\right) - A\left(z^* - \frac{\mathrm{d}z}{2}\right)P\left(z^* - \frac{\mathrm{d}z}{2}\right)\\ &\quad + \int_{\Gamma_{\mathcal{P}}^w} Pn_z.\end{aligned} \tag{20.11}$$

Since

$$\int_{\partial\mathcal{P}} n_z = 0,$$

we may write that

$$\begin{aligned}\int_{\Gamma_{\mathcal{P}}^w} Pn_z &= P(z^*)\int_{\Gamma_{\mathcal{P}}^w} n_z + \mathrm{o}(\mathrm{d}z) = -P(z^*)\int_{\partial\mathcal{P}\setminus\Gamma_{\mathcal{P}}^w} n_z + \mathrm{o}(\mathrm{d}z)\\ &= -P(z^*)\left(A\left(z^* + \frac{\mathrm{d}z}{2}\right) - A\left(z^* - \frac{\mathrm{d}z}{2}\right)\right) + \mathrm{o}(\mathrm{d}z).\end{aligned}$$

By substituting the last result into (20.11), we have

$$\begin{aligned}\int_{\mathcal{P}} \frac{\partial P}{\partial z} &= A\left(z^* + \frac{\mathrm{d}z}{2}\right)P\left(z^* + \frac{\mathrm{d}z}{2}\right) - A\left(z^* - \frac{\mathrm{d}z}{2}\right)P\left(z^* - \frac{\mathrm{d}z}{2}\right) \\ &\quad - P(z^*)\left[A\left(z^* + \frac{\mathrm{d}z}{2}\right) - A\left(z^* - \frac{\mathrm{d}z}{2}\right)\right] + \mathrm{o}(\mathrm{d}z) \\ &= \frac{\partial(AP)}{\partial z}(z^*)\,\mathrm{d}z - P(z^*)\frac{\partial A}{\partial z}(z^*)\,\mathrm{d}z + \mathrm{o}(\mathrm{d}z) = A\frac{\partial P}{\partial z}(z^*)\,\mathrm{d}z + \mathrm{o}(\mathrm{d}z).\end{aligned}$$

We finally consider the viscous term,

$$\int_{\mathcal{P}} \Delta u_z = \int_{\partial\mathcal{P}} \nabla u_z \cdot \mathbf{n} = -\int_{\mathcal{S}^-} \frac{\partial u_z}{\partial z} + \int_{\mathcal{S}^+} \frac{\partial u_z}{\partial z} + \int_{\Gamma_{\mathcal{P}}^w} \nabla u_z \cdot \mathbf{n}.$$

We neglect $\partial u_z/\partial z$ by assuming that its variation along z is small compared to the other terms. Moreover, we split $\mathbf{n}$ into two vector components, the radial component $\mathbf{n}_r = n_r\mathbf{e}_r$ and $\mathbf{n}_z = \mathbf{n} - \mathbf{n}_r$. Owing to the cylindrical geometry, $\mathbf{n}$ has no component along the circumferential coordinate and, consequently, $\mathbf{n}_z$ is indeed oriented along z. We may thus write

$$\int_{\mathcal{P}} \Delta u_z = \int_{\Gamma_{\mathcal{P}}^w} (\nabla u_z \cdot \mathbf{n}_z + \nabla u_z \cdot \mathbf{e}_r n_r)\,\mathrm{d}\sigma.$$

Again, we neglect the term $\nabla u_z \cdot \mathbf{n}_z$, which is proportional to $\partial u_z/\partial z$. We recall relation (20.1) to write

$$\int_{\mathcal{P}} \Delta u_z = \int_{\Gamma_{\mathcal{P}}^w} n_r \nabla u_z \cdot \mathbf{e}_r\,\mathrm{d}\sigma = \int_{\Gamma_{\mathcal{P}}^w} \bar{u}R^{-1}s'(1)\mathbf{n}\cdot\mathbf{e}_r\,\mathrm{d}\sigma = 2\pi\int_{z^*-\mathrm{d}z/2}^{z^*+\mathrm{d}z/2} \bar{u}s'(1)\,\mathrm{d}z,$$

where we have used the relation $n_r\,\mathrm{d}\sigma = 2\pi R\,\mathrm{d}z$ and indicated by s' the first derivative of s.

Then,

$$\int_{\mathcal{P}} \Delta u_z \approx 2\pi\bar{u}(z^*)s'(1)\,\mathrm{d}z.$$

By substituting all results into (20.4a), dividing all terms by $\mathrm{d}z$ and passing to the limit as $\mathrm{d}z \to 0$, we may finally write the momentum equation of our one-dimensional model as follows:

$$\frac{\partial Q}{\partial t} + \frac{\partial(\alpha A\bar{u}^2)}{\partial z} + \frac{A}{\rho}\frac{\partial P}{\partial z} + K_r\bar{u} = 0,$$

where

$$K_r = -2\pi\nu s'(1)$$

is a *friction parameter*, which depends on the type of profile chosen, i.e., on the choice of the function s in (20.1). For a profile law given by $s(y) = \gamma^{-1}(\gamma+2)(1-y^\gamma)$, we have $K_r = 2\pi\nu(\gamma+2)$. In particular, for a parabolic profile $K_r = 8\pi\nu$, while for $\gamma = 9$ we obtain $K_r = 22\pi\nu$.

To conclude, the final system of equations reads

$$\frac{\partial A}{\partial t} + \frac{\partial Q}{\partial z} = 0, \quad z \in (0, L),\ t \in I, \tag{20.12a}$$

$$\frac{\partial Q}{\partial t} + \alpha \frac{\partial}{\partial z}\left(\frac{Q^2}{A}\right) + \frac{A}{\rho}\frac{\partial P}{\partial z} + K_r\left(\frac{Q}{A}\right) = 0, \quad z \in (0, L),\ t \in I, \tag{20.12b}$$

where the unknowns are A, Q and P and α is here taken constant.

20.2. Accounting for the vessel wall displacement

In order to close system (20.12), we provide a relation for the pressure. A possibility is to resort to an algebraic relation linking pressure to the wall deformation and consequently to the vessel section A.

More generally, we may assume that the pressure satisfies a relation like

$$P(t, z) - P_{\text{ext}} = \psi\big(A(t, z); A_0(z), \boldsymbol{\beta}(z)\big), \tag{20.13}$$

where we have outlined that the pressure will in general depend also on $A_0 = \pi R_0^2$ and on a set of coefficients $\boldsymbol{\beta} = (\beta_0, \beta_1, \ldots, \beta_p)$, related to physical and mechanical properties, that are, in general, *given* functions of z. Here P_{ext} indicates, as in Section 15, the external pressure. We require that ψ be (at least) a C^1 function of all its arguments and be defined for all $A > 0$ and $A_0 > 0$, while the range of variation of $\boldsymbol{\beta}$ will depend by the particular mechanical model chosen for the vessel wall.

Furthermore, we require that for all allowable values of A, A_0 and $\boldsymbol{\beta}$,

$$\frac{\partial \psi}{\partial A} > 0 \quad \text{and} \quad \psi(A_0; A_0, \boldsymbol{\beta}) = 0. \tag{20.14}$$

By exploiting the linear elastic law provided in (16.16), with the additional simplifying assumption (16.15), and using the fact that

$$\eta = \big(\sqrt{A} - \sqrt{A_0}\big)/\sqrt{\pi}, \tag{20.15}$$

we can obtain the following expression for ψ:

$$\psi(A; A_0, \beta_0) = \beta_0 \frac{\sqrt{A} - \sqrt{A_0}}{A_0}. \tag{20.16}$$

We have identified $\boldsymbol{\beta}$ with the single parameter

$$\beta_0 = \frac{\sqrt{\pi} h_0 E}{1 - \xi^2}.$$

The latter depends on z only in those cases where the Young modulus E or the vessels thickness h_0 are not constant.

For ease of notation, the dependence of A, A_0 and $\boldsymbol{\beta}$ from their arguments will be understood. It is immediate to verify that all the requirements in (20.14) are indeed satisfied.

Another commonly used expression for the pressure-area relationship is given by HAYASHI, HANDA, NAGASAWA and OKUMURA [1980], SMITH, PULLAN and

HUNTER [2003]:

$$\psi(A;A_0,\boldsymbol{\beta})=\beta_0\left[\left(\frac{A}{A_0}\right)^{\beta_1}-1\right].$$

In this case, $\boldsymbol{\beta}=(\beta_0,\beta_1)$, where $\beta_0>0$ is an elastic coefficient while $\beta_1>0$ is normally obtained by fitting the stress-strain response curves obtained by experiments.

Another alternative formulation (LANGEWOUTERS, WESSELING and GOEDHARD [1984]) is

$$\psi(A;A_0,\boldsymbol{\beta})=\beta_0\tan\left[\pi\left(\frac{A-A_0}{2A_0}\right)\right],$$

where again the coefficients vector $\boldsymbol{\beta}$ reduces to the single coefficient β_0.

In the following, whenever not strictly necessary we will omit to indicate the dependence of the various quantities on A_0 and $\boldsymbol{\beta}$, which is however always understood.

20.3. *The final model*

By exploiting relation, (20.12), we may eliminate the pressure P from the momentum equation. To that purpose we will indicate by $c_1=c_1(A;A_0,\boldsymbol{\beta})$ the following quantity:

$$c_1=\sqrt{\frac{A}{\rho}\frac{\partial\psi}{\partial A}}, \tag{20.17}$$

which has the dimension of a velocity and, as we will see later on, is related to the speed of propagation of simple waves along the tube.

By simple manipulations (20.12) may be written in *quasi-linear* form as follows:

$$\frac{\partial}{\partial t}U+\mathbf{H}(U)\frac{\partial U}{\partial z}+B(U)=\mathbf{0},\qquad z\in(0,L),\ t\in I \tag{20.18}$$

where,

$$U=\begin{bmatrix}A\\Q\end{bmatrix},$$

$$\mathbf{H}(U)=\begin{bmatrix}0 & 1\\ \dfrac{A}{\rho}\dfrac{\partial\psi}{\partial A}-\alpha\bar{u}^2 & 2\alpha\bar{u}\end{bmatrix}=\begin{bmatrix}0 & 1\\ c_1^2-\alpha\left(\dfrac{Q}{A}\right)^2 & 2\alpha\dfrac{Q}{A}\end{bmatrix}, \tag{20.19}$$

and

$$B(U)=\begin{bmatrix}0\\ K_R\left(\dfrac{Q}{A}\right)+\dfrac{A}{\rho}\dfrac{\partial\psi}{\partial A_0}\dfrac{\mathrm{d}A_0}{\mathrm{d}z}+\dfrac{A}{\rho}\dfrac{\partial\psi}{\partial\boldsymbol{\beta}}\dfrac{\mathrm{d}\boldsymbol{\beta}}{\mathrm{d}z}\end{bmatrix}.$$

Clearly, if A_0 and $\boldsymbol{\beta}$ are constant the expression for B becomes simpler. A *conservation form* for (20.18) may be found as well and reads

$$\frac{\partial U}{\partial t}+\frac{\partial}{\partial z}\left[F(U)\right]+S(U)=0, \tag{20.20}$$

where

$$F(U) = \begin{bmatrix} Q \\ \alpha \dfrac{Q^2}{A} + C_1 \end{bmatrix}$$

is the vector of fluxes,

$$S(U) = B(U) - \begin{bmatrix} 0 \\ \dfrac{\partial C_1}{\partial A_0}\dfrac{\mathrm{d}A_0}{\mathrm{d}z} + \dfrac{\partial C_1}{\partial \boldsymbol{\beta}}\dfrac{\mathrm{d}\boldsymbol{\beta}}{\mathrm{d}z}, \end{bmatrix}$$

and C_1 is a primitive of c_1^2 with respect to A, given by

$$C_1(A; A_0, \boldsymbol{\beta}) = \int_{A_0}^{A} c_1^2(\tau; A_0, \boldsymbol{\beta})\,\mathrm{d}\tau.$$

Again, if A_0 and $\boldsymbol{\beta}$ are constant, the source term S simplifies and becomes $S = B$. System (20.20) allows to identify the vector U as the *conservation variables* of our problem.

REMARK 20.1. In the case we use relation (20.16), we have

$$c_1 = \sqrt{\frac{\beta_0}{2\rho A_0}} A^{1/4}, \qquad C_1 = \frac{\beta_0}{3\rho A_0} A^{3/2}. \tag{20.21}$$

LEMMA 20.2. *If $A \geqslant 0$, the matrix* **H** *possesses two real eigenvalues. Furthermore, if $A > 0$ the two eigenvalues are distinct and* (20.18) *is a strictly hyperbolic system of partial differential equations.*

PROOF. By straightforward computations, we have the following expression for the eigenvalues of **H**:

$$\lambda_{1,2} = \alpha \bar{u} \pm c_\alpha, \tag{20.22}$$

where

$$c_\alpha = \sqrt{c_1^2 + \bar{u}^2 \alpha(\alpha - 1)}.$$

Since $\alpha \geqslant 1$, c_α is a real number. If $c_\alpha > 0$ the two eigenvalues are distinct. A sufficient condition to have $c_\alpha > 0$ is $c_1 > 0$ and, thanks to the definition of c_1 and (20.14), this is always true if $A > 0$. If $\alpha = 1$, this condition is also necessary.

The existence of a complete set of (right and left) eigenvectors is an immediate consequence of **H** having distinct eigenvalues. □

REMARK 20.2. System (20.12) shares many analogies with the 1D compressible Euler equations, after identifying the section area A with the density. The equivalence is not complete since the term $\partial P/\partial z$ in the Euler equations is here replaced by $A\partial P/\partial z$.

20.3.1. Characteristics analysis

Let $(\mathbf{l}_1, \mathbf{l}_2)$ and $(\mathbf{r}_1, \mathbf{r}_2)$ be two couples of left and right eigenvectors of the matrix $\mathbf{H}$ in (20.19), respectively. The matrices $\mathbf{L}$, $\mathbf{R}$ and $\mathbf{\Lambda}$ are defined as

$$\mathbf{L} = \begin{bmatrix} \mathbf{l}_1^{\mathrm{T}} \\ \mathbf{l}_2^{\mathrm{T}} \end{bmatrix}, \qquad \mathbf{R} = [\,\mathbf{r}_1 \quad \mathbf{r}_2\,], \qquad \mathbf{\Lambda} = \mathrm{diag}(\lambda_1, \lambda_2) = \begin{bmatrix} \lambda_1 & 0 \\ 0 & \lambda_2 \end{bmatrix}. \tag{20.23}$$

Since right and left eigenvectors are mutually orthogonal, without loss of generality we choose them so that $\mathbf{LR} = \mathbf{I}$. Matrix $\mathbf{H}$ may then be decomposed as

$$\mathbf{H} = \mathbf{R\Lambda L}, \tag{20.24}$$

and system (20.18) written in the equivalent form

$$\mathbf{L}\frac{\partial U}{\partial t} + \mathbf{\Lambda L}\frac{\partial U}{\partial z} + \mathbf{L}B(U) = \mathbf{0}, \quad z \in (0, L),\ t \in I. \tag{20.25}$$

If there exist two quantities W_1 and W_2 which satisfy

$$\frac{\partial W_1}{\partial U} = \mathbf{l}_1, \qquad \frac{\partial W_2}{\partial U} = \mathbf{l}_2, \tag{20.26}$$

we will call them *characteristic variables* of our hyperbolic system. We point out that in the case where the coefficients A_0 and $\boldsymbol{\beta}$ are not constant, W_1 and W_2 are not autonomous functions of U.

By setting $W = [W_1, W_2]^{\mathrm{T}}$, system (20.25) may be elaborated into

$$\frac{\partial W}{\partial t} + \mathbf{\Lambda}\frac{\partial W}{\partial z} + G = \mathbf{0}, \quad z \in (0, L),\ t \in I, \tag{20.27}$$

where

$$G = \mathbf{L}B - \frac{\partial W}{\partial A_0}\frac{\mathrm{d}A_0}{\mathrm{d}z} - \frac{\partial W}{\partial \boldsymbol{\beta}}\frac{\mathrm{d}\boldsymbol{\beta}}{\mathrm{d}z}. \tag{20.28}$$

We note that the extra terms on the right-hand side are a consequence of the fact that the characteristic variables depend parametrically on the coefficient A_0 and β, which may by a function of z, and thus

$$\frac{\partial W}{\partial z} = \frac{\partial W}{\partial U}\frac{\partial U}{\partial z} + \frac{\partial W}{\partial A_0}\frac{\mathrm{d}A_0}{\mathrm{d}z} + \frac{\partial W}{\partial \boldsymbol{\beta}}\frac{\mathrm{d}\boldsymbol{\beta}}{\mathrm{d}z} = \mathbf{L}\frac{\partial U}{\partial z} + \frac{\partial W}{\partial A_0}\frac{\mathrm{d}A_0}{\mathrm{d}z} + \frac{\partial W}{\partial \boldsymbol{\beta}}\frac{\mathrm{d}\boldsymbol{\beta}}{\mathrm{d}z}.$$

In the case where $B = 0$ and the coefficients A_0 and $\boldsymbol{\beta}$ are constant, (20.27) takes the simpler form

$$\frac{\partial W}{\partial t} + \mathbf{\Lambda}\frac{\partial W}{\partial z} = \mathbf{0}, \quad z \in (0, L),\ t \in I, \tag{20.29}$$

which component-wise reads

$$\frac{\partial W_i}{\partial t} + \lambda_i\frac{\partial W_i}{\partial z} = 0, \quad z \in (0, L),\ t \in I,\ i = 1, 2. \tag{20.30}$$

REMARK 20.3. From definition (20.26) and the fact that the left and right eigenvectors $\mathbf{l}_i$ and $\mathbf{r}_i$ are mutually orthogonal it follows that

$$\frac{\partial W_1}{\partial U}(U)\cdot \mathbf{r}_2(U)=0,$$

thus W_1 is a 2-*Riemann invariant* of our hyperbolic system (GODLEWSKI and RAVIART [1996]). Analogously, one may show that W_2 is a 1-Riemann invariant.

From (20.30) we have that W_1 and W_2 are constant along the two *characteristic curves* in the (z,t) plane described by the differential equations

$$\frac{\mathrm{d}z}{\mathrm{d}t}=\lambda_1 \quad\text{and}\quad \frac{\mathrm{d}z}{\mathrm{d}t}=\lambda_2,$$

respectively. In the more general case (20.27) we may easily show that W_1 and W_2 satisfy a coupled system of ordinary differential equations.

The expression for the left eigenvectors $\mathbf{l}_1$ and $\mathbf{l}_2$ is given by

$$\mathbf{l}_1=\zeta\begin{bmatrix} c_\alpha-\alpha\bar{u}\\ 1\end{bmatrix},\qquad \mathbf{l}_2=\zeta\begin{bmatrix} -c_\alpha-\alpha\bar{u}\\ 1\end{bmatrix},$$

where $\zeta=\zeta(A,\bar{u})$ is any arbitrary smooth function of its arguments with $\zeta>0$. Here we have expressed $\mathbf{l}_1$ and $\mathbf{l}_2$ as functions of $(A,\bar{u})$ instead of (A,Q) as is more convenient for the next developments. Thus, relations (20.26) become

$$\frac{\partial W_1}{\partial A}=\zeta\big[c_\alpha-\bar{u}(\alpha-1)\big],\qquad \frac{\partial W_1}{\partial \bar{u}}=\zeta A, \tag{20.31a}$$

$$\frac{\partial W_2}{\partial A}=\zeta\big[-c_\alpha-\bar{u}(\alpha-1)\big],\qquad \frac{\partial W_2}{\partial \bar{u}}=\zeta A. \tag{20.31b}$$

For a hyperbolic system of two equations it is always possible to find the characteristic variables (or, equivalently, the Riemann invariants) locally, that is in a sufficiently small neighbourhood of any point U (GODLEWSKI and RAVIART [1996], LAX [1973]), yet the existence of global characteristic variables is not in general guaranteed. However, in the special case $\alpha=1$, (20.31) takes the much simpler form

$$\frac{\partial W_1}{\partial A}=\zeta c_1,\qquad \frac{\partial W_1}{\partial \bar{u}}=\zeta A,$$

$$\frac{\partial W_2}{\partial A}=-\zeta c_1,\qquad \frac{\partial W_2}{\partial \bar{u}}=\zeta A.$$

Let us show that a set of global characteristic variables for our problem does exist in this case. We remind that the characteristic variable W_1 exists if and only if

$$\frac{\partial^2 W_1}{\partial A\,\partial\bar{u}}=\frac{\partial^2 W_1}{\partial\bar{u}\,\partial A},$$

for all allowable values of A and $\bar{u}$. Since now c_1 does not depend on $\bar{u}$, the above condition yields

$$c_1\frac{\partial\zeta}{\partial\bar{u}}=\zeta+A\frac{\partial\zeta}{\partial A}.$$

In order to satisfy this relation, it is sufficient to take $\zeta = \zeta(A)$ such that $\zeta = -A(\partial\zeta/\partial A)$. A possible instance is $\zeta = A^{-1}$. The resulting differential form is

$$\partial W_1 = \frac{c_1}{A}\partial A + \partial \bar{u},$$

and by proceeding in the same way for W_2, we have

$$\partial W_2 = -\frac{c_1}{A}\partial A + \partial \bar{u},$$

To integrate it in the $(A, \bar{u})$ plane, we need to fix the value at a reference state, for instance $W_1 = W_2 = 0$ for $(A, \bar{u}) = (A_0, 0)$. We finally obtain

$$W_1 = \bar{u} + \int_{A_0}^{A} \frac{c_1(\tau)}{\tau}\,\mathrm{d}\tau, \qquad W_2 = \bar{u} - \int_{A_0}^{A} \frac{c_1(\tau)}{\tau}\,\mathrm{d}\tau. \tag{20.32}$$

REMARK 20.4. If we adopt relation (20.16) and use the expression for c_1 given in (20.21), after simple computations we have

$$W_1 = \bar{u} + 4(c_1 - c_{1,0}), \qquad W_2 = \bar{u} - 4(c_1 - c_{1,0}), \tag{20.33}$$

where $c_{1,0}$ is the value of c_1 corresponding to the reference vessel area A_0.

Under physiological conditions, typical values of the flow velocity and mechanical characteristics of the vessel wall are such that $c_\alpha \gg \alpha\bar{u}$. Consequently, $\lambda_1 > 0$ and $\lambda_2 < 0$, i.e., the flow is sub-critical everywhere. Furthermore, the flow is smooth. Discontinuities, which would normally appear when treating a non-linear hyperbolic system, do not have indeed the time to form in our context because of the pulsatility of the boundary conditions. It may be shown (CANIC and KIM [2003]) that, for the typical values of the mechanical and geometric parameters in physiological conditions and the typical vessel lengths in the arterial tree, the solution of our hyperbolic system remains smooth, in accordance to what happens in the actual physical problem (which is however dissipative, a feature which has been neglected in our one-dimensional model).

In the light of the previous considerations, from now on we will always assume sub-critical regime and smooth solutions.

20.3.2. Boundary conditions

System (20.12) must be supplemented by proper boundary conditions. The number of conditions to apply at each end equals the number of characteristics entering the domain through that boundary. Since we are only considering sub-critical flows, we need to impose exactly one boundary condition at both $z = 0$ and $z = L$.

An important class of boundary conditions, called non-reflecting or 'absorbing', are those that allow the simple wave associated to the outgoing characteristic to exit the computational domain with no reflections. Following THOMPSON [1987], HEDSTROM [1979], non-reflecting boundary conditions for one-dimensional systems of non-linear hyperbolic equation like (20.20) may be written as

$$\mathbf{l}_1\left(\frac{\partial U}{\partial t} + S(U)\right) = 0 \quad \text{at } z = 0, \qquad \mathbf{l}_2\left(\frac{\partial U}{\partial t} + S(U)\right) = 0 \quad \text{at } z = L,$$

for all $t \in I$. When $S = 0$ these conditions are equivalent to impose a constant value (typically set to zero) to the incoming characteristic variable. When $S \neq 0$ they take into account the "natural variation" of the characteristic variables due to the presence of the source term. A boundary condition of this type is quite convenient at the outlet section.

At the inlet instead one usually desires to impose values of pressure or mass flux derived from measurements or other means. Let us suppose that $z = 0$ is an inlet section (the following discussion may be readily extended to the boundary $z = L$). Whenever an explicit formulation of the characteristic variables is available, the boundary condition may be expressed directly in terms of the entering characteristic variable W_1, i.e., for all $t \in I$,

$$W_1(t) = g_1(t) \quad \text{at } z = 0, \tag{20.34}$$

g_1 being a given function. However, seldom one has directly g_1 at disposal, as the available boundary data is normally given in terms of physical variables. Let us suppose that we know the time variation of both pressure and mass flux at that boundary (for instance, taken from measurements). We may derive the corresponding value of g_1 using directly the definition of the characteristic variable W_1. If $P_m = P_m(t)$ and $Q_m = Q_m(t)$ are the measured average pressure and mass flux at $z = 0$ for $t \in I$ and $W_1(A, Q)$ indicates the characteristic variable W_1 as function of A and Q, we may pose

$$g_1(t) = W_1\big(\psi^{-1}\big(P_m(t) - P_{\text{ext}}\big), Q_m(t)\big), \quad t \in I,$$

in (20.34). This means that P_m and Q_m are not imposed exactly at $z = 0$ (this would not be possible since our system accounts for only one boundary condition at each end of the computational domain), yet we require that at all times t the value of A and Q at $z = 0$ lies on the curve in the (A, Q) plane defined by

$$W_1(A, Q) - W_1\big(\psi^{-1}\big(P_m(t) - P_{\text{ext}}\big), Q_m(t)\big) = 0.$$

If instead one has at disposal the time history $q(t)$ of a just one physical variable $\phi = \phi(A, Q)$, the boundary condition

$$\phi\big(A(t), Q(t)\big) = q(t), \quad \forall t \in I, \quad \text{at } z = 0,$$

is admissible under certain restrictions (QUARTERONI and VALLI [1994]), which in our case reduce to exclude the case where ϕ may be expressed as function of only W_2. In particular, it may be found that for the problem at hand the imposition of either average pressure or mass flux are both admissible.

REMARK 20.5. If the integration of (20.26) is not feasible (as, for instance, in the case $\alpha \neq 1$), one may resort to the *pseudo-characteristic* variables (QUARTERONI and VALLI [1994]), $Z = [Z_1, Z_2]^{\mathrm{T}}$, defined by linearising (20.26) around an appropriately chosen reference state. One obtains

$$Z = \overline{Z} + \mathbf{L}(\overline{U})(U - \overline{U}), \tag{20.35}$$

where $\overline{U}$ is the chosen reference state and $\overline{Z}$ the corresponding value for $\mathbf{Z}$. One may then use the pseudo-characteristic variables instead of W, by imposing

$$Z_1(t) = g_1(t) \quad \text{at } z = 0, \qquad Z_2(t) = 0 \quad \text{at } z = L.$$

In the context of a time advancing scheme for the numerical solution of (20.20) the pseudo-characteristics are normally computed linearising around the solution computed at the previous time step.

REMARK 20.6. When considering the numerical discretisation, we need in general to provide an additional equation at each end point in order to close the resulting algebraic system. Typically, this extra relation is provided by the so-called *compatibility conditions* (QUARTERONI and VALLI [1994]), which read as follows:

$$\mathbf{l}_2^{\mathrm{T}}\left(\frac{\partial}{\partial t}U+\mathbf{H}\frac{\partial U}{\partial z}+B\right)=0, \quad z=0,\ t\in I, \tag{20.36a}$$

$$\mathbf{l}_1^{\mathrm{T}}\left(\frac{\partial}{\partial t}U+\mathbf{H}\frac{\partial U}{\partial z}+B\right)=0, \quad z=L,\ t\in I. \tag{20.36b}$$

20.3.3. Energy conservation for the 1D model

Most of the results presented in this section are taken from FORMAGGIA, GERBEAU, NOBILE and QUARTERONI [2001] and CANIC and KIM [2003].

LEMMA 20.3. *Let us consider the hyperbolic problem* (20.18) *and assume that the initial and boundary conditions are such that* $\forall z\in(0,L)$,

$$A(0,z)>0 \quad \textit{and} \quad A(t,0)>0, \quad A(t,L)>0, \quad \forall t\in I,$$

and that the solution U is smooth for all $(t,z)\in I\times(0,L)$. Then $A(t,z)>0$ for all $(t,z)\in I\times(0,L)$.

PROOF. Let us suppose that we have $A(t^*,z^*)=0$ at a generic point $(t^*,z^*)\in I\times(0,L)$. From the definition of λ_1 and λ_2 the line $l=\{(t,z)\colon z=z_u(t)\}$ satisfying

$$\frac{\mathrm{d}z_u}{\mathrm{d}t}(t)=\bar{u}\big(t,z_u(t)\big)$$

and ending at the point (t^*,z^*), lies between the two characteristic curves passing through the same point. Therefore, it completely lies inside the domain of dependence of (t^*,z^*) and either intersects the segment $z\in(0,L)$ at $t=0$ or one of the two semi-lines $z=0$ or $z=L$ at $t\geqslant 0$. We indicate this intersection point by $(\bar{t},\bar{z})$. The corresponding value of A, call it $\overline{A}$, is positive by hypothesis. From the continuity equation, A satisfies along the line l the following ordinary differential equation:

$$\frac{\mathrm{d}A}{\mathrm{d}t}=-A\frac{\partial\bar{u}}{\partial z},$$

where here the $\mathrm{d}A/\mathrm{d}t$ indicates the directional derivative along l. Therefore,

$$A(t^*,z^*)=\overline{A}\int_{\bar{t}}^{t^*}\frac{\partial\bar{u}}{\partial z}\big(\tau,z_u(\tau)\big)\,\mathrm{d}\tau>0,$$

in contradiction with the hypothesis. Therefore, we must have $A>0$. □

Here we derive now an *a priori* estimate for the solution of system (20.18) under the hypotheses of $\alpha = 1$, sub-critical smooth flow, and $A > 0$. We will consider the following initial and boundary conditions:

$$\text{initial conditions} \qquad A(0,z) = A^0(z), \quad Q(0,z) = Q^0(z), \quad z \in (0,L) \tag{20.37}$$

$$\text{boundary conditions} \qquad W_1(t,0) = g_1(t), \quad t \in I, \qquad W_2(t,L) = g_2(t), \quad t \in I. \tag{20.38}$$

Let the quantity e be defined as

$$e = \frac{\rho}{2} A\bar{u}^2 + \Psi, \tag{20.39}$$

where $\Psi = \Psi(A)$ is given by

$$\Psi(A) = \int_{A_0}^{A} \psi(\zeta)\,\mathrm{d}\zeta. \tag{20.40}$$

Here and in what follows we omit to indicate the dependence of ψ on A_0 and $\boldsymbol{\beta}$, since it is not relevant to obtain the desired result, which can be however extended also to the general case where the coefficients A_0 and $\boldsymbol{\beta}$ depend on z.

An energy of the 1D model is given by

$$\mathcal{E}(t) = \int_0^L e(t,z)\,\mathrm{d}z, \quad t \in I. \tag{20.41}$$

Indeed, owing to the assumptions we have made on ψ in (20.14), we may observe that ψ attains a minimum at $A = A_0$, since

$$\Psi(A_0) = \Psi'(A_0) = 0 \quad \text{and} \quad \Psi''(A) > 0, \quad \forall A > 0.$$

It follows that $\Psi(A) \geqslant 0$, $\forall A > 0$. Consequently, $\mathcal{E}(t)$ is a positive function for all Q and $A > 0$ and, moreover,

$$\mathcal{E}(t) = 0 \quad \text{iff} \quad \big(A(t,z), Q(t,z)\big) = (A_0, 0), \quad \forall z \in (0,L).$$

The following lemma holds.

LEMMA 20.4. *In the special case $\alpha = 1$, system (20.12), supplied with an algebraic pressure-area relationship of the form (20.13) and under conditions (20.14), satisfies the following conservation property, $\forall t \in I$:*

$$\mathcal{E}(t) + \rho K_R \int_{t_0}^{t}\int_0^L \bar{u}^2\,\mathrm{d}z\,\mathrm{d}\tau + \int_{t_0}^{t} Q(P_{\text{tot}} - P_{\text{ext}})\Big|_0^L \mathrm{d}\tau = \mathcal{E}(0), \tag{20.42}$$

where $\mathcal{E}(0)$ depends only on the initial data A^0 and Q^0, while $P_{\text{tot}} = P + \frac{1}{2}\rho\bar{u}^2$ is the fluid total pressure.

PROOF. Let us multiply the second equation of (20.12) by $\bar{u}$ and integrate over $(0,L)$. We will analyse separately the four terms that are obtained.

- First term:

$$
\begin{aligned}
I_1 = \int_0^L \frac{\partial (A\bar{u})}{\partial t}\bar{u}\,\mathrm{d}z &= \frac{1}{2}\int_0^L A\frac{\partial \bar{u}^2}{\partial t}\,\mathrm{d}z + \int_0^L \frac{\partial A}{\partial t}\bar{u}^2\,\mathrm{d}z \\
&= \frac{1}{2}\frac{\mathrm{d}}{\mathrm{d}t}\int_0^L A\bar{u}^2\,\mathrm{d}z + \frac{1}{2}\int_0^L \bar{u}^2\frac{\partial A}{\partial t}\,\mathrm{d}z.
\end{aligned} \tag{20.43}
$$

- Second term:

$$
\begin{aligned}
I_2 &= \alpha\int_0^L \frac{\partial (A\bar{u}^2)}{\partial z}\bar{u}\,\mathrm{d}z = \alpha\left[\int_0^L \frac{\partial (A\bar{u})}{\partial z}\bar{u}^2\,\mathrm{d}z + \int_0^L A\bar{u}^2\frac{\partial \bar{u}}{\partial z}\,\mathrm{d}z\right] \\
&= \alpha\left[\frac{1}{2}\int_0^L \frac{\partial (A\bar{u})}{\partial z}\bar{u}^2\,\mathrm{d}z + \frac{1}{2}\int_0^L \frac{\partial A}{\partial z}\bar{u}^3\,\mathrm{d}z + \frac{3}{2}\int_0^L A\bar{u}^2\frac{\partial \bar{u}}{\partial z}\,\mathrm{d}z\right] \\
&= \alpha\left[\frac{1}{2}\int_0^L \frac{\partial Q}{\partial z}\bar{u}^2\,\mathrm{d}z + \frac{1}{2}\int_0^L \frac{\partial (A\bar{u}^3)}{\partial z}\,\mathrm{d}z\right].
\end{aligned} \tag{20.44}
$$

Now, using the continuity equation, we obtain

$$
I_2 = \frac{\alpha}{2}\left[-\int_0^L \frac{\partial A}{\partial t}\bar{u}^2\,\mathrm{d}z + \left(A\bar{u}^3\right)\Big|_0^L\right]. \tag{20.45}
$$

- Third term:

$$
\begin{aligned}
I_3 &= \int_0^L \frac{A}{\rho}\frac{\partial P}{\partial z}\bar{u}\,\mathrm{d}z = \frac{1}{\rho}\int_0^L A\frac{\partial}{\partial z}(P - P_{\mathrm{ext}})\bar{u}\,\mathrm{d}z \\
&= \frac{1}{\rho}\left[-\int_0^L \frac{\partial Q}{\partial z}\psi(A)\,\mathrm{d}z + (P - P_{\mathrm{ext}})Q\Big|_0^L\right].
\end{aligned} \tag{20.46}
$$

Again, using the first of (20.12), we have

$$
\begin{aligned}
I_3 &= \frac{1}{\rho}\left[\int_0^L \frac{\partial A}{\partial t}\psi(A)\,\mathrm{d}z + (P - P_{\mathrm{ext}})Q\Big|_0^L\right] \\
&= \frac{1}{\rho}\left[\frac{\mathrm{d}}{\mathrm{d}t}\int_0^L \Psi(A)\,\mathrm{d}z + (P - P_{\mathrm{ext}})Q\Big|_0^L\right].
\end{aligned}
$$

- Fourth term:

$$
I_4 = \int_0^L K_r\frac{Q}{A}\bar{u}\,\mathrm{d}z = K_r\int_0^L \bar{u}^2\,\mathrm{d}z. \tag{20.47}
$$

By summing the four terms and multiplying by ρ, we obtain the following equality when $\alpha = 1$:

$$
\frac{1}{2}\rho\frac{\mathrm{d}}{\mathrm{d}t}\int_0^L A\bar{u}^2\,\mathrm{d}z + \frac{\mathrm{d}}{\mathrm{d}t}\int_0^L \Psi(A)\,\mathrm{d}z + \rho K_r\int_0^L \bar{u}^2\,\mathrm{d}z + Q(P_{\mathrm{tot}} - P_{\mathrm{ext}})\Big|_0^L = 0. \tag{20.48}
$$

Integrating Eq. (20.48) in time between t_0 and t leads to the desired result. □

In order to draw an energy inequality from (20.42), we need to investigate the sign of the last term on the left-hand side. With this aim, let us first analyse the homogeneous case $g_1 = g_2 = 0$.

We will rewrite the boundary term in (20.42) as a function of A, $\psi(A)$ and c_1 (which, in its turn, depends on A, see (20.17)).

If $g_1 = g_2 = 0$ in (20.38), then

$$\text{at } z=0, \quad W_1 = \bar{u} + \int_{A_0}^{A} \frac{c_1(\zeta)}{\zeta}\,\mathrm{d}\zeta = 0 \quad \Longrightarrow \quad \bar{u}(t,0) = -\int_{A_0}^{A} \frac{c_1(\zeta)}{\zeta}\,\mathrm{d}\zeta,$$

$$\text{at } z=L, \quad W_2 = \bar{u} - \int_{A_0}^{A} \frac{c_1(\zeta)}{\zeta}\,\mathrm{d}\zeta = 0 \quad \Longrightarrow \quad \bar{u}(t,L) = \int_{A_0}^{A} \frac{c_1(\zeta)}{\zeta}\,\mathrm{d}\zeta$$

and thus

$$Q(P_{\text{tot}} - P_{\text{ext}})\Big|_0^L = F\big(A(t,0)\big) + F\big(A(t,L)\big), \tag{20.49}$$

where

$$F(A) = A\int_0^A \frac{c_1(\zeta)}{\zeta}\,\mathrm{d}\zeta\left[\psi(A) + \frac{1}{2}\rho\left(\int_{A_0}^{A} \frac{c_1(\zeta)}{\zeta}\,\mathrm{d}\zeta\right)^2\right]. \tag{20.50}$$

From our assumption of sub-critical flow we have $|\bar{u}| < c_1$ which implies that at $z=0$ and $z=L$ we have

$$\left|\int_{A_0}^{A} \frac{c_1(\zeta)}{\zeta}\,\mathrm{d}\zeta\right| < c_1(A). \tag{20.51}$$

We are now in the position to conclude with the following result.

LEMMA 20.5. *If the function pressure-area relationship $P = \psi(A)$ is such that $F(A) > 0$ whenever* (20.51) *is satisfied, then inequality*

$$\mathcal{E}(t) + \rho K_r \int_{t_0}^{t}\int_0^L \bar{u}^2\,\mathrm{d}z\,\mathrm{d}\tau \leqslant \mathcal{E}(0) \tag{20.52}$$

holds for system (20.12), *provided homogeneous conditions on the characteristic variables, $W_1 = 0$ and $W_2 = 0$, are imposed at $z=0$ and $z=L$, respectively.*

PROOF. It is an immediate consequence of (20.42), (20.49) and (20.50). □

By straightforward computations, one may verify that the pressure–area relationship given in (20.16) satisfies the hypotheses of Lemma 20.5 (see FORMAGGIA, GERBEAU, NOBILE and QUARTERONI [2001]). Therefore, in that case the 1D model satisfies the energy inequality (20.52).

Under relation (20.16), we can prove a more general energy estimate, valid also in the case of non homogeneous boundary conditions. We state the following result.

LEMMA 20.6. *If the pressure–area relationship is given by* (20.16), *and the boundary data satisfy*

$$g_1(t) > -4c_{1,0}(0) \quad \textit{and} \quad g_2(t) < 4c_{1,0}(L), \quad \forall t \in I, \tag{20.53}$$

where

$$c_{1,0}(z) = \sqrt{\frac{\beta_0(z)}{\rho} A_0(z)^{-1/4}}$$

is the value of c_1 at the reference vessel area, then there exists a positive quantity $G(t)$ which continuously depends on the boundary data $g_1(t)$ and $g_2(t)$, as well as on the values of the coefficients A_0 and $\boldsymbol{\beta}$, at $z=0$ and $z=L$, such that, for all $t \in I$,

$$\mathcal{E}(t) + \rho K_r \int_{t_0}^{t} \int_0^L \bar{u}^2 \, \mathrm{d}z \, \mathrm{d}\tau \leqslant \mathcal{E}(0) + \int_0^t G(t) \, \mathrm{d}t. \tag{20.54}$$

PROOF. We will consider only the case where $g_1 \neq 0$ and $g_2 = 0$, since the most general case may be derived in a similar fashion. We recall that relationship (20.16) together with the assumption of sub-critical flow, complies with the conditions stated for $F(A)$ in Lemma 20.5. Then from (20.48) we obtain the following inequality:

$$\begin{aligned} \frac{\mathrm{d}}{\mathrm{d}t}\mathcal{E} + \rho K_r \int_0^L \bar{u}^2 \, \mathrm{d}z &\leqslant Q(P_{\text{tot}} - P_{\text{ext}})\Big|_{z=0} \\ &\leqslant \left(A|\bar{u}|\big|\psi(A)\big| + \frac{1}{2}\rho A |\bar{u}|^3 \right)\Big|_{z=0}. \end{aligned} \tag{20.55}$$

At $z=0$, we have from (20.33) that

$$\bar{u} + 4(c_1 - c_{1,0}) = g_1.$$

On the other hand, the condition $\lambda_1 = \bar{u} + c_1 > 0$ gives

$$c_1 < \frac{1}{3}(g_1 + 4c_{1,0}). \tag{20.56}$$

Since c_1 is a non-negative quantity, we must necessarily have $g_1 > -4c_{1,0}$. We now note that from (20.16) and the definition of $c_{1,0}$ we may write

$$\psi(A) = 2\rho\big(c_1^2(A) - c_{1,0}^2\big),$$

which together with (20.56) and the fact that $c_{1,0}$ is a positive function, allows us to state that, at $z=0$,

$$\psi(A) \leqslant \frac{2\rho}{9}\big(g_1^2 + 15c_{1,0}^2 + 8g_1c_{1,0}\big) \equiv f_1(g_1), \tag{20.57}$$

where f_1 is a positive continuous function depending parametrically on the values of A_0 and β_0 at $z=0$. Furthermore, condition $|\bar{u}| < c_1$ together with inequality (20.56) imply that

$$|\bar{u}| \leqslant f_2(g_1),$$

being f_2 another positive and continuous function. Finally, from the definition of ψ and $c_{1,0}$ we have

$$A = \frac{A_0^2}{\beta_0^2}\left(\psi(A) + \sqrt{A_0}\right)^2 \leqslant \frac{2A_0^2}{\beta_0^2}\left(\psi^2(A) + A_0\right)$$
$$\leqslant \frac{2A_0^2}{\beta_0^2}\left(f_1^2(g1) + A_0\right) \equiv f_3(g_1),$$

where we have exploited (20.57). By combining all previous inequalities, we deduce that the right-hand side in (20.55) may be bounded by a positive and continuous function of the boundary data g_1 that depend parametrically on the value of A_0 and β_0 at $z = 0$. By repeating a similar argument for the boundary conditions at $z = L$, we then obtain the desired stability inequality. □

20.3.4. *Weak form*

We consider the hyperbolic system (20.20) with initial condition $U = U_0$, at $t = t_0$, and appropriate boundary conditions at $z = 0$ and $z = L$. We indicate by $C_0^1((0, L) \times [t_0, t_1))$ the set of functions which are the restriction to $(0, L) \times [t_0, t_1)$ of C^1 functions with compact support in $(0, L) \times (-\infty, t_1)$. We will assume that U_0 is a bounded measurable function in $(0, L)$.

A function $U \in [L^\infty((0, L) \times [t_0, t_1))]^2$ is a weak solution of the equation in conservation form (20.20) if for all $\boldsymbol{\phi} \in [C_0^1((0, L) \times [t_0, t_1))]^2$ we have

$$\int_{t_0}^{t^1}\int_0^L \left(U \cdot \frac{\partial \boldsymbol{\phi}}{\partial t} + F(U) \cdot \frac{\partial \boldsymbol{\phi}}{\partial z} - S(U) \cdot \boldsymbol{\phi}\right) \mathrm{d}z\, \mathrm{d}t + \int_0^L U_0 \cdot \boldsymbol{\phi}|_{t=0} = 0. \qquad (20.58)$$

Moreover, we will require that U complies given boundary conditions.

A solution of (20.58) is called a weak solution of our hyperbolic system. Clearly, "classical" smooth solutions of (20.20) are also weak solutions. Conversely, it may be shown that a smooth weak solution, i.e., belonging to $[C^1((0, L) \times [t_0, t_1))]^2$, is also solution of (20.20) in a classical sense. However, the weak form accommodates also for less regular U. In particular, weak solutions of our hyperbolic problem may be discontinuous. The weak form is furthermore the basis of a class of numerical schemes, in particular, the finite element method, as already seen for the Navier–Stokes equations.

REMARK 20.7. The conservation formulation (20.20) accounts also for mechanical properties which vary smoothly along z. However, there are some fundamental difficulties in extending it to the case of discontinuous mechanical characteristics (e.g., discontinuous $\boldsymbol{\beta}$). On the other hand, this situation has a certain practical relevance, for instance in stented arteries or in the presence of a vascular prosthesis. A stent is a metal meshed wire structure inserted into a stenotic artery (typically a coronary) by angioplasty, in order to restore the original lumen dimension. Vascular prostheses are used to treat degenerative pathologies, such as aneurysms, or when angioplasty is not possible.

A possibility (FORMAGGIA, NOBILE and QUARTERONI [2002]) is to model the sharp variation of the Young modulus at the interface between the artery and the prosthesis by a regular function. Fig 20.2 illustrates a possible description of the change in the

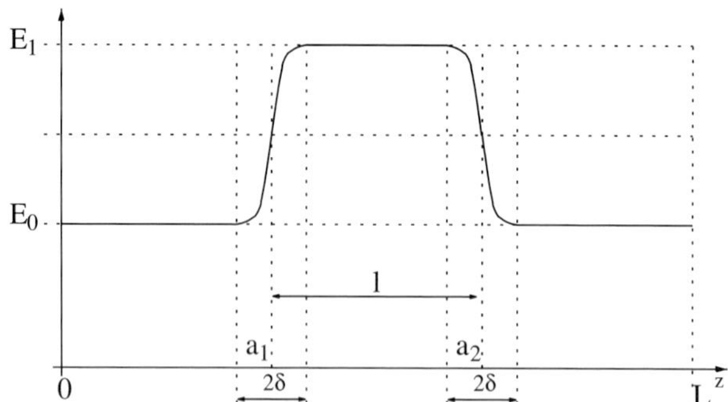

FIG. 20.2. The sharp variation of the Young modulus E from the value E_0 to the value E_1, due to the presence of a prosthesis, is modeled by a smooth function. One may argue what would happen when the parameter δ in figure tends to zero.

Young modulus due to the presence of a prosthesis. One may argue what would happen when the parameter δ in figure tends to zero. Numerical experiments have shown that the solution remains bounded although it becomes discontinuous at the location of the discontinuity in the Young modulus. This fact has been recently investigated in CANIC [2002] where an expression for the jump of mass flow and area across the discontinuity is derived by computing a particular limit of weak solutions of a regularised problem. More details are found in the cited reference.

20.3.5. *An entropy function for the 1D model*

Let us consider the hyperbolic system written in quasi-linear form (20.18). A pair of functions $e:\mathbb{R}^2 \to \mathbb{R}$ and $F_e:\mathbb{R}^2 \to \mathbb{R}$ is called *entropy pair* for the system if e is a convex function of U (called *entropy*) and the following condition is satisfied:

$$\left(\frac{\mathrm{d}e}{\mathrm{d}U}\right)^{\mathrm{T}} \mathbf{H}(U) = \frac{\partial F_e}{\partial U} \tag{20.59}$$

for all admissible values of U.

F_e is the *entropy flux* associated to the entropy e. If the hyperbolic system admits an entropy pair then the entropy function satisfies a conservation law of the form

$$\frac{\partial e}{\partial t} + \frac{\partial F_e}{\partial z} + B_e(U) = 0,$$

where

$$B_e(U) = \frac{\mathrm{d}e}{\mathrm{d}U} \cdot B(U) - \frac{\partial F_e}{\partial A_0}\frac{\mathrm{d}A_0}{\mathrm{d}z} - \frac{\partial F_e}{\partial \boldsymbol{\beta}}\frac{\mathrm{d}\boldsymbol{\beta}}{\mathrm{d}z}$$

is a source term. The last two terms in the previous expression account for the possible dependence of the coefficients A_0 and $\boldsymbol{\beta}$ on z.

The existence of an entropy pair is of a certain importance when studying the weak solution of the hyperbolic problem and, in particular, discontinuous solutions (more details in LAX [1973] and GODLEWSKI and RAVIART [1996]). Although we have here considered only smooth solutions, the identification of an entropy for our problem is important to set the basis for the extension of the model to more general situations.

In the case $\alpha = 1$,

$$e = \frac{1}{2}\rho A\bar{u}^2 + \Psi(A) = \frac{1}{2}\rho\frac{Q^2}{A} + \Psi(A)$$

is indeed an entropy for the problem at hand, with associated flux

$$F_e = Q\left(\psi(A) + \frac{1}{2}\rho\bar{u}^2\right) = Q(P_{\text{tot}} - P_{\text{ext}}).$$

Indeed, we have

$$\frac{\partial e}{\partial U} = \begin{bmatrix} -\frac{\rho\bar{u}^2}{2} + \psi(A) \\ \rho\bar{u} \end{bmatrix}, \qquad \frac{\partial F_e}{\partial U} = \begin{bmatrix} Q\frac{\partial\psi}{\partial A}(A) - \rho\bar{u}^3 \\ \psi(A) + \frac{3}{2}\rho\bar{u}^2 \end{bmatrix}$$

and we may directly verify condition (20.59) by recalling (20.19). Furthermore, $B_e = \rho K_r \bar{u}^2$ and the entropy balance equation thus read

$$\frac{\partial}{\partial t}\left(\frac{1}{2}\rho A\bar{u}^2 + \Psi(A)\right) + \frac{\partial}{\partial z}\left[Q\left(\psi(A) + \frac{1}{2}\rho\bar{u}^2\right)\right] + \rho K_r\bar{u}^2 = 0. \tag{20.60}$$

It is valid for any smooth solution of our hyperbolic model. Furthermore, the following lemma ensures the convexity of e.

LEMMA 20.7. *The entropy*

$$e(A, Q) = \frac{\rho}{2}\frac{Q^2}{A} + \Psi(A)$$

is convex for all $A > 0$.

PROOF. By a straightforward calculation one finds that the Hessian of e is given by

$$H_e = \begin{bmatrix} \dfrac{\partial^2 e}{\partial A^2} & \dfrac{\partial^2 e}{\partial A\partial Q} \\ \dfrac{\partial^2 e}{\partial A\partial Q} & \dfrac{\partial^2 e}{\partial Q^2} \end{bmatrix} = \frac{\rho}{A}\begin{bmatrix} \bar{u}^2 + c_1^2 & -\bar{u} \\ -\bar{u} & 1 \end{bmatrix}.$$

Its eigenvalues are

$$\lambda_{1,2}(H_e) = \rho\frac{c_1^2 + \bar{u}^2 + 1 \pm \sqrt{(c_1^2 + \bar{u}^2 + 1)^2 - 4c_1^2}}{2A}.$$

The condition for the discriminant to be positive is

$$4c_1^2 \leqslant \left(c_1^2 + \bar{u}^2 + 1\right)^2.$$

Since $c_1 > 0$ whenever $A > 0$, this inequality is equivalent to impose that

$$c_1^2 + \bar{u}^2 + 1 - 2c_1 = (c_1 - 1)^2 + \bar{u}^2 \geqslant 0,$$

which it is always true. Therefore, the two eigenvalues are strictly positive for all $A > 0$. This completes our proof. □

20.4. More complex wall laws that account for inertia and viscoelasticity

The algebraic relation (20.13) assumes that the wall is instantaneously in equilibrium with the pressure forces acting on it. Indeed, this approach correspond to the independent ring model introduced in Section 15.

At the price of some approximations it is possible to maintain the simple structure of a two-equations system while introducing effects, such as the inertia, which depend on the time derivative of the wall displacement.

We will consider as starting point relation (16.17) where we account for the inertia term and we model the viscoelastic property of the wall by adding a term proportional to the displacement rate, while we will still use the approximation (16.15) for the forcing term. We may thus write

$$P - P_{\text{ext}} = \gamma_0 \frac{\partial^2 \eta}{\partial t^2} + \gamma_1 \frac{\partial \eta}{\partial t} + \psi(A; A_0, \boldsymbol{\beta}), \tag{20.61}$$

where $\gamma_0 = \rho_w h_0$, $\gamma_1 = \gamma / R_0^2$ and the last term is the elastic response, modelled is the same way as done before. Here γ is the same viscoelasticity coefficient of (16.17) and η is the wall displacement, linked to A by (20.2).

In the following, we indicate by $\dot{A}$ and $\ddot{A}$ the first and second time derivative of A. We will substitute the following identities:

$$\frac{\partial \eta}{\partial t} = \frac{1}{2\sqrt{\pi A}} \dot{A}, \qquad \frac{\partial^2 \eta}{\partial t^2} = \pi^{-1/2} \left(\frac{1}{2\sqrt{A}} \ddot{A} - \frac{1}{4\sqrt{A^3}} \dot{A}^2 \right),$$

that are derived from (20.2), into (20.61) to obtain a relation that links the pressure also to the time derivatives of A, which we write in all generality as

$$P - P_{\text{ext}} = \tilde{\psi}(A, \dot{A}, \ddot{A}; A_0) + \psi(A; A_0, \boldsymbol{\beta}),$$

where $\tilde{\psi}$ is a non-linear function which derives from the treatment of the terms containing the time derivative of η. Since it may be assumed that the contribution to the pressure is in fact dominated by the term ψ, we will simplify this relationship by linearising $\tilde{\psi}$ around the state $A = A_0$, $\dot{A} = \ddot{A} = 0$. By doing that, after some simple algebraic manipulations, one finds

$$P - P_{\text{ext}} = \frac{\gamma_0}{2\sqrt{\pi A_0}} \ddot{A} + \frac{\gamma_1}{2\sqrt{\pi A_0}} \dot{A} + \psi(A; A_0, \boldsymbol{\beta}). \tag{20.62}$$

Replacing this expression for the pressure in the momentum equation requires to compute the term

$$\frac{A}{\rho} \frac{\partial P}{\partial z} = \frac{\gamma_0 A}{2\rho\sqrt{\pi A_0}} \frac{\partial^3 A}{\partial z \partial t^2} + \frac{\gamma_1 A}{2\rho\sqrt{\pi A_0}} \frac{\partial^2 A}{\partial z \partial t} + \frac{A}{\rho} \frac{\partial \psi}{\partial z}.$$

The last term in this equality may be treated as previously, while the first two terms may be further elaborated by exploiting the continuity equation. Indeed, we have

$$\frac{\partial^2 A}{\partial z \partial t} = -\frac{\partial^2 Q}{\partial z^2}, \qquad \frac{\partial^3 A}{\partial z \partial t^2} = -\frac{\partial^3 Q}{\partial t \partial z^2}.$$

Therefore, the momentum equation with the additional terms deriving from inertia and viscoelastic forces becomes

$$\frac{\partial Q}{\partial t}+\frac{\partial F_2}{\partial z}-\frac{\gamma_0 A}{2\rho\sqrt{\pi A_0}}\frac{\partial^3 Q}{\partial t\partial z^2}-\frac{\gamma_1 A}{2\rho\sqrt{\pi A_0}}\frac{\partial^2 Q}{\partial z^2}+S_2=0, \tag{20.63}$$

where with F_2 and S_2 we have indicated the second component of F and S, respectively.

REMARK 20.8. This analysis puts into evidence that the wall inertia introduces a *dispersive* term into the momentum equation, while the viscoelasticity has a *diffusion* effect.

20.5. Some further extensions

More general one-dimensional models may be derived by accounting for vessel curvature. This may be accomplished by enriching the description of the velocity field on each vessel section to allow asymmetries of the velocity profile to develop.

Another enhancement of the model is to account for vessel branching. By employing domain decomposition techniques, each branch is simulated by a separate one-dimensional model and interface conditions are used to account for the appropriate "transfer" of mass and momentum across the branching point. All these aspects are not covered in these notes. They are subject of current research and preliminary results may be found in FORMAGGIA, LAMPONI and QUARTERONI [2003].

Beside providing valuable information about average pressure and mass flux along an arterial segment, a one-dimensional model of blood flow may be used in the context of a multiscale/multimodel description of the cardiovascular system. In the multiscale framework, models of different level of complexity of the various cardiovascular elements are coupled together with the objective of simulating the whole cardiovascular system. Only the elements of major interest for the problem under study will be simulated at the highest level of detail (e.g., by employing a three-dimensional fluid–structure interaction model), while reduced models are adopted in the remaining parts. This technique allows us to account (at least partially) for the complex feedback mechanisms of the complete cardiovascular system, while keeping the overall computational costs at a reasonable level. More details on this technique may be found in FORMAGGIA, NOBILE, QUARTERONI and VENEZIANI [1999], FORMAGGIA, GERBEAU, NOBILE and QUARTERONI [2001], QUARTERONI, RAGNI and VENEZIANI [2001] while in PIETRABISSA, QUARTERONI, DUBINI, VENEZIANI, MIGLIAVACCA and RAGNI [2000] a first example on the use of this multiscale approach for a realistic clinical application is presented.

CHAPTER VII

21. Some numerical results

We provide some numerical results to illustrate applications of the techniques discussed in the previous sections. The aim here is to show the potential of the numerical modelling to reproduce realistic flow fields relevant for medical investigations. Many of the results here presented are substantially taken from previous works of the authors, in particular from QUARTERONI, TUVERI and VENEZIANI [2000], FORMAGGIA, GERBEAU, NOBILE and QUARTERONI [2001] and FORMAGGIA, NOBILE and QUARTERONI [2002]. More details and other examples may be found in the cited references.

21.1. Compliant pipe

Here we consider two examples of a fluid–structure interaction problem like the one presented in Section 19, namely a 2D and a 3D computation of a pressure wave in a compliant tube.

In the 2D case, we have considered a rectangular domain of height 1 cm and length $L = 6$ cm. The fluid is initially at rest and an over pressure of 15 mmHg ($2 \cdot 10^4$ dynes/cm^2) has been imposed at the inlet for 0.005 seconds. The viscosity of the fluid is equal to 0.035 poise, its density is 1 g/cm^3, the Young modulus of the structure is equal to $0.75 \cdot 10^6$ dynes/cm^2, its Poisson coefficient is 0.5, its density is 1.1 g/cm^3 and its thickness is 0.1 cm.

In the 3D case, our computation has been made on a cylindrical domain of radius $R_0 = 0.5$ cm and length $L = 5$ cm, with the following physical parameters: fluid viscosity: 0.03 poise, fluid density: 1 g/cm^3, Young modulus of the structure: $3 \cdot 10^6$ dynes/cm^2, Poisson coefficient: 0.3 and structure density: 1.2 g/cm^3. Again, an over-pressure of 10 mmHg ($1.3332 \cdot 10^4$ dynes/cm^2) is imposed at the inlet for 0.005 seconds.

The fluid equations are solved using the ALE approach, with a piece-wise linear finite element space discretisation. More precisely, for the 2D case the pressure is piece-wise linear on triangular elements and the velocity is linear over each of the four sub-triangles obtained by joining the midpoints of the edges of each pressure triangle (this is the so called P1isoP2–P1 discretisation). We have employed the Yosida technique illustrated in Section 14.5.2. For the 3D case we have used a stabilised scheme (HUGHES, FRANCA and BALESTRA [1986]) and piece-wise linear elements for both velocity and pressure.

For the 2D case, the equation for the structure displacement (18.13) has been solved using a P^1 finite element space discretisation, with nodes coincident with the ones of the pressure discretisation. In the 3D case, we have used a shell-type formulation (SIMO and FOX [1989], SIMO, FOX and RIFAI [1989]) to describe the dynamics of the wall structure. In both cases, the coupling scheme adopts a sub-iterations strategy of the type illustrated in Section 19.

In order to reduce spurious wave reflections at the outlet, we have coupled the fluid–structure interaction problem with a one-dimensional system of the type described in Section 20. For more details on this technique see FORMAGGIA, GERBEAU, NOBILE and QUARTERONI [2001], as well as FORMAGGIA, GERBEAU, NOBILE and QUARTERONI [2002].

Figs. 21.1 and 21.2 show the fluid pressure and the domain deformation in the 2D and the 3D case, respectively. For the sake of clarity, the displacements shown in Fig. 21.2 are magnified by a factor 10.

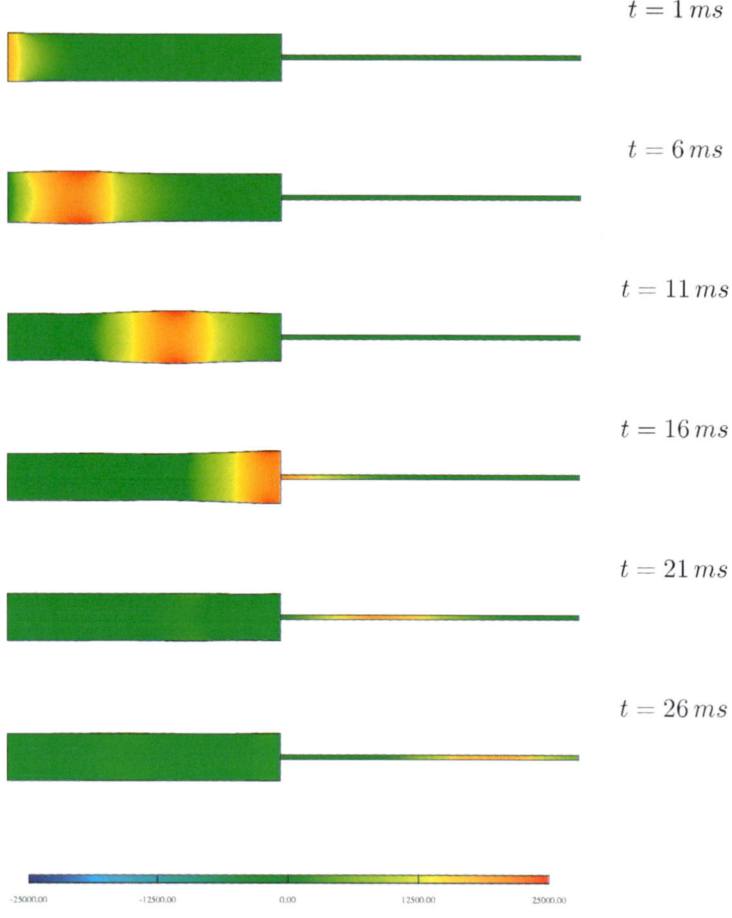

FIG. 21.1. Pressure pulse entering at the *inflow*. A non-reflecting boundary condition at the outlet has been obtained by the coupling with a 1D hyperbolic model. Solutions every 5 ms.

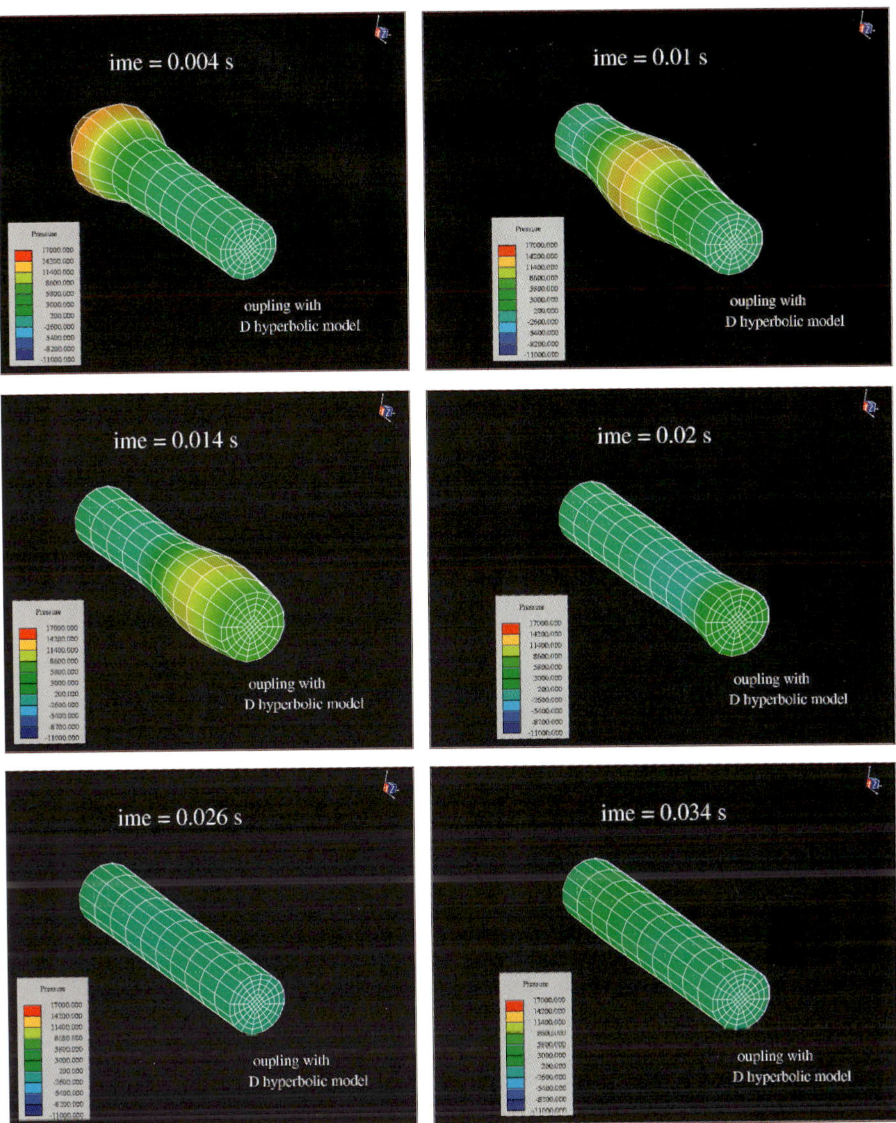

FIG. 21.2. A pressure pulse traveling in a 3D compliant vessel. The displacement of the structure has been magnified by a factor 10. A non-reflecting boundary condition at the outlet has been obtained by the coupling with a 1D hyperbolic model (not shown in the picture).

21.2. Anastomosis models

Anastomosis is the a surgical operation by which the functionality of a blocked artery (typically a coronary) is restored thanks to by-pass. The flow condition when the blood in the by-pass re-joins the main artery may be critical. If we have a large recirculation

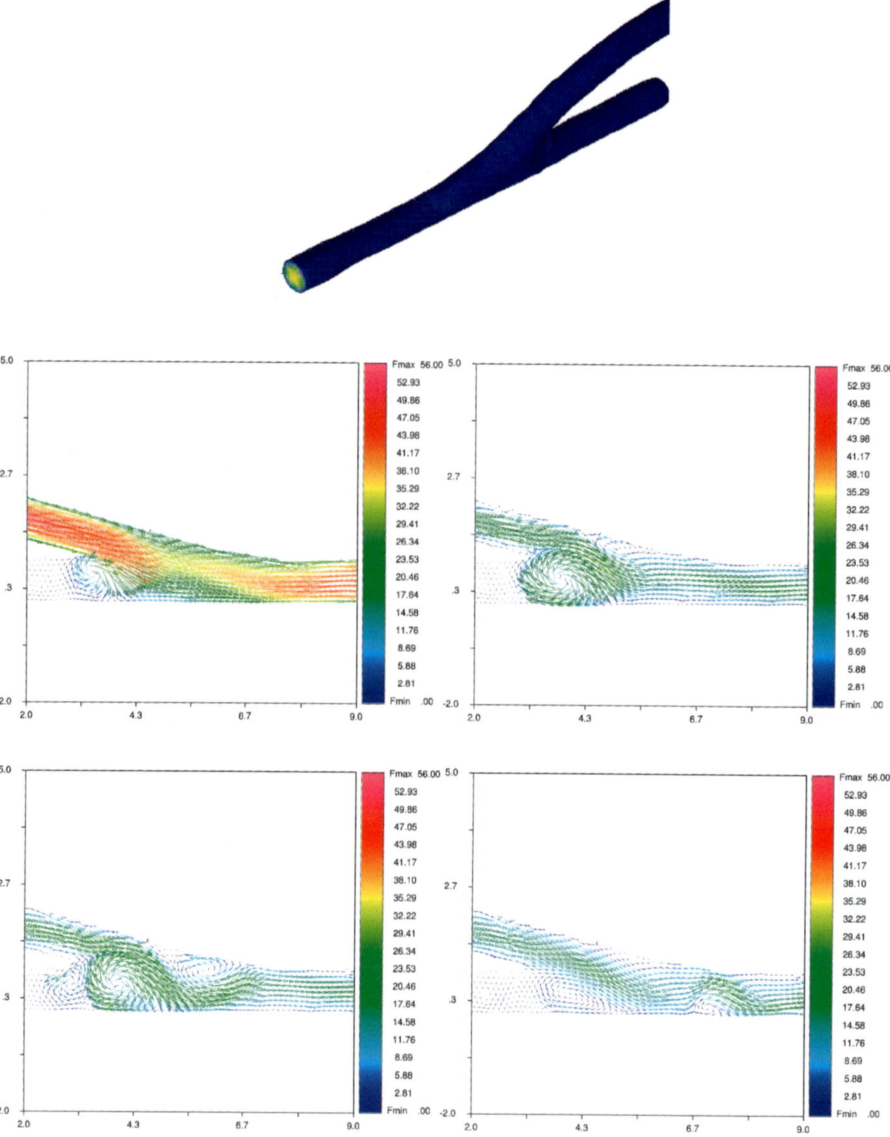

FIG. 21.3. A model of a coronary by-pass anastomosis (top) and the velocity vector field on the median plane at four different instants of the heart beat. Flow at systole (top, left), initial deceleration phase (top, right), beginning of diastole (bottom, left) and end of diastole (bottom, right). The recirculation regions upstream and downstream of the junction are evident.

area, the higher latency time of blood particles there may favor plaque growing and cause a new blockage further downstream.

The simulations here presented aim at highlight the problem. We illustrate the flow in the median plane of a 3D model of an anastomosis.[7] The junction angle is 15 degrees. The diameter of the occluded branch (below) is 1 cm, and the one of the by-pass (above) is 0.96 cm. The simulations have been carried out setting the dynamic viscosity $\mu = 0.04\ \mathrm{g\,cm^{-1}\,s^{-1}}$ and the density $\rho = 1\ \mathrm{g\,cm^{-3}}$. In this simulation the vessel wall has been assumed fixed and the boundary conditions prescribe null velocity on the walls and on the upstream section of the stenotic branch (100% stenosis), while a parabolic velocity profile has been prescribed at the inlet section with a peak velocity of $56\ \mathrm{cm\,s^{-1}}$, corresponding to a flow rate of $1320\ \mathrm{ml\,min^{-1}}$. On the downstream section a Neumann-type condition has been assigned.

Fig. 21.3 clearly illustrates the appearance and the evolution of the flow recirculation zones during the different phases of the heart beat.

21.3. *Pressure wave modification caused by a prosthesis*

Here we present a numerical simulation obtained using the one-dimensional model (20.12) to investigate the effect of a prosthesis in an artery, in particular with respect to the alteration of the pressure wave pattern. To that purpose we have considered the portion of an artery of length L and a prosthesis of length l (see Fig. 21.4) and a Young modulus varying as already illustrated in Fig. 20.2.

In order to assess the effect of the changes in vessel wall elastic characteristic on the pressure pattern, we have devised several numerical experiments. Two types of pressure input have been imposed at $z = 0$, namely an impulse input, that is a single sine wave with a small time period and a single sine wave with a more realistic time period (see Fig. 21.5). The impulse has been used to better highlight the reflections induced by the vascular prosthesis.

The part that simulates the presence of the prosthesis or stent of length L is comprised between coordinates a_1 and a_2. The corresponding Young's modulus has been taken as a multiple of the basis Young's modulus E_0 associated to the physiological tissue.

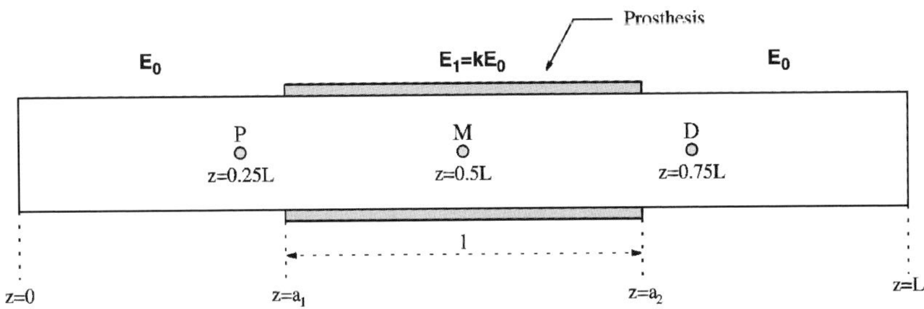

FIG. 21.4. The layout of our numerical experiment. The points P, M and D are used as 'monitoring stations' to assess the modifications on the pressure wave caused by the prosthesis.

[7]The model geometry has been provided by the Vascular Surgery Skejby Sygheus of the Aahrus University Hospital in Denmark.

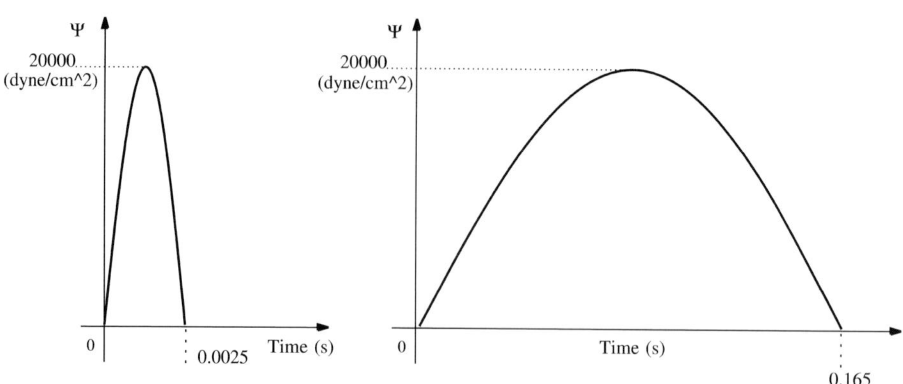

FIG. 21.5. The two types of pressure input profiles used in the numerical experiments: an impulse (left) and a more realistic sine wave (right).

TABLE 21.1
Data used in the numerical experiments

	Parameters	Value
Fluid	Input pressure amplitude	20×10^3 dyne/cm^2
	Viscosity, ν	0.035 poise
	Density, ρ	1 g/cm^3
Structure	Young's modulus, E_0	3×10^6 dyne/cm^2
	Wall thickness, h	0.05 cm
	Reference radius, R_0	0.5 cm

Three locations along the vessel have been identified and indicated by the letters D (distal), M (medium) and P (proximal). They will be taken as monitoring point for the pressure variation. Different prosthesis length L have been considered; in all cases points P and D are located outside the region occupied by the prosthesis. Table 21.1 indicates the basic data which have been used in all numerical experiments. In this numerical experiment we have considered the conservation form (20.20) setting the friction term K_r to zero. The numerical scheme adopted is a second order Taylor–Galerkin (DONEA, GIULIANI, LAVAL and QUARTAPELLE [1984]). A time step $\Delta t = 2 \times 10^{-6}$ s and the initial values $A = A_0$ and $Q = 0$ have been used throughout.

At the outlet boundary $z = L$ we have kept W_2 constant and equal to its initial value (non-reflecting boundary condition). At the inlet boundary we have imposed the chosen pressure input in an approximate fashion, following a technique of the type illustrated in Section 20.3.2.

21.3.1. *Case of an impulsive pressure wave*

In Fig. 21.6 we show the results obtained for the case of a pressure impulse. We compare the results obtained with uniform Young modulus E_0 and the corresponding solution when $E_1 = 100E_0$, $l = 5$ cm and the transition zone between healthy artery and pros-

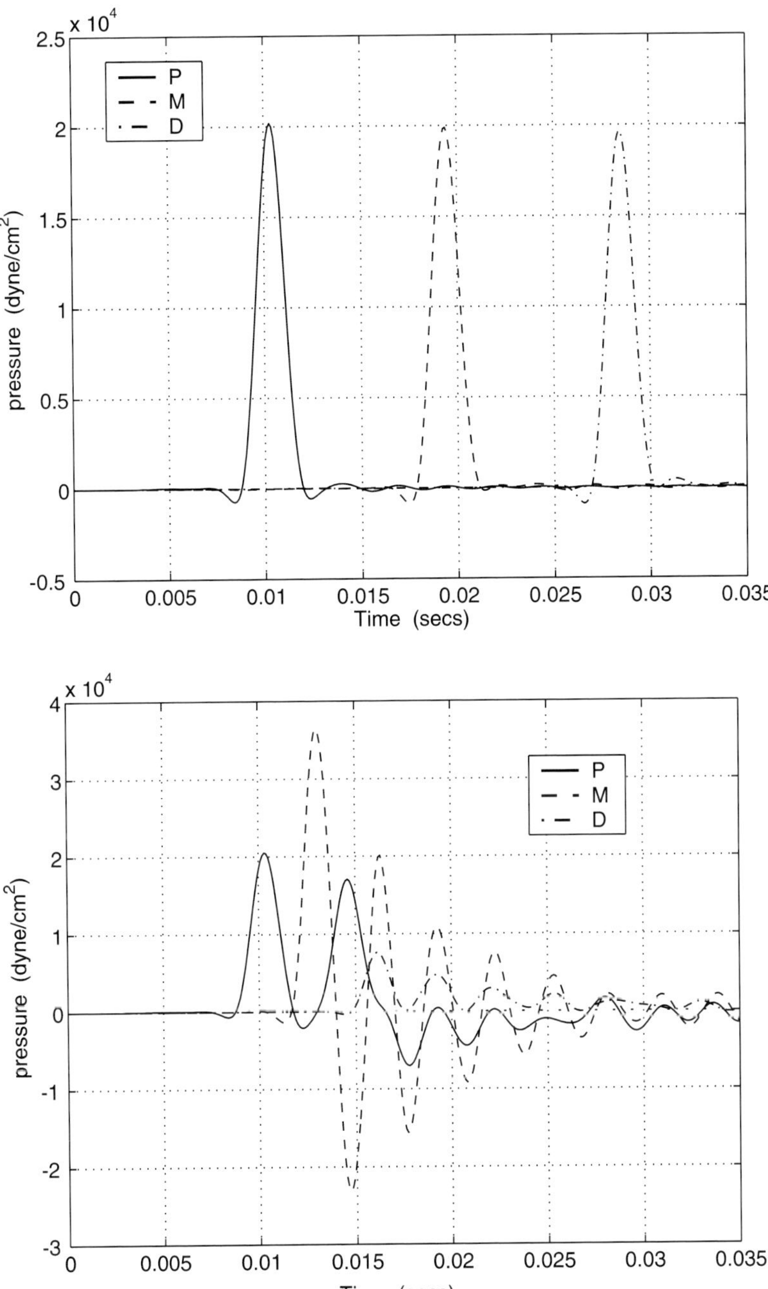

FIG. 21.6. Pressure history at points P, M and D of Fig. 21.4, for an impulsive input pressure, in the case of constant (upper) and variable (lower) E.

thesis is $\delta = 0.5$ cm. We have taken $L = 15$ cm and a non-uniform mesh of 105 finite elements, refined around the points a_1 and a_2. When the Young modulus is uniform, the impulse travels along the tube undisturbed. The numerical solution shows a little dissipation and dispersion due to the numerical scheme. In the case of variable E the situation changes dramatically. Indeed, as soon as the wave enters the region at higher Young's modulus it gets partially reflected (the reflection is registered by the positive pressure value at point P and $t \approx 0.015$ s) and it accelerates. Another reflection occurs at the exit of the 'prosthesis', when E returns to its reference value E_0. The point M indeed registers an oscillatory pressure which corresponds to the waves that are reflected back and forth between the two ends of the prosthesis. The wave at point D is much weaker, because part of the energy has been reflected back and part of it has been 'captured' inside the prosthesis itself.

21.3.2. Case of a sine wave

Now, we present the case of the pressure input given by the sine wave with a larger period shown in Fig. 21.5, which describes a situation closer to reality than the impulse. We present again the results for both cases of a constant and a variable E. All other problem data have been left unchanged from the previous simulation. Now, the interaction among the reflected waves is more complex and eventually results in a less oscillatory solution (see Fig. 21.7). The major effect of the presence of the stent is a pressure increase at the proximal point P, where the maximum pressure is approximately 2500 dynes/cm^2 higher than in the constant case. At a closer inspection one may note that the interaction between the incoming and reflected waves shows up in discontinuities in the slope, particularly for the pressure history at point P. In addition, the wave is clearly accelerated inside the region where E is larger.

In Table 21.2 we show the effect of a change in the length of the prosthesis by comparing the maximum pressure value recorded for a prosthesis of 4, 14 and 24 cm, respectively. The values shown are the maximal values in the whole vessel, over one period. Here, we have taken $L = 60$ cm, $\delta = 1$ cm, a mesh of 240 elements and we have positioned in the three cases the prosthesis in the middle of the model. The maximum value is always reached at a point upstream the prosthesis. In the table we give the normalised distance between the upstream prosthesis section and of the point where the pressure attains its maximum.

Finally, we have investigated the variation of the pressure pattern due to an increase of $k = E/E_0$. Fig. 21.8 shows the result corresponding to $L = 20$ cm and $\delta = 1$ cm and various values for k. The numerical result confirms the fact that a stiffer prosthesis

TABLE 21.2
Maximum pressure value for prosthesis of different length

Prosthesis length (cm)	Maximal pressure (dyne/cm^2)	Maximum location z_{max}/l
4	23.5×10^3	0.16
14	27.8×10^3	0.11
24	30.0×10^3	0.09

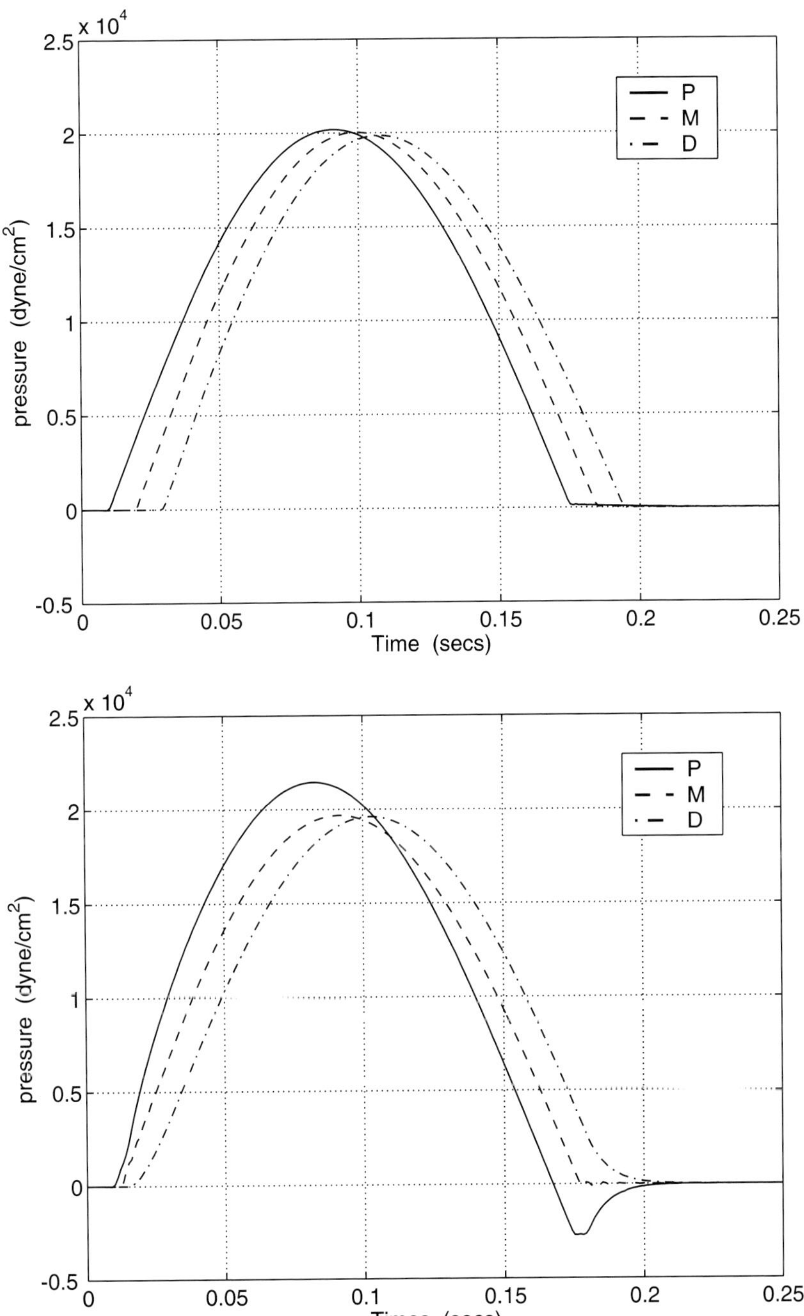

FIG. 21.7. Pressure history at points P, M and D of Fig. 21.4, for a sine wave input pressure, in the case of constant (upper) and variable (lower) E.

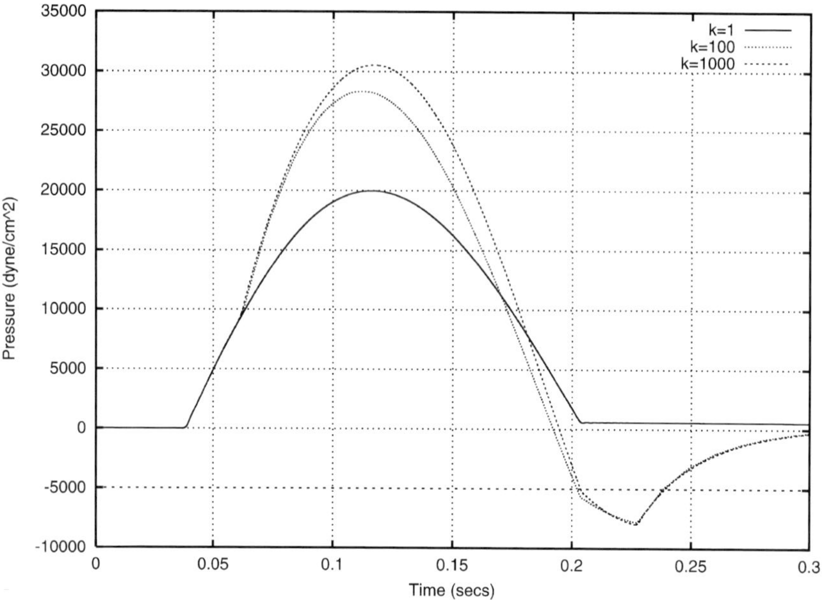

FIG. 21.8. Pressure history at point P of Fig. 21.4, for a sine wave input pressure and different Young's moduli $E = kE_0$.

causes a higher excess pressure in the proximal region, a fact that may have negative effects on the heart.

21.4. *Some examples of the geometrical multiscale approach*

We end this section by giving some examples of the geometrical multiscale approach, where models of different geometrical complexity are coupled together to provide the simulation of the global cardiovascular system, at different level of detail.

Fig. 21.9 shows an example of the simulation of a by-pass, with the interplay between three-dimensional, one-dimensional and lumped parameters models. A detailed description of the flow in the by-pass is obtained by solving the fluid–structure interaction problem (here using a two-dimensional model). The presence of the global cardiovascular system is provided by a system of algebraic and ordinary differential equations (ODE) for average mass flow and pressure. This system is here illustrated by means of an electrical analog, where voltage plays the role of average pressure and the current that of mass flow. A transition between the two models is provided by the use of the one-dimensional description detailed in the previous section.

A simpler example of this coupling strategy, yet on a realistic three-dimensional geometry, is shown in Fig. 21.10. A three-dimensional model of the modified Blalock–Taussig shunt a surgical operation meant to cure the consequences of a severe cardiac malformation, has been devised with the intent of finding the optimal design for the shunt. The three-dimensional model (on a fixed geometry) has been coupled with

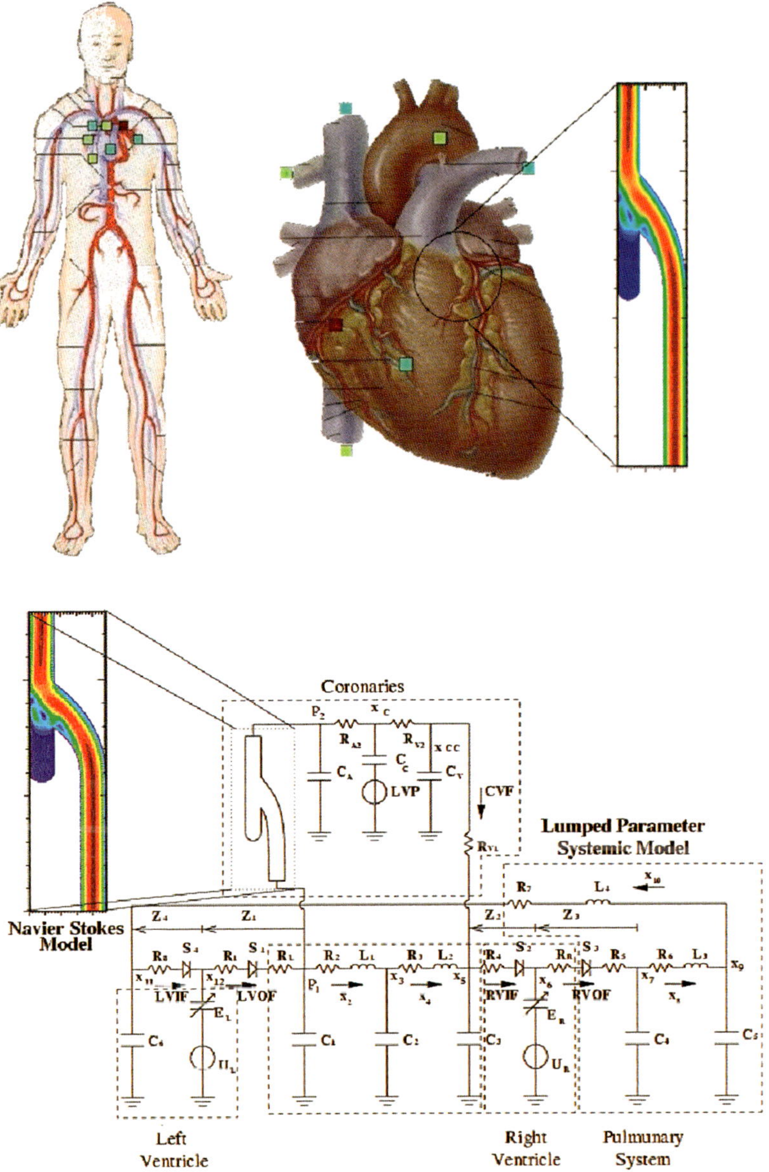

FIG. 21.9. On top we show a global model of the circulatory system where a coronary by-pass is being simulated by a Navier–Stokes fluid–structure interaction model. The rest of the circulatory system is described by means of a lumped parameter model, based on the solution of a system of ODEs, is here represented by an electrical circuit analog in the bottom part of the figure.

the systemic lumped parameter model, which provides the boundary conditions for the Navier–Stokes equations at the inlet and outlet sections. Thanks to this multiscale approach it has been possible to compute velocity profiles and flow patterns which are

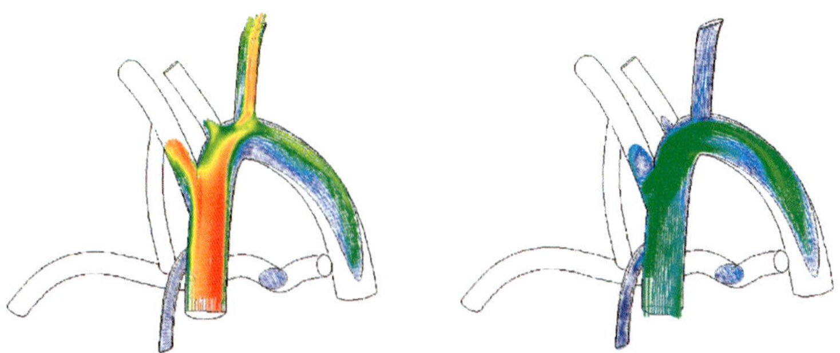

FIG. 21.10. Simulation of the haemodynamics in the modified Blalock–Taussig shunt obtained using a geometrical multiscale approach. Velocity field in the ascending aorta at two different times during the cardiac cycle.

closer to reality than those obtained by using more standard boundary conditions. An analysis of the technique is found in QUARTERONI and VENEZIANI [2003], while more details on this and other test cases may be found in PIETRABISSA, QUARTERONI, DUBINI, VENEZIANI, MIGLIAVACCA and RAGNI [2000], MIGLIAVACCA, LAGANÁ, PENNATI, DE LEVAL, BOVE and DUBINI [2004].

22. Conclusions

The development of mathematical models, algorithms and numerical simulation tools for the investigation of the human cardiovascular system has received a great impulse in the last years. These notes intended to cover just a few of the relevant issues. There are however other important aspects which require the use of sophisticated mathematical and numerical tools. We here mention just a few, namely the reconstruction of geometries from medical data; the transport of biochemicals in blood and vessel wall tissue; the heart dynamics; blood rheology. Besides, the need of validating the models calls for development of accurate in-vivo measurement techniques.

The number and complexity of the mathematical, numerical and technological problems involved makes the development of tools for accurate, reliable and efficient simulations of the human cardiovascular system one of the challenges of the next decades.

Acknowledgements

The authors thank Prof. Alessandro Veneziani and Dr. Fabio Nobile for their valuable contributions during the preparation of these notes and for having provided most of the numerical results here presented. We thank Dr. G. Dubini and Dr. F. Migliavacca for the availability of the numerical results for the modified Blalock–Taussig shunt.

Our research activity on the mathematical modelling of the cardiovascular system has been partially supported by grants from various research agencies, which we gratefully acknowledge. In particular, grants 21-54139.98, 21-59230.99 and 20-61862.00 from the Swiss National Science Foundation, the project of Politecnico di Milano "LSC-

Multiscale Computing in Biofluiddynamics", the project "Agenzia-2000" by the Italian CNR, titled "Modeling the fluid structure interaction in the arterial system", and a research contract "Cofin-2000" by the Italian Ministry of Education (MURST) titled "Scientific Computing: Innovative Models and Numerical Methods". Finally, the authors acknowledge the support by the European Union through the Research Training and Network project "HaeMOdel", contract number HPRN-CT-2002-002670.

References

ACHDOU, Y., GUERMOND, J.L. (2000). Convergence analysis of a finite element projection/Lagrange–Galerkin method for the incompressible Navier–Stokes equations. *SIAM J. Numer. Anal.* **37** (3), 799–826.

ARIS, R. (1962). *Vectors, Tensors and the Basic Equations of Fluid Mechanics* (Prentice Hall, New York).

BARNARD, A.C.L., HUNT, W.A., TIMLAKE, W.P., VARLEY, E. (1966). A theory of fluid flow in compliant tubes. *Biophys. J.* **6**, 717–724.

BEIRÃO DA VEIGA, H. (2004). On the existence of strong solutions to a coupled fluid–structure evolution problem. *J. Math. Fluid Mech.* **6**, 21–52.

BOUKIR, K., MADAY, Y., MÉTIVET, B., RAZAFINDRAKOTO, E. (1997). A high-order characteristics/finite element method for the incompressible Navier–Stokes equations. *Internat. J. Numer. Methods Fluids* **25** (12), 1421–1454.

BREZIS, H. (1983). *Analyse Fonctionnelle* (Masson, Paris).

BREZZI, F., FORTIN, M. (1991). *Mixed and Hybrid Finite Elements*, Springer Ser. Comput. Math. **5** (Springer-Verlag, Berlin).

CANIC, S. (2002). Blood flow through compliant vessels after endovascular repair: wall deformations induced by the discontinuous wall properties. *Comput. Visual. Sci.* **4** (3), 147–155.

CANIC, S., KIM, E. (2003). Mathematical analysis of the quasilinear effects in a hyperbolic model of blood flow through compliant axi-symmetric vessels. *Math. Methods Appl. Sci.* **26** (14), 1161–1186.

CHORIN, A.J., MARSDEN, J.E. (1990). *A Mathematical Introduction to Fluid Mechanics*, third ed., Texts Appl. Math. **4** (Springer-Verlag, New York).

CIARLET, P.G. (1988). *Mathematical Elasticity, Volume I: Three-Dimensional Elasticity*, Stud. Math. Appl. **20** (North-Holland, Amsterdam).

CIARLET, P.G. (1998). Introduction to Linear Shell Theory (Gauthier–Villars, Paris).

CIARLET, P.G. (2000). *Mathematical Elasticity, Volume III: Theory of Shells* (North-Holland, Amsterdam).

COKELET, G.R. (1987). The rheology and tube flow of blood. In: Skalak, R., Chen, S. (eds.), *Handbook of Bioengineering* (McGraw–Hill, New York).

DONEA, J., GIULIANI, S., LAVAL, H., QUARTAPELLE, L. (1984). Time-accurate solutions of advection–diffusion problems by finite elements. *Comput. Methods Appl. Mech. Engrg.* **45**, 123–145.

DUVAUT, G., LIONS, J.-L. (1976). *Inequalities in Mechanics and Physics* (Springer-Verlag, Berlin).

FARHAT, C., LESOINNE, M. (2000). Two efficient staggered algorithms for the serial and parallel solution of three-dimensional nonlinear transient aeroelastic problems. *Comput. Methods Appl. Mech. Engrg.* **182**, 499–515.

FARHAT, C., LESOINNE, M., MAMAN, N. (1995). Mixed explicit/implicit time integration of coupled aeroelastic problems: three-field formulation, geometry conservation and distributed solution. *Internat. J. Numer. Methods Fluids* **21**, 807–835.

FORMAGGIA, L., GERBEAU, J.-F., NOBILE, F., QUARTERONI, A. (2001). On the coupling of 3D and 1D Navier–Stokes equations for flow problems in compliant vessels. *Comput. Methods Appl. Mech. Engrg.* **191**, 561–582.

FORMAGGIA, L., GERBEAU, J.-F., NOBILE, F., QUARTERONI, A. (2002). Numerical treatment of defective boundary conditions for Navier–Stokes equations. *SIAM J. Numer. Anal.* **40** (1), 376–401.

FORMAGGIA, L., LAMPONI, D., QUARTERONI, A. (2003). One-dimensional models for blood flow in arteries. *J. Engrg. Math.* **47**, 251–276.

FORMAGGIA, L., NOBILE, F. (1999). A stability analysis for the Arbitrary Lagrangian Eulerian formulation with finite elements. *East–West J. Numer. Math.* **7**, 105–131.

FORMAGGIA, L., NOBILE, F., QUARTERONI, A. (2002). A one-dimensional model for blood flow: application to vascular prosthesis. In: Babuska, I., Miyoshi, T., Ciarlet, P.G. (eds.), *Mathematical Modeling and Numerical Simulation in Continuum Mechanics*. In: Lect. Notes Comput. Sci. Eng. **19** (Springer-Verlag, Berlin), pp. 137–153.

FORMAGGIA, L., NOBILE, F., QUARTERONI, A., VENEZIANI, A. (1999). Multiscale modelling of the circulatory system: a preliminary analysis. *Comput. Visual. Sci.* **2**, 75–83.

FUNG, Y.C. (1984). *Biodynamics: Circulation* (Springer-Verlag, New York).

FUNG, Y.C. (1993). *Biomechanics: Mechanical Properties of Living Tissues* (Springer-Verlag, New York).

GASTALDI, L. (2001). A priori error estimates for the arbitrary Lagrangian Eulerian formulation with finite elements. *East–West J. Numer. Math.* **9** (2), 123–156.

GIRAULT, V., RAVIART, P.-A. (1986). *Finite Element Methods for Navier–Stokes Equations, Theory and Algorithms*, Springer Ser. Comput. Math. **5** (Springer-Verlag, Berlin).

GODLEWSKI, E., RAVIART, P.-A. (1996). *Numerical Approximation of Hyperbolic Systems of Conservation Laws*, Appl. Math. Sci. **118** (Springer-Verlag, New York).

GRANDMONT, C., GUIMET, V., MADAY, Y. (2001). Numerical analysis of some decoupling techniques for the approximation of the unsteady fluid structure interaction. *Math. Models Methods Appl. Sci.* **11** (8), 1349–1377.

GRANDMONT, C., MADAY, Y. (1998). Nonconforming grids for the simulation of fluid–structure interaction. In: *Domain Decomposition Methods 10*, Boulder, CO, 1997. In: Contemp. Math. **218** (Amer. Math. Soc., Providence, RI), pp. 262–270.

GRANDMONT, C., MADAY, Y. (2000). Fluid structure interaction: a theoretical point of view. In: Dervieux, A. (ed.), Revue européenne des éléments finis **9** (Hermes Science), pp. 633–653.

GUERMOND, J.-L. (1999). Un résultat de convergence d'ordre deux en temps pour l'approximation des équations de Navier–Stokes par une technique de projection incrémentale. *M2AN Math. Model. Numer. Anal.* **33** (1), 169–189.

GUILLARD, H., FARHAT, C. (2000). On the significance of the geometric conservation law for flow computations on moving meshes. *Comput. Methods Appl. Mech. Engrg.* **190** (11–12), 1467–1482.

HAYASHI, K., HANDA, K., NAGASAWA, S., OKUMURA, A. (1980). Stiffness and elastic behaviour of human intracranial and extracranial arteries. *J. Biomech.* **13**, 175–184.

HEDSTROM, G.W. (1979). Nonreflecting boundary conditions for nonlinear hyperbolic systems. *J. Comput. Phys.* **30**, 222–237.

HOLZAPFEL, G.A., GASSER, T.C., OGDEN, R.W. (2000). A new constitutive framework for arterial wall mechanics and a comparative study of material models. *J. Elasticity* **61**, 1–48.

HUGHES, T.J., FRANCA, L.P., BALESTRA, M. (1986). A new finite element formulation for computational fluid dynamics: V. Circumventing the Babuska Brezzi condition: a stable Petrov–Galerkin formulation of the Stokes problem accommodating equal-order interpolation. *Comput. Methods Appl. Mech. Engrg.* **59**, 85–99.

LANGEWOUTERS, G.L., WESSELING, K.H., GOEDHARD, W.J.A. (1984). The elastic properties of 45 human thoracic and 20 abdominal aortas *in vitro* and the parameters of a new model. *J. Biomech.* **17**, 425–435.

LAX, P.D. (1973). *Hyperbolic Systems of Conservation Laws and the Mathematical Theory of Shock Waves*, Ser. Appl. Math. **11** (SIAM, Philadelphia, PA).

LE TALLEC, P., MOURO, J. (2001). Fluid structure interaction with large structural displacements. *Comput. Methods Appl. Mech. Engrg.* **190**, 3039–3067.

LIONS, J.L., MAGENES, E. (1968). *Problèmes aux Limites non Homogènes et Applications, 1* (Dunod, Paris).

MEYER, C.D. (2000). *Matrix Analysis and Applied Linear Algebra* (SIAM, Philadelphia, PA).

MIGLIAVACCA, F., LAGANÁ, K., PENNATI, G., DE LEVAL, M., BOVE, E., DUBINI, G. (2004). Global mathematical modeling of the norwood circulation: a multiscale approach for the study of pulmonary and coronary perfusions. *Cardiology in the Young*. In press.

NOBILE, F. (2001). Numerical approximation of fluid–structure interaction problems with application to hemodynamics. PhD thesis, École Polytechnique Fédérale de Lausanne (EPFL), thesis N. 2458.

PEROT, B. (1993). An analysis of the fractional step method. *J. Comput. Phys.* **108**, 51–58.

PIETRABISSA, R., QUARTERONI, A., DUBINI, G., VENEZIANI, A., MIGLIAVACCA, F., RAGNI, S. (2000). From the global cardiovascular hemodynamics down to the local blood motion: preliminary applications of a multiscale approach. In: Oñate, E., et al. (eds.), *ECCOMAS 2000*, Barcelona.

PIPERNO, S., FARHAT, C. (2001). Partitione procedures for the transient solution of coupled aeroelastic problems. Part ii: energy transfer and three-dimensional applications. *Comput. Methods Appl. Mech. Engrg.* **190**, 3147–3170.

PROHL, A. (1997). *Projection and Quasi-compressibility Methods for Solving the Incompressible Navier–Stokes Equations* (Teubner, Stuttgart).

QUARTERONI, A., RAGNI, S., VENEZIANI, A. (2001). Coupling between lumped and distributed models for blood problems. *Comput. Visual. Sci.* **4**, 111–124.

QUARTERONI, A., SALERI, F., VENEZIANI, A. (1999). Analysis of the Yosida method for the incompressible Navier–Stokes equations. *J. Math. Pure Appl.* **78**, 473–503.

QUARTERONI, A., SALERI, F., VENEZIANI, A. (2000). Factorization methods for the numerical approximation of the incompressible Navier–Stokes equations. *Comput. Methods Appl. Mech. Engrg.* **188**, 505–526.

QUARTERONI, A., TUVERI, M., VENEZIANI, A. (2000). Computational vascular fluid dynamics: Problems, models and methods. *Comput. Visual. Sci.* **2**, 163–197.

QUARTERONI, A., VALLI, A. (1994). *Numerical Approximation of Partial Differential Equations* (Springer-Verlag, Berlin).

QUARTERONI, A., VALLI, A. (1999). *Domain Decomposition Methods for Partial Differential Equations* (Oxford Univ. Press, New York).

QUARTERONI, A., VENEZIANI, A. (2003). Analysis of a geometrical multiscale model based on the coupling of ODE's and PDE's for blood flow simulations. *Multiscale Model. Simul.* **1** (2), 173–195.

QUARTERONI, A., VENEZIANI, A., ZUNINO, P. (2002). Mathematical and numerical modelling of solute dynamics in blood flow and arterial walls. *SIAM J. Numer. Anal.* **39** (5), 1488–1511.

RAJAGOPAL, K.R. (1993). Mechanics of non-Newtonian fluids. In: Galdi, G., Necas, J. (eds.), *Recent Developments in Theoretical Fluid Mechanics*. In: Pitman Res. Notes Math. Ser. **291** (Longman, Harlow).

RAPPITSCH, G., PERKTOLD, K. (1996). Pulsatile albumin transport in large arteries: a numerical simulation study. *ASME J. Biomech. Eng.* **118**, 511–519.

REDDY, B.D. (1998). *Introductory Functional Analysis. With Applications to Boundary Value Problems and Finite Elements* (Springer-Verlag, New York).

SEGEL, L.A. (1987). *Mathematics Applied to Continuum Mechanics* (Dover, New York).

SERRIN, J. (1959). Mathematical principles of classical fluid mechanics. In: Flugge, S., Truesdell, C. (eds.), *Handbuch der Physik, VIII/1* (Springer-Verlag, Berlin).

SIMO, J.C., FOX, D.D. (1989). On a stress resultant geometrically exact shell model, Part I: formulation and optimal parametrization. *Comput. Methods Appl. Mech. Engrg.* **72**, 267–304.

SIMO, J.C., FOX, D.D., RIFAI, M.S. (1989). On a stress resultant geometrically exact shell model, Part II: the linear theory; computational aspects. *Comput. Methods Appl. Mech. Engrg.* **73**, 53–92.

SMITH, N., PULLAN, A., HUNTER, P. (2003). An anatomically based model of coronary blood flow and myocardial mechanics. *SIAM J. Appl. Math.* **62** (3), 990–1018.

TAYLOR, C.A., DRANEY, M.T., KU, J.P., PARKER, D., STEELE, B.N., WANG, K., ZARINS, C.K. (1999). Predictive medicine: Computational techniques in therapeutic decision-making. *Comput. Aided Surgery* **4** (5), 231–247.

TEMAM, R. (1984). *Navier–Stokes Equations, Theory and Numerical Analysis*, second ed. (North-Holland, Amsterdam).

THOMPSON, K.W. (1987). Time dependent boundary conditions for hyperbolic systems. *J. Comput. Phys.* **68**, 1–24.

VENEZIANI, A. (1998). Mathematical and numerical modelling of blood flow problems, PhD thesis, Politecnico di Milano, Italy.

WOMERSLEY, J.R. (1955). Method for the calculation of velocity, rate of flow and viscous drag in arteries when the pressure gradient is known. *J. Physiol.* **127**, 553–563.

Computational Methods for Cardiac Electrophysiology

Mary E. Belik, Taras P. Usyk, Andrew D. McCulloch

Department of Bioengineering, University of California, San Diego, 9500 Gilman Drive, Mail Code 0412, La Jolla, CA 92093-0412, USA
E-mail addresses: mellen@bioeng.ucsd.edu (M.E. Belik), tusyk@ucsd.edu (T.P. Usyk), amcculloch@bioeng.ucsd.edu (A.D. McCulloch)

Abstract

Computational methods for tissue biomechanics, electrophysiology, and cellular physiology separately provide frameworks for modeling functions of cardiac tissue. We review strategies currently available for meeting the goal of structurally and functionally integrated models of cardiac electromechanical function that combine data-intensive cellular systems models with compute-intensive anatomically detailed multiscale simulations.

1. Background

A fundamental goal of physiology is to identify how the cellular and molecular structure of tissues and organs gives rise to their function *in vivo*. Correspondingly, a key goal of *in silico* physiology is to develop *computational models* that can predict physiological function from quantitative measurements of tissue, cellular, or molecular structure. Computational modeling provides a potentially powerful way to *integrate* structural properties measured *in vitro* to physiological functions measured *in vivo*. It also provides a mechanism for integrating biophysical theory with experimental observation.

In this chapter, we are interested in cardiac electromechanical function, i.e., how the cellular and extracellular organization and function of myocardial tissue is integrated into the electromechanically coupled activation and pumping function of the whole

Computational Models for the Human Body
Special Volume (N. Ayache, Guest Editor) of
HANDBOOK OF NUMERICAL ANALYSIS, VOL. XII
P.G. Ciarlet (Editor)

ISSN 1570-8659
DOI 10.1016/S1570-8659(03)12002-9

heart. For example, how does myocardial fiber architecture influence the relation between the biophysics of action potential propagation and the three-dimensional mechanics of the ventricular chambers?

The physics of the heart and other organs are complex. Geometry, structure, and boundary conditions are often irregular, three-dimensional, non-homogeneous, and time varying. Constitutive properties and reaction kinetics are typically nonlinear and time dependent. Fundamental physiological functions include mechanical responses and electrical, chemical, thermal, and transport processes in cells and tissues. Therefore, computational methods are needed to realistically model many of these diverse and multidisciplinary processes and their integrated interactions encountered in electrophysiology, biomechanics, and tissue engineering.

Structural models are usually based on *in vitro* measurements of anatomy, tissue architecture and material properties, and cell biophysics. Their results must be validated with measurements from experiments conducted *in vivo* or in the whole isolated organ. This iteration between model and experiment also provides the opportunity for numerical hypothesis testing and *in vivo* constitutive parameter estimation. Once validated, the computational models have multidisciplinary applications to problems in medicine, surgery, and bioengineering like diagnostic imaging, surgical planning and intervention, medical therapy, and biomedical engineering design for tissue engineering or medical devices.

In addition to *structural* integration across scales of tissue organization from muscle and cell to organ and system, computational models also provide a foundation for *functional* integration across interacting biological processes. Computational models have been developed for a variety of physiological processes that can be coupled for more accurate modeling of the heart. These include biomechanics, ionic currents and action potential propagation, contractile dynamics, energy metabolism, and cell signaling. By developing a comprehensive model of cardiac electromechanics, we will also have a framework for developing integrated models of functional interactions such as excitation–contraction coupling, mechanoelectric feedback, mechanoenergetics, and mechanotransduction. This development goal of integration is common and has been considered by others in references such as HUNTER, ROBBINS and NOBLE [2002], GIMA and RUDY [2002], KOHL, HUNTER and NOBLE [1999], NOBLE [2001], NOBLE [2002], RUDY [2000], SUNDNES, LINES and TVEITO [2001], WINSLOW, SCOLLAN, HOLMES, YUNG, ZHANG and JAFRI [2000].

For many applications in cardiac physiology, the dynamic biophysical processes within the cell and their functional interactions can be expressed by systems models typically consisting of coupled sets of nonlinear ordinary differential equations (ODEs) such as the common pool ionic models of myocyte electrophysiology. Similarly lumped parameter ODE models have been developed of other biophysical processes such as energy metabolism (JAFRI, DUDYCHA and O'ROURKE [2001]) and crossbridge dynamics (LANDESBERG and SIDEMAN [1994]). This opens the prospect of functionally coupled cellular models such as models of excitation–contraction coupling (MICHAILOVA and SPASSOV [1992], MICHAILOVA and SPASSOV [1997]) or mechanoenergetics (TAYLOR and SUGA [1993]) because most of the functional coupling in cardiac physiology originates within the cell.

These cellular processes are spatially coupled at the tissue and organ scales, and physico-chemical principles such as mass or momentum conservation have been used to derive continuous field models of the resulting spatially heterogeneous behavior. These field equations are partial differential equations that also incorporate additional dynamical state variables governed by the empirical systems of ODEs described above. Thus the ODEs describe local cellular biophysical properties and the PDEs provide a means for structural integration from cell to organ as a three-dimensional continuum.

The geometry of the heart and other tissues and organs is complex and three-dimensional with nonhomogeneous boundary conditions and anisotropic microstructures. The solution of coupled nonlinear PDEs and ODEs on these domains invariably requires the use of numerical methods.

In the following section, we illustrate these concepts by deriving first some sets of ODEs used to model the action potential in a single cardiac myocyte in terms of voltage dependent transmembrane ionic currents, and second the reaction–diffusion equations used to model the spread of the electrical impulse through the myocardium modeled as a bidomain continuum. Then we introduce methods for modeling the geometry and anatomical structure of the heart, and finally we discuss numerical methods suitable for solving these equations and some of their present and future applications.

2. Cell biophysics

2.1. Cellular electrophysiology

The properties of the cell membrane allow for the existence of an imbalance of total ionic charge between the intracellular and extracellular spaces. This potential difference, the resting membrane potential, is mainly due to differing concentrations of ions across the membrane, namely Na^+, K^+, Mg^{2+}, Ca^{2+}, H^+, and Cl^-. The membrane potential varies as ion concentrations change under various conditions and stimuli. Fig. 2.1 shows the general shape of an action potential which occurs when a stimulus, for example an injection of current from an electrode, causes the membrane potential to rise until a threshold level of voltage is reached (region 1 on the schematic). This is followed by depolarization (region 2 to 3) caused mainly by an influx of Na^+ ions. In reaction,

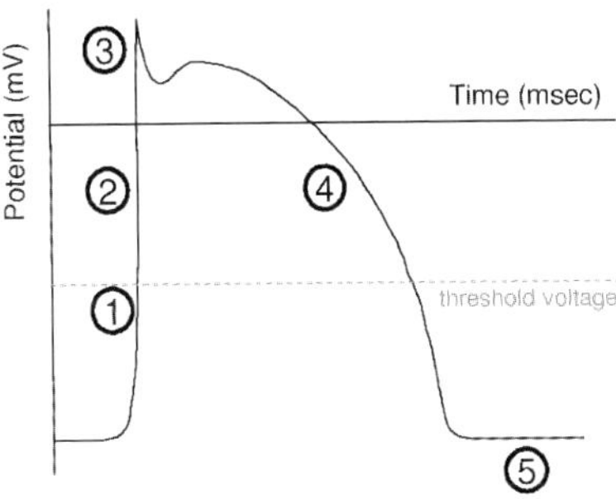

FIG. 2.1. Schematic of an action potential.

fluxes of other ions, mainly K^+ and Ca^{2+}, occur repolarizing the cell (region 4) which eventually returns to a resting equilibrium state (region 5).

Ions may cross the semi-permeable cell membrane by diffusion or by active transport. By Fick's law, ions will diffuse from regions of higher concentration to regions of lower concentration. For diffusion of charge carrying ions, a potential difference between regions may result which introduces a second driving force due to the presence of a non-zero electrical field. The electrochemical equilibrium between the forces due to a concentration gradient and forces due to a potential gradient for a particular ion is described by the Nernst equation (2.1) where R, F, and T are the gas constant, Faraday's constant, and temperature, respectively. V is the Nernst equilibrium potential, z is the valence of the ion in question, and C_o and C_i are the concentrations outside and inside the cell membrane.

$$V = \frac{RT}{zF} \ln \frac{[C_o]}{[C_i]}. \tag{2.1}$$

The Goldman–Hodgkin–Katz equation (2.2) accounts for electrochemical equilibrium of all ions present in a system; where P_{ion} is the permeability of the membrane to a particular ion, and V_m is the membrane potential:

$$V_m = \frac{RT}{zF} \ln \frac{\sum_{ion} P_{ion}[C_o]_{ion}}{\sum_{ion} P_{ion}[C_i]_{ion}}. \tag{2.2}$$

Existence of electrochemical gradients across the membrane causes facilitated diffusion through passive ion channels. A passive channel, when in an open state, allows passage of a particular ion across the cell membrane with a direction and rate determined by the forces of electrochemical equilibrium. In excitable cells such as myocytes, voltage-gating is an important determinant of the state of a channel. A closed channel at rest will change conformation to an open state when depolarization causes the threshold membrane potential to be reached. Various channels have other stable states other than merely open or closed such as open but inactivated. Active transport is accomplished by ion channels that use the energy of ATP hydrolysis to transport ions against the electrochemical gradient.

Differences in ion channel expression and resulting action potential morphology can be found between species. In addition, action potentials vary due to cell type, which is determined by the types of ion channels possessed and their mode of expression and varies with the myocardial region from which the cells originate. Mammalian ventricular tissue is thought to be composed of layers with different ionic properties and thus different action potential morphologies (YAN, SHIMIZU and ANTZELEVITCH [1998]). The ionic channels implicated in distinguishing transmural cell types are I_{Ks}, I_{Na}, and I_{to}. I_{Ks} has been found to be smaller in midmyocardial cells as compared to endocardial and epicardial cells, while I_{Na} and I_{to} are found to be larger (WOLK, COBBE, HICKS and KANE [1999], VISWANATHAN, SHAW and RUDY [1999], ZYGMUNT, EDDLESTONE, THOMAS, NESTERENKO and ANTZELEVITCH [2001]). In guinea pigs, it has been found that significantly larger differences in action potential duration (APD) exist on the epicardium from base to apex than exist between the endocardium and epicardium (LAURITA, GIROUARD and ROSENBAUM [1996]). This suggests that base to

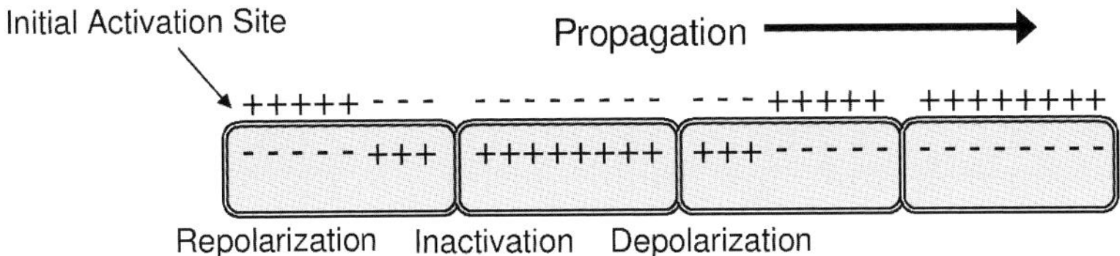

FIG. 2.2. Propagation travels from cell to cell spreading in all directions from the initial site of activation.

apex regional differences in cells may contribute more significantly to action potential heterogeneity than transmural differences in cells. In addition, it was found that an appropriately timed premature stimulus could reverse repolarization patterns in whole tissue even when fiber structure and propagation direction remain the same (LAURITA, GIROUARD and ROSENBAUM [1996]). This suggests that base to apex regional differences in cells may contribute more significantly to action potential duration heterogeneity than architectural fiber and sheet arrangement. Components of I_K, I_{Kr} and I_{Ks}, are implicated in base to apex action potential heterogeneity (CHENG, KAMIYA, LIU, TSUJI, TOYAMA and KODAMA [1999]). In a given cell, the character of a particular action potential is also highly dependent on the initial state (ROSENBAUM, KAPLAN, KANAI, JACKSON, GARAN, COHEN and SALAMA [1991]).

An action potential is a local event describing the variation of membrane potential with time at a particular location. Depolarization at an initial site spreads in all directions to neighboring sites through action potential propagation (Fig. 2.2). In the three-dimensional environment of the tissue, communication of action potentials from cell to cell is affected by local tissue architecture and direct cell coupling via gap junctions. Action potentials measured *in vitro* may differ from *in situ* due to differences in cell-to-cell coupling conditions (ANYUKHOVSKY, SOSUNOV and ROSEN [1996]). Propagation in whole ventricular tissue is further discussed in Section 3.

2.2. *Ion channels*

Early mathematical models of ion transport through cell membrane channels were developed by HODGKIN and HUXLEY [1952] based on experimental measurements made on the squid axon. Measurements were carried out under voltage-clamp conditions where the membrane potential is held at a constant voltage. Voltage clamping is accomplished by injecting current while monitoring membrane potential. Because the voltage is constant, no capacitive currents exist making this technique useful for isolating the contributions of ionic currents which have an electric circuit analog of a resistive component. According to Ohm's law, current through a resistor is equal to the product of conductance and the voltage drop. For the current describing the flow of a particular ion across the cell membrane, the driving voltage drop is theoretically the difference between the Nernst equilibrium potential of the ion and the actual membrane potential. Hence, an ionic current is described by the following equation, where g_{ion} is the conductance of the ion, and V_{ion} is the Nernst equilibrium potential for the ion:

$$I_{\text{ion}} = g_{\text{ion}}(V_{\text{m}} - V_{\text{ion}}). \tag{2.3}$$

For ion currents that pass through voltage-gated channels, g_{ion} is not a constant but a function of voltage and time. For example, a potassium channel: the protein in the cell membrane that forms this channel is composed of four identical subunits (gates) which each have an equal probability of being in an open or closed state (ALBERTS, BRAY, LEWIS, RAFF, ROBERTS and WATSON [1994]). Thus,

$$g_{\mathrm{K}} = \overline{g_{\mathrm{K}}} n^4, \tag{2.4}$$

where $\overline{g_{\mathrm{K}}}$ is the maximal channel conductance, n is the probability that one of the gates is open, and n^4 is the probability that all four gates are open. In the original Hodgkin–Huxley experiments the existence of four gates was found empirically. The open probability, which equals the fraction of open gates $n = open/(open + closed)$, is calculated by the law of mass action, which states that the change in the probability that a gate is open with respect to time equals the difference between the rate of closed gates opening and the rate of open gates closing:

$$\frac{\mathrm{d}n}{\mathrm{d}t} = \alpha(1-n) - \beta n, \tag{2.5}$$

where α and β are the rates of opening and closing, respectively. The rates, α and β, are voltage dependent and are found by empirically fitting experimental data.

By rewriting Eq. (2.5), physical meaning can be derived in terms of α and β,

$$\frac{\mathrm{d}n}{\mathrm{d}t} = \frac{(n_\infty - n)}{n_\tau}, \tag{2.6}$$

where

$$n_\infty = \frac{\alpha}{\alpha+\beta} \quad \text{and} \quad n_\tau = \frac{1}{\alpha+\beta}. \tag{2.7}$$

The steady state solution of Eq. (2.6) is

$$n(t) = n_\infty + (n_0 - n_\infty)\mathrm{e}^{-t/n_\tau}, \tag{2.8}$$

where n_0 is the initial probability that a gate is open. This solution is only approximate due to the changes of α and β with voltage which render n_∞ and n_τ non-constant. n_τ represents the time constant that determines the rate at which $n(t)$ approaches n_∞. When the time constant is sufficiently small so that the steady state is reached quickly after an initial change, $n(t)$ is approximately n_∞.

The Hodgkin and Huxley axon model included three currents: a constant conductance chloride current, the potassium current described above, and a sodium current. The sodium current was found to have a more complicated behavior including an inactivated state as well as the open and closed states. This resulted in Eq. (2.9) for sodium conductance with two gating variables: m for activation of the channel and h for inactivation. These gating variables are governed by empirically fit equations of the form of Eq. (2.5),

$$g_{\mathrm{Na}} = \overline{g_{\mathrm{Na}}} m^3 h. \tag{2.9}$$

Advances in technology since the time of Hodgkin and Huxley have significantly advanced knowledge of channels and their function. Modes of ion channel measurement include patch clamping techniques, where small patches of membrane rather than whole cells are studied under voltage clamp conditions. This method has the advantages of allowing precise control over the ionic contents of the spaces on both sides of the membrane as well as allowing measurements of individual ion channels. Many channels have been and continue to be discovered, some with much more complex behavior. These channel models combined appropriately with models of other myocyte functions are used to build models of whole myocytes.

2.3. *Modeling cellular kinetics*

2.3.1. *Basic ionic models*

Simulations of whole myocytes are typically derived from a statement of conservation of current. This includes the resistive terms that represent ion channels and a capacitive term. The cell membrane acts as a capacitor, a non-conductive dielectric, that separates the conductive extracellular space and the conductive cytoplasm. By the definition of capacitance,

$$q = C_{\mathrm{m}} V_{\mathrm{m}}, \tag{2.10}$$

where q is charge, and C_{m} is membrane capacitance (farad/cm^2). Changes in capacitive charge of the membrane over time result in a capacitive current (Fig. 2.3). By the definition of current, the capacitive current, I_{c}, is

$$I_{\mathrm{c}} = \frac{\mathrm{d}q}{\mathrm{d}t} = C_{\mathrm{m}} \frac{\mathrm{d}V_{\mathrm{m}}}{\mathrm{d}t}. \tag{2.11}$$

Thus the equation for the Hodgkin–Huxley nerve cell with sodium, potassium, and leakage ion channels is

$$\begin{aligned} -C_{\mathrm{m}} \frac{\mathrm{d}V_{\mathrm{m}}}{\mathrm{d}t} &= I_{\mathrm{Na}} + I_{\mathrm{K}} + I_{\mathrm{L}} \\ &= g_{\mathrm{Na}}(V_{\mathrm{m}} - V_{\mathrm{Na}}) + g_{\mathrm{K}}(V_{\mathrm{m}} - V_{\mathrm{K}}) + g_{\mathrm{L}}(V_{\mathrm{m}} - V_{\mathrm{L}}). \end{aligned} \tag{2.12}$$

A system of ordinary differential equations (ODEs) consisting of Eq. (2.12) and the ODEs for the gating variables, n, m, and h (equations of the form of (2.5)) is solved simultaneously for the membrane potential as it varies with time.

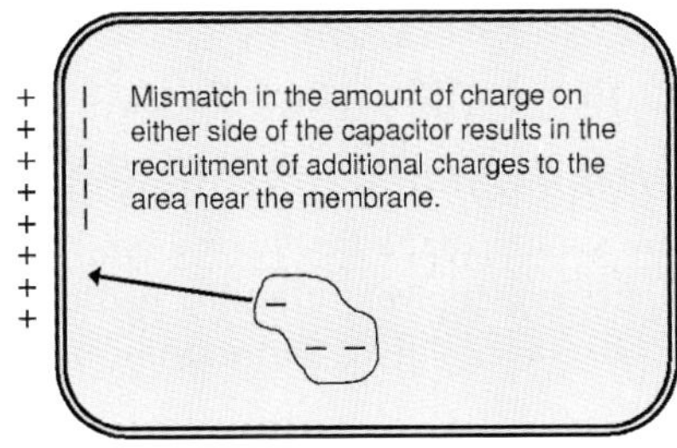

FIG. 2.3. Capacitive current is created by charge recruitment near the cell membrane.

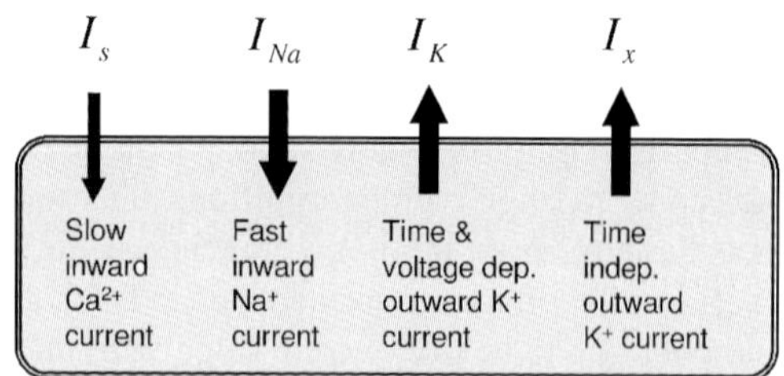

FIG. 2.4. A Beeler–Reuter ionic cell model. The size of the arrows represents relative current size.

The first ionic model adapted for mammalian cardiac cells was developed by BEELER and REUTER [1977]. This model requires solution of eight ODEs: the membrane potential, the myoplasmic calcium concentration, and six Hodgkin–Huxley type gating variables. The gating constants are associated with four ion channels: a fast inward sodium current, a slow inward current primarily carried by Ca^{2+} ions, an outward potassium current, and a voltage dependent outward current primarily carried by potassium ions. Fig. 2.4 shows a schematic of a Beeler–Reuter cell.

$$\text{state variables} = \{V_\mathrm{m}, [\mathrm{Ca}^{2+}]_\mathrm{i}, m, h, j, d, f, x\},$$

$$-C_\mathrm{m}\frac{\mathrm{d}V_\mathrm{m}}{\mathrm{d}t} = I_\mathrm{Na} + I_\mathrm{s} + I_\mathrm{K} + I_x - I_\mathrm{ext}. \tag{2.13}$$

The external current, I_ext, represents an externally applied stimulation. The ODE describing the change of calcium concentration with respect to time is a function of the slow inward current, I_s. The gating variables are governed by equations of the form of Eq. (2.5), and the twelve associated opening and closing rates are exponential functions of voltage fitted to measured data. Refer to BEELER and REUTER [1977] for complete information on these equations and parameters.

2.3.2. *Second generation models*

Complex interacting ionic processes not limited to ion channels give rise to action potentials. Second generation models incorporate intracellular processes and the resulting effects on intracellular concentrations of various ions. Various researchers have developed models of cardiac cellular kinetics adapted for the features particular to various species and cell types. The Luo–Rudy model (Fig. 2.5) is based on measurements from cell voltage clamp studies in guinea pig, and uses nine ODEs (LUO and RUDY [1991], LUO and RUDY [1994]).

$$\begin{aligned}\text{state variables} = \{&V_\mathrm{m}, [\mathrm{Na}^+]_\mathrm{i}, [\mathrm{K}^+]_\mathrm{i}, [\mathrm{Ca}^{2+}]_\mathrm{i}, [\mathrm{Ca}^{2+}]_\mathrm{JSR}, [\mathrm{Ca}^{2+}]_\mathrm{NSR},\\ &m, h, j, d, f, x\}.\end{aligned} \tag{2.14}$$

The six gating variables are similar to the Beeler–Reuter, Hodgkin–Huxley based, gating variables although the parameters were fit to the particular data of the guinea pig. The membrane potential ODE consists of the sum of three main currents: sodium, potassium, and calcium,

$$-C_\mathrm{m}\frac{\mathrm{d}V_\mathrm{m}}{\mathrm{d}t} = I_\mathrm{Na,tot} + I_\mathrm{K,tot} + I_\mathrm{Ca,tot}. \tag{2.14a}$$

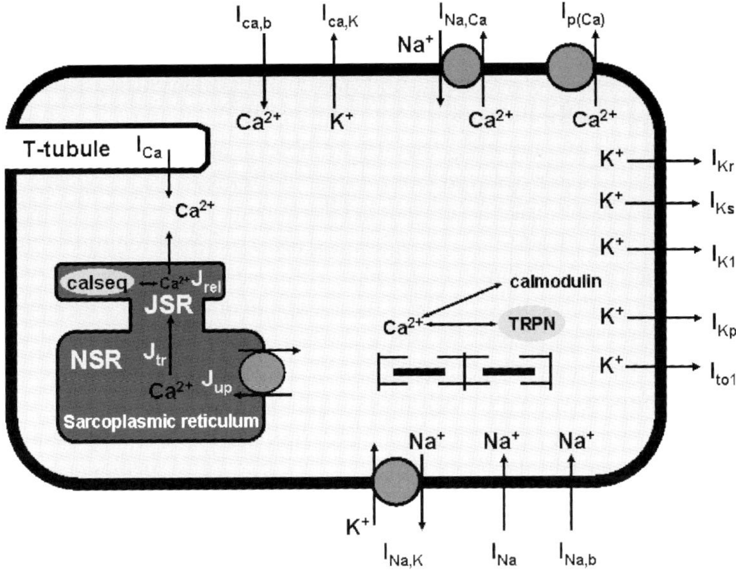

FIG. 2.5. Schematic of a Luo–Rudy ionic model.

However, each of these currents represents a sum of currents that relate to the particular ion:

$$I_{\mathrm{Na,tot}} = I_{\mathrm{Na}} + 3I_{\mathrm{NaCa}} + 3I_{\mathrm{NaK}} + I_{\mathrm{ns,Na}} + I_{\mathrm{Na,b}} + I_{\mathrm{CaNa}}, \tag{2.15}$$

$$I_{\mathrm{K,tot}} = I_{\mathrm{K}} + I_{\mathrm{K1}} + I_{\mathrm{Kp}} - 2I_{\mathrm{NaK}} + I_{\mathrm{ns,K}} + I_{\mathrm{CaK}}, \tag{2.16}$$

$$I_{\mathrm{Ca,tot}} = I_{\mathrm{Ca}} - 2I_{\mathrm{NaCa}} + I_{\mathrm{p(Ca)}} + I_{\mathrm{Ca,b}}. \tag{2.17}$$

The components of the total sodium current are the fast inward sodium current, the sodium calcium exchanger (pumping at a ratio of 3:2 sodium to calcium ions), the sodium potassium pump (3:2 ratio of sodium to potassium), a sodium current that passes through a non-specific calcium activated channel, a sodium background leakage current, and the sodium component of the current passing through the L-type calcium channel. Similar sums can be written to account for total potassium and calcium currents. Potassium currents are the time-dependent and time-independent potassium currents, a plateau potassium current, the sodium potassium pump, and potassium currents that flow through non-specific calcium activated channels and through the L-type calcium channels. The calcium currents are the calcium currents through the L-type and T-type channels, the sodium calcium exchanger, the sarcoplasmic pump current, and a calcium background leakage current. Of these currents, some are new in the Luo–Rudy model, others are taken from previous models with significant structural changes, and many were taken from earlier models with only their parameters adjusted to fit the experimental data used.

In addition to intracellular calcium concentration, intracellular potassium and sodium concentrations were included. The ODEs describing these three concentrations are dependent on the total current of the ion in question. The Luo–Rudy model also offers a more detailed description of calcium handling. The sarcoplasmic reticulum (SR) is represented by three intracellular currents and two concentrations. The SR is modeled as having two compartments: the network SR representing the longitudinal tubules where calcium is taken up from the myoplasm by an uptake current, and the junctional SR representing the cisternae where calcium is released back into the myoplasm. The compartments have a calcium transfer current between them, and each is described with a calcium concentration. In addition, myoplasmic calcium is buffered by calmodulin and troponin, and junctional calcium is buffered by calsequestrin.

The Luo–Rudy model is often the basis for models in other species such as the model of Winslow et al. for dog (WINSLOW, RICE, JAFRI, MARBAN and O'ROURKE [1999]) and the Puglisi–Bers model for rabbit (PUGLISI and BERS [2001]). These models extend the Luo–Rudy model by refitting the parameters that describe specific ion channel functions, and combining or exchanging new models of particular ion channels or ionic processes for the original Luo–Rudy kinetics in order to build species specific models. The main extension added by the Winslow et al. model besides fitting parameters to canine experimental data concerns the calcium dynamics. A complicated L-type channel model and a restricted subspace was added to better describe calcium induced calcium release. The subspace of the myoplasm is postulated to lie between the junctional SR and the cell membrane and allows close contact of L-type channels to ryanodine receptors on the junctional SR membrane. It is postulated that due to geometry, the subspace calcium concentration differs from the bulk myoplasm and is involved in triggering SR calcium release. The Puglisi–Bers model similarly was based on the Luo–Rudy model with the parameters rescaled for rabbit data. The main modifications in addition to rescaling were the addition of a rescaled Winslow transient outward potassium current, the addition of a calcium activated chloride current, and modifications of the T-type calcium (a component of I_{Ca}) and delayed rectifier potassium (a component of I_{K}) currents.

2.3.3. Other approaches

In the interest of creating integrated models across biological scales, many tissue level models rely on phenomenological models of cellular behavior in order to have computationally tractable problem formulations. The most basic phenomenological model was originally developed by FitzHugh and Nagumo (FHN) (FITZHUGH [1961], NAGUMO, ARIMOTO and YOSHIZAWA [1962]) for a general excitable media. It consists of only two variables, has cubic nonlinearity, and has an on–off behavior that results in an action potential-like depolarization and repolarization,

$$\frac{\mathrm{d}u}{\mathrm{d}t} = u(u-a)(1-u)c_1 - c_2 v, \tag{2.18}$$

$$\frac{\mathrm{d}v}{\mathrm{d}t} = bu - b\,\mathrm{d}v. \tag{2.19}$$

While the FHN model and its modifications (ROGERS and MCCULLOCH [1994], KOGAN, KARPLUS, BILLETT, PANG, KARAGUEUZIAN AND KHAN [1991]) can provide a qualitative action potential that is relatively computationally efficient, FHN models have two major disadvantages. First, they do not accurately reproduce features of the cardiac action potential that are of the most interest such as the rapid upstroke of depolarization. Second, they are not biophysically based, limiting their usefulness. For example, the character of the action potential computed can be changed through the various model parameters, but these cannot be related to specific biophysical mechanisms so that specific channels or interventions are not possible to simulate.

On the other hand, the biophysically detailed models present two main computational hurdles. These suffer from inherent instabilities (ENDRESEN and SKARLAND [2000]) which involve drift of ion concentrations and eventual equilibrium disruption. In addition, the growing number of variables included in these models that must be solved simultaneously are increasingly computationally inefficient. These significant drawbacks have led to the development of a range of intermediate models that provide better efficiency without retaining all the known ionic details (FENTON and KARMA [1998], DUCKETT and BARKLEY [2000], BERNUS, WILDERS, ZEMLIN, VERSCHELDE and PANFILOV [2002]). For example, Bernus et al. (BERNUS, WILDERS, ZEMLIN, VERSCHELDE and PANFILOV [2002]) have reduced a Priebe–Beuckelmann human myocyte model (PRIEBE and BEUCKELMANN [1998]) (an extension of a Luo–Rudy myocyte) to a six variable intermediate model. This model retains the fast kinetics of the sodium channel for accurate depolarization, but approximates other fast kinetics that do not have large effects on action potential shape.

2.4. *Sarcomere dynamics*

Potentially, ionic models may be linked to models of sarcomere dynamics through contractile activation and crossbridge mechanics. The basic unit of contraction is the sarcomere. Sarcomere dynamics, namely length–tension relations, play an important role in active force development. The length–tension relation in muscle arises from changes in the overlap of thick and thin filaments in the sarcomere. At the level of a single myocyte, contractile activation occurs shortly after depolarization of the cell as the ionic contents shift. The time course of cytosolic calcium concentration is also central in determining the contractile force and has been modeled using various force–calcium relations. Finally, mechanisms of crossbridge mechanics, namely crossbridge recruitment and length sensing, determine the forces developed by the sarcomere.

Long single fiber preparations have been valuable test specimens for studying the mechanisms of skeletal muscle mechanics. The lack of these ideal test specimens is one main reason that cardiac muscle mechanics testing is far more difficult than skeletal muscle testing. Moreover, under physiological conditions, cardiac muscle cannot be stimulated to produce sustained tetanic contractions due to the absolute refractory period of the myocyte cell membrane. Cardiac muscle also exhibits a mechanical property analogous to the relative refractory period of excitation. After a single isometric contraction, some recovery time is required before another contraction of equal amplitude can be activated (Fig. 2.6).

Unlike skeletal muscle, in which maximal active force generation occurs at a sarcomere length that optimizes myofilament overlap (~2.1 μm), the isometric twitch tension developed by isolated cardiac muscle continues to rise with increased sarcomere length

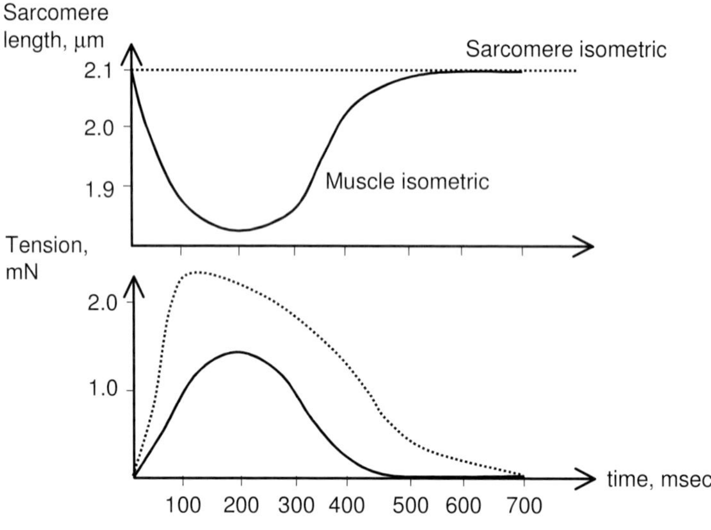

FIG. 2.6. Isometric testing.

in the physiological range (1.6–2.4 µm). Early evidence for a descending limb of the cardiac muscle isometric length–tension curve was found to be caused by shortening in the central region of the isolated muscle at the expense of stretching at the damaged ends where the specimen was tethered to the test apparatus. If muscle length is controlled so that sarcomere length in the undamaged part of the muscle is indeed constant, or if the developed tension is plotted against the instantaneous sarcomere length rather than the muscle length, the descending limb is eliminated (TER KEURS, RIJNSBURGER, VAN HEUNINGEN and NAGELSMIT [1980]). Thus, the increase with chamber volume of end-systolic pressure and stroke work is reflected in isolated muscle as a monotonic increase in peak isometric tension with sarcomere length. The increase in slope of the end-systolic pressure–volume relation (ESPVR) associated with increased contractility is mirrored by the effects of increased calcium concentration in the length–tension relation. The duration as well as the tension developed in the active cardiac twitch also increases substantially with sarcomere length.

The relation between cytosolic calcium concentration and isometric muscle tension has mainly been investigated in muscle preparations in which the sarcolemma has been chemically permeabilized. Because there is evidence that this chemical "skinning" alters the calcium sensitivity of myofilament interaction, recent studies have also investigated myofilament calcium sensitivity in intact muscles tetanized by high frequency stimulation in the presence of a compound such as ryanodine that opens calcium release sites in the sarcoplasmic reticulum. Intracellular calcium concentration was estimated using calcium-sensitive optical indicators such as fura. Myofilaments were activated in a graded manner by microMolar concentrations of calcium, which binds to troponin C according to a sigmoidal relation (RUEGG [1988]). Half-maximal tension in cardiac muscle was developed at intracellular calcium concentrations of 10^{-6} to 10^{-5} M (the $[Ca]_{50}$) depending on factors such as species and temperature (BERS [1991]). Hence, relative isometric tension $T_0/T_{\max}$ may be modeled using (TOZEREN [1985], HUNTER, MCCULLOCH, NIELSEN and SMAILL [1988])

$$\frac{T_0}{T_{\max}} = \frac{[\mathrm{Ca}]^n}{[\mathrm{Ca}]^n + [\mathrm{Ca}]_{50}^n}. \tag{2.20}$$

The Hill coefficient (n) governs the steepness of the sigmoidal curve. A wide variety of values have been reported but most have been in the range of 3 to 6 (KENTISH, TER KEURS, RICCIARDI, BUCX and NOBLE [1986], BACKX, GAO, AZAN-BACKX and MARBAN [1995]). The steepness of the isometric length–tension relation, compared with that of skeletal muscle is due to length-dependent calcium sensitivity. That is, the Ca_{50} (M), and perhaps n as well, change with sarcomere length, L.

The isotonic force–velocity relation of cardiac muscle is similar to that of skeletal muscle, and A.V. Hill's well-known hyperbolic relation is a good approximation except at larger forces greater than about 85% of the isometric value (Fig. 2.7a):

$$\frac{V}{V_{\max}} = \frac{1 - T/T_0}{1 + cT/T_0}. \tag{2.21}$$

The maximal (unloaded) velocity of shortening is essentially independent of preload, but does change with time during the cardiac twitch and is affected by factors that affect contractile ATPase activity and hence crossbridge cycling rates. De Tombe and colleagues (DE TOMBE and TER KEURS [1992]) using sarcomere length-controlled isovelocity release experiments found that viscous forces impose a significant internal load opposing sarcomere shortening. If the isotonic shortening response is adjusted for the confounding effects of passive viscoelasticity, the underlying crossbridge force–velocity relation is found to be essentially linear (Fig. 2.7b).

Cardiac muscle contraction also exhibits other significant length–history-dependent properties. An important example is "deactivation" associated with length transients. The isometric twitch tension redeveloped following a brief length transient that dissociates crossbridges reaches the original isometric value when the transient is imposed early in the twitch before the peak tension is reached. However, following transients

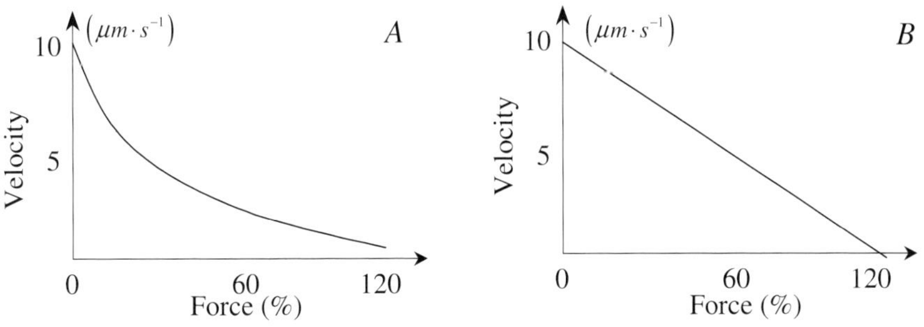

FIG. 2.7. Isotonic testing: (a) the results of an isovelocity release experiment conducted during a twitch; (b) cardiac muscle force–velocity relation corrected for viscous forces of passive cardiac muscle which reduce shortening velocity.

applied at times after the peak twitch tension has occurred, the fraction of tension redeveloped declines progressively since the activator calcium has fallen to levels below that necessary for all crossbridges to reattach (TER KEURS, RIJNSBURGER, VAN HEUNINGEN and NAGELSMIT [1980]).

A number of models of active tension development in cardiac muscle have been proposed. They may be grouped into three categories:

(1) time-varying elastance models include the essential dependence of cardiac active force development on muscle length and time (ARTS, RENEMAN and VEENSTRA [1979], CHADWICK [1982], TABER [1991]),
(2) "Hill" models, in which the active fiber stress development is modified by shortening or lengthening according to the force–velocity relation, so that fiber tension is reduced by increased shortening velocity (ARTS, VEENSTRA and RENEMAN [1982], NEVO and LANIR [1989]), and
(3) fully history-dependent models that are usually based on A.F. Huxley's crossbridge theory (PANERAI [1980], LANDESBERG and SIDEMAN [1994], LANDESBERG, MARKHASIN, BEYAR and SIDEMAN [1996]) which yields a system of partial differential equations as functions of time and crossbridge position.

Many of the early models were based on skeletal muscle models of Hill (HILL [1938], HILL [1970]). However, Hill's model considers tetanic contraction only and hence is inappropriate for describing cardiac muscle mechanics (FUNG [1981]). According to PANERAI [1980], Wong (WONG [1971], WONG [1972]) was the first to employ the sliding filament theory to model the mechanics of cardiac muscle. Wong generalized Huxley's model (HUXLEY [1957]) of the skeletal muscle crossbridge to partial and length-dependent activation. Panerai, using Huxley's original model, incorporated length-dependent activation in a first order kinetic equation describing Ca^{2+}-troponin C interaction. Instead of considering individual myofilaments, TOZEREN [1985] proposed a "continuum" model of cardiac muscle contraction. Tozeren generalized Hill's equation to partial activation to describe active fiber tension as a function of fiber strain, strain rate and time after onset of contraction. In these studies, model predictions were validated by experimental length–tension or force–velocity relations during contractions in which overall muscle length was controlled. Panerai accounted for the appreciable internal shortening that occurs during isometric contractions at the expense of lengthening in the damaged muscle at the clamped ends (KRUGER and POLLACK [1975], TER KEURS, RIJNSBURGER, VAN HEUNINGEN and NAGELSMIT [1980]).

Continuum models typically compute the active tension developed by a cardiac muscle fiber from the peak intracellular Ca^{2+} ion concentration, the time after onset of contraction and the sarcomere length–history (GUCCIONE and MCCULLOCH [1993]). Model contraction is driven by a free calcium transient that is independent of length. The number of actin sites available to react with myosin is determined from the total number of actin sites (available and inhibited), the free calcium, the length–history-dependent association and dissociation rates of Ca^{2+} ions to troponin binding sites and the troponin concentration.

HUNTER, MCCULLOCH and TER KEURS [1998] described intracellular calcium concentration as a function of time ($\mathrm{Ca_i}(t)$) as follows :

$$\mathrm{Ca}_i(t) = \mathrm{Ca}_0 + (\mathrm{Ca}_{\max} - \mathrm{Ca}_0)\frac{t}{\tau_{\mathrm{Ca}}}\mathrm{e}^{1-t/\tau_{\mathrm{Ca}}}, \tag{2.22}$$

where Ca_0 is the resting concentration of intracellular calcium and $\mathrm{Ca}_{\max}$ is the peak intracellular calcium concentration which occurs at $t = \tau_{\mathrm{Ca}}$.

Panerai described Ca^{2+}–troponin C interaction using a rate equation derived from classical chemical kinetics. This equation determines the concentration of actin that is free to react with myosin ($A_\mathrm{c}(t)$) from the kinetics of the binding of two calcium ions to independent sites on the troponin molecule,

$$\frac{\mathrm{d}A_\mathrm{c}(t)}{\mathrm{d}t} = c_1\mathrm{Ca}_i^2(t)\big[A_{\mathrm{co}} - A_\mathrm{c}(t)\big] - c_2A_\mathrm{c}(t), \tag{2.23}$$

where A_{co} is a constant reflecting the total amount of actin present in the muscle, and c_1 and c_2 are the association and dissociation rate constants, respectively.

Ca^{2+}–troponin C binding results in tropomyosin movement, which in turn controls the availability of actin binding sites. For modeling tropomyosin kinetics, HUNTER, MCCULLOCH and TER KEURS [1998] introduced a non-dimensional parameter z, ($0 \leqslant z \leqslant 1$), to represent the proportion of actin sites available for crossbridge binding. They proposed a model for z reflecting an exponential increase in tension with a first order rate constant dependent on calcium activation,

$$\frac{\mathrm{d}z}{\mathrm{d}t} = \alpha_0\left[\left(\frac{\mathrm{Ca}}{\mathrm{Ca}_{50}}\right)^n(1-z) - z\right], \tag{2.24}$$

where α_0 is the rate constant of tropomyosin movement. Ca_{50} and n are Hill parameters determined for a particular sarcomere length by fitting the equilibrium relation between z and [Ca]. Under an assumption of proportionality between steady state force and available actin binding sites, a sigmoidal response curve (Hill-type equation) is used to describe the steady state developed tension as a function of calcium concentration (HUNTER, MCCULLOCH and TER KEURS [1998]),

$$z_{\mathrm{ss}} = \frac{[\mathrm{Ca}]^n}{[\mathrm{Ca}]^n + [\mathrm{Ca}]_{50}^n} \tag{2.25}$$

where z_{ss} takes a value between 0 and 1, $[\mathrm{Ca}]_{50}$ is the calcium concentration required to produce 50% of peak contraction, and n is the Hill parameter describing the shape of the sigmoidal curve. Note that Eq. (2.25) is also known as a "Hill" equation, not to be confused with Hill's equation for the force–velocity relation (2.21).

Sarcomere length dependence is included by describing developed maximum tension as a function of λ, the extension ratio of sarcomeres (l_0/l_{ref}, where l_{ref} is resting sarcomere length):

$$T_0 = T_{\mathrm{ref}}\big(1 + \beta_0(\lambda - 1)\big). \tag{2.26}$$

From experimental observations in rats, $T_{\mathrm{ref}} = 125$ kPa (the reference tension when $\lambda = 1$) (HUNTER, MCCULLOCH and TER KEURS [1998]). The term β_0 describes myofilament cooperativity,

$$\beta_0 = \frac{1}{T_{\text{ref}}} \frac{\mathrm{d}T_0}{\mathrm{d}\lambda} = 1.45. \tag{2.27}$$

For less than full activation, the dependence on calcium under isometric conditions may be approximated by

$$T_0 = T_{\text{ref}}\big(1 + \beta_0(\lambda - 1)\big) z_{ss}. \tag{2.28}$$

Length-dependence of the parameters n and $[\mathrm{Ca}]_{50}$ is approximated in a similar form:

$$n = n_{\text{ref}}\big(1 + \beta_1(\lambda - 1)\big), \tag{2.29}$$

$$pC_{50} = pC_{50\text{ref}}\big(1 + \beta_2(\lambda - 1)\big), \tag{2.30}$$

$$[\mathrm{Ca}]_{50} = 10^{6 - pC_{50}} \quad \text{in } \mu\text{M}. \tag{2.31}$$

In order to fit the experimental data that KENTISH, TER KEURS, RICCIARDI, BUCX and NOBLE [1986] obtained from skinned rat right ventricular muscle, the parameters of the model were chosen as: $n_{\text{ref}} = 4.25$; $\beta_0 = 1.45$; $\beta_1 = 1.95$; $\beta_2 = 0.31$; $pC_{50\text{ref}} = 5.33$; $T_{\text{ref}} = 125$ kPa; $\mathrm{Ca}_{\max} = 4.3$ μM, the maximal intracellular calcium concentration.

Sarcomere length can affect generated force through crossbridge recruitment and through crossbridge distortion. Recruitment affects generated force by altering the numbers of attached crossbridges. Although this is the basis for Starling's law, the mechanisms are not yet clearly understood. Distortion is internal stretch of a crossbridge structure. Again mechanisms for sensing stretch are not fully understood. If crossbridges are modeled as having constant, linear elastic material properties, the total force generated in one half of a myosin and actin filament pair is as follows:

$$F_{\text{total}} = K\left(\frac{l_0}{l_{\text{ref}}}\right)N, \tag{2.32}$$

where K is a spring constant describing stretch of a single crossbridge, l_0/l_{ref} is the stretch of the crossbridge, and N is the number of attached crossbridges. Cooperativity mechanisms describe possible ways that crossbridge interactions sense and modulate force. Examples include sensing based on lateral spacing of actin and myosin, memory based mechanisms where recruitment depends on initial length, or existence of a crossbridge state that is attached but not yet generating force thereby accounting for exceedingly rapid recruitment rates that fall off during late stages of contraction (HUNTER [1999], RICE, WINSLOW and HUNTER [1999]).

3. Impulse propagation dynamics

3.1. *Physiology of propagation*

The nervous system directs the behavior of the heart; however, cardiac tissues have intrinsic rhythmicity (or automaticity) meaning cardiac tissues can initiate beats without nervous stimulus. This pacemaking activity may serve as a safety mechanism and is sufficient for successful cardiac function, for example, in the case of completely denervated mammalian hearts used for cardiac transplant. The cardiac conduction system in mammals begins with the sinoatrial node (SA node) located where the superior vena cava joins the right atria. The SA node may be referred to as the natural pacemaker of the heart, and it normally generates impulses at a higher frequency than other cardiac tissues. A cardiac impulse is transmitted via various internodal pathways from the SA node, across the atria to the atrioventricular node (AV node). The AV node is also capable of producing pacemaking impulses for the whole heart, at a somewhat lower frequency than the SA node. The impulse continues along the fibers of the AV node as they transition into fibers belonging to the bundle of His. In normal conditions this is the only pathway by which an impulse reaches the ventricles (see BERNE and LEVY [1997], KATZ [2001]).

The bundle of His represents the upper portion of the ventricular conduction system. It divides into left and right bundle branches one leading subendocardially to each of the ventricles. The bundle branches continue to subdivide into the complex Purkinje fiber network. The Purkinje fibers also possess automaticity but can only generate impulses at a significantly slow rate. Impulse conduction through the Purkinje fibers is the fastest of any cardiac tissue enabling rapid spread of the impulse throughout the ventricular endocardium. The Purkinje network has been mapped in various species. In the sheep, Purkinje fibers were found to be surrounded by a perifascicular sheath which may help to direct conduction along the fiber network rather than into the surrounding tissue or may provide protection from friction during contraction (ANSARI, HO and ANDERSON [1999]). In the rabbit, while the left ventricular Purkinje system follows a branching pattern similar to that in the dog, the right ventricle was found to have a denser web-like arrangement (CATES, SMITH, IDEKER and POLLARD [2001]). The connections to the myocardium at the ends of the Purkinje network consist of transitional cells that have also been characterized (CATES, SMITH, IDEKER and POLLARD [2001], TRANUM-JENSEN, WILDE, VERMEULEN and JANSE [1991]). An impulse first excites the septum as it spreads along the Purkinje network along the endocardium. From the ends of the Purkinje fibers, the excitation in the ventricular walls spreads through the myocardial tissue from endocardium to epicardium at a slower rate than the conduction through the Purkinje system. Apical and central regions of the ventricular free walls are excited first and activation proceeds toward the base. Spread of activation within the ventricular myocardium itself is affected by local tissue characteristics.

The microstructure of ventricular architecture is important in determining three-dimensional conduction patterns. As discussed in Section 5.4, ventricular myocardium is arranged in fibers which are in turn arranged in laminar sheets (LEGRICE, SMAILL, CHAI, EDGAR, GAVIN and HUNTER [1995]). This structural arrangement in addition to its affects on cardiac mechanics influences propagation. In pacing studies where stimulating electrodes were introduced into the ventricular wall in order to elicit activation at varying wall depths, clear anisotropic propagation was seen. Patterns of depolarization were found to align with the fiber architecture as demonstrated by helical rotation of the pattern as stimulation varied with depth (EFIMOV, ERMENTROUT, HUANG and SALAMA [1996]). Various modeling efforts have also confirmed the importance of fiber architecture to electrical propagation (FRANZONE, GUERRI, PENNACCHIO and TACCARDI [1998], FRANZONE, GUERRI, PENNACCHIO and TACCARDI [2000], MUZIKANT and HENRIQUEZ [1997], MUZIKANT and HENRIQUEZ [1998], KEENER and PANFILOV [1997], SAXBERG, GRUMBACH and COHEN [1985], ROGERS and MCCULLOCH [1994], VETTER and MCCULLOCH [2001]). Measurements indicating dependence on sheet architecture are virtually nonexistent. This is most likely due to the difficulties associated with obtaining three-dimensional recordings, particularly inside the ventricular wall. Conduction is known to be faster in the fiber direction compared to the sheet and sheet normal directions (KANAI and SALAMA [1995]), but little is known experimentally about the ratio of sheet to sheet normal conduction. A recent study of the effects of tissue microstructure on propagation supports the reasonability of modeling conduction as orthotropic rather than transversely isotropic (HOOKS, TOMLINSON, MARSDEN, LEGRICE, SMAILL, PULLAN and HUNTER [2002]).

In addition to structural heterogeneities, functional heterogeneities exist regionally within ventricular tissue. Studies show that the myocytes that compose endocardial and epicardial layers of the ventricle walls have different electrical properties from those in the mid-wall (ANTZELEVITCH, SHIMIZU, YAN, SICOURI, WEISSENBURGER, NESTERENKO, BURASHNIKOV, DI DIEGO, SAFFITZ and THOMAS [1999], SICOURI and ANTZELEVITCH [1995]). Changes in ion concentrations have been shown to result in differential regional effects supporting the variation of cell type as the underlying mechanism (WOLK, KANE, COBBE and HICKS [1998]). These midmyocardial cells (M cells) are characterized by the ability to disproportionately prolong action potential duration compared to other myocardial cells in response to slowing of stimulation rate or APD prolonging agents (ANTZELEVITCH, SHIMIZU, YAN, SICOURI, WEISSENBURGER, NESTERENKO, BURASHNIKOV, DI DIEGO, SAFFITZ and THOMAS [1999]). This arrangement may be the basis for the T-wave morphology of electrocardiograms. The T-wave characterizes repolarization. The peak of the T-wave has been found to correspond with epicardial repolarization and the end with M cell repolarization (YAN, SHIMIZU and ANTZELEVITCH [1998]). Very recent evidence suggests that the arrangement of cell types may not consist of transmural layers. Instead M cells may be located in islands in various regions of the ventricle wall (AKAR, YAN, ANTZELEVITCH and ROSENBAUM [2002]).

In the whole ventricle repolarization depends on tissue heterogeneity (WOLK, COBBE, HICKS and KANE [1999]). Endocardial myocytes depolarize before epicardial cells during normal sinus rhythm, while repolarization has been measured in the canine to occur first in the epicardium followed by the endocardium and finally the midmyocardium (YAN, SHIMIZU and ANTZELEVITCH [1998]). Repolarization patterns have been found to be much more sensitive to test conditions than depolarization patterns in transmural optical mapping studies in the canine (AKAR, YAN, ANTZELEVITCH and ROSENBAUM [2002]). Evidence for the significance of regional heterogeneities of cell types versus architectural fiber orientation in determining repolarization sequence seems to be conflicting between studies in dog and guinea pig. In guinea pig repolarization seemed to spread from apex to base regardless of pacing site while in dog repolarization was highly dependent on activation sequence (KANAI and SALAMA [1995]). Corroborating these findings, timed premature stimuli have been found to reverse the pattern of repolarization even though fiber structure and propagation direction remained the same in the guinea pig (LAURITA, GIROUARD and ROSENBAUM [1996]). Comparison of measurements of repolarization in canine and guinea pig have led to the conclusion that the smaller size of the guinea pig heart requires less time for complete depolarization so that intrinsic regional heterogeneities dominate the pattern of repolarization (KANAI and SALAMA [1995], EFIMOV, ERMENTROUT, HUANG and SALAMA [1996], LAURITA, GIROUARD and ROSENBAUM [1996]).

Regional variations in action potential morphology and ion channel expression have also been observed from base to apex. In rabbit myocytes, the density of I_{Kr} and I_{Ks} ion channels varied in basal and apical samples. I_{Kr} was expressed in higher density in the apex than the base. In addition, the ratio of I_{Ks} to I_{Kr} was larger in the base than in the apex. Action potential duration was found to be longer in the apex than base (CHENG, KAMIYA, LIU, TSUJI, TOYAMA and KODAMA [1999]). This study did not rule out the possibility of detecting transmural variations in the apical or basal sections; however, other observations of whole ventricle repolarization patterns also suggest base to apex variation of cell types. The actual three-dimensional variation of cell types remains debatable, and other regional variations are possible that have not yet been examined. For example, typically the left and right ventricular walls have been treated as consisting of similar cell type distributions; however, it is possible and perhaps likely that regional variations also exist in this case (WOLK, COBBE, HICKS and KANE [1999]).

3.2. *Cable theory*

Cable theory was developed for power transmission through uniform conducting cables surrounded by an insulating medium and was first applied to cardiac tissue by WEIDMANN [1970]. A muscle fiber has a conductive interior consisting of the myoplasm and is surrounded by the insulating cell membrane. The myoplasm has a material property, **D**, that describes its three-dimensional conductivity (mS/cm).

An electric field vector (mV/cm), **E**, is defined as a potential drop maintained spatially in a material,

$$\mathbf{E} = -\nabla \Phi_{\mathrm{i}}, \tag{3.1}$$

where

$$\nabla = \frac{\partial}{\partial x_1}\hat{i} + \frac{\partial}{\partial x_2}\hat{j} + \frac{\partial}{\partial x_3}\hat{k} \tag{3.2}$$

and Φ_{i} is the scalar function for potential inside the cable (or the intracellular potential, hence the subscript i). By Ohm's law, the flux vector ($\mu\mathrm{A/cm}^2$), **J**, which represents the current density inside the cable, is proportional to the electric field vector,

$$\mathbf{J} = \mathbf{D}\mathbf{E} = -\mathbf{D}\nabla\Phi_{\mathrm{i}}. \tag{3.3}$$

Physically, Eq. (3.3) states that current flux in a cable occurs in the direction of the greatest potential drop.

Considering a differentially small section of cable with volume, $\mathrm{d}\Omega$, and surface area, $\mathrm{d}\Gamma = \mathrm{d}\Gamma_{\mathrm{c}} + \mathrm{d}\Gamma_{\mathrm{m}}$, the total current that enters the section of cable must equal the current that leaves as dictated by the conservation of charge (see Fig. 3.1). Since charge

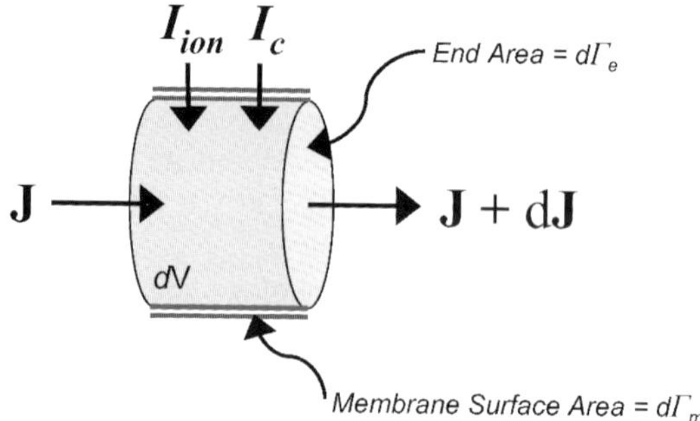

FIG. 3.1. Current in a section of cable. Inward currents are positive, and outward currents are negative.

cannot accumulate within the cable, the flux in through area $\mathrm{d}\Gamma_e$ minus flux out through $\mathrm{d}\Gamma_e$ (fluxes through the cable ends to neighboring conductive volumes) and the sum of membrane currents (per unit area) that cross the membrane through $\mathrm{d}\Gamma_m$ (positive membrane currents defined as flowing from inside to out) must balance:

$$\left[\mathbf{J}-(\mathbf{J}+\mathrm{d}\mathbf{J})\right]\mathrm{d}\Gamma_e-(I_c+I_{ion})\,\mathrm{d}\Gamma_m=0, \tag{3.4}$$

where the total differential of flux over the length of cable, $\mathrm{d}J$, in three dimensions is

$$\mathrm{d}\mathbf{J}=\frac{\partial J_{x_1}}{\partial x_1}\mathrm{d}x_1+\frac{\partial J_{x_2}}{\partial x_2}\mathrm{d}x_2+\frac{\partial J_{x_3}}{\partial x_3}\mathrm{d}x_3, \tag{3.5}$$

and

$$I_c=C_m\frac{\mathrm{d}V_m}{\mathrm{d}t}, \tag{3.6}$$

where I_c is the capacitive membrane current. I_{ion} is the sum of the currents that cross the membrane through ion channels with units ($\mu A/cm^2$). Then the units of the right-hand side of Eq. (3.4) are $(\mu A/cm^2)(cm^2)=\mu A$.

Rearranging Eq. (3.4),

$$-(\mathrm{d}\mathbf{J})\,\mathrm{d}\Gamma_e=(I_c+I_{ion})\,\mathrm{d}\Gamma_m. \tag{3.7}$$

Examining the terms of the right-hand side, the change in flux in the x_1 direction, for example, is

$$-\left(\frac{\partial J_{x_1}}{\partial x_1}\mathrm{d}x_1\right)\mathrm{d}x_2\,\mathrm{d}x_3=-\frac{\partial J_{x_1}}{\partial x_1}\mathrm{d}x_1\,\mathrm{d}x_2\,\mathrm{d}x_3, \tag{3.8}$$

where $\mathrm{d}x_2\,\mathrm{d}x_3=\mathrm{d}\Gamma_e$, and $\mathrm{d}x_1\,\mathrm{d}x_2\,\mathrm{d}x_3=\mathrm{d}\Omega$. In the general 3D case, the total change in flux is

$$-(\mathrm{d}\mathbf{J})\,\mathrm{d}\Gamma_e=-\left(\frac{\partial J_{x_1}}{\partial x_1}+\frac{\partial J_{x_2}}{\partial x_2}+\frac{\partial J_{x_3}}{\partial x_3}\right)\mathrm{d}\Omega=-(\nabla\cdot\mathbf{J})\,\mathrm{d}\Omega. \tag{3.9}$$

Substituting Eq. (3.9) into (3.7) results in the following statement of the conservation of charge:

$$-(\nabla\cdot\mathbf{J})\,\mathrm{d}\Omega=\left(C_m\frac{\mathrm{d}V_m}{\mathrm{d}t}+I_{ion}\right)\mathrm{d}\Gamma_m. \tag{3.10}$$

By setting $d\Gamma_m/d\Omega = S_v$, the cell membrane surface area to volume ratio, and expressing flux in terms of potential from Eq. (3.3), Eq. (3.10) results in the *three-dimensional cable equation*:

$$\frac{1}{S_v C_m}\nabla\cdot\mathbf{D}\nabla\Phi_i = \frac{dV_m}{dt} + \frac{1}{C_m}I_{ion}. \tag{3.11}$$

It should also be noted that the conductivity tensor, **D**, is often written with units of $cm^2/msec$ by combining **D** with S_v and C_m. Thus the cable equation may also be written as

$$\nabla\cdot\mathbf{D}\nabla\Phi_i = \frac{dV_m}{dt} + \frac{1}{C_m}I_{ion}. \tag{3.12}$$

The cable equation can also be derived by considering integral forms of the statement of charge conservation rather than the differential form presented above. Considering an arbitrary volume of cable, Ω, with surface, $\Gamma = \Gamma_e + \Gamma_m$, and outward normal **n**, an equivalent statement of Eq. (3.4) is

$$\int_{\Gamma_e}\mathbf{n}\cdot\mathbf{J}\,d\Gamma + \int_{\Gamma_m}\left(C_m\frac{dV_m}{dt} + I_{ion}\right)d\Gamma = 0, \tag{3.13}$$

where outward fluxes and currents are positive. Applying the Gauss theorem directly results in

$$\int_{\Gamma}\mathbf{n}\cdot\mathbf{J}\,d\Gamma = \int_{\Omega}\frac{\partial J_i}{\partial x_i}\,d\Omega = \int_{\Omega}\nabla\cdot\mathbf{J}\,d\Omega, \tag{3.14}$$

and Eq. (3.13) becomes

$$\int_{\Omega}\nabla\cdot\mathbf{J}\,d\Omega = -\int_{\Gamma_m}\left(C_m\frac{dV_m}{dt} + I_{ion}\right)d\Gamma. \tag{3.15}$$

Since this must hold for any arbitrary volume and associated surface area of the cable

$$\nabla\cdot\mathbf{J} = -S_v\left(C_m\frac{dV_m}{dt} + I_{ion}\right), \tag{3.16}$$

where S_v accounts for the fact that the membrane currents flow only through a proportionally sized portion of the surface area of the arbitrary volume. Applying the results of Ohm's law from Eq. (3.3), Eq. (3.16) also leads to the cable equation (Eq. (3.11)).

3.3. Governing equations of cardiac propagation

The bidomain model for describing the electrophysiology of cardiac tissue was introduced by GESELOWITZ and MILLER 3RD [1983]. This model treats cardiac tissue as two separate, continuous domains that both occupy the space occupied by the tissue. These two domains represent the intracellular and extracellular spaces. u_i and u_e are the electrical potentials in each domain. According to the conservation of current, any communication of current between the two domains is considered to have crossed the cell membrane. Current flowing strictly within either domain can be assumed to be purely resistive since in the actual microstructure of intra- and extracellular domains, that occupy mutually exclusive regions of space, continuous pathways exist. In the extracellular space current can flow continuously between the cells, and in the intracellular space current can flow continuously through cells since their myoplasm is connected via gap junctions. The three-dimensional resistance to current flow in each domain is described by a tensor of conductivity parameters.

Applying Ohm's law, Eq. (3.3), to these two domains results in

$$\mathbf{J}_{\mathrm{i}} = -\mathbf{D}_{\mathrm{i}} \nabla u_{\mathrm{i}}, \tag{3.17}$$

$$\mathbf{J}_{\mathrm{e}} = -\mathbf{D}_{\mathrm{e}} \nabla u_{\mathrm{e}}, \tag{3.18}$$

where $\mathbf{J}_{\mathrm{i}}$ and $\mathbf{J}_{\mathrm{e}}$ are the intra- and extracellular current densities, and $\mathbf{D}_{\mathrm{i}}$ and $\mathbf{D}_{\mathrm{e}}$ are anisotropic conductivity tensors. If an arbitrary volume domain of the tissue, Ω, with a surface Γ is defined, conservation of charge requires that the flux out of one domain must equal the flux into the other,

$$\int_{\Gamma} \mathbf{J}_{\mathrm{i}} \cdot \mathbf{n}\, \mathrm{d}\Gamma = -\int_{\Gamma} \mathbf{J}_{\mathrm{e}} \cdot \mathbf{n}\, \mathrm{d}\Gamma, \tag{3.19}$$

where $\mathbf{n}$ represents the unit outward normal to the arbitrary volume. Approximating the behavior of the intracellular domain as a cable, as in the above derivation, application of the Gauss theorem leads to

$$-\nabla \cdot \mathbf{J}_{\mathrm{i}} = S_{\mathrm{v}} \left(C_{\mathrm{m}} \frac{\mathrm{d}V_{\mathrm{m}}}{\mathrm{d}t} + I_{\mathrm{ion}} \right), \tag{3.20}$$

$$-\nabla \cdot \mathbf{J}_{\mathrm{e}} = -S_{\mathrm{v}} \left(C_{\mathrm{m}} \frac{\mathrm{d}V_{\mathrm{m}}}{\mathrm{d}t} + I_{\mathrm{ion}} \right), \tag{3.21}$$

where the opposite signs indicate that the same membrane currents flow out of the extracellular domain and into the intracellular domain.

The membrane potential is defined as the difference between the intra- and extracellular potentials,

$$V_{\mathrm{m}} = u_{\mathrm{i}} - u_{\mathrm{e}}. \tag{3.22}$$

Substitution of Eqs. (3.17) and (3.18) into Eqs. (3.20) and (3.21) results in

$$\nabla \cdot \mathbf{D}_{\mathrm{i}} \nabla u_{\mathrm{i}} = S_{\mathrm{v}} \left(C_{\mathrm{m}} \frac{\mathrm{d}V_{\mathrm{m}}}{\mathrm{d}t} + I_{\mathrm{ion}} \right), \tag{3.23}$$

$$\nabla \cdot \mathbf{D}_{\mathrm{e}} \nabla u_{\mathrm{e}} = -S_{\mathrm{v}} \left(C_{\mathrm{m}} \frac{\mathrm{d}V_{\mathrm{m}}}{\mathrm{d}t} + I_{\mathrm{ion}} \right). \tag{3.24}$$

Writing u_{i} in terms of V_{m} and u_{e},

$$\nabla \cdot \mathbf{D}_{\mathrm{i}} (\nabla V_{\mathrm{m}} + \nabla u_{\mathrm{e}}) = S_{\mathrm{v}} \left(C_{\mathrm{m}} \frac{\mathrm{d}V_{\mathrm{m}}}{\mathrm{d}t} + I_{\mathrm{ion}} \right), \tag{3.25}$$

$$\nabla \cdot \mathbf{D}_{\mathrm{e}} \nabla u_{\mathrm{e}} = -S_{\mathrm{v}} \left(C_{\mathrm{m}} \frac{\mathrm{d}V_{\mathrm{m}}}{\mathrm{d}t} + I_{\mathrm{ion}} \right). \tag{3.26}$$

By writing Eq. (3.26) in terms of Eq. (3.25), Eq. (3.26) becomes

$$\nabla \cdot \mathbf{D}_{\mathrm{e}} \nabla u_{\mathrm{e}} = -\nabla \cdot \mathbf{D}_{\mathrm{i}} (\nabla V_{\mathrm{m}} + \nabla u_{\mathrm{e}}). \tag{3.27}$$

Finally, rearranging both equations:

$$\nabla \cdot (\mathbf{D}_i \nabla V_m) + \nabla \cdot (\mathbf{D}_i \nabla u_e) = S_v \left(C_m \frac{dV_m}{dt} + I_{ion} \right), \tag{3.28}$$

$$\nabla \cdot \big((\mathbf{D}_i + \mathbf{D}_e) \nabla u_e \big) = -\nabla \cdot (\mathbf{D}_i \nabla V_m). \tag{3.29}$$

These are the governing equations for the bidomain treatment of ventricular tissue. The unknowns V_m and u_e vary spatially within the domain of the problem, in this case the ventricles, and are thus partial differential equations. I_{ion}, a function of V_m, is the collection of membrane currents detailed in Section 2.3 that describe the local ionic state of the cells in the tissue as a function of time through a set of ordinary differential equations.

The boundary conditions needed to complete the formulation describe the situation that on the surface of the heart, the flux in Eq. (3.21) from the extracellular space to the intracellular space does not exist since the intracellular space does not exist outside of the heart. Eq. (3.29) is an equivalent statement to (3.21), so the boundary condition can be derived as

$$\nabla \cdot \big((\mathbf{D}_i + \mathbf{D}_e) \nabla u_e \big) = -\nabla \cdot (\mathbf{D}_i \nabla V_m) = 0 \quad \text{on } \Gamma, \tag{3.30}$$

or

$$\mathbf{n} \cdot (\mathbf{D}_i \nabla V_m) = 0 \quad \text{on } \Gamma. \tag{3.31}$$

Finite elements can be used to solve these governing equations with the associated boundary conditions. We return to the solution of the bidomain field equations for cardiac electrophysiology after a derivation of some applicable finite element methods in Section 4.

4. Finite element methods

The finite element method is a popular computational approach to these problems that has applications in diverse areas of cardiovascular biophysics, such as

- blood flow in arteries;
- stress and strain distributions in the myocardium of the beating heart;
- bioheat transfer in myocardial tissue during laser, cryo or radio-frequency ablation;
- multicellular action potential propagation;
- shock and defibrillation;
- strain analysis from cardiac magnetic resonance imaging;
- the inverse electrocardiographic problem.

In this section we derive finite element equations useful for both constructing models of complex domains such as the heart and for solving the bidomain equations for action potential propagation. More detailed texts on finite element methods can be found in CIARLET and LIONS [1990], CAREY and ODEN [1983], LANGTANGEN [1999], LIONS and MAGENES [1972] and elsewhere. Results from this section are applied to the construction of models of cardiac geometry and structure that provide the domain on which the bidomain equations may be solved in Section 5. Then in Section 6, results from this section are applied to the bidomain equations themselves.

4.1. Formulation of FE equations

In the general case, an unknown function u satisfies a certain partial differential equation (PDE) represented by the differential operator $\mathcal{L}$ such that $\mathcal{L}(u)$ is satisfied on a domain, Ω,

$$\mathcal{L}(u)=0 \quad \text{on } \Omega, \tag{4.1}$$

subject to appropriate boundary conditions on the boundaries, Γ,

$$\mathcal{B}(u)=0 \quad \text{on } \Gamma. \tag{4.2}$$

In general, boundaries can be divided into two non-intersecting sections: one where conditions are prescribed on the derivative of the solution, u, and one where conditions are prescribed on u itself. In the following discussion, the boundary, Γ, is the portion of the boundary where conditions are prescribed on the derivative of u because all approximations to the solution are always chosen such that prescribed conditions on u are explicitly satisfied.

Eqs. (4.1) and (4.2) are the *strong formulation* since they require an exact solution u everywhere. The finite element method seeks an approximate solution $\hat{u}$ in the form

$$u \simeq \hat{u} = \sum_{i=1}^{M} \Psi_i u_i, \tag{4.3}$$

where $\boldsymbol{\Psi}$ are *basis functions* prescribed in terms of independent variables (such as spatial coordinates x, y, z), and some or all of the coefficients u_i are unknown. In determining $\mathbf{u}$ the goal is to minimize the error between $\hat{u}$ and u.

In general, because $\hat{u}$ is not the exact solution, substitution of the approximation into the differential equation results in

$$\mathcal{L}(\hat{u}) \neq 0, \tag{4.4}$$

$$\mathcal{B}(\hat{u}) \neq 0. \tag{4.5}$$

Since the exact solution is generally unknown, the differences between $\mathcal{L}(u)$ and $\mathcal{L}(\hat{u})$ and $\mathcal{B}(u)$ and $\mathcal{B}(\hat{u})$ are used as a measure of error. By definition, $\mathcal{L}(u)$ and $\mathcal{B}(u)$ equal zero, so $\mathcal{L}(\hat{u})$ and $\mathcal{B}(\hat{u})$ become the measure of error, the *residual*,

$$\mathcal{L}(\hat{u}) + \mathcal{B}(\hat{u}) = R. \tag{4.6}$$

Various procedures for minimizing R give rise to finite element methods including weighted residuals, collocation, and least squares.

Because the residual, R, is defined over a domain, Ω, it varies over the independent variables (for the present problem, it varies spatially over the domain of the ventricles). In order to find a minimum, a scheme for integrating R over the domain will be used to find an average measure of the error in the approximation $\mathcal{L}(\hat{u})$ and $\mathcal{B}(\hat{u})$ due to $\hat{u} \neq u$. Since the PDE in Eq. (4.1) must be zero at each point in the domain, Ω, it follows that

$$\int_{\Omega} \omega \mathcal{L}(u)\, \mathrm{d}\Omega \equiv 0, \tag{4.7}$$

where ω is an arbitrary weighting function. The same integral equation can be written for the boundary conditions

$$\int_{\Gamma} \overline{\omega}\mathcal{B}(u)\,\mathrm{d}\Gamma \equiv 0, \tag{4.8}$$

for any function $\overline{\omega}$. The integral statement that

$$\int_{\Omega} \omega\mathcal{L}(u)\,\mathrm{d}\Omega + \int_{\Gamma} \overline{\omega}\mathcal{B}(u)\,\mathrm{d}\Gamma = 0, \tag{4.9}$$

is satisfied for all ω and $\overline{\omega}$ and is equivalent to the PDE in Eq. (4.1) and the boundary conditions in Eq. (4.2).

If the unknown function is approximated by the expansion in Eq. (4.3), the result is the *weighted residual formulation*,

$$\int_{\Omega} \omega_i\mathcal{L}(\hat{u})\,\mathrm{d}\Omega + \int_{\Gamma} \overline{\omega_j}\mathcal{B}(\hat{u})\,\mathrm{d}\Gamma = 0,$$
$$\text{where } i = 1, \ldots, m,\ \ j = m+1, \ldots, M. \tag{4.10}$$

m is the number of unknowns, u_i, and $M - m$ is the number of unknowns on the boundary. In solving Eq. (4.10) for the approximate solution $\hat{u}$, the PDE is satisfied only in an average sense, so the weighted residual is a *weak formulation*. Certain choices of the weight functions result in the point collocation and least squares methods.

4.1.1. Point collocation method

Using the Dirac delta function (which equals x when $x = c$, but equals zero otherwise) in choosing the weight functions results in the collocation method:

$$\omega(x) = \sum_{i=1}^{m} \delta(x - c_i), \tag{4.11}$$

$$\overline{\omega}(x) = \rho^2 \sum_{i-m+1}^{M} \delta(x - c_i). \tag{4.12}$$

The constant ρ is introduced for consistency of units. The following property of the Dirac delta function

$$\int_{\Omega} f(x)\delta(x - c_i)\,\mathrm{d}\Omega = f(c_i), \tag{4.13}$$

results in

$$\mathcal{L}\big(\hat{u}(c_i)\big) = 0 \quad \text{for } i = 1, \ldots, m \text{ on } \Omega, \tag{4.14}$$

$$\mathcal{B}\big(\hat{u}(c_i)\big) = 0 \quad \text{for } i = m+1, \ldots, M \text{ on } \Gamma, \tag{4.15}$$

when these weight functions are applied to the weighted residual equation (Eq. (4.10)). The approximate solution is forced to satisfy the PDE only at M points in the body and on the boundary where the dependent variable x is equal to c_i.

4.1.2. Least squares method

The least squares method arises from seeking the minimum of the averaged square of the residual. The averaged square of the residual is

$$\int_{\Omega} \mathcal{L}(\hat{u})^2\,\mathrm{d}\Omega + \rho^2 \int_{\Gamma} \mathcal{B}(\hat{u})^2\,\mathrm{d}\Gamma. \tag{4.16}$$

The constant ρ is again introduced for consistency of units. Take the first variation to find the minimum

$$\frac{\partial}{\partial u_i}\left[\int_{\Omega} \mathcal{L}(\hat{u})^2\,\mathrm{d}\Omega + \rho^2 \int_{\Gamma} \mathcal{B}(\hat{u})^2\,\mathrm{d}\Gamma\right] = 0, \tag{4.17}$$

$$\int_{\Omega} 2\mathcal{L}(\hat{u})\frac{\partial \mathcal{L}(\hat{u})}{\partial u_i}\,\mathrm{d}\Omega + \rho^2 \int_{\Gamma} 2\mathcal{B}(\hat{u})\frac{\partial \mathcal{B}(\hat{u})}{\partial u_i}\,\mathrm{d}\Gamma = 0, \quad \text{where } i = 1, \ldots, M. \tag{4.18}$$

This is the same as choosing the weight functions as follows:

$$\omega(x) = 2\frac{\partial \mathcal{L}(\hat{u})}{\partial u_i} \quad \text{where } i = 1, \ldots, m \text{ on } \Omega, \tag{4.19}$$

$$\overline{\omega}(x) = 2\rho^2\frac{\partial \mathcal{B}(\hat{u})}{\partial u_i} \quad \text{where } i = m+1, \ldots, M \text{ on } \Gamma. \tag{4.20}$$

4.2. Boundary conditions

The following generic boundary value problem can be used to illustrate the treatment of boundary conditions:

$$-\nabla \cdot \left[k(\mathbf{X})\nabla u(\mathbf{X})\right] = f(\mathbf{X}) \quad \text{on } \Omega, \tag{4.21}$$

$$-\mathbf{n} \cdot k(\mathbf{X})\nabla u(\mathbf{X}) = g(\mathbf{X}) \quad \text{on } \Gamma_N, \tag{4.22}$$

$$u(\mathbf{X}) = \phi(\mathbf{X}) \quad \text{on } \Gamma_E, \tag{4.23}$$

where the boundary has been divided into two sections. It is important to note that this model has the same form as the bidomain equations and its boundary conditions, so the result here is directly applicable and will be described in Section 6.

A term added to the approximation in Eq. (4.3) with a requirement that $\Psi_i = 0$ on Γ_E causes the essential boundary conditions prescribed on Γ_E to be satisfied automatically by the choice of the following form for $\hat{u}$:

$$\hat{u} = \phi + \sum_{i=1}^{M} \Psi_i u_i. \tag{4.24}$$

Applying the weighted residual formulation of Eq. (4.10):

$$-\int_{\Omega} \left(\nabla \cdot [k\nabla\hat{u}] + f\right)\omega_i\,\mathrm{d}\Omega - \int_{\Gamma_N} (\mathbf{n} \cdot k\nabla\hat{u} + g)\overline{\omega_i}\,\mathrm{d}\Gamma = 0,$$
$$\text{where } i = 1, \ldots, M \tag{4.25}$$

for N linearly independent weight functions. Integrating the first term in the first integrand by parts:

$$\int_{\Omega} \nabla\omega_i \cdot k\nabla\hat{u}\,\mathrm{d}\Omega - \int_{\Gamma_N} (\mathbf{n}\cdot k\nabla\hat{u})\omega_i\,\mathrm{d}\Gamma - \int_{\Omega} f\omega_i\,\mathrm{d}\Omega$$
$$- \int_{\Gamma_N} (\mathbf{n}\cdot k\nabla\hat{u} + g)\overline{\omega_i}\,\mathrm{d}\Gamma = 0. \tag{4.26}$$

If the weight functions are chosen so that $\omega_i = \Psi_i$ and $\overline{\omega_i} = -\Psi_i$ (the Galerkin procedure), then the terms concerning the boundary conditions prescribed on the natural boundary "naturally" cancel from the equation, leaving

$$\int_{\Omega} \nabla\omega_i \cdot k\nabla\hat{u}\,\mathrm{d}\Omega - \int_{\Omega} f\omega_i\,\mathrm{d}\Omega - \int_{\Gamma_N} g\overline{\omega_i}\,\mathrm{d}\Gamma = 0. \tag{4.27}$$

Inserting the approximate solution, $\hat{u}$, and the chosen weighting functions:

$$\sum_{j=1}^{M}\left(\int_{\Omega} \nabla\Psi_i \cdot k\nabla\Psi_j\,\mathrm{d}\Omega\right)u_j = \int_{\Omega} f\Psi_i\,\mathrm{d}\Omega - \int_{\Gamma_N} g\Psi_i\,\mathrm{d}\Gamma$$
$$\text{where } i = 1, \ldots, M. \tag{4.28}$$

This yields a set of M linear equations in the form

$$\mathbf{Ku} = \mathbf{f}, \tag{4.29}$$

where $\mathbf{K}$ and $\mathbf{f}$ are known and $\mathbf{u} = u_i$ can be found using a linear solver.

4.3. Domain discretization

Because the problem of finding an approximation of u to satisfy $\mathcal{L}(u)$ has been expressed in an integral form over the domain Ω in Eq. (4.9), the properties of integration can be taken advantage of in order to make a continuous, intractable problem tractable with a finite number of unknowns. The key property is that of summation: an integral over an arbitrary domain is the same as the sum of integrals over a set of arbitrary non-overlapping subdomains whose union is the original domain. Domain discretization permits the approximation to be obtained element by element and an assembly to be achieved by summation.

In the case of the finite element method, the approximate solution, $\hat{u}$, becomes a set of piece-wise functions, $\hat{\mathbf{u}}$, spanning separate subdomains (the elements),

$$\mathbf{u} \simeq \hat{\mathbf{u}} = \sum_{n(e)=1}^{n_{\max}} \sum_{r=1}^{r_{\max}} \Psi_{n(e)}^{r} u_{n(e)}^{r}, \tag{4.30}$$

where $M = n_{\max} \times \gamma_{\max}$ terms make up each component of $\hat{\mathbf{u}}$ which has one component for each element. Recasting the original PDE in Eqs. (4.1) and (4.2) into subdomains over a domain Ω, subject to appropriate boundary conditions on the boundaries Γ:

$$\mathbf{A}(\mathbf{u}) = \begin{Bmatrix} \mathbf{A}_1(\mathbf{u}) \\ \mathbf{A}_2(\mathbf{u}) \\ \vdots \end{Bmatrix} = \mathbf{0}, \tag{4.31}$$

$$\mathbf{B}(\mathbf{u}) = \begin{Bmatrix} \mathbf{B}_1(\mathbf{u}) \\ \mathbf{B}_2(\mathbf{u}) \\ \vdots \end{Bmatrix} = \mathbf{0}, \tag{4.32}$$

where **A** is a set of piece-wise PDE operators on the subdomains with **B** on the boundary subdomains equivalent to the original PDE, $\mathcal{L}$, on the whole domain with $\mathcal{B}$ on the boundaries. The weak formulation can then be written in the following form:

$$\int_{\Omega} \mathbf{F}_j(\hat{\mathbf{u}})\,\mathrm{d}\Omega + \int_{\Gamma} \mathbf{f}_j(\hat{\mathbf{u}})\,\mathrm{d}\Gamma = \sum_{e=1}^{e_{\max}} \left(\int_{\Omega^e} \mathbf{F}_j(\hat{\mathbf{u}})\,\mathrm{d}\Omega^e + \int_{\Gamma^e} \mathbf{f}_j(\hat{\mathbf{u}})\,\mathrm{d}\Gamma^e \right), \tag{4.33}$$

where Ω^e is the domain of each element and Γ^e is its part of the boundary. $\mathbf{F}_j$ and $\mathbf{f}_j$ prescribe known functions or operators.

Considering linear partial differential equations and boundary conditions:

$$\mathbf{A}(\mathbf{u}) = \mathbf{L}\mathbf{u} - \mathbf{f} = \mathbf{0} \quad \text{in } \Omega, \tag{4.34}$$

$$\mathbf{B}(\mathbf{u}) = \mathbf{M}\mathbf{u} - \mathbf{p} = \mathbf{0} \quad \text{on } \Gamma. \tag{4.35}$$

The approximation scheme (4.33) will again yield a set of linear equations in the form

$$\mathbf{K}\mathbf{u} + \mathbf{f} = \mathbf{0}. \tag{4.36}$$

4.4. *Two uses of basis functions*

The set of basis functions, $\boldsymbol{\Psi}$, is generally used for two purposes: the approximation of the unknown dependent variable (as in Eq. (4.3)) and the parameterization of the dependent variables defining the problem domain into the finite elements. In the case that the same set of basis functions is used for both purposes, the result is an *isoparametric element interpolation* with an *isoparametric mapping* of element to global coordinates, i.e., an isoparametric interpolation for a field variable, $u(x)$, is created by defining the geometric (dependent) variable x as an interpolation of nodal parameters using the same basis functions:

$$\left.\begin{aligned} u(\xi) &= \textstyle\sum_{i=1}^{M} \Psi_i(\xi) u_i \\ x(\xi) &= \textstyle\sum_{i=1}^{M} \Psi_i(\xi) x_i \end{aligned}\right\} \quad \Longrightarrow \quad u = u(x). \tag{4.37}$$

4.4.1. *Dependent variable interpolation*

To approximate a set of points $\{x_k; u_k(x_k)\}$ by a continuous function, a convenient and popular method is to use a polynomial expression such as: $u(x) = a + bx + cx^2 + dx^3 + \cdots$ and then to estimate the monomial coefficients $a, b, c, d, \ldots$ to obtain a best approximation to the field variable $u(x)$. Since high-order polynomials, such as quartics

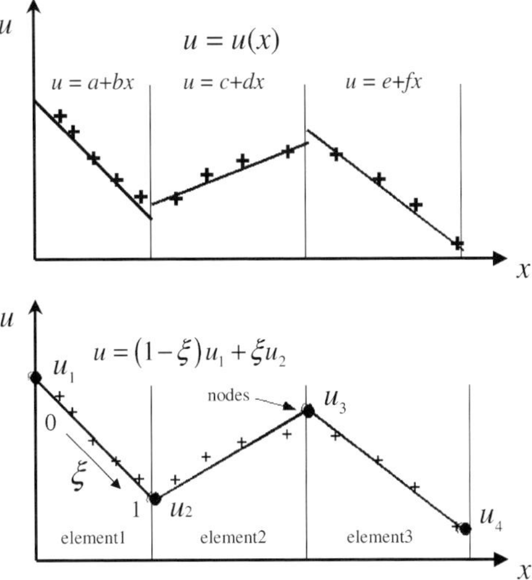

FIG. 4.1. Parameterizing piece-wise polynomials in terms of shared nodal parameters automatically ensures continuity of u across the element boundaries.

and quintics tend to oscillate unphysically, it is helpful to divide a large or complex domain into smaller subdomains and use low-order *piece-wise* polynomials over each of them – the subdomains again are the finite elements.

For example, a field variable $u(x)$, may be represented by several linear elements as illustrated in Fig. 4.1. It is generally necessary to impose constraints to ensure continuity of u across the element boundaries. Reparameterizing the linear function from monomial coefficients a and b in one of the elements in terms of the nodal values of u at each end of the element (u_1 and u_2), we write for one element

$$u(\xi) = u_1(1-\xi) + u_2\xi, \tag{4.38}$$

where $\xi \in (0, 1)$ is a normalized measure of distance along the one-dimensional element (Fig. 4.1).

Adjacent elements share global nodal parameters, U_Δ (see Fig. 4.2), defined at each global node, Δ. Thus, it is necessary to map global nodal parameters, U_Δ, defined at global node Δ, onto local node n of element e by use of a *connectivity* matrix, $\Delta(n, e)$,

$$u_n = U_{\Delta(n,e)}. \tag{4.39}$$

It is often desirable to use an interpolation that preserves continuity of the derivative of a field variable u with respect to ξ across element boundaries not only u itself. Then because neighboring elements may not have the same size in the global coordinate system, it is more accurate to define the global nodal derivative parameters as $(\partial u/\partial s)_n$ where s is arc length in global units, and then to compute the local basis function parameters $(\partial u/\partial \xi)_n$, for each element with respect to its own local coordinates ξ,

$$\left(\frac{\partial u}{\partial \xi}\right)_n = \left(\frac{\partial u}{\partial s}\right)_{\Delta(n,e)} \cdot \left(\frac{\partial s}{\partial \xi}\right)_n. \tag{4.40}$$

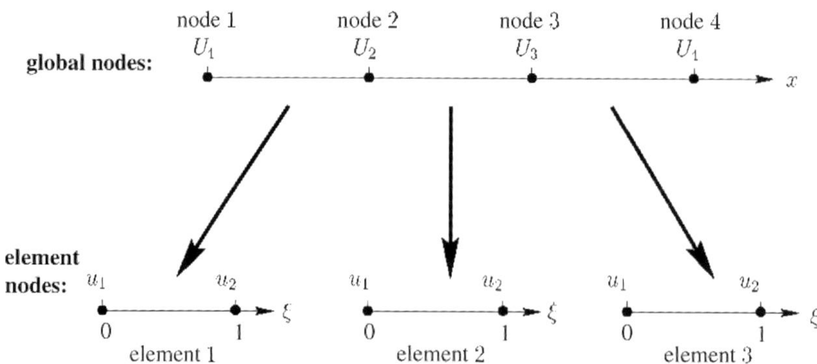

FIG. 4.2. The finite element method: relationship between global nodal parameters U_i $(i = 1, \ldots, 4)$ and local nodal parameters u_j $(j = 1, 2)$.

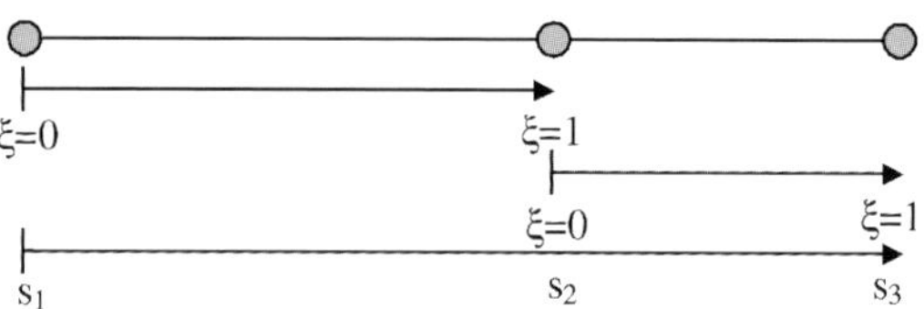

FIG. 4.3. Scaling factors.

The term $(\partial s/\partial \xi)_n$ is an element *scale factor*, which scales the arc length derivative of global node Δ to the local derivative of element node n required for the local interpolation (Fig. 4.3).

4.4.2. *Element parameterization*

Schemes for discretizing a domain into a finite number of elements involve choosing points within the domain that define each element and result in a mesh of nodes. (Various approaches and rules for creating meshes can be found elsewhere.) Since each finite element will be treated separately, similar calculations are repeated for each element. To make the process more convenient each element is mapped to a reference element, a basic change of coordinates. For example, the geometry of an element that is shaped like a thin plate of arbitrary size that may be warped out of plane with curved and skewed boundaries in actual space can be mapped to a simple square with sides one unit long (Fig. 4.4).

Similar to interpolation of a field variable, the geometry of Ω is then interpolated over $e_{\max}$ elements as

$$\mathbf{X} = \sum_{e=1}^{e_{\max}} \sum_{m=1}^{M(e)} [\boldsymbol{\Psi}^m][\mathbf{X}^m]\mathbf{e}, \tag{4.41}$$

where $M(e) = n_{\max}(e) \times \gamma_{\max}$, the number of nodes defining the eth element times the number of parameters per node. $[\boldsymbol{\Psi}^m]$ is matrix of interpolation functions defined

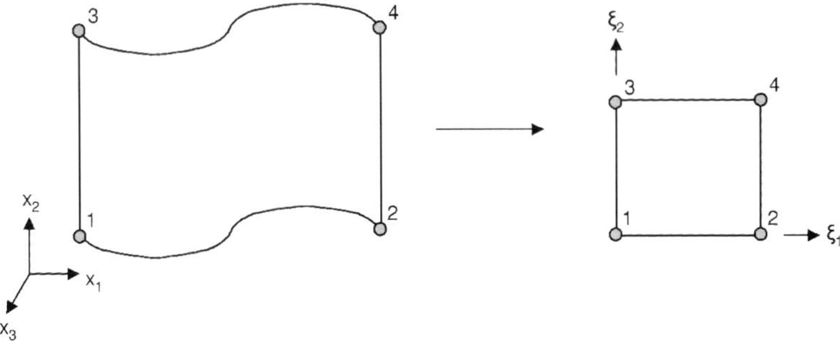

FIG. 4.4. Mapping of a geometric variable from the global to element coordinate system.

as products of independent Lagrange and Hermite polynomials in each ξ_k direction for each node; $\mathbf{e} = (\mathbf{e}_1, \mathbf{e}_2, \mathbf{e}_3)$ are unit vectors, which describe a curvilinear coordinate system; and $[\mathbf{X}^m]$ is a matrix-vector of nodal coordinates and generalized derivatives. Generalized derivatives are included in $[\mathbf{X}^m]$, as in the case of the interpolation of a field variable, for continuity of derivatives. Derivatives of the dependent variables may then appear in an integrand, so they must also be transformed from the global to the local coordinate system.

A scaling factor $[\mathbf{S}^m]$ matrix is also defined for each element, so that equations may be written

$$\mathbf{X} = \sum_{e=1}^{e_{\max}} \sum_{n=1}^{M(e)} [\boldsymbol{\Psi}^m][\mathbf{S}^m][\mathbf{X}^{\Delta(m,e)}]\mathbf{e}, \tag{4.42}$$

where $[\mathbf{X}^m] = [\mathbf{S}^m][\mathbf{X}^{\Delta(m,e)}]$.

Thus, both the geometric or dependent field variables of an element may in general be interpolated in two or three dimensions by equations of the form

$$\hat{u}_{(e)} = \sum_{n(e)=1}^{n_{\max}} \sum_{\gamma=1}^{\gamma_{\max}} \Psi^{\gamma}_{n(e)}(\xi_1, \xi_2, \xi_3) \cdot u^{\gamma}_{n(e)}, \tag{4.43}$$

where $\Psi^{\gamma}_{n_{(e)}}(\xi_1, \xi_2, \xi_3)$ are piece-wise interpolation functions, $u^{\gamma}_{n_{(e)}}$ are nodal parameters at local node n of element (e), and the index γ identifies each of the $\gamma_{\max}$ nodal parameters used to interpolate $u_{(e)}$. For three-dimensional elements, $n_{\max} = 8$. The interpolation functions $\Psi^{\gamma}_{n_{(e)}}(\xi_1, \xi_2, \xi_3)$ may be constructed as tensor products of separate polynomials in each ξ_k direction. For three-dimensional rectangular finite elements, these equations may be written as follows:

$$\Psi^{\gamma}_{n(e)}(\xi_1, \xi_2, \xi_3) = \varphi^{\gamma}_{n(e)_1}(\xi_1)\, \varphi^{\gamma}_{n(e)_2}(\xi_2)\, \varphi^{\gamma}_{n(e)_3}(\xi_3). \tag{4.44}$$

A full description of several Lagrange and Hermite interpolation functions and their combinations in two and three dimensions can be found in Section 8 (Appendix).

5. Anatomical models

5.1. Coordinate systems

The geometry of a region, over which a finite element solution is sought, is defined with respect to a coordinate system. The choice of coordinate system depends on the particular problem. Fig. 5.1 shows the relation of Cartesian coordinates to coordinates

A) Rectangular Cartesian Coordinates: $\{\Theta_A\}=(X,Y,Z)$

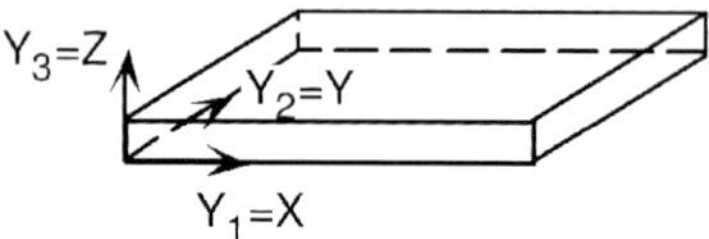

B) Cylindrical Polar Coordinates: $\{\Theta_A\}=(R,\Theta,Z)$

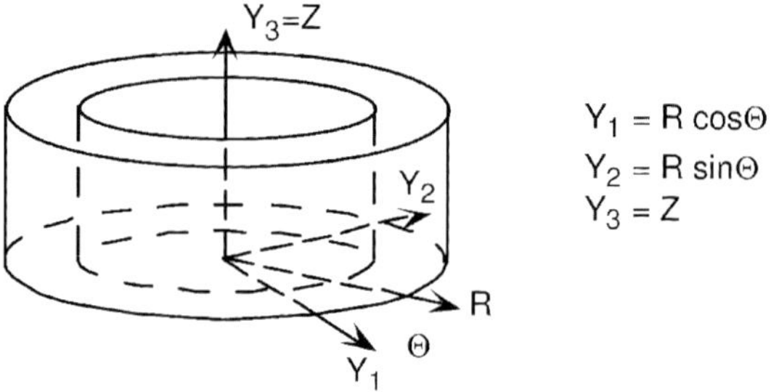

C) Spherical Polar Coordinates: $\{\Theta_A\}=(R,\Theta,\Phi)$

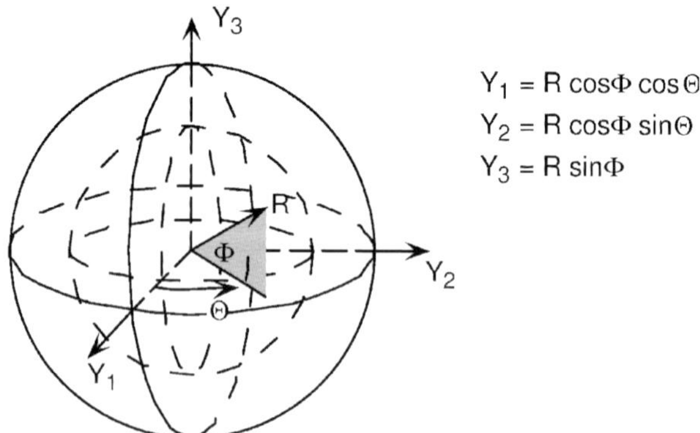

FIG. 5.1. The relationship between (A) rectangular Cartesian reference coordinates, Y_i, and curvilinear world coordinates, Θ_i, for two orthogonal coordinate systems that may be used to formulate the finite element equations: (B) cylindrical polar; and (C) spherical polar. Reprinted with permission of the ASME from COSTA, HUNTER, ROGERS, GUCCIONE, WALDMAN and MCCULLOCH [1996] ASME J. of Biomech. Eng., 118:452–463.

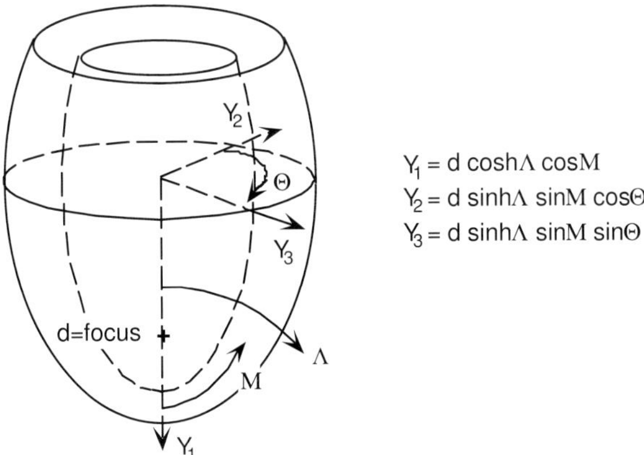

FIG. 5.2. The relationship between rectangular Cartesian reference coordinates, Y_i, and prolate spheroidal coordinates, $\{\Theta_A\} = \{\Lambda, M, \Theta\}$ used here to describe a thick-walled confocal ellipsoidal shell bounded by inner and outer surfaces of constant Λ (the dimensionless transmural coordinate) and truncated at $M = 120°$. Dimensional scaling is determined by the focal length, d. Reprinted with permission of the ASME from COSTA, HUNTER, WAYNE, WALDMAN, GUCCIONE and MCCULLOCH [1996] ASME J. of Biomech. Eng., 118:464–472.

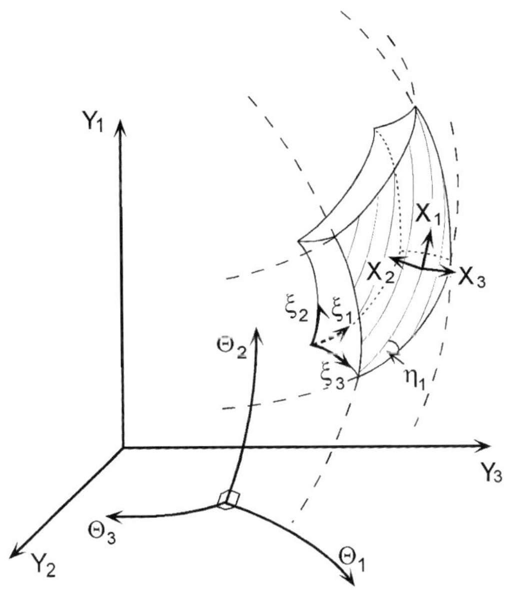

FIG. 5.3. Four coordinate systems are used in our finite element method. A rectangular Cartesian global reference coordinate system $\{Y_1, Y_2, Y_3\}$ and orthogonal curvilinear coordinate systems $\{\Theta_1, \Theta_2, \Theta_3\}$ are used to describe the geometry. Curvilinear local finite element coordinates are $\{\xi_1, \xi_2, \xi_3\}$, and locally orthonormal convecting body/fiber coordinates are $\{X_1, X_2, X_3\}$. Reprinted with permission of the ASME from COSTA, HUNTER, ROGERS, GUCCIONE, WALDMAN and MCCULLOCH [1996] ASME J. of Biomech Eng., 118:452–463.

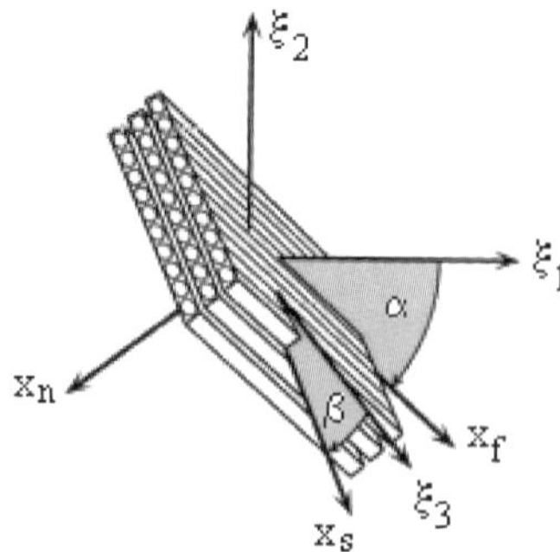

FIG. 5.4. The fibrous structure of the anisotropic myocardium can be defined using locally orthonormal coordinates defined by the tissue fiber-sheet microstructure and derived in terms of the local finite element coordinates $\{\xi_A\}$.

of other common coordinate systems. It is sometimes more efficient and convenient to use an orthogonal curvilinear coordinate system. The prolate spheroidal coordinate system shown in Fig. 5.2 is useful for finite element modeling of the ventricles.

In the case of the ventricles, position vectors $\mathbf{R} = Y^R \mathbf{e}_R$ are defined with respect to the global rectangular Cartesian reference coordinate system, Y_A, or a suitable curvilinear system of world coordinates, Θ_A. The geometry of the body is discretized, and nodal geometric variables are interpolated using polynomial functions of normalized finite element coordinates ξ_k (Fig. 5.3).

The fibrous structure of the anisotropic myocardium is defined using locally orthonormal body/fiber coordinates X_i in which X_f is aligned with the local muscle fiber axis and lies in the epicardial tangent coordinate plane (the ξ_1–ξ_2 plane). X_s lies in the laminar sheet coordinate plane, and X_n is orthogonal to the sheet plane (Fig. 5.4). The relationship between these coordinate systems is described by a transformation matrix [**M**], as shown by USYK, MAZHARI and MCCULLOCH [2000].

5.2. *Least squares fitting for nodal geometric parameters*

To apply the general theory of least squares fitting to fit finite element models to anatomical measurements, we introduce the objective function

$$F(\mathbf{X}) = \sum_{d=1}^{D} \gamma_d \left\| \mathbf{X}(\xi_d) - \mathbf{X}_d \right\|^2, \tag{5.1}$$

where $\mathbf{X}_d$ is the dth measured coordinate or field variable, $\mathbf{X}(\xi_d)$ is the interpolated value at ξ_d, which is defined by the projection of the measured point onto a surface, and γ_d is the corresponding weight applied to the data point. The objective function represents the error between the coordinate of a measured anatomical surface point and the corresponding coordinate projected on the element surface. Weight parameters are all set equal to one when all measurements can be assumed to be equally accurate. As shown above, we can define $\mathbf{X}$ as

$$\mathbf{X} = \sum_{e=1}^{e_{\max}} \sum_{m=1}^{M(e)} [\boldsymbol{\Psi}^m][\mathbf{S}^m][\mathbf{X}^{\Delta(m,e)}]\mathbf{e}, \tag{5.2}$$

so

$$F(\mathbf{X}) = \sum_{d=1}^{D} \sum_{e=1}^{e_{\max}} \sum_{m=1}^{M(e)} \gamma_d \left\| [\boldsymbol{\Psi}^m][\mathbf{S}^m][\mathbf{X}^{\Delta(m,e)}]\mathbf{e} - \mathbf{X}_d \right\|^2. \tag{5.3}$$

A least squares fit minimizes the objective function

$$\frac{\partial F}{\partial X_j^{\Delta(m,e)}} = 0. \tag{5.4}$$

The following sections describe various anatomical features that have been incorporated into ventricular models using least squares fitting techniques.

5.3. Ventricular geometry

The mammalian heart consists of four pumping chambers, the left and right atria and ventricles communicating through the atrioventricular (mitral and tricuspid) valves, which are structurally connected by chordae tendineae to papillary muscles that extend from the anterior and posterior aspects of the right and left ventricular lumens. The muscular cardiac wall is perfused via the coronary vessels that originate at the left and right coronary ostia located in the sinuses of Valsalva immediately distal to the aortic valve leaflets. Surrounding the whole heart is the collagenous parietal pericardium that fuses with the diaphragm and great vessels.

From the perspective of engineering mechanics, the ventricles are three-dimensional thick-walled pressure vessels with substantial variations in wall thickness and principal curvatures both regionally and temporally through the cardiac cycle. The ventricular walls in the normal heart are thickest at the equator and base of the left ventricle and thinnest at the left ventricular apex and right ventricular free wall. There are also variations in the principal dimensions of the left ventricle with species, age, phase of the cardiac cycle, and disease.

Ventricular geometry has been studied in most quantitative detail in the dog heart (NIELSEN, LE GRICE, SMAILL and HUNTER [1991]). Geometric models have been very useful in the analysis, especially the use of confocal and nonconfocal ellipses of revolution to describe the epicardial and endocardial surfaces of the left and right ventricular walls. The canine left ventricle is reasonably modeled by a thick ellipsoid of revolution truncated at the base. The crescentic right ventricle wraps about 180 degrees around the heart wall circumstantially and extends longitudinally about two thirds of the distance from the base to the apex. Using a truncated ellipsoidal model, left ventricular geometry in the dog can be defined by the major and minor radii of two surfaces, the left ventricular endocardium, and a surface defining the free wall epicardium and the septal endocardium of the right ventricle. STREETER JR. and HANNA [1973] described the position of the basal plane using a truncation factor f_b defined as the ratio between the longitudinal distances from equator-to-base and equator-to-apex. Hence, the overall longitudinal distance from base to apex is $(1 + f_b)$ times the major radius of the ellipse.

Since variations in f_b between diastole and systole are relatively small (0.45 to 0.51), they suggested a constant value of 0.5.

The focal length d of an ellipsoid is defined from the major and minor radii (a and b) by $d^2 = a^2 - b^2$, and varies only slightly in the dog from endocardium to epicardium between end-diastole (37.3 to 37.9 mm) and end-systole (37.7 to 37.1 mm) (STREETER JR. and HANNA [1973]). Hence, within measurement accuracy, the boundaries of the left ventricular wall can be treated as ellipsoids of revolution, and the assumption that the ellipsoids are confocal appears to be a good one. This has motivated the choice of prolate spheroidal (elliptic–hyperbolic–polar) coordinates (λ, μ, θ) as described earlier (NIELSEN, LE GRICE, SMAILL and HUNTER [1991], YOUNG and AXEL [1992]). Here, the focal length d defines a family of coordinate systems that vary from spherical polar when $d = 0$ to cylindrical polar in the limit when $d \to \infty$. A surface of constant transmural coordinate λ is an ellipse of revolution with major radius $a = d\cosh\lambda$ and minor radius $b = d\sinh\lambda$. In an ellipsoidal model with a truncation factor of 0.5, the longitudinal coordinate μ varies from 0° at the apex to 120° at the base. Integrating the Jacobian in prolate spheroidal coordinates gives the volume of the wall or cavity

$$
\begin{aligned}
&d^3 \int_0^{2\pi} \int_0^{\mu} \int_{\lambda_1}^{\lambda_2} \big((\sinh^2\lambda + \sin^2\mu)\sinh\lambda \sin\mu\big)\,\mathrm{d}\lambda\,\mathrm{d}\mu\,\mathrm{d}\theta \\
&\quad = \frac{2\pi d^3}{3} \big|(1-\cos\mu)\cosh^3\lambda - \big(1-\cos^3\mu\big)\cosh\lambda\big|_{\lambda_1}^{\lambda_2}.
\end{aligned}
\tag{5.5}
$$

Using a truncated ellipsoidal model (Fig. 5.5) as an initial approximation and using the finite element least squares fitting approach described above, VETTER and MCCULLOCH [1998] built a realistic anatomical model of the geometry of the right and left ventricles of the rabbit heart (Fig. 5.6). By using prolate spheroidal coordinates to

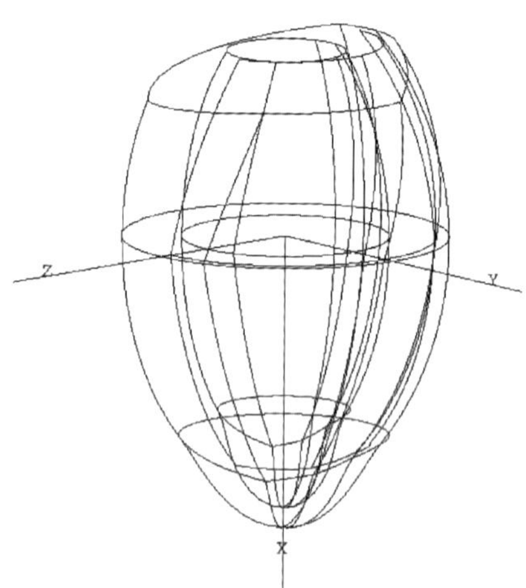

FIG. 5.5. Initial unfitted prolated spheroidal meshes for the epicardial surface and the left and right ventricular endocardium. The mesh represents the volume occupied by the left and right endocardial free walls and the septal wall.

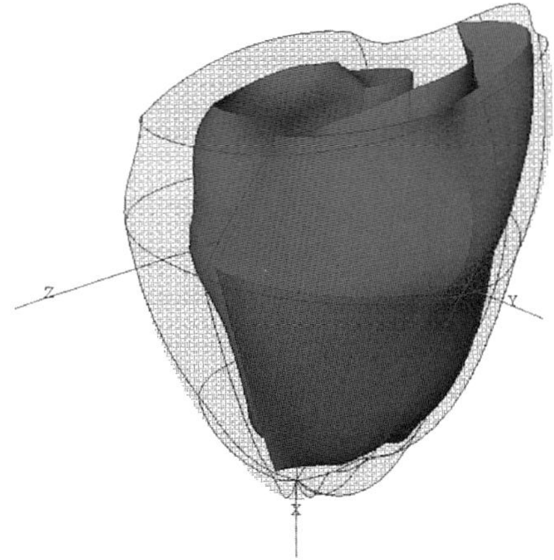

FIG. 5.6. Fitted model of the rabbit heart with epicardial and endocardial surfaces of the left and right ventricles rendered (from VETTER and MCCULLOCH [1998]).

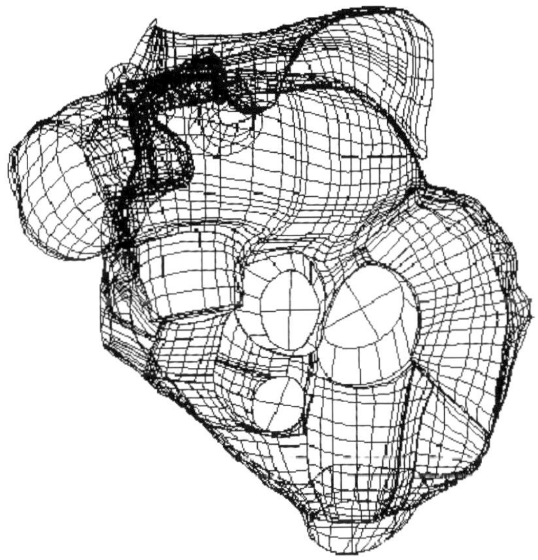

FIG. 5.7. Mesh fitted to porcine atrial anatomy.

construct surfaces as initial estimates for the left and right ventricular epicardia and endocardia (Fig. 5.5), VETTER and MCCULLOCH [1998] reduced the problem to a one-dimensional least squares fit of the λ coordinate alone, which was approximated on each surface using bicubic Hermite interpolation after the original work of NIELSEN, LE GRICE, SMAILL and HUNTER [1991].

Using the same general techniques, it is possible to fit anatomical models to measurements of other three-dimensional structures such as the atria and blood vessels, etc. Fig. 5.7 shows an anatomical model of the porcine atria.

5.4. *Fiber architecture*

The cardiac ventricles have a complex three-dimensional muscle fiber architecture (for a comprehensive review see STREETER JR. [1979]). Although the myocytes are relatively short, they are connected such that at any point in the normal heart wall there is a clear predominant fiber axis that is approximately tangent with the wall (within 3–5° in most regions, except near the apex and papillary muscle insertions). Each ventricular myocyte is connected via gap junctions at intercalated disks to an average of 11.3 neighbors, 5.3 on the sides and 6.0 at the ends (SAFFITZ, KANTER, GREEN, TOLLEY and BEYER [1994]). The classical anatomists dissected discrete bundles of fibrous swirls, though later investigations showed that the ventricular myocardium could be unwrapped by blunt dissection into a single continuous muscle "band" (TORRENT-GUASP [1973]). However, more modern histological techniques showed that in the plane of the wall, the muscle fiber angle makes a smooth transmural transition from epicardium to endocardium. Similar patterns have been described for humans, dogs, baboons, macaques, pigs, guinea pigs, and rats. In the human or dog left ventricle, the muscle fiber angle typically varies continuously from about $-60°$ (i.e., 60° clockwise from the circumferential axis) at the epicardium to about $+70°$ at the endocardium. The rate of change of fiber angle is usually greatest at the epicardium, so that circumferential (0°) fibers are found in the outer half of the wall, and the rate of angle change begins to slow transmurally approaching the inner third of the wall near the trabeculata–compacta interface. There are also small increases in fiber orientation from end-diastole to systole (7–19°), with the greatest changes at the epicardium and apex (STREETER JR., SPOTNITZ, PATEL, ROSS JR. and SONNENBLICK [1969]).

A detailed description of the morphogenesis of the muscle fiber system in the developing heart is not available, but there is evidence of an organized myofiber pattern by day 12 in the fetal mouse heart that is similar to that seen at birth (day 20) (MCLEAN, ROSS and PROTHERO [1989]). Abnormalities of cardiac muscle fiber patterns have been described in some disease conditions. In hypertrophic cardiomyopathy, which is often familial, there is substantial myofiber disarray, typically in the interventricular septum (MARON, BONOW, CANNON 3RD, LEON and EPSTEIN [1987]).

Regional variations in ventricular myofiber orientations are generally smooth except at the junction between the right ventricular free wall and septum. A detailed study in the dog that mapped fiber angles throughout the entire right and left ventricles described the same general transmural pattern in all regions including the septum and right ventricular free wall, but with definite regional variations (NIELSEN, LE GRICE, SMAILL and HUNTER [1991]). Transmural differences in fiber angle were about 120–140° in the left ventricular free wall, larger in the septum (160–180°), and smaller in the right ventricular free wall (100–120°). A similar study of fiber angle distributions in the rabbit left and right ventricles has recently been reported (VETTER and MCCULLOCH [1998]). Fiber angles in the rabbit heart were generally very similar to those in the dog, except for on the anterior wall, where average fiber orientations in the rabbit were 20–30° counterclockwise of those in the dog.

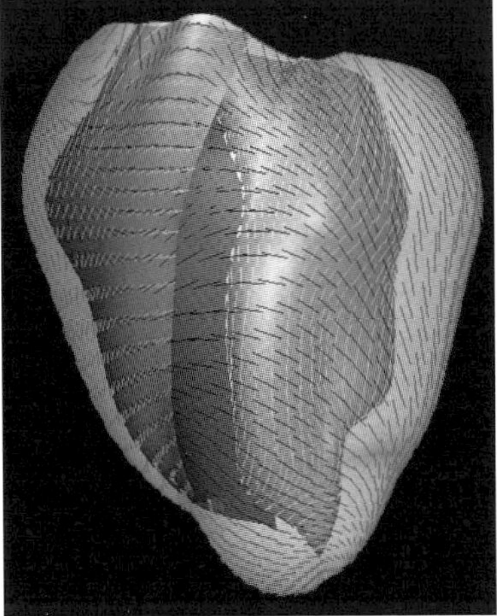

FIG. 5.8. Anatomical model of the rabbit left and right ventricles. 8,351 geometric points and 14,368 fiber angles were fitted using 36 high-order finite element elements.

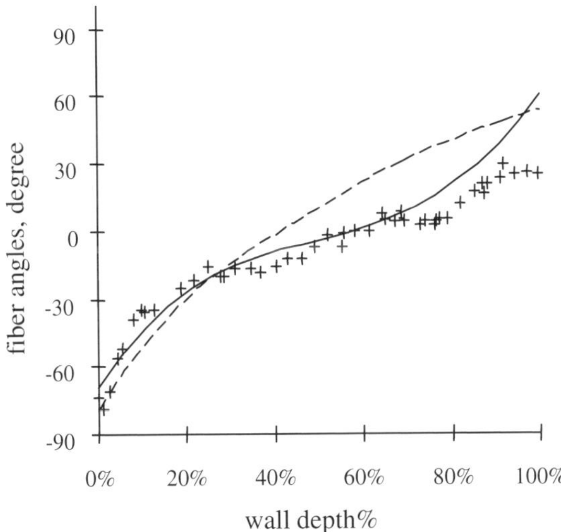

FIG. 5.9. Fitted fiber angles in the lateral left ventricular wall: + experimental measurements (rabbit); — fitted data (rabbit) (VETTER and MCCULLOCH [1998]); – – fitted data (dog) (NIELSEN, LE GRICE, SMAILL and HUNTER [1991]).

Using the same least squares method that was used to fit the ventricular geometry described in the previous section, VETTER and MCCULLOCH [1998] also fitted a model of fiber architecture into the anatomical model of the rabbit heart. The fitted model was based on about 14,000 histologically measured angles (Fig. 5.8). Figs. 5.9 and 5.10 show experimental measurements and model values for fiber angles in different

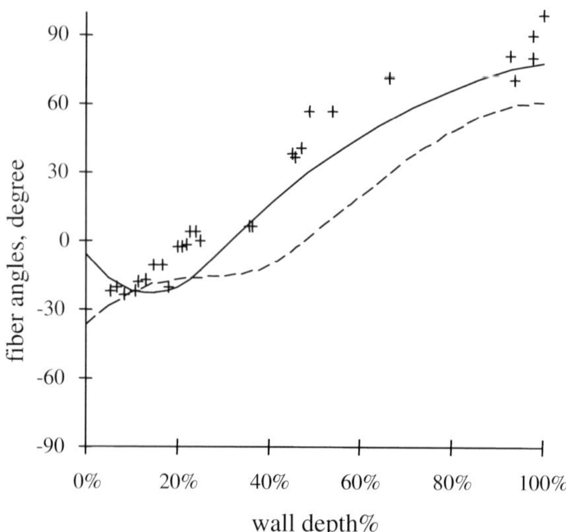

FIG. 5.10. Fitted fiber angles in the anterior wall: + local experimental measurements (rabbit); — fitted data (rabbit) (VETTER and MCCULLOCH [1998]); – – fitted data (dog) (NIELSEN, LE GRICE, SMAILL and HUNTER [1991]).

regions of the left ventricular wall from a dog model (NIELSEN, LE GRICE, SMAILL and HUNTER [1991]) and a rabbit model (VETTER and MCCULLOCH [1998]). Typical root-mean-squared fitting errors were less than 15–20° which is in the range of measurement error.

5.5. *Sheet architecture*

The fibrous architecture of the myocardium has motivated models of myocardial material symmetry as transversely isotropic. The recognition by LEGRICE, SMAILL, CHAI, EDGAR, GAVIN and HUNTER [1995] that planes of *cleavage* observed in transverse myocardial sections correspond to parallel, branching laminar *sheets* several myocytes thick are the best structural evidence for material orthotropy and have motivated the development of models describing the variation of fiber, sheet, and sheet-normal axes throughout the ventricular wall (LEGRICE, HUNTER and SMAILL [1997]). This also led to the hypothesis that the laminar architecture of ventricular myocardium is related to significant transverse shear strains (WALDMAN, FUNG and COVELL [1985]) and myofiber rearrangement (SPOTNITZ, SPOTNITZ, COTTRELL, SPIRO and SONNENBLICK [1974]) observed in the intact heart during systole. By measuring three-dimensional distributions of strain across the wall thickness using biplane radiography of radiopaque markers, LeGrice and colleagues (LEGRICE, TAKAYAMA and COVELL [1995]) found that the cleavage planes coincide closely with the planes of maximum shearing during ejection, and that the consequent reorientation of the myocytes may contribute 50% or more of normal systolic wall thickening. For a discussion of the implications of sheet organization on electrical propagation see Section 3.1.

5.6. Conductivity

Conductivity in the intracellular and extracellular spaces is represented by the tensors $\mathbf{D}_\mathrm{i}$ and $\mathbf{D}_\mathrm{e}$. The use of tensors allows directions of preferential conductivity to be defined. Since the tissue is known to have higher conductivity in the direction of the fibers than in other directions and is hypothesized to also have higher conduction in the direction of sheets than in the sheet-normal direction, the conductivity tensor is defined relative to the fiber–sheet coordinate system (see Fig. 5.4). When in this coordinate system, the conductivity tensor is diagonalized. Its highest eigenvalue corresponds to the eigenvector in the direction of the fiber coordinate direction. Because fiber and sheet angles vary regionally in the tissue, the directions of fastest conductivity will also vary in the same way. Conductivity can then be defined as a constant ratio of fiber to sheet to sheet normal magnitudes with only the orientation of the tensors varying regionally.

Recently least squares interpolation methods have also been used to incorporate a Purkinje fiber network into an electrophysiological model of the dog ventricles (USYK, LEGRICE and MCCULLOCH [2002]). Measurements of the geometry of the Purkinje fibers on the endocardium were used to fit a field representing the local angle of the Purkinje network. This field was defined for the whole endocardium even though Purkinje fibers are found only in discrete locations. A second field was fitted to represent the faster conduction that occurs along the Purkinje fibers. This field is defined with the constraint that its components are non-zero only in locations corresponding to positions where a fiber can be found. These two fields together define an additional conductivity tensor that can be included by superposition.

6. Solution implementation

The bidomain equations with the associated natural boundary conditions are summarized as

$$\nabla \cdot \mathbf{D}_\mathrm{i}\nabla V_\mathrm{m} + \nabla \cdot \mathbf{D}_\mathrm{i}\nabla u_\mathrm{e} = S_\mathrm{v} C_\mathrm{m}\frac{\mathrm{d}V_\mathrm{m}}{\mathrm{d}t} + S_\mathrm{v} I_\mathrm{ion} \quad \text{in } \Omega, \tag{6.1}$$

$$\nabla \cdot (\mathbf{D}_\mathrm{i} + \mathbf{D}_\mathrm{e})\nabla u_\mathrm{e} = -\nabla \cdot \mathbf{D}_\mathrm{i}\nabla V_\mathrm{m} \quad \text{in } \Omega, \tag{6.2}$$

$$\mathbf{n} \cdot \mathbf{D}_\mathrm{i}\nabla V_\mathrm{m} = 0 \quad \text{on } \Gamma_E. \tag{6.3}$$

Eqs. (6.1) and (6.2) must be solved simultaneously, so that one of each of the two unknowns, V_m and u_e, can be found from each of the equations. Collecting terms involving V_m on the left-hand side of the first equation:

$$S_\mathrm{v} C_\mathrm{m}\frac{\mathrm{d}V_\mathrm{m}}{\mathrm{d}t} - \nabla \cdot \mathbf{D}_\mathrm{i}\nabla V_\mathrm{m} = \nabla \cdot \mathbf{D}_\mathrm{i}\nabla u_\mathrm{e} - S_\mathrm{v} I_\mathrm{ion}. \tag{6.4}$$

Applying weighted residuals as in Eq. (4.28):

$$\begin{aligned} &\int_\Omega \Psi_k S_\mathrm{v} C_\mathrm{m}\frac{\mathrm{d}V_\mathrm{m}}{\mathrm{d}t}\,\mathrm{d}\Omega - \int_\Omega \nabla\Psi_k \cdot \mathbf{D}_\mathrm{i}\nabla V_\mathrm{m}\,\mathrm{d}\Omega \\ &\quad = \int_\Omega \Psi_k(\nabla \cdot \mathbf{D}_\mathrm{i}\nabla u_\mathrm{e} - S_\mathrm{v} I_\mathrm{ion})\,\mathrm{d}\Omega, \quad k = 1, \ldots, M, \end{aligned} \tag{6.5}$$

where the boundary conditions have been used to cancel out the surface integral term from the integration by parts of the second term on the left-hand side. Using integration by parts again on the first term on the right-hand side prevents any second derivative terms from entering the equation:

$$\sum_{j=1}^{M}\left[\int_{\Omega}\Psi_k S_{\mathrm{v}} C_{\mathrm{m}}\frac{\mathrm{d}(\Psi_j V_{\mathrm{m}j})}{\mathrm{d}t}\,\mathrm{d}\Omega-\left(\int_{\Omega}\nabla\Psi_k\cdot\mathbf{D}_{\mathrm{i}}\nabla\Psi_j\,\mathrm{d}\Omega\right)V_{\mathrm{m}j}\right]$$
$$=\int_{\Gamma}\Psi_k\mathbf{D}_{\mathrm{i}}\nabla u_{\mathrm{e}}\cdot\mathbf{n}\,\mathrm{d}\Gamma-\int_{\Omega}\nabla\Psi_k\cdot\mathbf{D}_{\mathrm{i}}\nabla u_{\mathrm{e}}\,\mathrm{d}\Omega-\int_{\Omega}\Psi_k S_{\mathrm{v}} I_{\mathrm{ion}}\,\mathrm{d}\Omega. \tag{6.6}$$

Eq. (6.2) is already written with terms involving u_{e} on the left-hand side. Similarly applying weighted residuals with natural boundary conditions:

$$\sum_{j=1}^{M}\left[\int_{\Omega}\nabla\Psi_k\cdot(\mathbf{D}_{\mathrm{i}}+\mathbf{D}_{\mathrm{e}})\nabla\Psi_j\,\mathrm{d}\Omega\right]u_{\mathrm{e}j}=-\int_{\Omega}\Psi_k(\nabla\cdot\mathbf{D}_{\mathrm{i}}\nabla V_{\mathrm{m}})\,\mathrm{d}\Omega. \tag{6.7}$$

In both equations $k=1,\ldots,M$.

Since the integral over the whole domain is the sum of integrals over each subdomain, the finite element method provides a mechanism not only for solving the numerical problem but for assembling it. Discretizing the domain into $e_{\max}$ elements, a pair of equations for each element subdomain can be written whose sum is the original Eqs. (6.6) and (6.7):

$$\sum_{j=1}^{M}\left[\int_{\Omega^e}\Psi_k S_{\mathrm{v}} C_{\mathrm{m}}\frac{\mathrm{d}(\Psi_j V_{\mathrm{m}j})}{\mathrm{d}t}\,\mathrm{d}\Omega^e-\left(\int_{\Omega^e}\nabla\Psi_k\cdot\mathbf{D}_{\mathrm{i}}\nabla\Psi_j\,\mathrm{d}\Omega^e\right)V_{\mathrm{m}j}\right]$$
$$=\int_{\Gamma^e}\Psi_k\mathbf{D}_{\mathrm{i}}\nabla u_{\mathrm{e}}\cdot\mathbf{n}\,\mathrm{d}\Gamma^e-\int_{\Omega^e}\nabla\Psi_k\cdot\mathbf{D}_{\mathrm{i}}\nabla u_{\mathrm{e}}\,\mathrm{d}\Omega^e-\int_{\Omega^e}\Psi_k S_{\mathrm{v}} I_{\mathrm{ion}}\,\mathrm{d}\Omega^e, \tag{6.8}$$

$$\sum_{j=1}^{M}\left[\int_{\Omega^e}\nabla\Psi_k\cdot(\mathbf{D}_{\mathrm{i}}+\mathbf{D}_{\mathrm{e}})\nabla\Psi_j\,\mathrm{d}\Omega^e\right]u_{\mathrm{e}j}=-\int_{\Omega^e}\Psi_k(\nabla\cdot\mathbf{D}_{\mathrm{i}}\nabla V_{\mathrm{m}})\,\mathrm{d}\Omega^e, \tag{6.9}$$

where an $M\times M$ system of equations can be constructed separately for each element. Because elements share global nodes, the system of equations for each element cannot be solved independently. Hence they are *assembled* into a single whole system for all the elements where there is one row for each global nodal parameter. The connectivity matrix dictates which local nodal parameters (rows in element $M\times M$ systems) correspond to the same global parameter and should thus be summed on a single row of the whole system.

At this stage, the anatomy of the ventricles is incorporated through the representation of the geometry of the elements in the domain. The structural properties of tissue architecture are incorporated through the combination of fields representing fiber geometry, sheet geometry, and conductivity in the terms $\mathbf{D}_{\mathrm{i}}$ and $\mathbf{D}_{\mathrm{e}}$. Local cellular characteristics are incorporated through the membrane currents calculated in the term I_{ion}. The next section describes methods that have been used to solve these equations.

6.1. Solution algorithms

The bidomain equations must be discretized in time as well as space. Finite difference schemes based on the θ-rule are commonly used. The θ-rule for a problem that is first order in time

$$\frac{du}{dt} = G, \tag{6.10}$$

where G is some spatially varying function of u, is written as

$$\frac{u^{n+1} - u^n}{\Delta t} = \theta G^{n+1} + (1 - \theta) G^n, \tag{6.11}$$

where values of u^n are known, u^{n+1} is the unknown quantity, and time has been discretized into steps of size Δt. Choices of θ lead to various methods, for example, $\theta = 0 \rightarrow$ forward Euler, $\theta = \frac{1}{2} \rightarrow$ Crank–Nicolson, $\theta = 1 \rightarrow$ backward Euler.

Various schemes have been used for discretizing the bidomain problems in time. For example, SUNDNES, LINES and TVEITO [2001] have recently used operator splitting methods to solve the bidomain equations simultaneously with PDEs representing the potential in the torso. Their calculations are useful for investigating the forward problem of electrophysiology, which refers to the calculation of body surface potentials from potentials originating in the heart as are measured clinically through electrocardiograms. Operator splitting methods consist of separating the ODE calculations from the PDE calculations by dividing a single time step. During the first part of the time step the ODEs are solved. Their solution is used to update the PDEs before solving them during the second half time step. In this way the I_{ion} term becomes a constant source term rather than a function of V_{m} during the solution of the PDEs with a value from the solution of the ODEs at the half time step.

In the discretized bidomain equations, Eq. (6.8) is first order in time due to the $\mathrm{d}V_{\text{m}}/\mathrm{d}t$ term on the left-hand side. This term can be discretized as follows:

$$\int_{\Omega^e} S_{\text{v}} C_{\text{m}} \Psi_k \frac{\mathrm{d}(\Psi_j V_{\text{m}j})}{\mathrm{d}t} \,\mathrm{d}\Omega^e = \int_{\Omega^e} \frac{S_{\text{v}} C_{\text{m}}}{\Delta t} \Psi_k \Psi_j \left(V_{\text{m}j}^{n+1} - V_{\text{m}}^{n} \right) \mathrm{d}\Omega^e, \tag{6.12}$$

where V_{m}^n is known and V_{m}^{n+1} is the variable for which an approximation is being sought. Employing the operator splitting method of Sundnes et al.: treating the I_{ion} term as known from a series of separate integrations of the ODEs over the global time interval from $t = n$ to $t = n + 1/2$, Eqs. (6.8) and (6.9) can be written as

$$\begin{aligned} &\sum_{j=1}^{M} \left[\int_{\Omega^e} \left[\frac{S_{\text{v}} C_{\text{m}}}{\Delta t} \Psi_k \Psi_j - \nabla \Psi_k \cdot \mathbf{D}_{\text{i}} \nabla \Psi_j \right] V_{\text{m}j}^{n+1} \,\mathrm{d}\Omega^e \right] \\ &\quad = \int_{\Omega^e} \left(\frac{S_{\text{v}} C_{\text{m}}}{\Delta t} \Psi_k \Psi_j V_{\text{m}}^{n} - \nabla \Psi_k \cdot \mathbf{D}_{\text{i}} \nabla u_{\text{e}}^{n} - \Psi_k S_{\text{v}} I_{\text{ion}}^{n+1/2} \right) \mathrm{d}\Omega^e \\ &\qquad + \int_{\Gamma^e} \Psi_k \mathbf{D}_{\text{i}} \nabla u_{\text{e}}^{n} \cdot \mathbf{n} \,\mathrm{d}\Gamma^e, \end{aligned} \tag{6.13}$$

$$\sum_{j=1}^{M}\left[\int_{\Omega^e} \nabla\Psi_k \cdot (\mathbf{D}_\mathrm{i} + \mathbf{D}_\mathrm{e})\nabla\Psi_j \,\mathrm{d}\Omega^e\right] u_{\mathrm{e}j}^{n+1} = -\int_{\Omega^e} \Psi_k \nabla \cdot \mathbf{D}_\mathrm{i} \nabla V_\mathrm{m}^{n+1} \,\mathrm{d}\Omega^e. \quad (6.14)$$

The steps for finding the solution proceed as follows. First, the I_ion term is found by separate integration of the ODEs at each integration point in the discretized mesh representing the ventricles using known quantities at time n as the initial conditions. Next, the results of the ODE integration at time $n + 1/2$ are combined with the known solutions of V_m and u_e at time n to find the terms in the right-hand side of the first bidomain equation. Following assembly, a solution of a linear system of the form

$$\mathbf{Ax} = \mathbf{b}, \quad (6.15)$$

is solved for V_m^{n+1} at each of the M global nodes. Finally, V_m^{n+1} is used to form the right-hand side of the second equation and a second assembly and linear system solution is performed to find u_e^{n+1} at the M global nodes.

6.2. *Implementation issues*

Although computer memory and speed continue to grow with advances in technology, simulations of cardiac electrophysiology problems remain large and time consuming and continue to rapidly grow more so with advances in biology. Parallel programming methods are useful in this situation. In incorporating local cellular processes into the simulation of tissue electrophysiology, a level of data parallelism can be achieved. If local values of membrane potential are known, the currents passing through the ion channels or within the intracellular spaces of a cell in one location can be calculated independently of those occurring within a cell in another location. In time integration schemes such as the method presented in the previous section, this is the exact situation. The result is that cellular information is data parallel. Calculations of local cellular processes can be distributed to a set of processors so that multiple calculations can be performed simultaneously. As the ODE calculation time grows with the sophistication of the ionic model, the solution of the linear systems do not change size or complexity. Then for complex ionic models, the data parallel ODE portion of the solution is also the portion where improvements in speed are most needed.

The nature of the propagation problem is that local areas of tissue near the wave front are the same areas experiencing fast ion kinetics. In locations far from the wave front, kinetics are relatively slow. Parallel programming that balances the calculation load could lead to even further speedups. Load balancing works by keeping processors busy, for example, while one processor calculates an ODE integration for a point near the wave front requiring many small time steps to cover the global time increment, another processor might calculate ODE integrations for several points away from the wave front that each require few local time steps to cover the same global time increment.

Adaptive meshing techniques involving element size can also be used for improving efficiency. These techniques use smaller mesh elements to discretize the domain in regions near the wave front and larger elements in areas away from the wave front. For adaptive meshing and load balancing techniques the location of the wave front must be identified. In addition, for adaptive meshing the wave front location must also be predicted so that it will not reach elements that are too large in a single time step leading to a divergent solution. Various adaptive meshing techniques applied to cardiac electrophysiology can be found in QU and GARFINKEL [1999], QUAN, EVANS and HASTINGS [1998], OTANI [2000] and elsewhere.

In examining the solution of cardiac propagation problems various major problem components can be identified each of which may be accomplished by significant separate software developments. For example, the ODE integration of the cellular ionic model may be carried out by an implicit Runge–Kutta solver suitable for stiff problems, but there are many equally effective choices of solvers. The best one to use may depend on exactly which ionic model has been chosen. Interchangeability of the ionic model itself may also be important since these models are evolving rapidly. Similarly, the solution of linear systems for the PDEs may be performed by a range of $\mathbf{Ax} = \mathbf{b}$ solvers. The best solver may depend on the number of mesh elements or the computational platform being utilized, i.e., some solvers use matrix free methods that can solve the linear system without assembling $\mathbf{A}$ resulting in less memory use while other solvers may be specialized for solving linear systems on a distributed memory multiprocessor system.

Several relatively new paradigms in computing can address these implementation requirements. Object-oriented programming treats software components as objects, black boxes that send and receive messages. For example, an ODE integrator object would be sent a message describing which equations to integrate, when to start, and when to stop integrating. All ODE objects need this same type of information, and all of them result in values of the dependent variables of the ODEs calculated over time. Every time a user has a new set of ODEs to integrate, it is not necessary to change the ODE integrator object itself in order to integrate the new set of equations.

The use of generic programming methods in software developments can help facilitate the adaptability and interoperability of separately developed software objects. The goal of generic programming is to express components of programs at the most general level possible without losing efficiency. This involves writing algorithms with minimum assumptions about the data to be processed, while also creating data structures with minimal assumptions about the algorithms to be used. This provides for maximum interoperability of separate components. With generic programming, components can be developed independently and combined arbitrarily as needed only requiring specified interfaces in order to communicate. Then treating the components as objects and gluing them together in a component-based environment, should lead to faster software development times where reuse of components and overall code modifiability can be maximized. In addition, generic programming helps to reduce the number of lines of code, thus reducing the possibility of bugs as well as lowering maintenance costs.

For the bidomain problem of electrophysiology various objects need to be used together in order to solve the problem. Component-based software design can be used for gluing objects together. Programming languages such as Python, a very high level object-oriented programming language, can be used for this implementation task. Python can interface code written in other languages and handle the conversion of data structures so that various independent pieces of code can function together as objects in an overall software package.

7. Integrated models

The electrophysiological function of the heart does not exist separately from its mechanical function. So far the integration across biophysical scales has been discussed in the context of incorporation of cellular ionic systems models into finite element simulations of wave propagation in tissues. However, the incorporation of mechanical function into such simulations is of fundamental interest. Contraction itself occurs within sarcomeres at the cellular level and depends on the intracellular calcium concentration (see Section 2.4).

Three-dimensional finite element stress analysis methods for large elastic deformations of nonlinear anisotropic materials can be applied to modeling the mechanics of the heart. For example, the steps needed to solve a mechanics problem may proceed as follows. Finite element equations are integrated using a Gaussian quadrature scheme, and the resulting system of nonlinear elliptic equations are solved for the unknown deformation and pressure using a Newton iterative method (ODEN [1972]). The non-symmetric element tangent stiffness matrix (Jacobian) may be approximated by forward differences or may be found analytically and updated at each full Newton iteration. Non-zero contributions to the constraint-reduced global tangent stiffness matrix are vectorized and solved using a general linear sparse solver with threshold pivoting. The iterative process is terminated when the sum of solution increments and the maximum unconstrained residual are both less than an acceptable threshold (USYK, LEGRICE and MCCULLOCH [2002]).

If the finite deformation stress analysis above incorporates cellular level information in a manner similar to the electrophysiology problem, it is conceivable that bidirectional influences of each of these large problems could be implemented through local interactions of cellular models. The applications of this type of integration are clear in measurable phenomena such as excitation–contraction coupling and mechanoelectric feedback. Other systems models would also be useful extensions to the electrophysiological problem. For example, signaling, metabolism, and energetics all represent fields where important components could be collected for integration into a single model. With these features, simulations of heart failure, ischemia, and other conditions, not to mention normal physiological function will be more complete.

The computational hurdles for creating an integrated model include schemes for synchronizing calculations and translating meshes since these problems are typically solved with very different time and space scales. Parallelism can be exploited in an additional layer in an integrated model where, for example, the electrical and mechanical portions of the problem are solved on different processors with carefully developed communication between these major problem objects. Issues of convergence and parameter sensitivity are also of increasing concern as the number of variables increases.

In summary, integration across biological scales in simulating cardiac electrophysiology is common in current cardiac models. This paves the way for integration of structurally and functionally integrated models of cardiac electromechanical function that combine data-intensive cellular systems models with compute-intensive anatomically detailed multiscale simulations.

8. Appendix: Lagrange and Hermite interpolations

8.1. *Linear interpolation*

For any variable u, a linear variation between two values, u_1 and u_2, may be described as

$$u(\xi) = (1 - \xi)u_1 + \xi u_2, \quad 0 \leqslant \xi \leqslant 1, \tag{8.1}$$

where the parameter ξ is a normalized measure of distance along the curve. (Notice that in previous sections u represented the solution over an entire domain, and $u_{(e)}$ represented the piece-wise solution on element subdomains:

$$u = \sum_{1}^{e_{\max}} u_{(e)}. \tag{8.2}$$

In this appendix, the (e) notation has been dropped for convenience and clarity so that u is $u_{(e)}$ used in other sections.) We define

$$\varphi_1(\xi) = 1 - \xi, \qquad \varphi_2(\xi) = \xi, \tag{8.3}$$

so that

$$u(\xi) = \varphi_1(\xi)u_1 + \varphi_2(\xi)u_2, \tag{8.4}$$

where $\varphi_1(\xi)$ and $\varphi_2(\xi)$ are the *linear Lagrange* basis functions associated with the nodal parameters u_1 and u_2 (Fig. 8.1).

8.2. *Quadratic interpolation*

A quadratic variation of u over a one-dimensional element requires three nodal parameters,

$$u(\xi) = \varphi_1(\xi)u_1 + \varphi_2(\xi)u_2 + \varphi_3(\xi)u_3. \tag{8.5}$$

The *quadratic Lagrange* basis functions are shown in Fig. 8.2. Notice that since $\varphi_1(\xi)$ must be zero at $\xi = 0.5$ (node 2), $\varphi_1(\xi)$ must have a factor $(\xi - 0.5)$ and since it is also zero at $\xi = 1$ (node 3), another factor is $(\xi - 1)$. Finally, since $\varphi_1(\xi)$ is 1 at $\xi = 0$

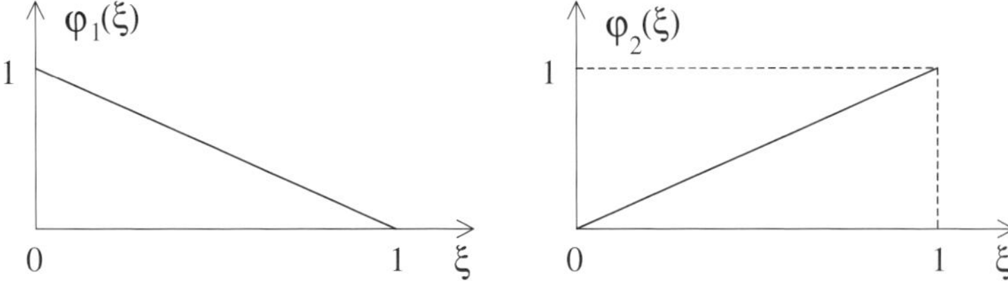

FIG. 8.1. The one-dimensional linear Lagrange basis functions.

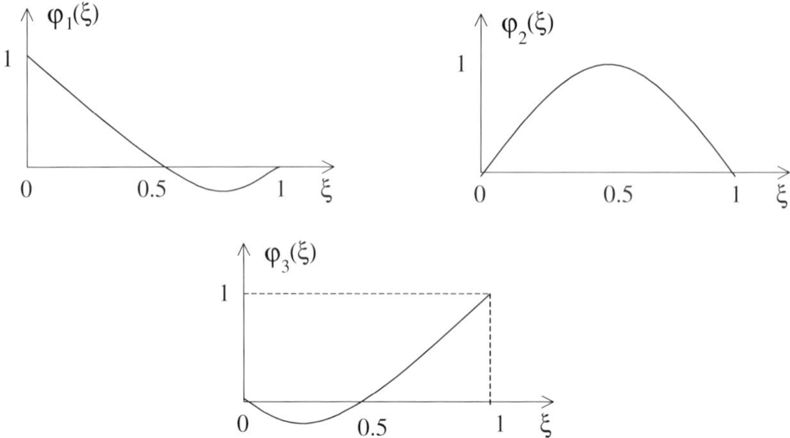

FIG. 8.2. The quadratic Lagrange basis functions.

(node 1), we have

$$\varphi_1(\xi) = 2(\xi - 0.5)(\xi - 1). \tag{8.6}$$

The other two quadratic Lagrange basis functions are found similarly, so the quadratic Lagrange basis functions are

$$\begin{aligned}
\varphi_1(\xi) &= 2(\xi - 0.5)(\xi - 1),\\
\varphi_2(\xi) &= -4\xi(\xi - 1),\\
\varphi_3(\xi) &= 2\xi(\xi - 0.5).
\end{aligned} \tag{8.7}$$

8.3. *Cubic Hermite interpolation*

All the basis functions mentioned thus far are Lagrange basis functions and provide C^0continuity of u across element boundaries but not higher-order continuity. In order to preserve continuity of the derivative of u with respect to ξ across element boundaries additional nodal parameters are included: the derivatives at node n, $(\frac{\partial u}{\partial \xi})_n$. The basis functions are chosen to ensure that

$$\left.\frac{\partial u}{\partial \xi}\right|_{\xi=0} = \left(\frac{\partial u}{\partial \xi}\right)_1 = u_1' \quad \text{and} \quad \left.\frac{\partial u}{\partial \xi}\right|_{\xi=1} = \left(\frac{\partial u}{\partial \xi}\right)_2 = u_2', \tag{8.8}$$

and since u is shared between adjacent elements, derivative continuity is ensured. The *cubic Hermite* basis functions are derived from

$$u(\xi) = a + b\xi + c\xi^2 + d\xi^3, \tag{8.9}$$

$$\frac{\partial u}{\partial \xi} = b + 2c\xi + 3d\xi^2, \tag{8.10}$$

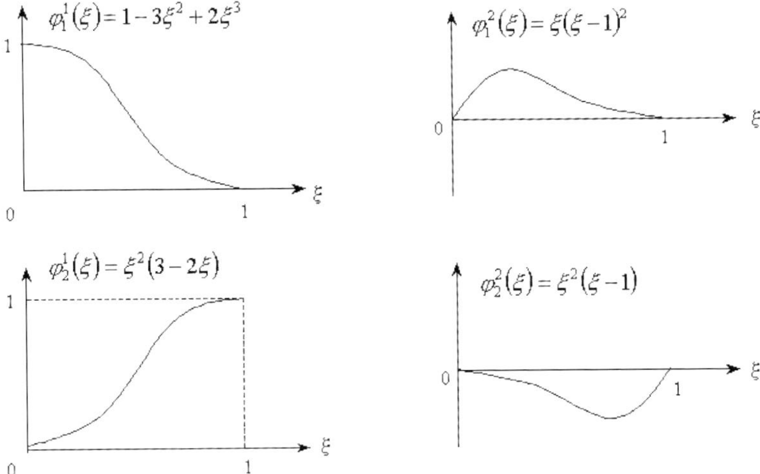

FIG. 8.3. Cubic Hermite basis functions.

subject to the constraints

$$\begin{aligned}
&u(0)=a=u_1,\\
&u(1)=a+b+c+d=u_2,\\
&\frac{\partial u}{\partial \xi}(0)=b=u_1', \qquad \frac{\partial u}{\partial \xi}(1)=b+2c+3d=u_2'.
\end{aligned} \tag{8.11}$$

Solving these equations, we get

$$u(\xi)=\varphi_1^1(\xi)u_1+\varphi_1^2(\xi)u_1'+\varphi_2^1(\xi)u_2+\varphi_2^2(\xi)u_2', \tag{8.12}$$

where the four cubic Hermite basis functions are sketched in Fig. 8.3, and are given by the following equations:

$$\begin{aligned}
&\varphi_1^1(\xi)=1-3\xi^2+2\xi^3, \qquad \varphi_1^2(\xi)=\xi(\xi-1)^2,\\
&\varphi_2^1(\xi)=\xi^2(3-2\xi), \qquad \varphi_2^2(\xi)=\xi^2(\xi-1).
\end{aligned} \tag{8.13}$$

8.4. Two-dimensional elements

Two-dimensional bilinear basis functions are readily constructed from the products of the above one-dimensional linear functions as follows:

$$u(\xi_1,\xi_2)=\sum_{n=1}^{4}\Psi_n(\xi_1,\xi_2)\cdot u_n, \tag{8.14}$$

where n are the four local nodes of a two-dimensional rectangular element (again the notation indicating that these terms belong to a single element has been dropped, so

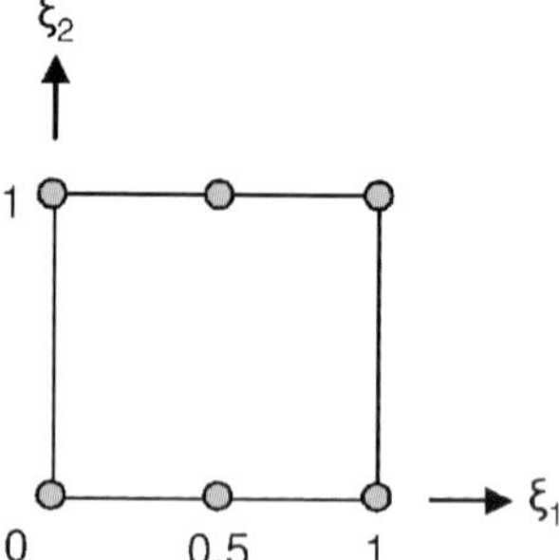

FIG. 8.4. Two-dimensional isoparametric element with linear Lagrange interpolation in one direction and quadratic Lagrange interpolation in the other.

$n = n(e)$ and $u = u_{(e)}$ of previous sections):

$$\begin{aligned}
&\Psi_1(\xi_1,\xi_2) = \varphi_1(\xi_1)\varphi_1(\xi_2), \qquad && \Psi_3(\xi_1,\xi_2) = \varphi_1(\xi_1)\varphi_2(\xi_2),\\
&\Psi_2(\xi_1,\xi_2) = \varphi_2(\xi_1)\varphi_1(\xi_2), \qquad && \Psi_4(\xi_1,\xi_2) = \varphi_2(\xi_1)\varphi_2(\xi_2),
\end{aligned} \tag{8.15}$$

and the functions $\varphi_i(\xi_k)$ $(i, k = 1, 2)$ are defined above by Eqs. (8.3).

Higher-order, two-dimensional parametric basis functions can be similarly constructed from products of the appropriate one-dimensional basis functions. For example, a six-noded (see Fig. 8.4) quadratic-linear element (quadratic in ξ_1 and linear in ξ_2) would be given by

$$u(\xi_1,\xi_2) = \sum_{n=1}^{6} \Psi_n(\xi_1,\xi_2) \cdot u_n, \tag{8.16}$$

$$\begin{aligned}
&\Psi_1(\xi_1,\xi_2) = \varphi_1^Q(\xi_1)\varphi_1^L(\xi_2), \qquad && \Psi_4(\xi_1,\xi_2) = \varphi_1^Q(\xi_1)\varphi_2^L(\xi_2),\\
&\Psi_2(\xi_1,\xi_2) = \varphi_2^Q(\xi_1)\varphi_1^L(\xi_2), \qquad && \Psi_5(\xi_1,\xi_2) = \varphi_2^Q(\xi_1)\varphi_2^L(\xi_2),\\
&\Psi_3(\xi_1,\xi_2) = \varphi_3^Q(\xi_1)\varphi_1^L(\xi_2), \qquad && \Psi_6(\xi_1,\xi_2) = \varphi_3^Q(\xi_1)\varphi_2^L(\xi_2),
\end{aligned} \tag{8.17}$$

where Q indicates quadratic basis functions as in Eqs. (8.7), and L indicates linear basis functions as in Eqs. (8.3).

A two-dimensional bicubic Hermite element requires four derivatives per node for a total of 16 parameters,

$$u, \quad \frac{\partial u}{\partial \xi_1}, \quad \frac{\partial u}{\partial \xi_2}, \quad \text{and} \quad \frac{\partial^2 u}{\partial \xi_1 \partial \xi_2}. \tag{8.18}$$

The need for the second order cross derivative term can be explained as follows: if u is cubic in ξ_1 and cubic in ξ_2 then $\frac{\partial u}{\partial \xi_1}$ is quadratic in ξ_1 and cubic in ξ_2, and $\frac{\partial u}{\partial \xi_2}$ is cubic in ξ_1 and quadratic in ξ_2. Now consider the 2–3 edge as shown in Fig. 8.5. The cubic variation of u with ξ_2 is specified by four nodal parameters: two at node 1, u_1, $\left(\frac{\partial u}{\partial \xi_2}\right)_1$ and two at node 3, u_3 and $\left(\frac{\partial u}{\partial \xi_2}\right)_3$. Since $\frac{\partial u}{\partial \xi_1}$ (the normal derivative) is also cubic in ξ_2 and is entirely independent of the four nodal parameters, four additional parameters are

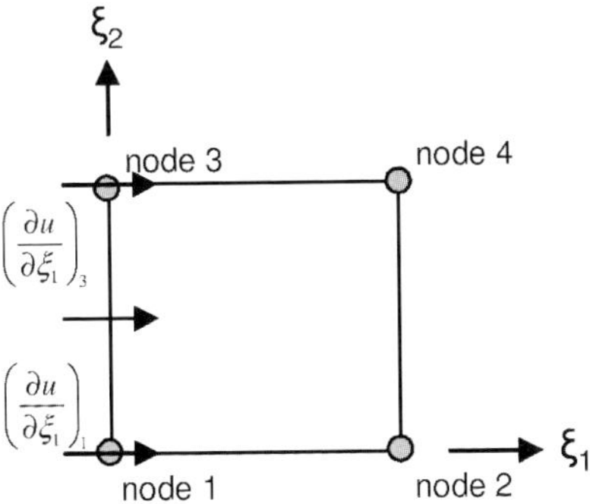

FIG. 8.5. Bicubic interpolation. Arrows on the 2–3 edge depict the direction of the normal derivatives, $\frac{\partial u}{\partial \xi_1}$.

required to specify that cubic shape. Two of these are specified by the normal derivatives at nodes 1 and 3, $\left(\frac{\partial u}{\partial \xi_1}\right)_1$ and $\left(\frac{\partial u}{\partial \xi_1}\right)_3$, and the remaining two by the variation of the normal derivatives with respect to ξ_2 at nodes 1 and 3, $\left(\frac{\partial^2 u}{\partial \xi_1 \partial \xi_2}\right)_1$ and $\left(\frac{\partial^2 u}{\partial \xi_1 \partial \xi_2}\right)_3$. The bicubic interpolation is thus given by four nodal parameters at each node,

$$u(\xi_1, \xi_2) = \sum_{n=1}^{4} \sum_{\gamma=1}^{4} \Psi_n^{\gamma}(\xi_1, \xi_2) \cdot u_n^{\gamma}, \tag{8.19}$$

where each node, n, of the element has the following nodal parameters:

$$\begin{aligned} u_n^1 &= u_n, & u_n^3 &= \left(\frac{\partial u}{\partial \xi_2}\right)_n, \\ u_n^2 &= \left(\frac{\partial u}{\partial \xi_1}\right)_n, & u_n^4 &= \left(\frac{\partial^2 u}{\partial \xi_1 \partial \xi_2}\right)_n, \end{aligned} \tag{8.20}$$

$$\begin{aligned}
\Psi_1^1(\xi_1, \xi_2) &= \varphi_1^1(\xi_1)\varphi_1^1(\xi_2), & \Psi_2^1(\xi_1, \xi_2) &- \varphi_2^1(\xi_1)\varphi_1^1(\xi_2), \\
\Psi_3^1(\xi_1, \xi_2) &= \varphi_1^1(\xi_1)\varphi_2^1(\xi_2), & \Psi_4^1(\xi_1, \xi_2) &= \varphi_2^1(\xi_1)\varphi_2^1(\xi_2), \\
\Psi_1^2(\xi_1, \xi_2) &= \varphi_1^2(\xi_1)\varphi_1^1(\xi_2), & \Psi_2^2(\xi_1, \xi_2) &= \varphi_2^2(\xi_1)\varphi_1^1(\xi_2), \\
\Psi_3^2(\xi_1, \xi_2) &= \varphi_1^2(\xi_1)\varphi_2^1(\xi_2), & \Psi_4^2(\xi_1, \xi_2) &= \varphi_2^2(\xi_1)\varphi_2^1(\xi_2), \\
\Psi_1^3(\xi_1, \xi_2) &= \varphi_1^1(\xi_1)\varphi_1^2(\xi_2), & \Psi_2^3(\xi_1, \xi_2) &= \varphi_2^1(\xi_1)\varphi_1^2(\xi_2), \\
\Psi_3^3(\xi_1, \xi_2) &= \varphi_1^1(\xi_1)\varphi_2^2(\xi_2), & \Psi_4^3(\xi_1, \xi_2) &= \varphi_2^1(\xi_1)\varphi_2^2(\xi_2), \\
\Psi_1^4(\xi_1, \xi_2) &= \varphi_1^2(\xi_1)\varphi_1^2(\xi_2), & \Psi_2^4(\xi_1, \xi_2) &= \varphi_2^2(\xi_1)\varphi_1^2(\xi_2), \\
\Psi_3^4(\xi_1, \xi_2) &= \varphi_1^2(\xi_1)\varphi_2^2(\xi_2), & \Psi_4^4(\xi_1, \xi_2) &= \varphi_2^2(\xi_1)\varphi_2^2(\xi_2),
\end{aligned} \tag{8.21}$$

and the functions $\varphi_i^j(\xi_k)$ $(i = 1, 2;\ j = 1, 2;\ k = 1, 2, 3)$ are defined above by Eqs. (8.13).

8.5. Three-dimensional elements

Three-dimensional trilinear Lagrange basis functions are similarly constructed from the products of the above one-dimensional linear functions as follows:

$$u(\xi_1, \xi_2, \xi_3) = \sum_{n=1}^{8} \Psi_n(\xi_1, \xi_2, \xi_3) \cdot u_n, \tag{8.22}$$

where

$$\begin{aligned}
&\Psi_1(\xi_1, \xi_2, \xi_3) = \varphi_1(\xi_1)\varphi_1(\xi_2)\varphi_1(\xi_3), && \Psi_5(\xi_1, \xi_2, \xi_3) = \varphi_1(\xi_1)\varphi_1(\xi_2)\varphi_2(\xi_3),\\
&\Psi_2(\xi_1, \xi_2, \xi_3) = \varphi_2(\xi_1)\varphi_1(\xi_2)\varphi_1(\xi_3), && \Psi_6(\xi_1, \xi_2, \xi_3) = \varphi_2(\xi_1)\varphi_1(\xi_2)\varphi_2(\xi_3),\\
&\Psi_3(\xi_1, \xi_2, \xi_3) = \varphi_1(\xi_1)\varphi_2(\xi_2)\varphi_1(\xi_3), && \Psi_7(\xi_1, \xi_2, \xi_3) = \varphi_1(\xi_1)\varphi_2(\xi_2)\varphi_2(\xi_3),\\
&\Psi_4(\xi_1, \xi_2, \xi_3) = \varphi_2(\xi_1)\varphi_2(\xi_2)\varphi_1(\xi_3), && \Psi_8(\xi_1, \xi_2, \xi_3) = \varphi_2(\xi_1)\varphi_2(\xi_2)\varphi_2(\xi_3)
\end{aligned} \tag{8.22a}$$

and the functions $\varphi_i(\xi_k)$ $(i = 1, 2;\ k = 1, 2, 3)$ are defined above by Eqs. (8.3). These eight basis functions correspond to the eight nodes of a trilinear brick element (Fig. 8.6).

A three-dimensional tricubic Hermite element requires eight derivatives per node,

$$u,\quad \frac{\partial u}{\partial \xi_1},\quad \frac{\partial u}{\partial \xi_2},\quad \frac{\partial u}{\partial \xi_3},\quad \frac{\partial^2 u}{\partial \xi_1 \partial \xi_2},\quad \frac{\partial^2 u}{\partial \xi_1 \partial \xi_3},\quad \frac{\partial^2 u}{\partial \xi_2 \partial \xi_3},\quad \text{and}\quad \frac{\partial^3 u}{\partial \xi_1 \partial \xi_2 \partial \xi_3}, \tag{8.23}$$

$$u(\xi_1, \xi_2, \xi_3) = \sum_{n=1}^{8} \sum_{\gamma=1}^{8} \Psi_n^{\gamma}(\xi_1, \xi_2, \xi_3) \cdot u_n^{\gamma}, \tag{8.24}$$

where

$$u_n^1 = u_n,\quad u_n^2 = \left(\frac{\partial u}{\partial \xi_1}\right)_n,\quad u_n^3 = \left(\frac{\partial u}{\partial \xi_2}\right)_n,$$

$$u_n^4 = \left(\frac{\partial^2 u}{\partial \xi_1 \partial \xi_2}\right)_n,\quad u_n^5 = \left(\frac{\partial u}{\partial \xi_3}\right)_n,\quad u_n^6 = \left(\frac{\partial^2 u}{\partial \xi_1 \partial \xi_3}\right)_n,$$

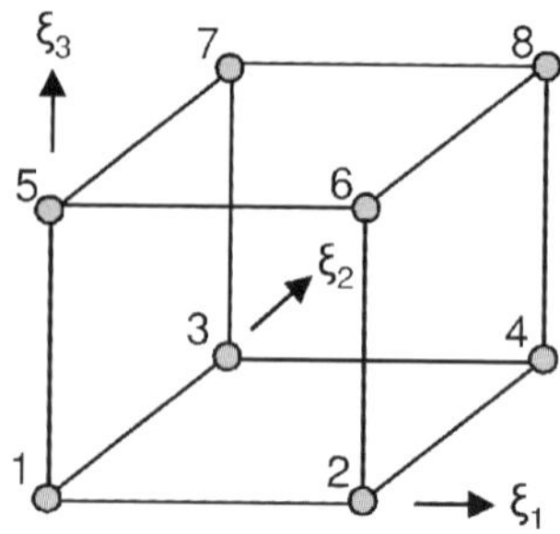

FIG. 8.6. An 8-noded three-dimensional isoparametric finite element.

$$u_n^7 = \left(\frac{\partial^2 u}{\partial \xi_2 \partial \xi_3} \right)_n, \qquad u_n^8 = \left(\frac{\partial^3 u}{\partial \xi_1 \partial \xi_2 \partial \xi_3} \right)_n, \tag{8.25}$$

$$\begin{aligned}
&\Psi_1^1(\xi_1,\xi_2,\xi_3) = \varphi_1^1(\xi_1)\varphi_1^1(\xi_2)\varphi_1^1(\xi_3), &&\Psi_2^1(\xi_1,\xi_2,\xi_3) = \varphi_2^1(\xi_1)\varphi_1^1(\xi_2)\varphi_1^1(\xi_3),\\
&\Psi_3^1(\xi_1,\xi_2,\xi_3) = \varphi_1^1(\xi_1)\varphi_2^1(\xi_2)\varphi_1^1(\xi_3), &&\Psi_4^1(\xi_1,\xi_2,\xi_3) = \varphi_2^1(\xi_1)\varphi_2^1(\xi_2)\varphi_1^1(\xi_3),\\
&\Psi_5^1(\xi_1,\xi_2,\xi_3) = \varphi_1^1(\xi_1)\varphi_1^1(\xi_2)\varphi_2^1(\xi_3), &&\Psi_6^1(\xi_1,\xi_2,\xi_3) = \varphi_2^1(\xi_1)\varphi_1^1(\xi_2)\varphi_2^1(\xi_3),\\
&\Psi_7^1(\xi_1,\xi_2,\xi_3) = \varphi_1^1(\xi_1)\varphi_2^1(\xi_2)\varphi_2^1(\xi_3), &&\Psi_8^1(\xi_1,\xi_2,\xi_3) = \varphi_2^1(\xi_1)\varphi_2^1(\xi_2)\varphi_2^1(\xi_3),
\end{aligned}$$

$$\begin{aligned}
&\Psi_1^2(\xi_1,\xi_2,\xi_3) = \varphi_1^2(\xi_1)\varphi_1^1(\xi_2)\varphi_1^1(\xi_3), &&\Psi_2^2(\xi_1,\xi_2,\xi_3) = \varphi_2^2(\xi_1)\varphi_1^1(\xi_2)\varphi_1^1(\xi_3),\\
&\Psi_3^2(\xi_1,\xi_2,\xi_3) = \varphi_1^2(\xi_1)\varphi_2^1(\xi_2)\varphi_1^1(\xi_3), &&\Psi_4^2(\xi_1,\xi_2,\xi_3) = \varphi_2^2(\xi_1)\varphi_2^1(\xi_2)\varphi_1^1(\xi_3),\\
&\Psi_5^2(\xi_1,\xi_2,\xi_3) = \varphi_1^2(\xi_1)\varphi_1^1(\xi_2)\varphi_2^1(\xi_3), &&\Psi_6^2(\xi_1,\xi_2,\xi_3) = \varphi_2^2(\xi_1)\varphi_1^1(\xi_2)\varphi_2^1(\xi_3),\\
&\Psi_7^2(\xi_1,\xi_2,\xi_3) = \varphi_1^2(\xi_1)\varphi_2^1(\xi_2)\varphi_2^1(\xi_3), &&\Psi_8^2(\xi_1,\xi_2,\xi_3) = \varphi_2^2(\xi_1)\varphi_2^1(\xi_2)\varphi_2^1(\xi_3),
\end{aligned}$$

$$\begin{aligned}
&\Psi_1^3(\xi_1,\xi_2,\xi_3) = \varphi_1^1(\xi_1)\varphi_1^2(\xi_2)\varphi_1^1(\xi_3), &&\Psi_2^3(\xi_1,\xi_2,\xi_3) = \varphi_2^1(\xi_1)\varphi_1^2(\xi_2)\varphi_1^1(\xi_3),\\
&\Psi_3^3(\xi_1,\xi_2,\xi_3) = \varphi_1^1(\xi_1)\varphi_2^2(\xi_2)\varphi_1^1(\xi_3), &&\Psi_4^3(\xi_1,\xi_2,\xi_3) = \varphi_2^1(\xi_1)\varphi_2^2(\xi_2)\varphi_1^1(\xi_3),\\
&\Psi_5^3(\xi_1,\xi_2,\xi_3) = \varphi_1^1(\xi_1)\varphi_1^2(\xi_2)\varphi_2^1(\xi_3), &&\Psi_6^3(\xi_1,\xi_2,\xi_3) = \varphi_2^1(\xi_1)\varphi_1^2(\xi_2)\varphi_2^1(\xi_3),\\
&\Psi_7^3(\xi_1,\xi_2,\xi_3) = \varphi_1^1(\xi_1)\varphi_2^2(\xi_2)\varphi_2^1(\xi_3), &&\Psi_8^3(\xi_1,\xi_2,\xi_3) = \varphi_2^1(\xi_1)\varphi_2^2(\xi_2)\varphi_2^1(\xi_3),
\end{aligned}$$

$$\begin{aligned}
&\Psi_1^4(\xi_1,\xi_2,\xi_3) = \varphi_1^2(\xi_1)\varphi_1^2(\xi_2)\varphi_1^1(\xi_3), &&\Psi_2^4(\xi_1,\xi_2,\xi_3) = \varphi_2^2(\xi_1)\varphi_1^2(\xi_2)\varphi_1^1(\xi_3),\\
&\Psi_3^4(\xi_1,\xi_2,\xi_3) = \varphi_1^2(\xi_1)\varphi_2^2(\xi_2)\varphi_1^1(\xi_3), &&\Psi_4^4(\xi_1,\xi_2,\xi_3) = \varphi_2^2(\xi_1)\varphi_2^2(\xi_2)\varphi_1^1(\xi_3),\\
&\Psi_5^4(\xi_1,\xi_2,\xi_3) = \varphi_1^2(\xi_1)\varphi_1^2(\xi_2)\varphi_2^1(\xi_3), &&\Psi_6^4(\xi_1,\xi_2,\xi_3) = \varphi_2^2(\xi_1)\varphi_1^2(\xi_2)\varphi_2^1(\xi_3),\\
&\Psi_7^4(\xi_1,\xi_2,\xi_3) = \varphi_1^2(\xi_1)\varphi_2^2(\xi_2)\varphi_2^1(\xi_3), &&\Psi_8^4(\xi_1,\xi_2,\xi_3) = \varphi_2^2(\xi_1)\varphi_2^2(\xi_2)\varphi_2^1(\xi_3),
\end{aligned}$$

$$\begin{aligned}
&\Psi_1^5(\xi_1,\xi_2,\xi_3) = \varphi_1^1(\xi_1)\varphi_1^1(\xi_2)\varphi_1^2(\xi_3), &&\Psi_2^5(\xi_1,\xi_2,\xi_3) = \varphi_2^1(\xi_1)\varphi_1^1(\xi_2)\varphi_1^2(\xi_3),\\
&\Psi_3^5(\xi_1,\xi_2,\xi_3) = \varphi_1^1(\xi_1)\varphi_2^1(\xi_2)\varphi_1^2(\xi_3), &&\Psi_4^5(\xi_1,\xi_2,\xi_3) = \varphi_2^1(\xi_1)\varphi_2^1(\xi_2)\varphi_1^2(\xi_3),\\
&\Psi_5^5(\xi_1,\xi_2,\xi_3) = \varphi_1^1(\xi_1)\varphi_1^1(\xi_2)\varphi_2^2(\xi_3), &&\Psi_6^5(\xi_1,\xi_2,\xi_3) = \varphi_2^1(\xi_1)\varphi_1^1(\xi_2)\varphi_2^2(\xi_3),\\
&\Psi_7^5(\xi_1,\xi_2,\xi_3) = \varphi_1^1(\xi_1)\varphi_2^1(\xi_2)\varphi_2^2(\xi_3), &&\Psi_8^5(\xi_1,\xi_2,\xi_3) = \varphi_2^1(\xi_1)\varphi_2^1(\xi_2)\varphi_2^2(\xi_3),
\end{aligned}$$

$$\begin{aligned}
&\Psi_1^6(\xi_1,\xi_2,\xi_3) = \varphi_1^2(\xi_1)\varphi_1^1(\xi_2)\varphi_1^2(\xi_3), &&\Psi_2^6(\xi_1,\xi_2,\xi_3) = \varphi_2^2(\xi_1)\varphi_1^1(\xi_2)\varphi_1^2(\xi_3),\\
&\Psi_3^6(\xi_1,\xi_2,\xi_3) = \varphi_1^2(\xi_1)\varphi_2^1(\xi_2)\varphi_1^2(\xi_3), &&\Psi_4^6(\xi_1,\xi_2,\xi_3) = \varphi_2^2(\xi_1)\varphi_2^1(\xi_2)\varphi_1^2(\xi_3),\\
&\Psi_5^6(\xi_1,\xi_2,\xi_3) = \varphi_1^2(\xi_1)\varphi_1^1(\xi_2)\varphi_2^2(\xi_3), &&\Psi_6^6(\xi_1,\xi_2,\xi_3) = \varphi_2^2(\xi_1)\varphi_1^1(\xi_2)\varphi_2^2(\xi_3),\\
&\Psi_7^6(\xi_1,\xi_2,\xi_3) = \varphi_1^2(\xi_1)\varphi_2^1(\xi_2)\varphi_2^2(\xi_3), &&\Psi_8^6(\xi_1,\xi_2,\xi_3) = \varphi_2^2(\xi_1)\varphi_2^1(\xi_2)\varphi_2^2(\xi_3),
\end{aligned}$$

$$\begin{aligned}
&\Psi_1^7(\xi_1,\xi_2,\xi_3) = \varphi_1^1(\xi_1)\varphi_1^2(\xi_2)\varphi_1^2(\xi_3), &&\Psi_2^7(\xi_1,\xi_2,\xi_3) = \varphi_2^1(\xi_1)\varphi_1^2(\xi_2)\varphi_1^2(\xi_3),\\
&\Psi_3^7(\xi_1,\xi_2,\xi_3) = \varphi_1^1(\xi_1)\varphi_2^2(\xi_2)\varphi_1^2(\xi_3), &&\Psi_4^7(\xi_1,\xi_2,\xi_3) = \varphi_2^1(\xi_1)\varphi_2^2(\xi_2)\varphi_1^2(\xi_3),\\
&\Psi_5^7(\xi_1,\xi_2,\xi_3) = \varphi_1^1(\xi_1)\varphi_1^2(\xi_2)\varphi_2^2(\xi_3), &&\Psi_6^7(\xi_1,\xi_2,\xi_3) = \varphi_2^1(\xi_1)\varphi_1^2(\xi_2)\varphi_2^2(\xi_3),\\
&\Psi_7^7(\xi_1,\xi_2,\xi_3) = \varphi_1^1(\xi_1)\varphi_2^2(\xi_2)\varphi_2^2(\xi_3), &&\Psi_8^7(\xi_1,\xi_2,\xi_3) = \varphi_2^1(\xi_1)\varphi_2^2(\xi_2)\varphi_2^2(\xi_3),
\end{aligned}$$

$$\Psi_1^8(\xi_1,\xi_2,\xi_3) = \varphi_1^2(\xi_1)\varphi_1^2(\xi_2)\varphi_1^2(\xi_3), \qquad \Psi_2^8(\xi_1,\xi_2,\xi_3) = \varphi_2^2(\xi_1)\varphi_1^2(\xi_2)\varphi_1^2(\xi_3),$$
$$\Psi_3^8(\xi_1,\xi_2,\xi_3) = \varphi_1^2(\xi_1)\varphi_2^2(\xi_2)\varphi_1^2(\xi_3), \qquad \Psi_4^8(\xi_1,\xi_2,\xi_3) = \varphi_2^2(\xi_1)\varphi_2^2(\xi_2)\varphi_1^2(\xi_3),$$
$$\Psi_5^8(\xi_1,\xi_2,\xi_3) = \varphi_1^2(\xi_1)\varphi_1^2(\xi_2)\varphi_2^2(\xi_3), \qquad \Psi_6^8(\xi_1,\xi_2,\xi_3) = \varphi_2^2(\xi_1)\varphi_1^2(\xi_2)\varphi_2^2(\xi_3),$$
$$\Psi_7^8(\xi_1,\xi_2,\xi_3) = \varphi_1^2(\xi_1)\varphi_2^2(\xi_2)\varphi_2^2(\xi_3), \qquad \Psi_8^8(\xi_1,\xi_2,\xi_3) = \varphi_2^2(\xi_1)\varphi_2^2(\xi_2)\varphi_2^2(\xi_3).$$

The functions $\varphi_i^j(\xi_k)$ $(i = 1, 2;\ j = 1, 2;\ k = 1, 2, 3)$ are defined above by Eqs. (8.13).

References

AKAR, F.G., YAN, G.X., ANTZELEVITCH, C., ROSENBAUM, D.S. (2002). Unique topographical distribution of M cells underlies reentrant mechanism of torsade de pointes in the long-QT syndrome. *Circulation* **105** (10), 1247–1253.

ALBERTS, B., BRAY, D., LEWIS, J., RAFF, M., ROBERTS, K., WATSON, J. (1994). *Molecular Biology of the Cell*, third ed. (Garland Publishing, Inc., New York).

ANSARI, A., HO, S.Y., ANDERSON, R.H. (1999). Distribution of the Purkinje fibres in the sheep heart. *Anat. Rec.* **254** (1), 92–97.

ANTZELEVITCH, C., SHIMIZU, W., YAN, G.X., SICOURI, S., WEISSENBURGER, J., NESTERENKO, V.V., BURASHNIKOV, A., DI DIEGO, J., SAFFITZ, J., THOMAS, G.P. (1999). The M cell: its contribution to the ECG and to normal and abnormal electrical function of the heart. *J. Cardiovasc. Electrophysiol.* **10** (8), 1124–1152.

ANYUKHOVSKY, E.P., SOSUNOV, E.A., ROSEN, M.R. (1996). Regional differences in electrophysiological properties of epicardium, midmyocardium, and endocardium. In vitro and in vivo correlations. *Circulation* **94** (8), 1981–1988.

ARTS, T., RENEMAN, R.S., VEENSTRA, P.C. (1979). A model of the mechanics of the left ventricle. *Ann. Biomed. Eng.* **7**, 299–318.

ARTS, T., VEENSTRA, P.C., RENEMAN, R.S. (1982). Epicardial deformation and left ventricular wall mechanics during ejection in the dog. *Am. J. Physiol. Heart Circ. Physiol.* **243**, H379–H390.

BACKX, P.H., GAO, W.D., AZAN-BACKX, M.D., MARBAN, E. (1995). The relationship between contractile force and intracellular [Ca2+] in intact rat cardiac trabeculae. *J. Gen. Physiol.* **105** (1), 1–19.

BEELER, G.W., REUTER, H. (1977). Reconstruction of the action potential of ventricular myocardial fibres. *J. Physiol.* **268** (1), 177–210.

BERNE, R.M., LEVY, M.N. (1997). *Cardiovascular Physiology*, seventh ed. (Mosby–Year Book, Inc., St Louis).

BERNUS, O., WILDERS, R., ZEMLIN, C.W., VERSCHELDE, H., PANFILOV, A.V. (2002). A computationally efficient electrophysiological model of human ventricular cells. *Am. J. Physiol. Heart Circ. Physiol.* **282** (6), H2296–H2308.

BERS, D.M. (1991). *Excitation–contraction Coupling and Cardiac Contractile Force* (Kluwer, Dordrecht).

CAREY, G.F., ODEN, J.T. (1983). *Finite Elements, A Second Course, vol. II* (Prentice Hall, Englewood Cliffs, NJ).

CATES, A.W., SMITH, W.M., IDEKER, R.E., POLLARD, A.E. (2001). Purkinje and ventricular contributions to endocardial activation sequence in perfused rabbit right ventricle. *Am. J. Physiol. Heart Circ. Physiol.* **281** (2), H490–H505.

CHADWICK, R.S. (1982). Mechanics of the left ventricle. *Biophys. J.* **39** (3), 279–288.

CHENG, J., KAMIYA, K., LIU, W., TSUJI, Y., TOYAMA, J., KODAMA, I. (1999). Heterogeneous distribution of the two components of delayed rectifier K^+ current: a potential mechanism of the proarrhythmic effects of methanesulfonanilideclass III agents. *Cardiovasc. Res.* **43** (1), 135–147.

CIARLET, P.G., LIONS, J.L. (eds.) (1990). *Finite Element Methods (Part 1)*. In: Handbook of Numerical Analysis **II** (North-Holland, Amsterdam).

COSTA, K.D., HUNTER, P.J., ROGERS, J.M., GUCCIONE, J.M., WALDMAN, L.K., MCCULLOCH, A.D. (1996). A three-dimensional finite element method for large elastic deformations of ventricular myocardium: Part I – Cylindrical and Spherical Polar Coordinates. *ASME J. Biomech. Eng.* **118**, 452–463.

COSTA, K.D., HUNTER, P.J., WAYNE, J.S., WALDMAN, L.K., GUCCIONE, J.M., MCCULLOCH, A.D. (1996). A three-dimensional finite element method for large elastic deformations of ventricular myocardium: Part II – Prolate spheroidal coordinates. *ASME J. Biomech. Eng.* **118**, 464–472.

DE TOMBE, P.P., TER KEURS, H.E. (1992). An internal viscous element limits unloaded velocity of sarcomere shortening in rat myocardium. *J. Physiol.* **454**, 619–642.

DUCKETT, G., BARKLEY, D. (2000). Modeling the dynamics of cardiac action potentials. *Phys. Rev. Lett.* **85** (4), 884–887.

EFIMOV, I.R., ERMENTROUT, B., HUANG, D.T., SALAMA, G. (1996). Activation and repolarization patterns are governed by different structural characteristics of ventricular myocardium: experimental study with voltage-sensitive dyes and numerical simulations. *J. Cardiovasc. Electrophysiol.* **7** (6), 512–530.

ENDRESEN, L.P., SKARLAND, N. (2000). Limit cycle oscillations in pacemaker cells. *IEEE Trans. Biomed. Eng.* **47** (8), 1134–1137.

FENTON, F., KARMA, A. (1998). Vortex dynamics in three-dimensional continuous myocardium with fiber rotation – filament instability and fibrillation. *Chaos* **8** (1), 20–47.

FITZHUGH, R. (1961). Impulses and physiological state in theoretical models of nerve membrane. *Biophys. J.* **1**, 445.

FRANZONE, P.C., GUERRI, L., PENNACCHIO, M., TACCARDI, B. (1998). Spread of excitation in 3-D models of the anisotropic cardiac tissue. II. Effects of fiber architecture and ventricular geometry. *Math. Biosci.* **147** (2), 131–171.

FRANZONE, P.C., GUERRI, L., PENNACCHIO, M., TACCARDI, B. (2000). Anisotropic mechanisms for multiphasic unipolar electrograms: simulation studies and experimental recordings. *Ann. Biomed. Eng.* **28** (11), 1326–1342.

FUNG, Y.C. (1981). *Biomechanics: Mechanical Properties of Living Tissues* (Springer-Verlag, New York).

GESELOWITZ, D.B., MILLER 3RD, W.T. (1983). A bidomain model for anisotropic cardiac muscle. *Ann. Biomed. Eng.* **11** (3-4), 191–206.

GIMA, K., RUDY, Y. (2002). Ionic current basis of electrocardiographic waveforms: a model study. *Circ. Res.* **90** (8), 889–896.

GUCCIONE, J.M., MCCULLOCH, A.D. (1993). Mechanics of active contraction in cardiac muscle: Part I – Constitutive relations for fiber stress that describe deactivation. *J. Biomech. Eng.* **115** (1), 72–81.

HILL, A.V. (1938). Time heart of shortening and the dynamic constants of muscle. *Proc. R. Soc.* **126**, 136–195.

HILL, A.V. (1970). *First and Last Experiments in Muscle Mechanics* (University Press, Cambridge).

HODGKIN, A.L., HUXLEY, A.F. (1952). A quantitative description of membrane current and its application to conduction and excitation in nerve. *J. Physiol.* **117**, 500–544.

HOOKS, D.A., TOMLINSON, K.A., MARSDEN, S.G., LEGRICE, I.J., SMAILL, B.H., PULLAN, A.J., HUNTER, P.J. (2002). Cardiac microstructure: Implications for electrical propagation and defibrillation in the heart. *Circ. Res.* **91** (4), 331–338.

HUNTER, P., ROBBINS, P., NOBLE, D. (2002). The IUPS human physiome project. *Pflugers Arch.* **445** (1), 1–9.

HUNTER, P.J., MCCULLOCH, A.D., NIELSEN, P.M.F., SMAILL, B.H. (1988). A finite element model of passive ventricular mechanics. In: Spilker, R.L., Simon, B.R. (eds.), Computational Methods in Bioengineering **9** (ASME, Chicago), pp. 387–397.

HUNTER, P.J., MCCULLOCH, A.D., TER KEURS, H.E. (1998). Modelling the mechanical properties of cardiac muscle. *Prog. Biophys. Mol. Biol.* **69** (2–3), 289–331.

HUNTER, W.C. (1999). Abstract: Making the heart beat: Dynamic models of myocardial sarcomere force generation. In: *1999 Physiome Symp. in Seattle: Integrated Biology of the Heart*, September 9–11.

HUXLEY, A.F. (1957). Muscle structure and theories of contraction. *Prog. Biophys. Chem.* **7**, 255–318.

JAFRI, M.S., DUDYCHA, S.J., O'ROURKE, B. (2001). Cardiac energy metabolism: models of cellular respiration. *Annu. Rev. Biomed. Eng.* **3**, 57–81.

KANAI, A., SALAMA, G. (1995). Optical mapping reveals that repolarization spreads anisotropically and is guided by fiber orientation in guinea pig hearts. *Circ. Res.* **77** (4), 784–802.

KATZ, A.M. (2001). *Physiology of the Heart*, third ed. (Lippincott Williams & Wilkins, Philadelphia).

KEENER, J.P., PANFILOV, A.V. (1997). The effects of geometry and fibre orientation on propagation and extracellular potentials in myocardium. In: Panfilov, A.V., Holden, A.V. (eds.), *Computational Biology of the Heart* (Wiley, New York), pp. 235–258.

KENTISH, J.C., TER KEURS, H.E., RICCIARDI, L., BUCX, J.J., NOBLE, M.I. (1986). Comparison between the sarcomere length-force relations of intact and skinned trabeculae from rat right ventricle. Influence of calcium concentrations on these relations. *Circ. Res.* **58** (6), 755–768.

KOGAN, B.Y., KARPLUS, W.J., BILLETT, B.S., PANG, A.T., KARAGUEUZIAN, H.S., KHAN, S.S. (1991). The simplified Fitzhugh–Nagumo model with action potential duration restitution – Effects on 2D-wave propagation. *Physica D* **50** (3), 327–340.

KOHL, P., HUNTER, P., NOBLE, D. (1999). Stretch-induced changes in heart rate and rhythm: clinical observations, experiments and mathematical models. *Prog. Biophys. Mol. Biol.* **71** (1), 91–138.

KRUGER, G.W., POLLACK, J.H. (1975). Myocardial sarcomere dynamics during isometric contraction. *J. Physiol.* **51**, 627–643.

LANDESBERG, A., MARKHASIN, V.S., BEYAR, R., SIDEMAN, S. (1996). Effect of cellular inhomogeneity on cardiac tissue mechanics based on intracellular control mechanisms. *Am. J. Physiol. Heart Circ. Physiol.* **270** (3 Pt 2), H1101–H1114.

LANDESBERG, A., SIDEMAN, S. (1994). Coupling calcium binding to troponin C and cross-bridge cycling in skinned cardiac cells. *Am. J. Physiol. Heart Circ. Physiol.* **266** (3 Pt 2), H1260–H1271.

LANGTANGEN, H.P. (1999). *Computational Partial Differential Equations Numerical Methods and Diffpack Programming* (Springer-Verlag, Berlin).

LAURITA, K.R., GIROUARD, S.D., ROSENBAUM, D.S. (1996). Modulation of ventricular repolarization by a premature stimulus Role of epicardial dispersion of repolarization kinetics demonstrated by optical mapping of the intact guinea pig heart. *Circ. Res.* **79** (3), 493–503.

LEGRICE, I.J., HUNTER, P.J., SMAILL, B.H. (1997). Laminar structure of the heart: A mathematical model. *Am. J. Physiol. Heart Circ. Physiol.* **272** (5 Pt 2), H2466–H2476.

LEGRICE, I.J., SMAILL, B.H., CHAI, L.Z., EDGAR, S.G., GAVIN, J.B., HUNTER, P.J. (1995). Laminar structure of the heart: ventricular myocyte arrangement and connective tissue architecture in the dog. *Am. J. Physiol. Heart Circ. Physiol.* **269** (2 Pt 2), H571–H582.

LEGRICE, I.J., TAKAYAMA, Y., COVELL, J.W. (1995). Transverse shear along myocardial cleavage planes provides a mechanism for normal systolic wall thickening. *Circ. Res.* **77**, 182–193.

LIONS, J.L., MAGENES, E. (1972). *Non-Homogeneous Boundary Value Problems and Applications, vol. 2* (Springer-Verlag, Berlin).

LUO, C.H., RUDY, Y. (1991). A model of the ventricular cardiac action potential depolarization, repolarization, and their interaction. *Circ. Res.* **68** (6), 1501–1526.

LUO, C.H., RUDY, Y. (1994). A dynamic model of the cardiac ventricular action potential. I. Simulations of ionic currents and concentration changes. *Circ. Res.* **74** (6), 1071–1096.

MARON, B.J., BONOW, R.O., CANNON, R.O. 3RD, LEON, M.B., EPSTEIN, S.E. (1987). Hypertrophic cardiomyopathy. Interrelations of clinical manifestations, pathophysiology, and therapy (1). *N. Engl. J. Med.* **316** (13), 780–789.

MCLEAN, M., ROSS, M.A., PROTHERO, J. (1989). Three-dimensional reconstruction of the myofiber pattern in the fetal and neonatal mouse heart. *Anat. Rec.* **224** (3), 392–406.

MICHAILOVA, A.P., SPASSOV, V.Z. (1992). Theoretical model and computer simulation of excitation–contraction coupling of mammalian cardiac muscle. *J. Mol. Cell. Cardiol.* **24** (1), 97–104.

MICHAILOVA, A.P., SPASSOV, V.Z. (1997). Computer simulation of excitation–contraction coupling in cardiac muscle A study of the regulatory role of calcium binding to troponin C. *Gen. Physiol. Biophys.* **16** (1), 29–38.

MUZIKANT, A.L., HENRIQUEZ, C.S. (1997). Paced activation mapping reveals organization of myocardial fibers: a simulation study. *J. Cardiovasc. Electrophysiol.* **8** (3), 281–294.

MUZIKANT, A.L., HENRIQUEZ, C.S. (1998). Validation of three-dimensional conduction models using experimental mapping: are we getting closer? *Prog. Biophys. Mol. Biol.* **69** (2–3), 205–223.

NAGUMO, J.S., ARIMOTO, S., YOSHIZAWA, S. (1962). *Proc. IRE* **50**, 2061.

NEVO, E., LANIR, Y. (1989). Structural finite deformation model of the left ventricle during diastole and systole. *J. Biomech. Eng.* **111** (4), 342–349.

NIELSEN, P.M., LE GRICE, I.J., SMAILL, B.H., HUNTER, P.J. (1991). Mathematical model of geometry and fibrous structure of the heart. *Am. J. Physiol. Heart Circ. Physiol.* **260** (4 Pt 2), H1365–H1378.

NOBLE, D. (2001). From genes to whole organs: connecting biochemistry to physiology. *Novartis Found Symp.* **239**, 111–123; discussion 123–128, 150–159.

NOBLE, D. (2002). Modeling the heart – from genes to cells to the whole organ. *Science* **295** (5560), 1678–1682.

ODEN, J.T. (1972). *Finite Elements of Nonlinear Continua* (McGraw–Hill, New York).

OTANI, N. (2000). Computer modeling in cardiac electrophysiology. *J. Comp. Phys.* **161**, 21–34.

PANERAI, R.B. (1980). A model of cardiac muscle mechanics and energetics. *J. Biomech.* **13** (11), 929–940.

PRIEBE, L., BEUCKELMANN, D.J. (1998). Simulation study of cellular electric properties in heart failure. *Circ. Res.* **82** (11), 1206–1223.

PUGLISI, J.L., BERS, D.M. (2001). LabHEART: an interactive computer model of rabbit ventricular myocyte ion channels and Ca transport. *Am. J. Physiol. Cell. Physiol.* **281** (6), C2049–C2060.

QU, Z., GARFINKEL, A. (1999). An advanced algorithm for solving partial differential equation in cardiac conduction. *IEEE Trans. Biomed. Eng.* **46** (9), 1166–1168.

QUAN, W., EVANS, S.J., HASTINGS, H.M. (1998). Efficient integration of a realistic two-dimensional cardiac tissue model by domain decomposition. *IEEE Trans. Biomed. Eng.* **45** (3), 372–385.

RICE, J.J., WINSLOW, R.L., HUNTER, W.C. (1999). Comparison of putative cooperative mechanisms in cardiac muscle: length dependence and dynamic responses. *Am. J. Physiol. Heart Circ. Physiol.* **276** (5 Pt 2), H1734–H1754.

ROGERS, J.M., MCCULLOCH, A.D. (1994). Nonuniform muscle fiber orientation causes spiral wave drift in a finite element model of cardiac action potential propagation. *J. Cardiovasc. Electrophysiol.* **5** (6), 496–509.

ROSENBAUM, D.S., KAPLAN, D.T., KANAI, A., JACKSON, L., GARAN, H., COHEN, R.J., SALAMA, G. (1991). Repolarization inhomogeneities in ventricular myocardium change dynamically with abrupt cycle length shortening. *Circulation* **84** (3), 1333–1345.

RUDY, Y. (2000). From genome to physiome: integrative models of cardiac excitation. *Ann. Biomed. Eng.* **28** (8), 945–950.

RUEGG, J.C. (1988). *Calcium in Muscle Activation: A Comparative Approach* (Springer-Verlag, Berlin).

SAFFITZ, J.E., KANTER, H.L., GREEN, K.G., TOLLEY, T.K., BEYER, E.C. (1994). Tissue-specific determinants of anisotropic conduction velocity in canine atrial and ventricular myocardium. *Circ. Res.* **74** (6), 1065–1070.

SAXBERG, B.E., GRUMBACH, M.P., COHEN, R.J. (1985). A time dependent anatomically detailed model of cardiac conduction. *Comput. Cardiol.* **12**, 401–404.

SICOURI, S., ANTZELEVITCH, C. (1995). Electrophysiologic characteristics of M cells in the canine left ventricular free wall. *J. Cardiovasc. Electrophysiol.* **6** (8), 591–603.

SPOTNITZ, H.M., SPOTNITZ, W.D., COTTRELL, T.S., SPIRO, D., SONNENBLICK, E.H. (1974). Cellular basis for volume related wall thickness changes in the rat left ventricle. *J. Mol. Cell Cardiol.* **6** (4), 317–331.

STREETER JR., D.D. (1979). Gross morphology and fiber geometry of the heart. In: Bethesda, M.D. (ed.), *Handbook of Physiology* (American Physiological Society), p. 61.

STREETER JR., D.D., HANNA, W.T. (1973). Engineering mechanics for successive states in canine left ventricular myocardium. I. Cavity and wall geometry. *Circ. Res.* **33** (6), 639–655.

STREETER JR., D.D., SPOTNITZ, H.M., PATEL, D.P., ROSS JR., J., SONNENBLICK, E.H. (1969). Fiber orientation in the canine left ventricle during diastole and systole. *Circ. Res.* **24** (3), 339–347.

SUNDNES, J., LINES, G.T., TVEITO, A. (2001). Efficient solution of ordinary differential equations modeling electrical activity in cardiac cells. *Math. Biosci.* **172** (2), 55–72.

TABER, L.A. (1991). On a nonlinear theory for muscle shells: Part II – Application to the beating left ventricle. *J. Biomech. Eng.* **113** (1), 63–71.

TAYLOR, T.W., SUGA, H. (1993). Variable crossbridge cycling-ATP coupling accounts for cardiac mechanoenergetics. *Adv. Exp. Med. Biol.* **332**, 775–782; discussion 782–783.

TER KEURS, H.E., RIJNSBURGER, W.H., VAN HEUNINGEN, R., NAGELSMIT, M.J. (1980). Tension development and sarcomere length in rat cardiac trabeculae. Evidence of length-dependent activation. *Circ. Res.* **46** (5), 703–714.

TORRENT-GUASP, F. (1973). *The Cardiac Muscle* (Juan March Foundation, Madrid).

TOZEREN, A. (1985). Continuum rheology of muscle contraction and its application to cardiac contractility. *Biophys. J.* **47** (3), 303–309.

TRANUM-JENSEN, J., WILDE, A.A., VERMEULEN, J.T., JANSE, M.J. (1991). Morphology of electrophysiologically identified junctions between Purkinje fibers and ventricular muscle in rabbit and pig hearts. *Circ. Res.* **69** (2), 429–437.

USYK, T.P., LEGRICE, I., MCCULLOCH, A.D. (2002). Computational model of three-dimensional cardiac electromechanics. *Comput. Visual. Sci.* **4** (4), 249–257.

USYK, T.P., MAZHARI, R., MCCULLOCH, A.D. (2000). Effect of laminar orthotropic myofiber architecture on regional stress and strain in the canine left ventricle. *J. Elasticity* **61** (1–3), 143–164.

VETTER, F.J., MCCULLOCH, A.D. (1998). Three-dimensional analysis of regional cardiac function: A model of the rabbit ventricular anatomy. *Prog. Biophys. Mol. Biol.* **69** (2–3), 157–183.

VETTER, F.J., MCCULLOCH, A.D. (2001). Mechanoelectric feedback in a model of the passively inflated left ventricle. *Ann. Biomed. Eng.* **29** (5), 414–426.

VISWANATHAN, P.C., SHAW, R.M., RUDY, Y. (1999). Effects of I_{Kr} and I_{Ks} heterogeneity on action potential duration and its rate dependence: a simulation study. *Circulation* **99** (18), 2466–2474.

WALDMAN, L.K., FUNG, Y.C., COVELL, J.W. (1985). Transmural myocardial deformation in the canine left ventricle. Normal in vivo three-dimensional finite strains. *Circ. Res.* **57** (1), 152–163.

WEIDMANN, S. (1970). Electrical constants of trabecular muscle from mammalian heart. *J. Physiol.* **210** (4), 1041–1054.

WINSLOW, R.L., RICE, J., JAFRI, S., MARBAN, E., O'ROURKE, B. (1999). Mechanisms of altered excitation–contraction coupling in canine tachycardia-induced heart failure, II: Model studies. *Circ. Res.* **84** (5), 571–586.

WINSLOW, R.L., SCOLLAN, D.F., HOLMES, A., YUNG, C.K., ZHANG, J., JAFRI, M.S. (2000). Electrophysiological modeling of cardiac ventricular function: from cell to organ. *Annu. Rev. Biomed. Eng.* **2**, 119–155.

WOLK, R., COBBE, S.M., HICKS, M.N., KANE, K.A. (1999). Functional, structural, and dynamic basis of electrical heterogeneity in healthy and diseased cardiac muscle: Implications for arrhythmogenesis and anti-arrhythmic drug therapy. *Pharmacol. Ther.* **84** (2), 207–231.

WOLK, R., KANE, K.A., COBBE, S.M., HICKS, M.N. (1998). Regional electrophysiological effects of hypokalaemia, hypomagnesaemia and hyponatraemia in isolated rabbit hearts in normal and ischaemic conditions. *Cardiovasc. Res.* **40** (3), 492–501.

WONG, A.Y. (1971). Mechanics of cardiac muscle, based on huxley's model: Mathematical stimulation of isometric contraction. *J. Biomech.* **4** (6), 529–540.

WONG, A.Y. (1972). Mechanics of cardiac muscle, based on Huxley's model: Simulation of active state and force–velocity relation. *J. Biomech.* **5** (1), 107–117.

YAN, G.-X., SHIMIZU, W., ANTZELEVITCH, C. (1998). Characteristics and distribution of M cells in arterially perfused canine left ventricular wedge preparations. *Circulation* **98** (18), 1921–1927.

YOUNG, A.A., AXEL, L. (1992). Three-dimensional motion and deformation of the heart wall: Estimation with spatial modulation of magnetization – a model-based approach. *Radiology* **185** (1), 241–247.

ZYGMUNT, A.C., EDDLESTONE, G.T., THOMAS, G.P., NESTERENKO, V.V., ANTZELEVITCH, C. (2001). Larger late sodium conductance in M cells contributes to electrical heterogeneity in canine ventricle. *Am. J. Physiol. Heart Circ. Physiol.* **281** (2), H689–H697.

Mathematical Analysis, Controllability and Numerical Simulation of a Simple Model of Avascular Tumor Growth

Jesús Ildefonso Díaz, José Ignacio Tello

Departamento de Matemática Aplicada, Universidad Complutense de Madrid, Avda Complutense, 28040 Madrid, Spain

Preface

Cancer is one of the most prevalent causes of natural death in the western world, and a high percentage of people develop some kind of this disease during their lives. For this reason medicine is one of the scientific fields which found significant interest not only within the scientific community, but also among the general population. The scientific community comprises medicine, but also other areas of research such us Biology, Chemistry, Mathematics, Pharmacy or Physics. This is evident from the huge number of research works and publications in the field and the great quantity of human and economical resources which have been devoted to cancer research in the last decades.

The development and growth of a tumor is a complicated phenomenon which involves many different aspects from the subcellular scale (gene mutation or secretion of substances) to the body scale (*metastasis*). This complexity is reflected by the different mathematical models given for each phase of the growth. The first phase is known as the *avascular* phase, previous to *vascularization*, and the second one, when *angiogenesis* occurs, is known as *vascular* phase.

The aim of this work is to present the study of the mathematical analysis, the controllability and a numerical simulation for a simple, avascular model of growth of a tumor. In Section 1, we describe the biological phenomenology of several processes which influence the growth and development of tumors. The mathematical modelling is

Computational Models for the Human Body
Special Volume (N. Ayache, Guest Editor) of
HANDBOOK OF NUMERICAL ANALYSIS, VOL. XII
P.G. Ciarlet (Editor)

ISSN 1570-8659
DOI 10.1016/S1570-8659(03)12003-0

presented by describing different models of partial differential equations (PDE). We focus our attention on a class of models proposed by GREENSPAN [1972] and BYRNE and CHAPLAIN [1995], BYRNE [1999a], BYRNE [1999b], BYRNE and CHAPLAIN [1996a], CHAPLAIN [1996], CHAPLAIN [1999], ORME and CHAPLAIN [1995], THOMPSON and BYRNE [1999], WARD and KING [1998], studied in FRIEDMAN and REITICH [1999], CUI and FRIEDMAN [1999], CUI and FRIEDMAN [2000], CUI and FRIEDMAN [2001], DÍAZ and TELLO [2004], DÍAZ and TELLO [2003] and by other authors. We prove the solvability of the model equations and establish uniqueness of solutions under additional conditions. In Section 6, we study the controllability of the growth of the tumor by a localized internal action of the inhibitor on a nonnecrotic tumor. It is obvious that this type of results has merely a mathematical interest and it does not suggest any special therapeutical strategy to inhibit tumor growth. Nevertheless our results show that there is not any *obstruction* to the controllability (as it appears, for instance, in some similar PDE's models: see DÍAZ and RAMOS [1995]). In a final section, we address the numerical simulation of the problem.

1. Phenomenology

A tumor originates from mutations of DNA inside cells. In order to create malignant cells, a sufficiently large number of such mutations has to occur. Factors for mutations can be external radiation, hereditary causes etc. Eventually, such gene mutations induce an uncontrolled reproduction, the onset of the formation of a malignant tumor. This process continues as long as the malignant cells find sufficient supply, and will generate a small spheroid of a few millimeters. During this time, called the *avascular* phase, nutrients (glucose and oxygen) arrive at the cells through diffusion. As the spheroid grows, the level of nutrients in the interior of the tumor decreases due to consumption by the outer cells. When the level of concentration of nutrients in the interior falls bellow a critical level, the cells cannot survive, a phenomenon called *necrosis*, and an inner region is formed in the center of the tumor by the dead cells, which decompose into simpler chemical compounds (mainly water). At this time, one can distinguish several regions in the tumor: a necrotic region in the center, an outer region, where *mitosis* (division of cells) occurs, and a region in between where the level of nutrients suffices for the cells to live, but not to proliferate. Until this moment, the tumor is a *multicell spheroid* whose radius is no more than a few millimeters.

The cells of the tumor secrete some chemical substances, known as *Tumor Angiogenesis Factors* (TAFs). These substances diffuse through the surrounding tissue. TAFs stimulate *endothelial* cells (ECs), located in neighboring blood vessels. Endothelial cells are thin cells which form the basement membrane of the blood vessels. When ECs are stimulated by TAFs, they destroy the membrane basement (by secretion of *proteases* and *collagenases*) and migrate towards the tumor forming capillary sprouts. These grow thanks to the proliferation of ECs and other substances located in the extracellular matrix (as fibronectin), forming a capillary network. Initially, the ECs move forming parallel vessels and as sprouts are closer to the tumor, the sprouts branch out and connect.

This process of formation of new vessels, known as *angiogenesis*, is one of the most decisive steps in the growth of a tumor. Angiogenesis is present in other contexts of life, as well, like in wound healing or in the formation of embryos.

The connection of the blood vessels to the tumor supplies nutrients to the malignant cells, aiding a faster proliferation of the tumor's cells. This phase of the tumor is known as *vasculature phase* and is characterized by an aggressive growth.

Finally, the cells of the tumor invade the surrounding tissue and metastasizing to other parts of the body. The circulatory and lymphatic systems are used by the malignant cells for transport to another sites. The process in which cells leave the tumor and enter into the vessels is known as *intravasation*. Cancer cells, which survive in the blood flow and escape from the circulatory system, arrive at a new site, where a new colony of cells may grow. Fortunately, less than 0.05 per cent of cells which were introduced in the circulation are able to create new colonies. Each tumor has a preference to metastasize to a specific organ.

During the growth of a tumor, the *immune system* competes with the malignant cells; it will be activated through the recognition of the cancer cells by the immune cells. *Macrophages* (Ms) are a type of white blood cells, which migrate into the tumor to the regions with low oxygen (*hypoxic* regions) in the interior of the tumor through the external layer of well nourished cells of the tumor. Ms move to the tumor (by chemotaxis) attracted by *macrophage chemoattractants*, which is secreted by the tumor. A *cytotoxic* substance is secreted into the tumor's cell which kills it. Ms may also help the growth of the tumor secreting other chemical substances which help angiogenesis.

It is the main strategy of all cancer therapies (apart from surgery) to inhibit the growth of tumors with tools adapted to the phase the tumor is in. E.g., chemotherapy or radiation therapy are intended to destroy cells of tumor, other treatments try to stimulate cells of the immune system. The first type of therapy is nonselective, destroying both, malignant cells and cells of the immune system. Another therapy based on genetic engineering is being studied. The idea is to insert a therapeutic gene into the cells of a patient and re-inject them back into the patient.

2. Mathematical modelling

Mathematical modelling of the growth of a tumor have been studied by several authors during the last thirty years in many different works.

Among the many different PDE models we can introduce (following FRIEDMAN [2002]) a rough classification into two classes: the mixed models, in which all the different population of cells are continuously present everywhere in the tumor, at all the times, and segregated models, perhaps less realistic but relevant for in vitro experiments, in which the different populations of cells are separated by unknown interfaces or free boundaries. Our analysis will be restricted to the second class of models (some references on mixed models can be found in BELLOMO and PREZIOSI [2000], DE ANGELIS and PREZIOSI [2000], CHAPLAIN and PREZIOSI [2002] and FRIEDMAN [2002]). Moreover, we shall consider spherical tumors (for other free boundary type tumors, without symmetrical shape, arising in tumoral masses growing around a blood vessel see, e.g., BERTUZZI, FASANO, GANDOLFI and MARANGI [2002] or BAZALIY and FRIEDMAN [2003]).

In this section, we describe different mathematical models for each phase. A first and simple model describing the avascular phase was presented in GREENSPAN [1972], assuming spherical symmetry in $\mathbb{R}^3$. The outer boundary delimiting the tumor is denoted by $R(t)$ and the concentration of nutrients and inhibitors by σ and β, respectively. According to principle of conservation of mass, the tumor mass is proportional to its volume $\frac{4}{3}\pi R^3(t)$, assuming the density of the cell mass is constant. The balance between the birth and death rate of cells is given as a function of the concentration of nutrients and inhibitors. Let $\widehat{S}$ be this balance, then after normalizing, we obtain the law

$$\frac{\mathrm{d}}{\mathrm{d}t}\left(\frac{4}{3}\pi R^3(t)\right) = \int_{\{|\tilde{x}|<R(t)\}} \widehat{S}\big(\sigma(\tilde{x},t), \beta(\tilde{x},t)\big)\,\mathrm{d}\tilde{x}.$$

Depending on the author, the function $\widehat{S}$ can be written in different ways. GREENSPAN [1972] studied the problem in the presence of an inhibitor, and the possibility that this affects mitosis, when the concentration of the inhibitor is greater than a critical level $\tilde{\beta}$. He proposed $\widehat{S}(\sigma,\beta) = sH(\sigma-\tilde{\sigma})H(\tilde{\beta}-\beta)$, where $H(\cdot)$ denotes the maximal monotone graph of $\mathbb{R}^2$ associate with the Heaviside function, i.e., $H(k)=0$ if $k<0$, $H(k)=1$ if $k>0$ and $H(0)=[0,1]$. BYRNE and CHAPLAIN [1996a] study the growth when the inhibitor affects the cell proliferation and propose $\widehat{S}(\sigma,\beta) = s(\sigma-\tilde{\sigma})(\tilde{\beta}-\beta)$ (for a positive constant s). In the absence of inhibitors or in case that the inhibitor does not affect mitosis, they choose $\widehat{S}(\sigma,\beta) = s\sigma(\sigma-\tilde{\sigma})$. FRIEDMAN and REITICH [1999] and CUI and FRIEDMAN [2000] study the asymptotic behavior of the radius, $R(t)$, with the cell proliferation rate free of the action of inhibitors. They assume that $\widehat{S} = s(\sigma-\tilde{\sigma})$, where $s\sigma$ is the cell birth-rate and the death-rate is given by $s\tilde{\sigma}$ (see also the survey SLEEMAN [1996]).

We assume that the tumor is composed of an homogeneous tissue and that the distribution of the concentration of nutrients σ is governed by a PDE in the spheroid. Assuming that there is no inhibitor, that the tumor has not necrotic core and that diffusion is high, we obtain the equation

$$d_1\Delta\sigma = \lambda\sigma, \quad |x|<R,$$

where $\lambda\sigma$ represents the nutrient consumption by cells and d_1 is the diffusion coefficient.

In necrotic tumors, an inner free boundary appears, which is denoted by $\rho(t)$. It separates the necrotic core (where σ falls below σ_n) from the remaining part. A model for necrotic tumors was presented in BYRNE [1997a], who proposes the equation

$$0 = \Delta\sigma - \lambda H\big(|x|-\rho(t)\big), \quad |x|<R(t),$$

where the effect of time-delay appears in the radial growth. In addition, asymptotic techniques are used to show the effect of the delay terms.

Several authors (ADAM [1986] and BRITTON and CHAPLAIN [1993]) studied a model proposed by SHYMKO and GLASS [1976] where cell proliferation is controlled by chemical substances *Growth inhibitor factor* (GIFs) as chalones. GIFs secreted by cells reduce the mitotic activity. Two different kinds of inhibitors appear, depending on the phase of the cell cycle stage at which inhibition occurs. The inhibitor can act before DNA synthesis (as epidermal chalon in Melanoma or granulocyte chalon in Leukemia) or before mitosis (see ATTALLAH [1976]). The concentration of GIF (denoted by C) is modeled by one PDE in a bounded domain Ω of $\mathbb{R}^3$,

$$\frac{\partial C}{\partial t} = d\Delta C + f(C) + S(x), \quad x \in \Omega,\ t > 0, \tag{2.1}$$

$$D\frac{\partial C}{\partial n} + PC = 0, \quad x \in \partial\Omega,\ t > 0,\ P \geqslant 0, \tag{2.2}$$

$$C(x,t) = C_0(x), \quad x \in \Omega, \tag{2.3}$$

where $S(x)$ is a source term and $f(C)$ represents the decay of GIF (see ADAM and BELLOMO [1997]).

In 1972, GREENSPAN [1972] proposed a radially symmetric model employing the Heaviside function H for modelling the necrotic part. The avascular model considers a chemical inhibitor β, which is produced in the necrotic core. The distribution of nutrients $\hat{\sigma}$ is given by the equation

$$\frac{\partial \sigma}{\partial t} - d_1\Delta\sigma = -\lambda(\sigma_B - \sigma)H\big(|x| - \rho\big)H\big(R - |x|\big), \tag{2.4}$$

where R is the outer boundary of the tumor and ρ is the radius of the necrotic core.

The chemical substance "β" (produced within the tumor) inhibits the mitosis of cancer cells without causing their death and satisfies the diffusion equation

$$\frac{\partial \beta}{\partial t} - d_2\Delta\beta = PH\big(|x| - \rho\big)H\big(R - |x|\big) - P_d H\big(\rho - |x|\big). \tag{2.5}$$

This model, proposed by Greenspan, has been studied by several authors in the last thirty years. We shall focus on the study of a similar model and detail the modelling and some mathematical results in the next section.

When asymmetric distribution of nutrients or displacement of cells produced by nonuniform density appears in the interior of the spheroid tumor, the internal forces may break the symmetry of the outer boundary. Several authors have studied, in different models, the symmetry breaking of the boundary. GREENSPAN [1976] studied a model where the pressure p of the cancer cells satisfies

$$\Delta p = S,$$

inside the tumor, where S is the rate of volume lost per unit volume (assumed constant). The distribution of nutrients σ satisfies a elliptic equation outside of the tumor. Using Darcy's law, (the velocity v of the boundary is proportional to the gradient of p) that is $v = \mu\nabla p$, with suitable boundary conditions for p and σ, Greenspan obtains nonsymmetric explicit solutions using spherical harmonics.

Darcy's law has been used in different models in order to describe the movement of the free boundary. BYRNE [1997b], BYRNE and CHAPLAIN [1996b] and BYRNE and MATTHEWS [2002] propose similar models improving GREENSPAN [1976]; they study the stability of radially symmetric solutions via perturbations with spherical harmonics. FRIEDMAN and REITICH [2001] study the bifurcation of non-symmetric solutions from any radially symmetric steady state. Bessel functions are used in FRIEDMAN and REITICH [2001] and also in FRIEDMAN, HU and VELÁZQUEZ [2001] in a protocell model.

LEVINE, SLEEMAN and NILSEN-HAMILTON [2000] and LEVINE, PAMUK, SLEEMAN and NILSEN-HAMILTON [in press] (see also HOLMES and SLEEMAN [2000]) developed models of angiogenesis based on analysis of the relevant biochemical processes and on the methodology of the reinforced random walk of OTHMER and STEVENS [1997]. A mathematical analysis of the model proposed in LEVINE, SLEEMAN and NILSEN-HAMILTON [2000] have been performed in FONTELOS, FRIEDMAN and HU [2002]. Their model involves several diffusing populations and several chemical species. Another model of angiogenesis with one diffusing population and two non-diffusing ones, was developed in ANDERSON and CHAPLAIN [1998] and CHAPLAIN and ANDERSON [1997]. They denote the density of the endothelial cells by p, the concentration of the tumor angiogenesis factor (secreted by the tumor) by c, and w represents the density of the fibronectin cells, then

$$\frac{\partial p}{\partial t} = \operatorname{div}\left(\nabla p - p\left(\frac{\alpha}{1+c}\nabla c + \rho \nabla w\right)\right), \qquad \frac{\partial w}{\partial t} = \gamma p(1-w),$$
$$\frac{\partial c}{\partial t} = -\mu p c,$$

where α, ρ, γ and μ are positives constants. The asymptotic behavior of the solutions has been studied for some values of the parameters and special initial data in FRIEDMAN and TELLO [2002]. A computational approach is used by VALENCIANO and CHAPLAIN [2003a], VALENCIANO and CHAPLAIN [2003b] to obtain numerical solutions for similar models. LEVINE and SLEEMAN [1997] study the chemotaxis equations developed in the context of reinforced random walks. They use the classification of the second order part of a modified equation in the "Hodograph plane" and study the existence of blow up of solutions in finite time.

Recently, BERTUZZI, FASANO, GANDOLFI and MARANGI [2002] have developed a model for the phase transition in tumor cells and their migration towards the periphery.

The macrophages cells are part of the response of the immune system to cancer; their movement has been modeled by different authors (see OWEN and SHERRATT [1999]).

3. A simple mathematical model

In this section we describe a simple mathematical model which will be studied throughout the remainder of this work. It belongs to a group of first generation cancer models with Greenspan's model (2.4), (2.5) being one of the earliest ones. Similar models have been proposed and studied by several authors (BYRNE and CHAPLAIN [1996a], FRIEDMAN and REITICH [1999], CUI and FRIEDMAN [2000], CUI and FRIEDMAN [2001] and DÍAZ and TELLO [2004], DÍAZ and TELLO [2003]). We assume that the density of live cells is proportional to the concentrations of the nutrients σ. The tumor occupies a ball in $\mathbb{R}^3$ of radius $R(t)$ which is unknown (which is reason why R is usually called the free boundary of the problem).

The tumor comprised a central necrotic core of dead cells, the necrotic core is covered with a layer (of living cells) resulting in a second free boundary denoted by $\rho(t)$ in GREENSPAN [1972].

The transfer of nutrients to the tumor through the vasculature occurs below a certain level σ_B, and it is done with a rate r_1. During the development of the tumor, the immune system secretes inhibitors as a immune response to the foreign body. The structure of inhibitor absorption is similar to the transference of nutrients (for a constant r_2). If we assume that the nutrient consumption rate is proportional to the concentrations of nutrients, the nutrient consumption rate is given by $\lambda\sigma$. Both processes, consumption and transference, occur simultaneously in the exterior of the necrotic core, where cells are inhibited by $\hat{\beta}$. We assume that the host tissue is homogeneous and that the diffusion coefficient, d_1, is constant. The reaction between nutrients and inhibitors can be globally modelled by introducing the Heaviside maximal monotone graph (as function of $\hat{\sigma}$) and some continuous functions $g_i(\hat{\sigma}, \hat{\beta})$. Then $\hat{\sigma}$ satisfies

$$\frac{\partial\sigma}{\partial t} - d_1\Delta\sigma \in \hat{r}_1\big((\sigma_B - \sigma) - \lambda_1\sigma - \lambda\beta\big)H(\sigma - \sigma_n) + \hat{g}_1(\sigma, \beta). \tag{3.1}$$

We also assume a constant diffusion coefficient for the inhibitor concentration $\hat{\beta}$, d_2. The model considers the permanent supply of inhibitors, modeled by $\tilde{f}$ and localized on a small region ω_0 inside the tumor. This term $\tilde{f}$ was introduced in DÍAZ and TELLO [2003] to control the growth of the tumor. Then β satisfies

$$\frac{\partial\beta}{\partial t} - d_2\Delta\beta \in -r_2(\beta - \beta_B)H(\sigma - \sigma_n) + \hat{g}_2(\sigma, \beta) + \tilde{f}\chi_{\omega_0}, \tag{3.2}$$

adding initial and boundary conditions, we obtain

$$\sigma(\tilde{x}, t) = \overline{\overline{\sigma}}, \qquad \beta(\tilde{x}, t) = \overline{\overline{\beta}}, \quad |\tilde{x}| = R(t), \tag{3.3}$$

$$\sigma(\tilde{x}, 0) = \sigma_0(\tilde{x}), \qquad \beta(\tilde{x}, 0) = \beta_0(\tilde{x}), \quad |\tilde{x}| < R_0. \tag{3.4}$$

In this formulation, the presence of the maximal monotone graph H is the reason why the symbol $\in$ appears in Eq. (3.2) instead of the equal sign (a precise notion of weak solution will be presented later). Different constants appears in the equations and boundary conditions which lead to a wide variety of special cases: σ_n is the level of concentration of nutrients above which the cells can live (below this level the cells die by *necrosis*), $\overline{\overline{\sigma}}$ and $\overline{\overline{\beta}}$ are the concentration of nutrients and inhibitors in the exterior of the tumor. The diffusion operator Δ is the Laplacian operator and χ_{ω_0} denotes the characteristic function of the set ω_0 (i.e., $\chi_{\omega_0}(\tilde{x}) = 1$, if $\tilde{x} \subset \omega_0$, and $\chi_{\omega_0}(\tilde{x}) = 0$, otherwise).

Notice that the above formulation is of global nature and that the inner free boundary $\rho(t)$ is defined implicitly as the boundary of the set $\{r \in [0, R(t))\colon\ \sigma \leqslant \sigma_n\}$. So, if for instance, the initial datum σ_0 satisfies $\sigma_0(\tilde{x}) = \sigma_n$ on $[0, \rho_0]$, for some $\rho_0 > 0$ and $\hat{g}_1(\sigma_n, \beta) \in [0, r_1(\sigma_B - \sigma_n) - \lambda\sigma_n]$ for any $\beta \geqslant 0$, the above formulation leads to the associate double free boundary formulation in which $\hat{\sigma}$ satisfies

$$\begin{cases} \dfrac{\partial\sigma}{\partial t} - d_1\Delta\sigma + \lambda_1\sigma = \hat{r}_1(\sigma_B - \sigma) + \hat{g}_1(\sigma, \beta), & \rho(t) < |\tilde{x}| < R(t), \\ \sigma(\tilde{x}, t) = \sigma_n, & |\tilde{x}| \leqslant \rho(t), \\ \sigma(\tilde{x}, t) = \overline{\overline{\sigma}}, & |\tilde{x}| = R(t), \\ R(0) = R_0,\ \rho(0) = \rho_0,\ \sigma(\tilde{x}, 0) = \sigma_0(\tilde{x}), & \rho_0 < |\tilde{x}| < R_0. \end{cases}$$

The free boundary $R(t)$ is described by the ODE presented in Section 2,

$$\frac{\mathrm{d}}{\mathrm{d}t}\left(\frac{4}{3}\pi R^3(t)\right) = \int_{\{|\tilde{x}|<R(t)\}} \widehat{S}\big(\sigma(\tilde{x},t), \beta(\tilde{x},t)\big)\,\mathrm{d}\tilde{x}, \quad R(0) = R_0. \tag{3.5}$$

4. Existence of solutions

In this section, we study the existence of solutions to (3.1)–(3.5) after introducing some structural assumptions on $\hat{g}_i$ and $\widehat{S}$. We also introduce some functional spaces and a useful change of variables. The existence result is presented in Theorem 4.1 and proved by using a Galerkin approximation based on a weak formulation of the problem.

We shall assume that the reaction terms $\hat{g}_i$ and the mass balance of the tumor $\widehat{S}$ satisfy:

$$\hat{g}_i \text{ are piecewise continuous,} \quad \big|\hat{g}_i(a,b)\big| \leqslant c_0 + c_1\big(|a| + |b|\big), \tag{4.1}$$

$$\widehat{S} \text{ is continuous and} \quad -\lambda_0 \leqslant \widehat{S}(a,b) \leqslant c_0 + c_1\big(|a|^2 + |b|^2\big) \tag{4.2}$$

for some positives constants λ_0, c_0, c_1.

The above assumptions ((4.1) and (4.2)) do not constitute biological restrictions, and previous models satisfy them provided σ and β are bounded. They are introduced in order to carry out the mathematical treatment, and its great generality allows us to handle all the special cases from the literature previously mentioned. They are relevant due to its generality. It is possible to show that the absence of one (or both) of the conditions implies the occurrence of very complicated mathematical pathologies, and much more sophisticated approaches would be needed for proving that the model admits a solution (in some very delicate sense).

We introduce the change of variables,

$$x = (x_1, x_2, x_3) = \frac{\tilde{x}}{R(t)}, \tag{4.3}$$

$$u(x,t) = \sigma\big(R(t)x, t\big) - \overline{\overline{\sigma}} \tag{4.4}$$

and

$$v(x,t) = \beta\big(R(t)x, t\big) - \overline{\overline{\beta}}. \tag{4.5}$$

Let the unit ball $\{x \in \mathbb{R}^3\colon\ |x| < 1\}$ be denoted by B and define functions from $\mathbb{R}^2$ to $2^{\mathbb{R}^2}$ by

$$\begin{cases} g_1\big(\sigma - \overline{\overline{\sigma}}, \beta - \overline{\overline{\beta}}\big) := \big(\hat{r}_1\big((\sigma_B - \sigma) - \lambda_1\sigma\big) - \lambda\beta\big)H(\sigma - \sigma_n) + \hat{g}_1(\sigma,\beta), \\ g_2\big(\sigma - \overline{\overline{\sigma}}, \beta - \overline{\overline{\beta}}\big) := -r_2(\beta - \beta_B)H(\sigma - \sigma_n) + \hat{g}_2(\sigma,\beta), \end{cases} \tag{4.6}$$

$$S\big(\sigma - \overline{\overline{\sigma}}, \beta - \overline{\overline{\beta}}\big) := \frac{4}{3\pi}\widehat{S}(\sigma,\beta) \tag{4.7}$$

and

$$f(x,t) := \tilde{f}\big(xR(t), t\big), \qquad \widetilde{\omega}_0^t = \big\{(x,t) \in B \times [0,T]\colon\ R(t)x \in \omega_0\big\}.$$

Problem (3.1)–(3.5) becomes

$$\begin{cases} \dfrac{\partial u}{\partial t} - \dfrac{d_1}{R(t)^2}\Delta u - \dfrac{R'(t)}{R(t)} x \cdot \nabla u \in g_1(u, v), & x \in B,\ t > 0, \\ \dfrac{\partial v}{\partial t} - \dfrac{d_2}{R(t)^2}\Delta v - \dfrac{R'(t)}{R(t)} x \cdot \nabla v \in g_2(u, v) + f\chi_{\tilde{\omega}_0^t}, & x \in B,\ t > 0, \\ R(t)^{-1}\dfrac{\mathrm{d}R(t)}{\mathrm{d}t} = \int_B S(u, v)\,\mathrm{d}x, & t > 0, \\ u(x, t) = v(x, t) = 0, & x \in \partial B,\ t > 0, \\ R(0) = R_0,\ u(x, 0) = u_0(x),\ v(x, 0) = v_0(x), & x \in B. \end{cases} \tag{4.8}$$

We introduce the Hilbert spaces

$$\mathbf{H}(B) := L^2(B)^2, \qquad \mathbf{V}(B) = H_0^1(B)^2$$

and define inner products by

$$\langle \Phi, \Psi \rangle_{\mathbf{H}(B)} = \int_B \Phi \cdot \Psi^t\,\mathrm{d}x, \qquad \langle \Phi, \Psi \rangle_{\mathbf{V}(B)} = \sum_{i=1,2} d_i \int_B (\nabla \Phi_i)^t \cdot \nabla \Psi_i\,\mathrm{d}x$$

for all $\Phi = (\Phi_1, \Phi_2)$, $\Psi = (\Psi_1, \Psi_2)$.

For the sake of notational simplicity we use $\mathbf{H} = \mathbf{H}(B)$ and $\mathbf{V} = \mathbf{V}(B)$. Given $T > 0$, we introduce $U = (u, v)$, $U_0 = (u_0, v_0)$ and define $G : \mathbb{R}^2 \to 2^{\mathbb{R}^2} \times 2^{\mathbb{R}^2}$ and $F : (0, T) \times B \to \mathbb{R}^2$ by

$$G(U) = \big(g_1(u, v), g_2(u, v)\big), \qquad F(t, x) = \big(0, f(t, x)\chi_{\tilde{\omega}_0^t}\big).$$

We have

$$\big|G(U)\big| = \big|g_1(u, v)\big| + \big|g_2(u, v)\big| \leqslant C_0 + C_1|U| = C_0 + C_1\big(|u| + |v|\big). \tag{4.9}$$

DEFINITION. $(U, R) \in L^2(0, T : \mathbf{V}) \times W^{1,\infty}(0, T : \mathbb{R})$ is a weak solution of the problem (4.8) if there exists $g^* = (g_1^*, g_2^*) \in L^2(0, T : \mathbf{H})$ with $g^*(x, t) \in G(U(x, t))$ a.e. $(x, t) \in B \times (0, T)$ satisfying

$$\int_0^T -\langle U, \Phi_t \rangle_{\mathbf{H}}\,\mathrm{d}t + \int_0^T \tilde{a}(t, U, \Phi)\,\mathrm{d}t = \int_0^T \langle g^*, \Phi \rangle_{\mathbf{H}}\,\mathrm{d}t + \langle U_0, \Phi(0) \rangle_{\mathbf{H}} + \int_0^T \langle F(t), \Phi \rangle_{\mathbf{H}}\,\mathrm{d}t,$$

$\forall \Phi \in L^2(0, T : \mathbf{V}) \cap H^1(0, T : \mathbf{H})$ with $\Phi(T) = 0$, where

$$\tilde{a}(t, U, \Phi) := \frac{1}{R^2(t)}\langle U, \Phi \rangle_{\mathbf{V}} - \frac{R'(t)}{R(t)}\langle x \cdot \nabla U, \Phi \rangle_{\mathbf{H}} \tag{4.10}$$

and $R(t)$ is strictly positive and given by

$$R(t)^{-1}\frac{\mathrm{d}R(t)}{\mathrm{d}t} = \int_B S\big(U(x, t)\big)\,\mathrm{d}x \quad \text{for } t \in (0, T).$$

DEFINITION. (σ, β, R) is a weak solution of (3.1)–(3.5) if

$$\sigma(\tilde{x}, t) = u\left(\frac{\tilde{x}}{R(t)}, t\right) + \overline{\overline{\sigma}} \quad \text{and} \quad \beta(\tilde{x}, t) = v\left(\frac{\tilde{x}}{R(t)}, t\right) + \overline{\overline{\beta}},$$

for $t \in (0, T)$ and $\tilde{x} \in \mathbb{R}^3$, $|\tilde{x}| \leqslant R(t)$, where $(U = (u, v), R)$ is a weak solution of (4.8) for any $T > 0$.

REMARK 4.1. The definition of weak solution and the structural assumptions on G imply that $\partial U/\partial t \in L^2(0, T : \mathbf{V}(B)')$ and the equation holds in $D'(B \times (0, T))$.

THEOREM 4.1. *Assume* (4.1), (4.2), $R_0 > 0$ *and* σ_0, $\beta_0 \in L^2(0, R_0)$, *then problem* (3.1)–(3.5) *has at least a weak solution for each* $T > 0$.

PROOF. We shall use a Galerkin method to construct a weak solution. Let $R(t) \in W^{1,\infty}(0, T : \mathbb{R})$ such that $R'(t)/R(t) \geqslant -\lambda_0$ a.e. $t \in (0, T)$. For fixed $t \in (0, T)$, we consider the operator $\mathbf{A}(t) \equiv \mathbf{A}(R(t)) : \mathbf{V} \to \mathbf{V}'$ defined by

$$\mathbf{A}\big(R(t)\big)(U) = \begin{pmatrix} -\frac{d_1}{R(t)^2}\Delta u - \frac{R'(t)}{R(t)}x \cdot \nabla u & 0 \\ 0 & -\frac{d_2}{R(t)^2}\Delta v - \frac{R'(t)}{R(t)}x \cdot \nabla v \end{pmatrix}.$$

$\mathbf{A}(t)$ defines a continuous, bilinear form on $\mathbf{V} \times \mathbf{V}$,

$$\tilde{a}(t : \cdot, \cdot) : \mathbf{V} \times \mathbf{V} \to \mathbb{R}$$

for a.e. $t \in (0, T)$ (see (4.10)). Since $R'(t)/R(t) \geqslant -\lambda_0$, $\tilde{a}$ satisfies

$$\begin{aligned} \tilde{a}(t, U, U) &= \frac{1}{R^2(t)}\langle U, U\rangle_{\mathbf{V}} - \frac{R'(t)}{R(t)}\langle x \cdot \nabla U, U\rangle_{\mathbf{H}} \\ &= \frac{1}{R^2(t)}\langle U, U\rangle_{\mathbf{V}} + \frac{R'(t)}{2R(t)}\langle U, U\rangle_{\mathbf{H}} \\ &\geqslant \Big(\max_{0<t<T}\big\{R(t)\big\}\Big)^{-2}\|U\|_{\mathbf{V}}^2 - \frac{\lambda_0}{2}\|U\|_{\mathbf{H}}^2. \end{aligned}$$

□

Now we establish some *a priori estimates* which will be used later. In fact, those estimates can be applied even for other existence methods, different from the Galerkin-type one, as, for instance, iterative methods, fixed point methods, etc. (see, for instance, SHOWALTER [1996]).

LEMMA 4.1.

$$\|U\|_{\mathbf{H}}^2 \leqslant C_0^2\big(\exp\{(\lambda_0 + 2C_1 + 1)T\} - 1\big) + \|F\|_{L^2(0,T:\mathbf{H})}^2 + \|U_0\|_{\mathbf{H}}^2.$$

PROOF. Inserting U^t as test function into the weak formulation of (4.8), one obtains

$$\frac{\mathrm{d}}{\mathrm{d}t}\int_B \frac{1}{2}U^2\,\mathrm{d}x + \tilde{a}(t, U, U) + \int_B g^*(U)U^t\,\mathrm{d}x = \int_B F \cdot U^t\,\mathrm{d}x$$

for some $g^* \in L^2((0,T)\times B)^2$ and $g^*(x,t) \in G(U(x,t))$ for a.e. $(x,t) \in B \times (0,T)$. The definition of $\tilde{a}$ yields

$$\frac{1}{2}\frac{\mathrm{d}}{\mathrm{d}t}\|U\|_{\mathbf{H}}^2 - \frac{\lambda_0}{2}\|U\|_{\mathbf{H}}^2 \leqslant \big(\|g^*\|_{\mathbf{H}} + \|F\|_{\mathbf{H}}\big)\|U\|_{\mathbf{H}}. \tag{4.11}$$

Thus by Young's inequality and (4.9) imply

$$\frac{1}{2}\frac{\mathrm{d}}{\mathrm{d}t}\|U\|_{\mathbf{H}}^2 - \left(\frac{\lambda_0}{2} + C_1 + \frac{1}{2}\right)\|U\|_{\mathbf{H}}^2 \leqslant \frac{1}{2}\big(C_0^2 + \|F\|_{\mathbf{H}}^2\big).$$

Integrating with respect to time, we get

$$\frac{1}{2}\|U\|_{\mathbf{H}}^2 - \frac{1}{2}\|U_0\|_{\mathbf{H}}^2 - \left(\frac{\lambda_0}{2} + C_1 + \frac{1}{2}\right)\|U\|_{L^2(0,T:\mathbf{H})}^2 \leqslant \frac{1}{2}\big(C_0^2 T + \|F\|_{L^2(0,T:\mathbf{H})}^2\big)$$

and by Gronwall's lemma,

$$\|U\|_{\mathbf{H}}^2 \leqslant C_0^2\big(\exp\big\{(\lambda_0 + 2C_1 + 1)T\big\} - 1\big) + \|F\|_{L^2(0,T:\mathbf{H})}^2 + \|U_0\|_{\mathbf{H}}^2 \leqslant C. \tag{4.12}$$

□

REMARK 4.2. Since U is bounded in $\mathbf{H}$ (by (4.12)), R satisfies

$$R(t) = R_0 \exp\left\{\int_0^t \int_0^1 S(U)\,\mathrm{d}x\,\mathrm{d}t\right\} \leqslant R_0 \mathrm{e}^{K_1 t} \tag{4.13}$$

and

$$R(t) \geqslant R_0 \exp\{-\lambda_0 t\}, \tag{4.14}$$

consequently, $R \in W^{1,\infty}(0,T)$.

LEMMA 4.2. $\|U\|_{L^2(0,T:\mathbf{V})} \leqslant K(T,F,G,U_0)$.

PROOF. Selecting U as test function in (4.8), we have

$$\begin{aligned}&\frac{D}{R_0^2 \mathrm{e}^{2K_1 T}}\|U\|_{L^2(0,T:\mathbf{V})}^2 - \frac{\lambda_0}{2}\|U\|_{L^2(0,T:\mathbf{H})}^2\\ &\quad \leqslant C_1\|U\|_{L^2(0,T:\mathbf{H})}^2 + \big(C_0 + \|F\|_{L^2(0,T:\mathbf{H})}\big)\|U\|_{L^2(0,T:\mathbf{H})}.\end{aligned}$$

By (4.12), we get

$$\|U\|_{L^2(0,T:V)} \leqslant K(F,G,U_0,T). \tag{4.15}$$

□

REMARK 4.3. By Lemma 4.2 and Remark 4.2, we get that

$$u_t - \frac{d_1}{R^2}\Delta u \in L^2\big(0,T:L^2(B)\big), \qquad v_t - \frac{d_2}{R^2}\Delta v \in L^2\big(0,T:L^2(B)\big)$$

and obtain the extra regularity

$$U_t, \quad \Delta U \in \big[L^2\big(0,T:L^2(B)\big)\big]^2. \tag{4.16}$$

Now, as previously in the proof of Theorem 4.1, we consider the approximate problem

$$\begin{cases} \dfrac{\partial U^\varepsilon}{\partial t} + A\big(R^\varepsilon(t)\big)U^\varepsilon = G^\varepsilon\big(U^\varepsilon\big) + F(t) \quad \text{on } B \times (0,T), \\ U^\varepsilon(0,x) = U_0, \qquad U^\varepsilon = 0 \quad \text{on } \partial B, \\ \dfrac{1}{R^\varepsilon}\dfrac{\mathrm{d}R^\varepsilon}{\mathrm{d}t} = \displaystyle\int_B S(U^\varepsilon)\,\mathrm{d}x, \end{cases} \tag{4.17}$$

where $G^\varepsilon = (g_1^\varepsilon, g_2^\varepsilon)$ is a Lipschitz continuous function such that

$$G^\varepsilon \to G \quad \text{when } \varepsilon \to 0 \text{ a.e. in } \mathbb{R}^2.$$

G^ε is obtained replacing H by

$$H^\varepsilon(s) = \begin{cases} 0 & \text{if } s < 0, \\ \dfrac{s}{\varepsilon} & \text{if } 0 \leqslant s \leqslant \dfrac{1}{\varepsilon}, \\ 1 & \text{if } s > \dfrac{1}{\varepsilon}. \end{cases}$$

Now, we apply the Galerkin method to the approximated problem. Let λ_n and $\phi_n \in H_0^1(B)$ for $n \in \mathbb{N}$ be the eigenvalues and eigenfunctions associated to $-\Delta$ satisfying

$$-\Delta\phi_n = \lambda_n\phi_n.$$

We consider V_m the finite-dimensional vector space spanned by $\{\phi_1, \dots, \phi_m\}$. We search for a solution $U_m^\varepsilon \in L^2(0, T : V_m)$ of the problem

$$\begin{cases} \dfrac{\mathrm{d}}{\mathrm{d}t}U_m^\varepsilon + A\big(R_m^\varepsilon(t)\big)U_m^\varepsilon = G^\varepsilon\big(U_m^\varepsilon\big) + F_m(t), \\ U_m^\varepsilon(0) = U_{0,m}^\varepsilon, \\ R_m^\varepsilon(t)^{-1}\dfrac{\mathrm{d}R_m^\varepsilon(t)}{\mathrm{d}t} = \displaystyle\int_B S\big(U_m^\varepsilon(x,t)\big)\,\mathrm{d}x, \end{cases} \tag{4.18}$$

where the initial conditions $U_{0,m}^\varepsilon = P_m(U_0)$ (where P_m is the orthogonal projection from $L^2(B)$ onto V_m) and $F_m = P_m(F)$. Then

$$R_m^\varepsilon(t) = R_0 \exp\left\{\int_0^t\int_B S\big(U_m^\varepsilon(x,s)\big)\,\mathrm{d}x\,\mathrm{d}s\right\}.$$

PROPOSITION 4.1. (4.18) *has a unique solution* U_m^ε *for any* $T < \infty$.

PROOF. Problem (4.18) can be written as a suitable nonlinear ordinary differential system. Let $U_m^\varepsilon = (u_m^\varepsilon, v_m^\varepsilon)$ be defined by

$$u_m^\varepsilon(t) = \sum_{n=1,\dots,m} a_n^{\varepsilon m}(t)\phi_n, \qquad v_m^\varepsilon(t) = \sum_{n=1,\dots,m} b_n^{\varepsilon m}(t)\phi_n$$

and denote

$$a^{\varepsilon m} = \left(a_1^{\varepsilon m}, a_2^{\varepsilon m}, \dots, a_m^{\varepsilon m}\right), \quad b^{\varepsilon m} = \left(b_1^{\varepsilon m}, b_2^{\varepsilon m}, \dots, b_m^{\varepsilon m}\right),$$
$$\lambda_a = \left(\lambda_1 a_1^{\varepsilon m}, \dots, \lambda_m a_m^{\varepsilon m}\right) \quad \text{and} \quad \lambda_b = \left(\lambda_1 b_1^{\varepsilon m}, \dots, \lambda_m b_m^{\varepsilon m}\right).$$

Then $a^{\varepsilon m}$, $b^{\varepsilon m}$ and R_m^ε satisfy

$$\dot a^{\varepsilon m} + \frac{\lambda_a}{(R_m^\varepsilon)^2} + \phi_\varepsilon\left(a^{\varepsilon m}, b^{\varepsilon m}\right) L_1^m\left(a^{\varepsilon m}, b^{\varepsilon m}\right) = g_1^m\left(a^{\varepsilon m}, b^{\varepsilon m}\right),$$
$$\dot b^{\varepsilon m} + \frac{\lambda_b}{(R_m^\varepsilon)^2} + \phi_\varepsilon\left(a^{\varepsilon m}, b^{\varepsilon m}\right) L_2^m\left(a^{\varepsilon m}, b^{\varepsilon m}\right) = g_2^m\left(a^{\varepsilon m}, b^{\varepsilon m}\right) + F^m(t),$$
$$\frac{\dot R_m^\varepsilon}{R_m^\varepsilon} = \phi_\varepsilon\left(a^{\varepsilon m}, b^{\varepsilon m}\right),$$

where

$$\phi_\varepsilon\left(a^{\varepsilon m}, b^{\varepsilon m}\right) = \int_B S\left(U_m^\varepsilon\right) \mathrm{d}x,$$
$$L_1^m\left(a^{\varepsilon m}, b^{\varepsilon m}\right) = \int_B x \cdot \nabla u_m^\varepsilon \phi_n \, \mathrm{d}x \quad \text{for } n = 1, \dots, m,$$
$$L_2^m\left(a^{\varepsilon m}, b^{\varepsilon m}\right) = \int_B x \cdot \nabla v_m^\varepsilon \phi_n \, \mathrm{d}x \quad \text{for } n = 1, \dots, m,$$
$$g_1^m\left(a^{\varepsilon m}, b^{\varepsilon m}\right) = \int_B g_1^\varepsilon(u_m^\varepsilon, v_m^\varepsilon)\phi_n \, \mathrm{d}x \quad \text{for } n = 1, \dots, m,$$
$$g_2^m\left(a^{\varepsilon m}, b^{\varepsilon m}\right) = \int_B g_2^\varepsilon\left(u_m^\varepsilon, v_m^\varepsilon\right)\phi_n \, \mathrm{d}x \quad \text{for } n = 1, \dots, m.$$

Since G_ε is a Lipschitz function, we obtain that there exists a unique solution $a^{\varepsilon m}, b^{\varepsilon m}, R^{\varepsilon m}$ to the system for T small enough. Moreover, (4.12) and (4.14) hold, and we get the existence of a solution of (4.18) for any $T < \infty$. By (4.15) and (4.16), $\{(U_m^\varepsilon, \frac{\mathrm{d}}{\mathrm{d}t} U_m^\varepsilon)\}_{m=1,\infty}$ is uniformly bounded in $L^2(0, T : \mathbf{V}) \times L^2(0, T : \mathbf{V}')$. So, there exists a subsequence $U_{mi}^\varepsilon \in L^2(0, T : \mathbf{V})$ with $\frac{\mathrm{d}}{\mathrm{d}t} U_{mi}^\varepsilon \in L^2(0, T : \mathbf{V}')$ such that

$$\left(U_{mi}^\varepsilon, \frac{\mathrm{d}}{\mathrm{d}t} U_{mi}^\varepsilon\right) \rightharpoonup \left(U^\varepsilon, \frac{\mathrm{d}}{\mathrm{d}t} U^\varepsilon\right) \quad \text{weakly in } L^2(0, T : \mathbf{V}) \times L^2(0, T : \mathbf{V}'),$$

and $R_{mi}^\varepsilon \rightharpoonup R^\varepsilon$ weakly in $W^{1,p}(0, T)$ for $p < \infty$. Taking limits when $mi \to \infty$, we get the existence of a weak solution to (4.17) for any $T < \infty$.

To end the proof of Theorem 4.1, we take limits in the equation when $\varepsilon \to 0$. We employ (4.12) and (4.14) and the compact embedding $\mathbf{H}_0^1(B) \subset \mathbf{L}^s(B)$ (for $s < 6$) in order to obtain the existence of a subsequence $U^{\varepsilon i}$ such that

$$U^{\varepsilon i} \to U \quad \text{in } L^2\left(0, T : \left[L^s(B)\right]^2\right)$$

and in particular

$$U^{\varepsilon i} \to U \quad \text{in } L^2(0, T : \mathbf{H})$$

(see, e.g., SIMON [1987]). Since

$$H^{\varepsilon}\big(u^{\varepsilon}+\overline{\overline{\sigma}}\big) \rightharpoonup h \in H(u+c) \quad \text{weakly in } L^2\big(0,T:L^s(B)\big)$$

and

$$v^{\varepsilon} \to v \quad \text{in } L^2\big(0,T:L^s(B)\big)$$

(see Lemma 3.4.1 of VRABIE [1995]), we have

$$G^{\varepsilon i}\big(U^{\varepsilon i}\big) \rightharpoonup g^* \in G(U) \quad \text{weakly in } L^1(0,T:\mathbf{H}).$$

Since $|R'| \leqslant C$, there exists a subsequence $R_{\varepsilon ij}$ such that

$$R_{\varepsilon ij} \rightharpoonup R \quad \text{weakly in } W^{1,p}(0,T),\ p<\infty,$$

and we deduce that $R_{\varepsilon ij} \to R$ in $C^0([0,T])$. Finally, taking limits in the weak formulation of the problem (4.17), we get

$$\int_0^T \langle U_t,\Phi\rangle_{\mathbf{H}}\,\mathrm{d}t + \int_0^T \tilde{a}\big(R(t),U,\Phi\big)\,\mathrm{d}t + \int_0^T \langle g^*,\Phi\rangle_{\mathbf{H}}\,\mathrm{d}t = \int_0^T \langle F,\Phi\rangle_{\mathbf{H}}\,\mathrm{d}t$$

for all $\Phi \in L^2(0,T:V)$ and, moreover,

$$R(t)^{-1}\frac{\mathrm{d}R(t)}{\mathrm{d}t} = \int_B S\big(U(x,t)\big)\,\mathrm{d}x.$$

Notice that

$$\int_0^T \frac{R'_{\varepsilon ij}}{R_{\varepsilon ij}} \int_B x\cdot\nabla u_{\varepsilon ij}\psi\,\mathrm{d}x\,\mathrm{d}t = \int_0^T \frac{R'_{\varepsilon ij}}{R_{\varepsilon ij}} \int_B u_{\varepsilon ij}\psi - u_{\varepsilon ij}x\cdot\nabla\psi\,\mathrm{d}x\,\mathrm{d}t$$

and

$$\int_0^T \frac{R'_{\varepsilon ij}}{R_{\varepsilon ij}} \int_B x\cdot\nabla v_{\varepsilon ij}\psi\,\mathrm{d}x\,\mathrm{d}t = \int_0^T \frac{R'_{\varepsilon ij}}{R_{\varepsilon ij}} \int_B v_{\varepsilon ij}\psi - v_{\varepsilon ij}x\cdot\nabla\psi\,\mathrm{d}x\,\mathrm{d}t.$$

We conclude that (σ,β,R) defined by

$$\sigma(t,\tilde{x}) = u\left(t,\frac{\tilde{x}}{R(t)}\right)+\overline{\overline{\sigma}} \quad \text{and} \quad \beta(t,\tilde{x}) = v\left(t,\frac{\tilde{x}}{R(t)}\right)+\overline{\overline{\beta}}$$

is a weak solution to (3.1)–(3.5). The additional regularity

$$\sigma_t - d_1\Delta\sigma \quad \text{and} \quad \beta_t - d_2\Delta\beta \in L^2\left(\bigcup_{t\in[0,T]}\big(0,R(t)\big)\times\{t\}\right)$$

follows from the fact that

$$\frac{\partial U}{\partial t}(t) + \mathbf{A}\big(R(t)\big)U(t) \in L^2\big(0,T:L^2(B)^2\big). \qquad \square$$

5. Uniqueness of solutions

We begin by pointing out that if, for instance,

$$\sigma_n \geqslant \frac{r_1 \sigma_B}{r_1 + \lambda}, \qquad r_1 \sigma_B > 0, \ \hat{g}_1(\hat{\sigma}, \hat{\beta})$$

is a decreasing function of $\hat{\sigma}$ and independent of $\hat{\beta}$ and the initial datum $\sigma_0(\tilde{x})$ is such that $\sigma_0'(\rho_0) = \sigma_0''(\rho_0) = 0$, then it is possible to adapt the arguments of DÍAZ and TELLO [1999] in order to construct more than one solution of problem (3.1)–(3.5). This and the presence of non-Lipschitz terms at both equations clarify that any possible uniqueness result will require an significant set of additional conditions.

In this section we prove the uniqueness of solution for two different cases. CUI and FRIEDMAN [2000] prove uniqueness of radial symmetric solutions without forcing term (i.e., $f = 0$).

5.1. 3-dimensional case with forcing term

When a tumor does not have a necrotic core, Eqs. (3.1) and (3.2) simplify such that reaction terms become linear, i.e., the nutrients concentration $\hat{\sigma}$ and the inhibitors concentration $\hat{\beta}$ satisfy

$$\frac{\partial \hat{\sigma}}{\partial t} - d_1 \Delta \hat{\sigma} - \hat{r}_1(\sigma_B - \hat{\sigma}) + \lambda_1 \hat{\sigma} + \lambda \hat{\beta} = 0, \quad |x| < R(t), \ t \in (0, T),$$

$$\frac{\partial \hat{\beta}}{\partial t} - d_2 \Delta \hat{\beta} - r_2(\beta_B - \hat{\beta}) = f \chi_{\omega_0}, \quad |x| < R(t), \ t \in (0, T).$$

For notational convenience we shall assume that the diffusion coefficients d_1 and d_2 are equal and constant $d_1 = d_2 = d$. Thus by normalizing the unknown densities

$$\sigma := \hat{\sigma} - \frac{\hat{r}_1 \sigma_B + \lambda \beta_B}{(\hat{r}_1 + \lambda_1)}, \qquad \beta := \hat{\beta} - \beta_B,$$

and setting

$$r_1 := \hat{r}_1 + \lambda_1, \qquad S(\sigma, \beta) := \frac{3}{4\pi} \widehat{S}(\hat{\sigma}, \hat{\beta}),$$

we arrive at the formulation

$$\frac{\partial \sigma}{\partial t} - d \Delta \sigma + r_1 \sigma + \lambda \beta = 0, \quad |x| < R(t), \ t \in (0, T), \tag{5.1}$$

$$\frac{\partial \beta}{\partial t} - d \Delta \beta + r_2 \beta = f \chi_{\omega_0}, \quad |x| < R(t), \ t \in (0, T), \tag{5.2}$$

$$R(t)^2 \frac{\mathrm{d}R(t)}{\mathrm{d}t} = \int_{|x| < R(t)} S(\sigma, \beta) \, \mathrm{d}x, \quad R(0) = R_0, \ t \in (0, T), \tag{5.3}$$

$$\sigma(x, 0) = \sigma_0(x), \qquad \beta(x, 0) = \beta_0(x), \quad |x| < R_0, \tag{5.4}$$

$$\sigma(x, t) = \overline{\overline{\sigma}}, \qquad \beta(x, t) = \overline{\overline{\beta}}, \quad |x| = R(t), \ t \in (0, T), \tag{5.5}$$

where $R_0 > 0$, the normalized nutrient and inhibitor densities at the exterior of the tumor $\overline{\overline{\sigma}}$, $\overline{\overline{\beta}}$ and the initial densities (σ_0, β_0) are known. We introduce again the changes of unknown and variables (4.3)–(4.5) and set

$$\tilde{t}(t) := \int_0^t R^{-2}(\rho)\,\mathrm{d}\rho. \tag{5.6}$$

Note that since R is a continuous function and $1/R^2(t) > 0$, we obtain that $\tilde{t}(t) \in C^1([0, \widetilde{T}])$ and employing the implicit function theorem, one derives the existence of the inverse function $t(\tilde{t}) \in C^1([0, T])$. Then, problem (5.1)–(5.5) reduces to

$$\frac{\partial u}{\partial \tilde{t}} + A(u) + R^2 r_1 u = R^2\big(r_1\overline{\overline{\sigma}} + \lambda\big(v + \overline{\overline{\beta}}\big)\big), \quad \tilde{x} \in B,\ \tilde{t} \in (0, \widetilde{T}), \tag{5.7}$$

$$\frac{\partial v}{\partial \tilde{t}} + A(v) + R^2 r_2 v = R^2 f \chi_{\widetilde{\omega}_0^{\tilde{t}}} - R^2 r_2 \overline{\overline{\beta}}, \quad \tilde{x} \in B,\ \tilde{t} \in (0, \widetilde{T}), \tag{5.8}$$

$$R(\tilde{t})\frac{\mathrm{d}}{\mathrm{d}\tilde{t}}R(\tilde{t}) = \int_B S\big(u(\tilde{x}, \tilde{t}) + \overline{\overline{\sigma}}, v(\tilde{x}, \tilde{t}) + \overline{\overline{\beta}}\big)\,\mathrm{d}\tilde{x}, \quad R(0) = R_0, \tag{5.9}$$

$$u(\tilde{x}, \tilde{t}) = v(\tilde{x}, \tilde{t}) = 0, \quad \tilde{x} \in \partial B,\ \tilde{t} \in (0, \widetilde{T}), \tag{5.10}$$

$$u(\tilde{x}, 0) = u_0(\tilde{x}) = \sigma_0(\tilde{x}R_0), \qquad v(\tilde{x}, 0) = v_0(\tilde{x}) = \beta_0(\tilde{x}R_0), \tag{5.11}$$

where $\widetilde{T} = \tilde{t}(T)$, $\widetilde{\omega}_0^{\tilde{t}} = \{\tilde{x} \in B\colon R(t(\tilde{t}))\tilde{x} \in \omega_0\}$, for any $\tilde{t} \in [0, \widetilde{T}]$ and

$$A(w) := -d\Delta w - R\dot{R}\tilde{x} \cdot \nabla w.$$

We assume that

$$\widehat{S} \in W^{1,\infty}(\mathbb{R}^2), \tag{5.12}$$

$$f\chi_{\widetilde{\omega}_0^{\tilde{t}}} \in L^p\big((0, T) \times \Omega\big), \quad p > 4, \tag{5.13}$$

$$(\sigma_0, \beta_0) \in W^{2,\infty}\big(B(R_0)\big)^2. \tag{5.14}$$

LEMMA 5.1. *Assume* (5.12)–(5.14), *then the solution* (u, v, R) *to the problem* (5.7)–(5.11) *satisfies*

$$u \in L^q\big(0, \widetilde{T} : W^{2,q}(B)\big) \cap W^{1,q}\big(0, \widetilde{T} : L^q(B)\big)$$

for all $1 < q < \infty$ *and*

$$v \in L^p\big(0, \widetilde{T} : W^{2,p}(B)\big) \cap W^{1,p}\big(0, \widetilde{T} : L^p(B)\big).$$

PROOF. By Theorem 4.1, we know that

$$(u, v, R) \in \big[L^2\big(0, \widetilde{T} : H^1(B)\big)\big]^2 \times W^{1,\infty}(0, \widetilde{T}).$$

Since $v_0 \in H^2(B)$ and $f \in L^p((0, T) \times B)$, we get

$$v \in W^{1,p}\big((0, \widetilde{T}) \times B\big) \cap L^p\big(0, \widetilde{T} : W^{2,p}(B)\big)$$

(see, e.g., LADYZENSKAJA, SOLONNIKOV and URALSEVA [1991], Theorem 9.1, Chapter IV). Since $p > 4$, $W^{1,p}((0, T) \times B) \subset L^\infty([0, \widetilde{T}] \times B)$, hence

$$u \in W^{1,q}\big((0,T)\times B\big) \cap L^q\big(0,T:W^{2,q}(B)\big),$$

for $q \leqslant \infty$. Consequently, we get $R \in W^{2,p}(0,T)$. □

One obtains from the lemma, in view of $W_0^{1,p}(B\times[0,\widetilde{T}]) \subset L^\infty(B\times[0,\widetilde{T}])$ (for $p>4$) the following corollary.

COROLLARY 5.1. $u, v \in L^\infty(B\times[0,\widetilde{T}])$.

Utilizing the continuous embedding

$$W^{1,q}\big((0,T)\times B\big) \cap L^q\big(0,T:W^{2,q}(B)\big) \subset L^2\big(0,T:W^{1,\infty}(B)\big),$$
$$W^{1,p}\big((0,\widetilde{T})\times B\big) \cap L^p\big(0,\widetilde{T}:W^{2,p}(B)\big) \subset L^2\big(0,T:W^{1,\infty}(B)\big),$$

and undoing the change of variables and unknown (4.3)–(4.5) and (5.17), we obtain

COROLLARY 5.2. *Under the assumptions of Theorem* 4.1, *we have*

$$\int_0^T \big(\|\sigma\|^2_{W^{1,\infty}(R(t))} + \|\beta\|^2_{W^{1,\infty}(R(t))}\big)\,\mathrm{d}t \leqslant k_0$$

for some $k_0 < \infty$.

The uniqueness of solutions is established in the next theorem.

THEOREM 5.1. *Let* $f \in L^p(\omega_0 \times (0,T))$ *with* $p>4$, *and* $(\sigma_0 - \overline{\overline{\sigma}}, \beta_0 - \overline{\overline{\beta}}) \in W^{2,s}(B(R_0)) \cap H_0^1(B(R_0))$, *for* $s>4$. *Then, there exists a unique solution to* (5.1)–(5.5).

PROOF. In arguing by contradiction, we assume that there exist two different solutions (σ_1, β_1, R_1) and (σ_2, β_2, R_2). Let

$$R(t) = \min\big\{R_1(t), R_2(t)\big\}, \qquad \sigma = \sigma_1 - \sigma_2, \qquad \beta = \beta_1 - \beta_2.$$

Then (σ, β, R) satisfies the problem,

$$\frac{\partial \sigma}{\partial t} - d\Delta\sigma + r_1\sigma + \lambda\beta = 0, \quad |x| < R(t),\ t \in (0,T), \tag{5.15}$$

$$\frac{\partial \beta}{\partial t} - d\Delta\beta + r_2\beta = 0, \quad |x| < R(t),\ t \in (0,T), \tag{5.16}$$

$$\sigma(x,0) = 0, \qquad \beta(x,0) = 0, \quad |x| < R_0, \tag{5.17}$$

$$\sigma(x,t) = \sigma_1(x,t) - \sigma_2(x,t), \quad |x| = R(t),\ t \in (0,T), \tag{5.18}$$

$$\beta(x,t) = \beta_1(x,t) - \beta_2(x,t), \quad |x| = R(t),\ t \in (0,T). \tag{5.19}$$

We introduce a new unknown defined by

$$z = k_1\sigma - k_2\beta,$$

with

$$k_1 = 1, \qquad k_2 = \frac{\lambda}{r_1 - r_2} \quad \text{if } r_1 \neq r_2,$$
$$k_1 = \frac{1}{2}, \qquad k_2 = \frac{\lambda}{r_1 - 2r_2} \quad \text{if } r_1 = r_2 \neq 0.$$

By construction of z, we have

$$\begin{cases} \dfrac{\partial z}{\partial t} - d\Delta z + r_1 z = 0, & |x| < R(t),\ t \in (0,T), \\ z(x,0) = 0, & |x| < R_0, \\ z = k_1\sigma - k_2\beta, & |x| = R(t),\ t \in (0,T). \end{cases} \tag{5.20}$$

We need the following preliminary result.

LEMMA 5.2. *Let z be the solution to the problem* (5.20) *and β the solution to* (5.16), (5.19), *then $e^{r_1 t} z$ and $e^{r_2 t}\beta$ take their maximum and minimum on $|x| = R(t)$.*

PROOF. Multiplying Eq. (5.20) by $e^{r_1 t}$, we obtain that $e^{r_1 t} z$ satisfies

$$\begin{cases} \dfrac{\partial}{\partial t}\left(e^{r_1 t} z\right) - d\Delta\left(e^{r_1 t} z\right) = 0, & |x| < R(t),\ t \in (0,T), \\ z(x,0) = 0, & |x| < R_0, \\ e^{r_1 t} z = e^{r_1 t}(k_1\sigma - k_2\beta), & |x| = R(t),\ t \in (0,T). \end{cases} \tag{5.21}$$

In the same way, $e^{r_2 t}\beta$ satisfies

$$\begin{cases} \dfrac{\partial}{\partial t}\left(e^{r_2 t}\beta\right) - d\Delta\left(e^{r_2 t}\beta\right) = 0, & |x| < R(t),\ t \in (0,T), \\ \beta(x,0) = 0, & |x| < R_0, \\ e^{r_2 t}\beta = e^{r_2 t}(\beta_1 - \beta_2), & |x| = R(t),\ t \in (0,T). \end{cases} \tag{5.22}$$

Applying Corollary 5.1, we obtain that $e^{r_1 t} z$ and $e^{r_2 t}\beta$ are bounded. Let

$$z^{**} = \max\left\{e^{r_1 t} z(x,t), t \in [0,T], x \in \partial B\left(R(t)\right)\right\},$$
$$z_{**} = \min\left\{e^{r_1 t} z(x,t), t \in [0,T], x \in \partial B\left(R(t)\right)\right\},$$
$$\beta^{**} = \max\left\{e^{r_2 t}\beta(x,t), t \in [0,T], x \in \partial B\left(R(t)\right)\right\},$$
$$\beta_{**} = \min\left\{e^{r_2 t}\beta(x,t), t \in [0,T], x \in \partial B\left(R(t)\right)\right\}.$$

Notice that $z^{**} \geqslant 0$, $\beta^{**} \geqslant 0$, $z_{**} \leqslant 0$ and $\beta_{**} \leqslant 0$. Let T_k and T^k be defined by

$$T_k(s) = \begin{cases} s, & \text{if } s > k, \\ k, & \text{if } s \leqslant k, \end{cases} \quad \text{and} \quad T^k(s) = \begin{cases} k, & \text{if } s \geqslant k, \\ s, & \text{if } s < k. \end{cases}$$

Taking $T_0(e^{r_1 t} z - z^{**})$ as test function in (5.21) and integrating by parts over $B(R(t))$, we arrive after some manipulations at

$$\frac{d}{dt}\int_{B(R(t))} \left[T_0\left(e^{r_1 t} z - z^{**}\right)\right]^2 dx \leqslant 0.$$

We deduce that $e^{r_1 t}z$ takes his maximum on $|x| = R(t)$. In the same way, taking $T^0(e^{r_1 t}z - z_{**})$ as test function, we obtain

$$z_{**} \leqslant e^{r_1 t} z \leqslant z^{**}. \tag{5.23}$$

The proof of

$$\beta_{**} \leqslant e^{r_2 t} \beta \leqslant \beta^{**}, \tag{5.24}$$

is analogous. □

END OF THE PROOF OF THEOREM 5.1. Given $t \in [0, T]$, we can assume, without lost of generality, that $R_1(t) \leqslant R_2(t)$. Consider

$$\begin{aligned} R_1^2(t)\dot{R}_1(t) - R_2^2(t)\dot{R}_2(t) = &\int_{B(R(t))} \big(S(\sigma_1, \beta_1) - S(\sigma_2, \beta_2)\big)\,dx \\ &- \int_{R_1(t)<|x|<R_2(t)} S(\sigma_2, \beta_2)\,dx. \end{aligned}$$

Since S is bounded, then

$$\left| \int_{R_1(t)<|x|<R_2(t)} S(\sigma_2, \beta_2)\,dx \right| \leqslant N\big|R_1^3(t) - R_2^3(t)\big| \leqslant M\big|R_1(t) - R_2(t)\big|,$$

where M depends only of $|S|_{L^\infty}$. Since S is Lipschitz continuous, integrating in time, it results

$$\begin{aligned} &\int_0^T \int_{B(R(t))} \big|S(\sigma_1, \beta_1) - S(\sigma_2, \beta_2)\big|\,dx\,dt \\ &\quad \leqslant \int_0^T \int_{B(R(t))} |S|_{W^{1,\infty}(\mathbb{R}^2)} \big(\sup|\sigma| + \sup|\beta|\big)\,dx\,dt \\ &\quad \leqslant \int_0^T \int_{B(R(t))} k_0 \left(\frac{1}{k_1} \sup|z + k_2\beta| + \sup|\beta| \right) dx\,dt \\ &\quad \leqslant \int_0^T \int_{B(R(t))} C\big(\sup|z| + \sup|\beta|\big)\,dx\,dt \\ &\quad \leqslant \int_0^T \int_{B(R(t))} C\big(\sup\big|e^{-r_1 t} e^{r_1 t} z\big| + \sup\big|e^{-r_2 t} e^{r_2 t} \beta\big|\big)\,dx\,dt \\ &\quad \leqslant \int_0^T \int_{B(R(t))} C\big(e^{|r_1|T} \sup\big|e^{r_1 t} z\big| + e^{|r_2|T} \sup\big|e^{r_2 t} \beta\big|\big)\,dx\,dt \\ &\quad \leqslant \int_0^T \int_{B(R(t))} k_3\big(\sup\big|e^{r_1 t} z\big| + \sup\big|e^{r_2 t} \beta\big|\big)\,dx\,dt. \end{aligned}$$

From Lemma 5.2, we know

$$\int_0^T \int_{B(R(t))} \sup\big|e^{r_1 t} z(x,t)\big|\,dx\,dt \leqslant e^{r_1 T} \frac{3\pi}{4} \int_0^T R^3(t) \sup_{|x|=R(t)} \big|z(x,t)\big|\,dt.$$

By Corollary 5.2, we deduce that

$$\int_0^T \left(\|\sigma_2\|^2_{W^{1,\infty}(B(R(t)))} + \|\beta_2\|^2_{W^{1,\infty}(B(R(t)))}\right) \mathrm{d}t \leqslant K_0,$$

and consequently,

$$\int_0^T \|z\|^2_{W^{1,\infty}(B(R(t)))} \,\mathrm{d}t \leqslant K.$$

Since

$$\mathrm{e}^{r_1 t} z(x,t) = \mathrm{e}^{r_1 t}\left(k_1\left(\sigma_2(x,t) - \overline{\overline{\sigma}}\right) - k_2\left(\beta_2(x,t) - \overline{\overline{\beta}}\right)\right), \quad \text{on } |x| = R(t),$$

we deduce

$$\begin{aligned}
&\mathrm{e}^{r_1 T} \frac{3\pi}{4} \int_0^T R^3(t) \sup_{|x|=R(t)} \left|z(x,t)\right| \mathrm{d}t \\
&\leqslant k_4 \int_0^T \|\sigma_2\|_{W^{1,\infty}(B(R_2(t)))} + \|\beta_2\|_{W^{1,\infty}(B(R_2(t)))} \left|R_1(t) - R_2(t)\right| \mathrm{d}t \\
&\leqslant k_4 \sup_{0<t<T} \left|R_1(t) - R_2(t)\right| T^{1/2} \int_0^T \left(\|\sigma_2\|^2_{W^{1,\infty}(B(R_2(t)))} \right. \\
&\qquad\qquad \left. + \|\sigma_2\|^2_{W^{1,\infty}(B(R_2(t)))}\right) \mathrm{d}t \\
&\leqslant k \sup_{0<t<T} \left|R_1(t) - R_2(t)\right| T^{1/2}.
\end{aligned}$$

In the same way,

$$\int_0^T \int_{B(R(t))} k_3 \sup |\beta| \,\mathrm{d}x\,\mathrm{d}t \leqslant k \sup_{0<t<T} \left|R_1(t) - R_2(t)\right| T^{1/2}.$$

Then

$$\int_0^T \left|R_1^2(t)\dot{R}_1(t) - R_2^2(t)\dot{R}_2(t)\right| \mathrm{d}t \leqslant C_0 \sup_{0<t<T} \left|R_1(t) - R_2(t)\right| \left(T + T^{1/2}\right). \tag{5.25}$$

Let $\delta = \max_{t\in[0,T]}\{R_1(t) - R_2(t)\}$ then

$$\left|R_1^3(t) - R_2^3(t)\right| \leqslant 3C_0\delta\left(T + T^{1/2}\right),$$

since $|R_1^3(t) - R_2^3(t)| \geqslant 3R_0^2|R_1(t) - R_2(t)|$, it follows $\delta \leqslant k_0\delta(T + T^{1/2})$. Furthermore, if $T < T_1 = \min\{1/4k_0^2, 1\}$, necessarily $R_1(t) = R_2(t)$. Since $\mathrm{e}^{r_1 t} z$ and $\mathrm{e}^{r_2 t}\beta$ take their maximum and minimum on $R(t) = R_1(t) = R_2(t)$, and $R(t)$ is zero, $\beta = 0$ and $z = 0$, and we deduce $\sigma = 0$. Repeating the process, starting now from T_1, we conclude the uniqueness of solutions for any $T > 0$ provided $R(T) > 0$. □

REMARK 5.1. Other qualitative properties of the solutions of this type of models have been studied in the literature by different authors. In particular, we mention the study of the asymptotic behavior, when $t \to +\infty$ (see, e.g., BYRNE and CHAPLAIN [1996a], FRIEDMAN and REITICH [1999], CUI and FRIEDMAN [2000], CUI and FRIEDMAN

[2001]) and the continuous dependence and bifurcation phenomena with respect to parameters (see, e.g., BYRNE and CHAPLAIN [1995], FRIEDMAN and REITICH [2001], FRIEDMAN, HU and VELÁZQUEZ [2001], among others).

5.2. *Uniqueness of solutions with radial symmetry*

Let $(\hat{\sigma}, \hat{\beta})$ be a solution of problem (3.1)–(3.5) without forcing term (i.e., $f = 0$). We assume the solution is radially symmetric and define $\sigma = \hat{\sigma} - \overline{\overline{\sigma}}$, $\beta = \hat{\beta} - \overline{\overline{\beta}}$ and $r = |x|$. Then (σ, β) verifies

$$\begin{cases} \dfrac{\partial \sigma}{\partial t} - \dfrac{d_1}{r^2}\dfrac{\partial}{\partial r}\left(r^2 \dfrac{\partial}{\partial r}\sigma\right) \in g_1(\sigma, \beta), & 0 < r < R(t),\ 0 < t < T, \\ \dfrac{\partial \beta}{\partial t} - \dfrac{d_2}{r^2}\dfrac{\partial}{\partial r}\left(r^2 \dfrac{\partial}{\partial r}\beta\right) = g_2(\sigma, \beta), & 0 < r < R(t),\ 0 < t < T, \\ R(t)^2 \dfrac{\mathrm{d}R(t)}{\mathrm{d}t} = \displaystyle\int_0^{R(t)} S(\sigma, \beta) r^2\,\mathrm{d}r, & 0 < t < T, \\ \dfrac{\partial \sigma}{\partial r}(0, t) = 0,\ \dfrac{\partial \beta}{\partial r}(0, t) = 0, & 0 < t < T, \\ \sigma\big(R(t), t\big) = 0,\ \beta\big(R(t), t\big) = 0, & 0 < t < T, \\ R(0) = R_0, & \\ \sigma(r, 0) = \sigma_0(r),\ \beta(r, 0) = \beta_0(r), & 0 < r < R_0, \end{cases} \tag{5.26}$$

where g_i are given by

$$g_1(\sigma, \beta) = -\big[(r_1 + \lambda)\big(\sigma + \overline{\overline{\sigma}}\big) - r_1\sigma_B + \big(\beta + \overline{\overline{\beta}}\big)\big] H\big(\sigma + \overline{\overline{\sigma}} - \sigma_n\big), \tag{5.27}$$

$$g_2(\sigma, \beta) = -r_2\big(\beta + \overline{\overline{\beta}}\big). \tag{5.28}$$

We will assume in this subsection that

$$S \in W^{1,\infty}_{\mathrm{loc}}\big(\mathbb{R}^2\big), \tag{5.29}$$

$$S \text{ is an increasing function in } \sigma \text{ and decreasing in } \beta, \tag{5.30}$$

$$\sigma_n \geqslant \frac{r_1\sigma_B - \overline{\overline{\beta}}}{r_1 + \lambda} \tag{5.31}$$

and the initial data $(\sigma_0 = \hat{\sigma} - \overline{\overline{\sigma}}, \beta_0 = \hat{\beta}_0 - \overline{\overline{\beta}})$ belong to $H^2(0, R_0)$ and satisfy

$$\frac{\partial \sigma_0}{\partial r}(0, t) = 0, \qquad \frac{\partial \beta}{\partial r}(0, t) = 0, \quad 0 < t < T, \tag{5.32}$$

$$\sigma\big(R(t), t\big) = 0, \qquad \beta\big(R(t), t\big) = 0, \quad 0 < t < T. \tag{5.33}$$

THEOREM 5.2. *There is, at most, one solution to* (5.26).

We will use some earlier results in the proof.

LEMMA 5.3. *Every solution* (σ, β) *of the problem* (5.26) *is bounded and satisfies* $\sigma_n \leqslant \sigma \leqslant \sigma_B$ *and* $-\overline{\overline{\beta}} \leqslant \beta \leqslant \max\{\beta_0\}$ *provided* $\sigma_n \leqslant \sigma_0 \leqslant \sigma_B$ *and* $-\overline{\overline{\beta}} \leqslant \beta_0$.

PROOF. By the "integrations by parts formula" (justifying the multiplication of the equation by $T_0(\sigma - \sigma_B)$ and posterior integrations in time and space, see ALT and LUCKHAUS [1983], Lemma 1.5), we have

$$\frac{1}{2}\int_0^{R(t)} \big[T_0(\sigma-\sigma_B)\big]^2 r^2\,\mathrm{d}r \leqslant \int_0^t\int_0^{R(s)} g_1(\sigma,\beta)T_0(\sigma-\sigma_B)r^2\,\mathrm{d}r\,\mathrm{d}s.$$

Since

$$\begin{aligned}
&-\big[(r_1+\lambda)\big(\sigma+\overline{\overline{\sigma}}\big) - r_1\sigma_B + \big(\beta+\overline{\overline{\beta}}\big)\big]H\big(\sigma+\overline{\overline{\sigma}}-\sigma_n\big)T_0(\sigma-\sigma_B)\\
&\quad = -(r_1+\lambda)T_0(\sigma-\sigma_B)^2 - \big[(r_1+\lambda)\big(\sigma_B+\overline{\overline{\sigma}}\big) - r_1\sigma_B + \big(\beta-\overline{\overline{\beta}}\big)\big]T_0(\sigma-\sigma_B)\\
&\quad \leqslant -\big[\lambda\sigma_B + (r_1+\lambda)\overline{\overline{\sigma}} + \big(\beta+\overline{\overline{\beta}}\big)\big]T_0(\sigma-\sigma_B)\\
&\quad \leqslant T^0\big(\beta+\overline{\overline{\beta}}\big)T_0(\sigma-\sigma_B) \leqslant \frac{1}{2}\Big(\big[T^0\big(\beta+\overline{\overline{\beta}}\big)\big]^2 + \big[T_0(\sigma-\sigma_B)\big]^2\Big),
\end{aligned}$$

we obtain

$$\int_0^{R(t)} T_0(\sigma-\sigma_B)^2 r^2\,\mathrm{d}r \leqslant \int_0^t\int_0^{R(s)} \big[T^0\big(\beta+\overline{\overline{\beta}}\big)^2 + T_0(\sigma-\sigma_B)^2\big]r^2\,\mathrm{d}r\,\mathrm{d}s. \tag{5.34}$$

In the same way, we consider $T^0(\beta+\overline{\overline{\beta}})$, and since

$$r_2\big(\beta+\overline{\overline{\beta}}\big)H\big(\sigma+\overline{\overline{\sigma}}-\sigma_n\big)T^0\big(\beta+\overline{\overline{\beta}}\big) \leqslant r_2\big[T^0\big(\beta+\overline{\overline{\beta}}\big)\big]^2,$$

it follows that

$$\int_0^{R(t)} \big[T^0\big(\beta+\overline{\overline{\beta}}\big)\big]^2 r^2\,\mathrm{d}r \leqslant \int_0^t\int_0^{R(s)} r_2 T^0\big(\beta+\overline{\overline{\beta}}\big)r^2\,\mathrm{d}r\,\mathrm{d}s. \tag{5.35}$$

Adding (5.34) and (5.35), we obtain thanks to Gronwall's lemma

$$\sigma \leqslant \sigma_B \quad \text{and} \quad \beta \geqslant -\overline{\overline{\beta}}.$$

Notice that $\beta \geqslant -\overline{\overline{\beta}}$ implies $\hat{\beta} \geqslant 0$.

Let us consider $\varepsilon > 0$ and take $T^0(\sigma - \sigma_n - \varepsilon)$ as test function in the weak formulation, then

$$\frac{1}{2}\int_0^{R(t)} \big[T^0(\sigma-\sigma_n-\varepsilon)\big]^2 r^2\,\mathrm{d}r \leqslant 0.$$

Now, taking limits as $\varepsilon \to 0$, one concludes

$$\frac{1}{2}\int_0^{R(t)} \big[T^0(\sigma-\sigma_n)\big]^2 r^2\,\mathrm{d}r \leqslant 0,$$

which proves $\sigma \geqslant \sigma_n$.

Knowing σ and R, β is well-defined as the unique solution of the equation

$$\frac{\partial\beta}{\partial t} - \frac{d_2}{r^2}\frac{\partial}{\partial r}\left(r^2\frac{\partial}{\partial r}\beta\right) = -r_2\left(\beta+\overline{\overline{\beta}}\right), \quad 0<r<R(t),\ 0<t<T,$$

$$\beta\left(R(t),t\right)=0, \qquad \frac{\partial\beta}{\partial r}=0 \quad \text{on } 0<t<T.$$

Since $\beta_0 \geqslant -\overline{\overline{\beta}}$, it follows that

$$\frac{\partial\beta}{\partial t} - \frac{d_2}{r^2}\frac{\partial}{\partial r}\left(r^2\frac{\partial}{\partial r}\beta\right) \leqslant 0,$$

and we obtain by maximum principle that $\beta \leqslant \max\{\beta_0\}$. □

COROLLARY 5.3. *There exists a positive constant M such that $R(t) \leqslant R_0 \mathrm{e}^{Mt}$ and $R'(t) \leqslant R_0 M \mathrm{e}^{MT}$.*

PROOF. The above result shows $(\sigma(r,t), \beta(r,t)) \in [\sigma_n, \sigma_B] \times [-\overline{\overline{\beta}}, \max\{\beta_0\}]$. Since S is a continuous function, it attains its maximum (denoted by $3M$) on that set. Thus,

$$R^2(t)\frac{\mathrm{d}R(t)}{\mathrm{d}t} \leqslant \int_0^{R(t)} 3Mr^2\,\mathrm{d}r.$$

Integrating the above equation, we have $\mathrm{d}R(t)/\mathrm{d}t \leqslant MR(t)$. Finally, the conclusion follows by Gronwall's lemma. □

REMARK 5.2. As in the previous subsection the solution (σ, β) of (5.26) satisfies

$$\int_0^T \left(\|\sigma\|^2_{W^{1,\infty}(\varepsilon,R(t))} + \|\beta\|^2_{W^{1,\infty}(\varepsilon,R(t))}\right)\mathrm{d}t \leqslant C_1$$

for all $\varepsilon > 0$.

PROOF OF THEOREM 5.2. We argue by contradiction and assume that (σ_1, β_1, R_1) and (σ_2, β_2, R_2) are two solutions of the problem. Let $R(t) := \min\{R_1(t), R_2(t)\}$, $\sigma := \sigma_1 - \sigma_2$ and $\beta := \beta_1 - \beta_2$ be the solution to

$$\begin{cases} \dfrac{\partial\sigma}{\partial t} - \dfrac{d_1}{r^2}\dfrac{\partial}{\partial r}\left(r^2\dfrac{\partial}{\partial r}\sigma\right) = g_1(\sigma_1,\beta_1) - g_1(\sigma_2,\beta_2), & r<R(t),\ 0<t<T, \\ \dfrac{\partial\beta}{\partial t} - \dfrac{d_2}{r^2}\dfrac{\partial}{\partial r}\left(r^2\dfrac{\partial}{\partial r}\beta\right) = g_2(\sigma_1,\beta_1) - g_2(\sigma_2,\beta_2), & r<R(t),\ 0<t<T, \\ \dfrac{\partial\sigma}{\partial r}(0,t)=0, \quad \dfrac{\partial\beta}{\partial r}(0,t)=0, & 0<t<T, \\ \sigma\left(R(t),t\right) = \sigma_1\left(R(t),t\right) - \sigma_2\left(R(t),t\right), & 0<t<T, \\ \beta\left(R(t),t\right) = \beta_1\left(R(t),t\right) - \beta_2\left(R(t),t\right), & 0<t<T, \\ \sigma(r,0)=0, \quad \beta(r,0)=0, & 0<r<R_0. \end{cases} \tag{5.36}$$

Now, we state a technical lemma.

LEMMA 5.4. *$|\beta|$ takes the maximum on the boundary $R(t)$ and σ satisfies*

$$\int_0^{R(t)} \left[T_0(\sigma-\sigma^*)\right]^2 r^2\,\mathrm{d}r \leqslant TC\Big[\max_{t\in[0,T]}\{\beta\}\Big]^2,$$

where

$$\sigma^* = \max_{t\in[0,T]}\big\{\sigma\big(R(t),t\big)\big\}.$$

PROOF. Let us consider $\beta_* = \min\{0, \beta(R(t),t)\}$ and

$$g_2(\beta_1) - g_2(\beta_2) = -r_2\big[\big(\beta_1 - \overline{\overline{\beta}}\big) - \big(\beta_2 - \overline{\overline{\beta}}\big)\big] = -r_2\beta,$$

then

$$\big(g_2(\beta_1) - g_2(\beta_2)\big)T^0(\beta-\beta_*) = -r_2\beta T^0(\beta-\beta_*) \leqslant 0.$$

Multiply the equation by $T^0(\beta-\beta_*)$, we get

$$\int_0^{R(t)} \left[T^0(\beta-\beta_*)\right]^2 r^2\,\mathrm{d}r \leqslant 0$$

and obtain $\beta \geqslant \beta_*$. In the same way, we prove that β takes its maximum on $R(t)$.

Let us consider

$$\begin{aligned}
&g_1(\sigma_1,\beta_1) - g_1(\sigma_2,\beta_2)\\
&\quad= -\Big(\big[(r_1+\lambda)\big(\sigma_1+\overline{\overline{\sigma}}\big) - r_1\sigma_B + \big(\beta_1+\overline{\overline{\beta}}\big)\big]H\big(\sigma_1+\overline{\overline{\sigma}}-\sigma_n\big)\\
&\qquad - \big[(r_1+\lambda)\big(\sigma_2+\overline{\overline{\sigma}}\big) - r_1\sigma_B + \big(\beta_2+\overline{\overline{\beta}}\big)\big]H\big(\sigma_2+\overline{\overline{\sigma}}-\sigma_n\big)\Big)\\
&\quad= (r_1+\lambda)\big[\big(\sigma_1+\overline{\overline{\sigma}}-\sigma_n\big)H\big(\sigma_1+\overline{\overline{\sigma}}-\sigma_n\big) - \big(\sigma_2+\overline{\overline{\sigma}}-\sigma_n\big)H\big(\sigma_2+\overline{\overline{\sigma}}-\sigma_n\big)\big]\\
&\qquad + \big(-(r_1+\lambda)\sigma_n + r_1\sigma_B - \overline{\overline{\beta}}\big)\big(H\big(\sigma_1+\overline{\overline{\sigma}}-\sigma_n\big) - H\big(\sigma_2+\overline{\overline{\sigma}}-\sigma_n\big)\big)\\
&\qquad - \big[\beta_1 H\big(\sigma_1+\overline{\overline{\sigma}}-\sigma_n\big) - \beta_2 H\big(\sigma_2+\overline{\overline{\sigma}}-\sigma_n\big)\big].
\end{aligned}$$

Since $(\sigma+\overline{\overline{\sigma}}-\sigma_n)H(\sigma+\overline{\overline{\sigma}}-\sigma_n)$ is an increasing function of σ, we obtain that

$$\begin{aligned}
&-\big[\big(\sigma_1+\overline{\overline{\sigma}}-\sigma_n\big)H\big(\sigma_1+\overline{\overline{\sigma}}-\sigma_n\big) - \big(\sigma_2+\overline{\overline{\sigma}}-\sigma_n\big)H\big(\sigma_2+\overline{\overline{\sigma}}-\sigma_n\big)\big]\\
&\quad\times T_0(\sigma_1-\sigma_2-\sigma^*) \leqslant 0.
\end{aligned}$$

Since $-(r_1+\lambda)\sigma_n + r_1\sigma_B - \overline{\overline{\beta}} \leqslant 0$, it follows that

$$\begin{aligned}
&\big(-(r_1+\lambda)\sigma_n + r_1\sigma_B - \overline{\overline{\beta}}\big)\big(H\big(\sigma_1+\overline{\overline{\sigma}}-\sigma_n\big) - H\big(\sigma_2+\overline{\overline{\sigma}}-\sigma_n\big)\big)\\
&\quad\times T_0(\sigma_1-\sigma_2-\sigma^*) \leqslant 0.
\end{aligned}$$

Then

$$\begin{aligned}
&\big[g_1(\sigma_1,\beta_1) - g_1(\sigma_2,\beta_2)\big]T_0(\sigma_1-\sigma_2-\sigma^*)\\
&\quad\leqslant -\big[\beta_1 H\big(\sigma_1+\overline{\overline{\sigma}}-\sigma_n\big) - \beta_2 H\big(\sigma_2+\overline{\overline{\sigma}}-\sigma_n\big)\big]T_0(\sigma_1-\sigma_2-\sigma^*)\\
&\quad\leqslant -(\beta_1-\beta_2)H\big(\sigma_2+\overline{\overline{\sigma}}-\sigma_n\big)T_0(\sigma_1-\sigma_2-\sigma^*)\\
&\quad\leqslant -T^0(\beta_1-\beta_2)T_0(\sigma_1-\sigma_2-\sigma^*) \leqslant -\beta_* T_0(\sigma_1-\sigma_2-\sigma^*).
\end{aligned}$$

Multiplying the equation, as before, by $T_0(\sigma-\sigma^*)$, we get

$$\int_0^{R(t)}\left[T_0(\sigma-\sigma^*)\right]^2 r^2\,\mathrm{d}r+\int_0^t\int_0^{R(s)}\left[\frac{\partial}{\partial r}T_0(\sigma-\sigma^*)\right]^2 r^2\,\mathrm{d}r\,\mathrm{d}s$$
$$=\int_0^t\int_0^{R(s)}\big(g_1(\sigma_1,\beta_1)-g_1(\sigma_2,\beta_2)\big)T_0(\sigma-\sigma^*)r^2\,\mathrm{d}r\,\mathrm{d}s$$
$$\leqslant-\int_0^t\int_0^{R(s)}\beta_* T_0(\sigma-\sigma^*)r^2\,\mathrm{d}r\,\mathrm{d}s$$
$$\leqslant\frac{T\widetilde{C}}{\lambda}\beta_*^2+\lambda\int_0^t\int_0^{R(s)}\left[T_0(\sigma_1-\sigma_2-\sigma^*)\right]^2 r^2\,\mathrm{d}r\,\mathrm{d}s.$$

Now, we choose λ such that

$$\lambda\int_0^{R(s)}\left[T_0(\sigma_1-\sigma_2-\sigma^*)\right]^2 r^2\,\mathrm{d}r$$
$$-\int_0^{R(s)}\left[\frac{\partial}{\partial r}T_0(\sigma-\sigma^*)\right]^2 r^2\,\mathrm{d}r\leqslant 0\quad\text{a.e. } t\in(0,T),$$

then,

$$\int_0^{R(t)}\left[T_0(\sigma-\sigma^*)\right]^2 r^2\,\mathrm{d}r\leqslant TC\beta_*^2$$

holds, which ends the proof. □

END OF THE PROOF OF THEOREM 5.2. Let us define

$$\delta=\max_{t\in[0,T]}\left\{\left|R_1(t)-R_2(t)\right|\right\}\geqslant 0,$$

and consider

$$R_1^2(t)R_1'(t)-R_2^2(t)R_2'(t)$$
$$=\int_0^{R(t)}\big(S(\sigma_1,\beta_1)-S(\sigma_2,\beta_2)\big)r^2\,\mathrm{d}r$$
$$+\int_{R(t)}^{R_1(t)}S(\sigma_1,\beta_1)r^2\,\mathrm{d}r-\int_{R(t)}^{R_2(t)}S(\sigma_2,\beta_2)r^2\,\mathrm{d}r.\tag{5.37}$$

By (5.29) and Lemma 5.3, we obtain

$$\left|\int_{R(t)}^{R_i(t)}S(\sigma_i,\beta_i)r^2\,\mathrm{d}r\right|\leqslant M\delta\quad(\text{for } i=1,2),\tag{5.38}$$

where

$$M=\max\left\{S(\sigma,\beta)\text{ for any }(\sigma,\beta)\in[\sigma_n,\sigma_B]\times\left[\overline{\overline{\beta}},\max\{\beta_0\}\right]\right\}.$$

(5.29) and (5.30) imply

$$\int_0^{R(t)}\big(S(\sigma_1,\beta_1)-S(\sigma_2,\beta_2)\big)r^2\,\mathrm{d}r\leqslant C\int_0^{R(t)}\big(T_0(\sigma)-T^0(\beta)\big)r^2\,\mathrm{d}r.$$

Since $T_0(\sigma) \leqslant T_0(\sigma - \sigma^*) + \sigma^*$ and $-T^0(\beta) \leqslant -\beta_*$, we obtain

$$\begin{aligned}\int_0^{R(t)} \big(S(\sigma_1, \beta_1) - S(\sigma_2, \beta_2)\big) r^2 \,\mathrm{d}r &\leqslant C \int_0^{R(t)} \big(T_0(\sigma - \sigma^*) + \sigma^* - \beta_*\big) r^2 \,\mathrm{d}r \\ &\leqslant C' \left(\left[\int_0^{R(t)} T_0(\sigma - \sigma^*)^2 r^2 \,\mathrm{d}r \right]^{1/2} + \sigma^* - \beta_* \right).\end{aligned}$$

By Lemma 5.4, it follows that

$$C' \left(\left[\int_0^{R(t)} T_0(\sigma - \sigma^*)^2 r^2 \,\mathrm{d}r \right]^{1/2} + \sigma^* - \beta_* \right) \leqslant C'' \big(\sigma^* - (T+1)\beta_*\big).$$

Since $\sigma_i(R_i(t), t) = 0$ (for $j = 1$ or 2), we obtain

$$\big|\sigma\big(R(t), t\big)\big| \leqslant \left(\sum_{i=1,2} \|\sigma_i\|_{W^{1,\infty}(R(t), R_i(t))} \right) \big|R_1(t) - R_2(t)\big|,$$

$$\big|\beta\big(R(t), t\big)\big| \leqslant \left(\sum_{i=1,2} \|\beta_i\|_{W^{1,\infty}(R(t), R_i(t))} \right) \big|R_1(t) - R_2(t)\big|$$

and then

$$\int_0^{R(t)} \big(S(\sigma_1, \beta_1) - S(\sigma_2, \beta_2)\big) r^2 \,\mathrm{d}r \leqslant C(T+2)\delta. \tag{5.39}$$

Integrating in time in (5.37), we get thanks to (5.38) and (5.39) that

$$R_1^3(t) - R_2^3(t) \leqslant TC(T+2)\delta + 2TM\delta. \tag{5.40}$$

On the other hand, one has

$$R_1^3(t) - R_2^3(t) = \big(R_1(t) - R_2(t)\big)\big(R_1^2 + R_1 R_2 + R_1^3\big).$$

We can assume without lost of generality that $\delta = R_1(t_0) - R_2(t_0)$ (for some $t_0 \in [0, T]$), hence

$$R_1^3(t) - R_2^3(t) \geqslant 4R^2\delta.$$

Substituting this into (5.40) leads to $\delta \leqslant k_0 \delta T$. Furthermore, taking $T_1 < 1/k_0$ necessitates $R_1(t) = R_2(t)$ for any $t \in [0, T_1]$. Since $|\beta|$ takes its maximum at $R(t) = R_1(t) = R_2(t)$ (and this maximum is 0), we get that $\beta = 0$. Substituting in (5.36) and taking σ as test function, we obtain

$$\int_0^{R(t)} \sigma^2 r^2 \,\mathrm{d}r \leqslant \int_0^t \int_0^{R(s)} \big(g_1(\sigma_1, \beta_1) - g_1(\sigma_2, \beta_2)\big) \sigma r^2 \,\mathrm{d}r \,\mathrm{d}s.$$

As in Lemma 5.4, since $(\sigma_i + \overline{\overline{\sigma}}_i - \sigma_n) H(\sigma_i + \overline{\overline{\sigma}} - \sigma_n)$ is a increasing function of σ, we obtain by (5.27) and Lemma 5.3 that $(g_1(\sigma_1, \beta_1) - g_1(\sigma_2, \beta_2))\sigma \leqslant 0$, which prove $\sigma = 0$.

Repeating the above process, starting now from T_1, we get the uniqueness of solutions for arbitrary $T > 0$, provided $R(T) > 0$. □

6. Approximate controllability

In this section we study the controllability of distribution of nutrients (in the usual weak sense of parabolic system) by the internal localized action of inhibitors. The main results of this section is the following theorem.

THEOREM 6.1. *Given $T > 0$, $\omega_0 \subset B(R_0 \exp\{-\|S\|_{L^\infty} T\})$, $\varepsilon > 0$, and $\hat{\sigma}^d \in L^p_{\mathrm{loc}}(\mathbb{R}^3)$, for some $p > 1$, there exists $f \in L^p((0,T) \times \omega_0)$ such that, if (σ, β, R) is the solution of the problem* (5.1)–(5.5), *then*

$$\|\sigma(T) - \sigma^d\|_{L^p(B(R(T)))} \leqslant \varepsilon, \tag{6.1}$$

where $\sigma^d := \hat{\sigma}^d \chi_{B(R(T))}$.

Due to some technical reasons, we shall prove the theorem firstly for $p > 4$. This assumption is a prerequisite in order to obtain the boundedness of the solution in the proof of Lemma 5.1 in view of the Sobolev compact embedding $W^{1,p}((0,T) \times B) \subset L^\infty((0,T) \times B)$. Finally, we prove the theorem for any $p > 1$ by Hölder inequality.

We shall establish the result in several steps. For $n \in \mathbb{N}$, we start by assuming $R_n(t)$ prescribed and look for a control f_n in ω_0 such that the solution (σ_n, β_n) of problem (5.1), (5.2), (5.4) and (5.5), satisfies (6.1). Then we obtain R_{n+1} and f_{n+1} from (σ_n, β_n) which allows us to find $(\sigma_{n+1}, \beta_{n+1})$. The proof of the theorem relies mostly on methods introduced in the study of approximate controllability (notion attributed to conclusions such as (6.1)) by different authors (see LIONS [1990], LIONS [1991], FABRE, PUEL and ZUAZUA [1995], GLOWINSKI and LIONS [1995] and DÍAZ and RAMOS [1995]). Iterating the process, we obtain a sequence $(R_n, f_n, \sigma_n, \beta_n)$ such as we shall show possesses a subsequence that converges to the searched control f and the associate solution of problem (5.1)–(5.5).

The next result shows the conclusion of Theorem 6.1 (the so-called approximate controllability in L^p) under some particular assumptions (mainly, $R(t)$ is a priori prescribed).

PROPOSITION 6.1. *Let $\omega_0 \subset B(R_0 \exp\{-\|S\|_{L^\infty} T\})$ and $\sigma_0 = \beta_0 - \overline{\overline{\sigma}} = \overline{\overline{\beta}} = 0$. Let $R \in W^{1,\infty}(0,T)$ a given function such that $R(0) = R_0$, $|\dot{R}| \leqslant \|S\|_{L^\infty} R_0 \exp\{|S|_{L^\infty} T\}$. Then, given $\ddot{\sigma}^d \in L^2_{\mathrm{loc}}(\mathbb{R}^3)$, there exists $f \in L^p(\omega_0 \times (0,T))$, with $p > 4$, such that, if (σ, β) is the solution of problem* (5.1), (5.2), (5.4) *and* (5.5), *then*

$$\|\sigma(T) - \sigma^d\|_{L^p(B(R(T)))} \leqslant \varepsilon,$$

where $\sigma^d = \hat{\sigma}^d|_{B(R(T))}$.

PROOF. Let $p' = p/(p-1)$ and consider the functional $J : L^{p'}(B(R(T))) \to \mathbb{R}$ defined by

$$J(\varphi^0) = \frac{1}{p'} \int_0^T \int_{\omega_0} |\psi(x,t)|^{p'} \,\mathrm{d}x \,\mathrm{d}t + \varepsilon \|\varphi^0\|_{L^{p'}(B(R(T)))} - \int_{B(R(T))} \sigma^d \varphi^0 \,\mathrm{d}x,$$

where $\varphi_0 \in L^{p'}(B(R(T)))$, and (φ, ψ) is the solution to the adjoint problem

$$-\frac{\partial \varphi}{\partial t} - d\Delta\varphi + r_1\varphi = 0, \quad |x| < R(t),\ t \in (0, T), \tag{6.2}$$

$$-\frac{\partial \psi}{\partial t} - d\Delta\psi + r_2\psi + \lambda\varphi = 0, \quad |x| < R(t),\ t \in (0, T), \tag{6.3}$$

$$\varphi(x, T) = \varphi_0(x), \qquad \psi(x, T) = 0, \quad |x| < R(T), \tag{6.4}$$

$$\varphi(x, t) = 0, \qquad \psi(x, t) = 0, \quad |x| = R(t),\ t \in (0, T). \tag{6.5}$$

We point out that the existence of a weak solution (φ, ψ) of (6.2)–(6.5) can be obtained as in Section 5, by employing (4.3)–(4.5) and (5.6).

In order to prove the uniqueness of solutions by contradiction, we assume that there exist two solutions (φ_1, ψ_1), (φ_2, ψ_2). Then $\varphi := \varphi_1 - \varphi_2$ satisfies (6.2) and taking $|\varphi|^{p'-2}\varphi$ as test function and integrating by parts it follows that

$$-\frac{\mathrm{d}}{\mathrm{d}t}\int_{B(R(t))} |\varphi|^{p'}\,\mathrm{d}x \leqslant r_1 \int_{B(R(t))} |\varphi|^{p'}\,\mathrm{d}x.$$

We obtain $\varphi = \varphi_1 - \varphi_2 = 0$ by Gronwall's lemma. Having proved $\varphi \equiv 0$, in the same way, $\psi := \psi_1 - \psi_2$ satisfies (6.3) and taking $|\psi|^{p'-2}\psi$ as test function, we obtain $\psi \equiv 0$, which proves the uniqueness.

Let us assume that J is convex, continuous and coercive (in the sense that $\liminf J \to \infty$ as $\|\varphi^0\|_{L^{p'}(B(R_0))} \to \infty$), facts, which shall be proved at the end of the proposition. Then J takes a minimum φ_0 (see BREZIS [1983], Corollary III.20). Moreover, if (ξ, ζ) is the solution of the problem (6.2)–(6.5) with datum $(\xi^0, 0)$, we have

$$\int_0^T \int_{\omega_0} |\psi|^{p'-2}\psi\zeta\,\mathrm{d}x\,\mathrm{d}t - \int_{B(R(T))} \sigma^d \xi^0\,\mathrm{d}x$$
$$+ \varepsilon \|\varphi^0\|^{1-p'}_{L^{p'}(B(R(T)))} \int_{B(R(T))} |\varphi^0|^{p'-2}\varphi^0\xi^0\,\mathrm{d}x = 0. \tag{6.6}$$

Multiplying (5.1), (5.2) by (ξ, ζ), integrating by parts and applying Leibnitz theorem, we arrive at

$$-\int_0^T \left\langle \sigma, \frac{\partial \xi}{\partial t}\right\rangle \mathrm{d}t - d\int_0^T \langle \sigma, \Delta\xi\rangle\,\mathrm{d}t + \int_0^T \int_{B(R(t))} r_1\sigma\xi\,\mathrm{d}x\,\mathrm{d}t$$
$$+ \int_0^T \int_{B(R(t))} \lambda\beta\xi\,\mathrm{d}x\,\mathrm{d}t - \int_0^T \left\langle \beta, \frac{\partial \zeta}{\partial t}\right\rangle \mathrm{d}t - d\int_0^T \langle \beta, \Delta\zeta\rangle\,\mathrm{d}t$$
$$+ \int_0^T \int_{B(R(t))} r_2\beta\zeta\,\mathrm{d}x\,\mathrm{d}t - \int_0^T \int_{\omega_0} f\zeta\,\mathrm{d}x\,\mathrm{d}t + \int_{B(R(t))} \sigma\xi\,\mathrm{d}x\Big]_0^T$$
$$+ \int_{B(R(t))} \beta\zeta\,\mathrm{d}x\Big]_0^T = 0,$$

where $\langle\ ,\ \rangle$ is the duality product $W_0^{1,p'}(B(R(t))) \times W^{-1,p'}(B(R(t)))$. We obtain from the choice of (ξ, ζ) and $\sigma(0, x) = \beta(0, x) = 0$ that

$$-\int_0^T \int_{\omega_0} f\zeta\,\mathrm{d}x\,\mathrm{d}t + \int_{B(R(T))} \sigma(T)\xi^0\,\mathrm{d}x = 0. \tag{6.7}$$

Let us take

$$f := |\psi|^{p'-2}\psi.$$

Substituting this into (6.7) and using (6.6), one has

$$\int_{B(R(T))} (\sigma(T) - \sigma^d)\xi^0 \,\mathrm{d}x + \varepsilon \|\varphi^0\|^{1-p'}_{L^{p'}(B(R(T)))} \int_{B(R(T))} |\varphi^0|^{p'-2}\varphi^0\xi^0 \,\mathrm{d}x = 0,$$

for all $\xi^0 \in L^{p'}(B(R(T)))$. Taking

$$\xi^0 = (\sigma(T) - \sigma^d)^{\frac{1}{p'-1}} \in L^{p'}(B(R(T))),$$

we obtain in view of $p = 1 + 1/(p'-1)$ that

$$\begin{aligned}
&\|\sigma(T) - \sigma^d\|^p_{L^p(B(R(T)))} \\
&\quad = \varepsilon \|\varphi^0\|^{1-p'}_{L^{p'}(B(R(T)))} \int_{B(R(T))} |\varphi^0|^{p'-2}\varphi^0 |\sigma(T) - \sigma^d|^{\frac{1}{p'-1}-1}(\sigma(T) - \sigma^d)\,\mathrm{d}x.
\end{aligned}$$

By Hölder inequality, we have

$$\begin{aligned}
&\|\varphi^0\|^{1-p'}_{L^{p'}(B(R(T)))} \int_{B(R(T))} |\varphi^0|^{p'-2}\varphi^0 |\sigma(T) - \sigma^d|^{\frac{1}{p'-1}-1}(\sigma(T) - \sigma^d)\,\mathrm{d}x \\
&\quad \leqslant \|\sigma(T) - \sigma^d\|^{p-1}_{L^p(B(R(T)))},
\end{aligned}$$

which leads to

$$\|\sigma(T) - \sigma^d\|_{L^p(B(R(T)))} \leqslant \varepsilon$$

and the conclusion holds.

So, it only remains to check the mentioned properties of J:

J is convex. We can write J as the sum of the functionals

$$J_1(\varphi^0) := -\int_{B(R(T))} \sigma^d\varphi^0 \,\mathrm{d}x, \qquad J_2(\varphi^0) := \varepsilon \|\varphi^0\|_{L^{p'}(B(R(T)))},$$

$$J_3(\varphi^0) := \frac{1}{p'} \int_0^T \int_{B(R(t))} |\psi|^{p'} \,\mathrm{d}x\,\mathrm{d}t.$$

First, we shall see that J_3 is convex. Let (φ_1, ψ_1) and (φ_2, ψ_2) be the solutions to (6.2)–(6.5) with datum φ_1^0, $\varphi_2^0 \in L^p(B(R(T)))$, respectively. Then, since the system is linear, we get, for $\alpha \in (0, 1)$,

$$J_3(\alpha\varphi_1^0 + (1-\alpha)\varphi_2^0) = \frac{1}{p'} \int_0^T \int_{B(R(t))} (|\alpha\psi_1 + (1-\alpha)\psi_2|^{p'})\,\mathrm{d}x\,\mathrm{d}t$$

and then

$$\begin{aligned}
&J_3(\alpha\varphi_1^0 + (1-\alpha)\varphi_2^0) - \alpha J_3(\varphi_1^0) - (1-\alpha)J_3(\varphi_2^0) \\
&\quad = \frac{1}{p'} \int_0^T \int_{B(R(t))} (|\alpha\psi_1 + (1-\alpha)\psi_2|^{p'} - \alpha|\psi_1|^{p'} - (1-\alpha)|\psi_2|^{p'})\,\mathrm{d}x\,\mathrm{d}t.
\end{aligned}$$

Since $p' > 1$, we obtain

$$\left|\alpha\psi_1 + (1-\alpha)\psi_2\right|^{p'} - \alpha|\psi_1|^{p'} - (1-\alpha)|\psi_2|^{p'} \leqslant 0,$$

and integrating, we have

$$\frac{1}{p'}\int_0^T \int_{B(R(t))} \left(\left|\alpha\psi_1 + (1-\alpha)\psi_2\right|^{p'} - \alpha|\psi_1|^{p'} - (1-\alpha)|\psi_2|^{p'}\right) \mathrm{d}x\, \mathrm{d}t \leqslant 0,$$

which proves the convexity of J_3. Finally, J_1 is linear and so convex and since $\|\cdot\|_{L^{p'}(B(R(T))}$ is convex, J_2 is also convex.

J is continuous. By construction, J_1 and J_2 are continuous. We are going to prove that J_3 is also continuous. Let $\varphi_n^0 \in L^{p'}(B(R(T)))$ such that $\varphi_n^0 \to \varphi^0$ and let (φ_n, ψ_n), (φ, ψ) be the solutions to (6.2)–(6.5) with datum φ_n^0 and φ^0. Subtracting both systems and taking

$$\left(p'|\varphi - \varphi_n|^{p'-2}(\varphi - \varphi_n), p'|\psi - \psi_n|^{p'-2}(\psi - \psi_n)\right)$$

as test function, using the integration by parts formula (see, e.g., ALT and LUCKHAUS [1983]) and Young's inequality, we arrive at

$$-\frac{\partial}{\partial t}\int_{B(R(t))} \left[|\varphi - \varphi_n|^{p'} + |\psi - \psi_n|^{p'}\right] \mathrm{d}x$$
$$+ \int_{B(R(t))} \left(r_1 p' - |\lambda|\right)|\varphi - \varphi_n|^{p'} \mathrm{d}x + \int_{B(R(t))} \left(r_2 p' - |\lambda|\right)|\psi - \psi_n|^{p'} \mathrm{d}x \leqslant 0.$$

Let X_n be defined by

$$X_n(t) = \|\varphi - \varphi_n\|^{p'}_{L^{p'}(B(R(t)))} + \|\psi - \psi_n\|^{p'}_{L^{p'}(B(R(t)))},$$

then,

$$-X_n'(t) \leqslant C X_n(t), \quad t \in (0,T), \qquad X_n(T) = \left\|\varphi_n^0 - \varphi^0\right\|^{p'}_{L^{p'}(B(R(T))}$$

are satisfied, where $C = \max\{-r_1 p' + |\lambda|, -r_2 p' + |\lambda|\}$. Thus, we obtain

$$0 \leqslant X_n(t) \leqslant \left|X_n(T)\right| \mathrm{e}^{-C(t-T)}.$$

Since

$$0 \leqslant \int_{\omega_0} |\psi - \psi_n|^{p'} \mathrm{d}x \leqslant X_n(t),$$

we conclude by integrating over $[0, T]$ and taking limits as $n \to \infty$ that

$$\int_0^T \int_{\omega_0} |\psi - \psi_n|^{p'} \mathrm{d}x\, \mathrm{d}t \leqslant \int_0^T X_n(t)\, \mathrm{d}t \to 0,$$

which proves the continuity of J_3.

J is coercive. Let $\varphi_n^0 \in L^{p'}(B(R(T)))$ such that $\|\varphi_n^0\|_{L^{p'}(B(R(T)))} \to \infty$, when $n \to \infty$. We claim

$$\liminf_{n\to\infty} \frac{J(\varphi_n^0)}{\|\varphi_n^0\|_{L^{p'}(B(R(T)))}} \geqslant \varepsilon.$$

Let

$$I := \liminf_{n\to\infty} \frac{J(\varphi_n^0)}{\|\varphi_n^0\|_{L^{p'}(B(R(T)))}} \geqslant -\left\|\sigma^d\right\|_{L^p(B(R(T)))}.$$

Then, there exists a minimizing subsequence (which we call again by φ_n^0) such that

$$\lim_{n\to\infty} \frac{J(\varphi_n^0)}{\|\varphi_n^0\|_{L^{p'}(B(R(T)))}} = I.$$

We define

$$\bar{\varphi}_n^0 := \frac{\varphi_n^0}{\|\varphi_n^0\|_{L^{p'}(B(R(T)))}},$$

and let $(\bar{\varphi}_n, \bar{\psi}_n)$ be the solution to (6.2)–(6.5) with data $(\bar{\varphi}_n^0, 0)$. Since the system is linear, we have

$$(\bar{\varphi}_n, \bar{\psi}_n) = \frac{1}{\|\varphi_n^0\|_{L^{p'}}} (\varphi_n, \psi_n).$$

Then

$$\frac{J(\varphi_n^0)}{\|\varphi_n^0\|_{L^{p'}(B(R(T)))}} = \left\|\varphi_n^0\right\|^{p'-1} \int_0^T \int_{\omega_0} |\bar{\psi}_n|^{p'} \,\mathrm{d}x\,\mathrm{d}t - \int_{B(R(T))} \sigma^d \bar{\varphi}_n^0 \,\mathrm{d}x + \varepsilon.$$

Now, it is clear that, if

$$\liminf_{n\to\infty} \int_0^T \int_{\omega_0} \bar{\psi}_n^{p'} \,\mathrm{d}x \geqslant \alpha_0, \tag{6.8}$$

for some positive α_0, then

$$\frac{J(\varphi_n^0)}{\|\varphi_n^0\|_{L^{p'}(B(R(T)))}} \geqslant \alpha_0 \left\|\varphi_n^0\right\|^{p'-1}_{L^{p'}(B(R(T)))} + \varepsilon - \left\|\sigma^d\right\|_{L^p(B(R(T)))} \to \infty$$

as $n \to \infty$, which proves the property. Let us assume that

$$\liminf \int_0^T \int_{\omega_0} |\bar{\psi}_n|^{p'} \,\mathrm{d}x = 0.$$

Then there exists a subsequence $\bar{\psi}_{n_i}$ such that

$$\int_0^T \int_{\omega_0} |\bar{\psi}_{n_i}|^{p'} \,\mathrm{d}x\,\mathrm{d}t \to 0,$$

therefore $\bar{\psi}_{n_i} \to 0$ in $L^{p'}(\omega_0 \times [0, T])$. Taking $(0, \zeta)$ as test function in (6.3), where $\zeta \in C_c^2((0, T) \times \omega_0)$, we obtain

$$\int_0^T \int_{\omega_0} \bar{\psi}_{n_i} \frac{\partial \zeta}{\partial t} \, dx \, dt - \int_0^T \int_{\omega_0} \bar{\psi}_{n_i} \Delta \zeta \, dx \, dt \\ - r_2 \int_0^T \int_{\omega_0} \bar{\psi}_{n_i} \zeta \, dx \, dt + \lambda \int_0^T \int_{\omega_0} \bar{\varphi}_{n_i} \zeta \, dx \, dt = 0.$$

Taking limits, we conclude that

$$\int_0^T \int_{\omega_0} \bar{\varphi}_{n_i} \zeta \, dx \, dt \to 0, \tag{6.9}$$

where $\bar{\varphi}_{n_i}$ is the solution to

$$\begin{cases} -\dfrac{\partial \bar{\varphi}_{n_i}}{\partial t} - d\Delta \bar{\varphi}_{n_i} - r_1 \bar{\varphi}_{n_i} = 0, & |x| < R(t), \ t \in (0, T), \\ \bar{\varphi}_{n_i}(t, x) = 0, & |x| = R(t), \ t \in (0, T), \\ \bar{\varphi}_{n_i}(T, x) = \bar{\varphi}^0, & |x| < R_0. \end{cases} \tag{6.10}$$

Repeating the change of (5.6) and introducing the unknown

$$\bar{u}_{n_i}(\tilde{x}, \tilde{t}) := \bar{\varphi}_{n_i}\big(R\big(t(\tilde{t})\big)\tilde{x}, t(\tilde{t})\big),$$

we obtain

$$\begin{cases} -\dfrac{\partial \bar{u}_{n_i}}{\partial \tilde{t}} - d\Delta \bar{u}_{n_i} - R^2 R' \tilde{x} \cdot \nabla \bar{u}_{n_i} + R^2 r_1 \bar{u}_{n_i} = 0, & B \times (0, \widetilde{T}), \\ \bar{u}_{n_i}(\tilde{x}, \tilde{t}) = 0, & \partial B \times (0, \widetilde{T}), \\ \bar{u}_{n_i}(\tilde{x}, \widetilde{T}) = \bar{u}^0_{n_i}(\tilde{x}) = \bar{\varphi}^0_{n_i}(\tilde{x} R_0), & \tilde{x} \in B. \end{cases} \tag{6.11}$$

Since $\bar{u}^0_{n_i} \rightharpoonup \bar{u}_0$ belongs to $L^{p'}(B)$, it follows that $\bar{u}_{n_i} \rightharpoonup \bar{u}$ (the solution of (6.11) with $\bar{u}_0 = \bar{\varphi}^0$). By (6.9), $\bar{u}_{n_i} \to 0$ weakly in $L^{p'}(B(\widehat{\omega}_0))$, where $\widehat{\omega}_0$ is an open subset of B such that $\widehat{\omega}_0 \subset \widetilde{\omega}_0$. Consequently, $\bar{u} \equiv 0$ on $\widetilde{\omega}_0$ for all $0 \leqslant \tilde{t} \leqslant \widetilde{T}$. By the unique continuation of the solution to Eq. (6.11) (see FRIEDMAN [1964], CHI-CHEUNG POON [1996], Theorem 1.1′), we deduce that $\bar{u} = 0$ in $B \times (0, \widetilde{T})$, which implies $\bar{u}_0 \equiv 0$ and $\bar{\varphi}^0 \equiv 0$ by uniqueness of (6.11). Furthermore,

$$-\int_{B(R(T))} \sigma^d \bar{\varphi}^0 \, dx = 0$$

and $I = \varepsilon$, which proves the coerciveness of J. □

PROOF OF THEOREM 6.1. We consider the function $\theta : C^1([0, T]) \to H^2(0, T)$, $\theta(R^*) = R$, where R is defined by

$$R^2(t)\dot{R}(t) = \int_{B(R^*(t))} S\big(\sigma + \sigma^s, \beta + \beta^s\big) \, dx, \qquad R^*(0) = R_0,$$

where (σ^s, β^s) is the solution to the problem (5.1), (5.2), (5.4) and (5.5), with $f \equiv 0$, and initial data $\sigma^s_{n-1}(x,0) = \sigma_0(x)$, $\beta^s(x,0) = \beta_0(x)$, and (σ, β) is the solution mentioned in Proposition 6.1. Since S is bounded, $R \in W^{1,\infty}(0,T)$. By Proposition 6.1, for each R^* there exists a minimum function φ^0_n which minimize the functional

$$J(\varphi^0) := \frac{1}{p'}\int_0^T \int_{\omega_0} |\psi|^{p'}\,\mathrm{d}x\,\mathrm{d}t + \varepsilon \|\varphi^0\|_{L^{p'}(B(R^*(T)))} - \int_{B(R^*(T))} \sigma^d \varphi^0\,\mathrm{d}x,$$

where $\sigma^d = \hat{\sigma}^d \chi_{B(R^*(T))}$. We are going to show that $\|\varphi^0\|_{L^{p'}(B(R^*(T)))}$ is uniformly bounded. To the contrary, we assume that there exists a sequence φ^0_n such that $\|\varphi^0_n\|_{L^{p'}(B(R^*(T)))} \to \infty$ and get

$$\begin{aligned}\frac{J(\varphi^0_n)}{\|\varphi^0_n\|_{L^{p'}}} &= \frac{1}{p'}\|\varphi^0_n\|^{p'-1}_{L^{p'}(B(R^*(T)))}\int_0^T\int_{\omega_0} |\bar{\psi}_n|^{p'}\,\mathrm{d}x\,\mathrm{d}t \\ &\quad + \varepsilon - \int_{B(R^*(T))} \sigma^d_n \bar{\varphi}^0_n\,\mathrm{d}x \leqslant 0 \end{aligned} \tag{6.12}$$

in view of $J_n(\varphi^0_n) \leqslant 0$. Since

$$\left|\int_{B(R^*(T))} \frac{\sigma^d_n \varphi^0_n}{\|\varphi^0_n\|_{L^{p'}(B(R^*(T)))}}\,\mathrm{d}x\right| \leqslant \|\sigma^d_n\|_{L^p(B(R^*(T)))} \leqslant \|\hat{\sigma}^d\|_{L^p(B(R_0\exp\{MT\}))},$$

it follows, by (6.12) that

$$\int_0^T\int_{\omega_0} |\bar{\psi}_n|^{p'}\,\mathrm{d}x\,\mathrm{d}t \to 0 \quad \text{when } n \to \infty.$$

Using the same argument as in the proof of coerciveness of J, we obtain

$$\bar{\varphi}^0_n \rightharpoonup 0 \quad \text{in } L^{p'}\big(B\big(R^*(T)\big)\big)$$

and

$$\liminf_{n\to\infty} \frac{J_n(\varphi^0_n)}{\|\varphi^0_n\|} \geqslant \varepsilon,$$

which contradicts (6.12). Consequently $\|\varphi^0_n\|_{L^{p'}(B(R^*(T)))}$ is uniformly bounded, hence $\|\varphi_n\|_{L^{p'}(B(R^*(T)))}$ is uniformly bounded. Furthermore, the set of controls is uniformly bounded. Performing the change of (4.3)–(4.5) and (5.6), applying Lemma 5.1, we obtain that θ is continuous and compact. Then, there exists a fixed point (σ, β, R) which satisfies (5.1)–(5.5) and condition (6.1). Thus the theorem is proved in the case $p > 4$.

In the case $p \leqslant 4$, we consider the control f for any $s > 4$, for instance $f \in L^5((0,T)\times\Omega)$, then

$$\begin{aligned}\|\sigma(T) - \sigma^d\|_{L^p(B(R(T)))} &\leqslant \left(\frac{3\pi}{4}\operatorname{meas}\big\{B\big(R(T)\big)\big\}\right)^{\frac{5}{p(5-p)}} \|\sigma(T) - \sigma^d\|_{L^5(B(R(T)))} \\ &\leqslant \varepsilon\left(\frac{3\pi}{4}\exp\big\{T\|S\|_{L^\infty}\big\}\right)^{\frac{5}{p(5-p)}},\end{aligned}$$

setting

$$\varepsilon = \varepsilon' \left(\frac{3\pi}{4} \exp\{T \|S\|_{L^\infty}\} \right)^{-\frac{p(5-p)}{5}},$$

we obtain the theorem. □

REMARK 6.1. Notice that the final observation is made regarding the density $\sigma(T, \cdot)$ and that once we have chosen the control to obtain (6.1). The free boundary, $R(t)$, and the inhibitor density $\beta(T, \cdot)$ are univocally determined.

REMARK 6.2. There exists a long literature on the application of Optimization and Control Theory to different mathematical tumor growth models. We refer the interested reader to the works by SWAM [1984], FISTER, LENHART and MCNALLY [1998], BELLOMO and PREZIOSI [2000] and the references therein.

7. Numerical analysis

In this section we establish a numerical solution to the problem (5.1)–(5.5) by employing a time discretization scheme which is implicit with respect to u and v and explicit for the free boundary R. We assume radial symmetry, no forcing terms (i.e., $f = 0$), and a nonnecrotic core. Let $x := r_1/R(t)$ and

$$u(x,t) = \sigma\bigl(xR(t),t\bigr) - \overline{\overline{\sigma}}, \qquad v(x,t) = \beta\bigl(xR(t),t\bigr) - \overline{\overline{\beta}}.$$

Then, problem (3.1)–(3.5) becomes

$$\begin{aligned}
&\frac{\partial u}{\partial t} = \frac{d_1}{x^2R^2}\frac{\partial}{\partial x}\left(x^2\frac{\partial}{\partial x}u\right) + x\frac{R'}{R}\frac{\partial u}{\partial x} - r_1 u - \lambda v + r_1\overline{\overline{\sigma}} + \lambda\overline{\overline{\beta}}, \quad (0,1)\times(0,T),\\
&\frac{\partial v}{\partial t} = \frac{d_2}{x^2R^2}\frac{\partial}{\partial x}\left(x^2\frac{\partial v}{\partial x}\right) + x\frac{R'}{R}\frac{\partial}{\partial x}v - r_2 v + r_2\overline{\overline{\beta}}, \quad (0,1)\times(0,T),\\
&R(t) = R_0 \exp\left\{\int_0^t \int_0^1 x^2 S(u,v)\,\mathrm{d}x\,\mathrm{d}t\right\}, \quad t>0,\\
&u_x(0,t) = v_x(0,t) = u(1,t) = v(1,t) = 0, \quad t>0,\\
&R(0) = R_0, \qquad u(x,0) = u_0(x), \qquad v(x,0) = v_0(x), \quad x\in(0,1).
\end{aligned}$$

7.1. *Time discretization*

Let $N \in \mathbb{N}$, $n = 1, \ldots, N$ and $t_n = n(T/N)$. We introduce the approximations

$$\begin{aligned}
&u^n(x) \approx u(x,t_n), \qquad v^n(x) \approx v(x,t_n), \qquad R^n \approx R(t_n),\\
&\dot{R}^n \approx \frac{\mathrm{d}R(t)}{\mathrm{d}t} \quad \text{in } t = t_n,
\end{aligned}$$

defined by the following algorithm:

Step 0:

$$(0.1)\quad \left(R^0, u^0, v^0\right) = (R_0, u_0, v_0),$$

$$(0.2)\quad R^{1/2} = \frac{1}{2}\left(R_0 + R_0 \mathrm{e}^{\Delta t \int_0^1 x^2 S(u^0, v^0)\,\mathrm{d}x}\right),$$

$$(0.3)\quad \dot{R}^0 = R_0 \int_0^1 x^2 S(u^0, v^0)\,\mathrm{d}x\, R_0 \mathrm{e}^{\Delta t \int_0^1 x^2 S(u^0, v^0)\,\mathrm{d}x}.$$

Now, for $1 < n \leqslant N$, assuming $(R^{n-1}, u^{n-1}, v^{n-1})$ be given, we calculate (R^n, u^n, v^n) as follows:

Step n:

(n.1)

$$\begin{cases} \dfrac{v^n - v^{n-1}}{\Delta t} = \dfrac{d_2}{(R^{n-1})^2} x^{-2} \dfrac{\partial}{\partial x}\left(x^2 \dfrac{\partial}{\partial x} v^n\right) + x \dfrac{\dot{R}^{n-1}}{R^{n-1}} \dfrac{\partial}{\partial x} v^{n-1} \\ \qquad - r_2 v^n + r_2 \overline{\overline{\beta}}, \quad \text{in } 0 < x < 1, \\ \dfrac{\partial v^n}{\partial x}(0) = v^n(1) = 0, \end{cases}$$

(for $n = 1$, we use $R^{1/2}$).

(n.2)

$$\begin{cases} \dfrac{u^n - u^{n-1}}{\Delta t} = \dfrac{d_1}{(R^{n-1})^2} x^{-2} \dfrac{\partial}{\partial x}\left(x^2 \dfrac{\partial}{\partial x} u^n\right) + x \dfrac{\dot{R}^{n-1}}{R^{n-1}} \dfrac{\partial}{\partial x} u^{n-1} \\ \qquad - r_1 u^n - \lambda v^n + r_1 \overline{\overline{\sigma}} + \lambda \overline{\overline{\beta}}, \quad \text{in } 0 < x < 1, \\ \dfrac{\partial u^n}{\partial x}(0) = u^n(1) = 0. \end{cases}$$

(n.3) We compute R^n by integrating according the compound trapezium rule

$$\begin{aligned} R^n &= R_0 \exp\left\{ \Delta t \sum_{j=0}^{n-1} \int_0^1 x^2 \frac{1}{2}\left(S(u^j, v^j) + S(u^{j+1}, v^{j+1})\right) \mathrm{d}x \right\} \\ &= R_0 \exp\left\{ \Delta t \int_0^1 x^2 \left[\frac{1}{2}\left(S(u^0, v^0) + S(u^n, v^n)\right) + \sum_{j=1}^{n-1} S(u^j, v^j) \right] \mathrm{d}x \right\}. \end{aligned}$$

(n.4)

$$\begin{aligned} \dot{R}^n = R_0 \int_0^1 x^2 S(u^n, v^n)\,\mathrm{d}x \exp\Bigg\{ \Delta t \sum_{j=0}^{n-1} \int_0^1 x^2 \frac{1}{2}\big(S(u^j, v^j)\,\mathrm{d}x \\ + S(u^{j+1}, v^{j+1})\,\mathrm{d}x\big) \Bigg\} \end{aligned}$$

$$= R_0 \int_0^1 x^2 S(u^n, v^n)\,\mathrm{d}x \exp\left\{\Delta T \int_0^1 x^2 \left[\frac{1}{2}\big(S(u^0, v^0) + S(u^n, v^n)\big) + \sum_{j=1}^{n-1} S(u^j, v^j)\right] \mathrm{d}x\right\}.$$

7.2. Full discretization

We approximate $H^1(0,1)$ by space V_h defined by

$$V_h := \{\phi \in C^0([0,1]):\ \phi|_{(x_{j-1},x_j)} \in P_1,\ \text{for } j = 1, s+1\},$$

where $x_j = j/(s+1)$ and P_1 is the space of those polynomials of degree 0 or 1. We approximate the above implicit–explicit scheme by the system

$$\frac{u_h^n - u_h^{n-1}}{\Delta T} = \frac{D_1}{(xR^{n-1})^2}\frac{\partial}{\partial x}\left(x^2 \frac{\partial}{\partial x} u_h^n\right) + x\frac{\dot{R}^{n-1}}{R^{n-1}}\frac{\partial}{\partial x} u_h^n - r_1 u_h^n - \lambda v_h^n + r_1\overline{\overline{\sigma}} + \lambda\overline{\overline{\beta}},$$
$$\text{in } 0 < x < 1,\ n = 1, \ldots, N,$$
$$\frac{v_h^n - v_h^{n-1}}{\Delta T} = \frac{D_2}{(xR^{n-1})^2}\frac{\partial}{\partial x}\left(x^2 \frac{\partial}{\partial x} v_h^n\right) + x\frac{\dot{R}^{n-1}}{R^{n-1}}\frac{\partial}{\partial x} v_h^n - r_2 v_h^n + r_2\overline{\overline{\beta}},$$
$$\text{in } 0 < x < 1,\ i = 1, \ldots, N,$$
$$u_h^n(1) = v_h^n(1) = 0, \qquad \frac{\partial u_h^n}{\partial x} = \frac{\partial v_h^n}{\partial x} = 0, \quad \text{on } x = 0,$$
$$R(0) = R_0, \qquad u_h^0(x) = u_{h,0}(x), \qquad v_h^0(x) = v_{h,0}(x),$$
$$R_h^n = R_0 \exp\left\{\Delta T \int_0^1 x^2 \left[\frac{1}{2}\big(S(u_h^0, v_h^0) + S(u_h^n, v_h^n)\big) + \sum_{j=1}^{n-1} S(u_h^j, v_h^j)\right] \mathrm{d}x\right\}.$$

7.3. Weak formulation of the discrete problem

Setting

$$b(\zeta, \varphi) = \int_0^1 x^2 \zeta\varphi\,\mathrm{d}x,$$

the weak formulation of the discrete problem is given by $(\forall\varphi \in V_h)$

$$(1 + \Delta T r_1) b(u_h^n, \varphi) + \frac{d_1 \Delta T}{(R^{n-1})^2} b\big((u_h^i)_x, \varphi_x\big) - \frac{\Delta T \dot{R}^{n-1}}{R^{n-1}} b\big(x(u_h^n)_x, \varphi\big)$$
$$= b\big(u_h^{n-1} - v_h^n + \Gamma_1\overline{\overline{\sigma}} + \overline{\overline{\beta}}, \varphi\big) = b\big(u_n^{n-1} + \Delta T(-\lambda v_h^n + r_1\overline{\overline{\sigma}} + \lambda\overline{\overline{\beta}}), \varphi\big),$$
$$(1 + \Delta T r_2) b(v_h^n, \varphi) + \frac{d_1 \Delta T}{(R^{n-1})^2} b\big((v_h^n)_x, \varphi_x\big) - \frac{\Delta T \dot{R}^{n-1}}{R^{n-1}} b\big(x(v_h^n)_x, \varphi\big)$$
$$= b\big(v_h^{n-1} + \Delta T r_2\overline{\overline{\beta}}, \varphi\big).$$

7.4. *Numerical experiments*

We consider the special case of $S(\sigma, \beta) = \sigma - \hat{\sigma}$, $T = 3$, $N = 501$, (i.e., $\Delta T = 3/500$) and $s = 20$ (i.e., $h = 1/20$) with the following choice of the parameters: $R_0 = 5$, $D_1 = D_2 = 1$, $\Gamma_1 = \Gamma_2 = \overline{\overline{\sigma}} = \overline{\overline{\beta}} = 1$. These values of the parameters have been taken merely with academical purpose. For other choices see, for instance, BYRNE and CHAPLAIN [1996a]. In Figs. 7.1, 7.5 and 7.9, we display the computed evolution of the radius of the tumor for experiments 1 ($\hat{\sigma} = 0.75$), 2 ($\hat{\sigma} = 1$) and 3 ($\hat{\sigma} = 1.5$). In Figs. 7.2, 7.6 and 7.10 we display visualized the computed evolution of the radius of the tumor in two

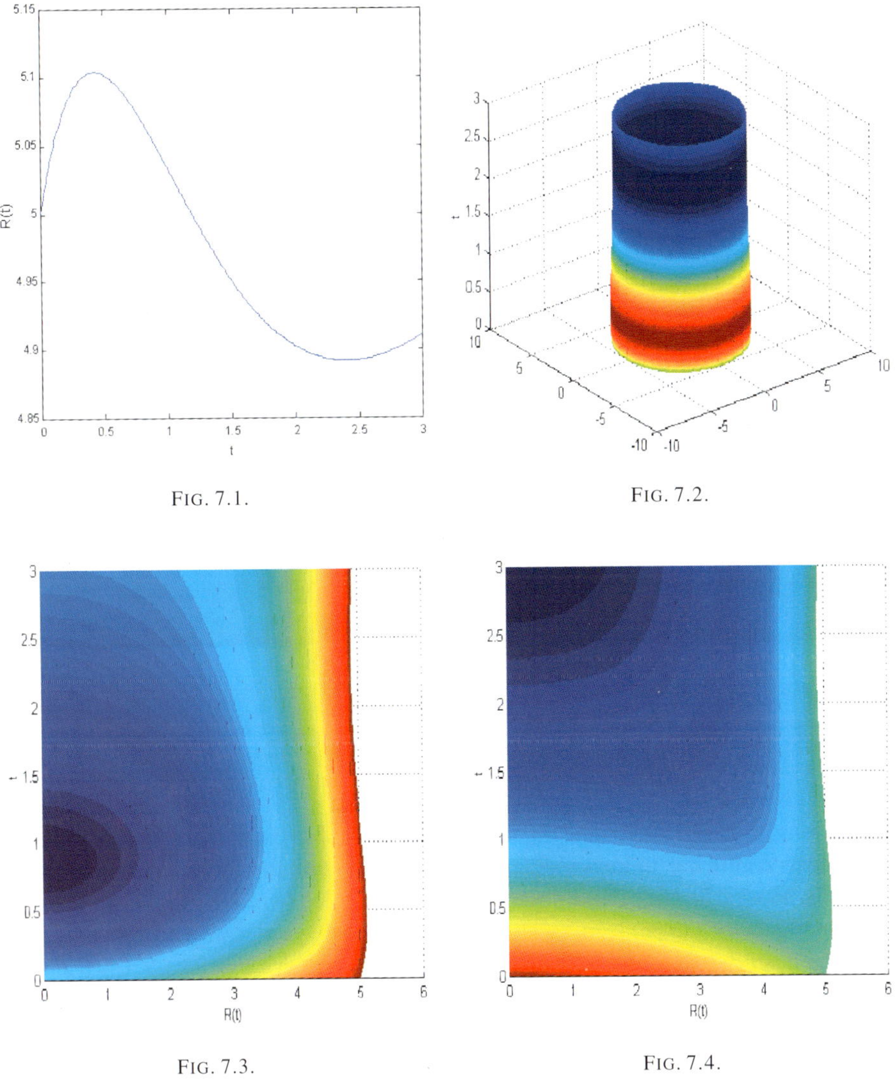

FIG. 7.1.

FIG. 7.2.

FIG. 7.3.

FIG. 7.4.

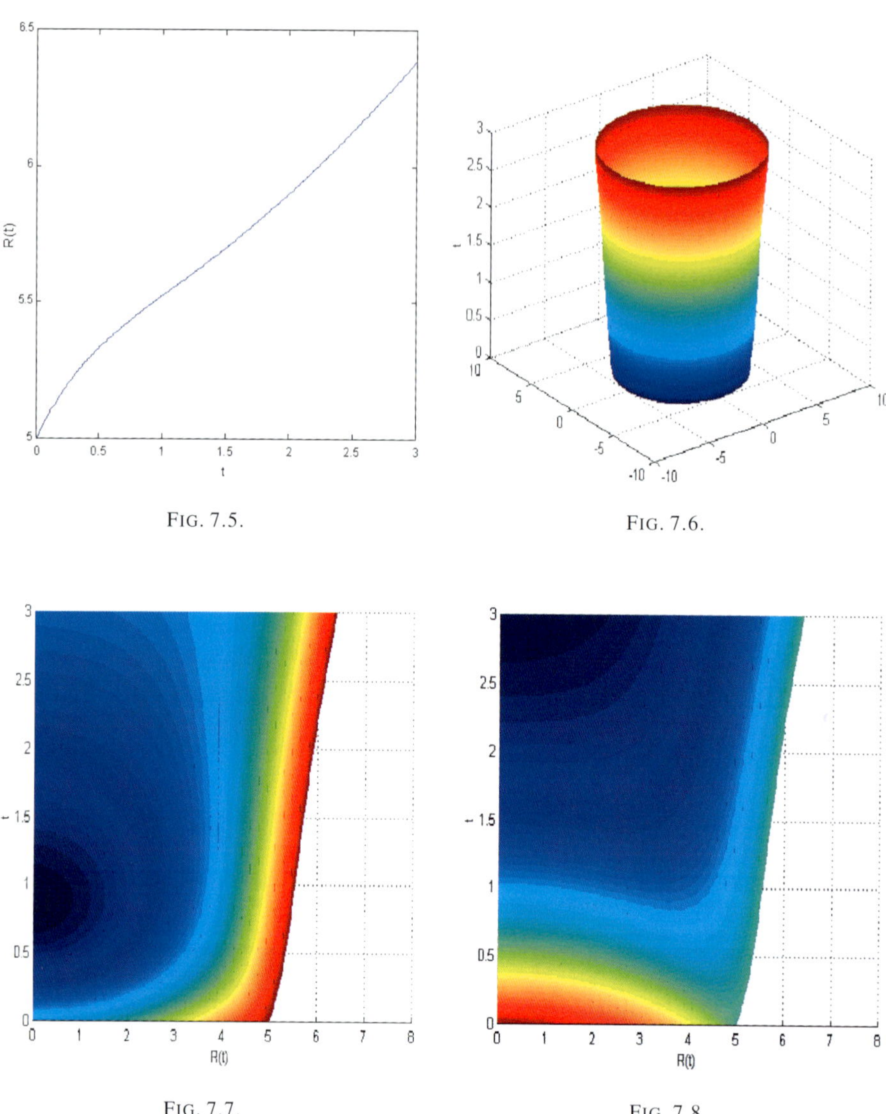

FIG. 7.5.

FIG. 7.6.

FIG. 7.7.

FIG. 7.8.

dimensions. Figs. 7.3, 7.7 and 7.11 show the computed evolution of the concentration of nutrients σ. Finally, in Figs. 7.4, 7.8 and 7.12 we exhibit the computed concentration of the inhibitors β. Numerical simulation of the model (when $S = \sigma - \tilde{\sigma}$) show us the importance of the parameter $\tilde{\sigma}$ in the behavior of the boundary. As it is expected, a smaller $\tilde{\sigma}$ produces a faster growth of the boundary. We can see in Figs. 7.1, 7.5 and 7.9 an initial concave growth of the radius that becomes convex after a time (which depends on $\tilde{\sigma}$). Among other different aspects it can be appreciated that the free boundary is not necessarily increasing in time (see Fig. 7.1).

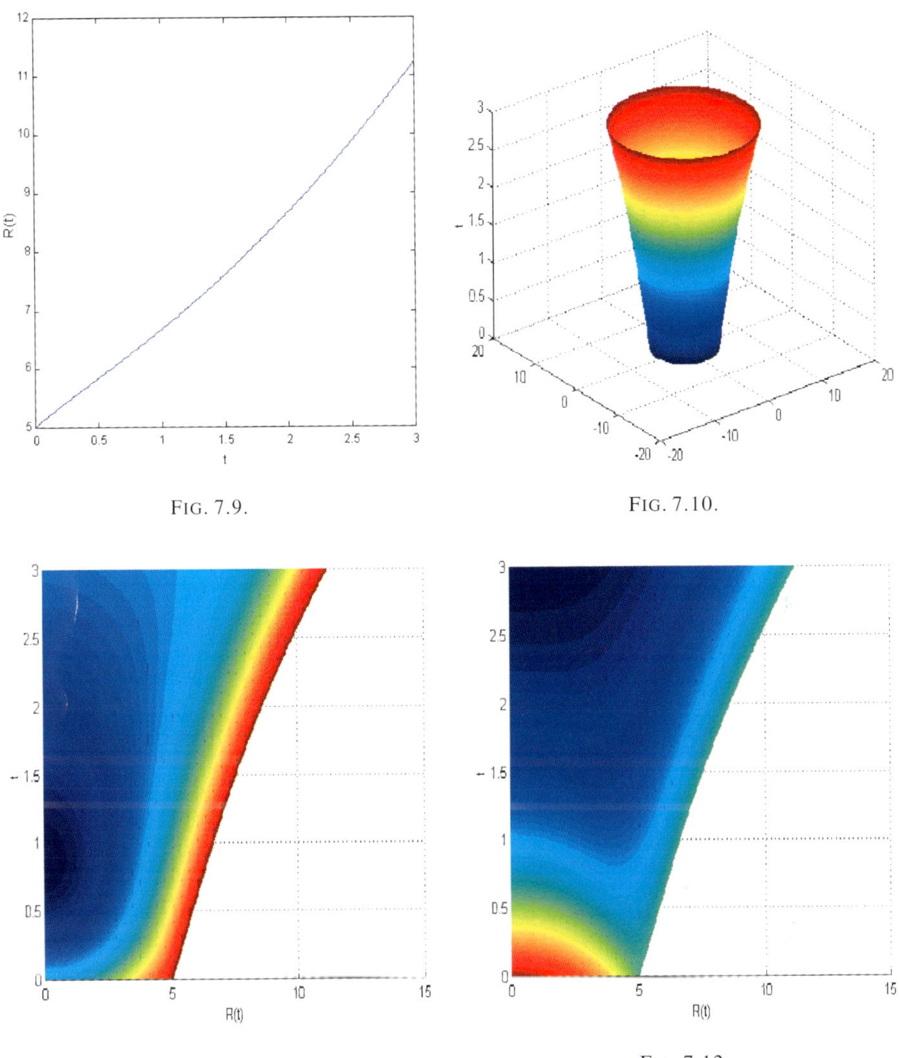

FIG. 7.9.

FIG. 7.10.

FIG. 7.11.

FIG. 7.12.

Acknowledgement

The work of first author was partially supported by the DGES (Spain) project REN2000/0766 and RTN HPRN-CT-2002-00274.

References

ADAM, J.A. (1986). A simplified mathematical model of tumor growth. *Math. Biosci.* **81**, 229–244.

ADAM, J.A., BELLOMO, N. (1997). *A Survey of Models for Tumor–Immune System Dynamics* (Birkhäuser, Boston).

ALT, H.W., LUCKHAUS, S. (1983). Quasi-linear elliptic–parabolic differential equations. *Math. Z.* **183**, 311–341.

ANDERSON, A.R.A., CHAPLAIN, M.A.J. (1998). Continuous and discrete mathematical models of tumor-induced angiogenesis. *Bull. Math. Biology* **60**, 857–899.

ATTALLAH, A.M. (1976). Regulation of cell growth in vitro and in vivo: point/counterpoint. In: Houck, J.C. (ed.), *Chalones* (North-Holland, Amsterdam), pp. 141–172.

BAZALIY, B.V., FRIEDMAN, A. (2003). A free boundary problem for an elliptic–parabolic system: application to a model of tumor growth. *Comm. Partial Differential Equations* **28**, 517–560.

BELLOMO, N., PREZIOSI, L. (2000). Modelling and mathematical problems related to tumor evolution and its interaction with the immune system. *Math. Comput. Modelling* **32**, 413–452.

BERTUZZI, A., FASANO, A., GANDOLFI, A., MARANGI, D. (2002). Cell kinetics in tumour cords studied by a model with variable cell cycle length. *Math. Biosci.* **177–178**, 103–125.

BREZIS, H. (1983). *Analyse Fonctionnelle* (Masson, Paris).

BRITTON, N.F., CHAPLAIN, M.A.J. (1993). A qualitative analysis of some models of tissue growth. *Math. Biosci.* **113**, 77–89.

BYRNE, H.M. (1997a). The effect of time delays on the dynamics of avascular tumor growth. *Math. Biosci.* **144**, 83–117.

BYRNE, H.M. (1997b). The importance of intercellular adhesion in the development of carciomas. *IMA J. Math. Appl. Med. Biol.* **14**, 305–323.

BYRNE, H.M. (1999a). A comparison of the roles of localised and nonlocalised factors in solid tumour growth. *Math. Models Methods Appl. Sci.* **9**, 541–568.

BYRNE, H.M. (1999b). A weakly nonlinear analysis of a model of avascular solid tumour growth. *J. Math. Biol.* **39**, 59–89.

BYRNE, H.M., CHAPLAIN, M.A.J. (1995). Growth of nonnecrotic tumors in the presence and absence of inhibitors. *Math. Biosci.* **130**, 151–181.

BYRNE, H.M., CHAPLAIN, M.A.J. (1996a). Growth of necrotic tumors in the presence and absence of inhibitors. *Math. Biosci.* **135**, 187–216.

BYRNE, H.M., CHAPLAIN, M.A.J. (1996b). Modelling the role of cell–cell adhesion in the growth and development of carciomas. *Math. Comput. Modelling* **12**, 1–17.

BYRNE, H.M., MATTHEWS, P. (2002). Asymmetric growth of avascular solid tumors: exploiting symmetries, in press.

CHAPLAIN, M.A.J. (1996). Avascular growth, angiogenesis and vascular growth in solid tumours: the mathematical modelling of the stages of tumour development. *Math. Comput. Modelling* **23** (6), 47–87.

CHAPLAIN, M.A.J. (1999). Mathematical models for the growth, development and treatment of tumours. *Math. Models Methods Appl. Sci.* **9** (4), 171–206 (special issue).

CHAPLAIN, M.A.J., ANDERSON, A.R.A. (1997). Mathematical modelling, simulation and prediction of tumour-induced Angiogenesis. *Invas. Metast.* **16**, 222–234.

CHAPLAIN, M.A.J., PREZIOSI, P. (2002). Macroscopic modelling of the growth and development of tumour masses, in press.

CHI-CHEUNG POON (1996). Unique continuation for parabolic equations. *Comm. Partial Differential Equations* **21**, 521–539.

CUI, S., FRIEDMAN, A. (1999). Analysis of a mathematical of protocell. *Math. Anal. Appl.* **236**, 171–206.

CUI, S., FRIEDMAN, A. (2000). Analysis of a mathematical model of effect of inhibitors on the growth of tumors. *Math. Biosci.* **164**, 103–137.

CUI, S., FRIEDMAN, A. (2001). Analysis of a mathematical model of the growth of the necrotic tumors. *Math. Anal. Appl.* **255**, 636–677.

DE ANGELIS, E., PREZIOSI, L. (2000). Advection–diffusion models for solid tumor evolution in vivo and related free boundary problem. *Math. Models Methods Appl. Sci.* **10**, 379–407.

DÍAZ, J.I., RAMOS, A.M. (1995). Positive and negative approximate controllability results for semilinear parabolic equations. *Rev. Real Acad. Cienc. Exact., Fís. Natur. Madrid* **LXXXIX**, 11–30.

DÍAZ, J.I., TELLO, L. (1999). A nonlinear parabolic problem on a Riemannian manifold without boundary arising in Climatology. *Collect. Math.* **50**, 19–51.

DÍAZ, J.I., TELLO, J.I. (2003). On the mathematical controllability in a mathematical in a simple growth tumors model by the internal localized action of inhibitors. *Nonlinear Anal.: Real World Appl.* **4**, 109–125.

DÍAZ, J.I., TELLO, J.I. (2004). Mathematical analysis of a simple model for the growth of necrotic tumors in presence of inhibitors. Int. J. Pure Appl. Math., in press.

FABRE, C., PUEL, J.P., ZUAZUA, E. (1995). Approximate controllability of the semilinear heat equation. *Proc. Roy. Soc. Edinburgh Sect. A* **125**, 31–61.

FISTER, K.R., LENHART, S., MCNALLY, J.S. (1998). Optimizing chemotherapy in HIV model. *Electron. J. Differential Equations* **32**, 1–12.

FONTELOS, M.A., FRIEDMAN, A., HU, B. (2002). Mathematical analysis of a model for the initiation of angiogenesis. *SIAM J. Math. Anal.* **33** (6), 1330–1355.

FRIEDMAN, A. (1964). *Partial Differential Equations of Parabolic Type* (Prentice Hall, New York).

FRIEDMAN, A. (2002). A hierarchy of cancer models and their mathematical challenges. In: *Lecture at the Workshop on Mathematical Models in Cancer*, Vanderbilt University, May 3–5.

FRIEDMAN, A., HU, B., VELÁZQUEZ, J.J.L. (2001). A Stefan problem for a protocell model with symmetry-breaking bifurcations of analytic solutions. *Interfaces and Free Boundaries* **3**, 143–199.

FRIEDMAN, A., REITICH, F. (1999). Analysis of a mathematical model for the growth of tumors. *J. Math. Biol.* **38**, 262–284.

FRIEDMAN, A., REITICH, F. (2001). Symmetry-breaking bifurcation of analytic solutions to free boundary problems: An applications to a model of tumor growth. *Trans. Amer. Math. Soc.* **353**, 1587–1634.

FRIEDMAN, A., TELLO, J.I. (2002). Stability of solutions of chemotaxis equations in reinforced random walks. *Math. Anal. Appl.* **272**, 138–163.

GLOWINSKI, R., LIONS, J.L. (1995). Exact and approximate controllability for distributed parameter systems, Part II. *Acta Numer.*, 157–333.

GREENSPAN, H.P. (1972). Models for the growth of solid tumor by diffusion. *Stud. Appl. Math.* **52**, 317–340.

GREENSPAN, H.P. (1976). On the growth and stability of cell cultures and solid tumors. *J. Theoret. Biol.* **56**, 229–242.

HOLMES, M.J., SLEEMAN, B.D. (2000). A mathematical model of tumour angiogenesis incorporating cellular traction and viscoelastic effects. *J. Theoret. Biol.* **202**, 95–112.

LADYZENSKAJA, O.H., SOLONNIKOV, V.A., URALSEVA, N.N. (1991). *Linear and Quasi-linear Equations of Parabolic Type*, Transl. Math. Monogr. **23** (Amer. Math. Soc., Providence, RI).

LEVINE, H.A., SLEEMAN, B.P. (1997). A system of reaction diffusion equations arising in the theory of reinforced random walks. *SIAM J. Appl. Math.* **57**, 683–730.

LEVINE, H.A., PAMUK, S., SLEEMAN, B.P., NILSEN-HAMILTON, M. Mathematical modeling of capillary formation and development in tumor angiogenesis: Penetration into the stroma, in press.

LEVINE, H.A., SLEEMAN, B.P., NILSEN-HAMILTON, M. (2000). A mathematical modeling for the roles of pericytes and macrophages in the initiation of angiogenesis I. The role of protease inhibitors in preventing angiogenesis. *Math. Biosci.* **168**, 75–115.

LIONS, J.L. (1990). Remarques sur la contrôllabilité approchée. In: *Actas de Jornadas Hispano–Francesas sobre Control de Sistemas Distribuidos*, Universidad de Malaga, pp. 77–88.

LIONS, J.L. (1991). Exact Controllability for distributed systems: Some trends ans some problems. *Appl. Indust. Math.*, 59–84.

ORME, M.E., CHAPLAIN, M.A.J. (1995). Travelling waves arising in mathematical models of tumour angiogenesis and tumour invasion. *Forma* **10**, 147–170.

OTHMER, H.G., STEVENS, A. (1997). Aggregation, blowup, and collapse: The ABC's of taxis in reinforced random walks. *SIAM J. Appl. Math.* **57**, 1044–1081.

OWEN, M.R., SHERRATT, J.A. (1999). Mathematical modeling of macrophage dynamics in tumors. *Math. Models Methods Appl. Sci.* **9**, 513–539.

SIMON, J. (1987). Compact sets in the space $L^p((0,T),B)$. *Ann. Mat. Pura Appl.* **CXLVI**, 65–96.

SLEEMAN, B.D. (1996). Solid tumor growth: A case study in mathematical biology. In: Aston, Ph.J. (ed.), *Nonlinear Mathematics and its Applications* (Cambridge Univ. Press, Cambridge), pp. 237–254.

SHOWALTER, R.E. (1996). *Monotone Operator in Banach Space and Nonlinear Equations* (Amer. Math. Soc., Philadelphia).

SHYMKO, R.M., GLASS, L. (1976). Cellular and geometric control of tissue growth and mitotic instability. *J. Theoret. Biol.* **63**, 355–374.

SWAM, G. (1984). *Applications of Optimal Control Theory in Biomedicine* (Dekker, New York).

THOMPSON, K.E., BYRNE, H.M. (1999). Modelling the internalisation of labelled cells in tumour spheroids. *Bull. Math. Biol.* **61**, 601–623.

VALENCIANO, J., CHAPLAIN, M.A.J. (2003a). Computing highly accurate solutions of a tumour angiogenesis model. *Math. Models Methods Appl. Sci.* **13**, 747–766.

VALENCIANO, J., CHAPLAIN, M.A.J. (2003b). An explicit subparametric spectral element method of lines applied to a tumour angiogenesis system of partial differential equations. Math. Models Methods Appl. Sci., in press.

VRABIE, I.I. (1995). *Compactness Methods for Nonlinear Evolutions*, second ed. (Longman, Essex).

WARD, J.P., KING, J.R. (1998). Mathematical modelling of avascular tumor growth II: Modeling growth saturation. *IMA J. Math. Appl. Med. Biol.* **15**, 1–42.

Human Models for Crash and Impact Simulation

Eberhard Haug

ESI Software S.A., 99, rue des Solets, BP 80112,
94513 Rungis Cedex, France
E-mail: eha@esi-group.com
URL: http://www.esi-group.com

Hyung-Yun Choi

Hong-Ik University, Seoul, South Korea
E-mail: hychoi@hongik.ac.kr

Stéphane Robin

LAB PSA-Renault, Paris, France
E-mail: stephane.robin@mpsa.com

Muriel Beaugonin

ESI Software S.A., Paris, France
E-mail: mbe@esi-group.com

Computational Models for the Human Body
Special Volume (N. Ayache, Guest Editor) of
HANDBOOK OF NUMERICAL ANALYSIS, VOL. XII
P.G. Ciarlet (Editor)

ISSN 1570-8659
DOI 10.1016/S1570-8659(03)12004-2

Contents

Preface

This article deals with the application of computational impact biomechanics to the consequences of real world passenger car accidents on human occupants, using computer models in numerical simulations with industrial crash codes. The corresponding developments are illustrated on the subject of safety simulations of human passenger car occupants. With some adaptations, the developed models apply equally well to the simulation of pedestrian accidents and to the design for occupant safety of motorbikes, trucks, railway vehicles, airborne vehicles, seagoing vessels and more.

The human models elaborated in this article belong to the class of finite element models. They can be adapted, specialized and packaged for other industrial applications in human ergonomics and comfort analysis and design, in situations where humans operate at their work place, as military combatants, or in sports and leisure activities and more. In the medical field, biomechanical human models can serve as a basis for the simulation and design of orthopedic prostheses, for bone fracture planning, physical rehabilitation analysis, the simulation of blood flow, artificial blood vessels, artificial heart valves, bypass operations, and heart muscle activity, virtual organ surgery, etc.

There exists indeed a large overlap, and a pressing urge and opportunity for creating a synergy of very diverse disciplines, which all deal with the simulation of the biomechanical response of the human body.

Most considerations of this article are related to the application of modern crash codes, which discretize space with the finite element method and which apply the explicit time integration scheme of the dynamic equations of motion to discrete numerical models. The reader is assumed to be familiar with the associated basic theory, needed for the use of such codes.

The article is structured as follows.

Chapter I provides an introduction on the interest, need and difficulties of using human models in occupant safety design and analysis. It contains a short overview on mechanical dummies, often used for the design of occupant safety of transport vehicles, and it summarizes some so far existing biomechanical human computer models.

Chapter II discusses "MB (multi-body)" or "HARB (Human Articulated Rigid Body)" or "ATB (Articulated Total Body)" models. These simplest human models consist in rigid body segments, joined at the locations of their skeletal articulations, which can provide gross overall kinematic responses of the human body to static and dynamic load scenarios. For more detailed investigations, they can serve as a basis for modular plug-in of more elaborate and deformable segment models, for making zooms on the detailed response of various body parts. The chapter closes with applications to occupant safety of HARB models, including the fifth percentile female and a six year old child model.

Chapter III discusses deformable human models. In a first section, the results of the first European HUMOS (Human Models for Safety) project (1999–2001) are summarized. The HUMOS-1 project was funded by the European Commission in the Industrial and Materials Technologies (IMT) program (Brite–EuRam III). In this project the geometry of a near fiftieth percentile human cadaver geometry was acquired in a passenger car driving posture and human models were derived from the anthropometrical, biomaterial and validation database, compiled and generated within this project.

In a second section, a systematic presentation of the generation of human models and sub-models is given and illustrated on the example of a deformable fiftieth percentile human model (H-Model). This section first outlines the HARB version of the model and then the deformable sub-models of the head, skull and brain, the neck and cervical spine, the torso with the rib cage, thoracic and abdominal organs, the upper extremity with the shoulder and arm complex, the lower extremity with the knee, thigh and hip complex and the ankle-foot complex. For each deformable sub-model the relevant anatomy, the main injury mechanisms, the basic model structure, its calibration and the basic validations of the models are outlined. A validation of an abdomen model is discussed in the first section. A final section outlines the emerging deformable models of the fifth percentile female.

Appendix A gives an overview on the basic theory of explicit solution and on contact treatment. Appendix B contains data on biomaterials. Appendix C outlines the Hill type muscle models. Appendix D discusses the numerical entities of air- and bio-bags, used to simulate protective airbags and hollow organs. Appendix E provides an insight into the management of the interaction of parts and organs in biomechanical simulation of the human body.

It is clear that this article can only be an incomplete outline over the fast growing, vast and stimulating subject of biomechanical (and biomedical) modeling techniques of the human body. The presented models and methodologies will undoubtedly be upgraded by the time this article is printed. The interested reader is therefore encouraged to keep a close watch on the corresponding web sites and the open literature.

CHAPTER I

Introduction

1. On the interest, need and difficulties of using human models in virtual passenger car crash tests

1.1. Crash design

Crash tests. In car design, standardized "legal laboratory crash tests" are made in order to assess the protective and life saving performance of the car body and its built-in passive occupant safety devices, such as airbags, protective paddings and seat belts. Conventionally, the response of car occupants under accidental conditions, as in frontal crashes, lateral side impacts, rollover accidents, etc., is studied, using re-usable "mechanical occupant surrogate devices", often called "mechanical dummies" or "legal crash dummies". According to existing regulations, passenger transport vehicles must be designed to pass standard crash tests safely to obtain legal certification for selling them to customers. The achieved safety levels are assessed through the mechanical responses of the used dummy devices, as recorded by instruments in physical crash tests carried out in crash laboratories. These recordings are correlated heuristically with human injury. Safe crash design methodologies have their widest use in passenger car design, but apply to all road, water and airborne passenger transport vehicles and working devices. Recent efforts towards "legal virtual testing" try to establish regulatory frameworks that can be used to replace laboratory tests for the purpose of legal certification of vehicles with simulation. While desirable for working with dummies, such process will be mandatory for working with human models since no real world tests can be made to back up the simulations.

Crash simulation. In numerical passenger car crash simulations, numerical models of the car structure, the passive safety restraints (seat belts, airbags, cushions) and the dummy devices are made, the latter as simpler multi-body models, or as more elaborate deformable finite element dummy models. Care must be taken that the passive safety devices are modeled with enough detail, so that their deployment, deformation and energy absorption capacities are well represented in the simulation of a car crash. The numerical dummy models are placed inside the numerical models of the vehicle structure, and their performance under an imposed crash scenario is evaluated. Models of passive safety devices, such as airbags, seat belts, knee bolsters, etc., will be designed and optimized to improve the car safety or crashworthiness performance with respect to the used dummy models. The safety of the car for human occupants is assessed through the simulated response output of the virtual dummy devices, which are modelled and "instrumented" to behave like the real world mechanical dummy devices. If human models were used instead of models of mechanical occupant surrogates, or dummies, a more direct access to human injury could be provided.

Crash codes overview. Numerical crash simulations are performed with specialized crash codes, which were conceived during the eighties of the last century (Pam–Crash, Radioss, Dyna3D), following an urgent need for economy, safety and speed of passenger car design. This need was expressed by the world's passenger car manufacturers. Since the standard safety regulations in all countries became more and more strict, the conventional methods to hand-make ever lighter new car prototype structures and to crash test them became increasingly uneconomical, time consuming and unsafe. The only answer to satisfy the pressure for crashworthiness, safety, quick time to market and economy of design lay in the emerging methodologies of virtual prototyping and design, using high performance computing. This is why several commercial crash codes have emerged, all based in essence on the dynamic explicit finite element method of structural analysis, which uses the proven finite element method for discretizing space, and the explicit direct integration scheme of the non-linear equations of motion to discretize time. One early account of the practical application of a commercial crash code is given by HAUG and ULRICH [1989].

The numerical models treated by these codes started with the car body-in-white (mostly steel structures), modeled with thin shells and contacts. Soon increasingly trade specific models of passive safety devices (airbags, seat belts, knee bolsters, etc.), modeled with cables, bars, joints, membranes, shells and solids followed. Within a few years, models of mechanical dummies, impact barriers and crash obstacles appeared. Today numerical models of human occupants are under active development, with worldwide active support of national agencies for traffic safety. Like always in numerical simulation, a trade-off between computational efficiency, robustness of execution and accuracy must be found. It is therefore legitimate to create numerical models of the human body at different levels of discretization, where the less discretized models execute faster to provide more approximate answers in early design stages, and the more elaborate models cost more computer time and resources, but provide more information and yield more accurate results for the final design.

The correct simulation of contact events or collisions is one of the most crucial features of crash codes. Collisions can occur between the structure of interest and objects in its environment, such as contact between a car and a rigid wall, car-to-car contact, or contact of an occupant with an airbag or seat belt. Contact can also occur between different parts of a crashed structure, such as between the engine and the car body, tire-to-wheel case, roof-to-steering wheel, occupant arm to occupant chest. Finally, self-contact can occur within a single car body component after buckling and wrinkling of its constituent thin sheet metal parts. The correct and efficient treatment of collision events is therefore of great importance, and crash codes have been conceived giving great attention to contact algorithms. Early accounts on the conception of such algorithms are found in HUGHES, TAYLOR, SACKMAN, CURNIER and KANOKNUKULCHAI [1976], HALLQUIST, GOUDREAU and BENSON [1985], and others.

1.2. Occupant safety design

Occupant surrogates. In real world crash tests, it is common practice to use mechanical dummies as surrogates for the human vehicle occupants. Mechanical dummies are instrumented biofidelic occupant surrogate devices, made of metallic, rubber, foam and plastic materials, that are widely used by car makers in real vehicle crash tests. The impact of car accidents on human occupants is inferred from the impact performance of the used mechanical dummies, expressed in terms of standard response measurements, such as head accelerations, chest deflections, femur loads, etc. These measurements can be correlated with human injury via so-called injury criteria. The latter give rough insight into the real injuries a human occupant might experience in each studied crash scenario.

The consistent use of dummies in crash tests is not ideal, because even the best crash dummies can only approximate the behavior of real humans in a crash. Humans undergo wider trajectories inside a vehicle than dummies. Therefore ever more advanced dummies are needed to provide more representative injury data. Nevertheless, dummies and dummy models enabled car manufacturers to very significantly increase the passive safety performances of their products. Most of the current safety devices were indeed developed with the well-known Hybrid III frontal dummy, or with the EuroSID 1 side impact dummy. Since humans cannot be used in real world crash tests, dummies are the only workable alternative, and their use is mandatory. Crash dummies are under continuous improvement, and next generation mechanical occupant surrogate devices are under development (e.g., the THOR dummy developed by NHTSA), often with the help of numerical simulations using human models.

Human subjects. The direct use of humans in the everyday safety design of transport vehicles is excluded due to obvious ethical and practical constraints. Some exceptional uses of live and dead humans (cadavers or PMHS for post mortem human subjects) for research purposes and for indirect studies of the response of the human body in crash situations are listed next. All tests involving human volunteers and human cadavers are subject to very rigorous screening procedures by competent ethics committees in all countries. Adult persons can dedicate their bodies in case of decease by an act of will to science. Children can not grasp such an act, and their parents can not, in general, decide for their children. Child cadavers can therefore hardly be used for destructive tests. Exceptions may exist in using body scan images of children when the parents give their consent. Modern non-destructive bone density measurements and similar existing or emerging techniques can be used to circumnavigate this ethical dilemma.

Human volunteers. Human volunteers can only exceptionally serve in experimental impact tests. One well known historical contribution were the human tests carried by US Air Force Colonel John P. Stapp, who studied from 1946 to 1958 the effects of deceleration on both humans and animals at the Edwards Air Force Base in California and at the Holloman Air Force Base in New Mexico. Stapp exposed belt restrained volunteers, including himself, to decelerations of up to 40 g, using rocket driven sleds. Since 1955 by now yearly Stapp Car Crash Conferences take place (46th by 2002). One recent example involving human volunteers is given by low energy rear and front impact crash tests, where the principal effect of neck "whiplash" motions is studied in purely research oriented projects under medically controlled conditions. ONO [1999] studies the relationship between localized spine deformation and cervical vertebral motions for low speed rear impacts using human volunteers. In such exceptional test setups, human volunteers are subjected to sub-injury rear or front impact equivalent acceleration levels. In particular, these studies employ X-ray cine-radiography, accelerometer recordings and electro-myographic recordings on the neck response in very low speed rear-end car impacts. In such recordings the activation level of the neck muscles can be monitored via their variable electrical characteristics. The resulting data are used to develop models to evaluate neck injuries caused by higher-speed rear-end impacts, and to improve the accuracy of conventional crash-test dummies.

Accidentological studies and accident reconstructions. Such studies can give insight into crash events after a real life accident has occurred. These investigations can determine what might have happened to the involved human occupants. Accidentological studies provide data about the ways the accidents occurred, the involved vehicles, vehicle trajectories and collision with obstacles, and data about the injuries and the medical consequences for the human occupants. Accident reconstruction studies often re-enact reported crashes in the laboratory, or use numerical simulation of the reported crashes. In such re-enactions and simulations, mechanical dummies and their models can be employed. In numerical simulations of the re-enacted crashes, the use of human models is of particular interest, since the regulations which prescribe the mandatory use of mechanical occupant surrogates in certification tests do not apply. Car companies re-enact reported crashes in order to better understand the causes of injury and to improve the car design.

Human cadaver tests. Tests with human cadavers (PMHS tests) can be carried out at the exceptional research level in experimental impact studies. Most cadaver tests study the basic biomechanical mechanisms that lead to injuries of the human body (e.g., SCHMIDT, KALLIERIS, BARZ, MATTERN, SCHULZ and SCHÜLER [1978]). In no case can cadaver tests be used in everyday car design. Only principal injury mechanisms can be deduced from cadaver tests, and each cadaver tends to be different. Average human response "corridors" can be derived from test campaigns which may involve many different cadavers, each subjected to the same test. In the past, the design of mechanical dummies was based largely on the knowledge derived from specific series of different types of cadaver tests. For example, KALLIERIS and SCHMIDT [1990] describe the neck response and injury assessment using cadavers and the US-SID side impact dummy for far-side lateral impacts of rear seat occupants with inboard-anchored shoulder belts.

Cadaver test results can produce valuable information for the construction of human numerical models, rather than to be of direct value in everyday car design. For example, cadaver test studies on the human skull and the mechanisms of brain injury can clarify the relationship between different types of impacts and the nature and extent of injury. Tests on the brain and the skull are carried out in order to improve two- and three-dimensional models of the head for computer simulations, to understand the mechanisms through which injuries develop in the brain and skull. Neck tests improve the knowledge about human neck injury tolerance and mechanisms. Pendulum impact tests on the thorax and pelvis shed light on the response of the skeleton and organs in frontal and side impacts. Impact tests on the abdomen can give insight in the action of lap belts on the visceral organs. Upper extremity impact tests yield information about aggression from side impact airbags. Cadaver test research into leg injuries typically involves examining intrusion of the fire wall of passenger cars into the occupant compartment, the sitting position and kinematics of the occupant, the effectiveness of knee bolsters, the position of the pedals, and the anatomical nature of these injuries.

Animal tests. Tests which involve life or dead animals are subject to ethics committee constraints, as are tests involving human cadavers or human volunteers. In order to discern the different behavior of body segments, organs and biomaterials of the live organism, animal tests have be performed on live anesthetized animals. Again, such tests can not serve as a basis for everyday car design, but are sometimes carried out in purely research type projects where the use of humans is excluded. For example, some brain injury mechanisms were studied in the past on primates by ONO, KIKUCHI, NAKAMURA, KOBAYASHI and NAKAMURA [1980]. Pigs were also used to study the consequences of chest impacts by KROELL, ALLEN, WARNER and PERL [1986].

Humans in crash tests. While humans cannot replace mechanical dummies in real world crash tests, this is the case in virtual crash simulations. It is therefore of great potential advantage to build human models, and to use them in accident simulations. By combining crash analysis and biomechanical analysis, it is possible to advance the understanding of how injuries occur. This is the most important step towards creating safer automobiles and safer roads. As a by-product, human models can be used for improving the design of mechanical occupant surrogate devices.

1.3. Injury and trauma

Humans vs. dummies. Human models represent "bone, soft tissue, flesh and organs" instead of "steel, rubber, plastic materials and foam", as it is the case with dummy models. Injury in the sense of biological damage does not exist in todays mechanical dummies, because dummy devices are designed for multiple re-use without repair. The danger of injury to humans is deduced indirectly from the instrument responses of the mechanical dummies (or their models), as obtained during a real world (or simulated) crash test.

Human bone, soft tissue, flesh and organ injury prediction is the primary goal of impact biomechanics. If injuries can be predicted directly and reliably, then cars can be designed safer. In impact biomechanics, two classes of human parts and organs may be distinguished from a purely structural point of view: first the ones who have an identifiable structural function, and second the ones who have not. Skeletal bones, for example, are "structural" elements in the sense that they must carry the body weight and mass, and their material resembles conventional engineering materials. The brain, on the other hand, has hardly any structural function, and it resembles a tofu-like material with a maze of reinforcing small and tiny blood vessels, not unlike a soft "composite" material. The structural response of skeletal bone can be modeled more readily with standard engineering procedures than the structural response of the brain, and the injury to skeletal bone can be inferred easily as fracture from its structural response, while neuronal brain injury is not easily derived from the structural response of the brain material.

Injury prediction. Injury of human parts, before any healing takes place, can either be defined as instantaneously irreversible mechanical damage, for example damaged articulations, broken bones, aorta rupture or soft tissue and organ laceration, or, as a reduction of the physiological functioning, for example of the neurological functions of the brain, sometimes without much visible physical damage.

Bone fracture, on the one hand, is largely characterized by the mechanical levels of stress, strain and rate of strain in the skeletal bones, as calculated readily from accurate mechanical models in the simulation of an impact event. Long bones (femur, ribs, humerus, etc.), short bones (calcaneous, wrist bones, etc.) and flat bones (skull, pelvis, scapula) can often be modeled using standard brittle material models for the harder cortical bone, and standard collapsible foam material models for the softer, spongy, trabecular or cancellous bone. Ligaments and tendons, and sometimes passive muscles, skin, etc., can be modeled fairly well using standard non-linear rubber-like hyper-visco-elastic materials.

Internal organs have physiological functions. Their structural attachment inside the body cavities is given by mutual sliding contact, by in and outgoing vessels, by ligaments and by sliding contact with the body cavity walls. Their structural response to impact is harder to calculate and the calculated mechanical response fields are hard to correlate with their physiological functioning or injury.

The heart can act like a structural vessel, for example, when it is compressed and shifted in a chest impact. Gross shifts may cause strain and rupture of the aorta, an event which can be modeled with advanced solid–fluid interaction simulation techniques. The mechanical simulation of this process requires a detailed model of the heart, the aorta walls, and of the way the heart and aorta are anchored inside the chest. The blood should then be modeled as a fluid medium.

The other internal organs are either solid (liver, spleen, kidneys, etc.), hollow (stomach, intestines, bladder, etc.), or spongeous (lungs). The solid organs respond with their bulk matter to mechanical aggression in crash events. For example, the liver might be lacerated by the action of a lap belt. However, the tender liver parenchyma is invested by tough-walled vessels which render the material heterogeneous and anisotropic. The hollow organs should be modeled as hollow cavities, with an adequate model of their contents, which might interact with the organ walls during a mechanical aggression.

For the brain, the mechanical stress and strain fields and their histories, once calculated, must yet be linked to neurological damage. After impact, the neurons are still there, but they may have ceased to function properly because they became disconnected

TABLE 1.1
AIS injury scale

AIS code	Description	General injury	Thorax injury (example)
0	No injury	–	–
1	Minor	Abrasions, sprains, cuts, bruises	–
2	Moderate	Extended abrasions and bruises; extended soft tissue wounds; mild brain concussions without loss of consciousness	Single rib fracture
3	Serious (not life threatening)	Open wounds with injuries of vessels and nerves; skull fractures; brain concussions with loss of consciousness (5–10 minutes)	2–3 rib fractures sternum fracture
4	Severe (life threatening; probability of survival)	Severe bleeding; multiple fractures with organ damage; brain concussion with neurological signs; amputations	>4 rib fractures 2–3 rib fracture with hemo/ pneumothorax
5	Critical (survival is uncertain)	Rupture of organs; severe skull and brain trauma; epidural and subdural hematoma; unconsciousness over 24 hours	>4 rib fractures with hemo/ pneumothorax
6	Maximum (treatment not possible; virtually unsurvivable)	Aorta rupture; collapse of thoracic cage; brain stem laceration; annular fracture of base of skull; separation of the trunk; destruction of the skull	Aorta laceration

at certain strain levels, or because these cells were asphyxiated from the pressure generated by hematomas, which may prevent proper blood supply to uninjured parts of the brain. While this may or may not create visible mechanical "material" damage, it will cause reduction or total loss of the brain functions, hence injury.

The definition of biological and medical injury to the internal organs and its correlation with mechanical output fields as obtained from impact biomechanics models remains an open field for intensive research.

Injury scales. Criteria for injury potential were proposed by GADD [1961], GADD [1966]. The most often used injury scale for impact accidents is the Abbreviated Injury Scale (AIS). Table 1.1 contains AIS scores and some associated injuries.

References on injury and trauma. Detailed descriptions and further extensive bibliographies of injury and trauma of the skull and facial bone, the brain, the head, the cervical spine, the thorax, the abdomen, the thoraco-lumbar spine and pelvis and the extremities can be found in the book by NAHUM and MELVIN (eds.) [1993] *Accidental Injury – Biomechanics and Prevention.*

In this book first general aspects related to impact biomechanics are discussed in the following chapters: Chapter 1: The Application of Biomechanics to the Understanding of Injury and Healing (FUNG [1993b]); Chapter 2: Instrumentation in Experimental Design (HARDY [1993]); Chapter 3: The Use of Public Crash Data in Biomechanical Research (COMPTON [1993]); Chapter 4: Anthropometric Test Devices (MERTZ [1993]); Chapter 5: Radiologic Analysis of Trauma (PATHRIA and RESNIK [1993]); Chapter 6: A Review of Mathematical Occupant Simulation Models (PRASAD and CHOU [1993]); Chapter 7: Development of Crash Injury Protection in Civil Aviation (CHANDLER [1993]); Chapter 8: Occupant Restraint Systems (EPPINGER [1993]); Chapter 9: Biomechanics of Bone (GOLDSTEIN, FRANKENBURG and KUHN [1993]); Chapter 10: Biomechanics of Soft Tissues (HAUT [1993]).

It is recommended to read these chapters for obtaining a good background for the following chapters, which are devoted to the trauma and injury of the individual body segments: ALLSOP [1993] Skull and Facial Bone Trauma: Experimental aspects (Chapter 11); MELVIN, LIGHTHALL and UENO [1993] Brain Injury Biomechanics (Chapter 12); NEWMAN [1993] Biomechanics of Head Trauma: Head Protection (Chapter 13); MCELHANEY and MYERS [1993] Biomechanical Aspects of Cervical Trauma (Chapter 14); CAVANAUGH [1993] The Biomechanics of Thoracic Trauma (Chapter 15); ROUHANA [1993] Biomechanics of Abdominal Trauma (Chapter 16); KING [1993] Injury to the Thoraco–Lumbar Spine and Pelvis (Chapter 17); LEVINE [1993] Injury to the Extremities (Chapter 18).

These chapters provide a broad overview and many references on injury and trauma of the human body parts, and most of the brief discussions of injury and trauma in this article are based on this book. The book further contains chapters on child passenger protection (Chapter 19), isolated tissue and cellular biomechanics (Chapter 20) and on vehicle interactions with pedestrians (Chapter 21).

1.4. Human models

Models of mechanical dummies simulate their metallic, rubber and plastic parts. Human models simulate the response of bone, flesh, muscles, and hollow and solid organs humans are made of. While humans cannot replace mechanical dummies in real world crash tests, numerical models of humans can readily replace numerical models of mechanical dummies in virtual crash simulations.

Generic and specific models. Depending on the application, human models can be conceived either as "generic" or as "specific" models.

"Generic" models describe the geometry and the physical properties of average size members of the population. They are needed for industrial design, whenever objects are designed for the "average" human user. The average size of the human body can be expressed in statistical "percentiles" of a given population, where the "nth height percentile" means that "n" percent of the population is smaller in height. For example, 40% of a population is smaller than its 40th height percentile specimen, while 60% is taller. Each average height and weight percentile specimen can still have different relative size distribution of its body segments, as well as different biomechanical properties. The variations around an average percentile specimen of a population are called its "stochastic variants".

"Specific" models describe the geometry and physical properties of given human subjects. They are needed, for example, for virtual surgery, where the surgeon wants to deal with the precise bone or organ of a given patient. In the case of generic models, the acquisition process of the geometry of the body may take time, whereas, for practical reasons, the time needed for the establishment of specific models must be short. Therefore, slow mechanical slicing techniques on cadavers can serve for the data acquisition of generic models, while fast X-ray and scanning techniques on patients are required to construct specific models.

Generic human models should be comprehensive in the sense that all body sizes, genders, ages, races and body morph-types are covered. To achieve this goal, great amounts of anthropometrical and biomechanical data must be acquired and collected in databases, including for children. In fact, human computer modeling and simulation created new demands for data which were not needed or collected before, and novel physical experiments are required. Concerning model validation, modern practice of simulation tends to reverse the role of physical experiments, or laboratory tests, which tend to back up model calibration and validation, rather than to yield primary results, now obtained by the simulations.

Scaling, morphing, aging. Generic models of any type and size should be made available in data bases and through mathematical scaling, morphing and aging techniques, which can generate any given percentile human model and its stochastic variants, with long or short trunks and extremities, thin or fat, older or younger, male or female. Experimental results of standardized validation test cases, together with simulation accuracy norms, must also be provided in such data bases, allowing the human modeler to judge the performance and the quality of the models under controlled conditions.

As almost none of these new requirements are met fully today, there is plenty of room for human model development work. In this article, only the fiftieth percentile "average" male human models for passenger car occupant safety analysis and design are discussed in detail. Female and child models are discussed more briefly, since they are less advanced and their modeling techniques resemble the average male models.

Omnidirectionality. Unlike the well-known existing families of mechanical passenger car occupant surrogates ("dummies"), which are widely used in standard real crash tests by the world's car manufacturers for distinct frontal, side and rear impacts, human models should not be specialized to certain types of crash. They should rather be modeled as "omni-directional" objects, to the image of their real counterparts, i.e., respond equally well for all conceivable types of crash scenarios, impact directions and locations.

1.5. Biomaterials

Biomaterials are "exotic" as compared to most conventional structural materials. A good starting point for their analysis is nevertheless the existing library of material models, offered in modern dynamic structural analysis codes, or crash codes. The theory of the available standard material descriptions can be found in the handbooks of the commercial crash codes and need not to be discussed here in their mathematical detail. Ongoing work will adapt and refine these existing models as new knowledge on the mechanical behavior of biomaterials emerges. A condensation of the abundant literature on biomaterials is added below in brief discussions on fundamental works on biomaterials.

Basic literature. The older book by Yamada (YAMADA [1970]) contains global information about basic material properties of most biological tissues, such as the average Young's modulus and the fracture strength of the tested parts and organs, which permits to get first rough ideas about the mechanical properties of human tissues and organs. The editor of Yamada's book, F.G. Evans, states in his 1969 preface:

> "... *It is a unique book in several respects. First, it contains more data on strength of more tissues from more individuals of different ages than any other study of which I am aware. Second, all of the material used in the study was fresh and unembalmed. Third, the tests were made with standard testing machines of known accuracy or with machines that, after consultation with the manufacturer, had been specifically modified for testing biological materials. Fourth, all of the human material was obtained from one ethnic group. Thus the strength characteristics and other mechanical properties of organs and tissues from Japanese can be easily compared with those from other racial groups. Fifth, data were included on the strength characteristics of organs and tissues from other mammals as well as birds, reptiles, amphibians, and fish.*"

These remarks clearly express not only the durable value of this introductory book, but contain the fundamental specifications for the structure and contents of a comprehensive data base of biological materials. Among the tests that were carried out in order to characterize the strengths of the materials were tests in tension, compression, bending, impact bending, impact snapping, torsion, expansion, bursting, tearing, cleavage, shearing, extraction, occlusion, abrasion, crushing and hardness. The book by Yamada next contains an impressive array of basic information about the mechanical properties for humans and animals of the loco-motor organs and tissues (bone, cartilage, ligaments, muscle and tendons); the circulatory organs and tissues (heart, arteries, veins and red blood cells); the respiratory and digestive organs and tissues (larynx, trachea, lungs, teeth, masticatory muscles, esophagus, stomach, small intestine, large intestine, liver and gall bladder); the uro-genital organs and tissues (kidney, ureter, urinary bladder, uterus, vagina, amnion membrane and umbilical cord); the nervous system, integument, sense organs and tissues (nerves, dura mater, skin, panniculus adiposus (fat), hair, nails, horn sheath, cornea and sclera (eye), auricle and tympanic membrane (ear)). The mechanical properties of certain organs such as the brain, the tongue, the spleen are missing, however. Then the book compares the mechanical properties of human organs and tissues according to their strength and with respect to other materials from industry and nature. Finally, varations (scatter), age effects and aging rates are discussed.

More recently, VIANO [1986] describes the biological structures, material properties and failure characteristics of bone, articular cartilage, ligament and tendon. In his article, the load-deformation of biological tissues is presented with particular reference to the microstructure of the material.

> *"Although many of the tissues have been characterized as linear, elastic and isotropic materials, they actually have a more complicated response to load, which includes stiffening with increasing strain, inelastic yield and strain rate sensitivity. Failure of compact and cancellous bone depends on the rate, type and direction of the loading. Soft biological tissues are visco-elastic and exhibit a higher load tolerance with an increasing rate of loading."*

Viano's paper includes a discussion on the basic principles of biomechanics and emphasizes material properties and failure characteristics of biological tissues subjected to impact loading. The author presents on more than 30 pages what should be known from an engineering point of view about biological tissues. He discusses what types of fibers (collagen), bulky tissue with visco-elastic properties, some of which can consolidate (hyaline cartilage), and crystals in bone tissue (calcium), are responsible for the cohesion of the skeleton (ligaments), the attachment of the muscles to the skeleton (tendons), the transmission of compression forces across the articular surfaces (articular cartilage) and for maintaining the overall shape of the skeleton (bones). For each discussed material the paper describes its biological microstructure and composition, it discusses laboratory setups for material testing, it gives typical stress-strain samples and it outlines possible mathematical models to describe the measured properties up to rupture and fracture.

The textbook by FUNG [1993a] describes "The Mechanical Properties of Living Tissues" (book title). The approach to the description of biomaterials chosen by Fung is the study of the morphology of the organism, the anatomy of the organ, the histology of the tissue, and the determination of the mechanical properties of the materials or tissues in the form of their constitutive equations. The book further deals with setting up the governing differential or integral equations of biomechanical processes, their boundary conditions, their calibration, solution and validation on experiments and predicted results. The constitutive behavior of biomaterials is identified and their equations are defined for the flow properties of blood, blood cells and their interaction with vessel walls, for bio-visco-elastic fluids, for bio-visco-elastic solids, for blood vessels, for skeletal muscle (with a description of Hill's active and passive muscle model), for the heart muscle, for smooth muscles and for bone, cartilage, tendons and ligaments, including the mechanical aspects of the remodeling or growth of certain tissues. The detailed derivation and the mathematical description of the constitutive equations of living tissues is the most distinguishing feature of Fung's textbook. For each treated subject, the book contains extensive lists of references that may be consulted for further reading.

Further collections of biomaterial properties can be found in more recent references, such as the handbook of biomaterial properties by BLACK and HASTINGS (eds.) [1998], which describes in its Part I, from the view point of surgical implants, the properties of cortical bone, cancellous bone, dentin and enamel, cartilage, fibro-cartilage, ligaments, tendons and fascia, skin and muscle, brain tissues, arteries, veins and lymphatic vessels, the intra-ocular lens, blood and related fluids and the vitreous humour. (Part II deals with the properties of surgical implant materials and Part III with the biocompatibility of such materials, not relevant in impact biomechanics.) The cortical bone material is described in its composition (organic, mineral), in its physical properties (density, electromechanical, other) and in its mechanical properties (dry, wet, scatter within the skeleton, stiffness, strength, strain rate and visco-elastic effects). At the end of each chapter additional readings and many references are indicated.

Papers that give detailed stress-strain behavior of biomaterials, including rate effects, are still scarce. Differences between dead and life tissue behavior are seldom described, and data are often inaccessible. The natural scatter between tissues from different individuals is sometimes discussed. The numerical analyst is still constrained to use approximate, incoherent or incomplete data. Many efforts are now undertaken to alleviate this lack of data, a need that was generated only recently by the desire to simulate the biomechanical response the human body using modern computer simulation tools.

Simplest descriptions for biomaterials. Fig. 1.1 shows a selection of some typical biomaterial response curves, as extracted from YANG [1998] (a report of the HUMOS-1 Project, funded by the European Commission under the Industrial and Materials Technologies program (Brite–EuRam III)). The well-known basic elastic, visco-elastic, and elasto-plastic material laws that exist in most dynamic codes and have been applied for biomaterial description. The elastic laws can be linear or nonlinear elastic, isotropic, orthotropic or hyper-elastic. The linear elastic materials are characterized by the elastic moduli, Poisson's ratios, the shear moduli and the mass density. The hyper-elastic materials are characterized by their respective strain energy functions (Mooney, Hart-Smith, etc.). The elastic-plastic material laws are typically defined with the additional hardening modulus, the yield strength, the ultimate strength and strain at failure. The visco-elastic materials need additional coefficients describing the damping, creep and relaxation behavior. The material laws provided with commercial codes are often sufficient to describe hard tissues, such as long bones.

The application to soft tissues is less evident and more research and tests are needed to characterize these materials. In particular, the difference between life and dead tissue behavior is more pronounced in soft than in hard tissues.

In many cases so-called "curve description options" for the standard material laws, as available in the commercial dynamic codes, can be used in order to encode the results directly as obtained from biomaterial tests. These options provide a maximum freedom for the analyst, beyond the usual mathematical descriptions of the materials.

Appendix B gives a summary on the mechanical properties of biomaterials as extracted mainly from YANG [1998].

Bone materials. Fig. 1.2 exemplifies the most frequently studied bone material. Inset (a) shows a cross section through the femur head, with the cortical outer shell of compact bone and the trabecular inner fill of spongeous bone clearly visible. Inset (b) shows the same basic structure in a cross section through the skull bone. Insets (c) and (d) (after RIETBERGEN [1996], RIETBERGEN, MÜLLER, ULRICH, RÜEGSEGGER and HUISKES [1998] and ULRICH [1998]) show so-called "voxel models" of the bony structure, where the trabecular structure of the bone is modelled directly in the optical voxel resolution of micro-scans of the bone. Inset (c), for example, uses several million simplified voxel finite elements to trace (red colour) the linear elastic force path through the trabeculae from an axial force loading. Inset (d) shows two different voxel densities, and inset (e) demonstrates that there is practically no visible difference between a real

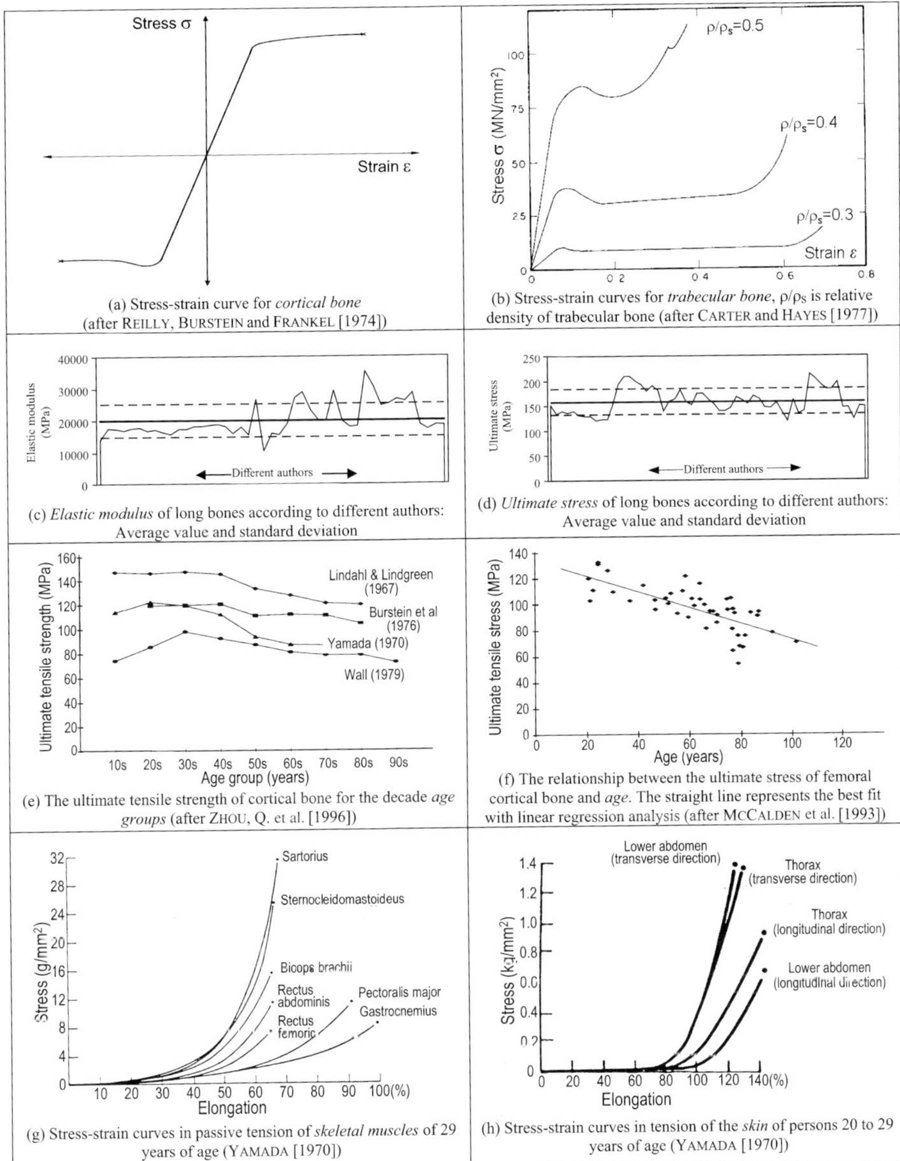

FIG. 1.1. Some typical biomaterial responses as a function of material type, inter-individual scatter, age, deformation rate (compiled by YANG [1998]). (Inset (a): Reproduced by permission of Elsevier Health Sciences Rights; Insets (b) and (f): Reproduced by permission of The Journal of Bone and Joint Surgery, Inc.; Insets (c) and (d): Reproduced by permission of Chalmers University of Technology; Inset (e): Reproduced by permission of The Stapp Association; Insets (g) and (h): Reproduced by permission of Lippincott, Williams and Wilkins.)

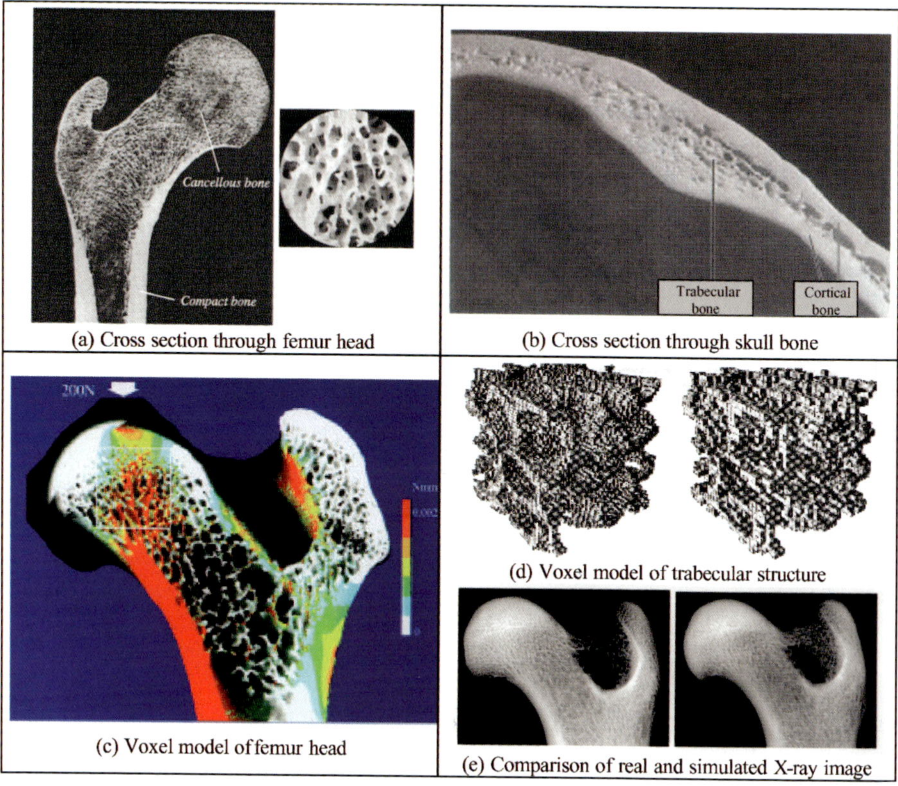

FIG. 1.2. Bone material structure and modeling (after RIETBERGEN, MÜLLER, ULRICH, RÜEGSEGGER and HUISKES [1998], ULRICH [1998]). (Insets (a)–(c), (e): Reproduced by permission of the Journal of Biomechanics; Inset (d): Courtesy Dr. Ulrich of ETH Zurich.)

X-ray picture of the bone and the simulated X-ray picture made from a voxel model of the bone.

This figure stands for the basic need for research concerning the modeling of biomaterials. The direct modeling of the fine structure of the bone material, used in the shown example, is certainly the best possible approach to model bone, since it disposes of uncertain macroscopic averaging processes, but remains in the realm of research and development. This approach may become common practice, once compute power will have increased to the required practical levels. Today, bones must be modeled with "macro" elements, the type of which must be chosen according to the type of bone (e.g., shells for cortical and solids for trabecular bone), and the material densities of which must be evaluated approximately from the average local density of the bone.

Similar remarks can be made for other types of biomaterials, which must be investigated indirectly using "smeared" properties and macro modeling techniques. The brain material, for example, is modeled as solids with homogenized gray or white matter, without taking into account the system of very fine blood vessels that it contains. If this system could be modeled in detail, injury to the vascular system of the brain could be accessed directly, see first attempts made in Fig. 3.3(e)–(h). Today brain injury is accessed indirectly by correlating mechanical field variables to injury through calibration.

References on biomaterial tests, laws, models and simulation. The literature on biomaterials and related subjects is relatively abundant. Appendix B contains references on the biomaterials While many of the indicated references deal with the experimental evaluation of biomaterial properties, others deal with the aspects of their modeling, the use of these materials in biomechanical models and the characterization of trauma and injury.

1.6. Human model validation

Segment and whole body validation. Provided human models can be built and the biomaterials can be calibrated, one of the greatest challenges remains their proper validation. A considerable number of tests on cadaver body segments, whole cadaver bodies and life volunteers were performed in the past, e.g., as listed in the report of the HUMOS-1 project: "Validation Data Base", ROBIN [1999]. As discussed in a later section, the European HUMOS-1 project (Human Models for Safety) produced a first near 50th percentile male European human model in a project funded by the European Commission (HUMOS-1, 1999–2001; HUMOS-2 is under way). The tests listed in this reference comprise the following topics, Fig. 1.3:

Head/neck complex: Frontal tests at 15 g, inset (a); lateral tests at 7 g, inset (b); oblique tests at 10 g, after EWING, THOMAS, LUSTICK, MUZZY III, WILLEMS and MAJEWSKI [1976] (not shown).

Thorax frontal impact: frontal impactor tests by INRETS at low and high velocity (not shown); frontal impactor tests by KROELL, SCHNEIDER and NAHUM [1971], KROELL, SCHNEIDER and NAHUM [1974] at 4.9 m/s, at 6.7 m/s and at 9.9 m/s, inset (c); at 7.0 m/s with seat back (not shown); frontal thorax impactor tests by STALNAKER, MCELHANEY, ROBERTS and TROLLOPE [1973] (not shown).

Thorax belt compression tests: by CESARI and BOUQUET [1990], CESARI and BOUQUET [1994], with a 22.4 kg mass at 2.9 m/s and 7.8 m/s impact velocities, with a 76.1 kg mass at 2.9 m/s impact velocity, inset (d).

Thorax lateral impact: lateral impactor tests by INRETS on the thorax at 3.3 m/s, at 5.9 m/s, inset (e).

Thorax oblique impact: oblique thorax impactor tests by VIANO [1989] at 4.42 m/s, at 6.52 m/s, at 9.32 m/s, inset (f).

Abdomen impact tests: frontal impactor tests on the abdomen by CAVANAUGH, NYQUIST, GOLDBERG and KING [1986], inset (g); oblique impactor tests by Cavanaugh at 31.4 kg and 6.9 m/s (not shown); oblique impactor tests by Viano on the abdomen at 4.8 m/s, at 6.8 m/s, at 9.4 m/s (not shown).

Pelvis impact tests: lateral tests on the pelvis by INRETS at 3.35 m/s, at 6.6 m/s, inset (h); lateral impact tests by Viano at 5.2 m/s, at 9.8 m/s (not shown) (BOUQUET, RAMET, BERMOND and CESARI [1994]).

It is clear that many more tests must be done in order to capture the biomechanical characteristics of humans, not only of the "average" subject (50th percentile male), but

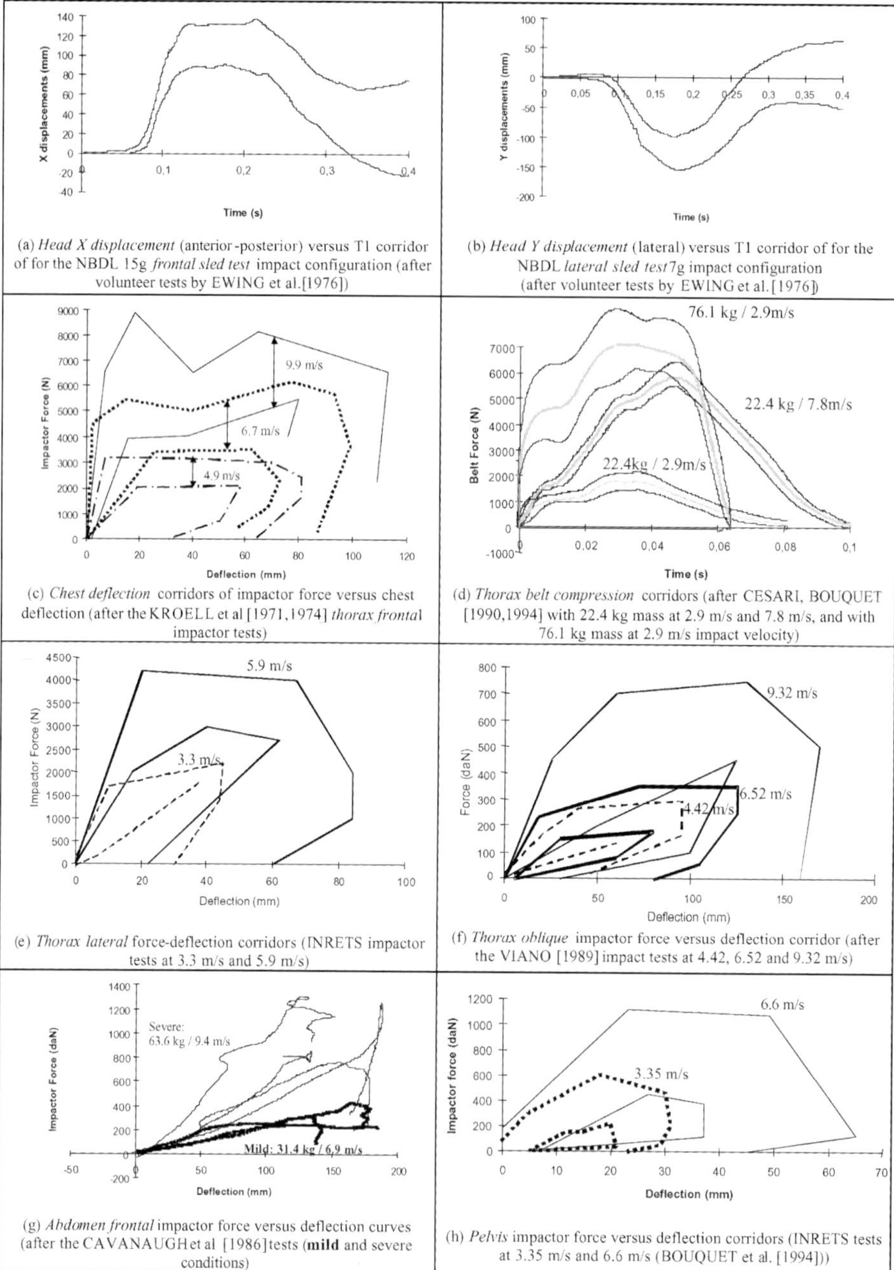

(a) *Head X displacement* (anterior-posterior) versus T1 corridor of for the NBDL 15g *frontal sled test* impact configuration (after volunteer tests by EWING et al.[1976])

(b) *Head Y displacement* (lateral) versus T1 corridor of for the NBDL *lateral sled test* 7g impact configuration (after volunteer tests by EWING et al.[1976])

(c) *Chest deflection* corridors of impactor force versus chest deflection (after the KROELL et al [1971,1974] *thorax frontal* impactor tests)

(d) *Thorax belt compression* corridors (after CESARI, BOUQUET [1990,1994] with 22.4 kg mass at 2.9 m/s and 7.8 m/s, and with 76.1 kg mass at 2.9 m/s impact velocity)

(e) *Thorax lateral* force-deflection corridors (INRETS impactor tests at 3.3 m/s and 5.9 m/s)

(f) *Thorax oblique* impactor force versus deflection corridor (after the VIANO [1989] impact tests at 4.42, 6.52 and 9.32 m/s)

(g) *Abdomen frontal* impactor force versus deflection curves (after the CAVANAUGH et al [1986] tests (**mild** and severe conditions)

(h) *Pelvis* impactor force versus deflection corridors (INRETS tests at 3.35 m/s and 6.6 m/s (BOUQUET et al. [1994]))

FIG. 1.3. Some typical test results for the validation of human models (compiled by ROBIN [1999]). (Insets (a)–(d), (f) and (g): Reproduced by permission of The Stapp Association; Inset (e): Reproduced by permission of INRETS; Inset (h): Material in the public domain by U.S. Department of Transportation.)

also of the inter-individual dispersions that distinguish humans. In order to arrive at "average" responses, and at their likely dispersions, multiple tests should be performed and the results collected in "corridors", which represent best the variable "response" of a given class of human individuals, such as, for example, the 50th percentile male. Some pertinent test results are summarized in Fig. 1.3.

In particular, whole body cadaver sled tests permit to assess the overall response of humans in car accident scenarios. Due to the fact that human models can now be built, calibrated, validated and used in crash simulation or in virtual crash testing, there is a pressing need for reliable data. The necessary tests on human volunteers and cadavers are subject to severe ethical control, which considerably restricts the frequency and number by which such tests can be performed.

2. Overview on mechanical dummies and models

This section may be skipped by readers not interested in mechanical dummies. The material is provided for to give an overview on the mechanical occupant surrogates or legal crash "dummies" as presently used by the auto industry for certification of new car models (KISIELEWICZ and ANDOH [1994]).

Mechanical dummies (occupant surrogates) and their numerical models are used heavily in crash tests and numerical simulations for safe car design. Due to the large number of car crashes each year, crash tests are administered by the National Highway Traffic Safety Administration (NHTSA), an agency within the United States Department of Transportation (DoT). About 35 new model cars have been tested every year since 1979 under the New Car Assessment Program (NCAP). The tests are to see how well different vehicles protect front-seat passengers in a car-to-car head-on collision at equal speeds. The head-on collision is used instead of a rear or a side collision because this is the collision that causes the most deaths and injuries.

The US federal law requires all cars to pass a 30 mph frontal rigid barrier test, so NCAP crash tests on fixed barriers (rigid walls) are performed at 35 mph (56.3 km/h), which corresponds to an impact of two identical cars colliding head-on at a relative velocity of 70 mph (112.6 km/h). These tests show the difference in protection in different car models. The results of the crashes are given on a one-to five star rating, with five being the highest level of protection.

These crash tests are all administered with dummies, the dummies are always wearing seat belts because they are standard equipment on cars today, air bags are used whenever they are available, and test results are only useful in comparing cars of similar weight (within 500 pounds (227 kg) of each other). Dummies heads and knees are painted before a test to see where these areas of the body make contact with the car.

The Hybrid III dummy family is used today in frontal impact tests. For side impact tests, the US DoT SID and the EuroSID special side impact dummies are used. Some of these mechanical occupant surrogates and typical protective measures are shown in Fig. 2.1.

The Hybrid III dummy family. The 50th percentile Hybrid III represents a man of average size. The European standard 50th percentile man is assumed 1.75 meters tall and having a body weight of 75.5 kilograms. Different "standard" sizes may exist in

(a) Hybrid III front impact dummy family: 5%, 50%, 95%; 3 and 6 year; CRABI – 6, 12, 18 months; pregnant woman 5% ; EuroSID, BioSID (FTSS)

(b) Airbags and seat belts (schematic models) upper: frontal impact driver and passenger airbags lower: side impact airbags

FIG. 2.1. Front impact dummies and passive safety systems. (Inset (a): Reproduced by permission of First Technology Safety Systems, Inc.; Inset (b) lower right-hand side: Courtesy Autoliv, BMW (FOGRASCHER [1998]); Inset (b) lower left-hand side: Courtesy AUDI.)

TABLE 2.1
Average body heights and weights (subject to variations)

Percentile age	Height		Weight	
	[cm]	[foot′in″]	[kg]	[lbs]
50th Hybrid III dummy	178	~ 5′10″	77.11	170
50th European adult male	175	~ 5′9″	75.5	166.45
5th adult female	152	~ 5′	49.89	110
95th adult male	188	~ 6′2″	101.15	223
6 year old child	113	~ $3'8\frac{1}{2}''$	21.32	47
3 year old child	99	~ 3′3″	14.97	33

different countries. Born in the USA in the labs of General Motors, the 50th percentile Hybrid III is the standard dummy used in frontal crash tests all over the world. It is called a hybrid, because it was created by combining parts of two different types of dummies. Beside the 50th percentile male, there are the 5th percentile female, the 95th percentile male and the 6 year-old and 3 year-old child dummies, Table 2.1.

Hybrid III dummy models. Fig. 2.2 shows numerical models of members of the Hybrid III dummy family (after FTSS/ESI Software). The models shown under Fig. 2.2(a) have 25 878 (50th percentile), 24 316 (5th percentile), 27 872 (95th percentile), 34 535 (6 year old child) and 13 345 (3 year old child) deformable finite elements, respectively. In (b) a Hybrid III dummy model is shown in a driver position. The 50th percentile male Hybrid III model is shown in Fig. 2.2(c) through (e). Insets (c) and (d) are finite element models, while inset (e) is a section through the simpler multi-body version. Most dummy models are made either as simpler multi-body models, or as more detailed finite element models.

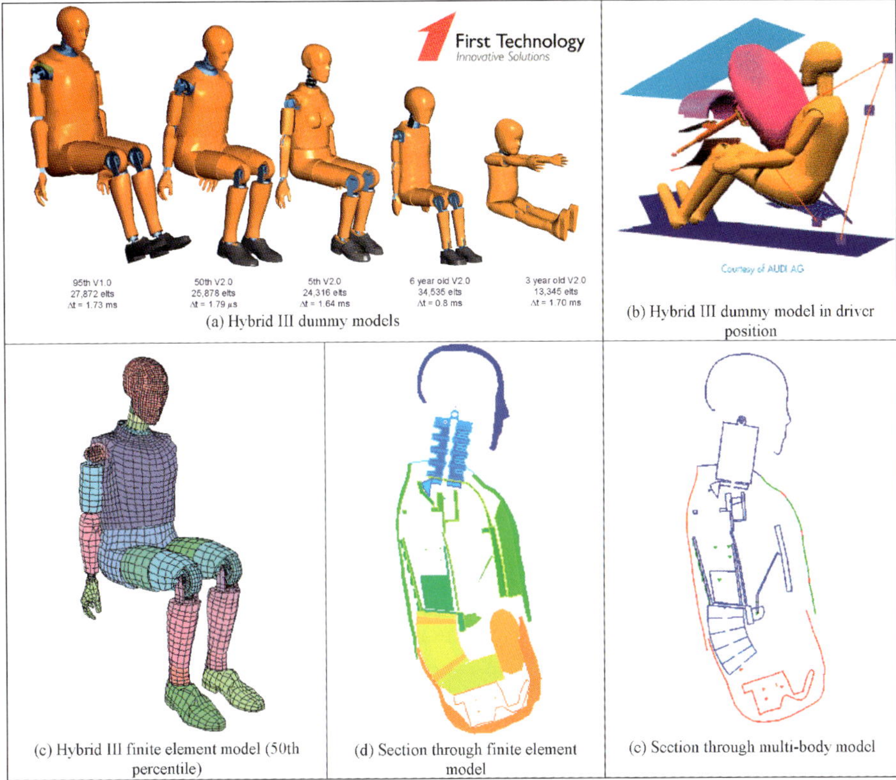

(a) Hybrid III dummy models (b) Hybrid III dummy model in driver position (c) Hybrid III finite element model (50th percentile) (d) Section through finite element model (e) Section through multi-body model

FIG. 2.2. Numerical models of the Hybrid III dummy family (FTSS/ESI Software). (Reproduced by permission of First Technology Safety Systems, Inc.)

Multi-body modelling techniques comprise linked rigid body tree structures, which contain relatively few deformable parts and which are linked together at the intersections of their anatomical segments. They execute faster but cannot yield detailed injury data. Finite element models are made of the usual standard library of finite elements of the used crash codes (solids, shells, membranes, beams, bars, springs, etc). They take more central processor unit (CPU) computer time, but can yield response data, which are more readily linked to human injuries. Typical solver codes used to analyse car crash scenarios execute the explicit time integration scheme for the set of non-linear equations of motion in the nodal degrees of freedom.

Standard injury criteria. The standard way of assessing injuries of vehicle occupants are heuristic injury coefficients that are calculated from injury criteria defined from the instrumented front or side impact dummy responses in crash tests.

Head injuries of occupants are assessed from the "Head Injury Coefficient" (HIC),

$$\mathrm{HIC} = \max_{t_1 < t_2}\left[(t_2 - t_1) \left\{ \frac{1}{t_2 - t_1} \int_{t_1}^{t_2} a(t)\, \mathrm{d}t \right\}^{2.5} \right],$$

where $\Delta t = t_2 \geqslant t \geqslant t_1$ is a normed time window (e.g., 22.5 ms) that is shifted along the Hybrid III head acceleration magnitude time history, $a(t)$, as recorded by the head accelerometers, to find the maximal value of the HIC coefficient over the duration of the crash event. If the calculated HIC-value is below critical (e.g., 1000), then it is assumed that no serious injury (skull fracture; neuro-vascular damage) occurs.

Neck injuries can be assessed (among other criteria) from the N_{ij} neck injury criterion,

$$N_{ij} = F_Z/F_{Z,\text{crit}} + M_Y/M_{Y,\text{crit}},$$

where F_Z is the recorded neck axial force (tension/compression), M_Y is the recorded sagittal neck bending moment (flexion/extension) and subscripts "crit" indicate the respective critical values, set such that $N_{ij} = 1.0$ corresponds to a 30% probability of injury.

Similar criteria exist for the *thorax* (chest acceleration, chest compression, viscous criterion, side impact dummy rib deflection, thoracic trauma index), for the *abdomen* (abdominal peak force, pelvis acceleration, pubic symphysis peak force) and for the *lower extremity* (femur load, tibia index).

SID, EuroSID, BioSID, SID II(s). Hybrid III dummies are designed to be used in frontal crash tests. For tests representing crashes in which a vehicle is struck on the side, a number of dedicated side-impact dummies have been created to measure injury risk to the ribs, spine, and internal organs, such as the liver and spleen, Fig. 2.3.

US DoT SID was the first side-impact dummy. It was developed in the late 1970s by the US National Highway Traffic Safety Administration (NHTSA) of the US Department of Transportation (DoT) and is used in US government-required side-impact testing of new cars.

EuroSID was developed by the European Experimental Vehicles Committee (EEVC) and is used to assess compliance with the European side-impact requirements.

BioSID is based on a General Motors design. It is more advanced than SID and EuroSID, but it is not specified as a test dummy to be used in legal tests.

SID, EuroSID, and BioSID are designed to represent 50th percentile or average-size men 5 feet 10 inches (1.78 m) tall and 170 pounds (77.11 kg).

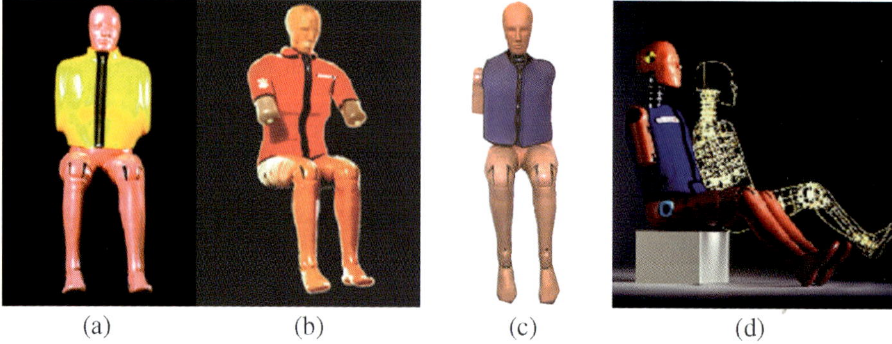

(a) (b) (c) (d)

FIG. 2.3. Side impact dummies (hardware). (a) US DoT SID; (b) EuroSID; (c) BioSID; (d) SID II(s) 5th percentile female. (Reproduced by permission of First Technology Safety Systems, Inc.)

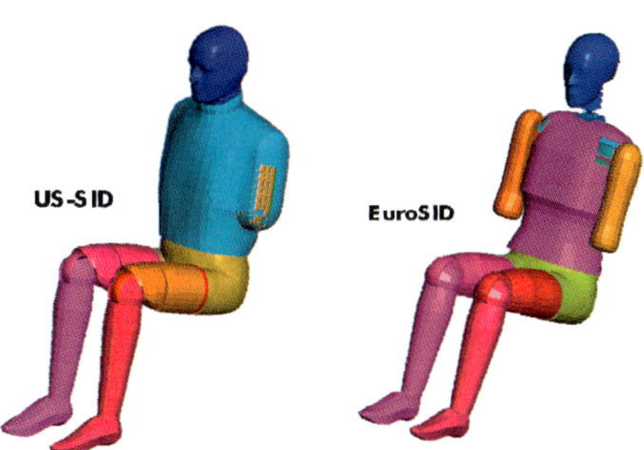

FIG. 2.4. US DoT SID and EuroSID side impact dummy models (FAT/ESI Software).

SID II(s) represents a 5th percentile small female who is 5 feet (1.52 m) tall and weighs 110 pounds (49.89 kg). SID II(s) was created by a research partnership of US automakers. It is the first in a family of technologically advanced side-impact dummies.

SID measures the acceleration of the spine and ribs. Acceleration is the rate of velocity change, and measuring it indicates the forces inflicted on the body during the crash. EuroSID, BioSID, and SID II(s) measure acceleration plus compression of the rib cage. Compression refers to the extent body regions are squeezed during the impact and is used as an indicator of injury to internal organs.

Side impact dummy models. In Fig. 2.4 finite element models of the US DoT SID and of the EuroSID dummy are shown. These models were elaborated on the basis of material, component and whole body tests, performed by the German car manufacturer consortium FAT. The models have about 38 000 deformable finite elements and a time step of 1.5 microseconds in explicit solver codes.

BioRID. A rear-impact dummy has been developed to measure the risk of minor neck injuries, sometimes called whiplash, in low-speed rear-end crashes, which is a big problem worldwide.

BioRID was developed in the late 1990s by a consortium of Chalmers University of Technology in Sweden, restraint manufacturer Autoliv, and automakers Saab and Volvo. It is designed to represent a 50th percentile or average-size man, 5 feet 10 inches tall and 170 pounds in weight, Fig. 2.5, DAVIDSSON, FLOGARD, LÖVSUND and SVENSSON [1999].

BioRID has been designed especially to study the relative motion of the head and torso. For tests representing crashes in which a vehicle is struck in the rear, BioRID can help researchers learn more about how seatbacks, head restraints, and other vehicle characteristics influence the likelihood of whiplash injury.

Unlike Hybrid III dummies, the BioRID spine is composed of 24 vertebra-like segments, so that in a rear-end crash BioRID interacts with vehicle seats and head restraints in a more humanlike way than the Hybrid III. The BioRID segmented neck can take on the same shapes observed in human necks during rear-end collisions, an important characteristic for measuring some risk factors associated with whiplash injury. Comparative cross sections through the human spine and BioRID are shown in Fig. 2.5.

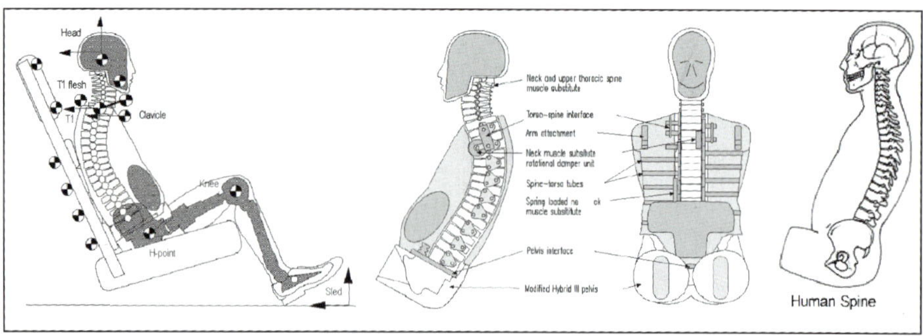

FIG. 2.5. BioRID Rear Impact Dummy and comparative human section. (Reproduced by permission of The Stapp Association.)

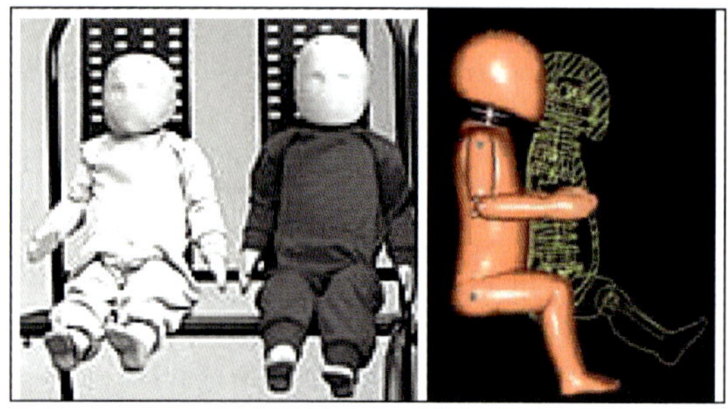

FIG. 2.6. Child Restraint Air Bag Interaction dummy (CRABI). (Reproduced by permission of First Technology Safety Systems, Inc.)

CRABI. The Child Restraint Air Bag Interaction dummy was developed by First Technology Safety Systems (FTSS) to represent children, Fig. 2.6. It is used to evaluate child restraint systems, including airbags. There are three sizes: 18 month-old, 12 month-old, and 6 month-old. These dummies have sensors in the head, neck, chest, back, and pelvis that measure forces and accelerations.

THOR. This advanced 50th percentile male dummy is being developed in the United States by NHTSA for use in frontal crash tests, Fig. 2.7 (http://www-nrd.nhtsa.dot.gov/departments/nrd-51/THORAdv/THORAdv.htm). THOR has more human-like features than Hybrid III, including a spine and pelvis that allow the dummy to assume various seating positions, such as slouching, for example, or sitting upright. THOR also has sensors in its face that measure forces so that the risk of facial injury can be assessed, which is not possible with current dummies. In fact, all of THORs standard sensors will provide more injury measurements than those available on Hybrid III.

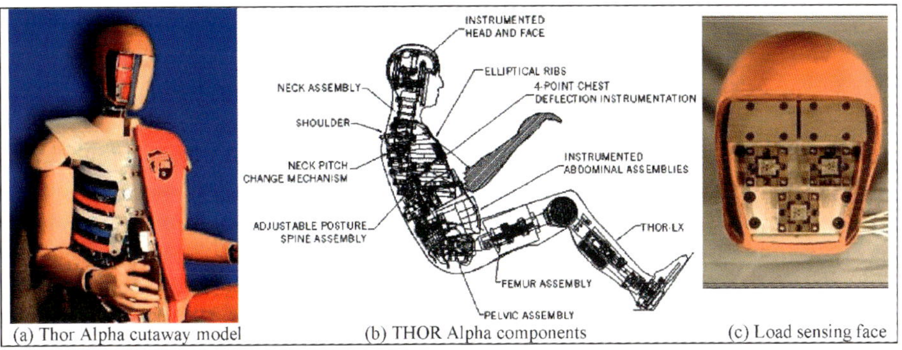

(a) Thor Alpha cutaway model (b) THOR Alpha components (c) Load sensing face

FIG. 2.7. Advanced THOR dummy. (Reproduced by permission of NHTSA, U.S. Government.)

3. Overview of existing human models for occupant safety

The following paragraphs briefly outline the present state of the art in human biomechanical modeling for occupant safety. Most of the information has been drawn from the indicated INTERNET web sites. The interested reader is invited to consult these sites, and more, to get up-to-date information of this rapidly expanding field. The selected examples demonstrate the extensive level of development of human models for impact biomechanics by research institutions and by private industry. The fact that car manufacturers invest actively in human models reflects the need for the use of human models in safe car design. At the same time a unification of these efforts is needed, as reflected, for example, by the pre-competitive European joint development project HUMOS, in which five car manufacturers, several equipment suppliers, several research institutes and three engineering software houses are active partners. While the official approval by the respective National Road Safety Administration authorities for the use of human models for safe car design in "legal virtual crash tests" is pending, new car models are being certified using mechanical dummies in real physical legal crash tests. An interesting concept on the way towards legal virtual crash tests with human models has been introduced recently by the National Highway Traffic Safety Administration (NHTSA), US Department of Transportation (DoT), which can be considered an encouraging step towards the increased use of human models in safe car design. This new concept is discussed first.

3.1. SIMon (Simulated Injury Monitor)

NHTSA experts have developed human models (http://www.nhtsa.dot.gov), which directly simulate bodily injury, unlike dummies or models of dummies, which access injury through equivalent measures. The developed models are meant to help new, advanced, mechanical dummy design (e.g., THOR) on the one hand, and, on the other hand, they serve as vehicles in the numerical interpretation of the enriched output data, harvested from the new generation dummies in crash tests.

SIMon-Head. The recently released first SIMon-Head model is discussed below and it provides the first step towards a new standardized analytical occupant safety analysis methodology, called SIMon (Simulated Injury Monitor). The objective of this particular research is to evaluate injury to the soft tissue of the human brain using finite element models of the brain together with dynamic load data from mechanical dummies, harvested in actual crash tests. The so far released SIMon-Head model consists in a CD Rom with an NT software package that can accept the measured output of nascent new generation mechanical dummies (e.g., THOR). The new dummy head is equipped with nine instead of three accelerometers, which permits to record translational, as well as rotational accelerations of the head. The SIMon-Head package uses these comprehensive acceleration time histories as an input to a built-in calibrated finite element model of the head and brain, to which it applies the recorded accelerations by running an explicit solver code, itself locked into the SIMon-Head package.

The package then analyzes the built-in head/brain model output data and it generates three new brain injury coefficients, namely the "Cumulated Strain Damage Measure (CSDM)", the "Dilatation Damage Measure (DDM)" and the "Relative Motion Damage Measure (RMDM)". The CSDM tells what cumulative volume fraction of the brain matter experienced at some time principal strains larger than a fixed threshold value (15%), known to cause Diffuse Axonal Injury (DAI). The DDM tells what instantaneous volume fraction of the brain matter experienced negative dynamic pressures that can cause vaporization of the cerebral fluids, and contusion. The RMDM tells the percentage of the bridging veins that have stretched beyond a limit curve in a calibrated strain vs. strain-rate diagram, each possibly causing Acute Subdural Hematoma (ASDH) through rupture. The bridging veins connect the soft brain tissues to the skull and may rupture through excessive shearing motions of the brain with respect to the skull.

This new concept will raise the level of precision for injury prediction by directly addressing different types of local injuries, rather than by comparing abstract global coefficients, such as the well-known Head Injury Coefficient (HIC), with calibrated threshold values (cf. NHTSA Federal Motor Vehicle Safety Standard (FMVSS)). The HIC is an integral of translational dummy head accelerations over a moving fixed time window, which tells the maximum acceleration the head experienced over the fixed time interval during the crash event. This measure can be linked to real brain injuries only in a purely heuristic fashion. The SIMon concept will be extended in the near future to further body segments (neck, thorax, femur, etc.), and it will undoubtedly accelerate the widespread use of human models in safe car design.

Fig. 3.1(a)–(h) show a typical (side) impact experiment in a crash laboratory (a), a model of the advanced (frontal) crash test dummy THOR (b), as well as several views of the SIMon Head model, labeled: a mid-coronal section view (c), a coronal-sagittal view, highlighting the boundary between the falx cerebri and the skull (d), a 3D view of the opened SIMon skull model (e), a 3D top view of the brain (f), a top view indicating the location of the parasagittal bridging veins (g), and a 3D view of a brain model (h).

SIMon is designed to provide head (Fig. 3.1), and later neck and thorax (Fig. 3.2), and lower extremity models of occupants, including women and children. These models of human parts can be driven by instrumentation data from advanced crash dummies.

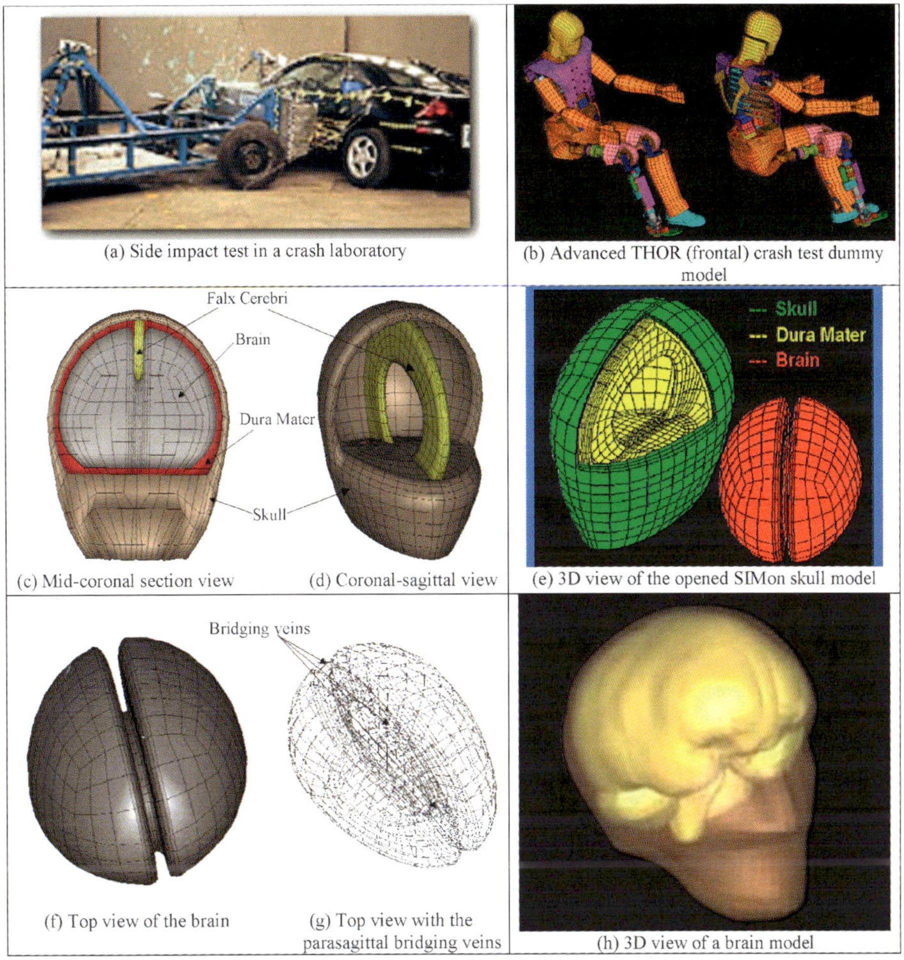

(a) Side impact test in a crash laboratory

(b) Advanced THOR (frontal) crash test dummy model

Falx Cerebri

Brain

Dura Mater

Skull

(c) Mid-coronal section view (d) Coronal-sagittal view

-- Skull

-- Dura Mater

-- Brain

(e) 3D view of the opened SIMon skull model

Bridging veins

(f) Top view of the brain (g) Top view with the parasagittal bridging veins

(h) 3D view of a brain model

FIG. 3.1. SIMon Head and Brain Models (http://www.nhtsa.dot.gov). (Reproduced by permission of NHTSA, U.S. Government.)

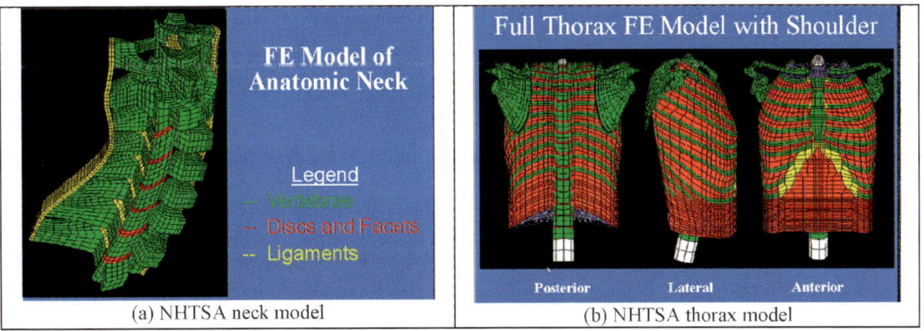

(a) NHTSA neck model (b) NHTSA thorax model

FIG. 3.2. NHTSA Neck and Chest Models (http://www.nhtsa.dot.gov). (Reproduced by permission of NHTSA, U.S. Government.)

References for SIMon-Head. The following references may be consulted on the subject of the SIMon-Head model: AL-BSHARAT, HARDY, YANG, KHALIL, TASHMAN and KING [1999] on brain/skull relative motions; BANDAK and EPPINGER [1994], BANDAK, TANNOUS, ZHANG, TORIDIS and EPPINGER [1996], BANDAK, TANNOUS, ZHANG, DIMASI, MASIELLO and EPPINGER [2001] on brain FE models and SIMon Head; FALLENSTEIN, HULCE and MELVIN [1970] on dynamic mechanical properties of human brain tissue; GENNARELLI and THIBAULT [1982], GENNARELLI, THIBAULT, TOMEI, WISER, GRAHAM and ADAMS [1987] on biomechanics of acute subdural hematoma (1982a) and on directional dependence of axonal brain injury (1987); LOWENHIELM [1974] on dynamic properties of bridging veins; MARGULIES and THIBAULT [1992] on diffuse axonal injury tolerance criteria; MEANY, SMITH, ROSS and GENNARELLI [1993] on diffuse axonal threshold injury animal tests; NUSHOLTZ, WILEY and GLASCOE [1995] on cavitation effects in head impact model; OMMAYA and HIRSCH [1971] on cerebral concussion tolerance in primates; ONO, KIKUCHI, NAKAMURA, KOBAYASHI and NAKAMURA [1980] on head injury tolerance for sagittal impact from tests.

3.2. *Wayne State University human models*

For over sixty years, the Wayne State University (WSU) Bioengineering Center has pioneered biomechanics research and issued injury tolerance thresholds. Over the past years, the center engaged in a continued activity of the development of models of the human body and its parts, in particular the WSU brain and head model, as shown below. These models are among the most advanced, and their validation is substantiated by experiments performed at the center itself. The WSU human models have served many workers and institutions as a basis for their own development and research (Ford, General Motors, Nissan, Toyota, ESI, Mecalog, etc.).

Fig. 3.3 gives an overview on the WSU human models. The reported information can be found on their web site http://ttb.eng.wayne.edu, as well as in publications by ZHANG, YANG, DWARAMPUDI, OMORI, LI, CHANG, HARDY, KHALIL and KING [2001] and HARDY, FOSTER, MASON, YANG, KING and TASHMAN [2001] for the WSU head injury model, Fig. 3.3(a)–(d); in ZHANG, BAE, HARDY, MONSON, MANLEY, GOLDSMITH, YANG and KING [2002] for the WSU vascular brain model, Fig. 3.3(e)–(h); in YANG, ZHU, LUAN, ZHAO and BEGEMAN [1998] for the WSU neck model, Fig. 3.3(i), (j); after SHAH, YANG, HARDY, WANG and KING [2001] for the WSU chest model, Fig. 3.3(k)–(m); in LEE and YANG [2001] for the WSU abdomen model, Fig. 3.3(n)–(q).

3.3. *THUMS (Total Human Model for Safety)*

The Total Human Model for Safety (THUMS) was recently assembled and tested by Toyota Research Company, see Fig. 3.4(a)–(f), cf. FURUKAWA, FURUSU and MIKI [2002], IWAMOTO, KISANUKI, WATANABE, FURUSU, MIKI and HASEGAWA [2002], KIMPARA, IWAMOTO and MIKI [2002], MAENO and HASEGAWA [2001], NAGASAKA, IWAMOTO, MIZUNO, MIKI and HASEGAWA [2002], OSHITA, OMORI,

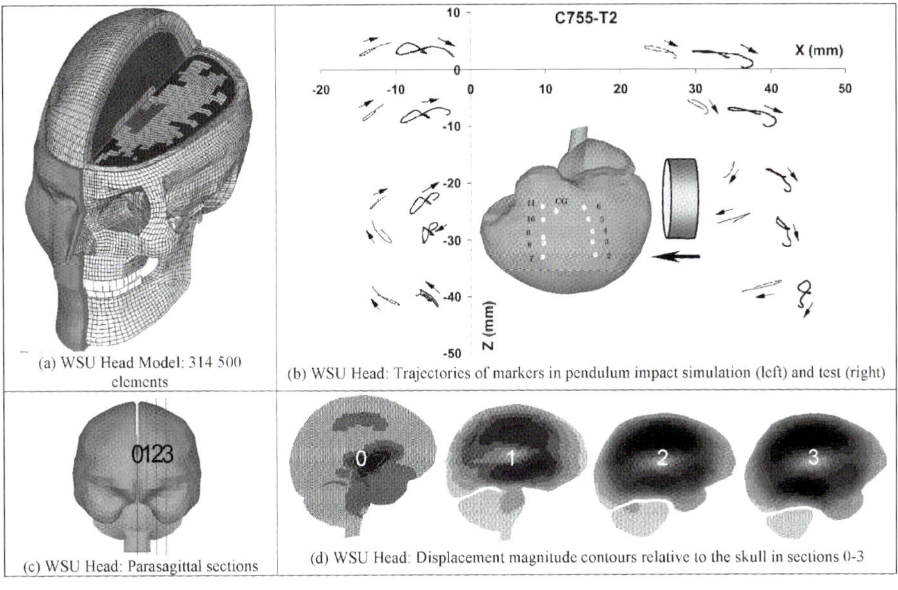

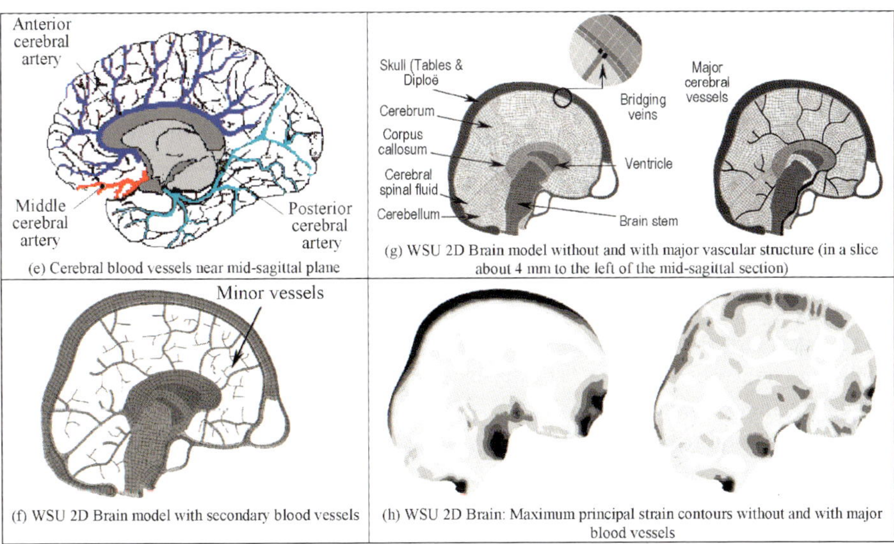

FIG. 3.3. The WSU human models. (Insets (a)–(h), (j) and (n)–(q): Reproduced by permission of The Stapp Association; Insets (i) and (k)–(m): Reproduced by permission of King H. Yang, Wayne State University Bio-engineering Center; Insets (a)–(d): WSUHIM Head Injury Model (ZHANG, YANG, DWARAMPUDI, OMORI, LI, CHANG, HARDY, KHALIL and KING [2001]) and response under occipital impact test C755-T2 (HARDY, FOSTER, MASON, YANG, KING and TASHMAN [2001]); Insets (e)–(h): WSU 2D Vasculated Brain Injury Model (ZHANG, BAE, HARDY, MONSON, MANLEY, GOLDSMITH, YANG and KING [2002]) and response under impact test nb.37 (NAHUM, SMITH and WARD [1977]); Insets (i) and (j): WSU Neck Model structure and response under whiplash conditions (YANG, ZHU, LUAN, ZHAO and BEGEMAN [1998]); Insets (k) to (q): WSU chest model (k), (l), (m) (SHAH, YANG, HARDY, WANG and KING [2001]), abdomen model (n), (o) (LEE and YANG [2001]) and response (p), (q) under lateral pendulum impact (VIANO [1989]).)

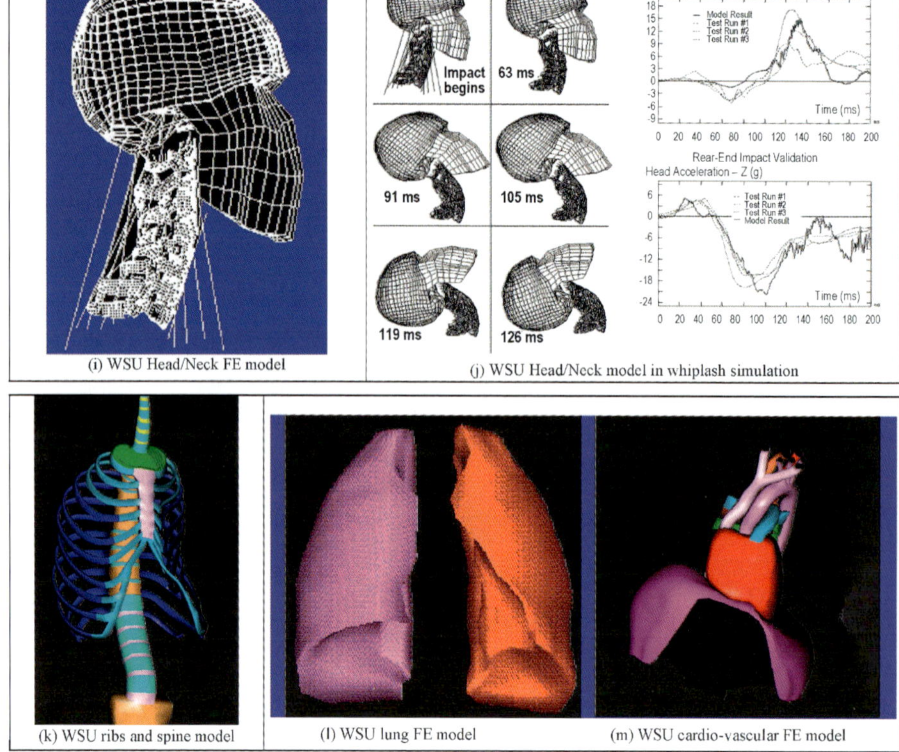

(i) WSU Head/Neck FE model

(j) WSU Head/Neck model in whiplash simulation

(k) WSU ribs and spine model

(l) WSU lung FE model

(m) WSU cardio-vascular FE model

FIG. 3.3. (*Continued.*)

NAKAHIRA and MIKI [2002] and WATANABE, ISHIHARA, FURUSU, KATO and MIKI [2001] and their web site http://www.tytlabs.co.jp/eindex.html. This considerable effort reflects the urgent need for car industry using human models for safe car design. A first 50th percentile human male model was completed in 2000, based on their own development and on models from Wayne State University. This model is relatively detailed, since it comprises more than 80 000 elements, which is about three times the density of the HUMOS-1 model, as discussed in a separate section of this article. Since conventional mechanical crash dummy models often have the same level of refinement, the model is suitable for running crash simulations.

THUMS is a family of models, which comprises the AM50 50th percentile male, Fig. 3.4(a), the AF05 5th percentile female and the 6 year old child, (b), and a pedestrian model, (c). The internal organs of the AM50 model are shown in inset (d). Insets (e) and (f) show the deformed shapes of the AM50 and the AF05 models, respectively, under the Kroell chest pendulum impact tests, with a pendulum mass of 23.4 kg, a diameter of 150 mm and an impact velocity of 7.29 m/s. The simulation results were compared with the Kroell tests (KROELL [1971] and KROELL, SCHNEIDER and NAHUM [1971]).

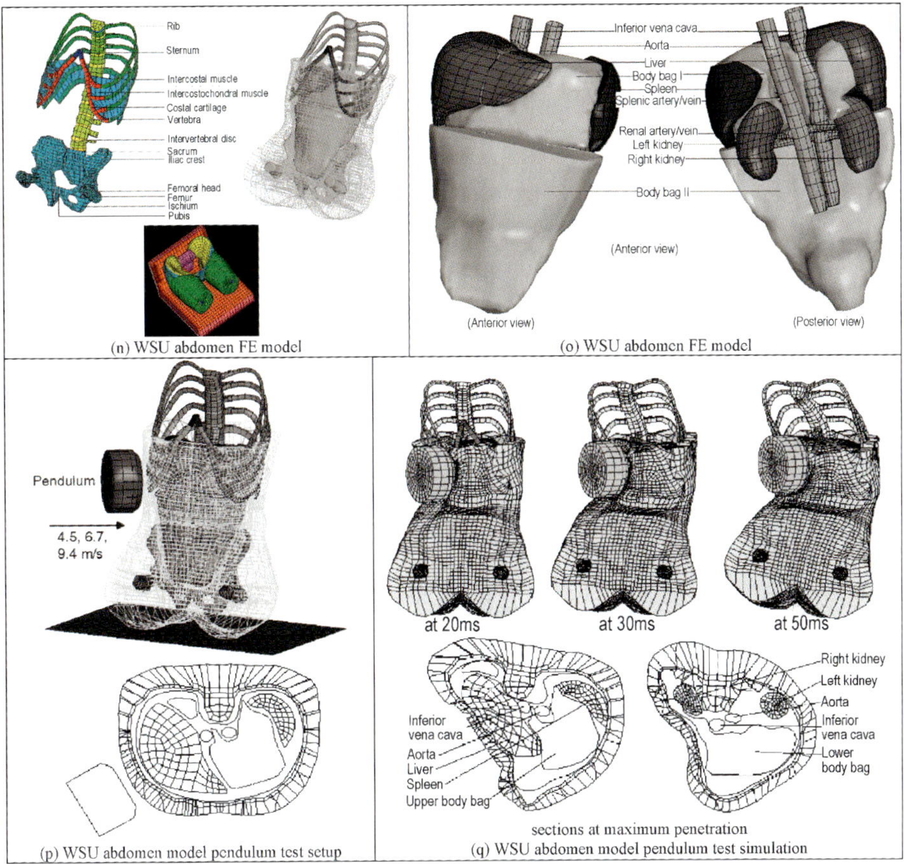

(n) WSU abdomen FE model

(o) WSU abdomen FE model

(p) WSU abdomen model pendulum test setup

(q) WSU abdomen model pendulum test simulation

FIG. 3.3. (*Continued.*)

The pedestrian simulations with 40 km/h bending and shearing tests, inset (c), were compared to tests by KAJZER, SCHROEDER, ISHIKAWA, MATSUI and BOSCH [1997].

3.4. LAB human model

The LAB (Laboratoire d'Accidentologie et de Biomécanique of PSA Peugeot, Citroën, RENAULT), in collaboration with CEESAR, ENSAM and INRETS, have developed a complete 50th percentile male human finite element model with 10 000 elements, Fig. 3.5(a)–(i).

The material properties were taken from the literature and a large data base of 30 test configurations and 120 test corridors was compiled and used to validate the model. Comparative studies were performed concerning the differences in the behavior of human models as compared to dummy models in frontal and lateral impact conditions. Fig. 3.5 gives an overview on this pioneering model and its comparisons with models of front impact (HYBRID III) and side impact (EuroSID) dummy models.

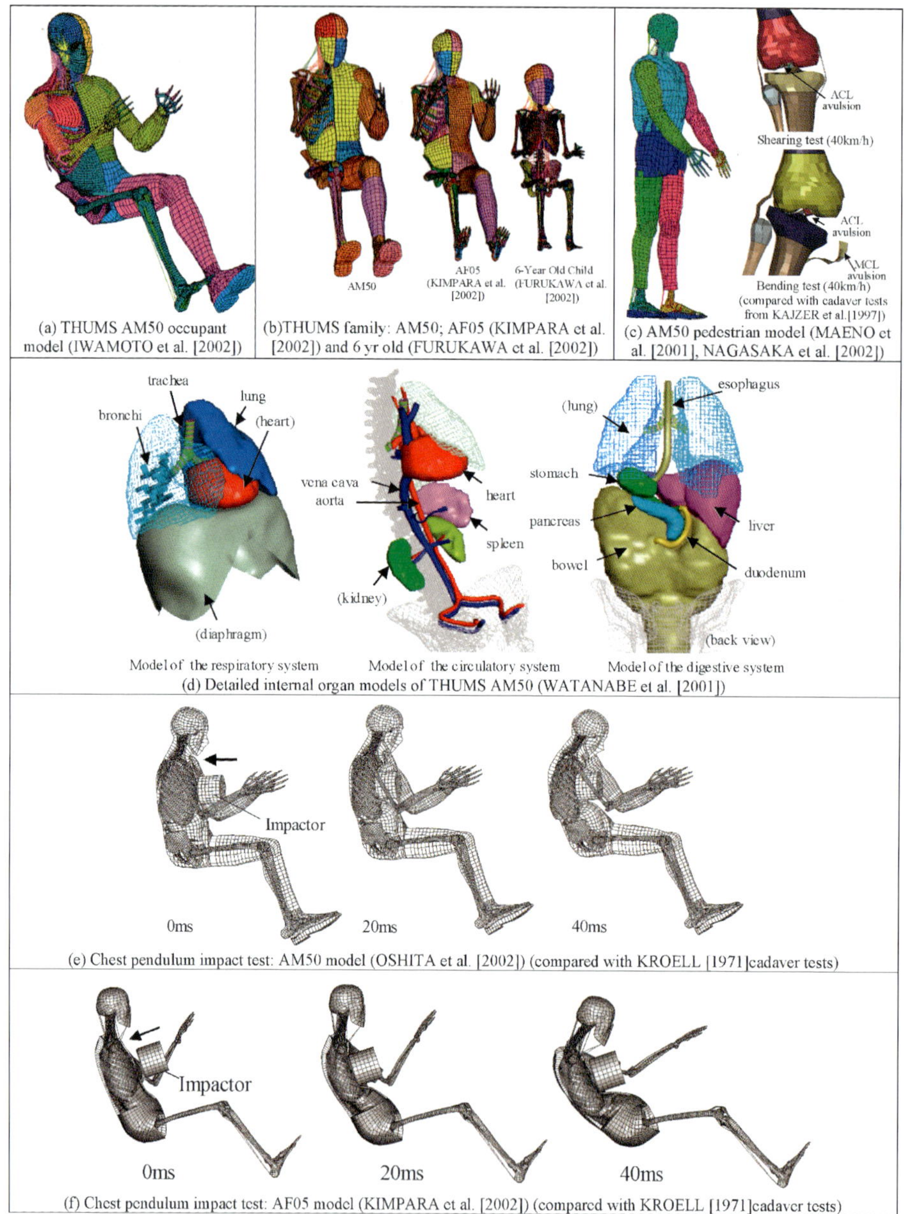

FIG. 3.4. TOYOTA's THUMS Total Model for Human Safety family. (Reproduced by permission of Kazuo Miki, Toyota CRDL.)

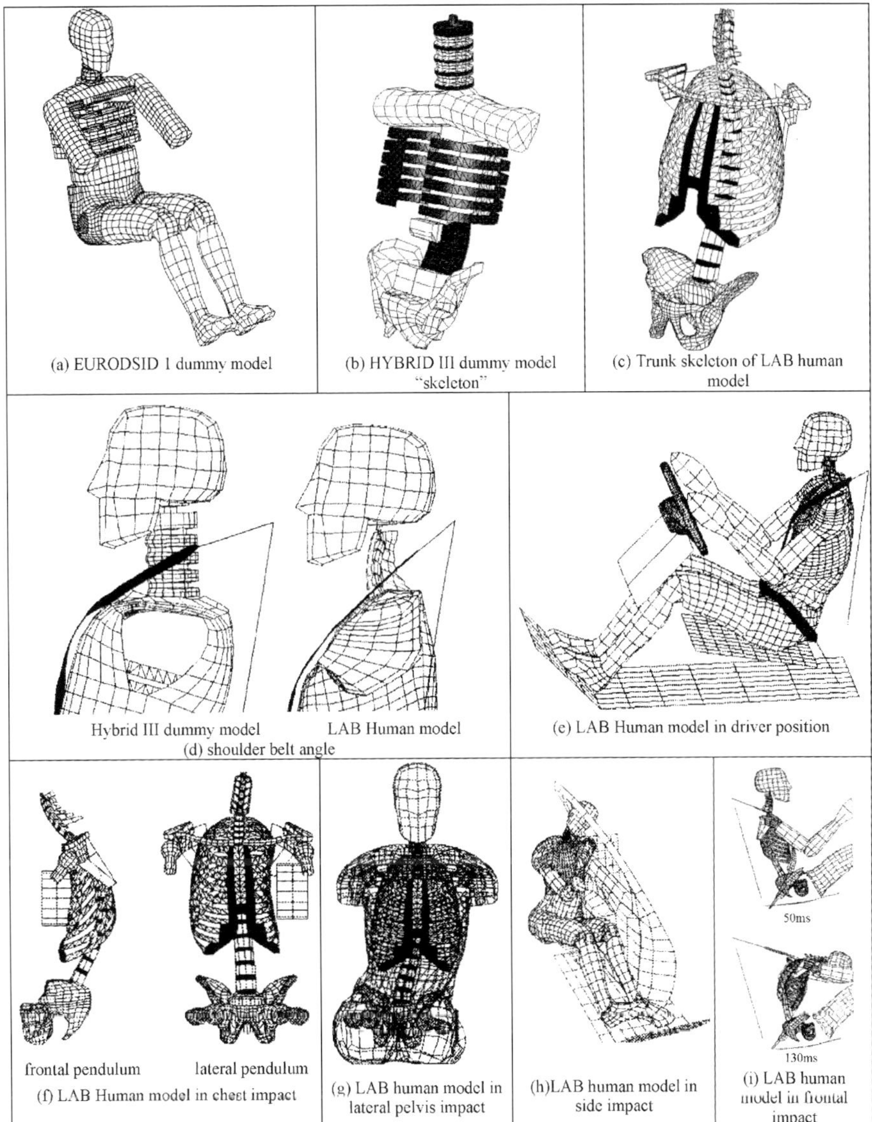

FIG. 3.5. The LAB human and dummy models (BAUDRIT, HAMON, SONG, ROBIN and LE COZ [1999], LIZEE, ROBIN, SONG, BERTHOLON, LECOZ, BESNAULT and LAVASTE [1998]). (Reproduced by permission of The Stapp Association.)

References for the LAB model. The LAB model has been published in the following papers: BAUDRIT, HAMON, SONG, ROBIN and LE COZ [1999] on comparing dummy and human models in frontal and lateral impacts; LIZEE, ROBIN, SONG, BERTHOLON, LECOZ, BESNAULT and LAVASTE [1998] on the development of a 3D FE human body

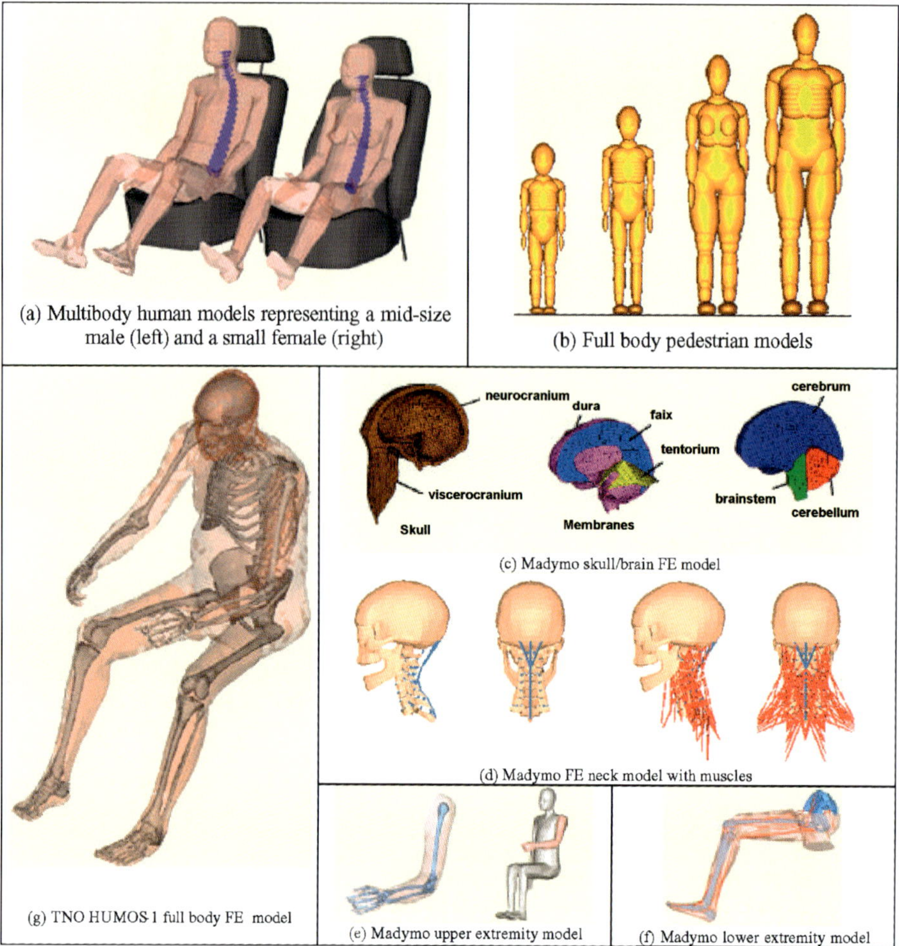

(a) Multibody human models representing a mid-size male (left) and a small female (right)

(b) Full body pedestrian models

(c) Madymo skull/brain FE model

(d) Madymo FE neck model with muscles

(e) Madymo upper extremity model

(f) Madymo lower extremity model

(g) TNO HUMOS-1 full body FE model

FIG. 3.6. Madymo multi-body and deformable finite element models of the human body. (Reproduced by permission of TNO Automotive.)

model; WILLINGER, KANG and DIAW [1999] on the validation of a 3D FE human model against experimental impacts.

3.5. MADYMO human models

MADYMO is a TNO Automotive engineering software tool that is used for the design of occupant safety systems. The following extract on their human models is drawn from their web site, http://www.madymo.com, Fig. 3.6(a)–(f).

Human body models have been developed for TNOs software program MADYMO, using a modular approach. Several combinations of detailed multi-body and finite element (FE) segment models are available. The models have been validated for impact

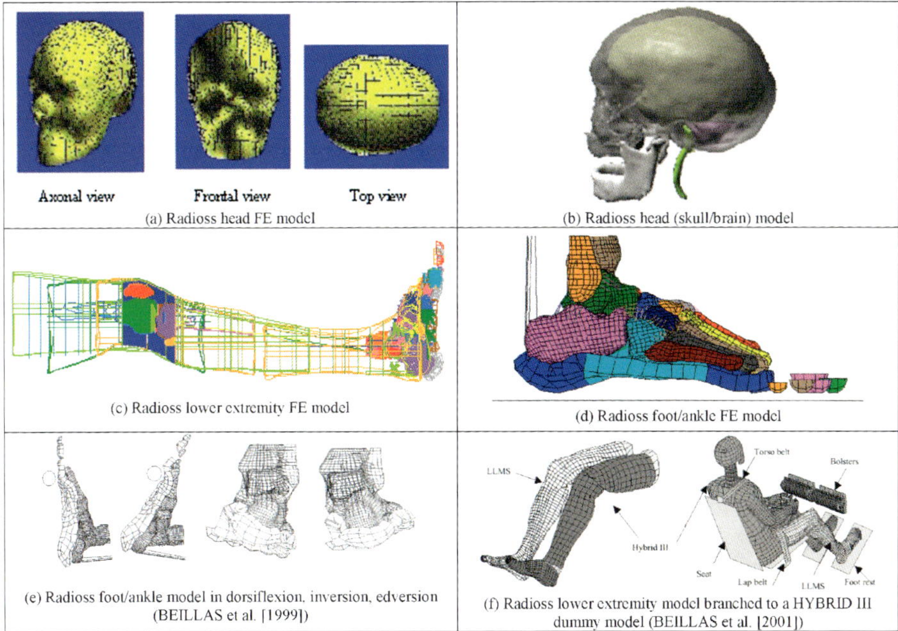

FIG. 3.7. Radioss human models (http://www.radioss.com). (Insets (a)–(d): Reproduced by permission of MECALOG Sarl; Insets (e) and (f): Reproduced by permission of The Stapp Association.)

loading. For their multi-body models, a combination of modeling techniques was applied using rigid bodies for most segments, but describing the thorax as a deformable structure.

A finite element mesh of the entire human body has been developed by TNO, based on the data produced in the EU project HUMOS, Fig. 3.6(g). The mesh was based on a European 50th percentile human in a seated driving position with a detailed 3-D numerical description of the subject's geometry. Apart of this full body model, TNO offers a series of deformable models of various body parts.

3.6. RADIOSS human models

As TNO and ESI Software, Mecalog is a partner in the European HUMOS projects. They have developed their own encrypted model from the common HUMOS-1 project data base. Further models of human parts were developed by Mecalog, as shown in Fig. 3.7(a)–(f). Their head and skull model was developed in collaboration with the University of Strassbourg (WILLINGER, KANG and DIAW [1999]). Their lower extremity model was developed in collaboration with WSU (BEILLAS, LAVASTE, NICOLOPOULOS, KAYVANTASH, YANG and ROBIN [1999], BEILLAS, BEGEMAN, YANG, KING, ARNOUX, KANG, KAYVANTASH, BRUNET, CAVALLERO and PRASAD [2001]). More information can be found on their web site http://www.radioss.com.

CHAPTER II

Human Articulated Multi-Body Models

The numerical models and materials presented in this chapter are based on work carried out at ESI Software and the University of West Bohemia (Robby family), and at IPS International and Hong-Ik University (H-Models).

4. Human Articulated Rigid Body (HARB) models

Open tree chain structure. The kinematics of the human body can be described to first order accuracy by the kinematics of a chain of articulated rigid bodies. Such models are computationally efficient, but they provide only limited information. Each member of the human body can be represented as a rigid body, while each skeletal joint is modeled by a corresponding numerical joint element or using non-linear springs. Modern crash codes have multi-body linkage algorithms with a comprehensive set of joint models, by which open-tree linked rigid body structures can be treated effectively. Details about these modeling techniques can be found in the handbooks of these codes. Human Articulated Rigid Body (HARB) models have therefore been introduced, which can describe the overall kinematics of the model under a crash load scenario. Fig. 4.1 shows a fiftieth percentile male HARB model, named "Robby", with its exterior skin, its tree link structure and its skeletal structure (ROBBY1 [1997], ROBBY2 [1998], HYNCIK [1997], HYNCIK [1999a], HYNCIK [2001a], HYNCIK [2002a], HAUG, BEAUGONIN, TRAMECON and HYNCIK [1999], BEAUGONIN, HAUG and HYNCIK [1998]).

Inset (a) of Fig. 4.1 defines the "sagittal", "coronal" (or "frontal") and "transverse" (or "horizontal") planes, used in anatomy to situate the parts of the body. To situate body parts relative to the body, directions are convened as follows: "anterior" pointing towards the front of the body, "posterior" towards the back, "medial" towards the midline, "lateral" away from the midline, "proximal" closer to the trunk and "distal" away from the trunk.

In inset (b), the upright HYBRID III 50th percentile mechanical dummy model is superimposed to the human model Inset (c) shows the upright skeleton with joints. Insets (d) to (f) show the model placed into the posture of a driver. Inset (g) details the articulated spine of the model.

Model geometry. The geometry of the shown fiftieth percentile male HARB model is based on the anatomical data sets available from DIGIMATION (DIGIMATION/VIEWPOINT CATALOG [2002]) and the VISIBLE HUMAN PROJECT [1994]. The exterior skin

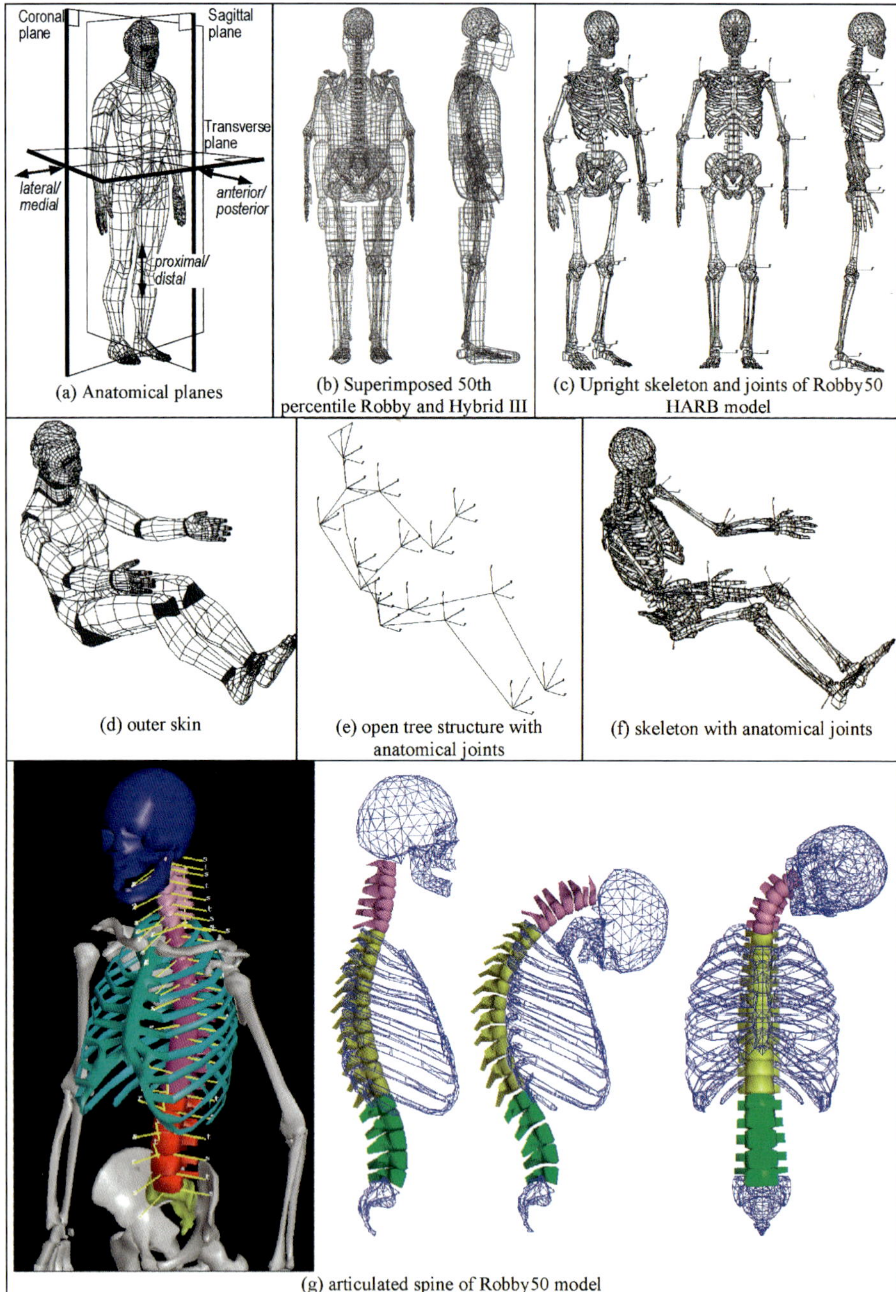

(a) Anatomical planes
(b) Superimposed 50th percentile Robby and Hybrid III
(c) Upright skeleton and joints of Robby50 HARB model
(d) outer skin
(e) open tree structure with anatomical joints
(f) skeleton with anatomical joints
(g) articulated spine of Robby50 model

FIG. 4.1. Human Articulated Rigid Body (HARB) model: 50 percentile male (ROBBY1 [1997], ROBBY2 [1998], HYNCIK [1997], HYNCIK [1999a], HYNCIK [2001a], HYNCIK [2002a], HAUG, BEAUGONIN, TRAMECON and HYNCIK [1999], BEAUGONIN, HAUG and HYNCIK [1998]).

and the skeleton of this particular HARB version are modeled with mostly triangular and quadrilateral rigid facets, while the anatomical joints are modeled with nonlinear joint elements that translate the major degrees of freedom and stiffness properties of the skeletal articulations. The body segments head, neck, thorax, upper arms, lower arms, hands, pelvis-abdomen, thighs, lower legs and feet represent each a rigid body made of its skin and skeletal parts (hand and foot articulations beyond the wrist and ankle joints are usually not modeled for crash simulations).

Mass and inertia properties. Each segment has the center of gravity, mass, principal inertias and directions of principal inertias of a fiftieth percentile male, as given in the pivotal UMTRI reports (ROBBINS [1983]), where the anthropometric, the mass and inertia and the basic joint biomechanical properties of the average fiftieth percentile American citizen and of the fifth percentile female and the ninety-fifth percentile male have been collected from an extensive study on human cadavers. The results of this report have led to the construction of the well known Hybrid III mechanical dummy family. One family of the discussed human HARB models is named after the reports by Robbins (Robby, Robina, Bobby).

Contact definition. The modeled body segments can interact with themselves and with the environmental structures and obstacles via their built in and user defined contact interface definitions. Despite the fact that the external skin is rigid, soft nonlinear contact penalty spring definitions permit a realistic treatment of the body contacts. Contact algorithms are described in the handbooks of modern solver codes (cf. Appendix A.3).

Joint modeling. The mechanical joint stiffnesses and maximum excursions of their rotational degrees of freedom are governed by user-specified nonlinear moment-rotation curves. This modeling technique can also allow for joint failure upon over-extension of the joints beyond the anatomical motion ranges. The stiffness and resistance properties are found in the literature (ex: ROBBINS [1983], ROBBINS, SCHNEIDER, SNYDER, PFLUG and HAFFNER [1983]) (see also Appendix B). Fig. 4.2 shows typical curves, based on the Robbins report, for example for the left knee joint.

This model is part of another HARB model, termed H-ARB (Section 9.2). The knee flexion–extension rotation about the joint r-axis has been modeled approximately by a moment-rotation curve with negligible stiffness and resistance in the range between $+1.0$ and -1.2 radians, and with steep slopes at the ends of the flexion–extension excursion range. The internal–external rotation of the knee joint about the t-axis is modeled in similar fashion with a reduced angular range. The varus–valgus rotation about the s-axis is penalized heavily by a stiff slope. The given curves are only approximations to the real motions of the human knee. In reality, the internal–external rotation range depends on the flexion–extension angle, there is a certain forward–backward mobility of the tibia with respect to the femur and the center of rotation of the flexion–extension rotation is mobile. These additional joint mobilities can be captured with more elaborate joint models, or when using detailed finite element models, where the articular surfaces and the stabilizing ligaments and soft tissues are modeled in sufficient detail, which is beyond the range of simple HARB models.

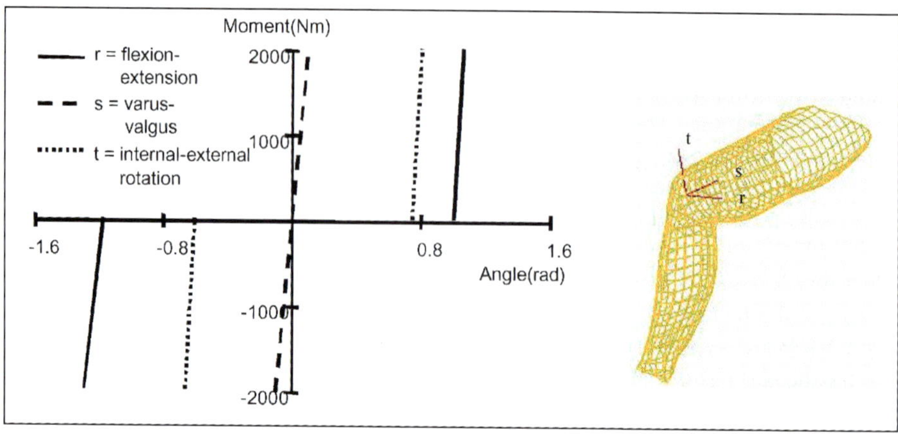

FIG. 4.2. Typical human joint moment-rotation curves (H-Model).

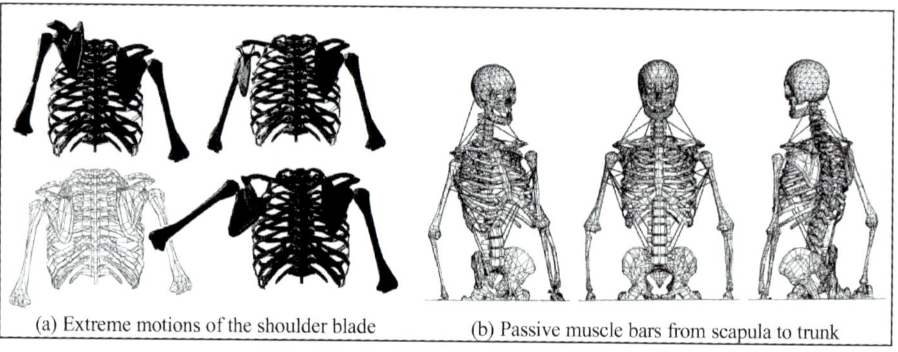

FIG. 4.3. The attachment of the arm–shoulder complex (ROBBY2 [1998], HYNCIK [1999a]).

Special attention must be given to the articulations of the arm–shoulder complex. The only points of the arm–shoulder complex that are fixed with respect to the trunk are the sterno-clavicular joints, which connect each clavicle to the sternum. The upper arm is connected to the scapula (shoulder blade) by the gleno-humeral joint, while the scapula connects to the clavicula by the acromio-clavicular joint. These joints can be modeled by kinematic joint elements, similar to the one described above. In addition, each scapula can slide about the outer surface of the rib cage inside pockets, which can be modeled as a sliding contact interface.

Fig. 4.3 shows the anatomical mobility of the shoulder blade with the extreme up-down and backward–forward translational motions and inside–outside rotation motions, inset (a), as well as a sub-system of passive Hill muscle bars that help confine these extreme motions by their passive stretch force reactions, inset (b) (HYNCIK [1999a]). If these muscles are absent, the motions of the shoulder blade and the clavicula will be stabilized only by their connecting ligaments, represented in the HARB models by rotational joint elements. A preliminary study by Ludek Hyncik (ROBBY2 [1998]) showed that the passive action of the stretched muscles at the ends of the scapular motion ranges is considerable (more than 30% in average). These muscles therefore help stabilizing the complex mobility of the arm–shoulder complex.

5. The Hill muscle model

The material described in this section is based on work carried out at ESI Software and the University of West Bohemia.

Hill-type muscle bars. The mechanical behaviour of the skeletal muscles in the directions of their fibers can be modeled to first order accuracy by Hill-type muscle bars (HILL [1970]). Each Hill-type muscle bar element is characterized by the physiological cross section area of the muscle, cut perpendicular to the fibers, and by the muscle fiber stretch and stretch velocity dependent active and passive mechanical properties of the Hill muscle model, described in Appendix C. The bars cannot, in general, transmit compressive forces.

These bar type finite elements can be arranged in the directions of the active muscle fibers. In that approach each skeletal muscle of interest is subdivided into a sufficient number of segments that can be approximated by bar like elements. For example, each segment of the biceps muscle of the arm can be approximated for most of its gross actions by a single bar element, fixed between its anatomical points of origin and insertion on the skeleton, provided the lever arm topology of the muscle is respected to sufficient precision.

For interactions with the skeleton and other tissues, and with the environment, further numerical modeling devices must be introduced, such as contact sliding interfaces and layers of deformable finite elements. For example, the trapezius muscle, being a flat surface muscle, can be subdivided into several parallel anatomical segments, each of which is represented with its tributary anatomical cross section area as one bar element. In order to facilitate contact of this muscle with the skeleton, each bar element can be further subdivided along its axis into a number of serial bar elements.

In successive further stages of refinement, skeletal muscles can be modeled as parallel assemblies of bars, Fig. 5.1(a), (b) (ROBBY-models, H-UE model: cf. Section 9.6); as bars attached to membrane finite elements which describe the resistance in the direction perpendicular to the muscle fibers, Fig. 5.1(c) (H-UE model); as heterogeneous two-dimensional fiber reinforced composite finite shell or membrane finite elements, where the composite fibers represent the mechanical properties attributed to the active muscle fibers, and where the composite matrix phase provides the passive in-surface stiffness and resistance of the muscle; as three-dimensional composite fiber reinforced solid finite elements, where the composite fiber phase represents the action of the muscle fibers and where the matrix phase provides the volume bulk response of the body of the muscle. It can be noted that the discussed muscle models use deformable finite elements, but the underlying model of the human body can still be a rigid multi-body model, as well as a deformable finite element model.

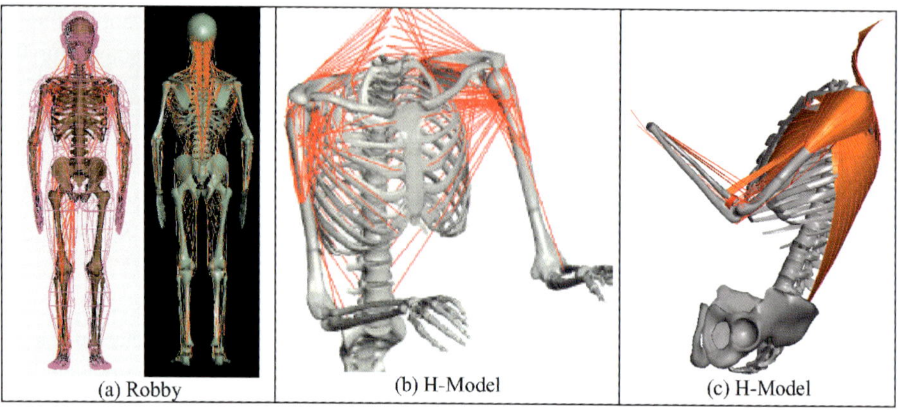

FIG. 5.1. Modeling of skeletal muscles with bars and membranes (ROBBY-models, H-UE model).

This underlines the fact that biomechanical models can be held mixed, incorporating deformable and rigid body models, as required by accuracy and computational efficiency.

The Hill muscle model. The Hill muscle model is one of the simplest phenomenological engineering models of the active and passive biomechanical behaviour of skeletal muscles. Its mathematical description is given in Appendix C.

6. Application of Hill muscle bars

Whiplash simulation. The neck models can be equipped with Hill muscle bars for the purpose of simulating the effect of the muscle forces on the head/neck displacements when, for example, a rear car impact victim is bracing the neck muscles in a voluntary or unvoluntary fashion. Fig. 6.1 shows a head/neck model with a number of Hill muscle bars added. The corresponding neck "whiplash" event simulation is discussed in the later section on the H-Neck model (cf. Section 9.4).

Static muscle force distribution. Recently Hill bar models have been used to calculate the distribution of the skeletal muscle forces of human subjects holding a given set of static loads at a given fixed posture (HAUG, TRAMECON and ALLAIN [2001]). To this end, the skeleton of a human model is equipped with Hill muscle bars and the static loads are applied. Since there are many more muscle segments that can be recruited to stabilize the kinematic degrees of freedom inherent in the articulated skeleton, the problem is hyperstatic, i.e., there are more equations than unknowns from which to calculate the equilibrium muscle forces. In fact there is an infinity of equilibrium force patterns for which static equilibrium can be achieved.

Optimization of muscle energy. One rational solution to this problem is given by the assumption that the human body will activate the muscles under a minimum expense

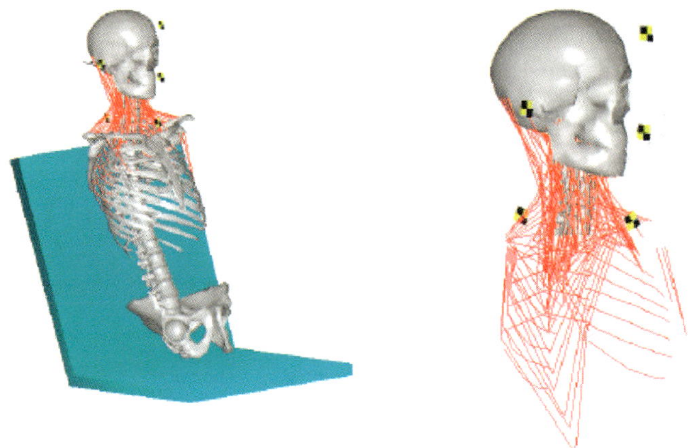

FIG. 6.1. Hill muscle bars added to the H-Neck model for whiplash studies (H-Neck).

of physiological energy. Physiological energy is spent when a muscle must activate to a constant force level over a given interval of time, even in the absence of any motion and external mechanical work ("isometric" conditions). This muscle energy is proportional to the product of force and time and the total energy is the sum of all muscle energies. One solution to the isometric muscle force problem can therefore be found by minimizing the total active muscle energy. This can be done by solving the associated *optimization* problem, where the simplest *objective function*, f, is given by the square root of the sum of the squares of the muscle forces, i.e.,

$$f = \left(\sum \gamma_i (\alpha_i - c)^2\right)^{1/2},$$

where the sum ranges over all participating muscle segments, i, activation level c is the given (average) voluntary level of muscle contraction before the load is applied (0–100%), α_i is the total activation level of the muscle segments that contribute to the task of carrying the load (0–100%) and γ_i is a switch that can have value "1" for the participating groups of muscles and "0", otherwise. This function can be thought to express the least possible overall level of muscle activation, or "energy", to be expended over a time with constant muscle forces. More complex objective functions have been proposed in the literature, for example, SEIREG and ARVIKAR [1989].

The *constraints* for the static optimization process are given by the fact that the accelerations of the links of the kinematic chain, constituted by the involved parts of the skeleton, must all be equal to zero in a position of static equilibrium. These accelerations can be calculated simply by performing an explicit analysis with a dynamic solver code, using the relevant muscled skeleton model, under the applied loads. In fact, one explicit solution time step at time = 0 is enough to determine whether or not the "structure" is in static equilibrium. At equilibrium, the internal muscle forces must balance the applied loads, and the accelerations, calculated by the solver at the centers of gravity of each rigid skeleton link, must be close to or equal to zero.

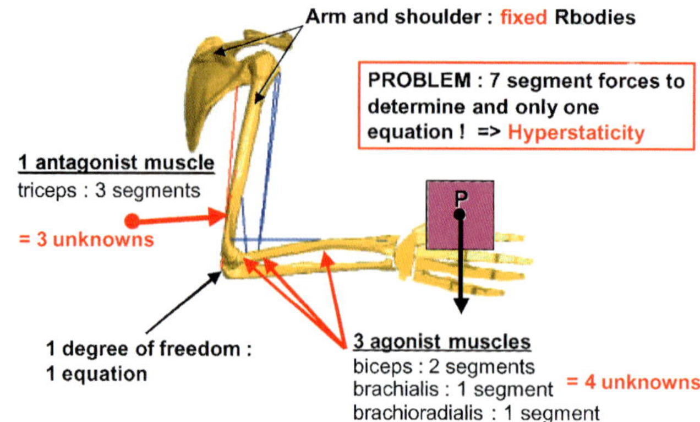

FIG. 6.2. Hill muscle bars activated at minimal energy under static loads (ROBBY models).

The *design parameters* of the optimization problem are given by the activation levels, α_i, of the participating muscle segments, which are the result of the optimization.

The *bounds* on the design parameters are given by $0 < \alpha_i < 100\%$, i.e., the activation level of a muscle cannot be less than zero and not greater than 100%. The outlined optimization procedure can be solved by standard optimization algorithms and it is applied to a simple one degree of freedom system.

Example. Fig. 6.2 shows a first example of an arm with fixed shoulder and upper arm, where a static load is held in place against one elbow rotation degree of freedom. In this example a minimum of seven muscle segments contributes to the task of supporting the load, and the only kinematic degree of freedom is the elbow flexion–extension rotation.

Fig. 6.3(a) shows the model of an upper torso with shoulders and arms, equipped with 132 Hill muscle bars. The model holds a constant static load in the left hand. The muscle forces are calculated for minimal energy. The colors of the muscle bars range from blue for zero activation to red for 100% activation.

In Fig. 6.3(b) a symmetric static load is applied to the same model (cubes), which increases linearly at equal increments. Inset (c) shows the calculated activation levels of some selected muscle bars at loads ranging from zero to 12 kg in each hand. In the diagram the agonist muscles (ex. biceps) are seen to quickly raise their levels of activation, initially in proportion to the load increase, while the less efficient muscles (ex. latissimus dorsi) initially contribute very little to the task. As soon as the first agonist muscles become saturated, i.e., reach 100% activation after about 3 kg loads, the less efficient muscles tend to increase their activation levels faster than linear. After the loads have reached about 8 kg, the inefficient muscles (ex. latissimus dorsi) must compensate for the saturation of the agonists and they raise their levels of activation dramatically. When all muscles reach saturation, the loads can no longer be increased and the limit load has been reached (2×12 kg). Limit loads for different postures are calculated in the examples shown in Appendix C, paragraph: ‘Applications’, Fig. C.3.

Fig. 6.3(d) and (e) show the model pulling the hand brakes in a passenger car and holding the steering wheel while driving. In both postures the needed muscle force

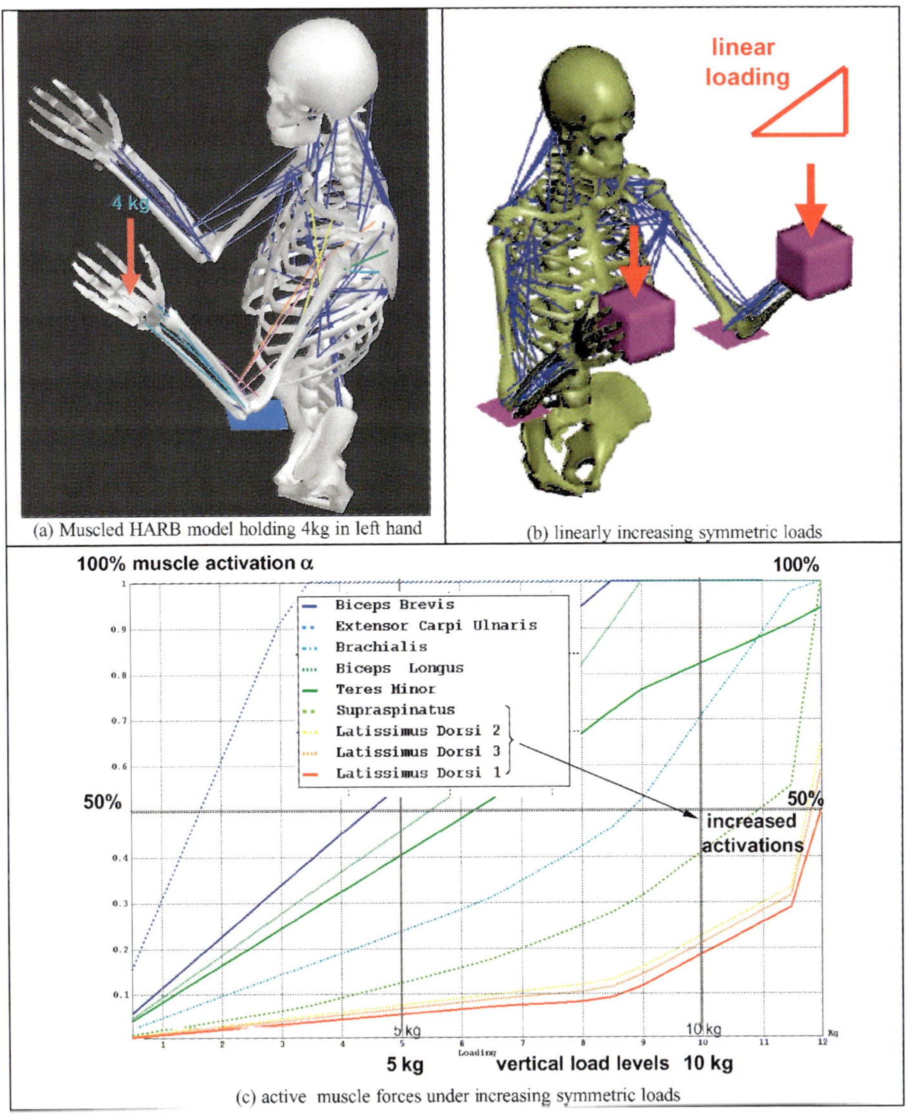

(a) Muscled HARB model holding 4kg in left hand

(b) linearly increasing symmetric loads

(c) active muscle forces under increasing symmetric loads

FIG. 6.3. HARB muscle forces under different static loads (ROBBY-models).

pattern was calculated for minimal energy. These examples show that articulated rigid body models can be used together with Hill muscle bars to calculate realistic muscle force distributions. These forces show the limit loads a subject can hold for a short time. For long term static muscle loads, such as in driving a car, the forces are less important, but will lead eventually to muscle fatigue. The levels of fatigue can best be assessed by knowing the forces that cause the fatigue.

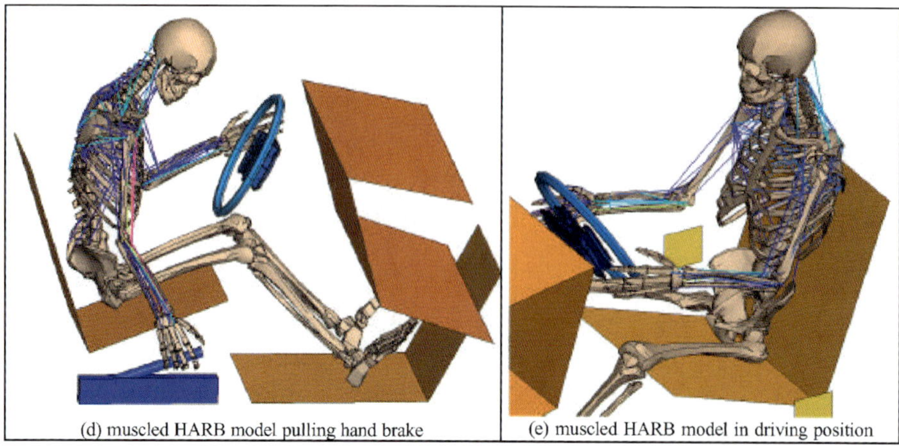

(d) muscled HARB model pulling hand brake (e) muscled HARB model in driving position

FIG. 6.3. (*Continued.*)

Fatigue. The Hill muscle model can be extended to add muscle fatigue. Under the isometric conditions, such as in the static scenarios described above, a given muscle will fatigue when forced to sustain an active muscle force over a prolonged period of time. If the active force output of a muscle must be constant over time, such as in the simple examples of holding a given load as shown in Fig. 6.2, then muscle fatigue will force the muscle to raise its level of activation in order to compensate for the fatigue-induced drop of force. If the activation level is kept constant, the muscle force will drop over time. Simple static muscle fatigue models are currently under investigation.

7. Application of HARB models

HARB model families. In Fig. 7.1 models of a fifth percentile female (ROBINA [1998]) and a six year old child (BOBBY – ESI Software) are shown. These models are built in a fashion which is analogous to the HARB model of the fiftieth percentile male, and their details are not discussed separately. It must be noted, however, that basic anthropometric data exist mostly for the adult humans, while such data for children are close to absent. Nevertheless, DIGIMATION (DIGIMATION/VIEWPOINT CATALOG [2002]) provide the external geometry, the skeleton geometry and the organ geometry of male and female adults, as well as some muscle surface geometry. The external geometry of pregnant woman and children is also supplied, Fig. 7.1(a). The skeletal geometry and the organ geometry of children is not available, however.

Fig. 7.1(b) shows the external and skeletal geometry of the 5th percentile female (Robina). Fig. 7.1(c) shows the external and skeletal geometry of a six year old child (Bobby). The external geometry (skin) was hand-scaled from an available eight year old, while the skeleton was hand-scaled from the adult geometry to the anthropometric dimension of the six year old child, where different body parts were scaled individually.

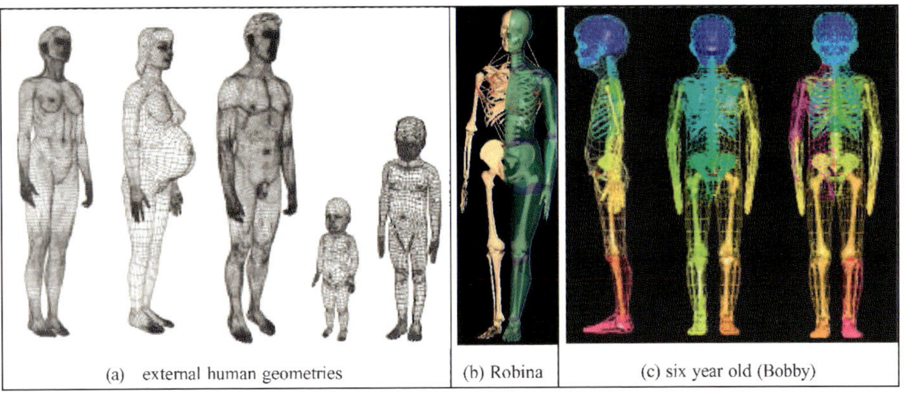

FIG. 7.1. Different HARB models (male, female, children). (Inset (a): Reproduced by permission of Digimation/Viewpoint, http://www.digimation.com.)

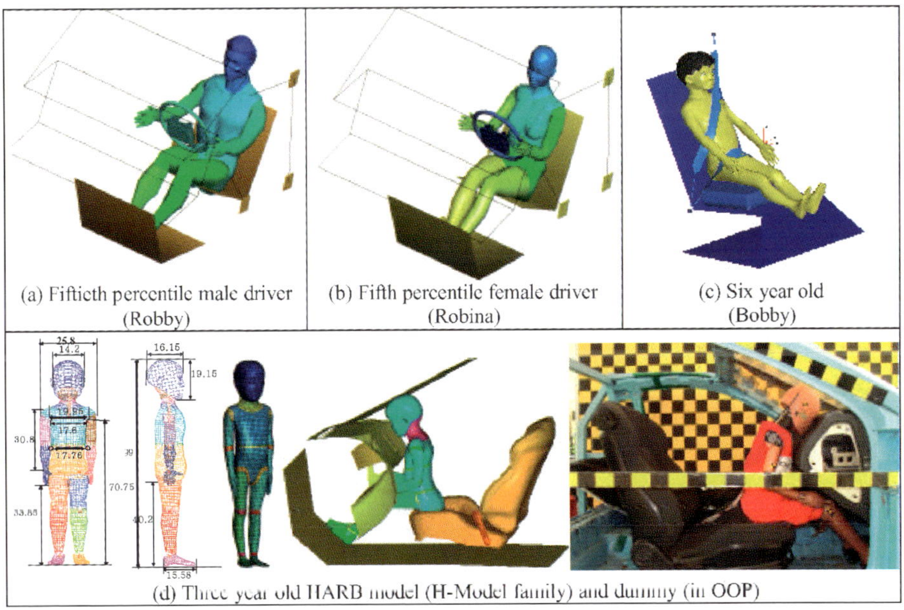

FIG. 7.2. HARB models in car environments (Robby and H-models). (Inset (d) right-hand side: Reproduced by permission of Prof. Kim, Kwangwon University.)

Crash test simulations with HARB models. Fig. 7.2 presents the different HARB models (Robby family) in a passenger car environment, with and without airbags in driver and passenger positions, including a three year old in passenger airbag out-of-position (OOP) posture (H-Model family).

Fig. 7.3(a) shows displaced shape snapshots of a frontal impact sled test simulation for the seat belted 50th percentile male (Robby) with a deploying driver airbag.

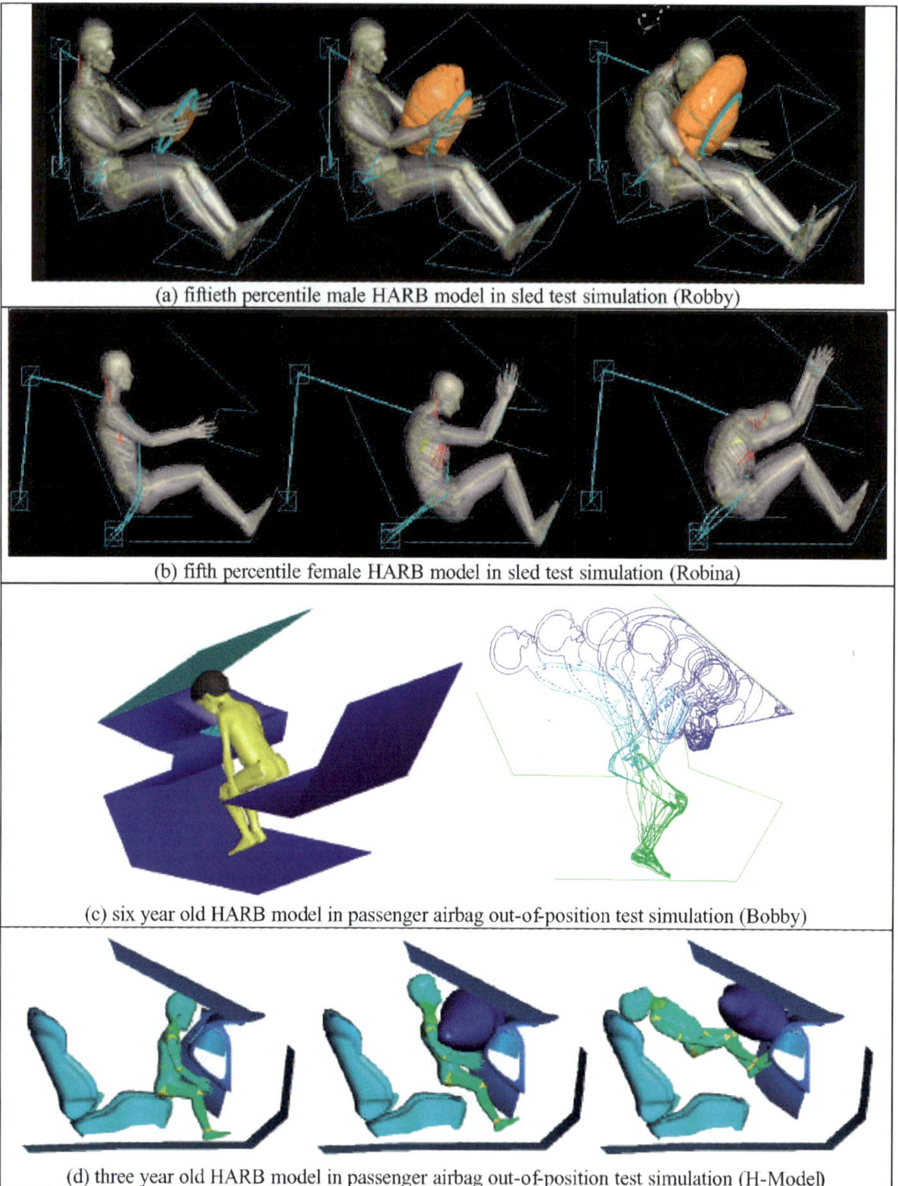

(a) fiftieth percentile male HARB model in sled test simulation (Robby)

(b) fifth percentile female HARB model in sled test simulation (Robina)

(c) six year old HARB model in passenger airbag out-of-position test simulation (Bobby)

(d) three year old HARB model in passenger airbag out-of-position test simulation (H-Model)

FIG. 7.3. HARB models in frontal crash test situations (Robby and H-Models).

Fig. 7.3(b) repeats this scenario for the belted 5th percentile female (ROBINA [1998]) who is not protected by an airbag. Fig. 7.3(c) shows the six year old HARB model (Bobby) in an OOP posture on the passenger seat. If the passenger side airbag fires at that moment, the subject is hit by the bag during its deployment phase, which can cause "bag slap" injuries and overextension and torsion of the neck.

Although the six year old should normally not be in the front seat, unbelted and leaning forward, car industry is interested in studying such events in an effort to render the airbag deployment phase as little aggressive to out-of-position subjects as possible, while keeping the airbag effective under standard conditions. Since the standard mechanical dummies are not biofidelic for such events, simulation with human models is recognized as a welcome alternative. Fig. 7.3(d), finally, shows a three year old HARB model (H-Model family) undergoing the same passenger airbag OOP scenario.

Validation of HARB models. Fig. 7.4 summarizes validations performed on the example of the model of the 50th percentile Robby model of the HARB family (ROBBY1 [1997], ROBBY2 [1998]). Validation of the other models were performed in a similar way.

Validation of biomechanical simulation techniques (algorithms, codes, models) in general can best be performed against physical test results. Directly exploitable physical results of crash events involving humans are rarely available. Access to physical results can be gained through accidentological studies, accident reconstruction analyses, cadaver test simulations and, rarely, volunteer test results for low energy impacts (rear impact).

The physical validation test results of Fig. 7.4 were obtained from an existing data base of cadaver tests performed at the University of Heidelberg (e.g., SCHMIDT, KALLIERIS, BARZ, MATTERN, SCHULZ and SCHÜLER [1978]). The subject selected from this data base was close to a 50th percentile male. The figure compares global head, thorax and pelvis accelerations with the corresponding cadaver test results.

In view of the large scatter expected from different human subjects, the comparison is considered satisfactory. In fact, validations should be made against experimental test corridors obtained from a sufficiently large subset of near 50th percentile human subjects. If the numerical simulation result curves remain within the given test corridors, the validation is successful. Moreover, the numerical results should not be performed on one "average" model of a given percentile slot, but simulations with stochastic variants of the average model should match the physical test corridors. This requirement is absent when working with mechanical dummies, because mechanical dummies of a given percentile category are all alike.

References on multi-body and muscle models. The following references may be consulted on the subjects discussed in this chapter: BAUDRIT, HAMON, SONG, ROBIN and LE COZ [1999] on dummy vs. human comparison; BEAUGONIN, HAUG and HYNCIK [1998] on the Robby family of articulated human models; CHOI and LEE [1999b] on a deformable human model; COHEN [1987] on the analysis of frontal impacts; DENG [1985], DENG and GOLDSMITH [1987] on the head/neck/upper torso response to dynamic loading; DIGIMATION/VIEWPOINT CATALOG [2002] contains CAD data of the human body; GRAY'S ANATOMY [1989]; HAUG, LASRY, GROENENBOOM, MUNCK, ROGER, SCHLOSSER and RÜCKERT [1993], HAUG [1995] and HAUG, BEAUGONIN, TRAMECON and HYNCIK [1999] on biomechanical models for vehicle accident simulation; HILL [1970] on muscle tests; HUANG, KING and CAVANAUGH [1994a], HUANG, KING and CAVANAUGH [1994b] on human models for

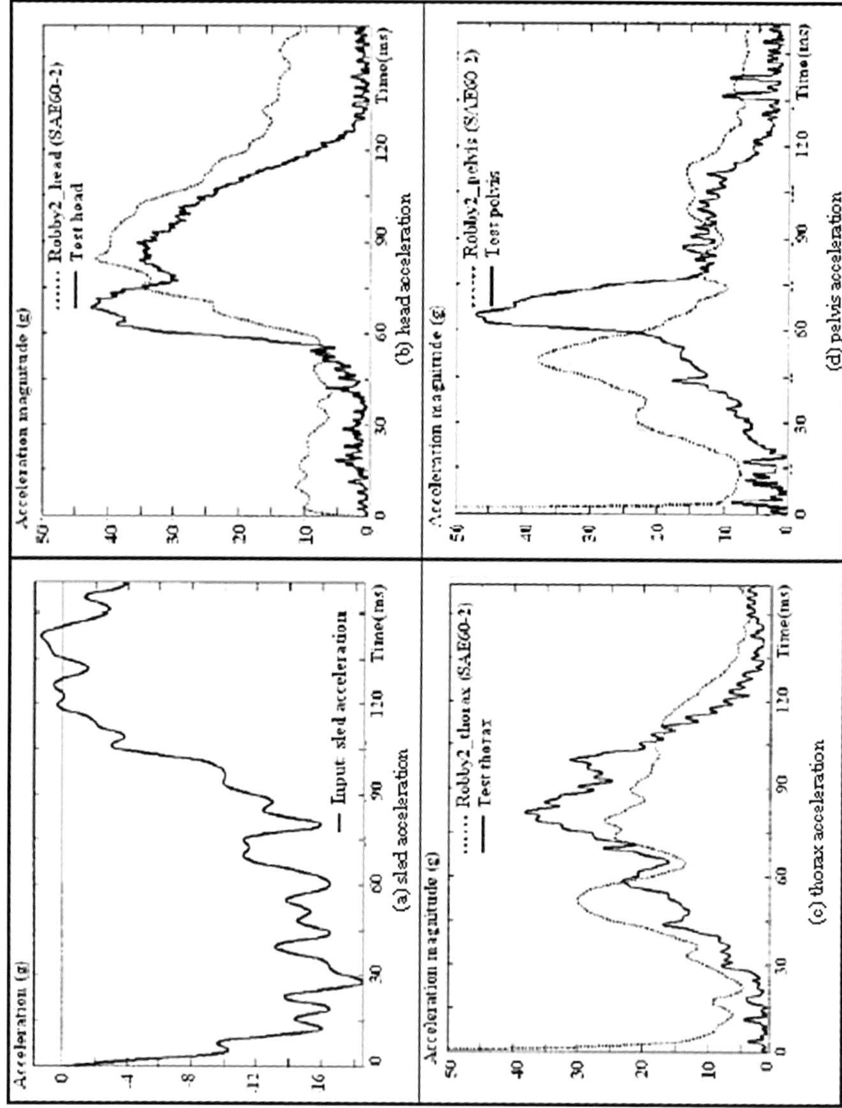

FIG. 7.4. 50th percentile HARB model validation with Heidelberg sled test results (Robby model).

side impact; HYNCIK [1997], HYNCIK [1999a], HYNCIK [1999b], HYNCIK [2000], HYNCIK [2001a], HYNCIK [2001b], HYNCIK [2002a], HYNCIK [2002b], HYNCIK [2002c] on human articulated rigid body models and deformable abdominal organ models; IRWIN and MERTZ [1997] on child dummies; KISIELEWICZ and ANDOH [1994] on critical issues in biomechanical tests and simulation; KROONENBERG, A. VAN DEN, THUNNISSEN and WISMANS [1997] on rear impact human models; LIZEE and SONG [1998], LIZEE, ROBIN, SONG, BERTHOLON, LECOZ, BESNAULT and LAVASTE [1998] on deformable human thorax and full body models; MA, OBERGEFELL and RIZER [1995] on human joint models; MAUREL [1998] on a model of the human upper limb; ROBBINS [1983], ROBBINS, SCHNEIDER, SNYDER, PFLUG and HAFFNER [1983] contains basic human percentile data; ROBBY1 [1997], and ROBBY2 [1998] on the 50th percentile Robby multi-body models; ROBIN [2001] on the HUMOS model; ROBINA [1998] on the 5th percentile female multi-body model of the Robby family; SCHMIDT, KALLIERIS, BARZ, MATTERN, SCHULZ and SCHÜLER [1978] on loadability limits of human vehicle occupant; SEIREG and ARVIKAR [1989] on muscle models; STÜRTZ [1980] on biomechanical data of children; VISIBLE HUMAN PROJECT [1994] is a CD ROM with anatomy data of the human body; WILLINGER, KANG and DIAW [1999] on 3D human model validation; WINTERS and STARK [1988] (muscle properties); WITTEK and KAJZER [1995], WITTEK and KAJZER [1997], WITTEK, HAUG and KAJZER [1999], WITTEK, KAJZER and HAUG [1999], WITTEK, ONO and KAJZER [1999], WITTEK, ONO, KAJZER, ÖRTENGREN and INAMI [2001] on active and passive muscle models.

CHAPTER III

Deformable Human Models

8. The HUMOS human models for safety

8.1. Introduction

International projects, like HUMOS-1, pave the way for standardizing the modeling methodology of the human body for impact biomechanics and occupant safety. The major foreground information produced by the first HUMOS project is therefore not specific computer models, but a database for the modeling of the human body with commercially available dynamic crash software. The encryption of the contents of the data base into specific versions of the individual commercial crash codes is a secondary result of the project. If models made with commercial software make use of the data published in the HUMOS database, if their modeling techniques are conform to the rules fixed in the base, and if they pass the standard tests laid down in this base, then the models could obtain the "HUMOS label". This label would certify the model to be conform to the standards laid down in the data base.

The HUMOS project embodies the fact that multi-disciplinary and multi-national efforts are needed to eventually achieve the goal of the industrial use of internationally standardized human models for safe car design. Putting humans into crash tests is only feasible in the computer. The methodology of "virtual testing" using compute models and codes, therefore, must be "certified" when it should complement or replace certain legal crash dummy tests. There is at present no officially binding certification procedure for simulation, neither for mechanical dummy nor for human models. The HUMOS projects can help to establish such procedures.

The following paragraphs are extracted from, and prepared after, reports of the HUMOS-1 project and the paper written by the HUMOS-1 project coordinator:

Robin, S. (2001), HUMOS: Human Model For Safety – A Joint Effort Towards the Development of Refined Human-Like Car Occupant Models, *17th ESV Conference*, Paper Number 297.

The selected paragraphs and material intend to briefly summarize the contents and major results of the first HUMOS project. The extracted material, figures and tables are reproduced by permission of the ESV Conference Organizers.

In the first European HUMOS program (1999–2001) the development of commonly accepted human car occupant models and computer methods was attacked. It was clear from the outset, that the results of the first HUMOS program must be extended into a second program, where other percentile human models and models with different human morphologies should be created by appropriate modeling, scaling and morphing techniques, and where the material description base should be enlarged.

Fourteen partners were involved, including six car manufacturers and several suppliers, software houses, public research organizations and universities. The program dealt with the synthesis and further development of the current knowledge of the human body in terms of geometry, cinematic behavior, injury threshold and risk, the implementation of this knowledge in new human body models, the development of the utilities for the design office use, and delivering of the models to be available for their integration in the car design process.

A wide bibliographical review supported those major goals. The geometry acquisition of a mid-sized 50th percentile male in a car driver seated position was achieved. The main human body structures were then reconstructed using a CAD method. The meshing of the different structures was based on the CAD definition and has led to models accounting for skin, bones, muscles as well as the main organs (lungs, heart, liver, kidneys, intestine etc.). The validation process was undertaken on a segment basis, each main part of the human body being confronted with the available literature results. The assembly of the whole model was the conclusion of this program.

The first problem that had to be solved was the acquisition of the inner and outer geometry of a human, seated in a car occupant posture (the driver posture was chosen). The publication made by ROBBINS, SCHNEIDER, SNYDER, PFLUG and HAFFNER [1983] on the seated posture of vehicle occupants served as a basis for many developments of human substitutes. This work mainly qualified the position of external anatomical landmarks on many different car occupants. But there were some limitations in this work, particularly concerning the relative position of the different bony structures and of the different organs, that had to be explored by the project.

8.2. Geometry acquisition

Selection of the subject. The geometry acquisition of a seated near 50th percentile adult male is one of the original achievements of the HUMOS-1 program. Commonly available human geometrical databases (e.g., GEBOD) furnish the main external dimensions of the human being. Little information is available concerning the geometry of the different organs and their relative positions of the different structures in a seated position. Table 8.1 summarizes the main characteristics of the selected HUMOS subject and those of the 50th percentile European male.

Subject frozen in driving position and physical slicing. The method consisted of placing the chosen cadaver in a driving position, Fig. 8.1.

Fig. 8.2 gives an overview on the applied methodology. The selected subject was installed in a standard full-size car cockpit with the hands placed on the steering wheel, and it was frozen in this position, Fig. 8.2, inset(1). A reference frame related to the cockpit was defined and the subject was embedded in a polymer block, Fig. 8.2(2). The method of physical slicing of a frozen cadaver was chosen, Fig. 8.2(3). Each slice was 5 mm thick, and the saw was 2.5 mm thick (268 slices were made). The different slices

TABLE 8.1
Main characteristics of the HUMOS subject and 50th percentile adult male (ROBIN [2001])

	Sitting height [mm]	Standing height [mm]	Weight [kg]
HUMOS subject	920	1730	80
50th percentile European male	915	1750	75.5

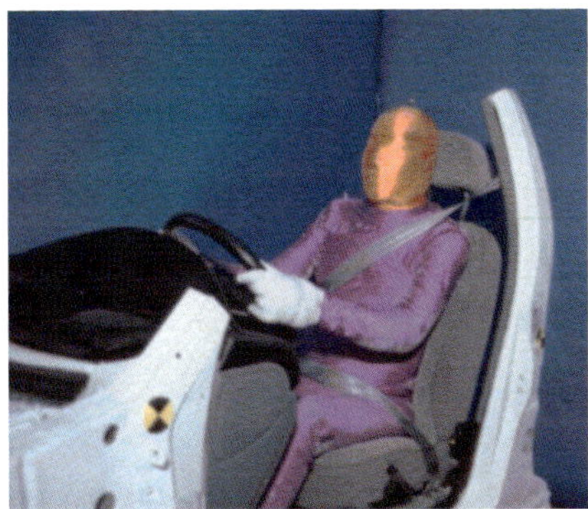

FIG. 8.1. HUMOS subject in its driving position. (Reproduced by permission of ESV Conference Organizers.)

were photographed on each side (491 images) and then each image was hand-contoured by anatomists, Fig. 8.2(4). This acquisition phase generated a set of about 13.000 files representing 300 different organs. Each file was composed of a set of points characterized by their 2D coordinates related to the slice, for example, Fig. 8.2(5). The position of each slice with regard to the reference frame was also available in the different files. A 3D visualization, Fig. 8.2(6), based on the nodes, was carried out in order to validate the acquisition process and, if need be, to modify some of the points describing the organs.

3D geometrical CAD reconstruction process. Results of the 3D CAD reconstruction process are shown on Fig. 8.3. During this process the point-by-point description files were transformed into CAD geometrical files. Standard commercially available CAD software was used. The main organs were reconstructed using available mathematical surface definitions. A back and forth process was set up between anatomists and CAD engineers in order to double-check at each step the shape of the reconstructed surfaces. When some assumptions had to be made during this phase, they were thoroughly discussed with the anatomists.

At this point it should be kept in mind that human occupant models for safety should be "generic", rather than "specific" in the sense that "average" people should be placed in cars during crash simulations. In reality, only specific subjects are available for data acquisition, with particular defaults and deviations from average. It is therefore legitimate to correct the acquired data for obvious deviations from the presumed anatomical

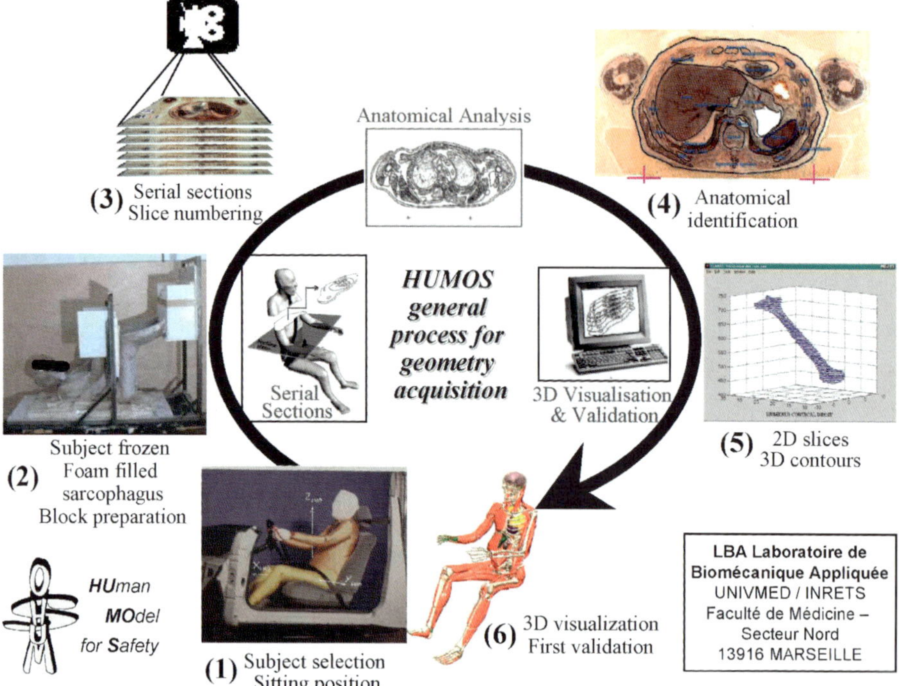

FIG. 8.2. Overview on the HUMOS-1 geometry acquisition process (HUMOS-1 project).

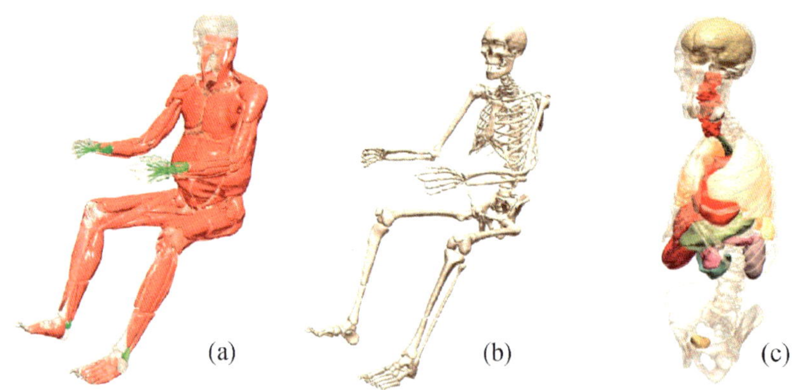

FIG. 8.3. Geometry acquisition result of (a) the whole body (without skin), (b) the skeleton and (c) the major organs. (Reproduced by permission of ESV Conference Organizers.)

standard. In fact, human models for safety should be as standardized as possible. Definitions of standard human anatomies have yet to be made. These goals must be achieved in follow-up projects.

8.3. Meshing process

During meshing, a close collaboration was maintained with the anatomists in order to double-check the shape of the reconstructed organs as well as the validity of the organ connectivity assumptions. Only half of the skeleton was meshed and sagittal symmetry was assumed. Further assumptions were made for some parts of the human body, which were not described accurately enough. For example, the posterior arcs of the different vertebrae were not always visible, and not all the ligaments and muscles were digitized during the acquisition process. Thus, some assumptions based on anatomical descriptions were made in order to add some muscle parts and some ligaments.

For reasons of computational efficiency it was decided to try to keep the number of deformable finite elements below 50 000, and requirements were issued on the final time step (cf. Appendix A) of the complete model, which put a lower limit on the size of the elements. Fig. 8.4 illustrates different segment meshes developed during this process.

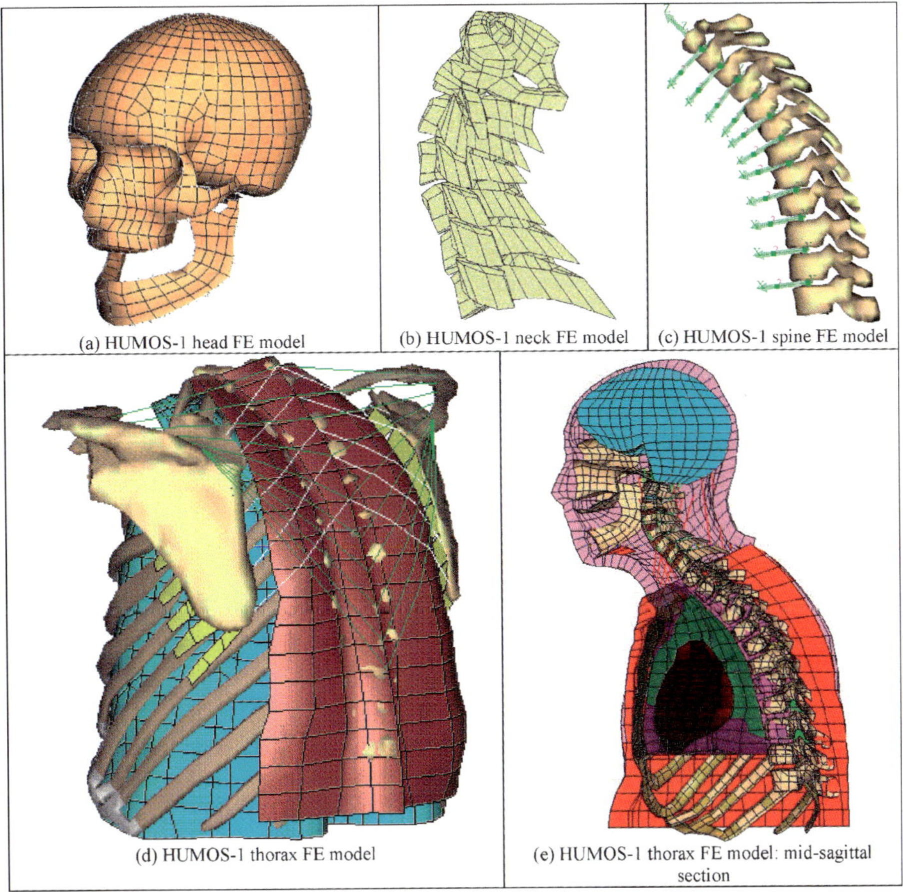

(a) HUMOS-1 head FE model
(b) HUMOS-1 neck FE model
(c) HUMOS-1 spine FE model
(d) HUMOS-1 thorax FE model
(e) HUMOS-1 thorax FE model: mid-sagittal section

FIG. 8.4. Mesh of skeleton and soft tissue parts. (Reproduced by permission of ESV Conference Organizers.)

It should be noted that the resulting mesh is relatively rough, and that finer meshes of parts might be substituted in the future for refined studies.

The bony part of the skull was modeled with standard thin shells. The cervical spine was modeled with shells and solids, some muscles and ligaments with bars. The remaining spinal column has rigid vertebrae and nonlinear joint elements for the inter-vertebral discs.

The final mesh of the whole thorax is a good example for the chosen mesh quality. The clavicle, sternum, ribs and connective cartilaginous structures were modeled using standard shell and solid finite elements. The scapula and the different vertebrae were modeled using shell elements for the definition of the shape of those structures. In the mechanical model, these bony parts are described as rigid elements. The muscular structures were meshed using both bar and solid elements. The flesh and connective tissues were modeled using solid elements.

The other body parts were meshed with comparable quality.

8.4. Material laws

A literature review enabled to gather the available knowledge, cf. Appendix B. Due to missing information, a limited number of experimental static and dynamic tests were performed within the program.

Ribs, clavicles, sternum and pelvic bone were tested extensively in order to gain specific knowledge about their mechanical properties. The results were associated with the definition of a parameterized material law (a power law was used).

Furthermore, static and dynamic muscle compression experiments were carried out. The strain-rate sensitivity was quantified for the soft tissues. Existing material laws were adapted and implemented in the different participating crash codes.

For the different organs, very few experimental results are available and there is an obvious lack of knowledge in this field. Available linear visco-elastic laws were chosen according to other modeling publications. Soft tissue characterization is still regarded as a largely uncovered area.

8.5. Segment validation process

Single rib model validation. The material properties were mainly derived from a literature review. The response corridors published by LIZEE and SONG [1998] and LIZEE, ROBIN, SONG, BERTHOLON, LECOZ, BESNAULT and LAVASTE [1998] were used for validation. Some material tests were carried out during the HUMOS-1 project. As an example, the rib models were first checked against static and dynamic tests carried out on isolated human ribs.

The ribs were modeled using shell elements for the cortical bone and solid elements for the trabecular bone. The thickness of the cortical bone was modeled using different shell thickness depending upon the location. In order to account for rib fracture, a failure plastic strain was introduced in the rib material law. The different values used for the material behavior definition of the ribs are reported in Table 8.2.

TABLE 8.2
Material properties of the rib cortical bones (ROBIN [2001])

Rib cortical bone material model: elastic-plastic behaviour with failure	
Elasticity modulus	14 GPa
Yield Stress	70 Mpa
Ultimate stress	70 Mpa
Maximum deformation	4%
Poisson's ratio	0.3
Density	6000 kg/m^3

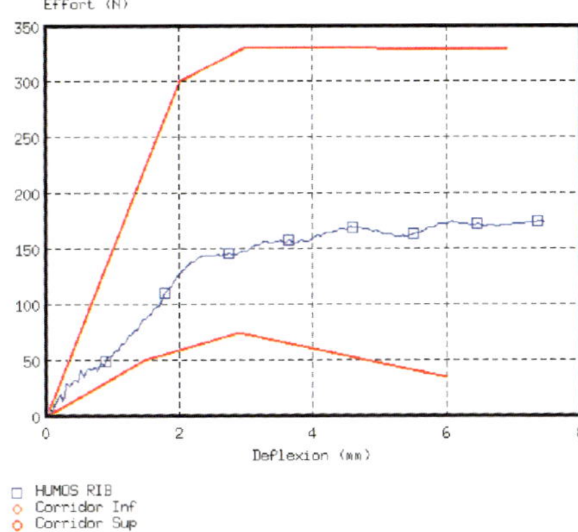

FIG. 8.5. Computer result compared with experimental corridor of an isolated rib subjected to a quasi-static loading. (Reproduced by permission of ESV Conference Organizers.)

These values were used for both the static and dynamic validation of the ribs. Fig. 8.5 illustrates the static behavior of the so-called "HUMOS rib" compared with the results obtained on an isolated human rib by the University of Heidelberg.

Thorax model validation. For the bones and cartilaginous parts of the thorax for which experimental data were available, the same process was used for the first description of the model. The assembled segment models were validated against the published experimental results. Figs. 8.6 and 8.7 represent the results obtained for the HUMOS-1 thorax model (ESI Software version) as compared to the experimental pendulum impact test results published by KROELL, SCHNEIDER and NAHUM [1971], KROELL, SCHNEIDER and NAHUM [1974] for frontal impact and by VIANO [1989] for lateral impact. In these experiments, the thorax was impacted with a 23.4 kg cylindrical impactor at 4.9 m/s and 4.6 m/s, in frontal and lateral directions, respectively.

The organs of the thorax were modeled with the airbag modeling technique (lungs) or as "bio-bags" (heart), cf. Appendix D. A bio-bag model is an adaptation of a standard airbag model, for which the mechanical properties of the enclosed gas are modified to be close to incompressible, like a fluid. This feature permits to approximately simulate

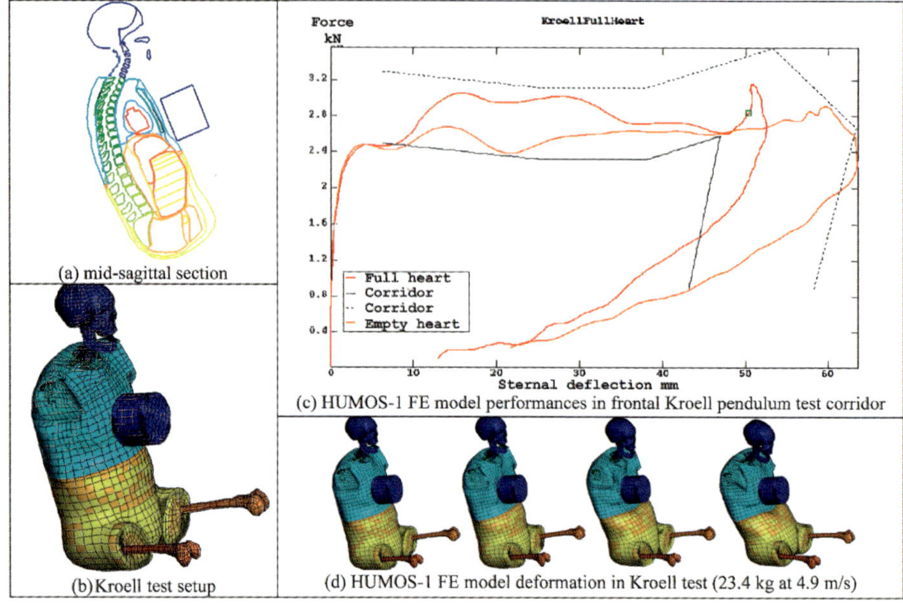

FIG. 8.6. HUMOS-1 FE model: frontal impact test (Kroell test) (HUMOS-1 project/ESI).

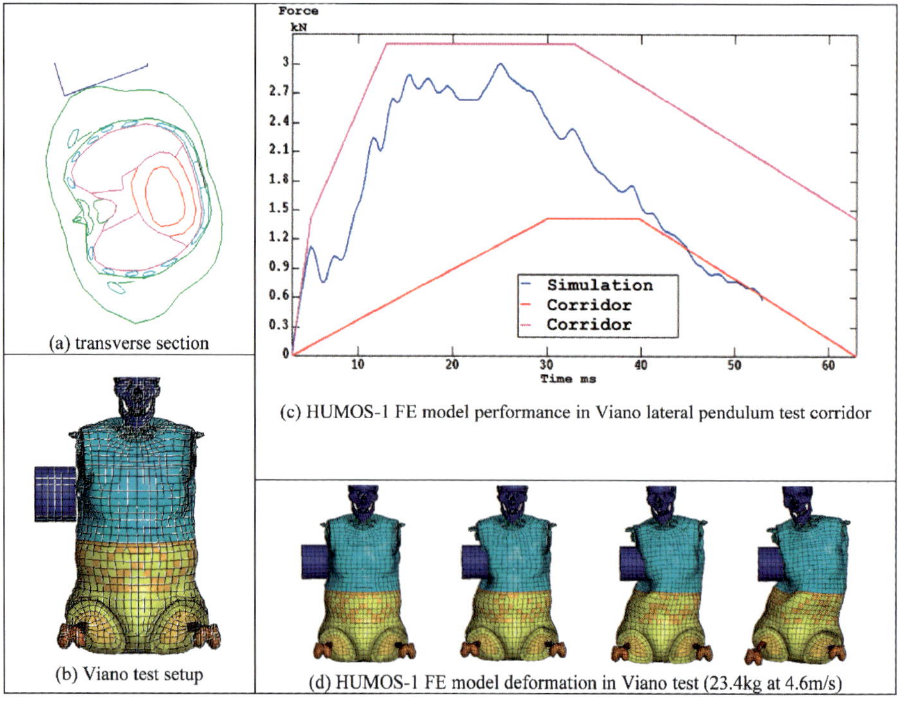

FIG. 8.7. Lateral impact test (Viano test) (HUMOS-1 project/ESI).

the impact response of hollow internal organs, filled partly or fully with air and liquid contents. It also allows organ contents to escape (vent out) under compression of the organ, when the standard venting options of the used airbag models are invoked. Standard airbag models are supported by all dynamic crash codes. They model a compressible gas contained in a flexible and extensible hull, made of membrane, shell or solid finite elements. More elaborate modeling of the gas or fluid contents of the hollow organs will require the techniques of fluid or particle dynamics, not discussed here.

The importance of at least approximate modeling of the blood contents of the heart is demonstrated in Fig. 8.6, where the impact force versus sternum deflection curves are shown both, for an empty and for a full heart.

Abdomen model validation. The test reported by CAVANAUGH, NYQUIST, GOLDBERG and KING [1986] was simulated with the HUMOS model (ESI Software version), Fig. 8.8. In this test, a bar (∅25 mm, length 381 mm) with total mass of 32 kg hits the abdomen with a velocity of 6.9 m/s. The simulation outcome depends largely on the way the abdominal cavity and the intestines are modeled, Fig. 8.8(a).

The abdominal cavity is treated like an equivalent bio-bag, closed by the abdominal walls, and filled with a fluid and with sub-models of the intestines and internal organs. The intestines can either be modeled as solids, Fig. 8.8(a)(1), as an inner simple (2) or "meandered" (3) bio-bag, each placed within the outer bio-bag of the abdominal cavity.

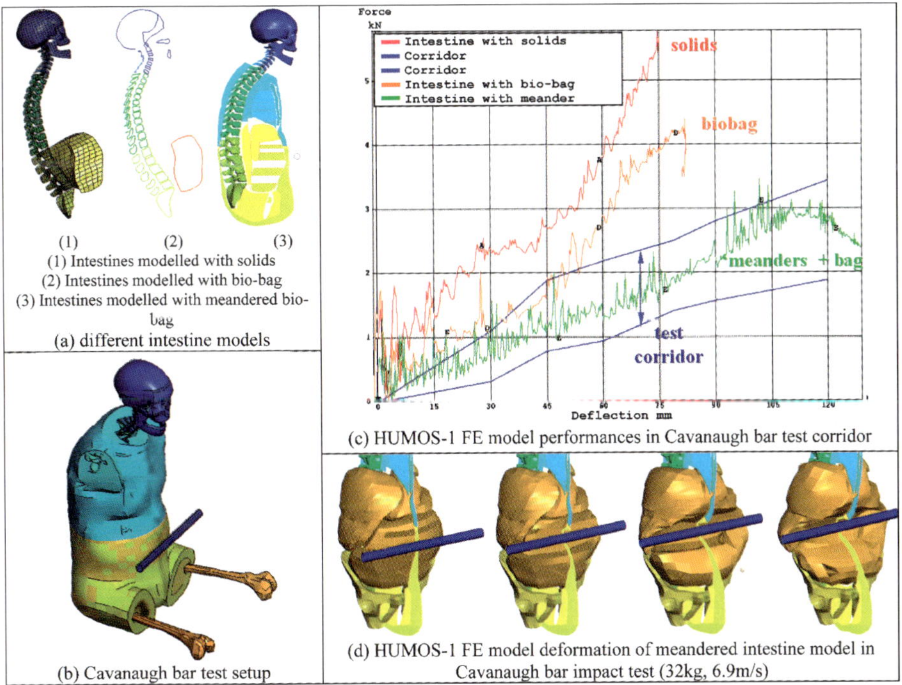

FIG. 8.8. Abdomen bar impact test (Cavanaugh test) (HUMOS-1 project/ESI).

The solid model of the intestines gave too much shear and compressive stiffness at large deformations, Fig. 8.8(c). The model with the bio-bag responded well initially, but developed too high internal fluid pressures at large deformations. The intestine model with a meandered bio-bag gave the best results and the response curve falls in between the experimental test corridor. The results show that even for a faithful reproduction of gross results, such as external load-deformation curves, the models should follow as closely as possible the anatomical realities, which allow for major shifts and relative displacements of the inner organs.

8.6. Discussion and conclusions

The HUMOS-1 program led to a first definition of a finite element model of the human body in a *seated driving posture*. In an effort to create a European standard human model, useful for car in industry, the model was implemented with three commercially available dynamic crash codes (Madymo, Pam–Crash and Radioss). The mesh of the model is shared by the different software packages, but the validation of the different models was carried out separately for the different codes.

From the beginning of this research work, it was foreseen that some major limitations would be encountered. First, the geometrical definition of the model comes from a unique human subject, i.e., is not generic. It is expected in a follow-up program to develop scaling techniques, which would enable to first define a generic 50th percentile model from the current reference mesh, and second 5th and 95th percentile occupant models. It is also expected to be able to derive from this first model some pedestrian models.

A validation database was built and used in order to validate the different segments of the model. The global validation of the whole model remains to be done. Some investigations need to be carried out on the muscle tonus contribution, especially for the low speed impact conditions that can be encountered in real field accident analysis. Furthermore, some limitations are due to the lack of knowledge of the injury mechanisms. The main currently used criteria were implemented in the model, but its injury prediction capabilities are limited with regard to its limited complexity.

8.7. Acknowledgements

The HUMOS project was funded by the European Commission under the Industrial and Materials Technologies program (Brite–EuRam III). The consortium partners were the LAB (Laboratory of Accidentology and Biomechanics PSA Peugeot, Citroën, Renault), who coordinated the work and was involved mainly in the meshing process of the thorax and in the validation database. The other participating car manufacturers were Volvo (meshing of the neck), BMW (meshing of the upper limbs), and VW (literature review of the existing models). Software developers were strongly involved in this program. ISAM/MECALOG (Radioss software) carried out the 3D CAD reconstruction of the model, the head mesh description, and the assembly of the final model with Radioss. ESI Software (Pam–Crash software) modeled the lower limbs and was in charge of the homogeneity of the different segment models. ESI Software also performed the assembly of the final model with Pam–Crash. TNO (Madymo Software) carried out modeling work and coordinated the soft tissue behavior activities. The supplier FAURECIA carried out the pelvis and abdomen modeling and coordinated the geometry acquisition process. INRETS LBMC (Laboratoire de Biomécanique et Mécanique des Chocs) carried out full-scale sled tests with human substitutes and contributed to the extension of the validation database. Marseille University was in charge of the geometrical acquisition of the seated human body, carried out by INRETS LBA (Laboratory of Applied Biomechanics). Athens University defined the physical material laws for the different human soft tissues and carried out experiments on some muscle properties. Heidelberg University was in charge of experimental investigations on different human tissues (bones and some cartilage structures) and produced a set of new experiments for the extension of the validation database. Chalmers University performed a bibliographical study on the current knowledge about human tissue behavior and identified the fields of missing knowledge. Chalmers University also contributed in the validation of the neck model.

8.8. References for the HUMOS model

The following references were found essential for the work reported in this section: CAVANAUGH, NYQUIST, GOLDBERG and KING [1986] on the impact response and tolerance of the human lower abdomen; KROELL, SCHNEIDER and NAHUM [1971], KROELL, SCHNEIDER and NAHUM [1974] on the impact response and tolerance of the human thorax; LIZEE and SONG [1998], LIZEE, ROBIN, SONG, BERTHOLON, LECOZ, BESNAULT and LAVASTE [1998] on 3D FE models of the thorax and whole body; MERTZ [1984] on a procedure for normalizing impact response data; ROBBINS, SCHNEIDER, SNYDER, PFLUG and HAFFNER [1983] on the seated posture of vehicle occupants; ROBIN [2001] on the HUMOS model for safety; VIANO [1989] on biomechanical responses and injuries in blunt lateral impact.

9. The fiftieth percentile male H-Model

9.1. Introduction

Preamble. This section gives an overview on the structure of the fiftieth percentile male H-Model of the human body. This family of models is presently distributed and developed by the private companies ESI Software (Paris) and IPS International (Seoul), and by Hong-Ik University (Seoul). The presently available basic model represents a 50th percentile male human body and it was conceived primarily to study injury mechanisms and to assess injuries of the human skeleton and organs, which result from car accidents, including pedestrian injuries, e.g., CHOI and LEE [1999b]. A fifth percentile female and a 3 and 6 year old child model are under development. The basic model permits a fast and ongoing absorption of the rapidly growing biomechanical research results. The model can assist the conception of improved crash protection measures and devices for the human driver and passenger. It allows the omni-directional analysis for different impact directions (front, side, rollover, etc.) with a single model with global and local responses to bags, belts, head restraints, knee bolsters, etc. The analysis of different body postures, and the effect of muscle activity an be studied using the model.

From the H-Models other models for the analysis of passenger sitting comfort and riding comfort can be derived. The models for sitting comfort permit the assessment of seat and backrest pressure and they are whole body models equipped with elaborate representations of the flesh. The models for riding comfort permit to assess the effect of seat vibrations and they have well adapted spines and organ masses to detect resonance.

An attempt is made to systematically present the material for each body part in paragraphs which discuss anatomy, injury, model structure and calibration, and validation. The discussed items are the whole body H-ARB models and models of the head, neck, torso, upper extremity, lower extremity and the foot/ankle complex.

Features. Major features of the H-Model are given by the possibility of modular assembly of the deformable external and internal components into an underlying multi-body H-ARB (Human Articulated Rigid Body) model. The H-ARB model assures a correct overall kinematic behavior in a simulated crash scenario. Modularity permits to study each body part of interest in detail. While assuring the good overall kinematics, detailed studies can so be made for the head in head impacts, the neck in whiplash events, the thorax and abdomen for belt injuries, the lower extremity in knee bolster impacts, the foot/ankle complex in toe panel intrusions and the upper extremity and shoulder complex in side impacts and airbag aggressions. Fig. 9.1 shows an overall view of the model with skin and skeleton.

Model structure. The H-Model is a recent combination of two previous biomechanical models of the human body, namely the "ROBBY" (ESI Software) and the "H-Dummy" (IPSI/Hong-Ik) model families. The overall anthropometric properties of both models were mainly extracted from the report edited by ROBBINS [1983], after which the ROBBY family had been named. The new model family is now co-developed by

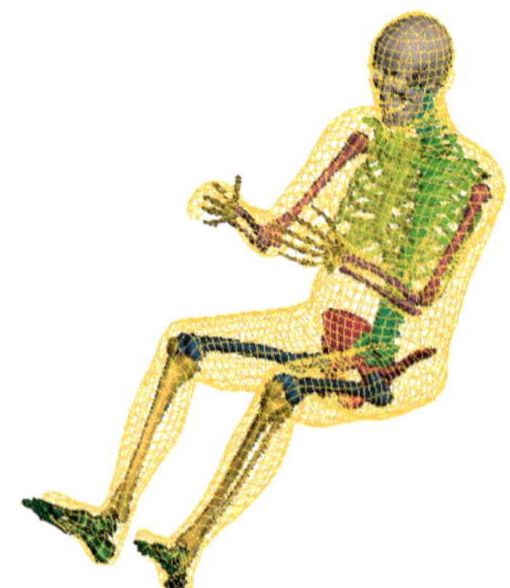

FIG. 9.1. H-Model with skin and skeleton (fiftieth percentile male).

the Seoul and Paris biomechanics teams. The development of the H-Model family was started prior to and is independent of the HUMOS-1 model described in Section 8 of this article.

The geometry of all listed models and sub-models has been derived mostly from the anatomy data found in the CAD geometry data base issued by DIGIMATION (DIGIMATION/VIEWPOINT CATALOG [2002]) and in the CD ROM issued by the VISIBLE HUMAN PROJECT [1994]. Further inspiration, concerning mainly the anatomical details of the ligaments, tendons, muscles, blood vessels and other soft tissues and internal organs was found in anatomical atlases, such as GRAY'S ANATOMY [1989]. Where necessary, the data were scaled to the presumed dimensions of the 50th percentile male. The Viewpoint CAD data were transformed into surface finite element meshes, and three-dimensional bulk FE meshes were generated in a compatible fashion, using commercially available FE mesh generator packages. The Visible Human CT data were digitized manually. The digitized anatomical cross sections were used to generate surface and bulk FE meshes.

The HUMOS-1 project (cf. Section 8) used a specific slicing technique, applied on a PMHS (post mortem human subject) in a sitting driver position. This procedure enabled the project to produce unique data of a seated subject. Since only one subject could be treated, scaling to the fiftieth percentile was necessary.

Calibration. All listed models and sub-models have been given material properties found in the open literature. A description of bio-materials is given in Section 1.5 and in Appendix B of this article.

Validation. All listed models and sub-models have been validated with available literature results. The basic validation tests and results are described in the available H-Model Reference Manuals. All validation test cases are relatively simple, representative, repeatable and well documented and controlled experiments, which facilitates modeling under well identified conditions.

9.2. The H-ARB Human Articulated Rigid Body model

Model structure and calibration. The H-ARB versions of the H-Models consist of articulated rigid body segments with flexible joints. The fiftieth percentile male model, H-ARB50, was built after ROBBINS [1983] and is documented in the H-Model Reference Manuals. It represents the basic platform for the modular assembly of detailed deformable and frangible skeleton components with soft tissues and internal organs. It serves for the evaluation of the overall kinematic and kinetic behavior of the occupant for omni-directional impacts. Fig. 9.2 shows the H-ARB50 version of the H-Model (76 kg) with the joint tripods (a) indicating the location and the orientations of the anatomical joints.

The outer geometry of the skin is taken from the clay model published in ROBBINS [1983], Fig. 9.2(b). The major dimensions of the outer surface of the model are shown in inset (c). The neck of the H-ARB model has several joints (skull-C1, C1-C2, etc.) as shown in inset (c), but no muscles. The various joints have rotational moment/angle

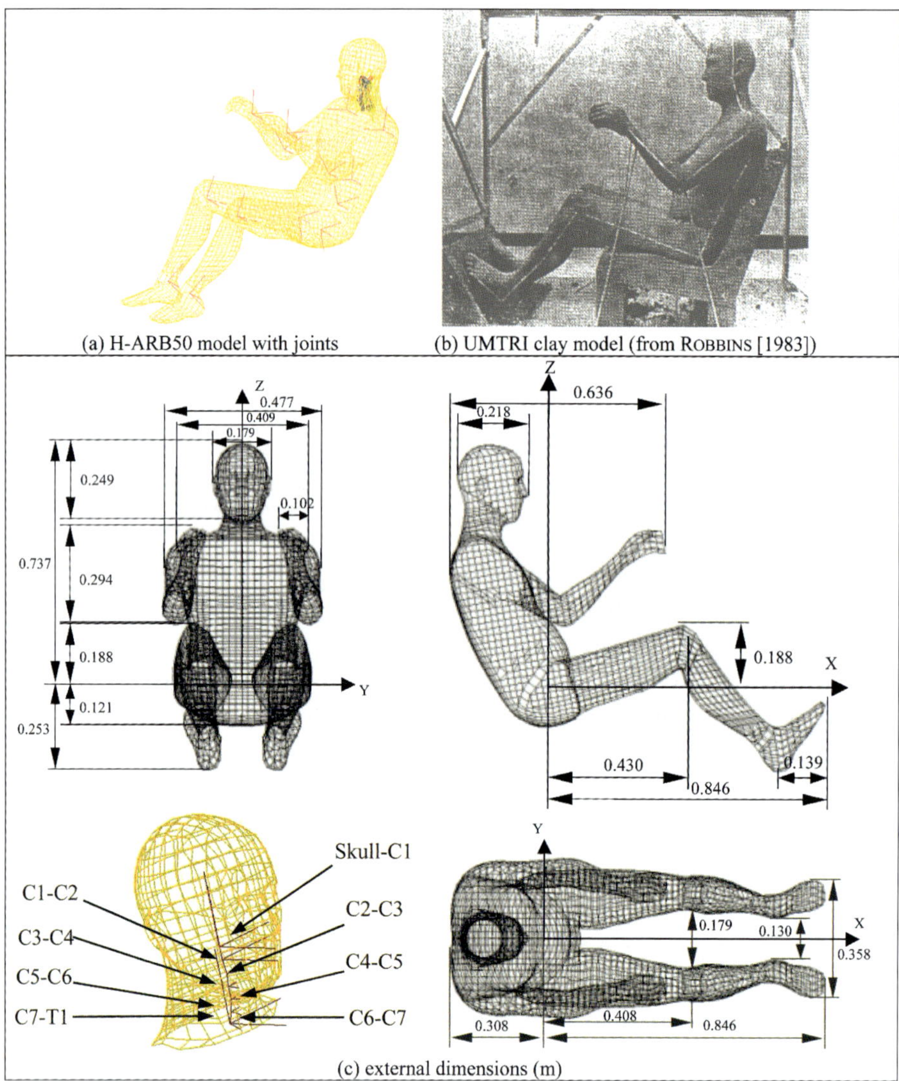

FIG. 9.2. H-ARB50 version of the H-Model (fiftieth percentile male). (Inset (b): from ROBBINS [1983] (UMTRI-83-53-2, US DoT NHTSA public domain report).)

relationships indicated in ROBBINS [1983] (see also Appendix B). Two typical curves are shown in Fig. 9.3. Some joints are given damping coefficients in order to stabilize their dynamic response. The joints are modeled with standard joint finite elements with linear and nonlinear moment-angle curves.

Accelerometers. Three accelerometers, which can be used to evaluate global motions of the H-ARB models, are defined using three additional nodal points. The locations of the accelerometers is adopted in analogy to the Hybrid III 50% male dummy model.

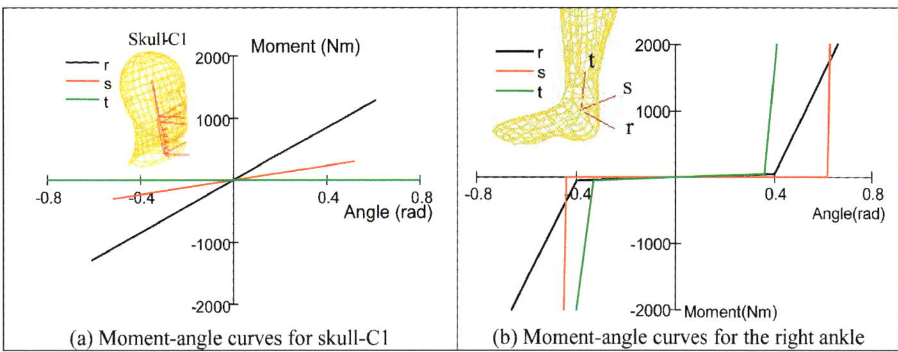

FIG. 9.3. H-Model H-ARB50 joint moment-angle curves.

The solver code will specifically output the acceleration time histories at these precise locations.

Contact. Contact surfaces in are defined between different segments of the H-Model, such as arms to thorax, between upper legs, etc. Note that in any application problem, additional contact interfaces involving the H-Models should be defined, such as H-Model to airbag, to belt, to car seat, etc. Standard contact algorithms are used, as implemented in the commercial crash codes (cf. Appendix A, Section A.3).

Validation and performance. First, the *extension motion* of the neck due to rear impact between two cars is presented. The geometry of the animated neck motion is shown in Fig. 9.4 at different times. The acceleration pulses of the striking and the hit car, the input acceleration pulse applied at T1 (first thorax vertebra), and the linear acceleration responses of the human volunteer and of the H-ARB50 head model are also shown in the figure. Any active muscle action is not considered in this simple H-ARB model. The passive muscle action, and the action of the discs and neck ligaments, is considered in a global fashion via the modeled standard joint finite elements with calibrated linear and nonlinear moment-angle curves.

Second, for the *lateral bending motion* of the neck due to the side impact, the animated neck motion for different times is shown in Fig. 9.5. The input acceleration pulse given at T1 (first thorax vertebra) is shown in the figure. The responses of the H-ARB50 neck model is compared with corridors of human volunteers.

Third, for a *frontal sled test simulation*, Fig. 9.6 represents the overall response of the H-ARB50 model to the medium intensity acceleration diagram indicated in inset (a). The H-ARB response is compared at different times to the response of a Hybrid III 50th percentile dummy model, inset (b). The head and upper torso linear acceleration versus time diagrams of both articulated rigid body models are compared in inset (c), where the differences in the responses reflect the differences between humans and crash dummies.

Finally, for an *arbitrary passenger out-of-position* (OOP) scenario, Fig. 9.7 shows the overall response of the H-ARB version of the H-Model.

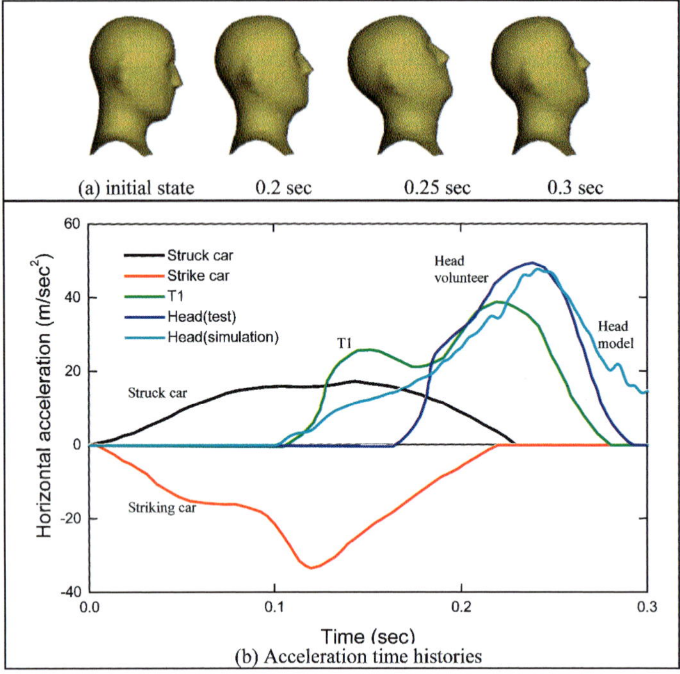

FIG. 9.4. H-ARB50 Validation: Horizontal acceleration histories due to rear impact.

Zooms with FE sub-models. For detailed studies ("zooms") of the impact response of the individual body parts, it is possible to substitute their detailed finite element models into the H-ARB versions of H-Model. The following sections outline such detailed sub-models. This modular approach saves CPU time while the overall model response correctly represents the kinematics of the body to generate the dynamic boundary conditions for the required zoom.

References on multi-body human models. To obtain further information on the shown HARB models, the following references may be consulted (in addition to the references listed at the end of Chapter II in the paragraph "References on multi-body and muscle models"): JAGER, SAUREN, THUNNISSEN and WISMANS [1994], JAGER [1996] on head/neck models; PRASAD [1990] on comparison between Hybrid dummy performances; SAE ENGINEERING AID 23 [1986] is the Hybrid III User's Manual; THUNNISSEN, WISMANS, EWING and THOMAS [1995] on human volunteers in whiplash loading; ZINK [1997] on 6 year old out-of-position passenger airbag simulation.

9.3. *H-Head: Skull and brain*

An overview on the H-Head model structure and application is given in CHOI [2001].

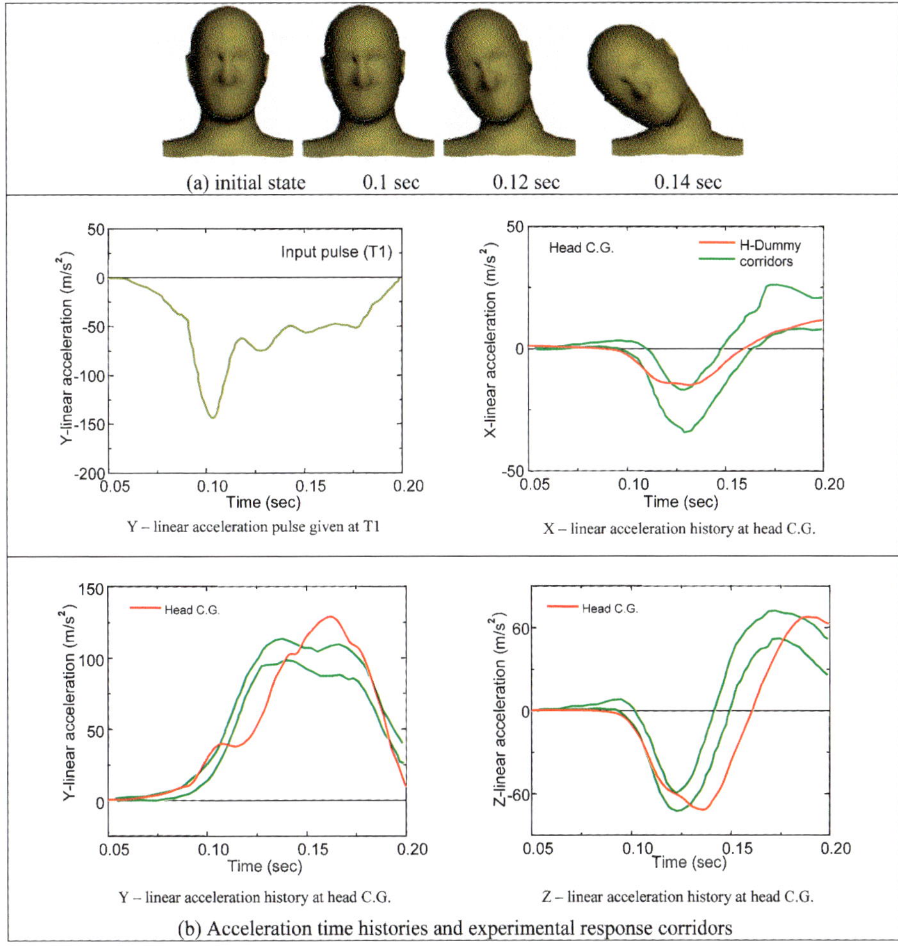

FIG. 9.5. H-ARB50 Validation: Neck motion due to side impact.

Anatomy. The complex anatomy of the skull and brain is summarized in Fig. 9.8. More details can be found in anatomical atlases and textbooks, e.g., GRAY'S ANATOMY [1989] and in the VISIBLE HUMAN PROJECT [1994]. The finite element model of the human head (H-Head) has been constructed by evaluating cross sections through the head. Three such sections (top-down) and the corresponding traces made of the brain matter are shown in Fig. 9.8(a) (SPITZER and WHITLOCK [1998]).

Cranial structure. The skull consists of three layers referred to as the outer table, diploe, and inner or vitreous table, Fig. 9.8(b) (PIKE [1990]) and (e) (PUTZ and PABST [2000]). The diploe consists of trabecular bone, and is located between the other two that are made up of compact (cortical) bone. The inside surface of the cranial cavity is lined by a layer of dense fibrous irregular connective tissue and enclosing venous sinuses. This layer adheres for the most part tightly to the cranial bones.

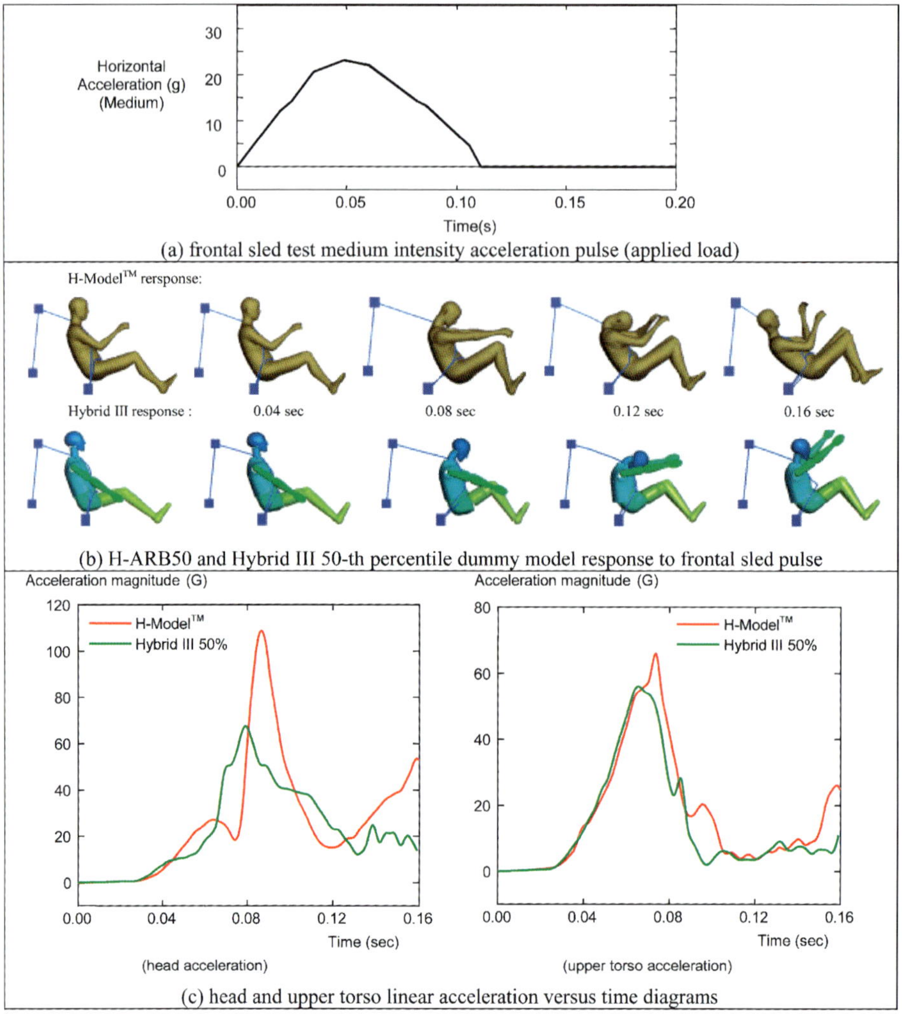

Horizontal Acceleration (g) (Medium)

30
20
10
0

0.00 0.05 0.10 0.15 0.20

Time(s)

(a) frontal sled test medium intensity acceleration pulse (applied load)

H-Model™ rersponse:

Hybrid III response : 0.04 sec 0.08 sec 0.12 sec 0.16 sec

(b) H-ARB50 and Hybrid III 50-th percentile dummy model response to frontal sled pulse

Acceleration magnitude (G)

120
100
80
60
40
20
0

H-Model™
Hybrid III 50%

0.00 0.04 0.08 0.12 0.16

Time (sec)

(head acceleration)

Acceleration magnitude (G)

80
60
40
20
0

H-Model™
Hybrid III 50%

0.00 0.04 0.08 0.12 0.16

Time (sec)

(upper torso acceleration)

(c) head and upper torso linear acceleration versus time diagrams

FIG. 9.6. H-ARB50 Validation: Sled test for frontal impact.

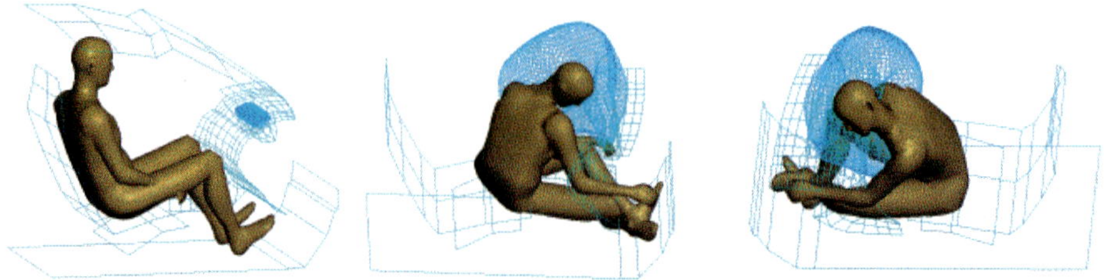

FIG. 9.7. OOP passenger airbag with H-ARB50.

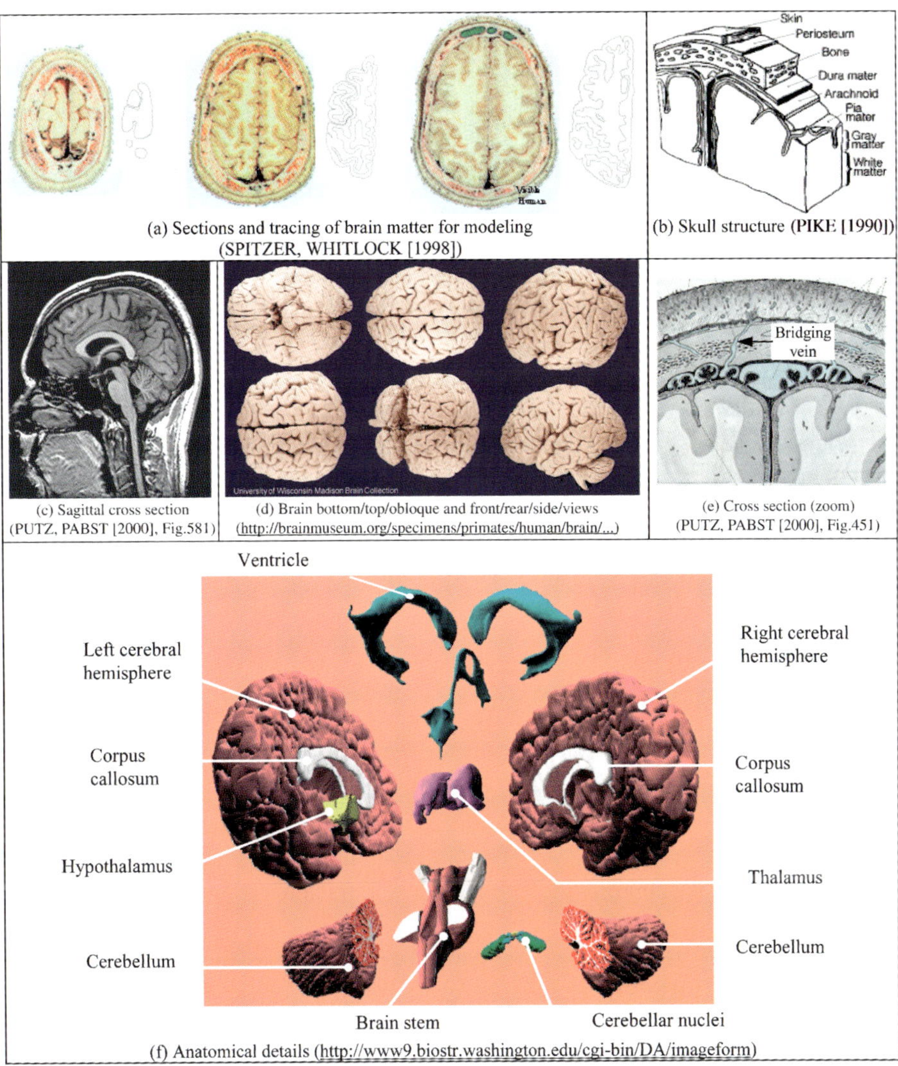

(a) Sections and tracing of brain matter for modeling (SPITZER, WHITLOCK [1998])

(b) Skull structure (PIKE [1990])

(c) Sagittal cross section (PUTZ, PABST [2000], Fig.581)

(d) Brain bottom/top/obloque and front/rear/side/views (http://brainmuseum.org/specimens/primates/human/brain/...)

(e) Cross section (zoom) (PUTZ, PABST [2000], Fig.451)

(f) Anatomical details (http://www9.biostr.washington.edu/cgi-bin/DA/imageform)

FIG. 9.8. Anatomy of the human head (overview). (Inset (a): Reproduced by permission of Jones & Bartlett Publishers Inc.; Inset (b): Reprinted with permission from "Automotove Safety" ©1990 SAE International; Insets (c) and (e): Reproduced by permission of Urban & Fischer Verlag; Inset (d): Reproduced by permission of Wally Welker at The Department of Neurophysiology, The University of Wisconsin; Inset (f): Graphic by John Sundsten, courtesy of the Structural Informatics Group at the University of Washington.)

Intra-cranial structure. The intra-cranial structure contains mainly the brain, the cerebrospinal fluid (CSF), the vascular structure, and the membranous coverings (dura mater, pia mater) and partition structures, Fig. 9.8(c)–(f) (PUTZ and PABST [2000] and web sites http://brainmuseum.org/specimens/primates/human/brain/human8sect6.jpg and http://www9.biostr.washington.edu/cgi-bin/DA/imageform). The brain occupies

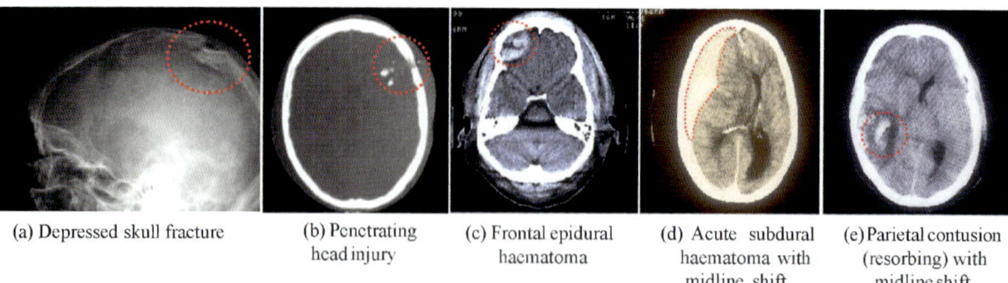

(a) Depressed skull fracture (b) Penetrating head injury (c) Frontal epidural haematoma (d) Acute subdural haematoma with midline shift (e) Parietal contusion (resorbing) with midline shift

FIG. 9.9. Brain injuries (http://www.trauma.org/imagebank/imagebank.html). (Reproduced by permission of Trauma.Org.)

most of the cranium and is composed of right and left hemispheres, separated by a fold of the dura mater called the falx cerebri. Another fold of the dura mater, called the tentorium cerebelli, separates the cerebrum and the cerebellum. The junction between the folds of the dura and the inner surface of the cranium forms some of the venous sinuses. These sinuses receive blood drained from the brain and reabsorb cerebrospinal fluid in regions knows as arachnoid granulations.

The next layers below the dura mater are the avascular arachnoid membrane and the delicate pia mater. The space between the arachnoid and the pia mater is called the subarachnoidal space, which is traversed by the arteries of the brain and cranial nerves and contains cerebrospinal fluid. The CSF circulates around the brain and spinal cord and through four ventricles (two lateral, the third ventricle at the mid-line below the lateral ventricles, and the fourth ventricle between the brain stem and the cerebellum).

Head injury. Head injuries can be caused by external or internal loads. External loads result from impacts with objects or obstacles, while internal dynamic loads result from motions of the rest of the body and are transmitted to the head through the neck (e.g., whiplash). Major head injuries are considered skull fractures (bone damage) and brain injuries (neural damage).

Skull fractures with increasing severity can be linear, comminuted (fragmented), depressed and basal fractures. Skull fractures correlate weakly with neural damage, i.e., neural damage may exist with and without skull fracture.

Neural damage can result from diffuse and focal brain injuries, Fig. 9.9. Diffuse brain injuries consist in brain swelling, concussion and diffuse axonal injury (DAI), while focal brain injuries lead to subdural hematomas (SDH), epidural hematomas (EDH), intra-cerebral hematomas (ICH), and contusions (from coup and countercoup pressure). For example, upon high angular head acceleration associated with an impact, the bridging veins at the top of the brain can rupture, causing a subdural hematoma or blood clot to develop. The clot exerts large pressures on the brain and if not quickly relieved can deprive the brain of its blood supply, causing brain death. The stretch in some of the bridging veins can reach 200%! (according to Viano: "humans are made to fall from trees").

Model structure and calibration. The detailed FE head model (H-Head) of the 50th percentile H-Model consists of deformable bone and brain components, Fig. 9.10. The modeling of the head comprises the scalp (can attenuate blunt impacts), the skull and

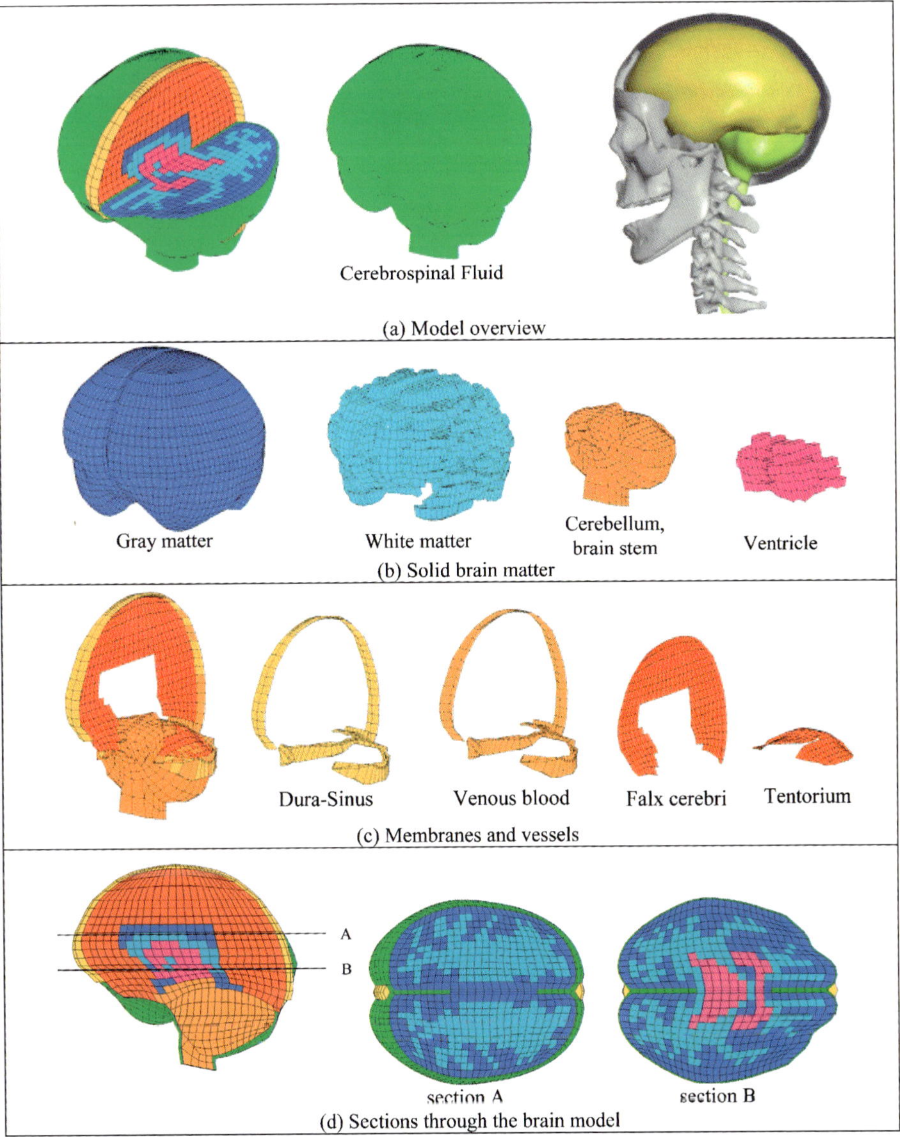

FIG. 9.10. H-Head model overview.

facial bone (the latter can be held more approximate), the cerebro-spinal fluid (CSF) layers and the ventricles, the brain membranes (dura mater), the falx, the tentorium, the gray and the white brain matter and the cerebellum. The CSF is presently modeled with the Murnaghan equation of state, $p = p_0 + B((\rho/\rho_0)^\gamma - 1)$, where p is the hydrostatic pressure, ρ is the mass density and B and γ are material constants. For the modeling of the vacuum cavity that might develop in the CSF at the interface between the skull

TABLE 9.1
Mechanical properties of the H-Head model

	Component	E [kPa]	K [kPa]	G [kPa]	ν	ρ [kg/m^3]
Skull	Outer table	7.3×10^6			0.22	3000
	Inner table	7.3×10^6			0.22	3000
	Diploe		2.02×10^5	1.39×10^5	0.22	1410
	Facial bone	7.3×10^6			0.22	2700
	Mandible	7.3×10^6			0.22	2700
	Dura mater	3.15×10^4			0.22	1133
	Venous sinus		1.0×10^5			1000
	CSF		1.0×10^5			1000
	Falx	3.15×10^4			0.45	1133
	Pia	3.15×10^4			0.45	1133
	Tentorium	3.15×10^4			0.45	1133
Brain	Gray matter		7.96×10^3	7.96×10^1	~ 0.499	1040
	White matter		1.27×10^4	1.27×10^2	~ 0.499	1040
	Ventricle		1.0×10^5			1000
	Cerebellum and stem		1.27×10^4	1.27×10^2	~ 0.499	1040

E = Young's modulus [kPa], K = Bulk modulus [kPa], G = Shear modulus [kPa], ν = Poisson's ratio, ρ = Mass density [kg/m^3].

and brain from negative countercoup pressures, ideal gas equations are used. The model is fully compatible with the H-Neck spinal cord model. Table 9.1 shows the material coefficients as calibrated for the H-Head model.

Validation. Fig. 9.11 summarizes a validation of the H-Head model carried out after tests performed by Nahum, using frontal pendulum impact. In Fig. 9.11(a) the test setup and the measured pressure time histories in coup, countercoup, parietal and occipital locations are shown. Fig. 9.11(b) contains the measured and calculated head accelerations. Fig. 9.11(c) gives an overview on the surface pressures as calculated with the H-Head model at different times. In Fig. 9.11(d)–(f) the pressure time histories in coup, countercoup and occipital locations are shown and comparison to Nahum's tests is seen to be satisfactory.

References on the H-Head model. The following references were considered relevant for the basic and detailed understanding, construction and validation of the head model: ABEL, GENNARELLI and SEGAWA [1978] on incidence and severity of cerebral concussion in rhesus monkeys from sagittal acceleration; BANDAK and EPPINGER [1994], BANDAK [1996] on brain FE models and impact traumatic brain injury; CHAPON, VERRIEST, DEDOYAN, TRAUCHESSEC and ARTRU [1983] on brain vulnerability from real accidents; CHOI and LEE [1999b] on deformable FE models of the human body; CLAESSENS [1997], CLAESSENS, SAUREN and WISMANS [1997] on FE modelling of the human head under impact conditions; COOPER [1982a], COOPER [1982b] on injury of the skull, brain and cerebro-spinal fluid related; DIMASI, MARCUS and EPPINGER [1991] on a 3D anatomical brain model for automobile crash loading;

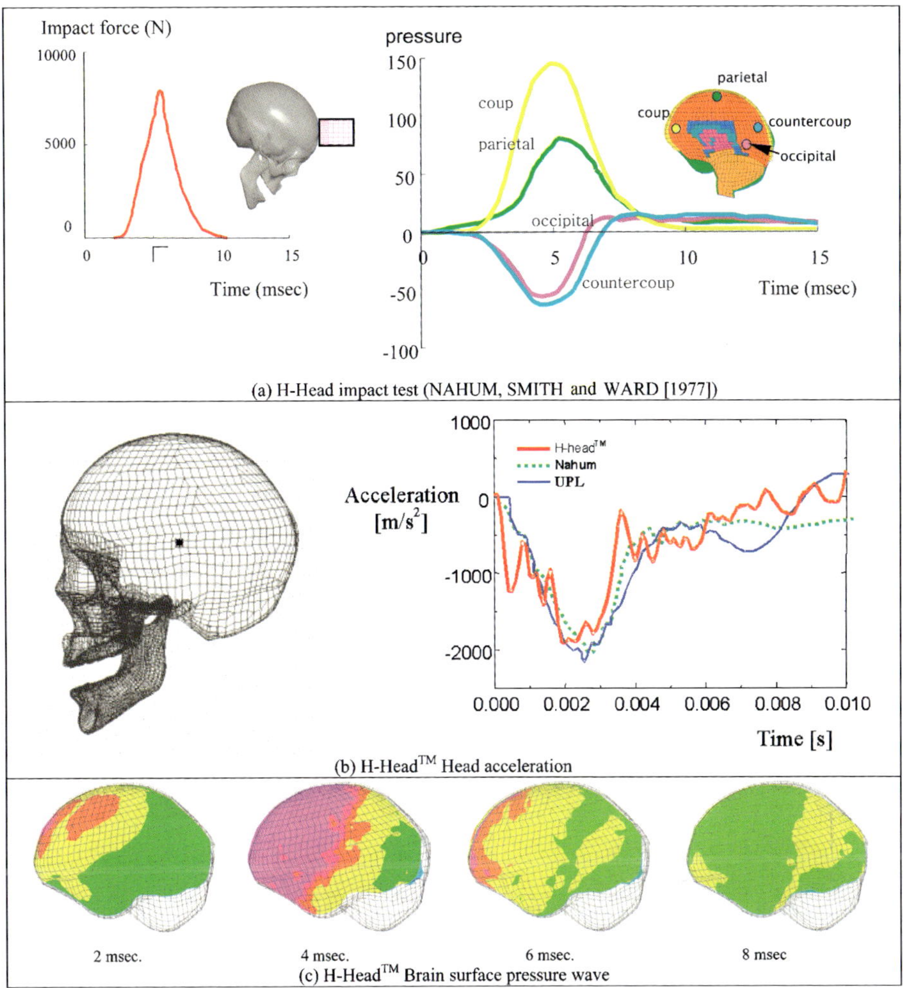

(a) H-Head impact test (NAHUM, SMITH and WARD [1977])

(b) H-Head™ Head acceleration

(c) H-Head™ Brain surface pressure wave

FIG. 9.11. H-Head model validations (tests: after NAHUM, SMITH and WARD [1977]).

DONNELLY and MEDIGE [1997] on shear properties of human brain tissues; EWING, THOMAS, LUSTICK, MUZZY III, WILLEMS and MAJEWSKI [1978] on the effect of the initial position on the head/neck response in sled tests; GENNARELLI [1980], GENNARELLI, THIBAULT, ADAMS, GRAHAM, THOMPSON and MARCINCIN [1982] on the analysis of head injury severity by AIS-80 (1980) and on diffuse axonal injury and traumatic coma in primates (1982); GRAY'S ANATOMY [1989]: Atlas of Anatomy; GURDJIAN and LISSNER [1944], GURDJIAN, WEBSTER and LISSNER [1955], GURDJIAN, ROBERTS and THOMAS [1966] on mechanisms of head injury and brain concussion and tolerance of acceleration and intra-cranial pressure; HOL-

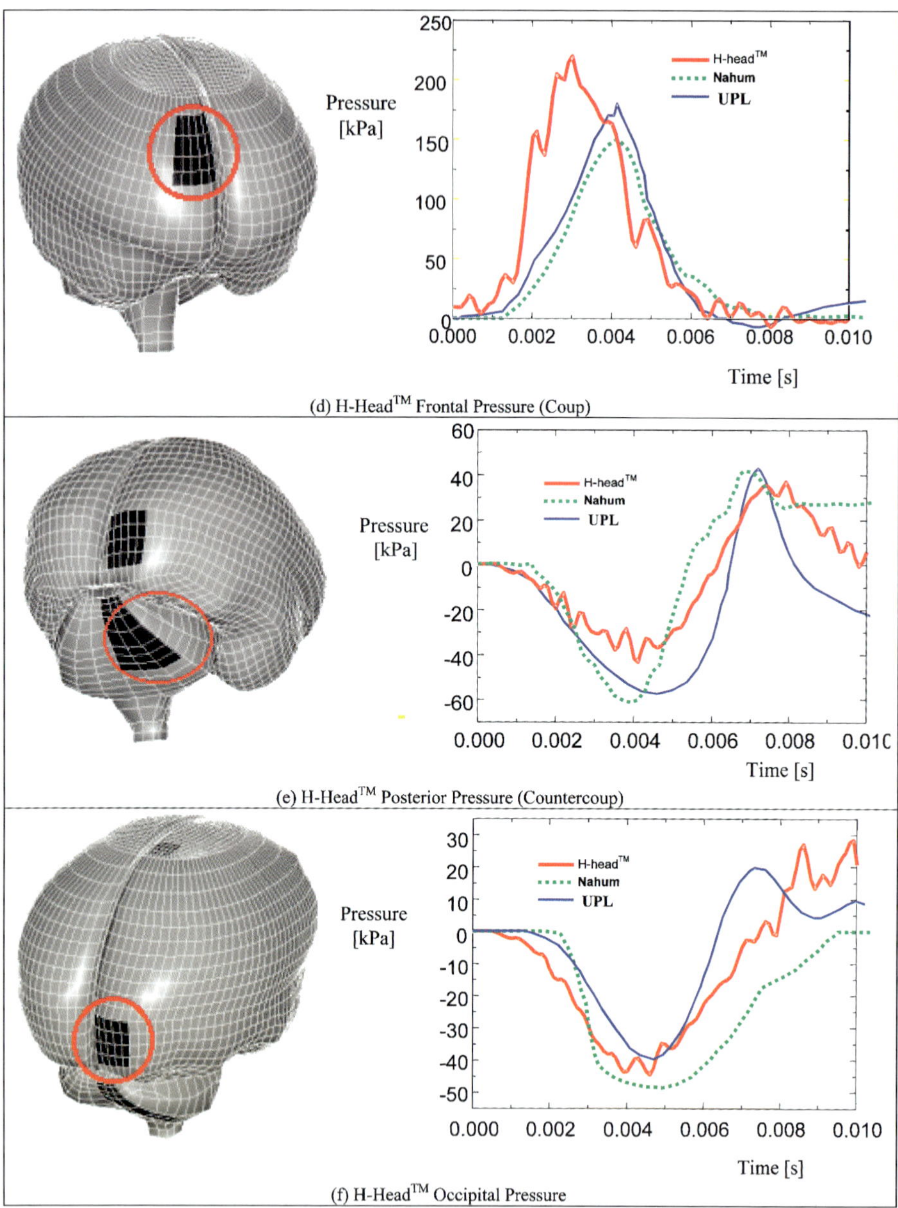

FIG. 9.11. *(Continued.)*

BOURN [1943] on the mechanics of head injury; KANG, WILLINGER, DIAW and CHINN [1997] on the validation of a 3D FE model of the human head in motorcycle accidents; LEE, MELVIN and UENO [1987], LEE and HAUT [1989] on FE analysis of subdural hematoma and bridging vein failure characteristics; LISSNER, LEBOW and

EVANS [1960] on experiments on intra-cranial pressure; MARGULIES, THIBAULT and GENNARELLI [1990] on modeling brain injury in primates; MILLER and CHINZEI [1997] on constitutive modelling of brain tissue; NAHUM, SMITH and WARD [1977] on intra-cranial pressure dynamics during head impact; NEWMAN [1993] on head protection; OMMAYA, HIRSCH, FLAMM and MAHONE [1966], OMMAYA and HIRSCH [1971], OMMAYA and GENNARELLI [1974] on cerebral concussions in the monkey (1966), on their tolerances (1971) and on their clinical/experimental correlation (1974); ONO, KIKUCHI, NAKAMURA, KOBAYASHI and NAKAMURA [1980], ONO [1999] on head injury tolerance for sagittal impact from tests (1980) and on spine deformation and on vertebral motion from whiplash test volunteers (1999); PENN and CLASEN [1982] on traumatic brain swelling and edema; PUTZ and PABST [2000]: Sobotta Atlas of Human Anatomy; RUAN, KHALIL and KING [1991] on human FE head model response in side impacts; RUAN and PRASAD [1994] on head injury assessment in frontal impacts by mathematical modelling; SANCES ET AL. [1982] on head and spine injuries; SCOTT [1981] on the epidemiology of motor cyclist head and neck trauma; SPITZER and WHITLOCK [1998]: Atlas of the Visible Human Male; TARRIERE [1981] on investigation of the brain with CT-scanners; TORG [1982], TORG and PAVLOV [1991] on athletic injury on the head, neck and face; TURQUIER, KANG, TROSSEILLE, WILLINGER, LAVASTE, TARRIERE and DÖMONT [1996] on the validation of a 3D FE head model against experiments; UENO, MELVIN, LUNDQUIST and LEE [1989] on 2D FE analysis of human brain under impact; VOO, KUMARESAN, PINTAR, YOGANANDAN and SANCES [1996] on a finite element model of the human head; WALKE, KOLLROS and CASE [1944] on the physiological basis of concussion; WILLINGER, KANG and DIAW [1999] on the validation of a 3D FE human model against experimental impacts; WISMANS ET AL. [1994] on injury biomechanics; ZEIDLER, STÜRTZ, BURG and RAU [1981] on injury mechanisms in head-on collisions; ZHOU, KHALIL and KING [1996] on the visco-elastic brain FE modeling for sagittal and lateral rotation acceleration. The following sites were used: http://www9.biostr.washington.edu/cgi-bin/DA/imageform; http://brainmuseum.org/specimens/primates/human/brain/human8sect6.jpg; http://www.trauma.org/imagebank/imagebank.html.

9.4. *H-Neck: Cervical spine with active muscles*

Overviews on the H-Neck model structure and applications are given in CHOI and EOM [1998], LEE and CHOI [2000] and CHOI, LEE and HAUG [2001b]. A study on the whiplash injury due to the low velocity rear-end collision is presented in CHOI, LEE, EOM and LEE [1999].

Anatomy. The anatomy of the cervical spine is summarized in Fig. 9.12 (after http://www.rad.washington.edu/RadAnat/Cspine.html and PUTZ and PABST [2000]). The human cervical spine is composed of 7 vertebrae, C1–C7, with five similar lower vertebrae (C3–C7) and two dissimilar upper vertebrae (C1 "atlas" and C2 "axis"). Approximately 47% of the bending and stretching of the cervical spine occurs between the head and C1, while over 50% of the rotational motions occur between C1 and C2. Between each pair of adjacent vertebrae except for between C1 and C2 exist a structure known as the disc of cervical spine. Each disc is composed of the nucleus pulposus,

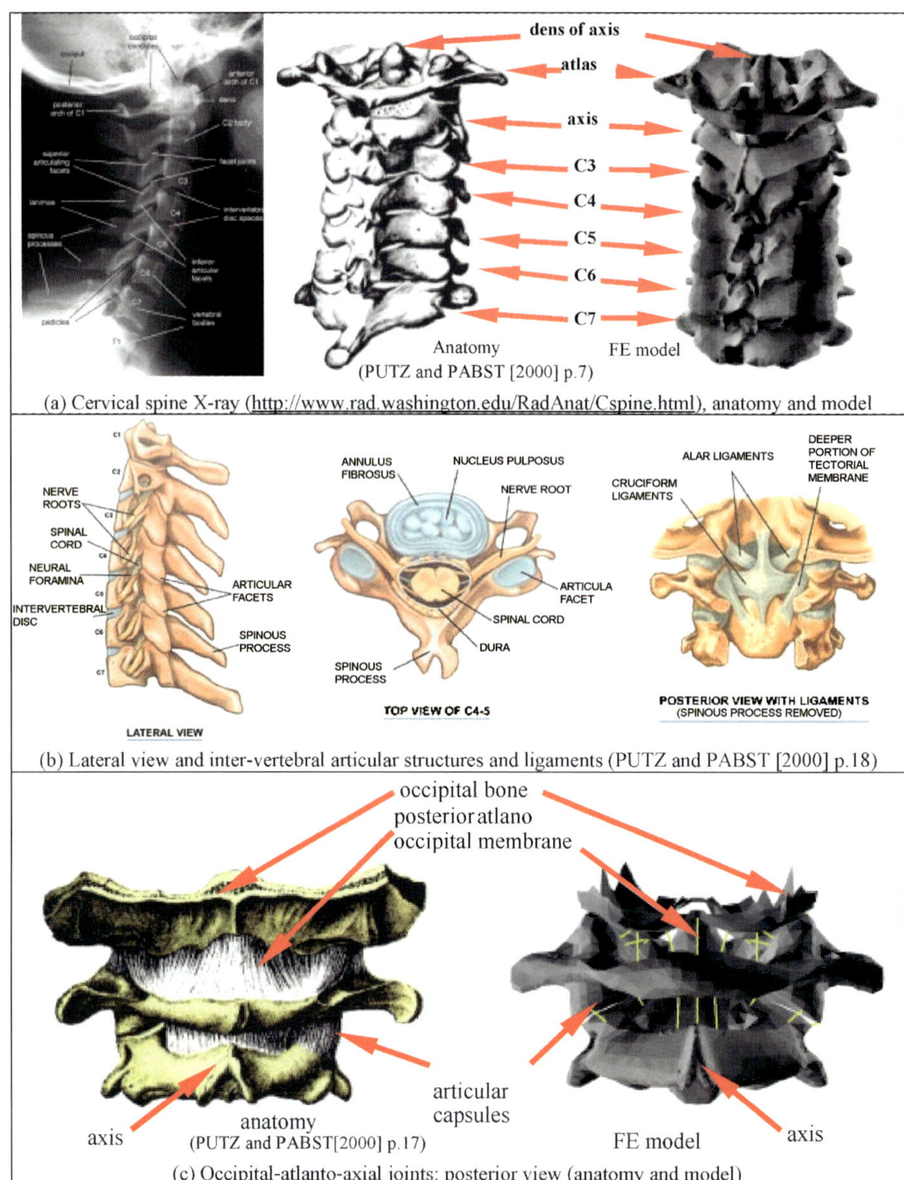

(a) Cervical spine X-ray (http://www.rad.washington.edu/RadAnat/Cspine.html), anatomy and model

(b) Lateral view and inter-vertebral articular structures and ligaments (PUTZ and PABST [2000] p.18)

(c) Occipital-atlanto-axial joints: posterior view (anatomy and model)

FIG. 9.12. Anatomy of the human cervical spine. (Inset (a) left: Reproduced by permission of Michael L. Richardson, University of Washington, Deparment of Radiology; Insets (a) center, (b) and (c) left: Reproduced by permission of Urban & Fischer Verlag.)

the annulus fibrosus and a cartilageous end-plate. Between 70 to 90% of the nucleus pulposus by weight is water, and takes up as much as 40 to 60% of the disc area.

The annulus fibrosus is a laminated and hence anisotropic structure composed of several layers in which each layer maintains a 30° angle of inclination from the horizontal plane. The inner boundary of the annulus fibrosus is attached to the cartilageous endplate, and the outer surface is directly connected to the vertebra body. The discs play a dominant role under compressive loads.

The discs show greater stiffness for the front/rearward motion than for the side/side motion. Spinal ligaments (anterior, posterior, ...) connect the vertebrae of the cervical spine. They ensure that the spinal motions occur within physiological bounds, thus preventing spinal code injury due to excessive spinal motion. The ligaments are usually situated between two adjacent vertebrae, and in some cases connects several vertebrae.

Neck injury. Neck injury from car accidents has been studied extensively and is summarized hereafter according to MCELHANEY and MYERS [1993] (eds. Nahum and Melvin). Fig. 9.13, inset (a), summarizes the anatomical head motions. Inset (b) shows the modes of loading on the neck. Inset (c) lists 10 distinct injury mechanisms and identifies a total of 25 different neck injuries. Inset (d) shows three tension-extension injury mechanisms, including the important whiplash injury (B). Inset (e) shows three flexion-compression neck injury mechanisms.

For example, the tension–extension injury (A) of inset (d) to Fig. 9.13 is described as "*Fixation of the head with continued forward displacement of the body. This occurs commonly in unbelted occupants hitting the windshield, and as a result of falls and dives*". The tension–extension whiplash injury (B) is described as "*Inertial loading of the neck following an abrupt forward acceleration of the torso as would occur in a rear-end collision*". The tension–extension injury (C) is described as "*Forceful loading below the chin directed postero-superiorly* (*as in a judicial hanging*).

The frequent whiplash injury occurs as a result of more or less mild rear-end collisions. It is considered a hyperextension injury and it may cause muscle stiffness and neck pain. Larger accelerations, in addition to producing whiplash symptoms, may produce disruption of the anterior longitudinal ligament and intervertebral disk, and horizontal fractures though the vertebra. The remaining neck injuries are described in the above reference. The survivable neck injuries from car accidents are often injuries of the soft connecting tissues, while bone fractures are less common.

Model structure and calibration. The modeling aspects of the human cervical spine are summarized in Figs. 9.14 and 9.15. The H-Neck model is fully compatible and interfaced with the H Head model. Based on 3D CAD data of 50% male cervical vertebrae supplied by DIGIMATION (DIGIMATION/VIEWPOINT CATALOG [2002]), the cervical spine in the current H-Neck model is built as a chain of articulated rigid vertebrae. This reduces computational cost, all the while soft tissue injury plays a more important role than bony fracture does in neck injury. Fully deformable vertebra bodies can easily be modeled, for example, as demonstrated earlier by NITSCHE, HAUG and KISIELEWICZ [1996].

The H-Neck model contains important elements such as vertebrae, discs, ligaments, and stabilizing muscles. The inter-vertebral contacts are modeled with sliding interfaces and contact stiffnesses. They can also be modeled by appropriate layers of finite elements. The vertebrae of the H-Neck model are interconnected by the neck ligaments,

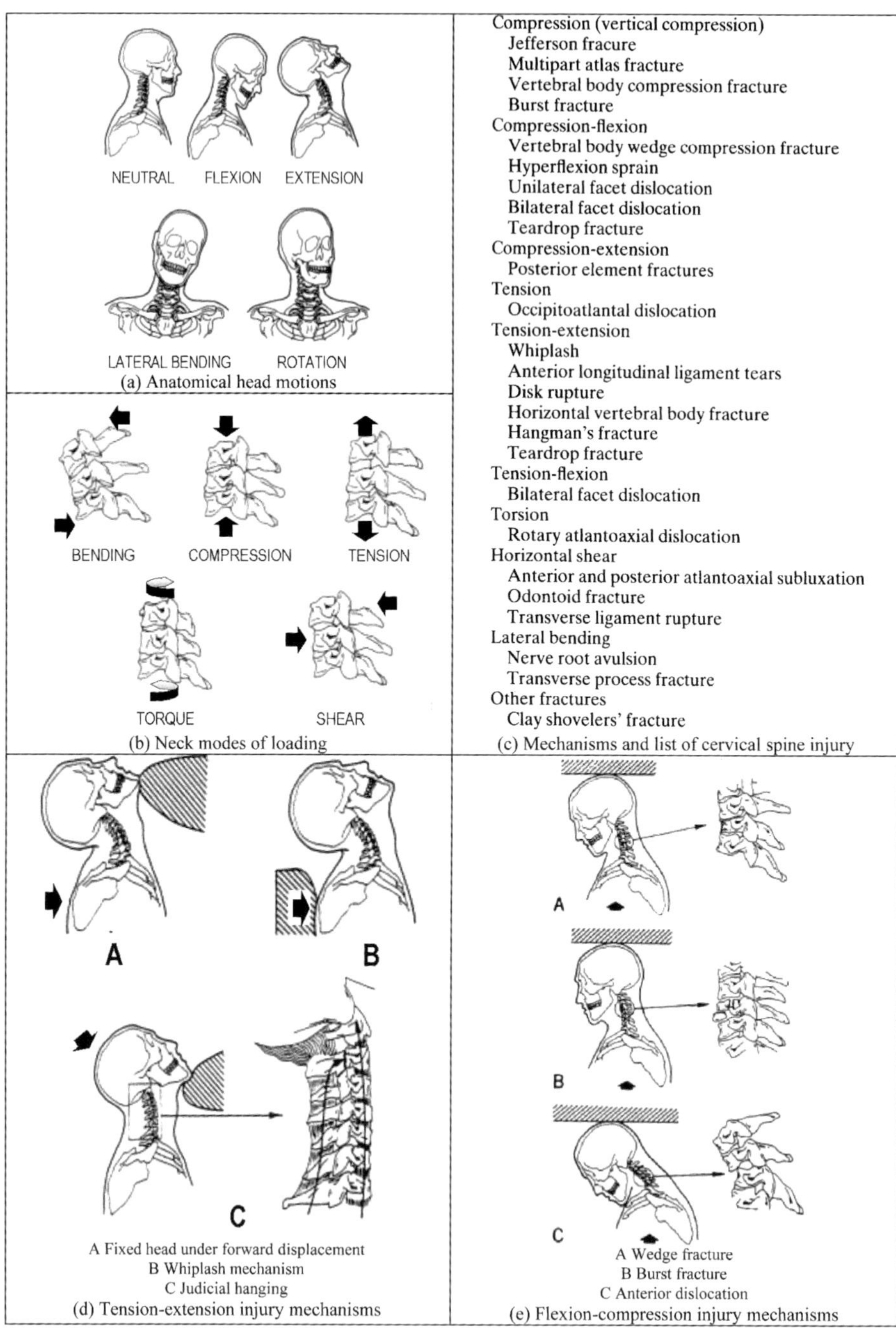

FIG. 9.13. Neck injury overview (after MCELHANEY and MYERS [1993]). (Reproduced by permission of Springer Verlag from: Nahum and Melvin eds., Accidental Injury, Biomechanics and Prevention: Insets (a) and (b): Figure 14.5, page 318; Inset (c): Table 14.5, page 319; Inset (d): Figure 14.10, page 324; Inset (e): Figure 14.8, page 320.)

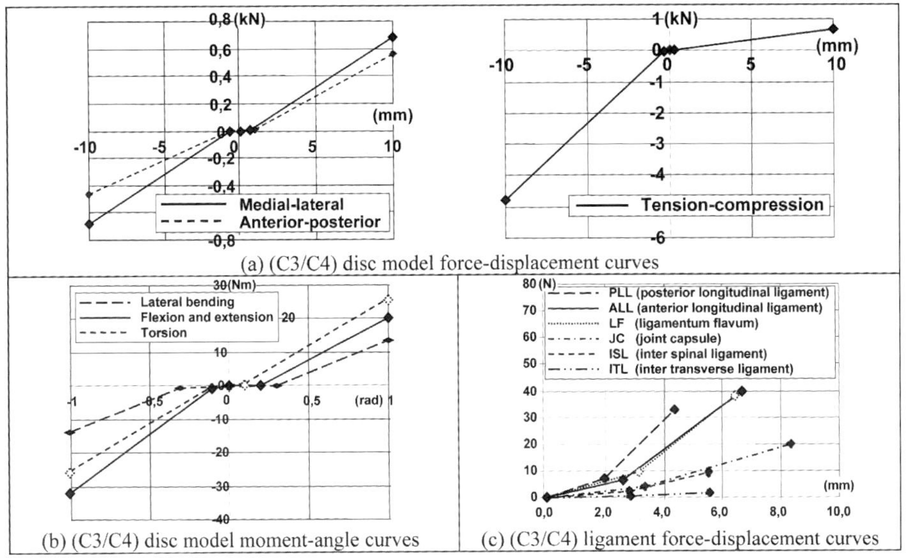

(a) (C3/C4) disc model force-displacement curves

(b) (C3/C4) disc model moment-angle curves

(c) (C3/C4) ligament force-displacement curves

FIG. 9.14. (C3–C4) disc 6 DOF kinematic joint element and ligament calibration (H-Neck).

modeled as bars or membranes. To account for the visco-elastic and anisotropic biomechanical properties of the ligaments, nonlinear one-dimensional tension-only bar elements are used. For cruciform, transverse, and tectorial ligaments in the atlanto-axial joint however, membrane elements are used to allow for surface contact (see also Fig. 9.12(c)). Each pair of adjacent vertebrae (except C1 and C2) is connected by the elastic discs of the cervical spine.

In the H-Neck model, the complex deformation of the discs is modeled by 6-DOF nonlinear spring elements with stiffness calibrated as shown in Fig. 9.14(a), (b). The ligament force-displacement calibrations are shown in Fig. 9.14(c).

These calibrations were made with test results found in the literature, for example: CHAZAL, TANGUY and BOURGES [1985] on biomechanical properties of the spinal ligaments; MCCLURE, SIEGLER and NOBILINI [1998] on in vivo 3D flexibility properties of the human cervical spine; MORONEY, SCHULTZ, MILLER and ANDERSSON [1988], MORONEY, SCHULTZ and MILLER [1988] on lower cervical spine properties and neck loads; MYKLEBUST and PINTAR [1988] on the tensile strength of spinal ligaments; NIGHTINGALE, WINKELSTEIN, KNAUB, RICHARDSON, LUCK and MYERS [2002] on upper/lower cervical spine strength comparison in flexion/extension; PANJABI, CRISCO, VASAVADA, ODA, CHOLEWICKI, NIBU and SHIN [2001] on cervical spine load-displacement curves; YOGANANDAN, SRIRANGAM and PINTAR [2001] on the biomechanics of the cervical spine.

Fig. 9.15(a) shows the anatomical details (PUTZ and PABST [2000]), the H-Neck ligaments and the assembled H-Head/H-Neck model, where the head can be modeled effectively as a rigid body for most whiplash simulations. The muscle forces obey the well known active and passive Hill muscle law, Fig. 9.15(b). The muscles are modeled by nonlinear bar elements, which can be curved in space, assuring a correct introduction

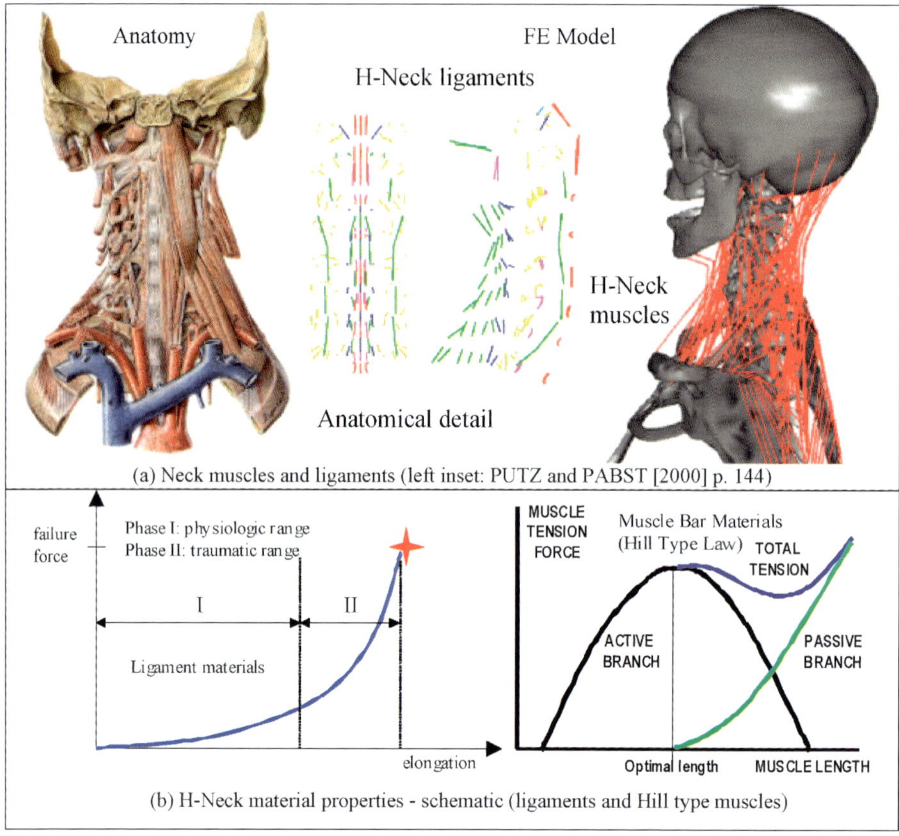

FIG. 9.15. H-Neck model overview. (Inset (a) left: Reproduced by permission of Urban & Fischer Verlag.)

of the muscle forces along the curved neck line. The active muscle force can contribute to conscious bracing and motion. The passive muscle forces contribute to the resistance of the connected body parts near the end of their respective anatomical motion ranges. It increases with muscle length in a highly nonlinear fashion (cf. Appendix C).

Validation. Fig. 9.16 shows in inset (a) the flexion response of the H-Neck model in a frontal passenger car impact simulation. In inset (b) the extension response in whiplash for a rear impact is shown. Inset (c) shows pictures of the deformed neck in lateral bending as from a side impact. The muscle bars are seen to follow the neck as it flexes and extends.

Fig. 9.17 contains several results of the H-Head/H-Neck response to a frontal car crash event, as compared to the response corridors obtained with volunteers in live tests (after THUNNISSEN, WISMANS, EWING and THOMAS [1995] and HORST, THUNNISSEN, HAPPEE, HAASTER and WISMANS [1997]).

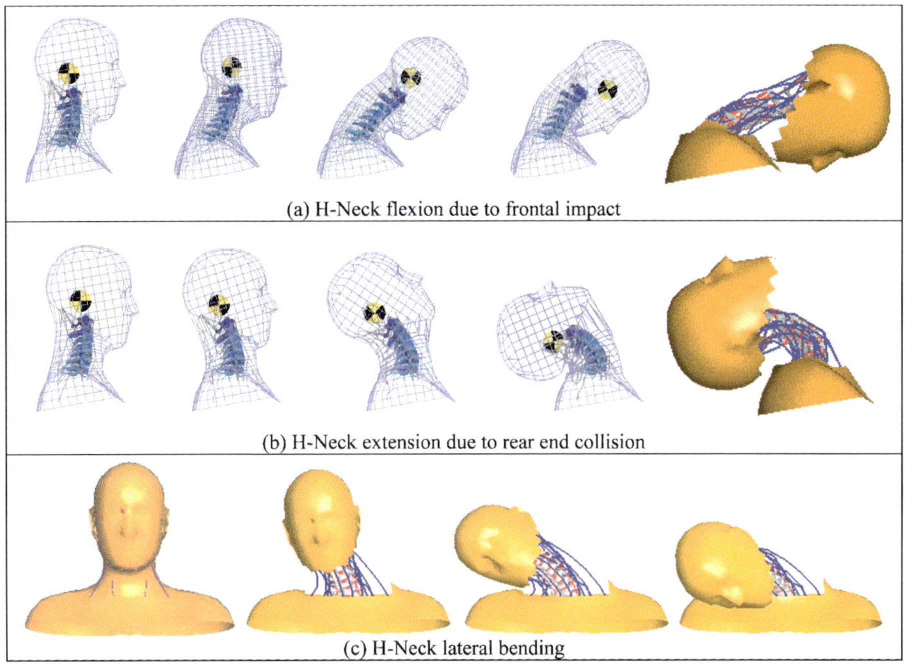

FIG. 9.16. H-Neck model validations: Flexion, extension and torsion motions.

Fig. 9.18 shows the trajectories of several points of the H-Neck model in extension from rear impacts (whiplash). The results compare well with corridors from tests performed with life volunteers (ONO, KANEOKA, WITTEK and KAJZER [1997]).

Fig. 9.19(a)–(c) compare the responses of the H-Neck model mounted on the H-ARB version in a rear impact sled test simulation, both, with and without a headrest. Fig. 9.19(d) compares the H-Head/H-Neck response to a rear impact with a headrest in a low and a high position. Fig. 9.20, finally, shows the head and the cervical spine skeleton response with a deformable headrest to such a rear impact. The results demonstrate the utility of such models to study the effects of preventive measures in the design of passive safety restraints of passenger cars.

JARI volunteer tests. Recently the H-Neck™ model is being validated against volunteer neck whiplash tests, Fig. 9.21, in which human volunteers underwent equivalent mild frontal and rear head accelerations in the sagittal plane, as they are typical in mild frontal and rear impacts of passenger cars, ONO, KANEOKA, SUN, TAKHOUNTS and EPPINGER [2001]. While the evaluations of the tests are still ongoing, some simulations of the scenarios have been made with an attempt to study the influence of voluntary and involuntary contraction of the neck muscles which counteract the accidental neck/head motions in the sagittal plane, Fig. 9.21(a). The neck muscles are divided into groups of neck flexors and neck extensors in the sagittal plane, insets (b) and (c), which are activated separately in a parametric study in different assumed reflex time scenarios to their

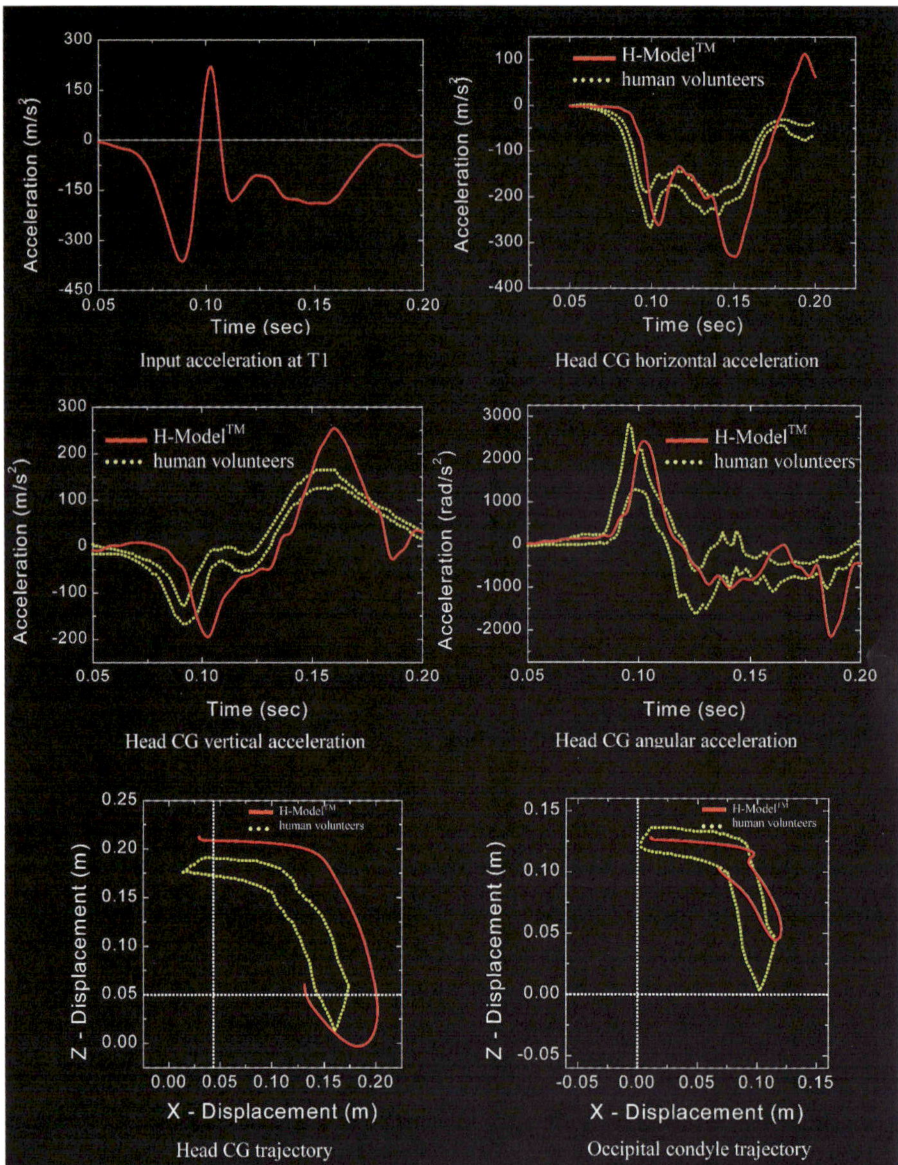

FIG. 9.17. H-Neck model validations: Volunteer tests in frontal crash (flexion). (Test results after Stapp papers THUNNISSEN, WISMANS, EWING and THOMAS [1995] and HORST, THUNNISSEN, HAPPEE, HAASTER and WISMANS [1997].)

highest level of activation (100%). The flexor group pulls the head forward (agonists) and the extensor group backward (antagonists).

For example, the subject's head is being suddenly pulled rearward with the help of a string attached to a mask, which contains accelerometers and optical gauges, as shown

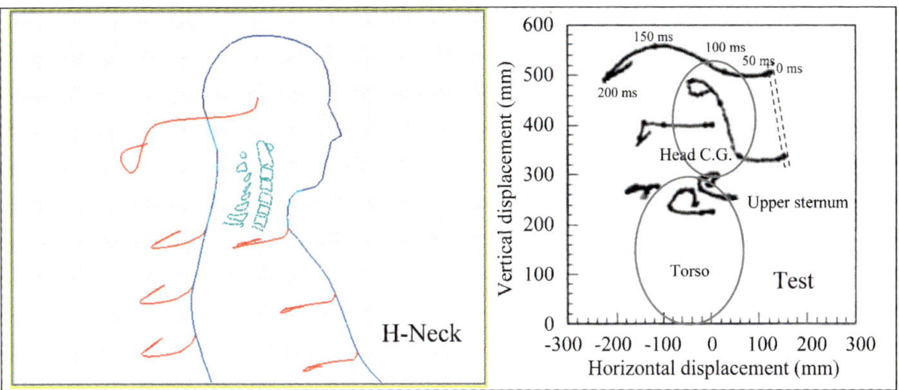

FIG. 9.18. H-Neck model validations: Trajectories in extension for a rear impact (whiplash). (Right inset: Reproduced by permission of The Stapp Association.)

in inset (a). The volunteer is either asked to relax, or to fully stiffen its neck muscles prior to the sudden pull. An intermediate case is when the subject involuntarily stiffens its neck muscles during the event (stretch reflex). In inset (d) deformed shape snapshots of a simulation of a relaxed subject with no activation of the active muscle forces are shown. In (e) the neck flexor muscles are assumed to be fully tensed, starting at 100 ms after the pull, followed by the extensor muscles, starting at 150 ms. As expected, the maximal neck flexion occurs for the relaxed subject, (d), with trajectories, (f), and this motion is reduced for the subject with the (arbitrarily) assumed pattern of full muscle activation, (g), (h). Since the exact pattern of involuntary muscle activation is unknown, such activation patterns can be found by back calculation from various assumed simple activation patterns and comparison with volunteer tests. Pattern (i) was not a good assumption, since the late activation of the flexors allowed for large neck flexion motions.

The active muscle forces were built up and applied according to Hill's law, using the physiological cross section area of each muscle. Although it is presently unknown in precisely what pattern a living subject activates the neck muscles in a whiplash event, such parametric studies can help clarify the open questions. Light can also be shed on the different loads and injuries of the cervical spine, which are influenced by the muscle actions.

References on the H-Neck model. The following references were considered useful for the preparation of the H-Neck model: BOKDUK and YOGANANDAN [2001] on minor injuries of the cervical spine; CUSICK and YOGANANDAN [2002] on major injuries of the cervical spine; DOHERTY, ESSES and HEGGENESS [1992] on odontoid fracture; DVORAK, HAYEK and ZEHNUDER [1987], DVORAK, PANJABI and FROEHLICH [1988] on rotary instability of the upper cervical spine; FOREMAN and CROFT [1995] (whiplash injuries); GRAY'S ANATOMY [1989]: Atlas of Anatomy; HIRSCH [1955] (disc compression reaction); HORST, THUNNISSEN, HAPPEE, HAASTER and WISMANS [1997] on the influence of muscle activity on head/neck impact response; HUELKE, NUSHOLTZ and KAIKER [1986] on an overview on cervical fracture and dislocation; JAGER [1996] on the mathematical modeling of the cervical spine (overview);

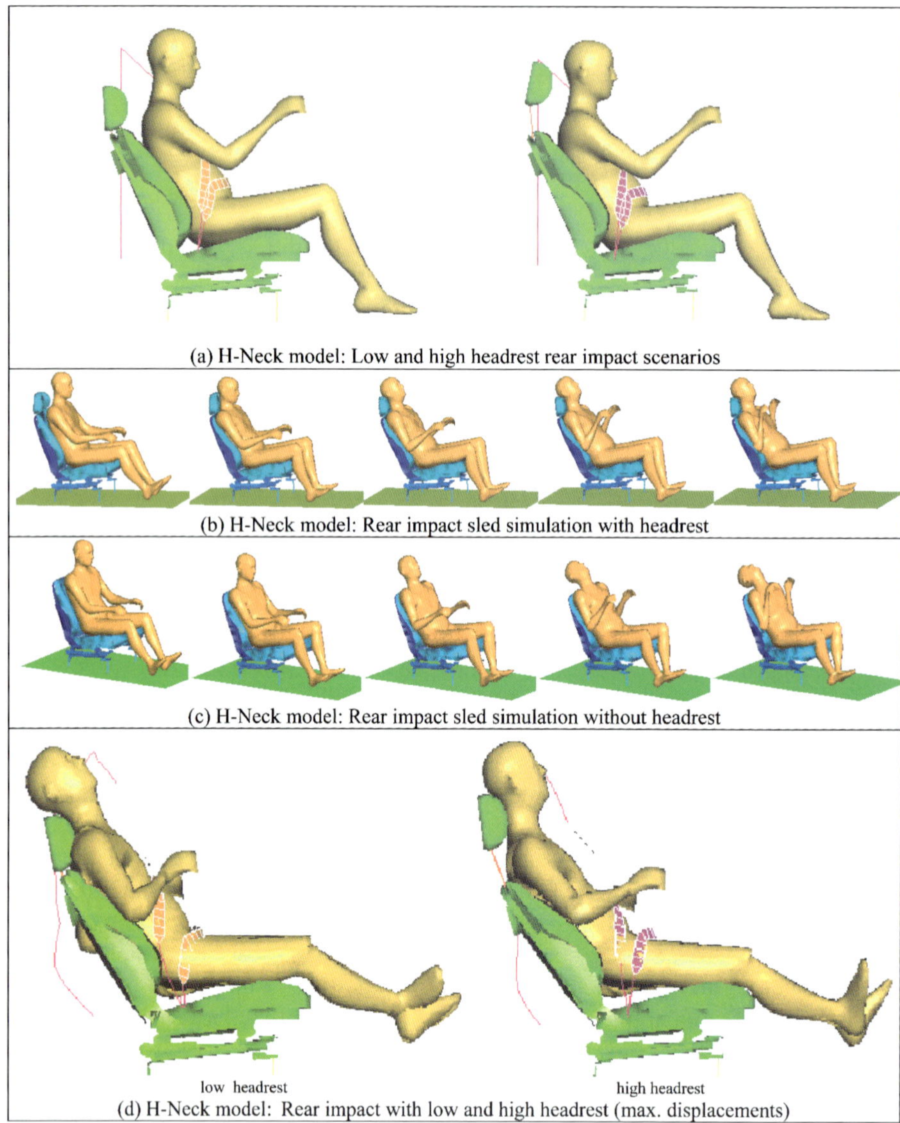

FIG. 9.19. H-Neck model validations: Low and high headrest scenarios.

JAGER, SAUREN, THUNNISSEN and WISMANS [1994] on head/neck models; KULAK, BELYTSCHKO, SCHULTZ and GALANTE [1976] on non-linear behaviour of discs under axial load; KALLIERIS and SCHMIDT [1990] on side impact neck response from cadaver tests; MARKOLF and MORRIS [1974] on components of intervertebral discs; MCELHANEY, PAVER, MCCRACKIN and MAXWELL [1983] on cervical spine compression response; MYERS, MCELHANEY, RICHARDSON, NIGHTINGALE and DOHERTY [1991], MYERS, MCELHANEY, DOHERTY, PAVER and GRAY [1991] on the influence of end conditions (a) and torsion (b) on cervical spine injury; NOVOTNY

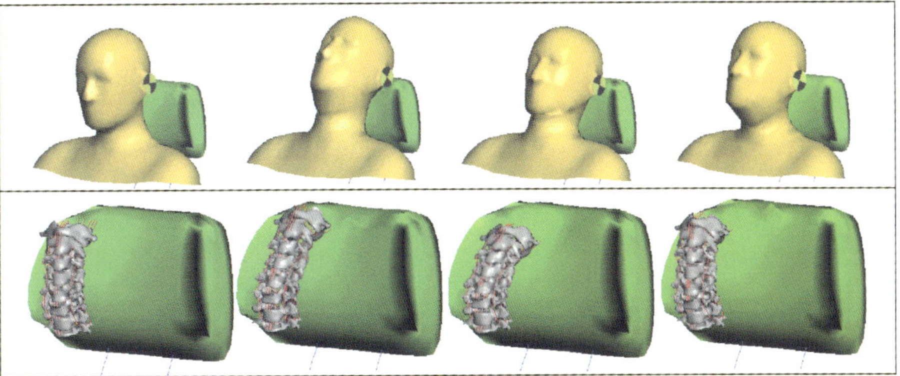

FIG. 9.20. H-Neck model validations: Rear impact whiplash response with deformable headrest.

[1993] on spinal biomechanics; ONO [1999] on spine deformation and vertebral motion from whiplash test volunteers; PANJABI, DVORAK and DURANCEAU [1988] on 3D movements of the upper cervical spine; PENNING [1979], PENNING and WILMARK [1987] on movements and rotations of the cervical spine; PUTZ and PABST [2000]: Sobotta Atlas of Human Anatomy; SONNERUP [1972] on cervical disc analysis in compression; SVENSSON and LÖVSUND [1992] on new dummy neck for rear end collision; TENNYSON and KING [1976a], TENNYSON and KING [1976b] on biodynamic model of the spinal column and measure of muscle action; THUNNISSEN, WISMANS, EWING and THOMAS [1995] on human volunteers in whiplash loading; VIRGIN and LUDHIANA [1951] on physical properties of intervertebral discs; WHITE and PANJABI [1990] on clinical biomechanics of the spine; WINTERS and STARK [1988] on muscle properties; WISMANS, VAN OORSCHOT and WOLTRING [1986] on omnidirectional human head-neck response; WITTEK and KAJZER [1995], WITTEK and KAJZER [1997], WITTEK, HAUG and KAJZER [1999], WITTEK, KAJZER and HAUG [1999], WITTEK, ONO and KAJZER [1999], WITTEK, ONO, KAJZER, ÖRTENGREN and INAMI [2001] on active and passive muscle models; YOGANANDAN, SANCES and PINTAR [1989b] on axial compression of human cadaver and manikin necks.

The following web sites were consulted: http://www.rad.washington.edu/RadAnat/Cspine.html.

9.5. H-Torso: Rib cage, spine and thoracic organs

An overview on the H-Torso model is given in CHOI and LEE [1999a].

Anatomy of the spine and torso. The anatomy of the human spine and thorax is summarized in Fig. 9.22(a)–(f), prepared with the web sites http://www9.biostr.washington.edu/cgi-bin/DA/imageform, (a) and (b), and http://www.nlm.nih.gov/research/visible/image/abdomen_mri.jpg, (c), of the National Library of Medicine (NLM).

The human spine consists in the flexible cervical spine (neck; seven vertebrae), the more rigid thoracic spine (twelve vertebrae with attached ribs), the flexible lumbar spine

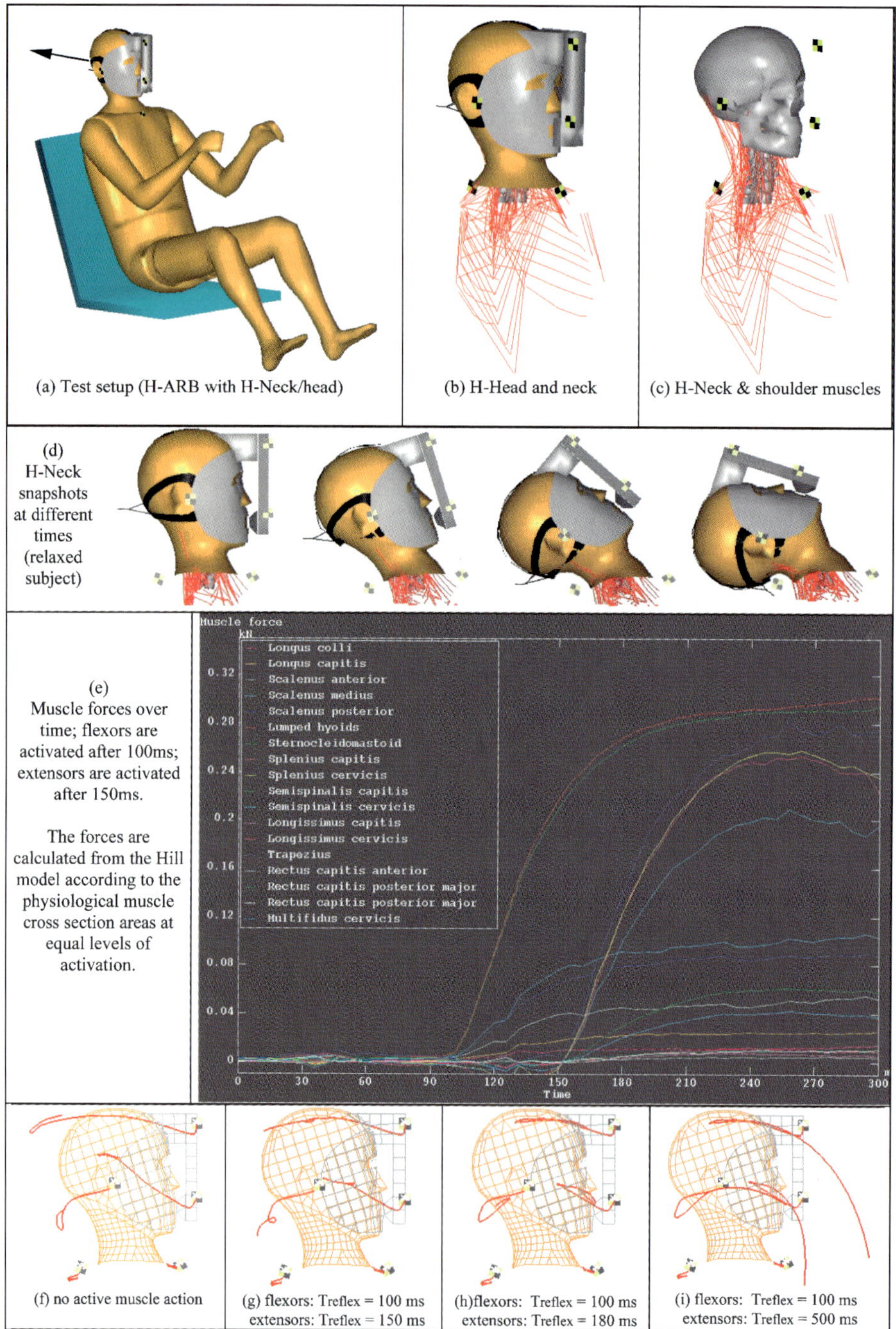

FIG. 9.21. Volunteer test simulation for sudden rearward pulling on subjects head mask (H-Neck).

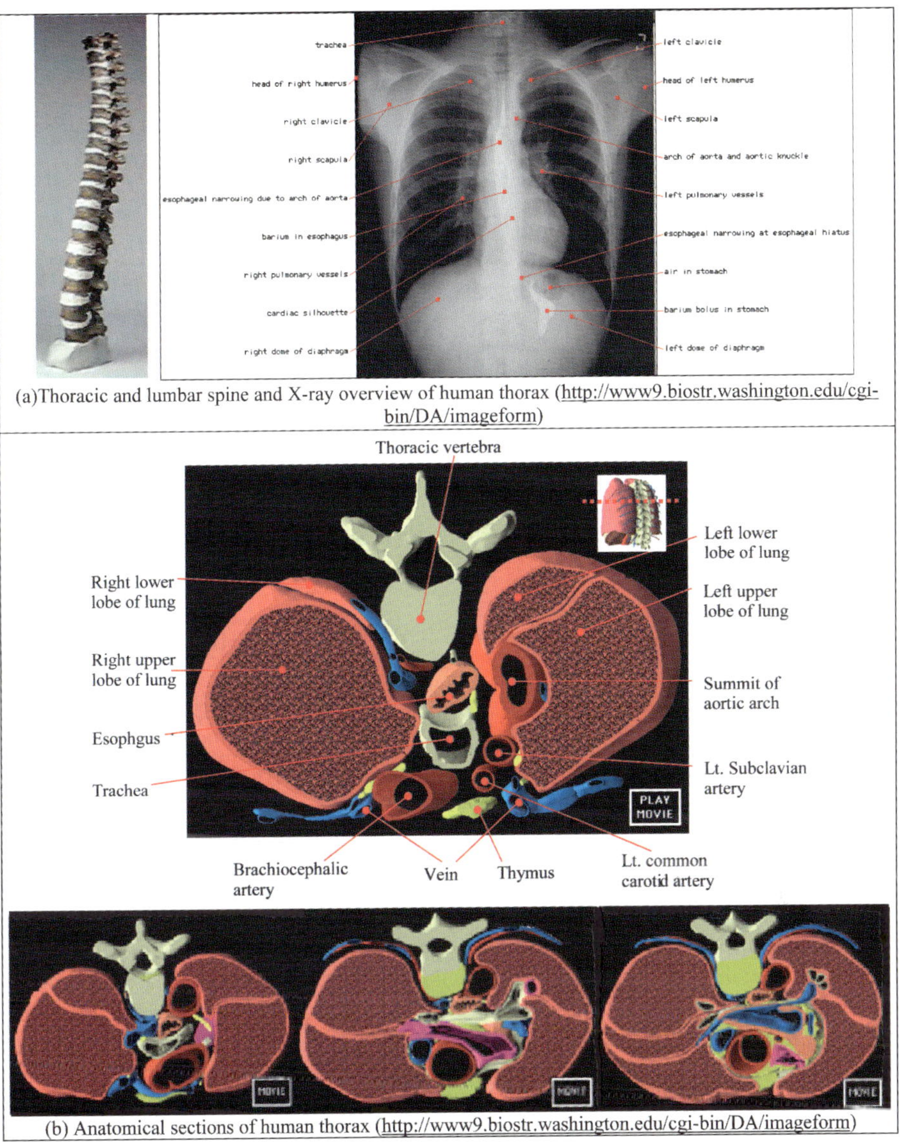

(a)Thoracic and lumbar spine and X-ray overview of human thorax (http://www9.biostr.washington.edu/cgi-bin/DA/imageform)

(b) Anatomical sections of human thorax (http://www9.biostr.washington.edu/cgi-bin/DA/imageform)

FIG. 9.22. Anatomy of the human torso. (Insets (a) right and (b): Reproduced by permission of the University of Washington Digital Anatomist; Inset (c): National Library of Medicine (NLM) public domain information.)

(five vertebrae), the rigid sacrum (five fused vertebrae) and the coccyx (tail bone; four fused segments of bone). The spine supports the upper body, it protects the spinal chord and it provides mobility via its twenty four movable and nine fixed vertebrae. The vertebrae are hinged together by their facet joints and by the inter-vertebral discs. The facet joints, discs and ligaments of the spine limit its overall deformation in bending, torsion and elongation.

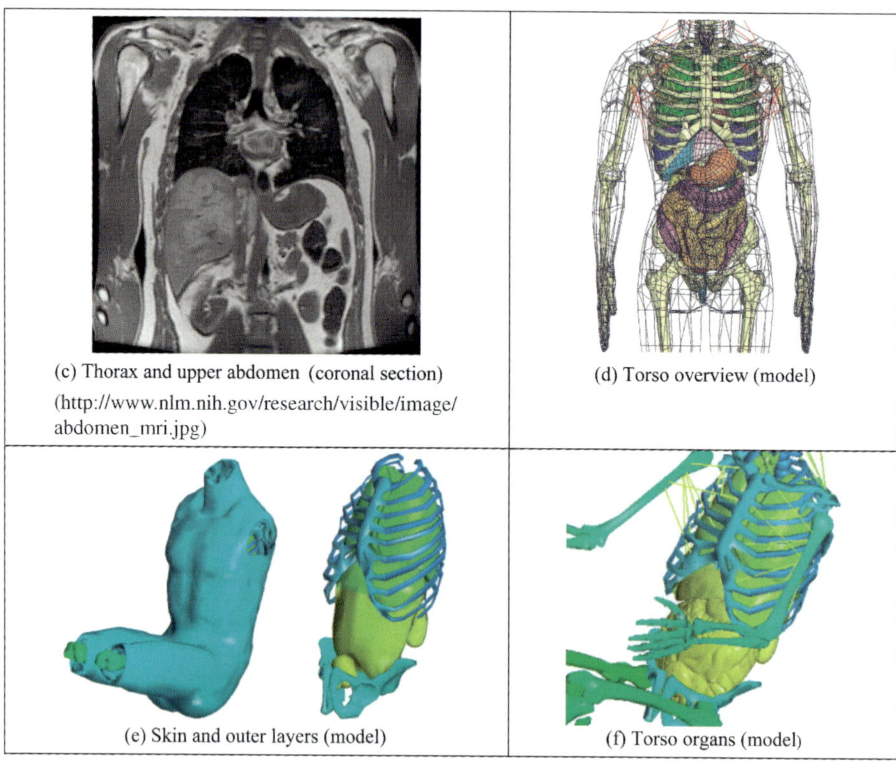

(c) Thorax and upper abdomen (coronal section) (http://www.nlm.nih.gov/research/visible/image/abdomen_mri.jpg)

(d) Torso overview (model)

(e) Skin and outer layers (model)

(f) Torso organs (model)

FIG. 9.22. (*Continued.*)

The human chest mainly consists of the ribs, the thoracic spine and the cardio-pulmonary organs. The upper ten ribs are directly or indirectly connected to the sternum via their costal cartilages, while the lowest two ribs are floating. The twelve ribs of each side articulate posterior with the twelve thoracic vertebrae. Three layers of intercostal muscles connect the ribs. The mediastinum forms the space between the cardio-pulmonary organs, consisting of trachea, lungs, aorta and the heart. The esophagus transverses the thorax. The diaphragm separates the thorax region from the upper abdomen. The thorax serves as a support for the shoulders and the upper extremities. The abdominal cavity, Fig. 9.22(c)–(f), is separated from the thorax by the diaphragm. The abdomen can be subdivided into the upper and lower abdomen. It contains the "solid" organs: liver, spleen, pancreas, kidneys, adrenal glands and ovaries; and the "hollow" organs: stomach, small and large intestines, urinary bladder and the uterus (ROUHANA [1993], HYNCIK [1999b], HYNCIK [2001b], HYNCIK [2002b], HYNCIK [2002c]). These organs are less well protected for impact than the organs of the thorax.

Injury to the thoraco-lumbar spine. According to KING [1993], injury to the bony part of the thoraco-lumbar spine is rare in frontal car accidents. Injury can be classified into anterior wedge fractures and burst fractures of vertebral bodies, dislocations and frac-

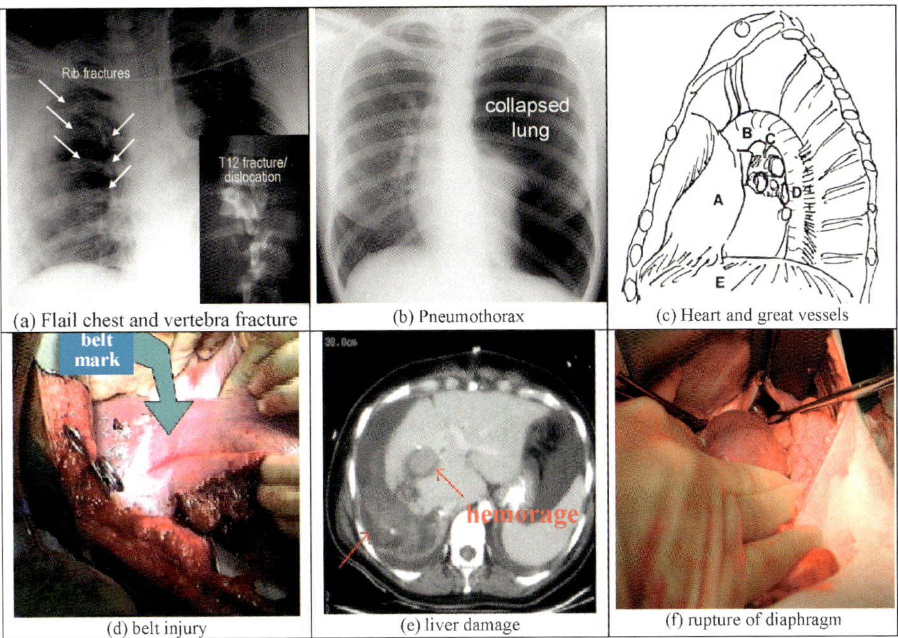

(a) Flail chest and vertebra fracture (b) Pneumothorax (c) Heart and great vessels
(d) belt injury (e) liver damage (f) rupture of diaphragm

FIG. 9.23. Chest and upper abdomen injury. (Insets (a) and (f): Reproduced by permission of Trauma.Org; Inset (b): Reproduced by permission of the Yale University School of Medicine; Inset (c): CAVANAUGH [1993], reproduced by permission of Springer Verlag from: Nahum and Melvin eds., Accidental Injury, Biomechanics and Prevention, Figure 15.4, page 367; Insets (d) and (e): from CIREN William Lehman Injury Research Center at the Ryder Trauma Center Case # 97-018 (public domain material, U.S. Department of Transportation, NHTSA), see NHTSA/CIREN [1997].)

ture dislocations, rotational injuries, Chance fractures, hyper-extension injuries and soft tissue injuries. A typical vertebra fracture is shown in the inset to Fig. 9.23(a). Anterior wedge fractures occur under combined axial compression and bending (pilot ejection) and consist in a lighter injury. Burst fractures threaten the integrity of the cord and occur at higher loads. Dislocations can produce misalignments of the facets between adjacent vertebrae and occur under combined flexion, rotation and postero anterior shear and can involve bone fracture and ligament rupture, threatening the cord. Rotational injuries come from excessive torsion of the spine under simultaneous axial and shear loads, with lateral wedge fractures and a danger of paraplegia. Chance fractures (CHANCE [1948]) occur by over-flexion of the lumbar spine when wearing only a lap belt in a frontal collision and it splits a lumbar vertebra in its transverse plane. If the upper body is properly fixed, excessive flexion over the restraining lap belt cannot occur. Hyper-extension injuries occur in pilot ejection and rarely in car accidents. They may produce bone fracture and ligament rupture. Soft tissue injuries involve the intervertebral discs, the ligaments, the joint capsule, the facet joints and the muscles and tendons attached to the spinal column.

Thorax/abdominal injury. For occupants of passenger cars, most crash injuries of the thorax are given by single and multiple rib fracture ("flail chest"), which result from the compression of the rib cage and the ensuing bending and breaking of the ribs, Fig. 9.23(a) (http://www.trauma.org). Flail chest prevents the rib cage from acting as a coherent protective cage and impairs respiration. Broken ribs can cause injury to the organs inside the rib cage, either from unrestrained large dynamic compression, or from punctures caused by the sharp ends of broken ribs. For the thorax region, pulmonary artery and lung injuries, and liver and heart injuries are frequently encountered. Lung contusions and laceration of lung tissue due to broken ribs can lead to hemothorax (bleeding of the lung tissue) and to pneumothorax (pneumatic collapse of one or both lungs) from punctures of the pleural sac between the lungs and the rib cage, Fig. 9.23(b), respectively (http://info.med.yale.edu/intmed/cardio/imaging/cases/pneumothorax_tension/graphics/rad1.gif).

The heart, Fig. 9.23(c) (from CAVANAUGH [1993]), letter A, can suffer contusion from compression and high rates of compression, and high compression of the sternum can lead to laceration. Rupture of the aorta, B, occurs frequently at its root, near the ligamentum arteriosum (isthmus), C, and at its insertion into the diaphragm, E. In lateral impacts, aorta rupture between its unsupported upper and its more firmly anchored descending branch, D, can result from lateral motions of the heart. The aorta walls act anisotropic in failure, with more frequent failure in transverse direction.

Trauma of the organs of the abdomen, Figs. 9.23(d) and (e) (http://www-nrd.nhtsa.dot.gov/departments/nrd-50/ciren/ciren1.html) and (f) (http://www.trauma.org/~imagebank/chest/images/chest0023c.jpg), may stem from penetrating objects, or blunt impacts, the latter being predominant in transport vehicle accidents, including injuries from the lap belt. The abdominal cavity is filled by the solid organs (liver, spleen, pancreas, kidneys, adrenal glands, ovaries) and by the hollow organs (stomach, large and small intestines, urinary bladder, uterus). These organs are only lightly tethered.

They can slide easily within their membrane envelopes (peritoneum) and they are held in place by their blood supply vessels (kidneys), or by peritoneal folds and ligaments. The large relative mobility of the abdominal organs makes injury sensitive to the attitude of the subject at the time of the accident. Blunt trauma affects the solid organs more than the hollow organs. Side impacts favor renal injury. The action of lap belts frequently leads to minor injuries, while preventing from serious injuries.

Model structure and calibration. The modeling aspects of the human torso are summarized in Figs. 9.24 and 9.25.

The *thoracic and lumbar spine* model is shown in Fig. 9.24.

The *thorax* of the H-Model mainly consists in deformable and damageable rib and rigid spine bones and in the thoracic organs and great vessels with internal air (lungs) or blood modeling, Fig. 9.25(a) (http://www.bartleby.com/107/illus490.html) and (b). The blood-filled vessels are simulated with an incompressible fluid model, while the lungs can compress like airbags. Both models can leak out fluid or compressible air ("bio-bags", cf. Appendix D). If impacts on the thorax are simulated, the abdominal cavity can be represented approximately as an incompressible bio-bag (see Fig. 9.27(b), (d)). In Section 8.5 on the HUMOS model validation, Fig. 8.6(c) demonstrates the effect of

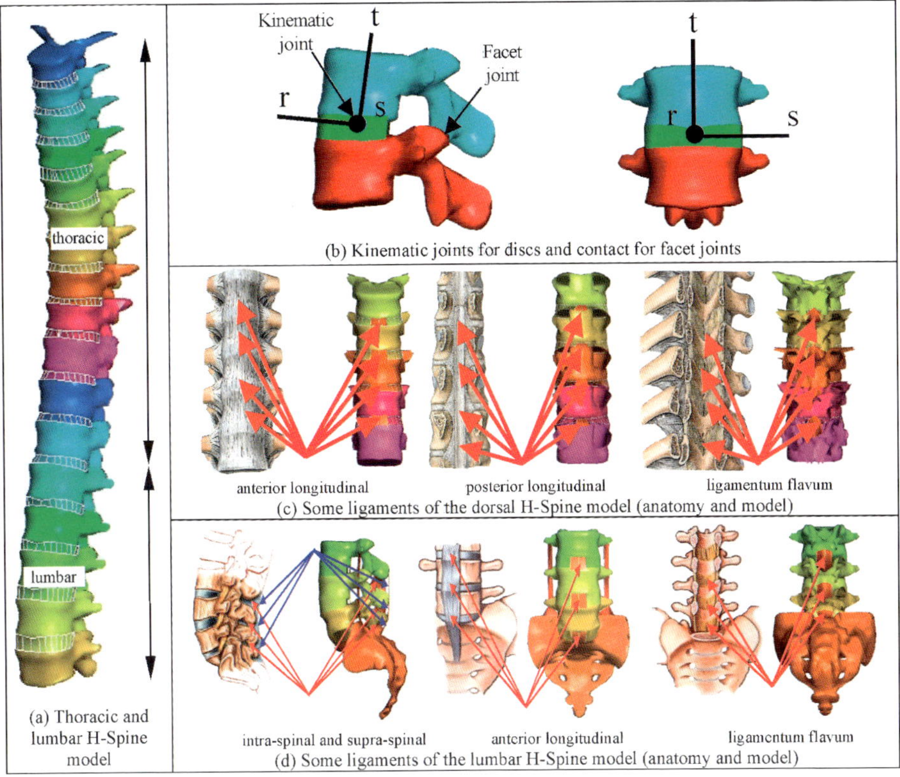

FIG. 9.24. H-Spine model: Thoracic and lumbar spine. (Insets (c) and (d): Anatomy after PUTZ and PABST [2000]: Reproduced by permission of Urban & Fischer Verlag.)

simulating the blood fill of the heart on the Kroell frontal pendulum impact test. If the heart was modeled as an empty bag with elastic skin, the resulting chest deformation was over-predicted.

Fig. 9.25(c) depicts the finite element models of the upper abdominal organs of the H-Model. The hollow organs (stomach, etc.) are modeled with bio-bags, while the solid organs (liver, etc.) are modeled with solid finite elements. An ad hoc modeling of the viscera was introduced in Section 8.5, Fig. 8.8(a), (d). Further investigation will clarify to which detail the visceral organs must be modeled for various purposes. If the upper abdominal organs are studied, the viscera can often be modeled approximately as bio-bags. Similar models were established in collaboration with the University of West Bohemia, Prof. Rosenberg, by HYNCIK [1999b], HYNCIK [2001b], HYNCIK [2002b], HYNCIK [2002c].

Fig. 9.26 summarizes the modeling and calibration aspects of the ribs of the H-Torso model. Fig. 9.26(a) shows the more complex thin shell solid models and the simpler tapered beam model of the ribs. The rib bone material properties, including for rib fracture, were calibrated against test results, Fig. 9.26(b) and (c).

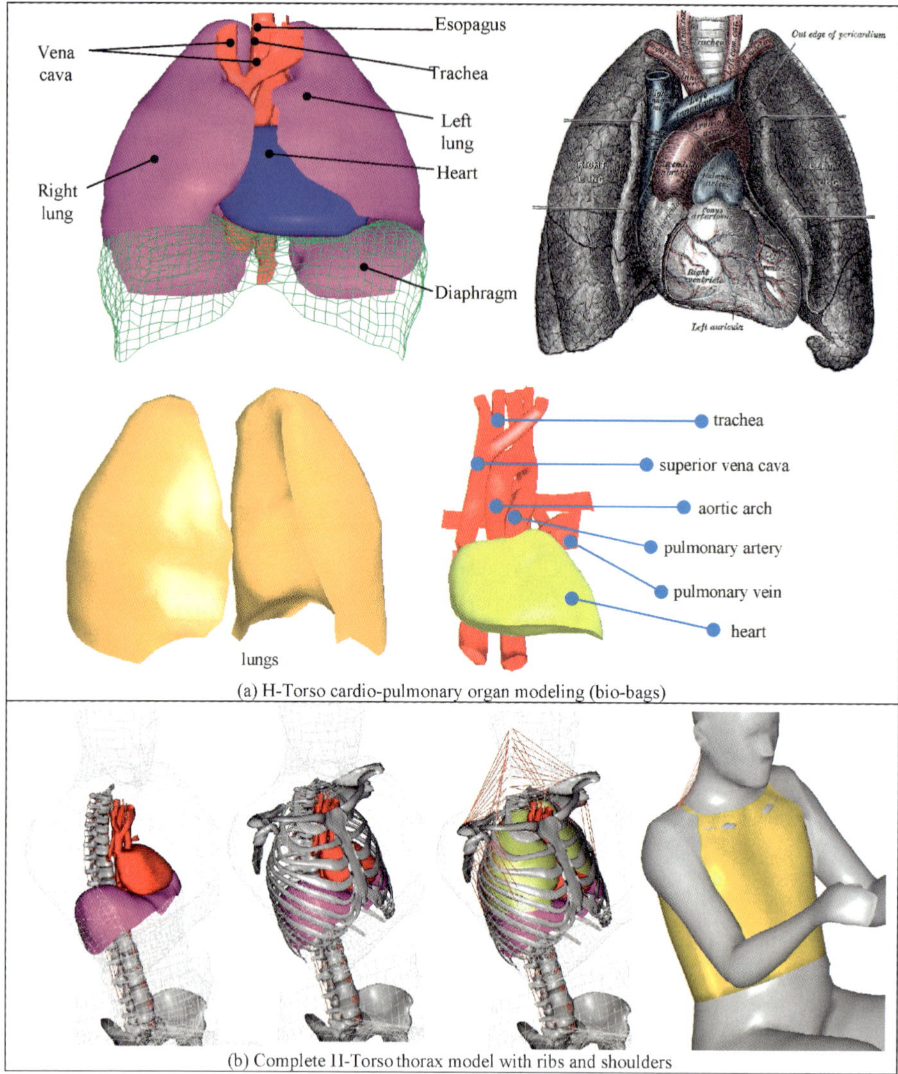

(a) H-Torso cardio-pulmonary organ modeling (bio-bags)

(b) Complete H-Torso thorax model with ribs and shoulders

FIG. 9.25. H-Torso model (thorax and upper abdomen). (Inset (a) anatomical drawing: Reproduced by permission of Bartleby.com, Inc.)

Validation. For a *frontal pendulum impact loading,* Fig. 9.27 shows the response of the H-Torso model after the cadaver (and animal) tests done by KROELL [1971], KROELL, SCHNEIDER and NAHUM [1971], KROELL, SCHNEIDER and NAHUM [1974], KROELL, ALLEN, WARNER and PERL [1986]. Fig. 9.27(a) represents the test setup and the resulting pendulum force-chest deflection corridor. Fig. 9.27(b)–(d) display the H-Torso model response to the applied frontal pendulum impact loading. In (c) and (d) the deformations of the heart, lungs and blood vessels are clearly visible.

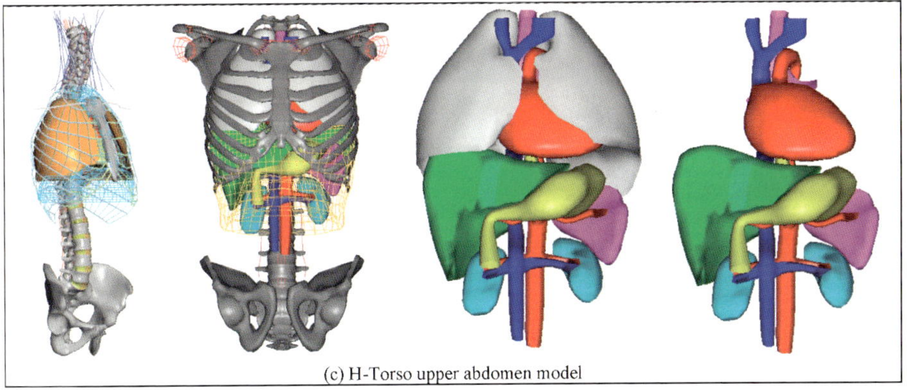

(c) H-Torso upper abdomen model

FIG. 9.25. (*Continued.*)

Fig. 9.27(e) shows the response of the H-Torso rib cage with von Mises surface stress contours added on the ribs and with the failure of some ribs ("flail chest"). Fig. 9.27(f), finally, shows the H-Torso model pendulum force versus thorax deflection curve to fall well into the response corridor obtained from many cadaver tests by Kroell. Similar good responses (not shown here) were obtained for lateral pendulum impacts (Viano tests), similar as shown in Section 8.5, Fig. 8.7, for the HUMOS model segment validation.

For *chest belt loading*, Fig. 9.28 summarizes the response obtained with the H-Torso model.

Fig. 9.28(a) shows the corresponding cadaver test setup (CESARI and BOUQUET [1990], CESARI and BOUQUET [1994]) with the measured cadaver and Hybrid III dummy chest band and thorax deflection versus impact energy results. The chest band device is an elastic steel band that is wrapped tightly around the chest. Strain gauges measure the flexural deformation from which the deflected shape of the tape, and hence the thorax, can be deduced. Fig. 9.28(b) shows the H-Torso model global deformation response to the applied belt loading. The response of the ribs, including fracture near maximum deformation, are shown in Fig. 9.28(c). Fig. 9.28(d) and (e) compare chest band results from the cadaver tests with the response of the H-Torso model, where the chest band results are approximated by plotting the external model section contours at the locations of the upper and lower chest bands, respectively. The agreement is deemed satisfactory, although the model should have duplicated the real chest bands with their true physical properties and contact with the chest for a closer representation.

For *abdomen bar impact tests*, validations of the H-Model similar as the ones carried out in Section 8.5 for the HUMOS model were performed (not repeated here).

References on the H-Torso model. The following references were consulted for the establishment of the H-Torso model (first authors): ALLAIN [1998] on thorax model calibration; ALLEN, FERGUSON, LEHMANN and O'BRIEN [1982] on fracture/dislocation of the lower cervical spine; BEGEMAN, KING and PRASAD [1973] on spinal loads from front/rear acceleration; BELYTSCHKO, KULAK, SCHULTZ and GALANTE [1972], BE-

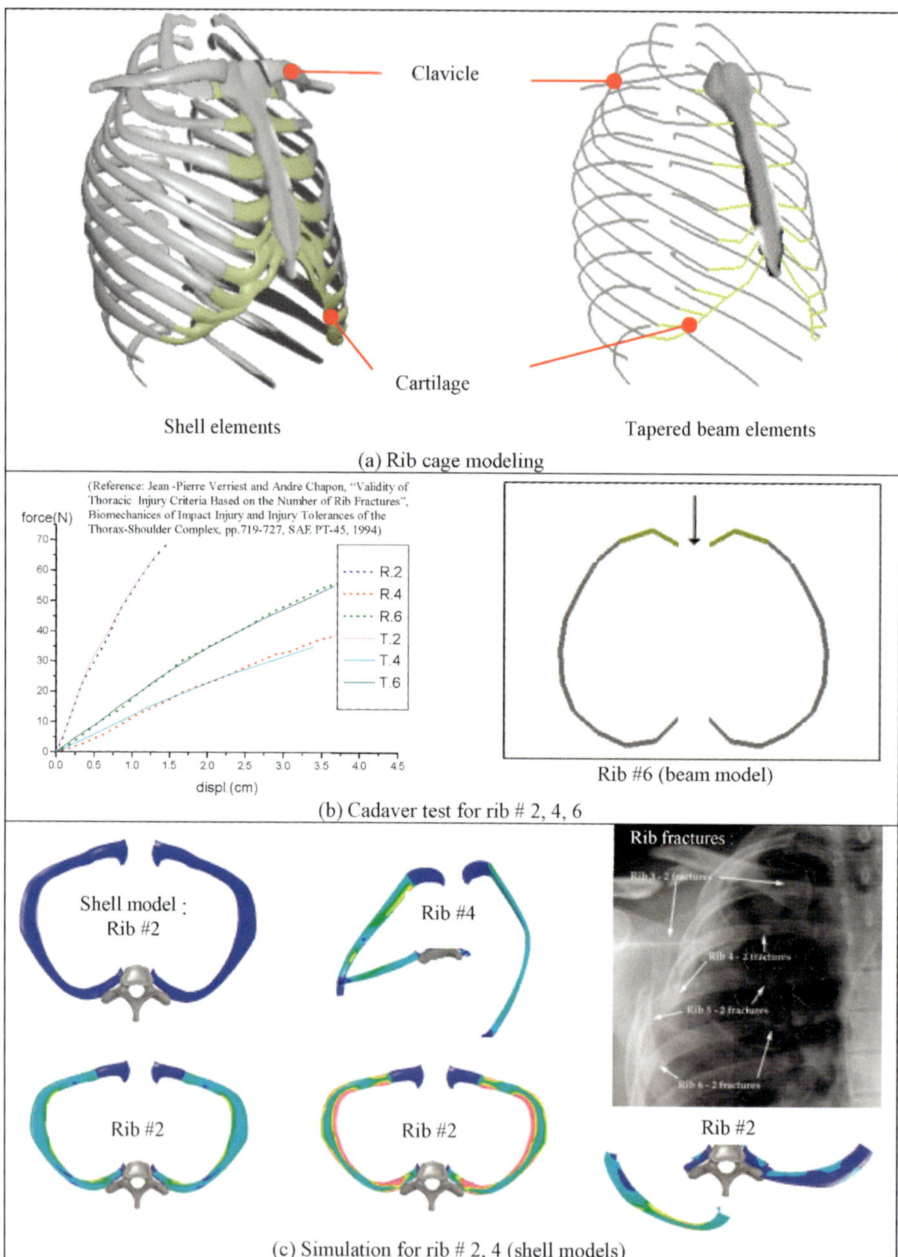

FIG. 9.26. H-Torso rib modeling and calibration. (Inset (b) left diagram: Reprinted with permission from SAE paper number 856027 © 1985 SAE International; Inset (c): X-ray picture: Reproduced by permission of Trauma.Org.)

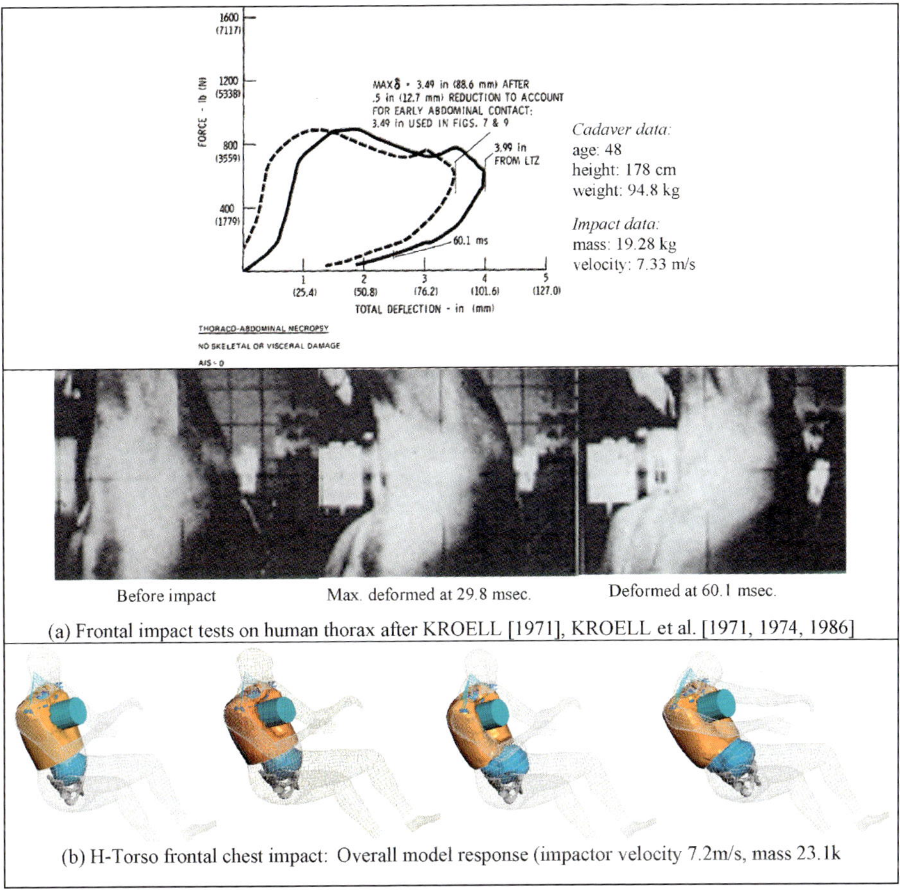

FIG. 9.27. H-Torso frontal chest pendulum impact validation. (Inset (a): Reproduced by permission of The Stapp Association.)

LYTSCHKO, KULAK, SCHULTZ and GALANTE [1974] on FE analysis of intervertebral discs; BOUQUET, RAMET, BERMOND and CESARI [1994] on thorax/pelvis response to impact; CESARI and BOUQUET [1990] and CESARI and BOUQUET [1994] on thoracic belt loading; CHAZAL, TANGUY and BOURGES [1985] on spinal ligament properties; CLEMENTE [1981] (anatomy atlas); COOPER, PEARCE, STAINER and MAYNARD [1982] on thorax trauma and cardiac injury; EPPINGER [1976], EPPINGER, MARCUS and MORGAN [1984] on thoracic injury and dummy development; FUNG and YEN [1984] on lung injury experiments; GOEL, GOYAL, CLARK, NISHIYAMA and NYE [1985] on lumbar spine kinematics; HUELKE, NUSHOLTZ and KAIKER [1986] on thoraco-abdominal biomechanics research; KAPANDJI [1974c] (joint physiology of the trunk and vertebral column); KAZARIAN, BEERS and HERNANDEZ [1979], KAZARIAN [1982] on the spine injuries; KEITHEL [1972] on thoraco-lumbar intervertebral joint deformation due to loads; KLEINBERGER, SUN, EPPINGER, KUPPA and SAUL [1998] on improved injury criteria; KULAK, BELYTSCHKO, SCHULTZ and GALANTE

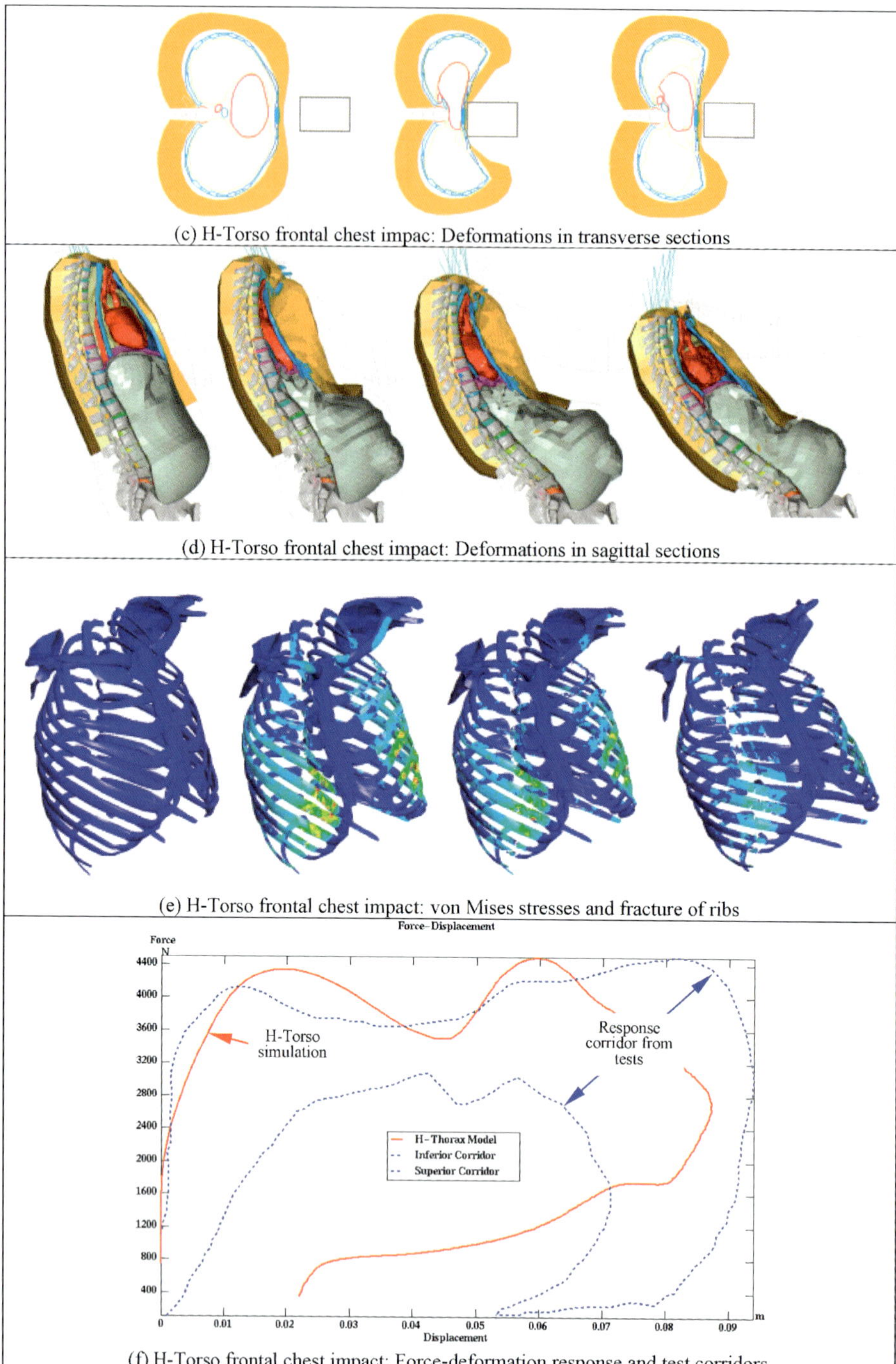

(c) H-Torso frontal chest impac: Deformations in transverse sections

(d) H-Torso frontal chest impact: Deformations in sagittal sections

(e) H-Torso frontal chest impact: von Mises stresses and fracture of ribs

(f) H-Torso frontal chest impact: Force-deformation response and test corridors

FIG. 9.27. (*Continued.*)

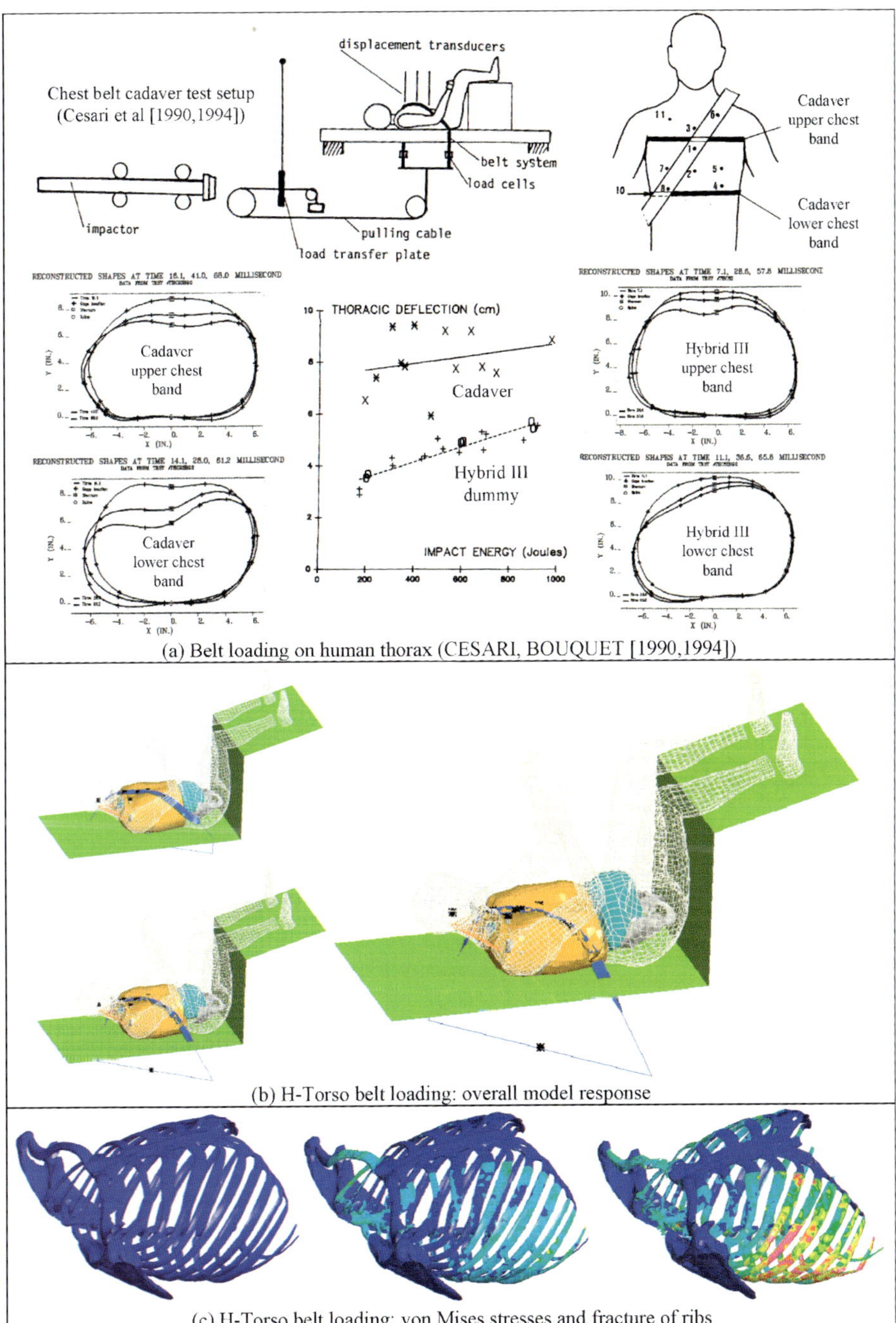

FIG. 9.28. H-Torso belt loading validation. (Inset (a), insets (d) and (e) test results: Reproduced by permission of The Stapp Association.)

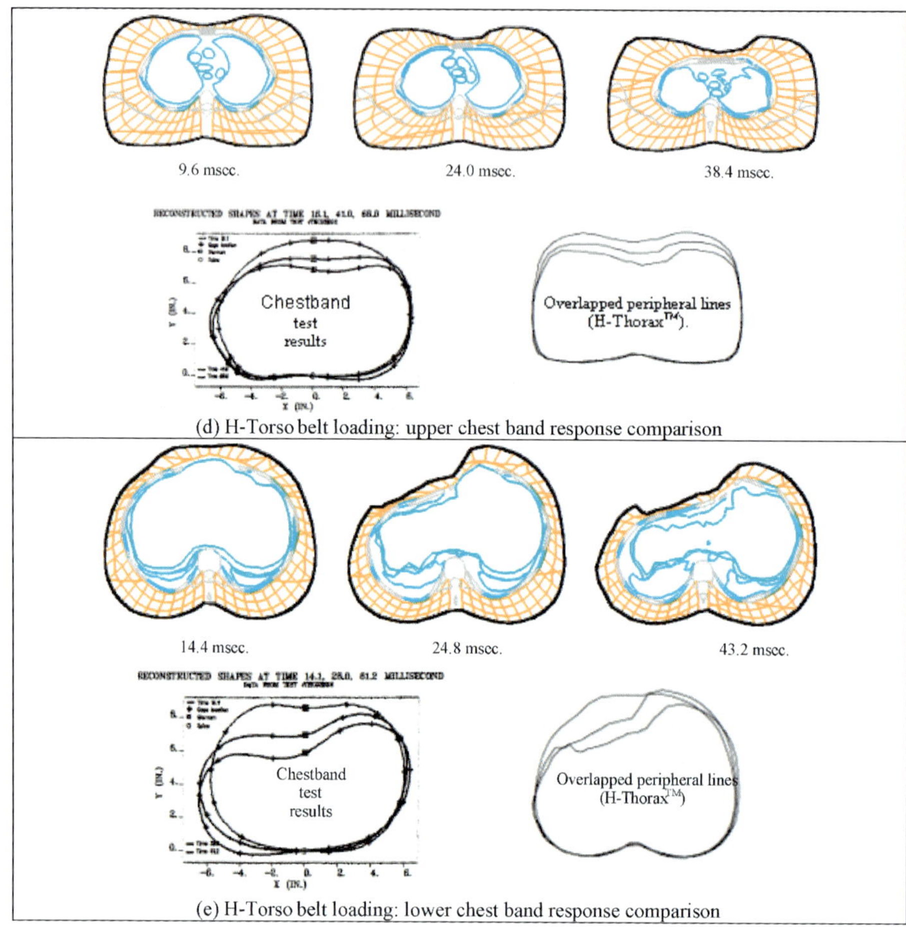

FIG. 9.28. (*Continued.*)

[1976] on non-linear behaviour of discs under axial load; LASKY, SIEGEL and NAHUM [1968] on automotive cardio-thoracic injuries; MAKHSOUS, HÖGFORS, SIEMIEN'SKI and PETERSON [1999] on shoulder strength in the scapular plane; MARKOLF [1972], MARKOLF and MORRIS [1974] on deformations under loads of thoraco-lumbar intervertebral joints and disc structure; MILLER, SCHULTZ, WARWICK and SPENCER [1986] on lumbar spine properties under large loads; MOFFATT, ADVANI and LIN [1971] on experiment and analysis of the human spine; MYKLEBUST and PINTAR [1988] on the tensile strength of spinal ligaments; NAHUM, GADD, SCHNEIDER and KROELL [1970], NAHUM, SCHNEIDER and KROELL [1975] on the response of the human thorax under blunt impact; NEUMANN, KELLER, EKSTROM, PERRY, HANSSON and SPENGLER [1992] on the properties of the human lumbar anterior ligament; NICOLL [1949] on fracture of the torso lumbar spine; NOVOTNY [1993] on spinal biomechanics; PANJABI and BRAND [1976] on the mechanical properties of the human thoracic spine; PLANK, KLEINBERGER and EPPINGER [1994] on FE analysis of thorax/restraint system interaction; POPE, KROELL, VIANO, WARNER and ALLEN [1979] on the postural influence on thoracic impact; PRASAD and KING [1994] on an experimentally validated dynamic spine model; ROAF [1960] on the mechanics of spinal injury; SCHNEIDER, KING and BEEBE [1990] on thorax-abdomen trauma assessment; STALNAKER and MOHAN [1974] on human chest impact protection criteria; STOCKIER, EPSTEIN and EPSTEIN [1969] on seat belt trauma to the lumbar spine; VERRIEST and CHAPON [1994] on thoracic injury criteria and rib fractures; VIANO and LAU [1983], VIANO [1989] on the influence of impact velocity and chest compression in thorax injury (1983) and on the response and injuries blunt lateral impact (1989); WHITE and PANJABI [1978], WHITE and PANJABI [1990] on spine biomechanics; YOGANANDAN, HAFFNER, MALMAN, NICHOLS, PINTAR, JENTZEN, WEINSHEL, LARSON and SANCES [1989], YOGANANDAN and PINTAR [1998] on the trauma of the human spine (1989a) and on the biomechanics of thoracic ribs (1998).

The following web sites were consulted: http://info.med.yale.edu./intmed/cardio/imaging/cases/pneumothorax_tension/graphics/rad1.gif; http://www.bartleby.com/107/illus490.html; http://www.bionetmed.com/; http://www.nlm.nih.gov/research/visible/image/abdomen_mri.jpg; http://www.trauma.org; http://www9.biostr.washington.edu/cgi-bin/DA/imageform.

9.6. *H-UE: Shoulder and arms*

An overview on the human upper extremity model structure and application is given in CHOI, LEE and HAUG [2001a].

Anatomy. The anatomy of the human upper extremity is summarized in Fig. 9.29. The upper extremity consists of the arm and the shoulder complex, Fig. 9.29(a) (after http://www.rad.washington.edu/RadAnat and PUTZ and PABST [2000]). The shoulder (scapula, clavicula) is connected to the thorax via the scapula-thoracic (sliding) joint and via the sterno-clavicular (ball) joint. The upper arm connects to the scapula by the gleno-humeral joint. The scapula connects to the clavicula by the acromio-clavicular joint.

The skeleton of the upper extremity is connected by several systems of muscles, Fig. 9.29(b) (after http://www.fitstep.com/Advanced/Anatomy/Shoulders.htm) and Fig. 9.29(c). These muscles connect the upper arm to the body (latissimus dorsi, pectoralis major), the upper arm to the scapula (supraspinatus, infraspinatus, teres, subscapularis, deltoid, biceps, triceps, coraco-brachialis, brachialis), the scapula to the thorax (trapezius, levator anguli, rhomboids, serratus, pectoralis minor), the upper arm to the clavicula (deltoid, pectoralis major) and the clavicula to the thorax (subclavius, trapezius). The bones of the upper extremity are further connected by ligaments, Fig. 9.29(d) (after http://eduserv.hscer.washington.edu/hubio553/atlas/shjointlig.html). The ligaments that are connecting the clavicula to the ribs are not shown.

Injury. The injury of the upper extremity is discussed in part in LEVINE [1993]. The upper extremity is most frequently hurt through side impacts between two vehicles,

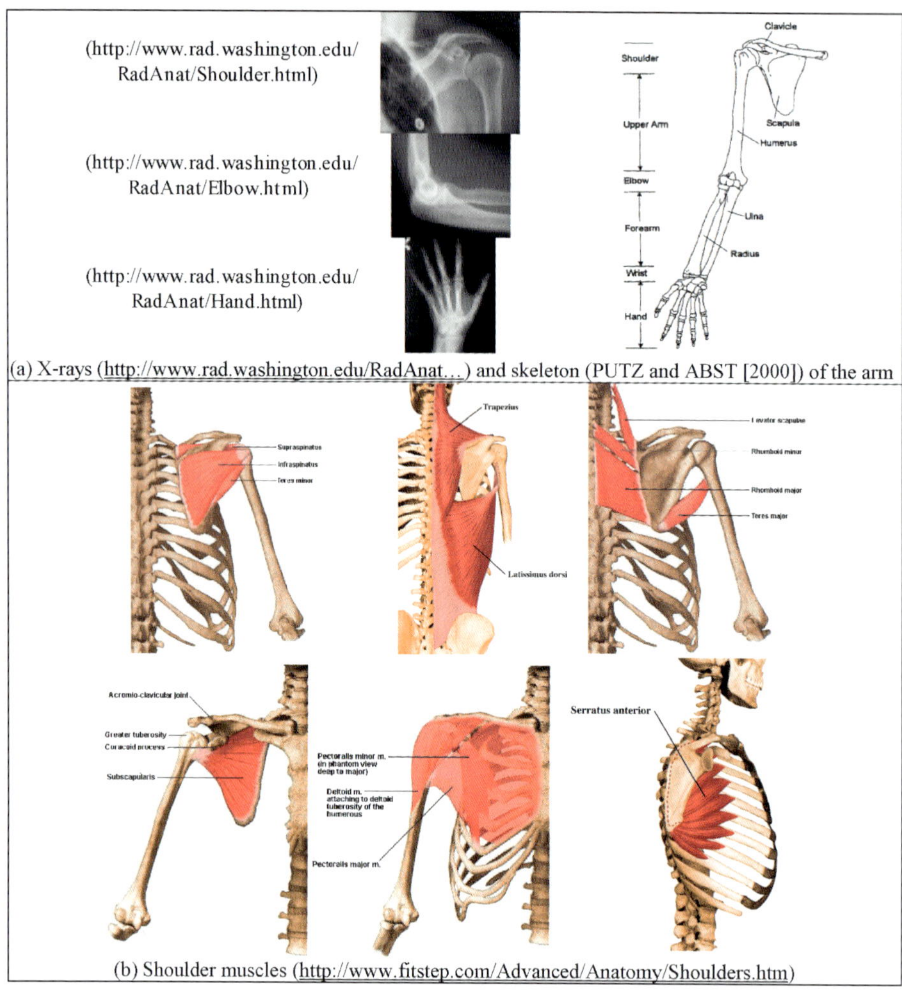

(a) X-rays (http://www.rad.washington.edu/RadAnat...) and skeleton (PUTZ and ABST [2000]) of the arm

(b) Shoulder muscles (http://www.fitstep.com/Advanced/Anatomy/Shoulders.htm)

FIG. 9.29. Anatomy of the human upper extremity. (Inset (a) X-ray pictures: Reproduced by permission of Michael L. Richardson, University of Washington Medical Center, Department of Radiology; Inset (a) anatomical drawing: Reproduced by permission of Urban & Fischer Verlag; Insets (b) and (c): Reproduced by permission of BetterU, Inc.; Inset (d): Reproduced by permission of Carol C. Teitz, University of Washington.)

where the occupant may hit the door in a lateral motion. The forearm may be affected when it is positioned over the steering wheel while the driver side airbag deploys. The upper extremity may also be injured from violent projection by side impact airbags. Because they must fire faster, side impact airbags turn out to be more aggressive than front impact airbags. Upon side impact, the clavicle with its acromio-clavicular and sterno-clavicular joints may be injured through fracture and dislocations. When the lower arm is placed over a driver side airbag deploying from the steering wheel, there may be fractures of the arm. Upon inflation of side airbags located in the back rest, the upper arm

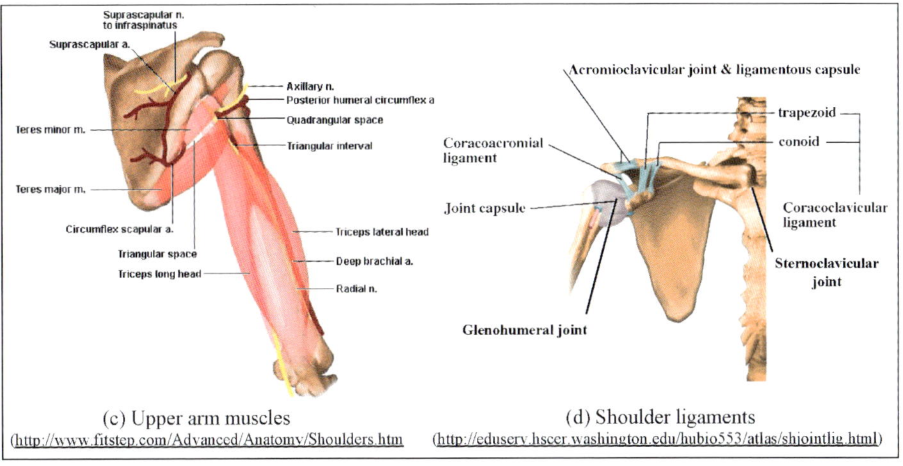

(c) Upper arm muscles
(http://www.fitstep.com/Advanced/Anatomy/Shoulders.htm

(d) Shoulder ligaments
(http://eduserv.hscer.washington.edu/hubio553/atlas/shjointlig.html)

FIG. 9.29. (*Continued.*)

may experience fracture and the gleno-humeral joint may become dislocated. The injuries can be more or less severe, ranging from dislocations to closed fractures to severe open fractures.

Model structure and calibration. The modeling aspects of the human upper extremity are summarized in Fig. 9.30. The upper extremity (H-UE) of the H-Model mainly consists of deformable and damageable bones and flesh padding. The flesh padding uses solid elements (not shown). Active and passive muscle forces are modeled with bar elements using the Hill muscle model. Nonlinear contact interfaces model the cartilage layers on the shoulder and elbow joints. Fig. 9.30(a) shows the H-UE model in its bodily context, while Fig. 9.30(b) shows the skeleton and the attached long muscle bars. Fig. 9.30(c) contains the modeling of the major joint ligaments, which are modeled with short nonlinear bars. Fig. 9.30(d) shows the modeling of surface muscles with two-dimensional finite elements and superimposed muscle bars. Here membrane elements are used as a “matrix” onto which a number of Hill type muscle bars are attached like “fibers” in series and in parallel, as required by the surface curvature and the surface area occupied by each muscle segment. The membranes represent the passive muscle material properties perpendicular to the fibers, while the bars represent the active and passive longitudinal muscle fiber properties. This way the curvature and the sliding contact of the muscles with the rib cage and other parts of the skeleton is well modeled. Fig. 9.30(e) shows the behavior of this model under imposed motions of the arm and shoulder. The next step might be to adapt “composite” materials, with the transverse and bulk properties of the muscles represented by the composite matrix, and the active and passive muscle fiber properties assigned to the fiber phase of the composite materials.

Fig. 9.31 shows, for example, the arm bone material calibration from *three point bending tests* on cadaver arms, after PINTAR, YOGANADAN and EPPINGER [1998] (public domain). The tests, Fig. 9.31(a), were simulated up to fracture, Fig. 9.31(b), and

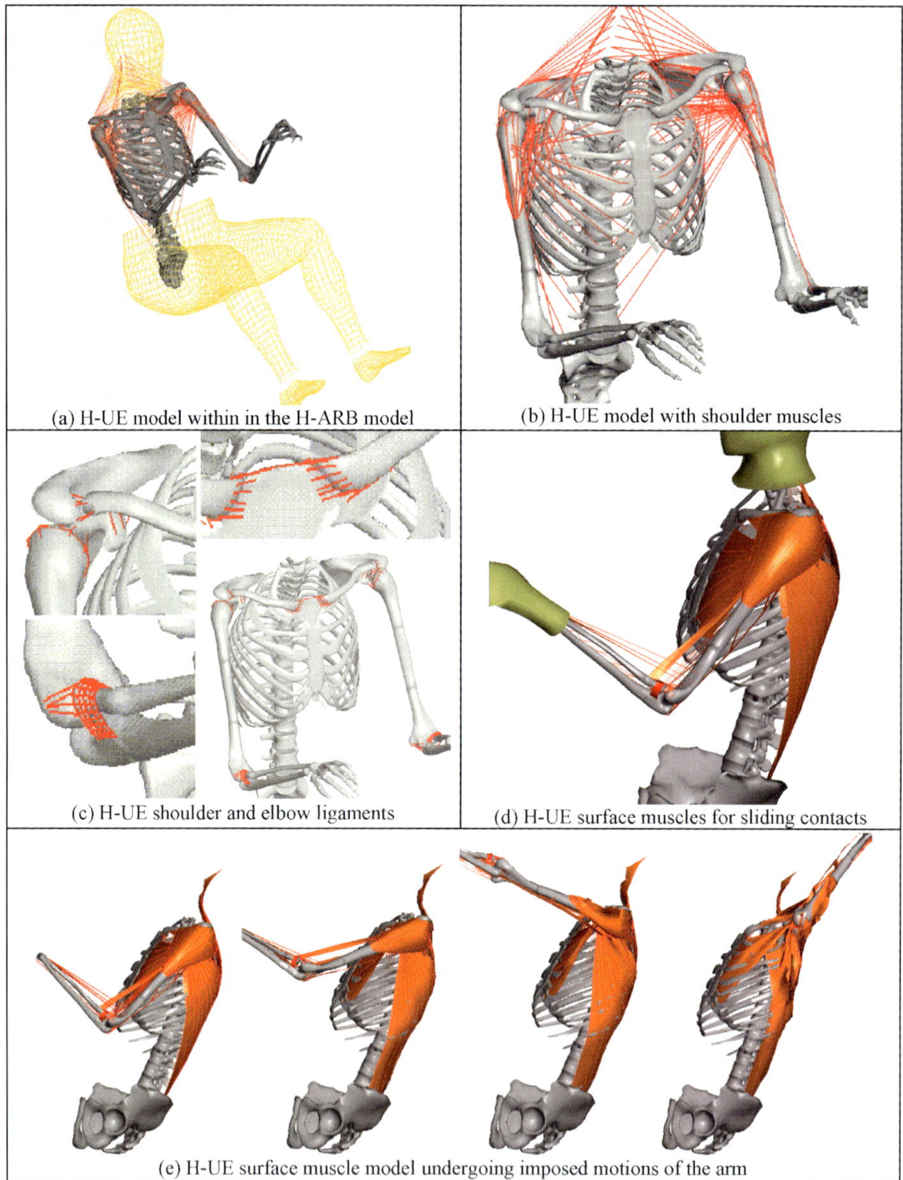

(a) H-UE model within in the H-ARB model

(b) H-UE model with shoulder muscles

(c) H-UE shoulder and elbow ligaments

(d) H-UE surface muscles for sliding contacts

(e) H-UE surface muscle model undergoing imposed motions of the arm

FIG. 9.30. H-UE modeling.

the calculated calibrated and the measured force-displacement curves are compared in Fig. 9.31(c).

Validation. Based on PALANIAPPAN, WIPASURAMONTON, BEGEMAN, TANAVDE and ZHU [1999], Fig. 9.32 shows the calibration of an earlier version of the model on

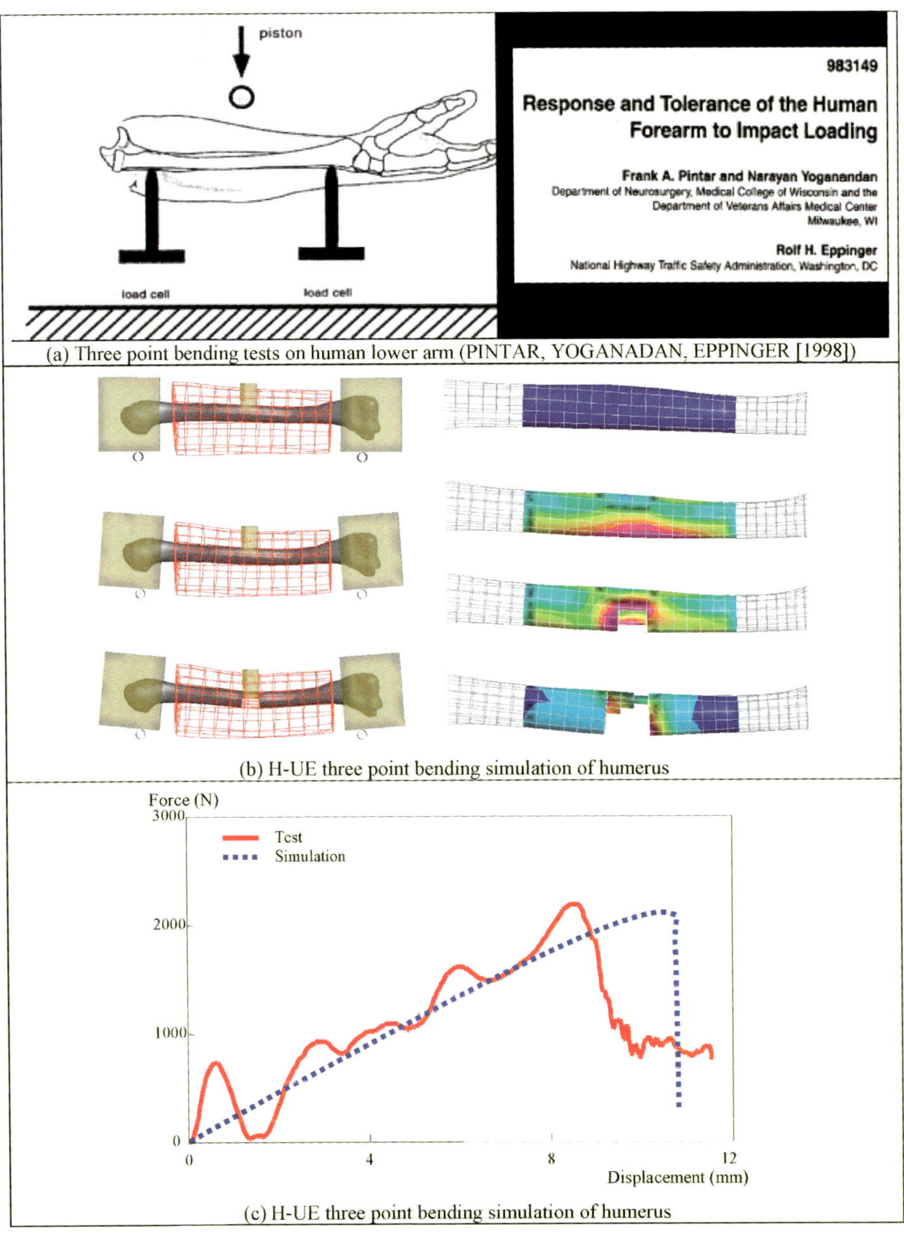

(a) Three point bending tests on human lower arm (PINTAR, YOGANADAN, EPPINGER [1998])

(b) H-UE three point bending simulation of humerus

(c) H-UE three point bending simulation of humerus

FIG. 9.31. H-UE arm bone material calibration.

a *pendulum impact* experiment. The hands were modeled rigid, the deformable bones were modeled with shells and the flesh was modeled with deformable solid elements. The flesh model was calibrated on the pendulum impact tests. The attenuation of the shocks through the deformable flesh was found to be an important element to achieve the required accuracy of the model.

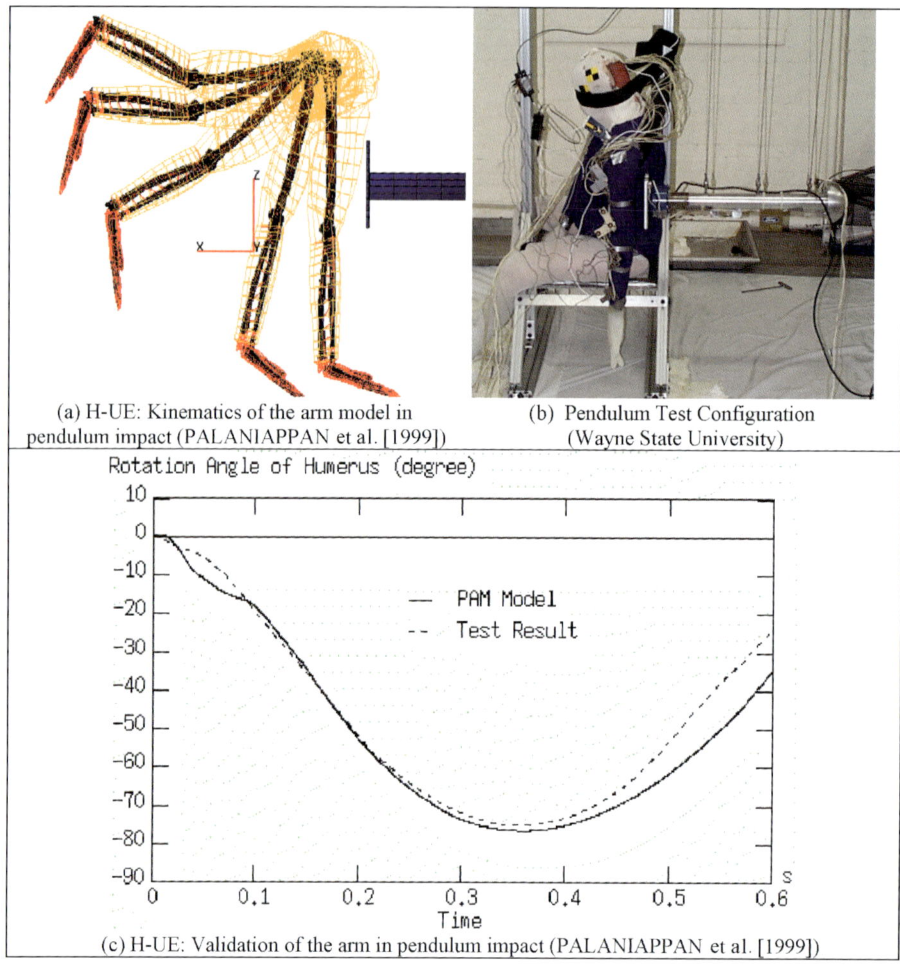

FIG. 9.32. H-UE pendulum impact validation (PALANIAPPAN, WIPASURAMONTON, BEGEMAN, TANAVDE and ZHU [1999]). (Reproduced by permission of The Stapp Association.)

From the same reference, Fig. 9.33 validates the H-UE model on a *side impact airbag out of position deployment*. The tests were performed at Wayne State University. In Fig. 9.33(a) the human arm model was mounted on a model of a fiftieth percentile Hybrid III dummy model. The left arm of the driver is out-of-position and it is projected forward by a deploying side impact airbag. This airbag is mounted in the backrest of the driver seat. In inset (b) the calculated and the measured arm displacement responses are compared.

References on the H-UE model. The following references were consulted when establishing the H-UE model (first authors): ENGIN [1979], ENGIN [1980], ENGIN [1983], ENGIN [1984], ENGIN and PEINDL [1987], ENGIN and CHEN [1989], ENGIN and T'MER [1989] on works on the biomechanics of the human shoul-

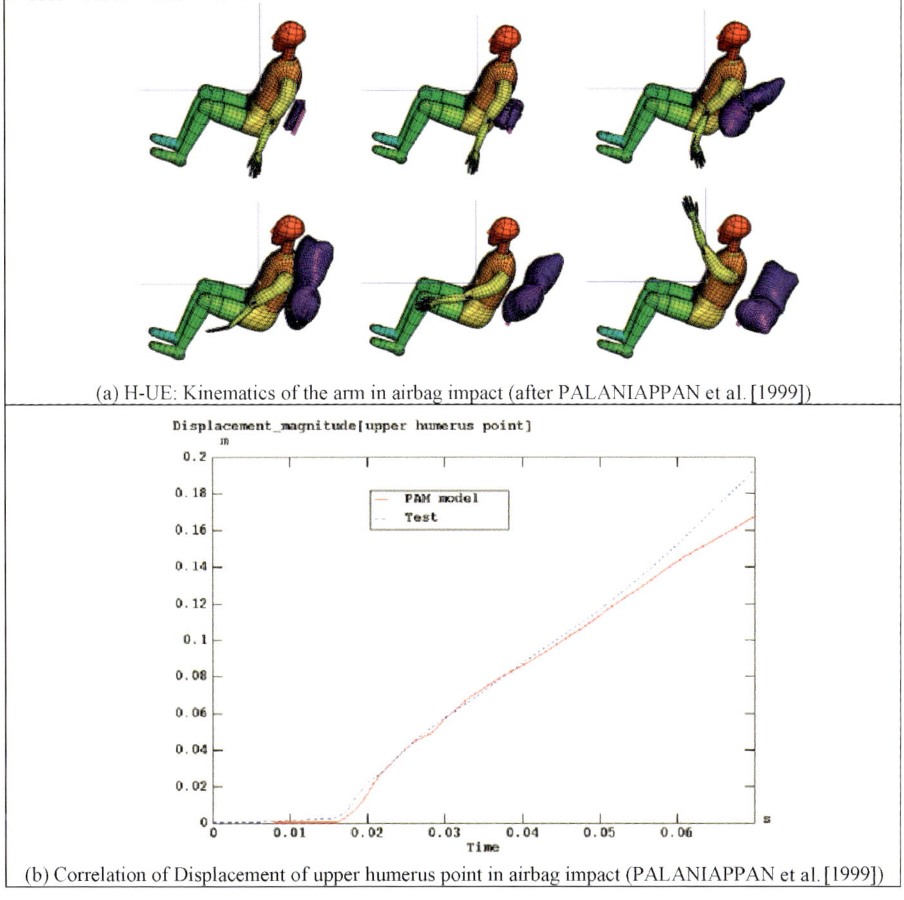

(a) H-UE: Kinematics of the arm in airbag impact (after PALANIAPPAN et al. [1999])

(b) Correlation of Displacement of upper humerus point in airbag impact (PALANIAPPAN et al. [1999])

FIG. 9.33. H-UE airbag impact validation (PALANIAPPAN, WIPASURAMONTON, BEGEMAN, TANAVDE and ZHU [1999]). (Reproduced by permission of The Stapp Association.)

der complex, bone and joint resistance; GOLDSTEIN, FRANKENBURG and KUHN [1993] on the biomechanics of bone; IRWIN [1994] on shoulder and thorax response cadaver tests and their analysis; KAPANDJI [1974a] (upper limb joint physiology); PEINDL and ENGIN [1987] on passive resistive properties of the shoulder complex; PRADAS and CALLEJA [1990] on the non-linear visco-elastic properties of human hand flexor tendon; PUTZ and PABST [2000]: Sobotta Atlas of Human Anatomy; T'MER and ENGIN [1989] on a 3D model of the shoulder complex. The following web sites were used: http://www.rad.washington.edu/RadAnat; http://www.fitstep.com/Advanced/Anatomy/Shoulders.htm; http://eduserv.hscer.washington.edu/hubio553/atlas/shjointlig.html.

9.7. H-LE: Knee-thigh-hip complex

An overview on the H-LE model and applications is provided in YOO and CHOI [1999].

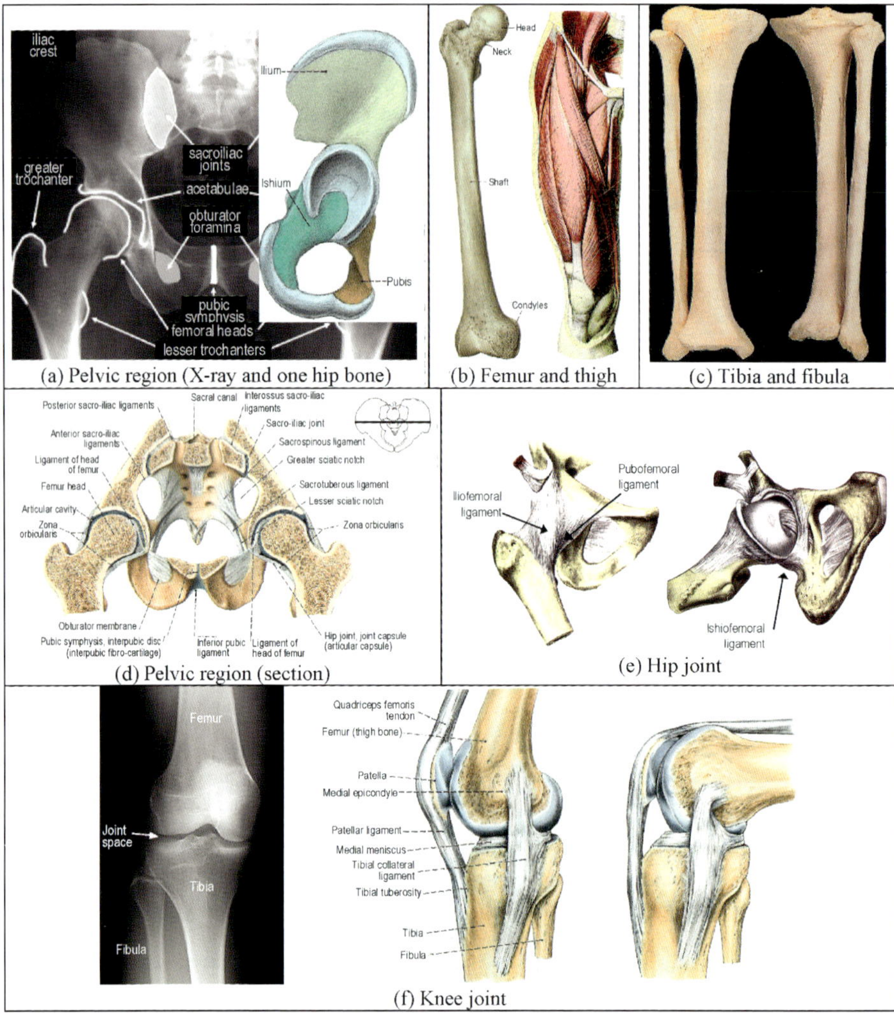

FIG. 9.34. Anatomy of the human lower extremity (without foot/ankle). (Insets (a) and (f) X-ray pictures: Reproduced by permission of Michael L. Richardson, University of Washington Medical Center, Department of Radiology; Insets (a) and (f), anatomical drawings, and insets (b) and (d): Reproduced by permission of Urban & Fischer Verlag; Insets (c) and (e): Figs. 7.16a, p. 205, 7.16d, p. 206 and 8.14c, p. 231 from HUMAN ANATOMY, 4th ed. by Frederic H. Martini, Michael J. Timmons and Robert H. Tallitsch, Copyright © 2003 by Frederic H. Martini, Inc. and Michael J. Timmons.)

Anatomy. The anatomy of the human lower extremity and pelvis is summarized in Fig. 9.34. It consists of hip, upper and lower legs, ankles and foot, articulated by the hip, knee, and ankle joints (the foot/ankle complex is discussed in a separate section). The pelvis supports the spinal column and it contains the sacrum, the coccyx, and the two hip bones, each made of the three fused ilium, ischium and pubic bones, Fig. 9.34(a) (http://www.rad.washington.edu/RadAnat/pelvis.html and PUTZ and PABST [2000]), Fig. 9.34(d) (PUTZ and PABST [2000]) and Fig. 9.34(e) (MARTINI, TIMMONS and TALLITSCH [2003]). These three pelvic bones converge at the acetabulum, the articulation for the head of the femur, (a). The femur (thigh bone), (b) (after PUTZ and PABST [2000]), is the longest bone in the body. Its lower end joins the tibia (shin) to form the knee joint. Its upper end is rounded into a ball (or "head" of the femur) that fits into a socket in the pelvis (the acetabulum) to form the hip joint, (d), (e).

The neck of the femur gives the hip joint a wide range of movement, but it is a point of weakness and a common site of fracture.

The tibia is the inner and thicker of the two long bones in the lower leg. The tibia runs parallel to the smaller and thinner fibula to which it is attached by ligaments. The upper end of the tibia joins the femur to form the knee joint. The lower ends of tibia and fibula form the ankle joint with the medial and lateral malleolus, Fig. 9.34(c) (MARTINI, TIMMONS and TALLITSCH [2003]).

The knee joint, Fig. 9.34(f) (http://www.rad.washington.edu/RadAnat/knee.html and PUTZ and PABST [2000]), is held together by flexible ligaments. The collateral ligaments run along the sides of the knee and limit sideways motion. The anterior cruciate ligament (ACL) limits rotation and relative forward motion of the tibia. The posterior cruciate ligament (PCL) limits relative backward motion of the tibia. The lateral meniscus and medial meniscus are pads of cartilage that cushion the joint, acting as shock absorbers between the bones. The patella is the roughly triangular-shaped bone at the front of the knee joint. It transmits redirecting forces from the quadriceps muscle to the knee joint, which it protects.

Injury. The injury of the lower extremity, Fig. 9.35, is discussed in part in LEVINE [1993]. The most frequent and severe injuries of the knee-thigh-hip complex are skeletal bone fractures and hip dislocations. Fractures of the skeletal bone can be classified roughly in shaft or diaphysis fracture of the long bones and in crushing or compression of the short bones and of the meta- and epiphysis (articular surface) near the articulations of the long bones. The injuries can range from dislocations to closed fractures to severe open fractures.

In front collision, knee or pelvic dislocation can occur if contact is made between the femur and the dash board, Fig. 9.35(a), (b), (d) (Profs. Choi, Poitout) and (e) (http://www.sicot.org/: Library: Online Report E006, February 2002, Figure 1 by: COSTA-PAZ, RANALLETTA, MAKINO, AYERZA and MUSCOLO [2002]). Because of crash forces transmitted along the axis of the femur to the hip joint, knee injuries, such as femoral condyle split and patella fracture, Fig. 9.35(d) and (e), would more likely occur with relatively harder surfaces of the dash board, while hip dislocation, (b), and femur shaft fracture, (c), occur with softer knee padding.

Hip dislocation, Fig. 9.35(b) and femoral neck and acetabulum fractures, (f), are typically seen in a collision of the pelvis with lateral components, causing the femoral head to punch through the acetabulum through direct force application via the greater trochanter (Fig. 9.34(a)). These injuries can occur as a result of direct contact with side structures. This produces a more complex injury pat-

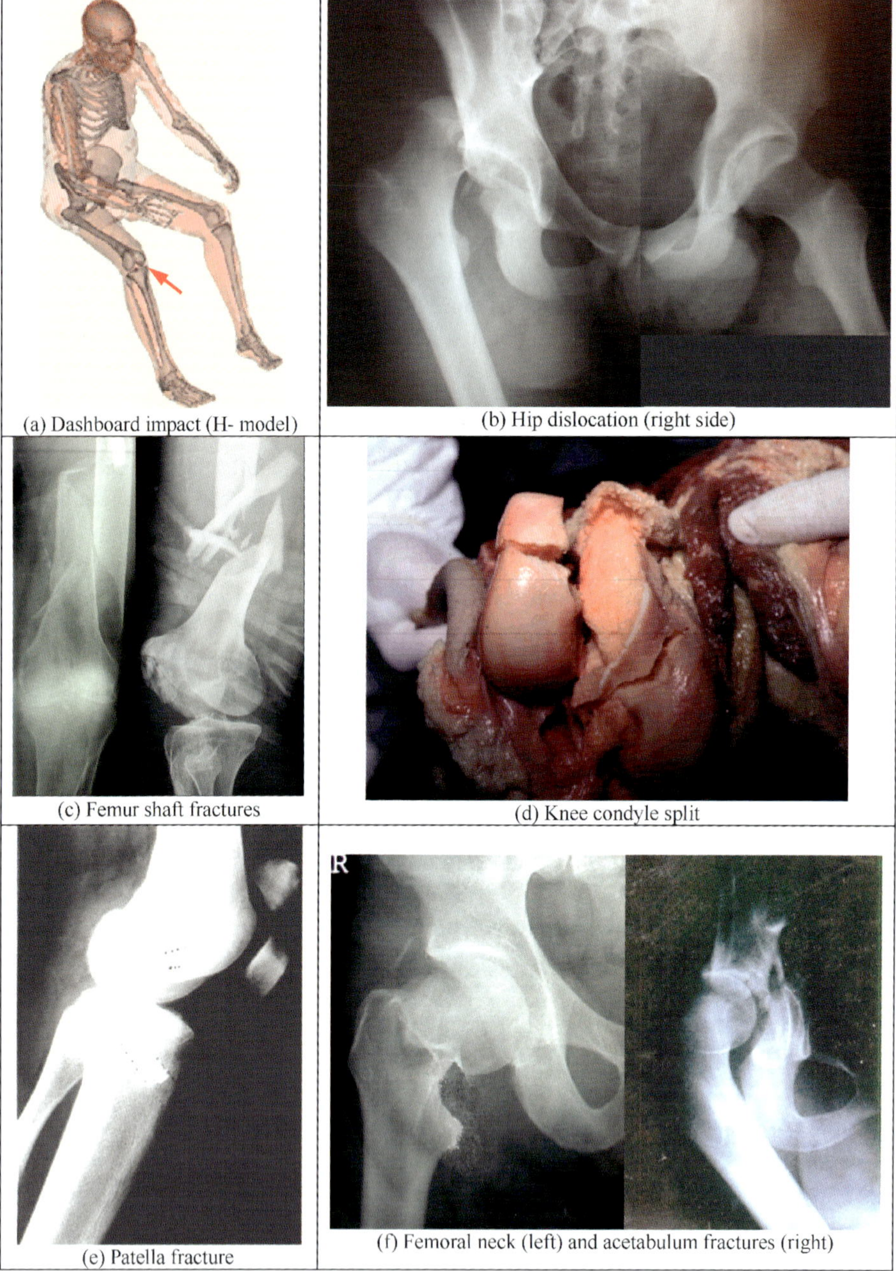

FIG. 9.35. Lower extremity injury. (Insets (b), (c) and (f) left-hand side: Courtesy Prof. D. Poitout, Service de Chirurgie Orthopédique et Traumatologique, Hôpital Nord, Marseille; Inset (d): Private photograph (Prof. H.Y. Choi); Inset (e): Reproduced by permission of SICOT Sociéte Internationale de Chirurgie Orthopedique et de Traumatologie; Inset (f) right-hand side: Reproduced by permission of Glacier Valley Medical Education.)

tern, as shown in Fig. 9.35(f) (Prof. D. Poitout and http://www.glaciermedicaled.com/bone/bonesc11p3.html), that not only involves the articular surfaces of the acetabulum, but also the iliac wing and the pubic rami (arch-like structures) of the pelvis. While bony disruption of the lower limb constitutes a severe injury, additional complications can result from the possibility of soft tissue injury.

Model structure and calibration. The modeling aspects of the human lower extremity are summarized in Fig. 9.36. The H-LE model mainly consists of deformable and damageable bones, including the pelvis bones, and flesh padding. Active and passive muscle forces are modeled with bar elements using the Hill muscle model. Nonlinear contact interfaces model the cartilage layers on hip and knee joints. Fig. 9.36(a) shows the H-LE model in its context, with the attached muscle bars and with the flesh paddings. Fig. 9.36(b) shows the modeling of the cartilage layers of the hip joint with nonlinear contact interfaces. Fig. 9.36(c) contains the modeling of the major hip joint ligaments, which are modeled with nonlinear bars, while Fig. 9.36(d) represents the major knee ligaments. The materials are chosen similar to the materials of the upper extremity.

Validations. From *knee bolster impact* in car frontal crash events, Fig. 9.37 summarizes an investigation made on the injury of the femur (HAYASHI, CHOI, LEVINE, YANG and KING [1996]). Fig. 9.37(a) shows the H-LE model response with distal femur fracture: ("condyle split") due to a hard knee bolster padding material. In that case, the impact force magnitude is high and leads to the observed injury mode. In Fig. 9.37(b) a soft knee bolster padding material was applied, which leads to the observed hip dislocation, because the femur axial force magnitude is too low to cause fracture, but too high over too much time for the hip joint to remain intact. Fig. 9.37(c) represents the same impact using a knee bolster padding material of intermediate stiffness, that leads to the observed femur shaft fracture. The latter injury is deemed to be the least damaging and best healing femur/hip injury. Fig. 9.37(d), finally, compares experimental and calculated impact force time histories.

For a *side impact* scenario, Fig. 9.38 shows the displacements and the deformations of the pelvis (upper pictures) and the pelvic bone with von Mises stress contours and fracture of the pelvic bone (lower pictures).

References on the H-LE model. The H-LE model was built under consultancy of the following references: BACH, HULL and PATTERSON [1997] on the measurement of strain in the anterior cruciate ligament; BEDEWI, MIYAMOTO, DIGGES and BEDEWI [1998] on the human femur FE impact and injury analysis; CAVANAUGH, WALILKO, MALHOTRA, ZHU and KING [1990] on the biomechanical response and injury tolerance of the pelvis in side impact tests; CESARI, BERMOND, BOUQUET and RAMET [1994] on testing and simulation of impacts on the human leg; DALSTRA and HUISKES [1995] on the load transfer across the pelvic bone; DOSTAL [1981] on a 3D biomechanical model of the hip musculature; ENGIN [1979], ENGIN and CHEN [1988a], ENGIN and CHEN [1988b] on the kinematics and passive resistances of the hip joint; FUKUBAYASHI and KUROSAWA [1980] on the contact area and the pressure distribution of the knee joint; HAYASHI, CHOI, LEVINE, YANG and KING [1996] on

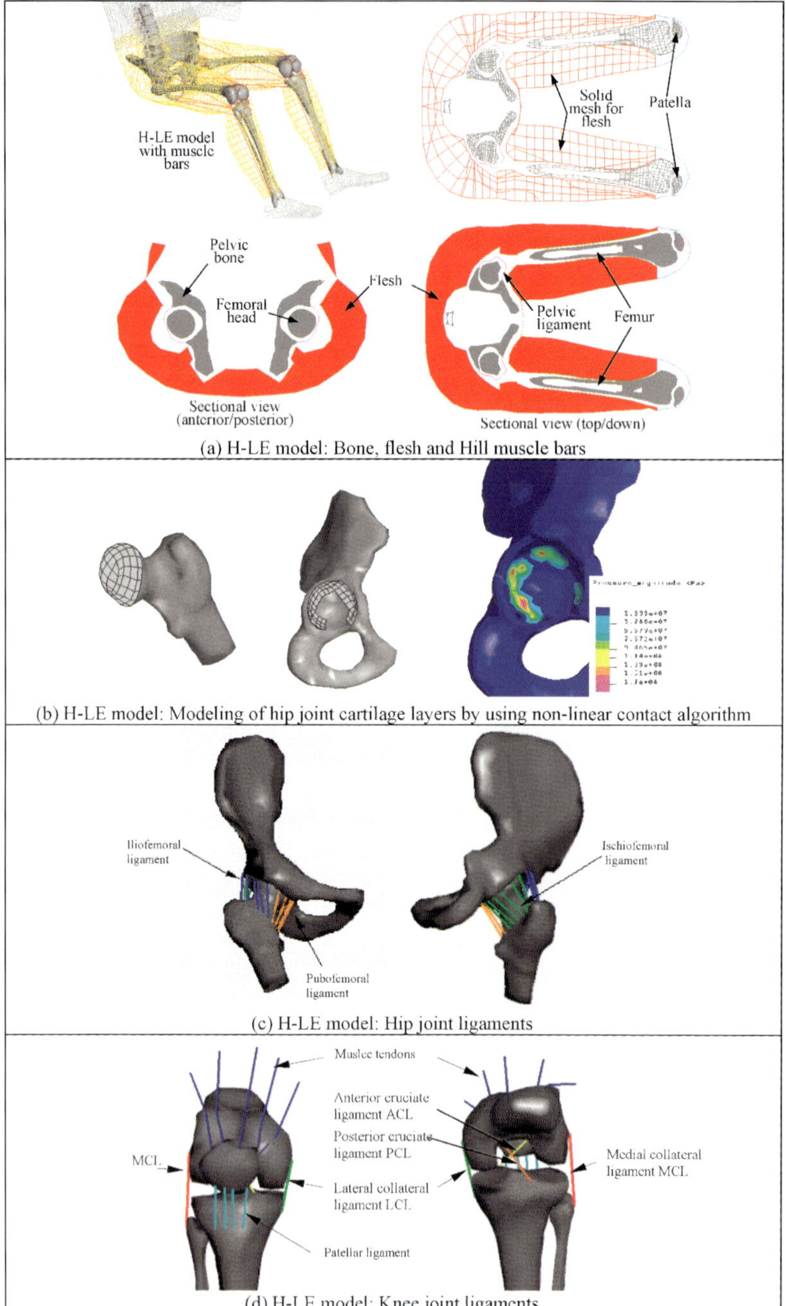

(a) H-LE model: Bone, flesh and Hill muscle bars

(b) H-LE model: Modeling of hip joint cartilage layers by using non-linear contact algorithm

(c) H-LE model: Hip joint ligaments

(d) H-LE model: Knee joint ligaments

FIG. 9.36. H-LE model overview.

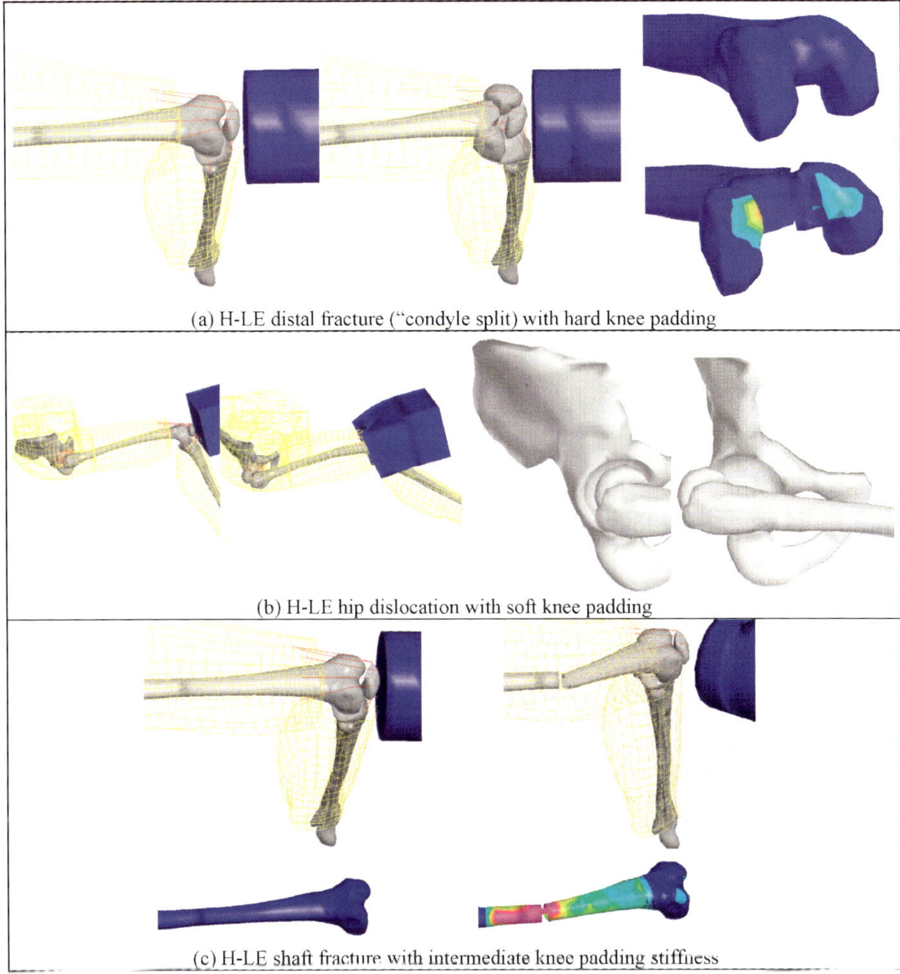

(a) H-LE distal fracture ("condyle split) with hard knee padding

(b) H-LE hip dislocation with soft knee padding

(c) H-LE shaft fracture with intermediate knee padding stiffness

Fig. 9.37. H-LE model knee bolster impact validations (test from Hayashi, Choi, Levine, Yang and King [1996]).

the experimental and analytical study of frontal knee impact; Kajzer [1991] on the impact biomechanics of knee injury; Kapandji [1974b] (lower limb joint physiology); King [1993] on the injury of the thoraco-lumbar spine and pelvis; Kress, Snider, Porta, Fuller, Wasserman and Tucker [1993] on the human femur response to impact loading; Levine [1993] on injury to the extremities; Martin and Thompson [1986] on Achilles tendon rupture; Martini, Timmons and Tallitsch [2003]: Human Anatomy; Momersteeg, Blaskevoort, Huiskes, Kooloos and Kauer [1996] on the mechanical behaviour of human knee ligaments; Nyquist, Cheng, El-Bohy and King [1985] on the tibia bending strength and response; Pattimore, Ward, Thomas and Bradford [1991] on the nature and causes of lower limb in-

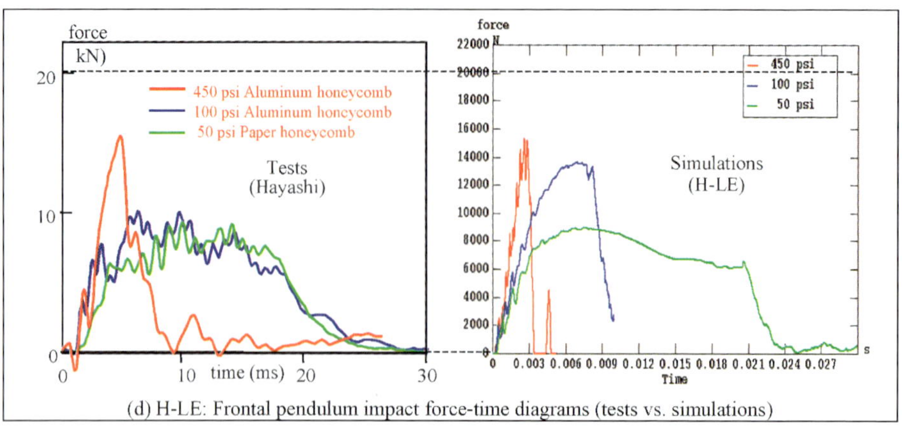

(d) H-LE: Frontal pendulum impact force-time diagrams (tests vs. simulations)

FIG. 9.37. (*Continued.*)

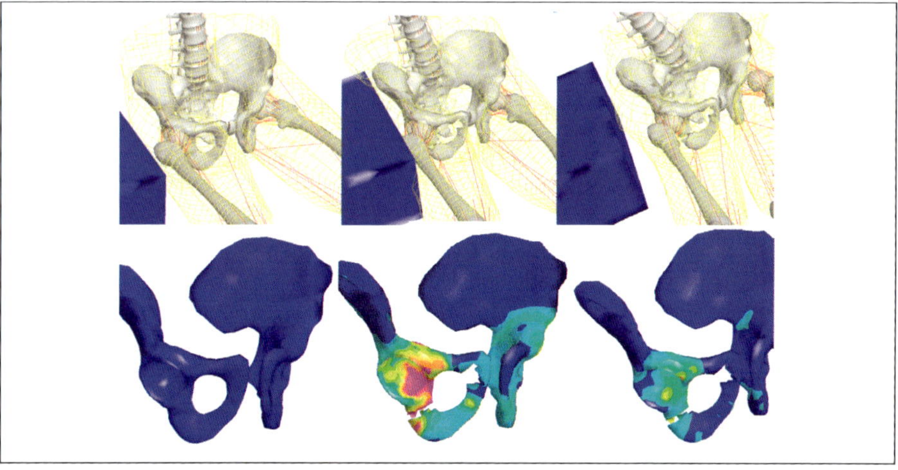

FIG. 9.38. H-LE model side impact validations (pelvis deformation and stress and fracture locations).

juries in car crashes; PORTIER, TROSSEILLE, LE COZ, LAVASTE and COLTAT [1993] on lower leg injuries in real-world frontal car accidents; PUTZ and PABST [2000]: Sobotta Atlas of Human Anatomy; RENAUDIN, GUILLEMOT, PÉCHEUX, LESAGE, LAVASTE and SKALLI [1993] on a 3D FE model of the pelvis in side impacts; ROHEN and YOKOCHI [1983]: Color Atlas of Anatomy; STATES [1986] on adult occupant injuries of the lower limb; WYKOWSKI, SINNHUBER and APPEL [1998] on a finite element model of the human lower extremity in frontal impact; YANG and LÖVSUND [1997] on a human model for pedestrian impact simulation. The following web sites were consulted: http://www.rad.washington.edu/RadAnat/pelvis.html; http://www.rad.washington.edu/RadAnat/knee.html; http://www.sicot.org/; http://www.glaciermedicaled.com/bone/bonesc11p3.html.

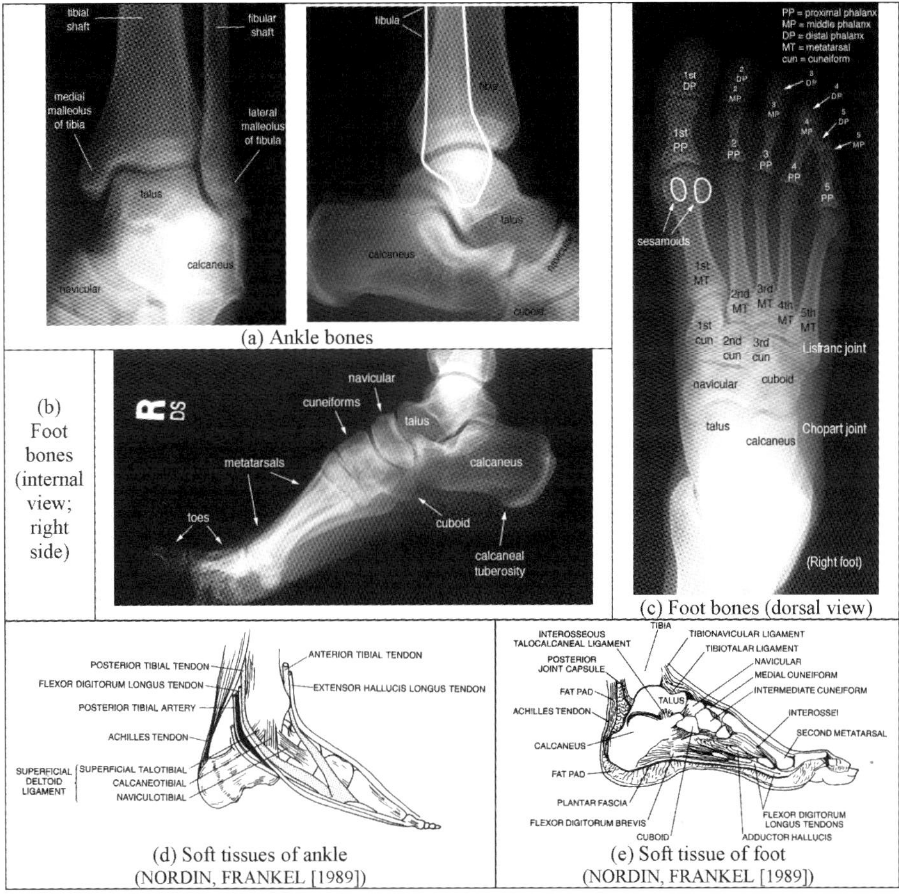

FIG. 9.39. Anatomy of the human foot/ankle complex. (Insets (a)–(c): Reproduced by permission of Michael L. Richardson, University of Washington Medical Center, Department of Radiology; Insets (d) and (e): Reproduced by permission of Lippincott, Williams & Wilkins.)

9.8. *H-Ankle&Foot*

Overviews on the H-Ankle&Foot model structure and applications are given in BEAUGONIN, HAUG, MUNCK and CESARI [1995], BEAUGONIN, HAUG and CESARI [1996], BEAUGONIN, HAUG, MUNCK and CESARI [1996], BEAUGONIN, HAUG and CESARI [1997].

Anatomy. The anatomy of the human foot/ankle complex is summarized in Fig. 9.39. The skeleton of the foot consists in the short bones of the tarsus (a) (http://www.rad.washington.edu/RadAnat/AnkleMortiseLabelled.html and http://www.rad.washington.edu/RadAnat/AnkleLaterallLabelled.html), namely the talus (or astragalus), which establishes the articulated connection between the foot and the leg (tibia, fibula), the calcaneous, the navicular (or scaphoid), the cuboid and the cuneiform bones,

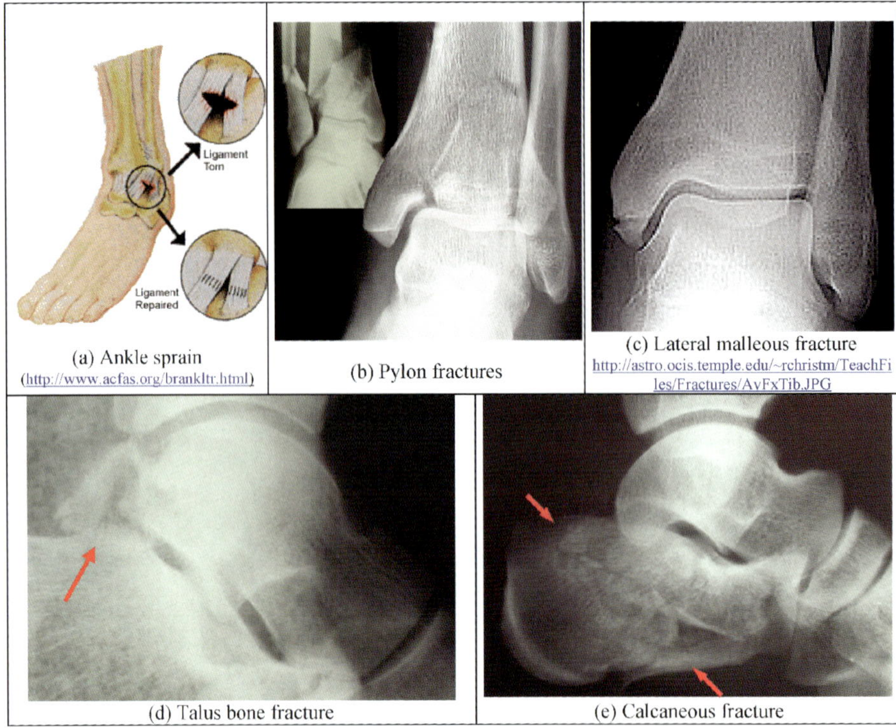

FIG. 9.40. Foot and ankle injury. (Inset (a): Reproduced by permission of ACFAS American College of Foot and Ankle Surgeons; Insets (b), (d) and (e): Courtesy Prof. D. Poitout, Service de Chirurgie Orthopédique et Traumatologique, Hôpital Nord, Marseille; Inset (c): Courtesy of Robert A. Christman, D.P.M., Philadelphia, PA.)

and of the metatarsal and the phalanx bones (b) (http://www.rad.washington.edu/RadAnat/FootLateralLabelled.html), (c) (http://www.rad.washington.edu/RadAnat/FootAPLabelled.html). Besides the major articulations of the talus bone, the Lisfranc and the Chopart joints provide minor mobility (c). These bones are connected by numerous ligaments and soft tissues, (d) and (e) (NORDIN and FRANKEL [1989]).

Injury. Injury of the foot/ankle complex, Fig. 9.40 (see indicated web sites), is discussed in part in LEVINE [1993]. As much as one third of all surviving vehicle crash victims sustain lower limb injury, where belt use does not alter significantly the risk. These injuries are not life threatening, but cause extensive health care cost and long periods of recovery. Ankle and foot injuries are mainly attributed to foot well intrusion in frontal crashes. They can be classified as *skeletal injury*, such as ankle fractures (malleolar or bimalleolar fractures; tibial pylon fractures; talar fractures) and foot fractures (e.g., the metatarsal bones; the calcaneus; the cuboid) and *internal injury,* such as joint injury (Lisfranc and Chopart joints), ligamentous injury (sprains, tears), tendon injury (Achilles tendon damage or rupture).

MORGAN, EPPINGER and HENNESSEY [1991] described six mechanisms which they consider the most frequent in vehicle crash (leg trapped between floor and instru-

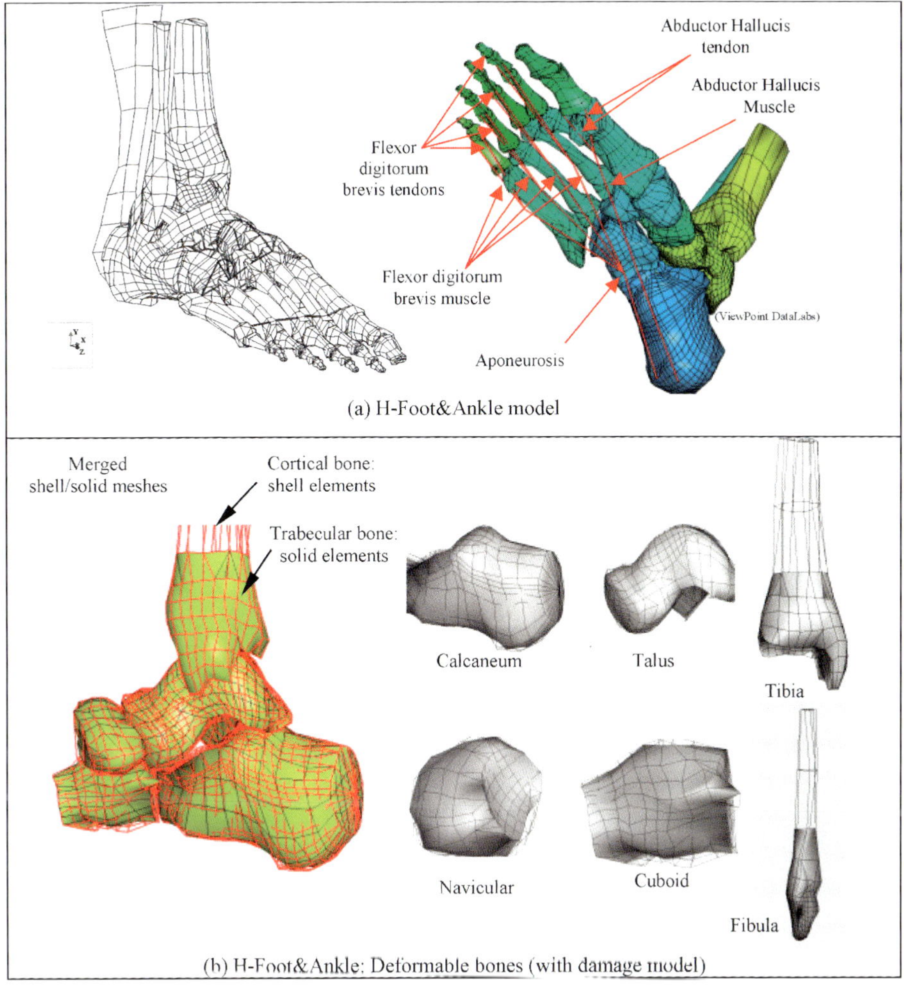

FIG. 9.41. H-Foot&Ankle model overview (BEAUGONIN, HAUG, MUNCK and CESARI [1995], BEAUGONIN, HAUG and CESARI [1996]). (Reproduced by permission of The Stapp Association.)

ment panel with pocketing of instrument panel; contact with foot controls; wheel well intrusion; contact with floor; collapse of leg compartment; foot trapped under pedals). These mechanisms correspond to a combination of ankle/foot simple movements like dorsiflexion, plantarflexion, pronation and supination. These movements can be associated with direct or indirect loading conditions.

Model structure and calibration. The foot/ankle complex (H-Ankle&Foot) of the H-Model mainly consists of deformable and damageable bone models with ligaments modeled with membranes and bars, Fig. 9.41 (BEAUGONIN, HAUG, MUNCK and CESARI [1995], BEAUGONIN, HAUG and CESARI [1996]).

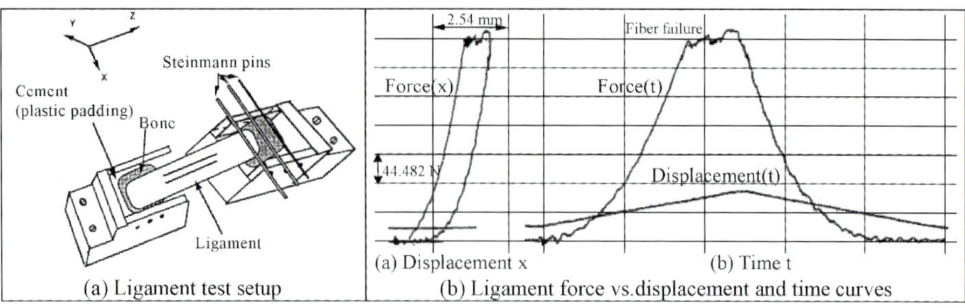

FIG. 9.42. H-Foot&Ankle model ligament calibration (after PARENTEAU, VIANO and PETIT [1996] and BEGEMAN and AEKBOTE [1996]). (Inset (a): Reproduced by permission of ASME; Inset (b): After material received from Prof. Begeman at Wayne State University, Bioengineering Center.)

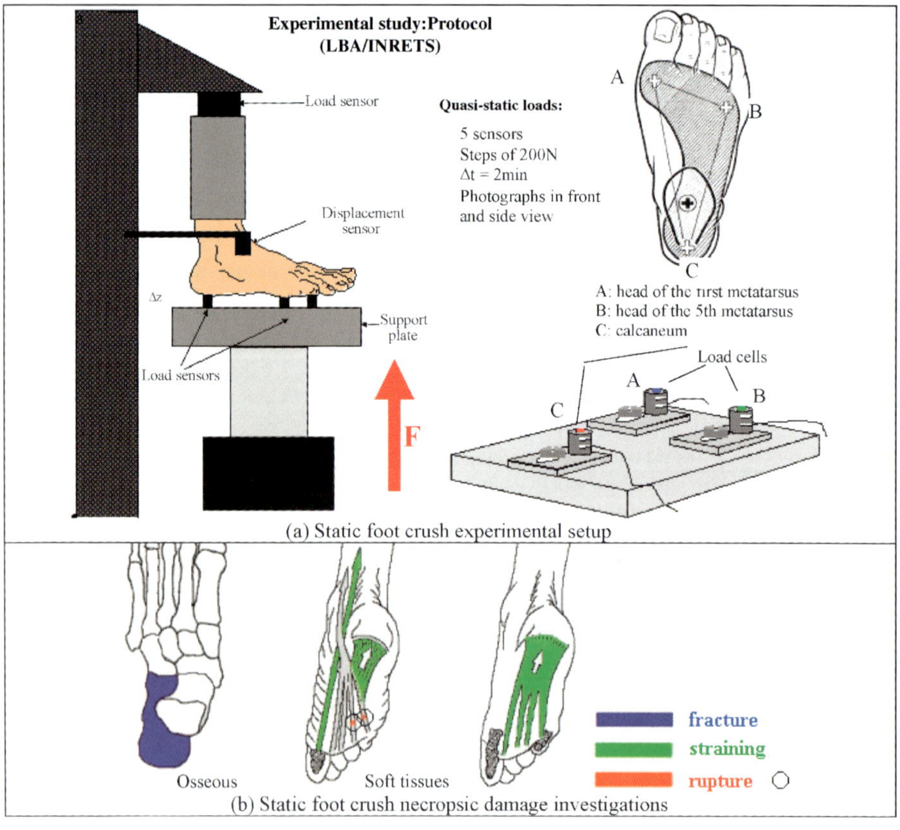

FIG. 9.43. H-Foot&Ankle validation: Static crush behaviour: Experimental tests and numerical simulations (MASSON, CESARI, BASILE, BEAUGONIN, TRAMECON, ALLAIN and HAUG [1999]). (Reproduced by permission of IRCOBI.)

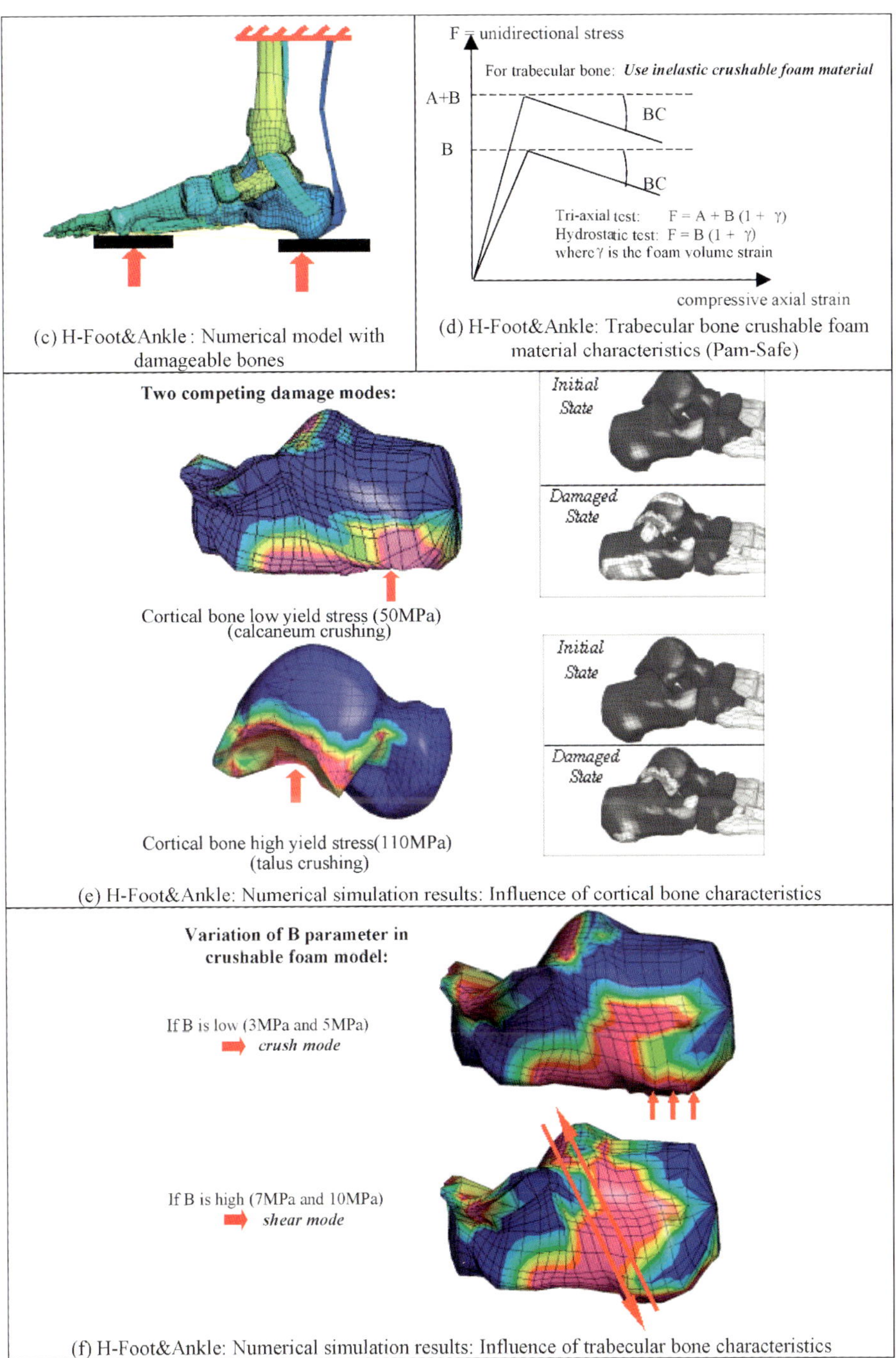

(c) H-Foot&Ankle: Numerical model with damageable bones

(d) H-Foot&Ankle: Trabecular bone crushable foam material characteristics (Pam-Safe)

(e) H-Foot&Ankle: Numerical simulation results: Influence of cortical bone characteristics

(f) H-Foot&Ankle: Numerical simulation results: Influence of trabecular bone characteristics

FIG. 9.43. (*Continued.*)

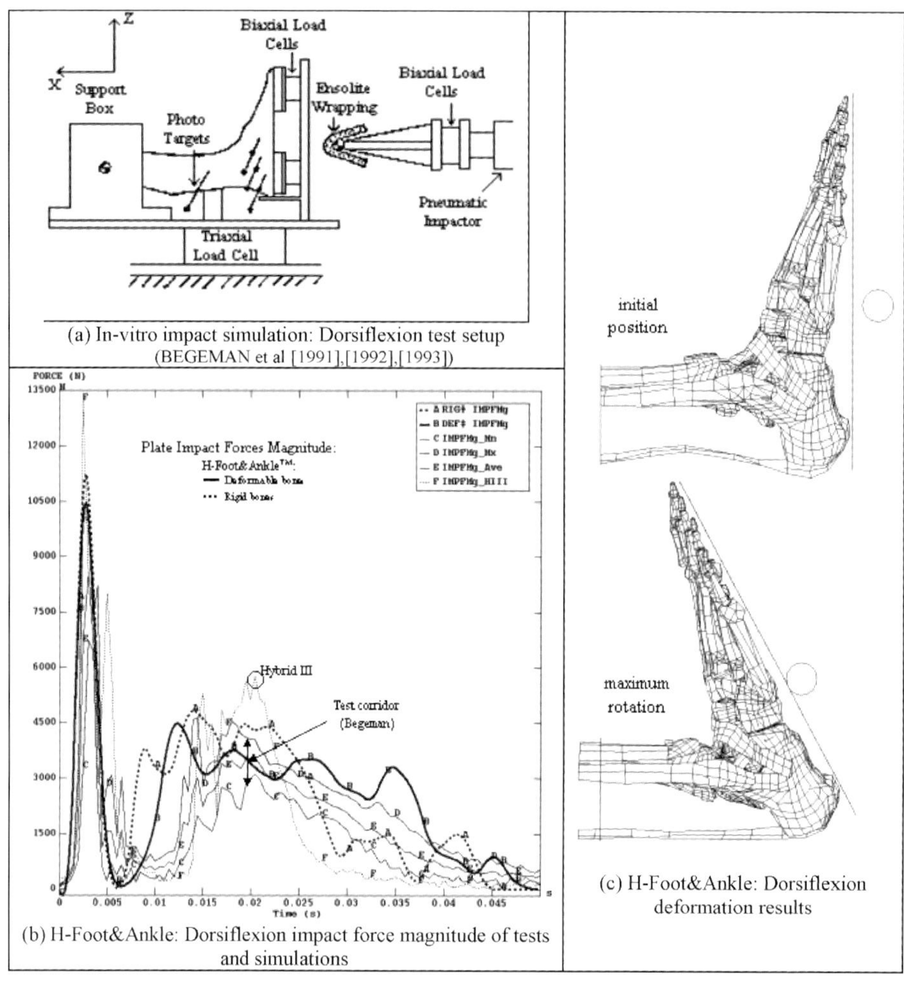

(a) In-vitro impact simulation: Dorsiflexion test setup (BEGEMAN et al [1991],[1992],[1993])

(b) H-Foot&Ankle: Dorsiflexion impact force magnitude of tests and simulations

(c) H-Foot&Ankle: Dorsiflexion deformation results

FIG. 9.44. H-Foot&Ankle validation: In vitro dynamic impacts: Experimental tests and numerical simulations (BEAUGONIN, HAUG and CESARI [1996], BEAUGONIN, HAUG, MUNCK and CESARI [1996], BEAUGONIN, HAUG and CESARI [1997], BEGEMAN and KOPACZ [1991], BEGEMAN, BALAKRISHNAN, LEVINE and KING [1992], BEGEMAN, BALAKRISHNAN, LEVINE and KING [1993]). (Insets (a)–(g) and (i): Reproduced by permission of The Stapp Association; Inset (h): ESI Software.)

Nonlinear contact interfaces model the cartilage layers of the major foot/ankle joints. Nonlinear joint elements connect the lesser bones. Fig. 9.41(a) and (b) show the H-Ankle&Foot model globally and with its major bones in detail. The latter are modeled with damageable thin shell elements for the cortical bone and with crushable foam solids for the trabecular bone.

Fig. 9.42 shows a typical ligament *calibration* test, where a ligament is isolated with its bony insertions, potted in grips and tested in tension (PARENTEAU, VIANO and PETIT [1996]). Typical force-displacement and force-time response curves are shown

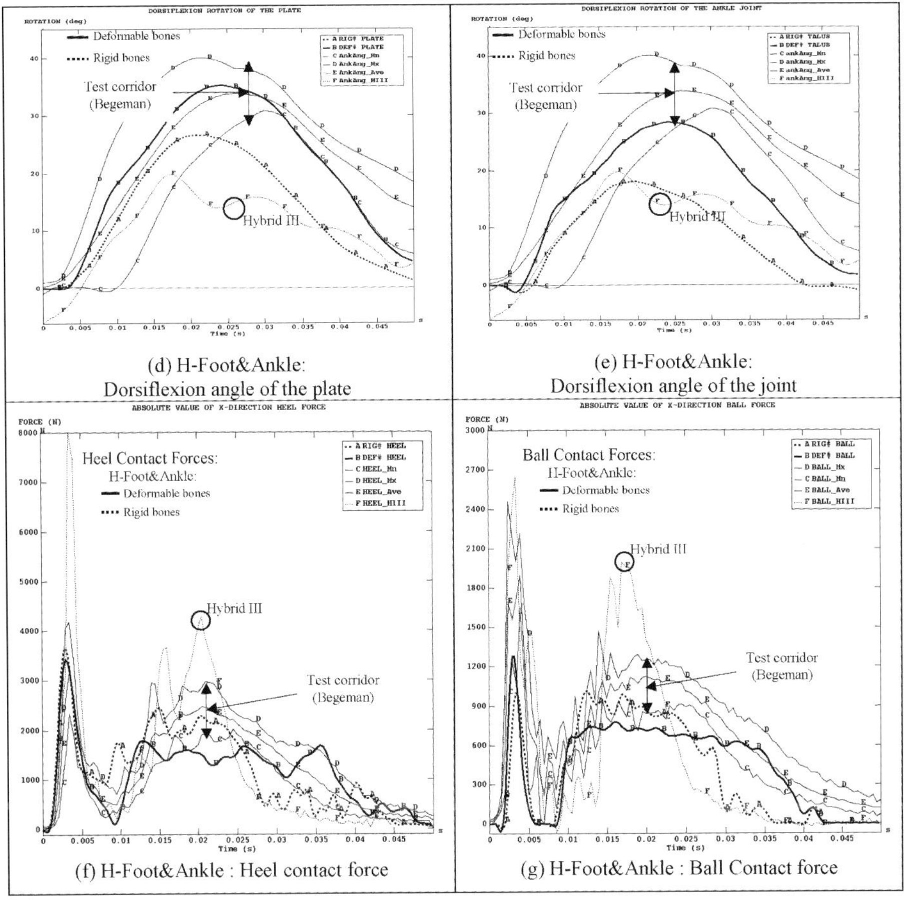

FIG. 9.44. (*Continued.*)

(BEGEMAN and KOPACZ [1991], BEGEMAN, BALAKRISHNAN, LEVINE and KING [1992], BEGEMAN, BALAKRISHNAN, LEVINE and KING [1993]), which exhibit un loading hysteresis and fiber failure. This behaviour is easily simulated using bars with nonlinear elastic-plastic material behaviour.

Validations. For *static crush behaviour* of the H-Foot&Ankle model, Fig. 9.43 shows experimental tests and numerical simulations, where different calcaneum and talus bone failure modes are obtained when the material properties of the cortical and trabecular bones are subject to parametric variations (MASSON, CESARI, BASILE, BEAUGONIN, TRAMECON, ALLAIN and HAUG [1999]).

For *dynamic plantar impacts*, Fig. 9.44 depicts the in vitro experimental tests (BEGEMAN and KOPACZ [1991], BEGEMAN, BALAKRISHNAN, LEVINE and KING [1992], BEGEMAN, BALAKRISHNAN, LEVINE and KING [1993]) and numerical simulations for the H-Foot&Ankle model validation (BEAUGONIN, HAUG, MUNCK and CESARI

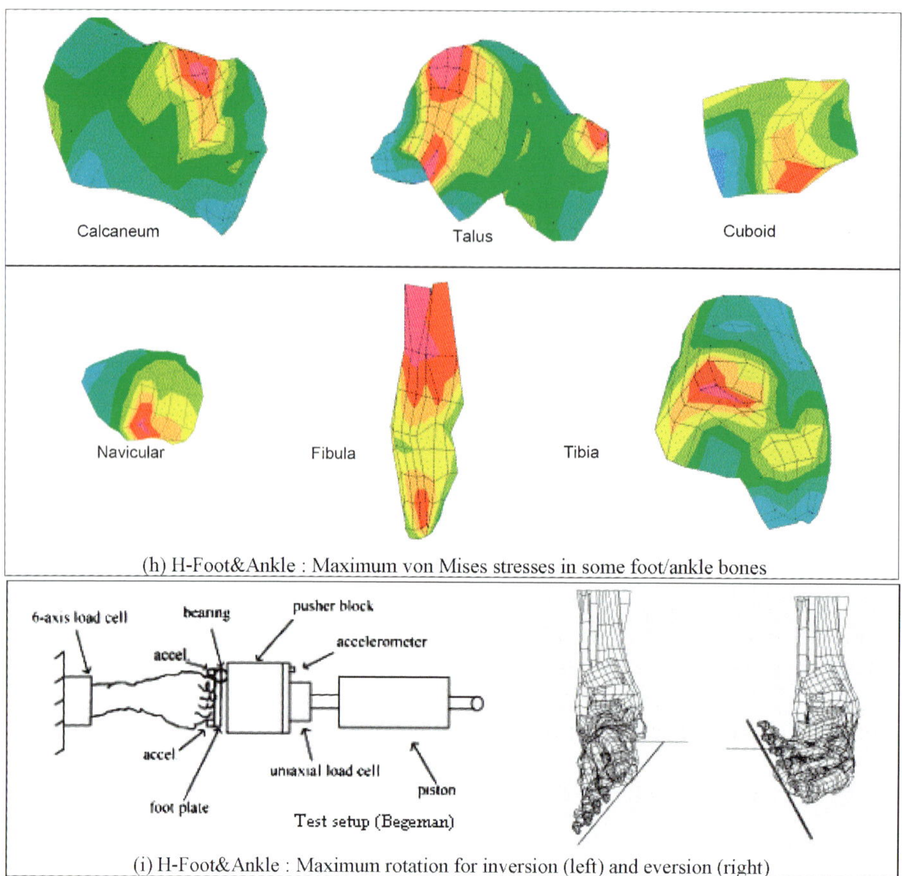

(h) H-Foot&Ankle : Maximum von Mises stresses in some foot/ankle bones

(i) H-Foot&Ankle : Maximum rotation for inversion (left) and eversion (right)

FIG. 9.44. (*Continued.*)

[1995], BEAUGONIN, HAUG and CESARI [1996], BEAUGONIN, HAUG, MUNCK and CESARI [1996], BEAUGONIN, HAUG and CESARI [1997]).

While Fig. 9.44(a)–(h) treat the dorsiflexion case, Fig. 9.44(i) treats the cases of inversion and eversion due to plantar impacts (BEGEMAN, BALAKRISHNAN, LEVINE and KING [1993], BEAUGONIN, HAUG and CESARI [1996]). The validations are described in more detail in the cited literature. The comparisons with the response curves of a Hybrid III mechanical crash dummy model, indicated in the diagrams, shows a considerable deviation from the response curves obtained with the human model, which demonstrates the limited biofidelity of crash dummies.

References on the H-Ankle&Foot model. The following references were consulted for the establishment of the H-Foot&Ankle model: ATTARIAN, MCCRACKIN, DEVITO, MCELHANEY and GARRETT [1985] on the biomechanical characteristics of the human ankle ligaments; BEGEMAN and KOPACZ [1991], BEGEMAN, BALAKRISHNAN,

LEVINE and KING [1992], BEGEMAN, BALAKRISHNAN, LEVINE and KING [1993] on the biomechanics of the human ankle impact in dorsiflexion (1991), on the human ankle dynamic response in dorsiflexion (1992) and for inversion/eversion (1993); CESARI, BERMOND, BOUQUET and RAMET [1994] on testing and simulation of impacts on the human leg; KAPANDJI [1974b] (lower limb joint physiology); LESTINA, KUHLMANN, KEATS and MAXWELL ALLEY [1992] on fracture mechanisms of the foot/ankle complex in car accidents; LUNDBERG, GOLDIE, KALIN and SELVIK [1989], LUNDBERG, SVENSSON, BYLUND, GOLDIE and SELVIK [1989] on the kinematics of the ankle/foot complex in dorsiflexion (a) and in pronation and supination (b); NAHUM, SIEGEL, HIGHT and BROOKS [1968] on lower extremity injuries in front seat occupants; NORDIN and FRANKEL [1989] on the Basic Biomechanics of the Musculoskeletal System; OTTE, VON RHEINBABEN and ZWIPP [1992] on the biomechanics of ankle/foot joint injury; PARENTEAU and VIANO [1996] on the kinematics and PARENTEAU, VIANO and PETIT [1996] on the biomechanical properties of the ankle subtalar joints in quasi-static loading up to failure; PATTIMORE, WARD, THOMAS and BRADFORD [1991] on the nature and causes of lower limb injuries in car crashes; PORTIER, TROSSEILLE, LE COZ, LAVASTE and COLTAT [1993] on lower leg injuries in real-world frontal car accidents; STATES [1986] on adult occupant injuries of the lower limb; WILSON-MACDONALD and WILLIAMSON [1988] on severy ankle joint injuries; WYKOWSKI, SINNHUBER and APPEL [1998] on a finite element model of the human lower extremity in frontal impact. The following web sites were consulted: http://www.rad.washington.edu/RadAnat; http://www.acfas.org/brankltr.html; http://astro.ocis.temple.edu/~rchristm/TeachFiles/Fractures/AvFxTib.JPG.

10. The fifth percentile female H-Model

Model structure and calibration. The fifth percentile female H-Model (weight 50 kg; height 1.52 m) is being built according to the same basic principles than the fiftieth percentile male H-Model (weight 75.5 kg; height 1.75 m), where the anatomy of the body was adjusted to the dimensions of the fifth percentile woman, Fig. 10.1. Inset (a) compares the shape of the model to the shape of the 5th percentile female Hybrid III mechanical dummy model. Inset (b) gives an overview of the model of the thorax. Inset (c) gives details of this model, concerning the deformable neck, breasts, rib cage, thoracic organs and the heart. All sub-models were built and calibrated in a way similar to the corresponding 50th percentile male sub-models, described earlier. The breasts were modeled as "bio-bags", Appendix D, which consist in a non-linear elastic outer skin with a quasi-incompressible filling in the volume created by the inner thorax lining and the outer skin. The modeling of the breasts is not considered final at that stage, and models using solid finite elements for the bulk (fatty tissue) and membranes for the outer envelope (skin) are built.

Calibration for thorax pendulum impact. Fig. 10.2(a) shows the Kroell pendulum impact test setup for female cadavers (KROELL, SCHNEIDER and NAHUM [1971], KROELL, SCHNEIDER and NAHUM [1974]). In inset (b) the measured, idealized, high velocity 6.71 m/s (solid lines) and low velocity 4.27 m/s (dashed lines) impact force

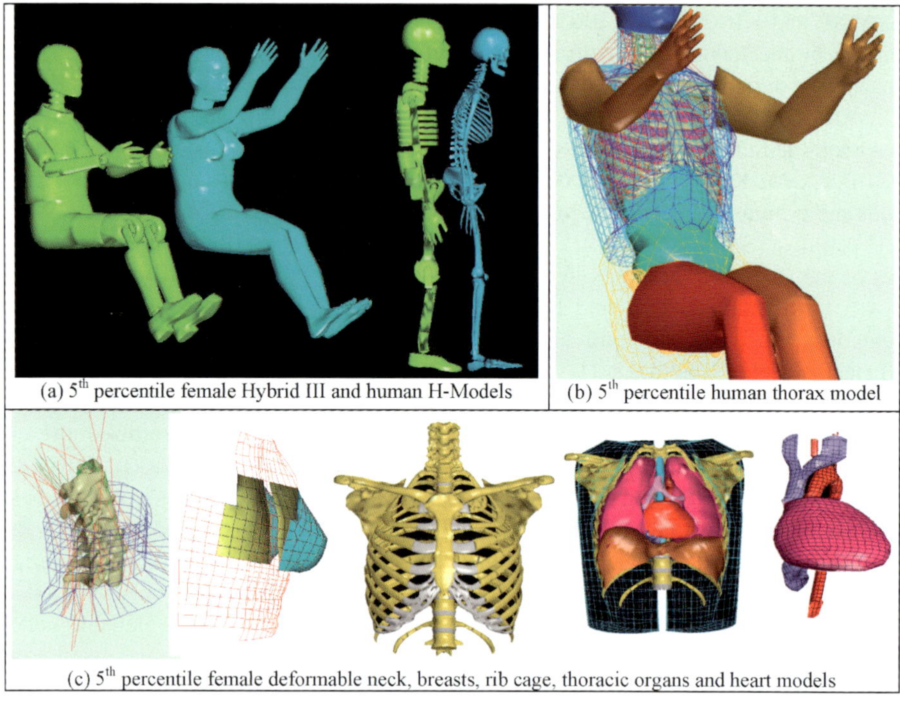
(a) 5th percentile female Hybrid III and human H-Models
(b) 5th percentile human thorax model
(c) 5th percentile female deformable neck, breasts, rib cage, thoracic organs and heart models

FIG. 10.1. Fifth percentile deformable female H-Model.

versus chest compression corridors for pendulum impacts of 23.15 kg mass are shown. Inset (c) demonstrates the final calibrated response of the female model at high impact velocity (6.71 m/s) to fall well within the test corridor. Inset (d) demonstrates at low impact velocity (4.27 m/s) the influence of various stiffness parameters, and notably of the presence and absence of the breasts during calibration.

Inset (e) contains sections at 10 millisecond intervals of the compression of the chest and organs, while inset (f) shows overall pictures and details on predicted rib fracture during this test.

During the calibration process, a first model used raw material data from the male model, leading to Fig. 10.2(d), curves (1a) and (1b). This model was generally too stiff. A preliminary evaluation of the modeling of the breasts with that model showed a considerable influence of their presence or absence. If breasts were modeled, curve (1a), the chest deformations were under-predicted, as compared to the case when the breasts were removed, curve (1b). After calibration of all material data, the response of the model with breasts was close to the measured test corridors, curve (2).

Validation for out-of-position airbag inflation. In Fig. 10.3(a) the fifth percentile female H-Model is exposed to a typical "out-of-position" (OOP) driver side airbag inflation scenario, cf. RUDOLF, FELLHAUER, SCHAUB, MARCA and BEAUGONIN [2002]. Contrary to a standard driver position (Fig. 10.1 insets (a), (b)), Fig. 10.3 shows the

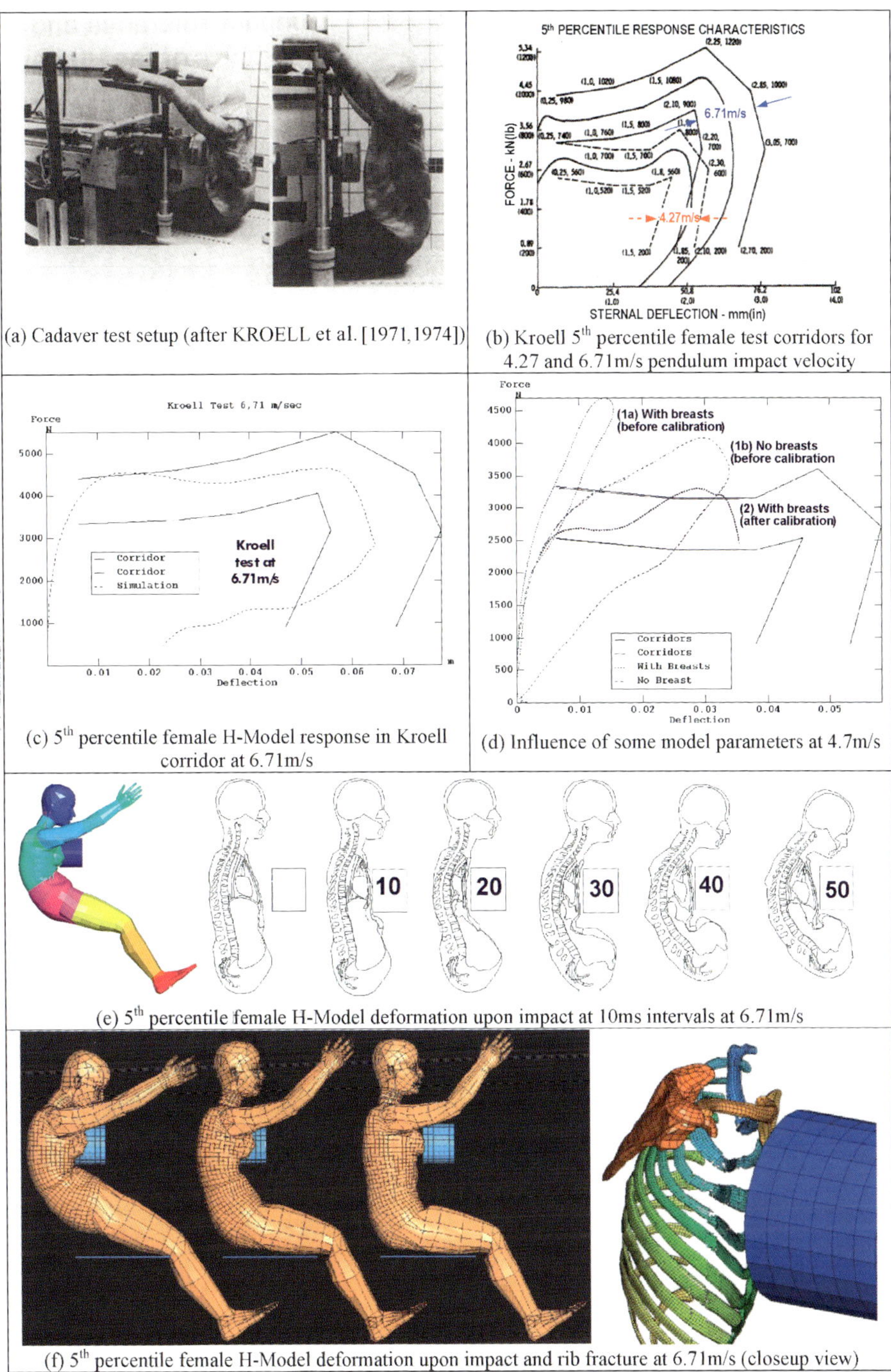

(a) Cadaver test setup (after KROELL et al. [1971,1974])

(b) Kroell 5th percentile female test corridors for 4.27 and 6.71m/s pendulum impact velocity

(c) 5th percentile female H-Model response in Kroell corridor at 6.71m/s

(d) Influence of some model parameters at 4.7m/s

(e) 5th percentile female H-Model deformation upon impact at 10ms intervals at 6.71m/s

(f) 5th percentile female H-Model deformation upon impact and rib fracture at 6.71m/s (closeup view)

FIG. 10.2. Calibration of the 5th percentile deformable female H-Thorax (tests after KROELL, SCHNEIDER and NAHUM [1971], KROELL, SCHNEIDER and NAHUM [1974]). (Insets (a), (b): Reproduced by permission of The Stapp Association.)

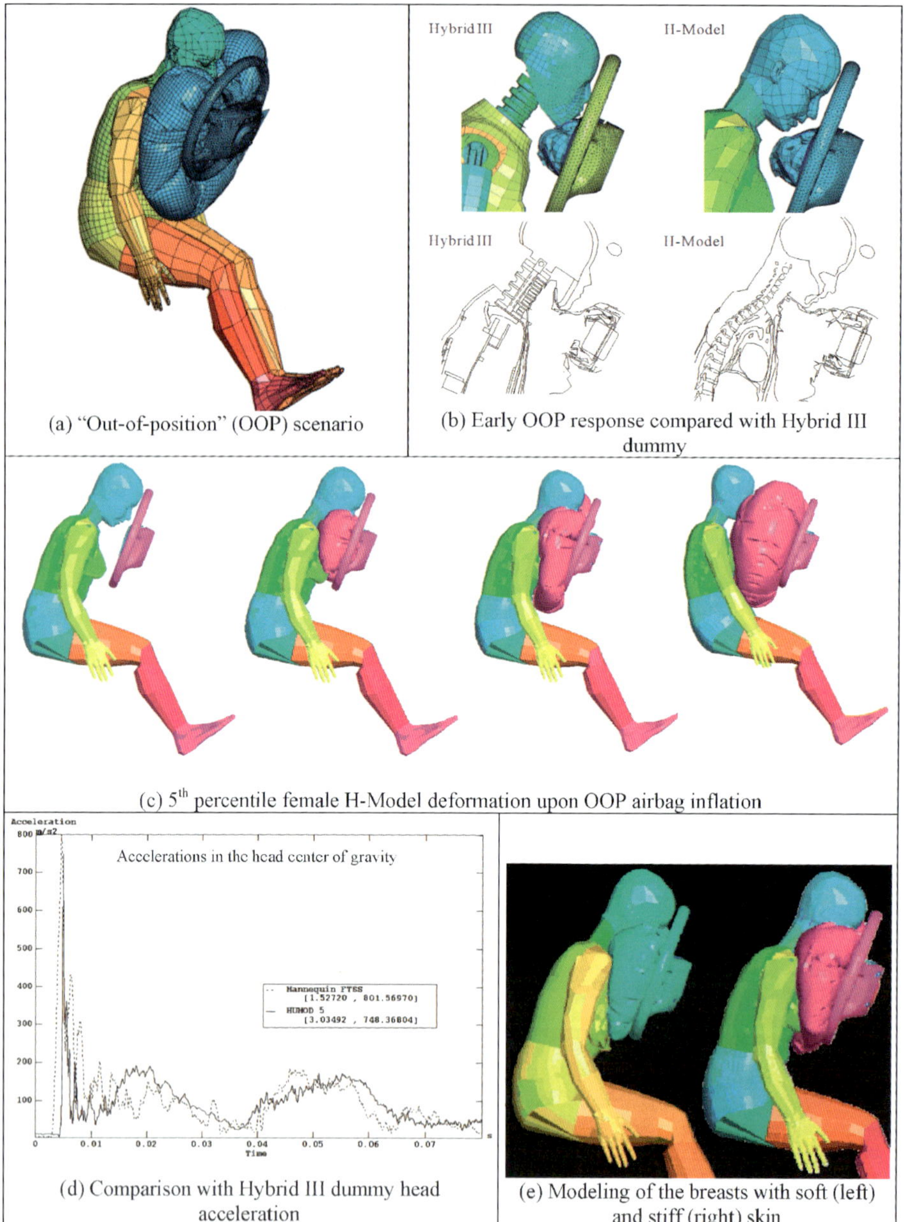

FIG. 10.3. Validation of the 5th percentile deformable female H-Model (OOP case) (RUDOLF, FELLHAUER, SCHAUB, MARCA and BEAUGONIN [2002]).

5th percentile female H-Model positioned close to the steering wheel, being rejected to the rear by the inflating driver side airbag, inset (c). Whereas in a normal position the occupant comes into contact with a fully inflated airbag, the OOP occupant is hit by the airbag while it inflates, which can case injury through the phenomenon called "bag-slap". Further, the inflating bag may be trapped under the chin, inset (b), and thus project the head and neck towards the rear with an over extension of the neck. The mid-sagittal section plots of inset (b) suggest that injury of the head/neck/thorax complex can be assessed much more readily from the human model than from the dummy model. Using human models and models of the standard 5th percentile female Hybrid III mechanical dummy, one can appreciate the differences between the response of the dummy and the human, inset (d). Inset (e), finally, gives an impression on the influence of the modeling if the outer skin of the breasts, which, when not correctly stiffening at large skin strain, may be reacting in an overly soft manner.

In the out-of-position airbag validation test case, the outer skin of the breasts was initially modeled with the isotropic material properties as calibrated from the pendulum impact tests. This led to large deformations in the OOP case, as observed in Fig. 10.3(e) (left). Such deformations were not seen previously in the pendulum impact simulations, because the pendulum impacted in a more concentrated fashion, between the breasts. Then the skin was re-modeled with an orthotropic fiber reinforced material, with stiffening fibers at larger strains, as it is characteristic of skin (YAMADA [1970] and Fig. 1.1(h)). If this stiffening effect was neglected, the skin was too soft and the final deformations were too large. It was found that the volume and the mechanical resistance of the breasts can play an important role in the energy absorption characteristics of the thorax from frontal impacts. It is therefore important to further evaluate their influence in future studies.

Further investigations on the small female dummy and cadaver tests under static airbag out-of-position deployment tests can be found, e.g., in CRANDALL, DUMA, BASS, PILKEY, KUPPA, KHAEWPONG and EPPINGER [1999].

APPENDIX A

Basic Theory of Crash Codes

The following paragraphs are based mainly on Pam–Crash documentation of ESI Software. See also HAUG, CLINCKEMAILLIE and ABERLENC [1989a], HAUG, CLINCKEMAILLIE and ABERLENC [1989b].

A.1. Overviews on solution methods and finite elements

Solution method overview. Modern crash and impact simulation codes are three-dimensional, Lagrangian, finite element, explicit and implicit, vectorized and multi-tasked *solid codes* for the non-linear dynamic and large deformation analysis of solid structures in the realm of computational structural mechanics (CSM). They analyze crash phenomena at discrete points in space and time.

Space discretization. Space is discretized with the most often used *finite element methodology*, which is based on the *displacement method* of structural analysis, where the discrete nodal displacements and rotations constitute the unknowns of the problem. A comprehensive account on the theory and application of the finite element method in engineering is given in the textbook by BATHE [1996]. Some solid codes are coupled to *flow codes*, which treat problems in the realm of computational fluid dynamics (CFD), in order to treat fluid-structure interaction (FSI) dominated problems. Most solid codes are provided with alternative spatial discretization schemes called *particle methods*, where a solid, fluid or gaseous medium is represented by *mesh-less* discrete (mass or integration) points, such as in the most common mass-point smooth particle hydrodynamics (SPH) scheme. This permits to solve certain classes of FSI problems with the solid codes without the necessity to couple with a separate flow code.

The solid codes allow to model 3D structures of arbitrary geometry using solid elements, membrane elements, plate and shell elements, beam and bar elements, and discrete spring and joint elements. In typical crashworthiness and impact simulations, plates and shells are used to model thin-walled metal or plastic components. Beams and bars are used for stiffening frames, wheel suspensions, shafts, special connections or secondary components. Solid elements may be used for modeling the bulk of crushable foams. In typical finite element simulations of the human body, plates and shells are used for the simulation of cortical bone, beams and bars are used for the modeling of long bones, tendons and muscles, and solid elements are used for the modeling of the bulk of soft tissues and for the spongeous bone.

Time discretization. The application of the displacement finite element method leads to the discretized, coupled and nonlinear equations of motion in each displacement and rotation degree of freedom. These equations can be integrated in the time domain by using either implicit or explicit methods. Both methods use time discretization operators which permit to solve for the unknown displacements, rotations, linear and angular velocities and accelerations of each degree of freedom, at a given discrete point in time, from the known states of the structure at previous points in time. An overview on the explicit and implicit solution methods can be found in HUGHES, PISTER and TAYLOR [1979].

Standard *implicit methods* require linearization of the set of nonlinear equations of motion and lead to sets of coupled algebraic equations which, for non-linear problems, must be solved at a considered point in time in an iterative fashion, in order to achieve dynamic equilibrium at that time. The degree of coupling of the equation set is measured by the "bandwidth" of the linearized system matrix, which envelopes the extent of non-zero matrix elements of each of its rows. In a "diagonal" equation matrix the bandwidth is minimal (one) while in a "full" matrix the bandwidth is maximal (number of equations). Implicit methods can be used together with explicit solution, when the explicit solution is costly, such as in quasi-static problems, which arise, for example, when seating an occupant model into a deformable car seat before simulating a crash scenario, or by calculating the elastic spring-back of stamped parts.

Standard *explicit methods* do not require repeated solution of linearized coupled equation systems, and lead to a set of uncoupled algebraic equations when a lumped (diagonal) mass matrix is assumed. Solution of diagonal equation systems is trivial and computer time per discrete solution time step is negligible as compared to the computer time needed to repeatedly solve the potentially huge coupled systems of algebraic equations of the implicit methods. To date, crash models with up to one million thin shell finite elements and more have been built and solved successfully on powerful computers with the explicit methods.

The time increment of explicit methods, however, is restricted for solution stability, while, in principle, the time increment in unconditionally stable implicit methods is not restricted in size. In typical crashworthiness and impact studies over relatively short durations and involving large distortions of the structural parts, this advantage of the implicit methods has no bearing, however, because the structural states must be known at many discrete points in time in order to allow for an accurate tracing of the complex physical phenomena, including "contact", and to account for material and geometrical non-linearity. Time increments of the order of one microsecond and less are typical in crash simulations with explicit time integration.

In crashworthiness and in higher velocity dynamic impact studies, therefore, the explicit time integration methods have proven computationally advantageous. Implicit solution algorithms, on the other hand, may be applied with advantage to quasi-static or slow vibration dynamic problems with limited non-linearity.

Finite elements overview. Standard finite elements are defined by a set of nodes and a connectivity array, which relates their geometric topology to these nodes. In nonlinear crash and impact analysis, experience shows that "higher order" elements are not beneficial. This is mainly due to the fact that sharp stationary or traveling plastic hinge folding lines of crushed thin shell structures, made of elastic-plastic material, are not well accommodated even by complex interpolation shape functions of higher order elements, and that plastic hinge lines therefore require a dense nodal point spacing in the finite element models of thin walled structures, even when computationally expensive higher order elements are used.

Similarly, high gradients of three-dimensional stress and of plastic deformation are better accommodated by dense finite element models of simple 8-node 3D solid elements, than by using higher order solids that, for computational competitiveness, must have coarser meshes. Similar remarks apply for the simulation of fracture, where the small scale of the physics of fracture requires even higher mesh densities, sometimes achieved by automatic adaptive local mesh refinement schemes. Recently, so-called "mesh-less" methods are applied to simulate fracture, such as the EFG (element-free Galerkin), DE (discrete elements), SPH (smoothed particle hydrodynamics) (MONAGHAN and GINGOLD [1983] and MONAGHAN [1988]), FPM (finite point methods), etc. These advanced space discretization methods are presently under development and they are implemented in commercial crash codes. One instance of recent trial applications in biomechanics of the mesh-less SPH methods is the simulation of coupled structure fluid interaction (FSI) problems, where the fluid is simulated with particles, enclosed in a deformable organ modeled with finite elements. Another emerging method is the coupling of solid codes and flow codes for the simulation of FSI events (LÖHNER [1990]), such as the rupture of the blood-filled aorta in a chest impact scenario.

In many structures, the most important crash simulation finite element is the thin shell element. The most used thin shell element is a bilinear four node quadrilateral element, based on the Mindlin–Reissner plate theory. One of the most efficient thin shell elements was originally developed by BELYTSCHKO and TSAY [1983] and BELYTSCHKO and LIN [1984]. The Mindlin–Reissner plate theory takes the transverse shear deformation of the plate into account and it presumes that lines normal to the plate mid-surface remain straight, but not necessarily normal.

In the classical Kirchhoff–Love plate theory the normal lines to the plate mid surface remain both, straight and normal, and the influence of the transverse shear deformation is neglected. The implementation of this theory requires slope-compatibility across element edges (C^1 continuity), resulting in a complex finite element formulation. On the other hand, Mindlin–Reissner theory requires only C^0-continuity in the shape functions for assuring complete inter-element deformation compatibility. This greatly simplifies the FE-formulation.

In the case of the Belytschko element, a reduced domain integration technique with one-point quadrature is applied for calculating the nodal forces contributed by the plate and shell elements. This technique avoids membrane-locking, but permits certain zero-energy or kinematic deformation modes of the elements, with which no resisting material stresses are associated. Such zero-energy modes are called hourglass-modes, which, if excited, can lead to numerical instability through uncontrolled spurious oscillations. In order to prevent hourglass modes, a built-in hourglass control algorithm is implemented, as developed by BELYTSCHKO, WONG, LIU and KENNEDY [1984].

It effectively avoids the numerical instability problem associated with one-point quadrature shell elements, by effectively damping out the zero-energy modes.

A.2. Explicit solution method outline

Three dimensional structures. In a three-dimensional solid model there are 3 translational degrees-of-freedom (DOF) per node. In a three-dimensional thin shell numerical model there are 6 degrees of freedom per nodal point, i.e., 3 translations and 3 rotations. In a three-dimensional thin plate numerical model there are 5 local degrees of freedom, i.e., 3 translations and 2 rotations. Crash models are mainly spatial models of elements that can be subdivided into layers with plane stress conditions (plate and thin shell theory).

Lagrangian discretization. This term refers to a choice of independent variables for the problem. In the Lagrangian formulation, each material particle is characterized by its initial coordinates (x_0, y_0, z_0) and its actual coordinates are chosen to be the dependent variables of the problem,

$$x = x(x_0, y_0, z_0, t),$$
$$y = y(x_0, y_0, z_0, t),$$
$$z = z(x_0, y_0, z_0, t).$$

In this formulation, the ordinary differential equations, obtained after spatial discretization, describe the dynamic equilibrium of the material particles which stand for the original continuum. For example,

$$\mathbf{M}\,\mathrm{d}^2\mathbf{x}/\mathrm{d}t^2 + \mathbf{C}\,\mathrm{d}\mathbf{x}/\mathrm{d}t + \mathbf{K}\mathbf{x} = \mathbf{F}_{\mathrm{ext}},$$

are the discretized equations of motion in the linear case, where $\mathbf{M}$ is the mass matrix, $\mathbf{C}$ is the damping matrix, $\mathbf{K}$ is the stiffness matrix and $\mathbf{F}_{\mathrm{ext}}$ is the vector of applied loads.

The acceleration of a particle is equal to the material derivative of the velocity, $\mathbf{v}$, and equal to the partial derivative of the velocity because x_0, y_0, z_0 are constants:

$$\frac{\mathrm{d}\mathbf{v}}{\mathrm{d}t} = \frac{\partial \mathbf{v}}{\partial t} + \frac{\partial \mathbf{v}}{\partial x_0}\frac{\partial x_0}{\partial t} + \frac{\partial \mathbf{v}}{\partial y_0}\frac{\partial y_0}{\partial t} + \frac{\partial \mathbf{v}}{\partial z_0}\frac{\partial z_0}{\partial t} = \frac{\partial \mathbf{v}}{\partial t}$$

The discrete finite element mesh points coincide with material points and have time dependent coordinates. The finite element mesh will thus deform with the material. This will engender distortions of the mesh, which can result in decreasing stable solution time steps and in solution inaccuracies due to excessive element distortions.

Finite elements. Finite element “unstructured” meshes permit to establish numerical models of a physical structure with realistic discretization and representation of the boundary conditions

Only the simplest finite elements are used in the PAM-Crash program because it is believed that fine meshes of simple elements give better results in highly distorting structures than coarse meshes of high-order elements. The most often used finite elements in crash codes are

- 4–8 node solids (1 point and fully integrated);
- 3–4 node bilinear shells (1 point and fully integrated);
- 3–4 node isoparametric membrane elements (fully integrated);
- 2 node beam, bar and joint elements.

The optimization of industrial crash codes has focused on the treatment of the most frequently used shell element.

Explicit integration scheme. A vibrating spring/dashpot-mass system consists of a mass, m, a dashpot with constant c, a spring with constant k, and an external load, $f(t)$, Fig. A.1.

Consider the second order ordinary differential equation which expresses dynamic equilibrium of this system, where x is the displacement, $\dot{x}$ is the velocity c is the damping coefficient, $\ddot{x}$ is the acceleration and m is the mass,

$$m\ddot{x} + c\dot{x} + kx = f(t) \quad \text{(1 DOF)}.$$

The *central difference scheme* considers the following time-axes around a discrete point in time, t_n, Fig. A.2.

The known quantities are the displacement x_n at time t_n, and the velocity $\dot{x}_{n-1/2}$ at the intermediate time $t_{n-1/2}$. The wanted quantities are the displacement x_{n+1} at time t_{n+1}, and the velocity $\dot{x}_{n+1/2}$ at time $t_{n+1/2}$. Dynamic equilibrium at time t_n is expressed

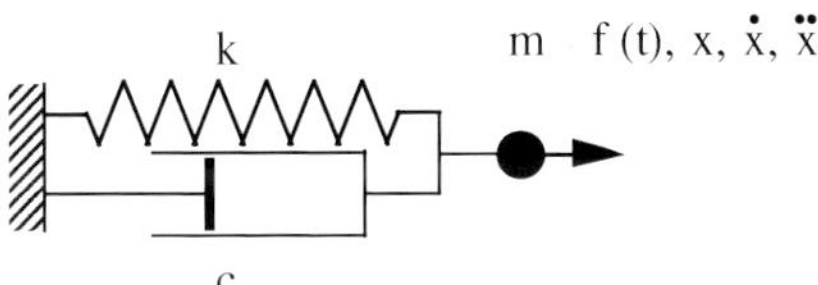

FIG. A.1. Spring/dashpot-mass system.

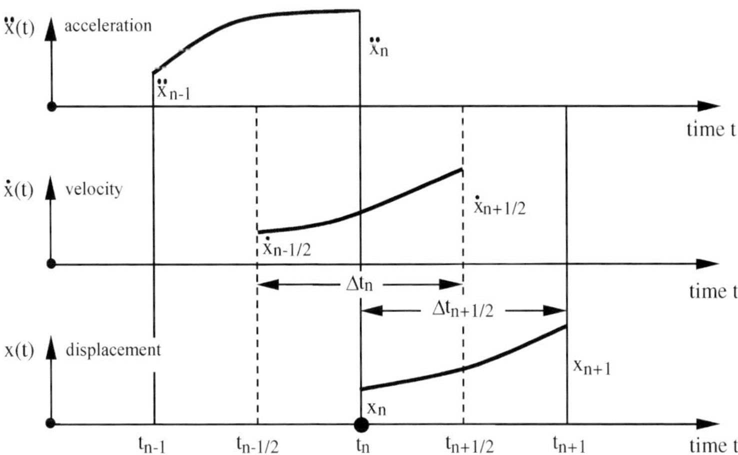

FIG. A.2. Central difference scheme.

as

$$m\ddot{x}_n = f_n - c\dot{x}_n - kx_n.$$

Since all terms on the right hand side are known, one can first solve for the acceleration at time t_n, $\ddot{x}_n$, and then apply the central difference time integration, Fig. A.2, and solve for the unknown quantities, $\dot{x}_{n+1/2}$ at time $t_{n+1/2}$ and x_{n+1} at time t_{n+1} as follows

$$\begin{aligned}\ddot{x}_n &= m^{-1}(f_n - c\dot{x}_n - kx_n),\\ \dot{x}_{n+1/2} &= \dot{x}_{n-1/2} + \Delta t_n \ddot{x}_n,\\ x_{n+1} &= x_n + \Delta t_{n+1/2}\dot{x}_{n+1/2}.\end{aligned}$$

Stable time step. The criterion for solution stability of the central difference explicit time integration can be derived formally as follows (BATHE [1996]). For a 1-DOF free-vibration system the equation of motion becomes

$$\ddot{x}_n + 2\xi\omega\dot{x}_n + \omega^2 x_n = 0 \quad \text{(free vibration)},$$

where $\xi = c/c_{\text{crit}}$ is the damping ratio, $c_{\text{crit}} = \sqrt{2km}$, is the critical damping, $\omega = \sqrt{k/m}$ is the circular frequency of vibration, and k and m are the spring stiffness and vibrating mass, respectively. The central difference scheme dictates, using constant time step, Δt, around time t_n,

$$\begin{aligned}\dot{x}_{n+1/2} &= (x_{n+1} - x_n)/\Delta t,\\ \dot{x}_{n-1/2} &= (x_n - x_{n-1})/\Delta t,\\ \ddot{x}_n &= (\dot{x}_{n+1/2} - \dot{x}_{n-1/2})/\Delta t = (x_{n+1} - 2x_n + x_{n-1})/\Delta t^2.\end{aligned}$$

Substitution of these expressions into the above equation of motion yields the following recursive system of equations

$$\begin{Bmatrix} x_{n+1} \\ x_n \end{Bmatrix} = \begin{bmatrix} \frac{2-\omega^2\Delta t^2}{1+\xi\omega\Delta t} & -\frac{1-\xi\omega\Delta t}{1+\xi\omega\Delta t} \\ 1 & 0 \end{bmatrix} \begin{Bmatrix} x_n \\ x_{n-1} \end{Bmatrix}.$$

For zero damping ($\xi = 0$) this becomes

$$\begin{Bmatrix} x_{n+1} \\ x_n \end{Bmatrix} = \begin{bmatrix} 2-\omega^2\Delta t^2 & -1 \\ 1 & 0 \end{bmatrix} \begin{Bmatrix} x_n \\ x_{n-1} \end{Bmatrix}, \quad \text{i.e.,}$$

$$\mathbf{x}_{n+1} = \mathbf{A}\mathbf{x}_n.$$

In general, one can write the equivalent recursive matrix relationship

$$\mathbf{x}_{n+1} = \mathbf{A}\mathbf{x}_n = \mathbf{A}^2\mathbf{x}_{n-1} = \cdots = \mathbf{A}^n\mathbf{x}_1,$$

where the operator matrix $\mathbf{A}$ can be seen to recur in powers of n. This relationship is stable only if $\mathbf{A}^n$ remains bounded for all values of n. This means that the absolute values of the eigenvalues of $\mathbf{A}$, or its "spectral radius", ρ, must be less than or equal to one, i.e.,

$$\rho(\mathbf{A}) = \max|\lambda_i| \leqslant 1 \quad \text{for stability.}$$

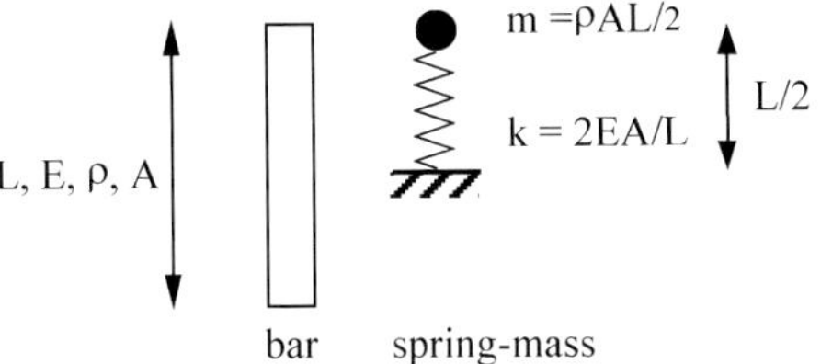

FIG. A.3. Free vibration bar element.

In case of zero damping, the eigenvalues of **A** follow from $\mathbf{Ax} = \lambda\mathbf{x}$, with solutions $\det(\mathbf{A} - \lambda\mathbf{I}) = 0$, where **I** is the 2×2 unit matrix. This leads to $\lambda_{1,2} = a \pm \sqrt{a^2 - 1}$, in which $a = (2 - \omega^2\Delta t^2)/2$. Enforcing $|\lambda_{1,2}| = 1$ one has $\lambda_{1,2}^2 = 1$, from which follows that $a^2 = 1$ satisfies the condition $|\lambda_{1,2}| = 1$ non-trivially. From this equation, the condition for stability of the time step is given by

$$\text{stable } \Delta t \leqslant 2/\omega.$$

For the case of the freely vibrating spring/mass system, stability is enforced if

$$\Delta t \leqslant 2\sqrt{m/k}.$$

EXAMPLE. Consider the free vibration 1 DOF system, made of a bar element with uniform mass and stiffness distribution, with half length, $L/2$, elastic modulus, E, mass density per unit volume, ρ, and cross section area, A, as shown in Fig. A.3.

The criterion for stability becomes in this case

$$\Delta t_n < 2\sqrt{\frac{m}{k}} = L\sqrt{\frac{\rho}{E}} = \frac{L}{\sqrt{E/\rho}} = \frac{L}{c},$$

where $c = \sqrt{E/\rho}$ is the speed of sound in the bar material. For steel one has roughly $c = 5$ km/s = 5 mm/microsecond. This means that for a shell element with dimensions of 5×5 mm the stable time step will be about 1 microsecond, which corresponds to the time it takes of an acoustic signal to travel across the element.

Implicit integration outline. Consider the second order differential equation ($c = 0$)

$$m\ddot{x} + kx = f$$

and a discretized time axis as follows,

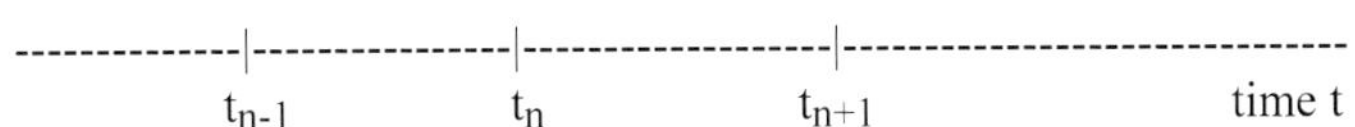

Let the known quantities at time t_n be x_n and $\dot{x}_n$. The wanted quantities are then $\dot{x}_{n+1}$ and x_{n+1}. Dynamic equilibrium at time t_{n+1} is expressed as

$$m\ddot{x}_{n+1} + kx_{n+1} = f_{n+1},$$

where x_{n+1} is unknown. Direct solution for $\ddot{x}_{n+1}$ or x_{n+1} is therefore impossible. One can now apply forward differences and substitute

$$\dot{x}_{n+1} = (x_{n+1} - x_n)/\Delta t,$$
$$\ddot{x}_{n+1} = (\dot{x}_{n+1} - \dot{x}_n)/\Delta t$$

and thus:

$$\ddot{x}_{n+1} = (x_{n+1} - 2x_n + x_{n-1})/(\Delta t)^2,$$

which after substitution yields

$$(m/\Delta t^2 + k)x_{n+1} = f_{n+1} - m/\Delta t^2(2x_n - x_{n-1}).$$

This equation can be solved for the unknown displacement at time t_{n+1},

$$x_{n+1} = (m/\Delta t^2 + k)^{-1}(f_{n+1} - m/\Delta t^2(2x_n - x_{n-1})).$$

The other variables are then obtained as

$$\dot{x}_{n+1} = (x_{n+1} - x_n)/\Delta t,$$
$$\ddot{x}_{n+1} = (\dot{x}_{n+1} - \dot{x}_n)/\Delta t.$$

This scheme works independently of the chosen value of Δt. It is said to be "unconditionally stable". It necessitates, however, to perform the solution of $(m/\Delta t^2 + k)^{-1}$, which is in general a costly operation because the stiffness matrix is not diagonal. For large time steps, however, the solution may remain stable, but one may accumulate period elongation and amplitude decay time integration errors in freely vibrating systems. For non-linear equations of motion, the linearized equations of dynamic equilibrium must be solved iteratively to ensure dynamic equilibrium at times t_n. If this is neglected, the solution, may become inaccurate and unstable.

Vectorized and multi-tasked codes. Explicit codes are well suited for vectorized super-computers because elements carry no dependencies, i.e., elements can be treated simultaneously, no K-matrix is stored and no I/O operations are required. This means that the elapsed time is practically equal to the CPU time in a computer run. For the same reasons, explicit codes parallelize well. Most industrial crash codes come with shared and distributed memory parallel architecture.

A.3. Contact treatment outline

Overview on contact algorithms. The successful and efficient treatment of contact events is of prime importance in crash and impact simulation. To identify potential contact surfaces of a given structure, the crash codes allow to define contact interface entities, which are most often given by collections of surface facets, identical to finite element surfaces. Actual contact can be defined a distance t_{contact} (contact thickness) away from the mid-surface defined by the facets represented by shell elements. Most contact algorithms are based on a node-to-segment treatment. After preliminary geometrical

TABLE A.1
Some typical contact options (overview)

Group	Type*	Action
Rigid walls		one-sided moving or fixed boundary condition
Nodal constraints		pairs of nodes are constrained to displace together
Special sliding and tied interfaces	Type 1	sliding without separation
	Type 2	3 DOF or 6 DOF tied with single failure
Internal solid element contact	Type 10	internal solid anti-collapse contact
Safety specific contacts	Type 7	multiply self-impacting contact for airbag
	Type 11	body-to-plane contact (force-deflection)
	Type 12	body-to-body contact (force-deflection)
	Type 21	body-to-multiplane contact (force-deflection or stress-strain)
	Type 37	enhanced self-impacting contact for airbag
Recommended contacts for crash simulation	Type 31	penalty-free node-to-segment contact
	Type 32	penalty or kinematic tied contact with distance and failure
	Type 33	segment-to-segment contact with edge treatment (3D bucket search)
	Type 34	node-to-segment contact with edge treatment (3D bucket search)
	Type 36	self-impacting contact with edge treatment (3D bucket search)
	Type 42	mesh independent spotweld parts
	Type 44	node-to-segment contact with smooth contact surface
	Type 46	edge-to-edge self-impacting contact

*PAM-Crash code.

proximity searches for potential contact events, the so identified candidate contact segments are fine-checked for mutual penetrations of their nodes and surfaces. Detected penetrations are limited ("penalized") or prevented by the algorithms.

In impact biomechanics several kinds of contact events are of importance. On the one hand, the biomechanical models may contact or collide with external objects (chest-to-seat belt, body-to-airbag, body-to-car interior, etc.). On the other hand, body parts and organs may be in mutual contact (arm-to-chest, chin-to-chest, leg-to-leg, internal organ-to-internal organ, organ-to-wall of body cavity, brain-to-skull, etc.).

Contacts may be sticking ("tied" options), sliding ("slide" options), multiple ("slide-and-void" options), failing (rupture options), and sliding can be with and without friction. Each contact event may require a particular treatment, with or without penalty algorithms. To demonstrate the importance editors of crash codes attribute to contact treatments, Table A.1 lists some often used contact options. The listed rigid wall and nodal constraint options are simpler contact treatments that can serve for detecting collisions with rigid obstacles (rigid walls) or that can be used to constrain specified pairs of nodes to move together (nodal constraints). The interested reader is

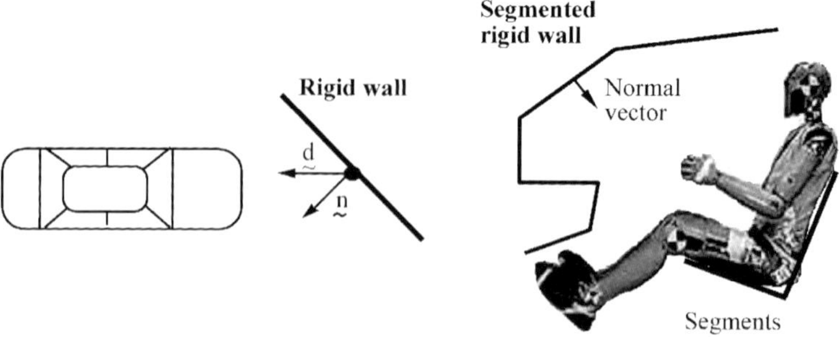

FIG. A.4. Rigid wall contact.

advised to consult the notes manuals of the commercial crash codes for more detail (see also: HUGHES, TAYLOR, SACKMAN, CURNIER and KANOKNUKULCHAI [1976], HALLQUIST, GOUDREAU and BENSON [1985] and HAUG, CLINCKEMAILLIE and ABERLENC [1989b]).

In simulations of impact biomechanics all listed options can be applied. Some details are provided next for the most popular rigid wall and penalty contact options. Contact types 11 and 12 are often used in the simple rigid multi-body models (HARB models) and they provide a "soft" penalization of penetrating rigid volumes according to user-defined penetration restoring force-deflection curves.

Rigid walls. Fig. A.4 shows the principle of rigid wall treatment, where a car may collide with an inclined rigid wall or where a car occupant may collide with a simplified model of the car interior, modeled with several rigid wall segments.

The rigid walls are impenetrable and the contacting bodies may slide along the walls with and without specified friction coefficients. The rigid walls may not be able to move, or they can be assigned some specified motion. Penetration of nodes into a rigid wall is prevented by an algorithm that sets the relative normal velocity to zero.

Nodal constraints. Nodal constraints are often used to tie together two non-matching finite element meshes, as it can arise when two internal organs are modelled independently and when the two organs have a common mesh interface. Instead of re-meshing the interfaces, one may simply identify pairs of nodes to move together.

Penalty methods. If penetration of a node into a segment is detected, it will be penalized in most contact algorithms by elastic restoring forces of fictitious (non-)linear springs, which are compressed by the amount of penetration. In dynamic contact events, dashpots can be added to dissipate energy and to prevent from spurious oscillation.

Fig. A.5 shows two potentially contacting surface segments. The algorithm checks if a slave node of the right segment touches the left master segment. Then the master and slave roles are reversed and the algorithm checks if a slave node of the left segment touches the right master segment. Inverting the master-slave roles prevents from contacts to be missed.

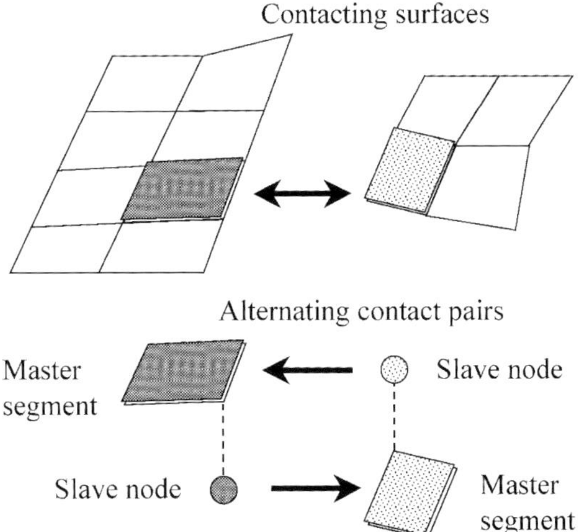

FIG. A.5. Contact interfaces.

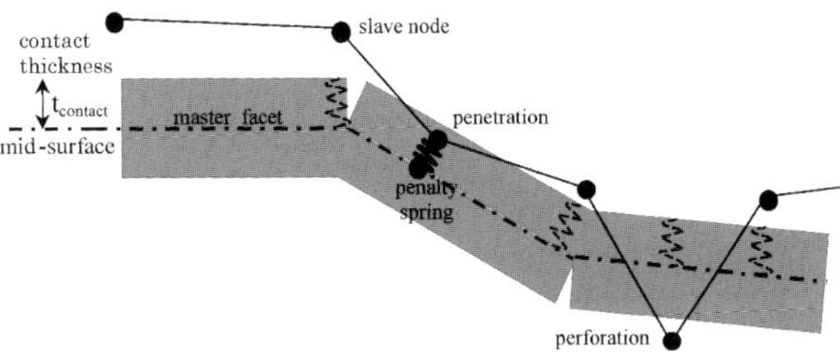

FIG. A.6. Penetration and perforation.

Fig. A.6 shows the application of a penalizing contact spring after "penetration" of a slave node into a master segment. If the slave node penetrates a distance beyond the contact thickness, $t_{contact}$, over the mid-surface of the master segment, then this node will have "perforated" the master segment and it will escape to the other side. Perforations can happen when the impact is too violent to be contained by the penalty springs.

APPENDIX B

Biomaterial Mechanical Properties

The following text and tables are extracted mainly from a report of the HUMOS-1 Project (YANG [1998]):

Yang, J. (1998) Bibliographic Study, Report 3CHA/980529/T1/DB, Chalmers University of Technology, SE-41296 Göteborg, Sweden.

The HUMOS project was funded by the European Commission under the Industrial and Materials Technologies program (Brite–EuRam III). This report has been prepared with the aim to provide input data for human impact models. It contains many references to the abundant literature on biomaterials, and the most important data are summarized hereunder. For simulations in impact biomechanics, the mechanical properties of the musculo–skeletal system are the most needed and known. The static and dynamic mechanical properties of the soft tissues and organs are the least known properties.

The mechanical properties of all living tissues undergo wide scatter through different ages, gender, biological, physical and loading conditions. Most of the listed properties delimit the observed scatter and can serve as guide lines and upper and lower bounds in human model calibration, and in stochastic analyses. Most biomaterials are non-linear, anisotropic, different in tension and compression and they are deformation rate dependent. The majority of the indicated values represent their average linear behavior. The properties of bone and soft tissue materials can depend much on the location in the human body, i.e., vary from member to member and within an individual part. Most properties were obtained from cadaver tests, which can only approximate their behavior in living bodies.

While the mechanical properties of the long bones (femur, tibia, fibula, humerus, ulna, radius) is a well studied subject the in biomechanics, data are much scarcer for pelvic bone, ribs, sternum, clavicle, scapula, short bones of the feet and other parts. For this reason the following data are concentrating on the properties of long bones.

B.1. Bone mechanical properties

Quasi-static properties of bone. Table B.1 contains the average quasi-isotropic properties of the most important skeletal bones, where N is the number of tested specimens, E is the linear isotropic Young's modulus, ν is Poisson's ratio, G is the shear modulus and σ_u is the ultimate stress.

Table B.2 shows the orthotropic stiffness matrix coefficients in directions 1, 2 and 3 of femoral cortical bone from 60 specimens (ASHMAN ET AL. [1984]), where axis 1 is

TABLE B.1
Elastic moduli and ultimate stress for bones (overview)

Authors	Bone	Test (N)	E [Gpa]	ν	G [Gpa]	σ_u [Mpa]
MCELHANEY [1966]	general cortical	quasi-static	20 ± 5		3.15 ± 0.25	16 ± 2.5
		strain rate 0.001	15.1			15.0
		strain rate 300	29.2			28.0
	general trabecular	compression	0.001–9.8			0.02–2.5
	vertebral bodies	compression				0.01–1.5
YAMADA [1970] SEDLIN ET AL. [1965] SEDLIN ET AL. [1966] REILLY, BURSTEIN and FRANKEL [1974] REILLY ET AL. [1975]	femur cortical	tensile	14–18			12–15
		compressive	15–19			14–21
		shear	15.5			7–8
		bending	3.28			16–18
		torsion				6.8
	tibia cortical	tensile	18–29			12–17
		compressive	25–35			18–21
		shear				7–8
YAMADA [1970]	femur	tensile	17.6			12.4
		(tensile dry)	(20.2)			(15.1)
	tibia	tensile	18.4			14.3
	fibula	tensile	18.9			14.9
	humerus	tensile	17.1			12.5
	radius	tensile	18.5			15.2
	ulna	tensile	18.8			15.1
	average	tensile	18.3			14.0
QUHAN [1989]	pelvis cortical	bending 63 yr (36)	5.26 ± 2.09			
		bending 23 yr (12)	3.76 ± 1.78			
	pelvis trabecular	bending 63 yr (29)	4.16 ± 2.02			
		bending 23 yr (13)	3.03 ± 1.63			
MCELHANEY [1970]	skull sandwich	radial comp.	2.4 ± 1.4	0.19 ± 0.08		740 ± 350
		tangential comp.	5.6 ± 3.0	0.22 ± 0.11		970 ± 360
		tangential tension	5.4 ± 2.9			430 ± 190
GRANIK and STEIN [1973]	ribs	bending	11.5 ± 2.1			11 ± 3
SACRESTE, BRUN-CASSAN, FAYON, TARRIERE, GOT and PATEL [1982]	ribs	bending	6.14 ± 4.26			8.6 ± 5.5

TABLE B.1
(*Continued*)

Authors	Bone	Test (N)	E [Gpa]	ν	G [Gpa]	σ_u [Mpa]
BURGHELE and SCHULLER [1968]	calcaneous	static (10 mm/min) intact bone				2100 [N] (mean force)
		dynamic (500 mm/min) intact bone				2620 [N] (mean force)
		cortical (largest over bone); cut specimens: 2 × 0.5 × 0.5 cm	2890–3200			2.87–3.79 (stress)
	talus	static (10 mm/min) intact bone				446 [N] (mean force)
		dynamic (500 mm/min) intact bone				468 [N] (mean force)

TABLE B.2
Orthotropic stiffness matrix [Gpa] for human femoral cortical bone

E_{11}	E_{22}	E_{33}	G_{12}	G_{13}	G_{23}	ν_{12}	ν_{13}	ν_{23}	ν_{21}	ν_{31}	ν_{32}
12.0	13.4	20.0	4.53	5.61	6.23	0.376	0.222	0.235	0.422	0.371	0.350

TABLE B.3
Transverse isotropic stiffness matrix for human femoral cortical bone

	E_{11}	E_{33}	ν_{31}	ν_{12}	G_{13}
Mean values [GPa]	11.5	17.0	0.46	0.58	3.28
Number of tests (N)	31	170	147	26	166
Standard deviation [%]	15–20% in tension 7–10% in compression		30%	30%	10%

the radial direction, axis 2 is the circumferential (or transverse) direction, axis 3 is the longitudinal direction of the cylindrical bone shaft, E_{ij} are the elastic modules, G_{ij} are the shear modules and ν_{ij} are Poisson's ratios.

Table B.3 contains transverse isotropic stiffness matrix coefficients for human femoral cortical bone ($E_{11} = E_{22}$) from a population over the age spans of 19–80 years (REILLY, BURSTEIN and FRANKEL [1974]), where N is the number of specimens, E_{11} is the elastic modulus for transverse or radial specimens, E_{33} is the elastic modulus for longitudinal specimens, ν_{12} is Poisson's ratio for transverse or radial specimens, ν_{31} is Poisson's ratio for longitudinal specimens and G_{13} is the shear modulus.

Table B.4 contains trabecular orthotropic stiffness properties found in the human proximal tibia from 3 males (ages 52, 55 and 67) (ASHMAN ET AL. [1986]).

TABLE B.4
Trabecular bone properties from human proximal tibias

Module	Average	Standard deviation	Range
E_{11} [MPa]	346.8	218	110–1230
E_{22} [MPa]	457.2	282	140–1750
E_{33} [MPa]	1107.1	634	340–3350
G_{12} [MPa]	98.3	66	30–380
G_{13} [MPa]	132.6	78	35–410
G_{23} [MPa]	165.3	94	45–460
Density [kg/m^3]	263.4	135	130–750

TABLE B.5
Trabecular bone properties from human pelvis

	E_{11}	E_{22}	E_{33}	G_{12}	G_{13}	G_{23}	ν_{12}	ν_{13}	ν_{23}	ν_{21}	ν_{31}	ν_{32}
Average [MPa]	59.8	57.3	43.2	26.0	22.6	22.6	0.18	0.24	0.21	0.17	0.16	0.14
Standard Dev.	44.9	44.6	39.9	19.1	17.1	17.2	0.11	0.14	0.16	0.10	0.07	0.09

Table B.5 lists the trabecular orthotropic stiffness properties of the pelvis bone (DALSTRA ET AL. [1993]).

Tables B.6 and B.7 list further trabecular bone mechanical properties, showing the great variety of these values.

Strain rate dependent properties of bone. Bone is strain rate dependent material. It exhibits stiffer and stronger behavior at high strain rate. Typical values found for cortical are listed in Table B.8.

Age dependent properties of bone. Mechanical properties of human bones change with age as a consequence of changes of density and mineral content. According to BURSTEIN ET AL. [1976], the elastic modulus and tensile strength of human bone decrease slowly after the age of about 45 years. CURREY [1975] showed that the elastic modulus and bending strength both increase with age until the age of about 30 years, but decrease thereafter. The decrease is associated with, and mainly caused by, the increased porosity and demineralization of bone. MCCALDEN ET AL. [1993] found that the change in porosity played a greater role in the reduction in strength than the change in mineral content. Both factors reduce the ability of bones to undergo plastic deformation before fracture starts. Table B.9 shows important trends of the age dependency of the mechanical properties of human bone with age. Table B.10 gives an example for the tensile and compression age dependent mechanical properties of the cortical bone of the femur. Similar trends with age are found in the other bones.

TABLE B.6
Compressive properties of trabecular bone [MPa]

(a) Proximal femur	Test piece	E	σ_u
BROWN [1980]	9.5 mm length	345	–
	5 mm cubes	–	1.2–3.1
CIARELLI ET AL. [1986], CIARELLI ET AL. [1991]	10 mm length	58–2248	2.1–16.2
	8 mm cubes	49–572	–
EVANS ET AL. [1961]	25 × 7.9 mm prisms	–	0.21–14.82
MARTENS ET AL. [1983]	8 mm diameter	1000–9800	0.45–15.6
SCHOENFELD ET AL. [1974]	7.9 mm cubes	20.68–965	–
	4.8 mm diameter	–	0.15–13.5
(b) Distal femur	**Test piece**	**E**	**σ_u**
BEHRENS [1974]	5 mm slab	–	2.25–66.2
CIARELLI ET AL. [1986], CIARELLI ET AL. [1991]	8 mm length	58.8–2942	–
	8 mm cubes	7.6–800	19
DUCHEYNE [1977]	5 mm diameter	–	0.98–22.5
(c) Proximal Tibia	**Test piece**	**E**	**σ_u**
BEHRENS [1974]	5 mm slab	–	1.8–63.6
	10.3 mm diameter	–	–
CARTER and HAYES [1977]	5 mm length	1.4–79	1.5–45
CIARELLI ET AL. [1986], CIARELLI ET AL. [1991]	8 mm cubes	–	0.52–11
GOLDSTEIN [1983]	7 mm diameter	8–457	1–13
	10 mm length		
HVID ET AL. [1985]	10 mm length	4–430	13.8–116.4
	5 mm slabs		
LINDAHL [1976]	14 × 9 mm		
	Male	34.6	3.9
	Female	23.1	2.2
LINDE [1989]	No constraint	113–853	
WILLIAMS ET AL. [1982]	5–6 mm cubes	10–500	1.5–6.7
(d) Vertebral bodies	**Test piece**	**E**	**σ_u**
ASHMAN ET AL. [1986]	5 mm diameter	158–378	
	10–15 mm length		
BARTLEY [1966]	Lumbar region	–	2.9
GALANTE ET AL. [1970]	7, 10 mm diameter	–	0.39–5.98
LINDAHL [1976]	10 × 9 × 14 mm		
	Male	55.6	4.6
	Female	35.1	2.7
MCELHANEY [1970]	10 mm length	151.7	4.13
ROCKOFF ET AL. [1969]	Lumbar region	–	0.69–6.9
STRUHL ET AL. [1987]	8&6 mm cubes	10–428	0.06–15
WEAVER ET AL. [1966]	10 mm cube	–	0.34–7.72
YAMADA [1970]	40–49 years	88.2	1.86
	60–69 years	68.6	1.37
(e) Pelvic trabecular	**Test piece**	**E**	**σ_u**
DALSTRA ET AL. [1993]	6.5 mm cubes	58.9	–

TABLE B.7
Shear properties of trabecular bone [MPa]

	Test piece	G	σ_u
ASHMAN ET AL. [1986]	5 mm diameter 10–15 mm length	58–89	

TABLE B.8
Cortical bone strain rate dependent properties

Cortical bone (MCELHANEY [1966])	Compression (human embalmed)	Compression (fresh cow bone)
Strain rate	from 0.001 to 300	from 0.001 to 300
E [GPa]	from 15.1 to 29.2	from 18.6 to 33
Strength [MPa]	from 150 to 280	from 175 to 280

TABLE B.9
Age dependency of bone properties

(a) *Cortical bone* ultimate tensile strength for the decade age groups (after ZHOU, Q. ET AL. [1996]).
(Reproduced by permission of The Stapp Association.)

Observations:
(1) The absolute strength values obtained by different authors are different, but the trend with age is similar.
(2) The maximum strength is reached around the 30–39 age group.
(3) The ultimate tensile strength of the elderly groups (70–89) drops by 15% to 25% compared to the maximum.

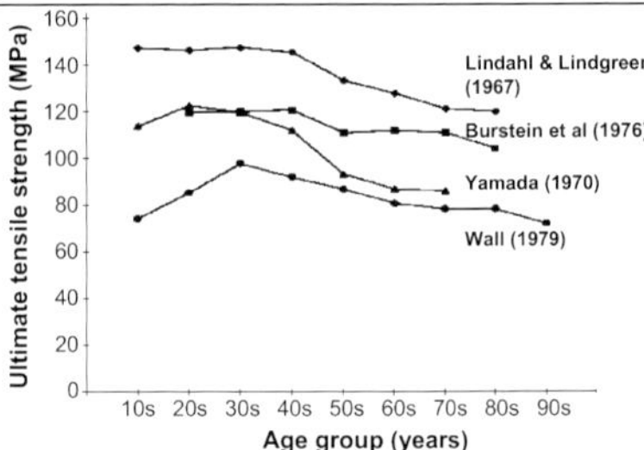

(b) *Femoral cortical bone* ultimate tensile stress for the decade age groups (after MCCALDEN ET AL. [1993]).
(Reproduced by permission of The Journal of Bone and Joint Surgery, Inc.)

The straight line represents the best fit with the use of linear regression analysis (from 253 specimens excised from 47 cadavers).

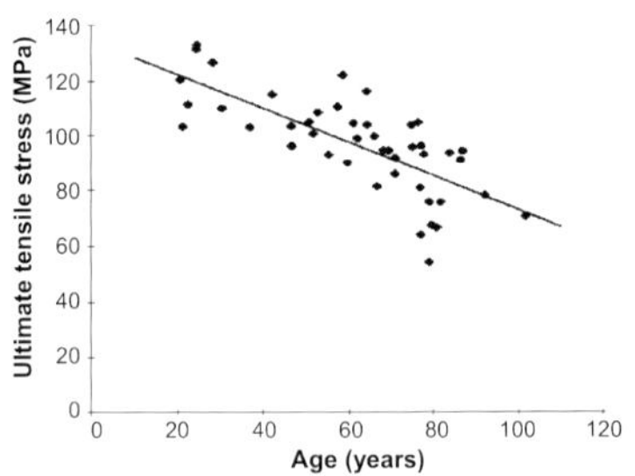

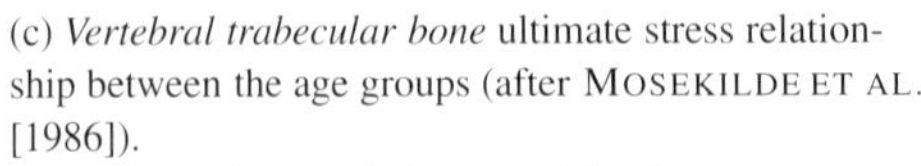

(c) *Vertebral trabecular bone* ultimate stress relationship between the age groups (after MOSEKILDE ET AL. [1986]).
(Reproduced by permission from Elsevier.)

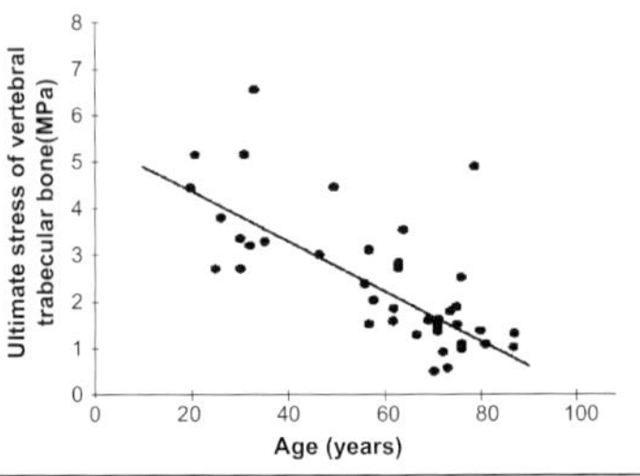

TABLE B.10
Age dependent femur cortical mechanical properties

(a) Tensile properties of femur		Elastic modulus [MPa]	Yield stress [MPa]	Ultimate stress [MPa]
BURSTEIN ET AL. [1976]	20–29 years	17000	120	140
	30–39 years	17600	120	136
	40–49 years	17700	121	139
	50–59 years	16600	111	131
	60–69 years	17100	112	129
	70–79 years	16300	111	129
	80–89 years	15600	104	120
(b) Compressive properties of femur		**Elastic modulus [MPa]**	**Yield stress [MPa]**	**Ultimate stress [MPa]**
BURSTEIN ET AL. [1976]	20–29 years	18100		209
	30–39 years	18600		209
	40–49 years	18700		200
	50–59 years	18200		192
	60–69 years	15900		179
	70–79 years	18000		190
	80–89 years	15400		180
(c) Bending properties of femur		**Elastic modulus [MPa]**	**Yield stress [MPa]**	**Ultimate stress [MPa]**
YAMADA [1970]	20–29 years			151
	30–39 years			174
	40–49 years			174
	50–59 years			162
	60–69 years			154
	70–79 years			139
	80–89 years			139
(d) Torsion properties of femur		**Elastic modulus [MPa]**	**Yield stress [MPa]**	**Ultimate stress [MPa]**
YAMADA [1970]	20–29 years	3430		57
	30–39 years	3430		57
	40–49 years	3140		52.7
	50–59 years	3140		52.7
	60–69 years	2940		48.6
	70–79 years	2940		48.6
	80–89 years	–		48.6

B.2. Ligament mechanical properties

Ligaments are fibrous materials with pronounced non-linear force-displacement response. The mechanical properties of some ligaments are summarized in Table B.11, where v_d is the deflection rate, D is the deflection at rupture, ε is the strain at rupture, E is the elastic modulus. For reasons of practical feasibility, ligaments are most often

TABLE B.11
Some ligament mechanical properties

Specimen type and conditions (source)	v_d [mm/ms]	D [mm]	F [kN]	ε [%]	E [kN/mm^2] (= GPa)	Other properties
(a) Collagen (VIIDIK [1987])						$\sigma_u = 0.045$–0.120 GPa
(b) Typical force-deformation curve for ligament for monotonic forcing (FRANK and SHRIVE [1994]) (Reproduced by permission of the University of Calgary.)	(I) = toe region; (II) = linear region; (III) = region of micro-failure; (IV) = failure region. At the top are schematic representations of fibers going from crimped (I) through recruitment (II) to progressive failure (III and IV)					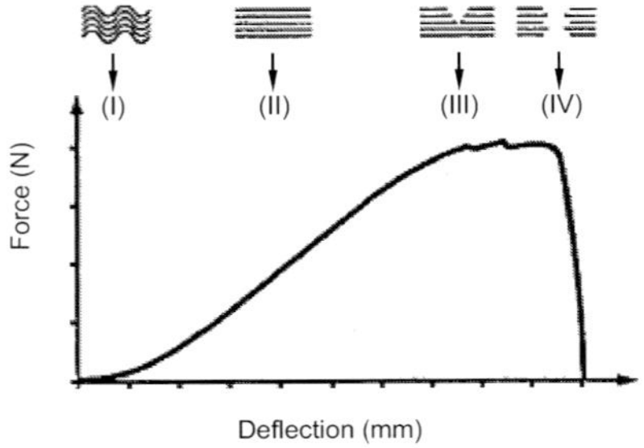
(c) Human cruciate ligament (NOYES and GROOD [1978]) Non-linear stress-strain behavior (Reproduced by permission of the Journal of Bone and Joint Surgery, Inc.)		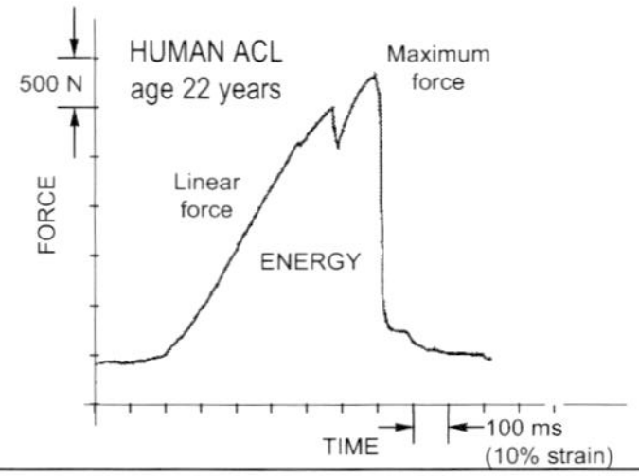				

TABLE B.11
(*Continued*)

Specimen type and conditions (source)	v_d [mm/ms]	D [mm]	F [kN]	ε [%]	E [kN/mm^2] (= GPa)	Other properties
(d) Human calcaneo-fibular and tibio-talar ligaments (BEGEMAN and AEKBOTE [1996]) Static and dynamic isolated bone-ligament-bone test pieces (Reproduced by permission of Prof. Begeman, Wayne State University, Bioengineering Center.)	maximum 0.4 mm/ms with cycle load					(viscoelastic relaxation)
(e) Human ankle/foot ligaments (ATTARIAN, MCCRACKIN, DEVITO, MCELHANEY and GARRETT [1985])						
(i) Anterior tibio-fibular ligament Cyclic load to isolated bone-ligament-bone test piece Effect of strain rate (Reproduced by permission of Foot & Ankle Intl.)	1.50E−04 1.50E−02 1.00E+00	3	0.016 (m) 0.02 (m) 0.042 (m)			

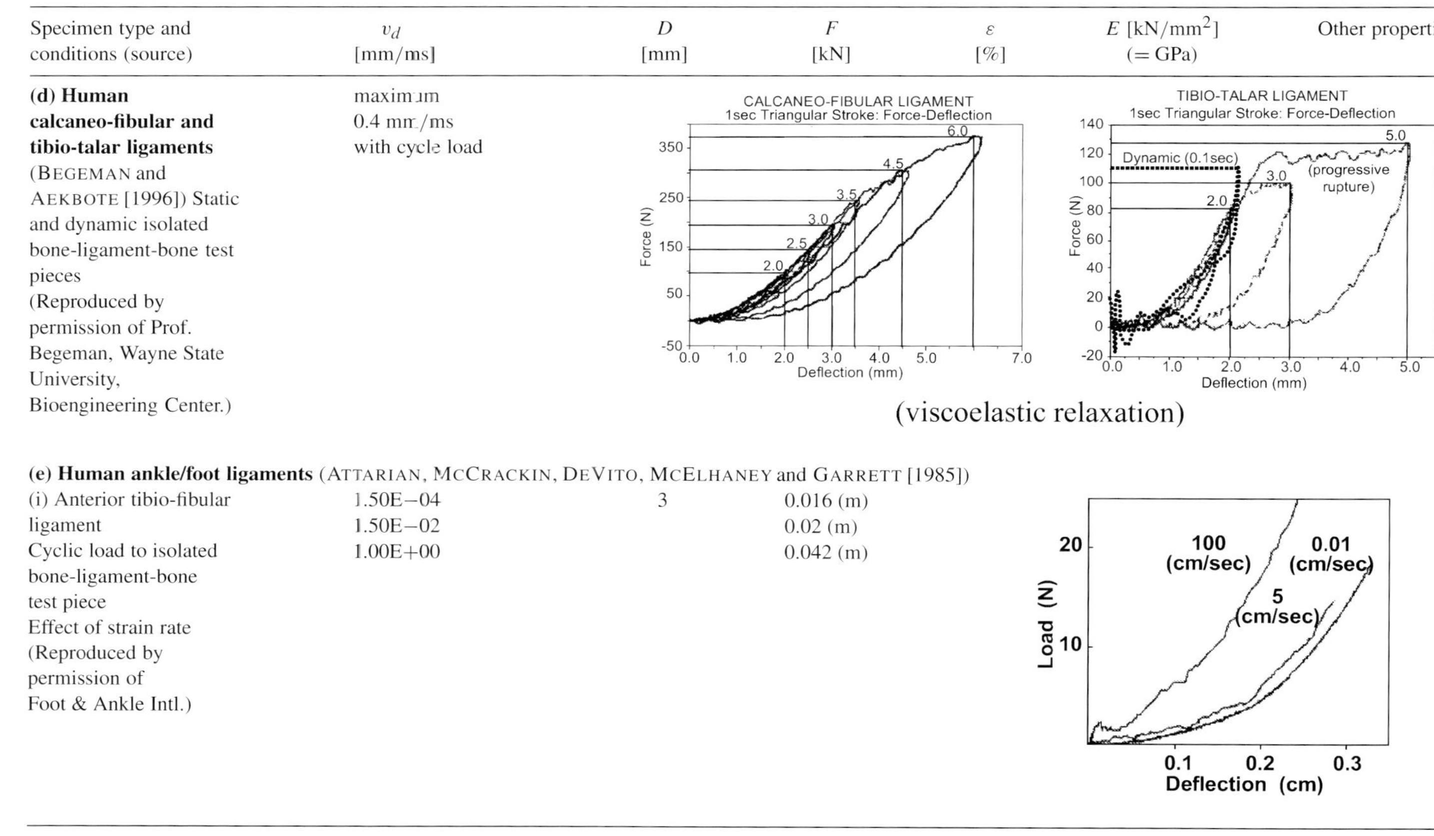

TABLE B.11
(Continued)

Specimen type and conditions (source)	v_d [mm/ms]	D [mm]	F [kN]	ε [%]	E [kN/mm^2] (= GPa)	Other properties
(ii) Calcaneo-fibular ligament Effect of strain rate (Reproduced by permission of Foot & Ankle Intl.)	1.50E−04 1.50E−02 1.00E+00	3	0.029 (m) 0.031 (m) 0.087 (m)		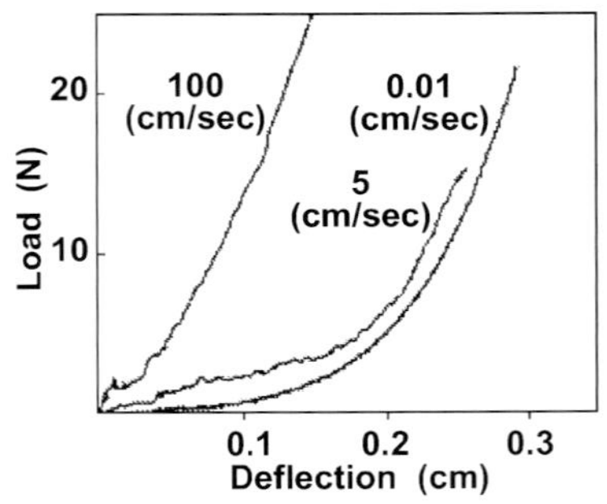	
(iii) Anterior tibio-fibular (12 specimens)	1.01 ± 0.07	5.1 ± 0.5 (f)	0.1389 ±0.0235 (max)	53 ±6 (f)	3.999E−02 ±0.854E−02	maximum load = force required to completely disrupt a ligament (grade III sprain)
(iv) Calcaneo-fibular (16 specimens)	1.06 ± 0.04	6.3 ± 0.5 (f)	0.3457 ±0.0552 (max)	38 ±3 (f)	7.051E−02 ±0.690E−02	
(v) Posterior talo-fibular (4 specimens)	0.82 ± 0.13	13.1 ± 1.6 (f)	0.2612 ±0.0324 (max)	100 ±15 (f)	3.975E−02 ±1.379E−02	
(vi) Tibiotalar ligament (6 specimens)	0.80 ± 0.13	10.5 ± 1.1 (f)	0.7138 ±0.0693 (max)	210 ±23 (f)	1.2282E−01 ±0.2504E−01	

TABLE B.11
(*Continued*)

Specimen type and conditions (source)	v_d [mm/ms]	D [mm]	F [kN]	ε [%]	E [kN/mm^2] (= GPa)	Other properties
(f) Anterior cruciate ligament (WAINWRIGHT, BIGGS and CURREY [1979] (Reproduced by permission of the Princeton University Press.)						
(g) Anterior cruciate ligament (WOO, PETERSON and OHLAND [1990] (Reproduced by permission of the Orthopaedic Research Society.)			force-elongation curves:			

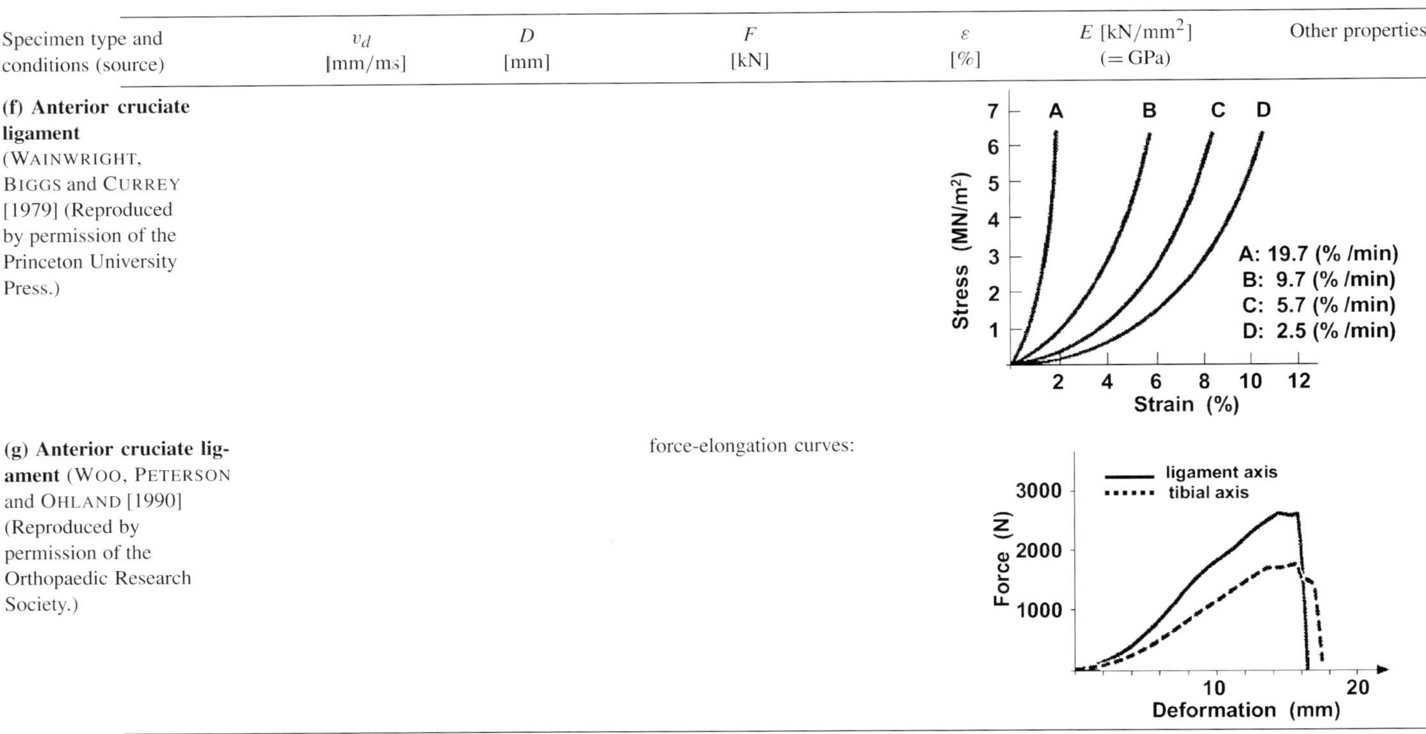

TABLE B.11
(*Continued*)

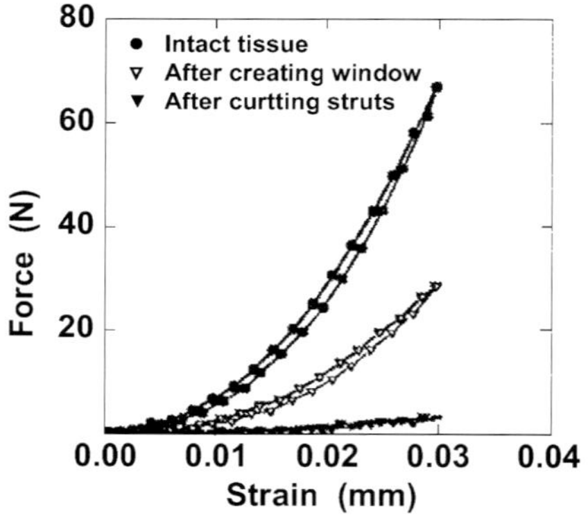

Specimen type and conditions (source)	v_d [mm/ms]	D [mm]	F [kN]	ε [%]	E [kN/mm^2] (= GPa)	Other properties
(h) Mediacollateral ligament						
			$\tau_{\max} = 4.0\text{e–}04\text{–}8.0\text{e–}04$Gpa			
shear stress-deformation (SHELTON, BUTLER and FEDER [1993]) (Reproduced by permission of ASME International.)			force-strain curves:			
(k) Human ACL in tension (HAUT [1993])						
(16–26 years)			1.73 ± 0.27 (u)	44.3 ± 8.5 (f)	0.111 ± 0.026	$\sigma_u = 37.8 \pm 9.3$ MPa (f)
(22–35 years)			2.16 ± 0.175 (u)			
(48–86 years)					0.065 ± 0.024	

TABLE B.11
(*Continued*)

Specimen type and conditions (source)	v_d [mm/ms]	D [mm]	F [kN]	ε [%]	E [kN/mm^2] (= GPa)	Other properties
(l) Spinal ligaments (PANJABI, JORNEUS and GREENSTEIN [1984]) (Reproduced by permission of the Orthopaedic Research Society.)			tensile response curves:			
(m) Human ankle/foot ligaments (PARENTEAU, VIANO and PETIT [1996])						
(i) force-strain curves (Reproduced by permission of ASME International.)			force-strain curves:			

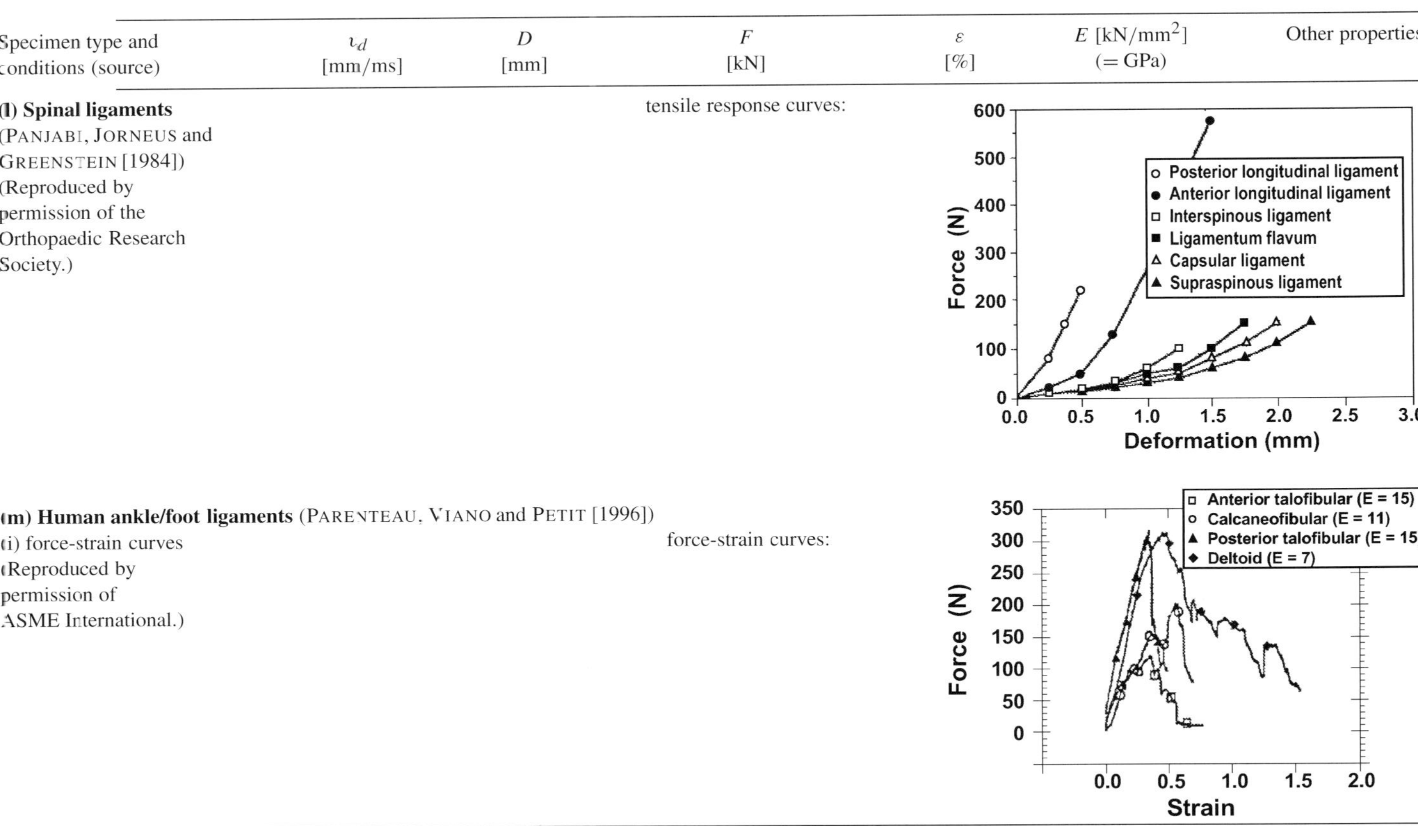

TABLE B.11
(*Continued*)

Specimen type and conditions (source)	v_d [mm/ms]	D [mm]	F [kN]	ε [%]	E [kN/mm^2] (= GPa)	Other properties
(ii) anterior talo-fibular ligament (5 specimens)			0.014–0.114 (y) 0.064–0.286 (u)	11–15 24–44		$k = 0.006$–0.023 kN/mm
(iii) calcaneo-fibular ligament (6 specimens)			0.053–0.259 (y) 0.120–0.290 (u)	21–39 30–84		$k = 0.018$-0.044 kN/mm
(iv) posterior talo-fibular ligament (1 specimen)			0.307 (y) 0.307 (u)	35 35		$k = 0.037$ kN/mm
(v) deltoid ligament (6 specimens)			0.119–0.355 (y) 0.239–0.507 (u)	10–50 45–79		$k = 0.026$–0.068 kN/mm
(vi) talo-calcaneal ligament (5 specimens)			0.061–0.117 (y) 0.078–0.156 (u)	13–33 13–33		$k = 0.021$–0.031 kN/mm
(vii) plantar ligament (6 specimens)			0.199–0.511 (y) 0.238–0.506 (u)	8–27 12–32		$k = 0.039$–0.495 kN/mm
(viii) metatarsal ligament (4 specimens)			0.075–0.220 (y) 0.103–0.247 (u)	11–40 0–54		$k = 0.016$–0.038 kN/mm
(n) Human ankle/foot ligaments (NIGG, SKARVAN and FRANK [1990])						
(i) anterior talo-fibular ligament			0.067–0.193	6–60		
(ii) calcaneo-fibular ligament			0.265–0.327	27–81		
(iii) deltoid ligament			0.173–0.315	34–58		

(f) = failure; (u) = ultimate; (m) = mean

tested as bone-ligament-bone test pieces, where the ligament and connected bones are isolated and where the bones are subjected to relative displacements in the direction of the fibers of the attached ligament. In the table, (f) means "failure", (u) means "ultimate" and (m) means "mean".

B.3. Brain mechanical properties

A synthesis of existing bibliography on the biomaterial behavior characterization of the brain is presented in Tables B.12 and B.13. A list of variables is given after Table B.14.

B.4. Joint mechanical properties

The physiological motions of the synovial joints of the human skeleton can be modeled approximately by computationally efficient point-like mechanical joint elements, with six more or less constrained motion degrees of freedom. Motion ranges, stiffness and resistance properties are found in the literature (ex: ROBBINS [1983], ROBBINS, SCHNEIDER, SNYDER, PFLUG and HAFFNER [1983]). Further data are given in Tables B.15 and B.16, in a coordinate system formed by the sagittal, coronal and horizontal planes. The rotation movements of joints are performed about the longitudinal axis of a body segment. The anatomically forbidden joint motions are often penalized by stiff springs, while the natural motions are hardly penalized within the admissible motion ranges and strongly penalized at the limits of these ranges. The corresponding stiffness is also shown in Tables B.15 and B.16. Fig. B.1 shows typical non-linear moment-rotation curves of the hip joint and of the subtalar joint of the ankle. The friction coefficients of the synovial joints of the human skeleton vary between 0.005 and 0.04.

In cases when more detail about the local motion and response under load is required, the joints can be modeled with their true articular surface geometry and their connecting ligaments. Detailed finite element models are also applied in orthopedic analysis. The gain in precision is obtained at a higher computational cost.

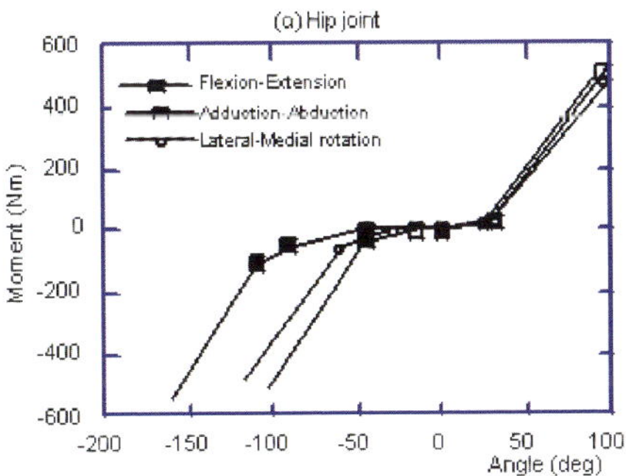

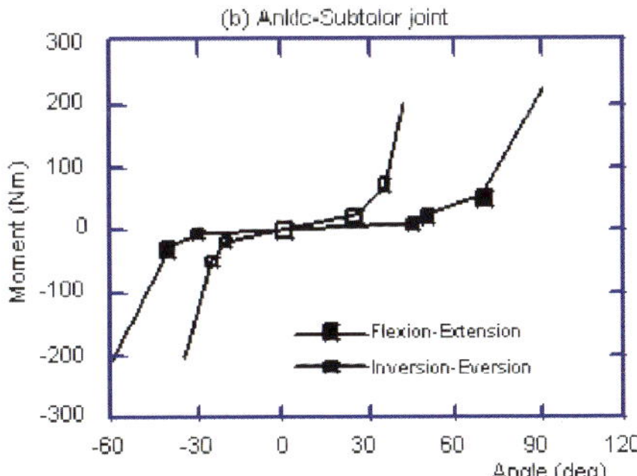

FIG. B.1. Typical moment-angle characteristics of the joints: (a) hip joint, and (b) ankle-subtalar joint (from YANG and LÖVSUND [1997] and PARENTEAU and VIANO [1996]). (Reproduced by permission of Chalmers University of Technology.)

TABLE B.12
Brain material properties

Authors	Experimental configuration	Characteristics					
	type and conditions	ρ [kg/mm^3]	E [GPa]	ν	K [GPa]	G [GPa]	Other
DIMASI, MARCUS and EPPINGER [1991]	linear visco elastic				6.895E−02	$G_s = 3.4474E{-}05$ $G_l = 1.723E{-}05$	$\beta = 0.100\ \mathrm{m\,s^{-1}}$
ESTES and MCELHANEY [1970]	incompressible with strain-rate independent bulk modulus (human, monkey)					2.070E+00	
FALLENSTEIN, HULCE and MELVIN [1970]	dynamic complex shear modulus by vibration tests at 10 Hz (human autopsy brain)					$G_1 = 6.00E{-}07$ to 1.10E−06 $G_2 = 3.00E{-}07$ to 6.50E−07	
FIROOZBAKSHK [1975]		1.000E−06				1.920E−04	
GALFORD and MCELHANEY [1970]	dynamic complex tensile modulus by vibration test at 34 Hz (human brain tissue)		$E_1 = 6.67E{-}05$ $E_2 = 2.62E{-}05$				
GOLDSMITH [1972]			7.800E−05		4.320E−10		
LEE, MELVIN and UENO [1987]		1.000E−06		0.49		8.000E−05	
MCELHANEY, MELVIN, ROBERTS and PORTNOY [1973]	dynamic complex shear modulus by vibration tests at 9–10 Hz					$G_1 = 4.30E{-}07$ to 9.50E−07 $G_2 = 3.50E{-}07$ to 6.00E−07	$G2/G1 = 0.72$

TABLE B.12
(Continued)

Authors	Experimental configuration type and conditions	Characteristics ρ [kg/mm^3]	E [GPa]	ν	K [GPa]	G [GPa]	Other
MARGULIES and THIBAULT [1989]		1.060E−06	4.000E−05	0.5	2.000E+00	1, 380E−05	
OMMAYA [1967]			8.00E−06 −1.5E−05				
ROSE and GORDON [1974]		1.050E−06			2.100E+00		
RUAN, KHALIL and KING [1991]		1.040E−06	6.670E−05	0.48–0.499	2.190E+00	1.680E−03	
RUAN, KHALIL and KING [1994]	linearly viscoelastic				1.279E−01	G_s = 5.28E−04 G_l = 1.68E−04	$\beta = 0.035\ \mathrm{m\,s^{-1}}$
SHUGAR [1975]		6.720E−07	1.030E−05	0.5	2.100E+00	3.450E−06	
SHUCK, HAYNES and FOGLE [1970]	dynamic complex shear modulus by vibration tests at 2–400 Hz					G_1 = 8.30E−07 to 1.38E−04 G_2 = 3.40E−07 to 8.27E−05	
SHUCK and ADVANI [1972]	dynamic complex shear modulus by vibration tests at 5–350 Hz					G_1 = 7.60E−06 to 3.39E−05 G_2 = 2.76E−06 to 8.16E−05	

TABLE B.12
(*Continued*)

Authors	Experimental configuration	Characteristics					
	type and conditions	ρ [kg/mm^3]	E [GPa]	ν	K [GPa]	G [GPa]	Other
THIBAULT and MARGULIES [1996]	Age effect on complex shear modulus in custom-designed oscillatory shear testing device, at shear strain 2.5% from 20–200 Hz, 25 °C, 100% humidity only in one location and one direction (neonatal (2–3 days) pigs: curve data)					$G = 7.500E{-}07$ to 1.5938E−06 $G_1 = 6.875E{-}07$ to 1.1875E−06 $G_2 = 1.875E{-}07$ to 1.00E−06	
TROSSEILLE, TARRIERE, LAVASTE, GUILLON and DOMONT [1992]		1.000E−06	2.400E−04	0.49–0.499			
UENO, MELVIN, LI and LIGHTHALL [1995]			2.400E−04	0.49	4.000E−03	8.000E−05	
WARD and THOMSON [1975]		1.040E−06	6.670E−05	0.48			
WARD [1982]			6.500E−04	0.48–0.499			
TURQUIER, KANG, TROSSEILLE, WILLINGER, LAVASTE, TARRIERE and DÖMONT [1996]		1.140E−6	6.750E−04	0.48	5.625E−03	$G_s = 5.28E{-}04$ $G_l = 1.68E{-}04$	$\beta = 0.035\ \mathrm{m\,s^{-1}}$

TABLE B.13
Brain material properties (white vs. gray matter and CSF)

Authors	Experimental configuration type and conditions	Characteristics					
		ρ [kg/mm^3]	E [GPa]	ν	K [GPa]	G [GPa]	Other
White matter:							
TADA and NAGASHIMA [1994]			5.000E−05	0.49			perm. = 1.0E−11 mm^2 poros. = 0.2
ZHOU, C. ET AL. [1996]		1.040E−06		0.4996	3.490E−01	2.680E−04	
ZHOU, C. ET AL. [1996]	SHUCK and ADVANI [1972]				2.190E−400	G_s = 4.10E−05 G_l = 7.6E−06	β = 0.70 ms^{-1}
Gray matter:							
TADA and NAGASHIMA [1994]			5.000E−04	0.49			perm. = 1.0E−14 mm^2 poros. = 0.2
ZHOU, C. ET AL. [1996]		1.040E−06		0.4996	2.190E−01	G = 1.68E−04	
ZHOU, C. ET AL. [1996]	SHUCK and ADVANI [1972]				2.190E+00	G_s = 3.40E−05 G_l = 6.3E−06	β = 0.70 ms^{-1}
CSF (cerebro-spinal fluid):							
RUAN, KHALIL and KING [1991]		1.040E−06		0.489	2.190E−02	5.000E−04	
RUAN, KHALIL and KING [1991]		1.000E−06			2.190E+00		
RUAN, KHALIL and KING [1994]		1.040E−06			2.190E−02	5.000E−05	
TADA and NAGASHIMA [1994]			9.930E−10	0.49			perm. = 1.0E−05 mm^2 poros. = 0.99

TABLE B.14
Brain material properties (other tissues)

Tissue & authors	Experimental configuration type and conditions	Characteristics					
		ρ [kg/mm^3]	E [GPa]	ν	K [GPa]	G [GPa]	Other
Cerebellum, brainstem RUAN, KHALIL and KING [1994]		1.040E−06		0.4996	2.19E−01	1.68E−04	
Pia mater ZHOU, Q. ET AL. [1996]		1.133E−06	1.150E−02	0.45			
Bridging veins ZHOU, Q. ET AL. [1996]		1.133E−06	1.100E−04	0.45			
Dura, falx & tentorium ZHOU, Q. ET AL. [1996]		1.133E−06	3.150E−02	0.45			
Dura/Falx MCELHANEY, MELVIN, ROBERTS and PORTNOY [1973]	tensile at frequency: 6.66E−05 msec^{-1} 6.66E−04 msec^{-1} 6.66E−03 msec^{-1}		4.157E−02 4.435E−02 6.069E−02				
MELVIN, MCELHANEY and ROBERTS [1970]			4.1382E−02 to 5.5176E−02				
Membrane RUAN, KHALIL and KING [1993]		1.133E−06	3.150E−02	0.45			
Dura Mater YAMADA [1970]	square 15 mm on each side expansive properties (curve Fig. 198 p. 222) (sample: cerebral rabbit)						$\sigma_u^e = 3.8\text{E}{-}02 \pm 0.0018$ [kg/mm^2] $\sigma_{\lim}^e = 0.2\sigma_u^e$
	rectangular strap 2 cm wide and 2.5 cm long						$T = 1.26$ [kg/mm]
	shearing properties (human adult average)						$\tau_u = 1.98$ [kg/mm^2]

TABLE

List of variables used in Tables B.11 through B.14

ρ = density	E_1 = storage tensile modulus
E = Young's modulus	E_2 = loss tensile modulus
ν = Poisson's ratio	β = decay factor
K = bulk modulus	σ_u^e = ultimate expansive strength
G = shear modulus	σ_{lim}^e = initial expansive strength
G_1 = storage shear modulus	T = shearing breaking load per unit width
G_2 = loss shear modulus	τ_u = ultimate shearing strength
G_e = equivalent shear modulus	perm. = permeability
G_l = long term shear modulus	poros. = porosity
G_s = short term shear modulus	

The shear relaxation behavior is described by the time dependent shear modulus $G(t) = G_l + (G_s - G_l)e^{-\beta t}$, where t is time in a relaxation test.

B.5. Inter-vertebral joint mechanical properties

Fig. B.2 indicates the average range of motion for rotations between the vertebrae, where C2–C7 are the vertebrae of the cerebral spine, T1–T12 of the thoracic spine, L1–L5 of the lumbar spine and S1 indicates the sacrum. The range of rotational motion between C1 (atlas) and C2 (axis) is about 35 degrees and accounts for about 50% of the rotation of the head. The range of motion also differs between individuals, sexes and is strongly age-dependent, decreasing by about 50% from youth to old age. (See Table B.17.)

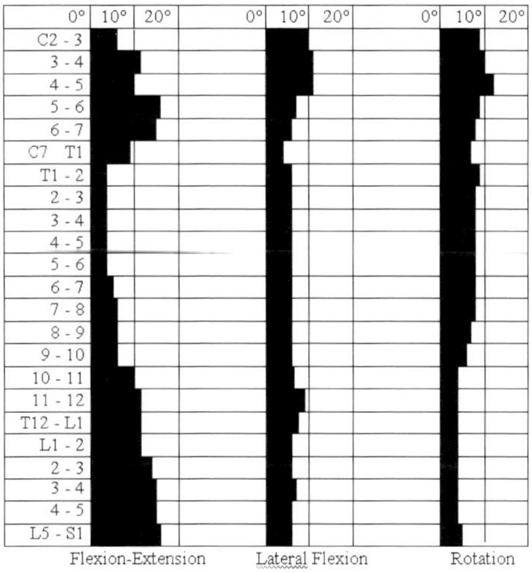

FIG. B.2. Inter-vertebral joint range of motion. (Reproduced by permission of Chalmers University of Technology.)

TABLE B.15
Upper limb joint mechanical properties

Joint	Authors	Experimental configuration	Characteristics			
			Physiological feature	Motion range [deg]	Stiffness [Nm/deg]	Mechanical feature
Shoulder	KAPANDJI [1974a]	Physiological motion	Flexion	180	0–0.3	Ball joint
			Extension	45–50	0–0.2	
			Abduction	140	0–0.3	
			Adduction	30–45	–	
	FRANKEL and NORDIN [1980]	Static load	Lateral rotation	80	0.3	
			Medial rotation	95	0.3	
Elbow	KAPANDJI [1974a]	Physiological motion	Flexion	145–160	0–0.2	Pin joint
			Extension	0	–	
			Abduction	0	–	
			Adduction	0	–	
			Lateral rotation	90	0.2	
			Medial rotation	85	0.2	
Wrist	KAPANDJI [1974a]	Physiological motion	Flexion	85		Cardan joint
			Extension	85		
			Abduction	15		
			Adduction	45		

TABLE B.16
Lower limb joint mechanical properties

Joint	Authors	Experimental configuration	Characteristics			
			Physiological feature	Motion range [deg]	Stiffness [Nm/deg]	Mechanical feature
Hip	KAPANDJI [1974b]	Physiological motion	Flexion	90–145	0–2.5	Ball joint
			Extension	20–30	1.2	
	FRANKEL and NORDIN [1980]		Abduction	45	0–1.2	
	MOW and HAYES [1991]	Static load	Adduction	30	0.8	
			Lateral rotation	30	0.6	
			Medial rotation	60	0.6	
Knee	KAPANDJI [1974b]	Physiological motion	Flexion	120–160	0–1.2	Pin joint
			Extension	5–10	2.0	
	FRANKEL and NORDIN [1980]		Abduction	0	–	
			Adduction	0	–	
			Lateral rotation	40	1.0	
			Medial rotation	30	1.0	
Ankle	KAPANDJI [1974b]	Physiological motion	Tibia-Talar:			Pin joint
			Dorsiflexion	20–30	0.5	
			Plantarflexion	30–50	0.3	
	PARENTEAU and VIANO [1996]	Quasi-static load	Subtalar:			Ball joint
			Inversion	15–20	1	
			Eversion	10–15	1.5	
			Abduction	30–40		
			Adduction	25–35		

TABLE B.17
Inter-vertebral joint mechanical properties

Joint	Authors	Experimental configuration	Characteristics			
			Physiological feature	Motion range [deg]	Stiffness [Nm/deg]	Other
Lumbar	KAPANDJI [1974a]	Physiological motion	Flexion	60	1.0–2.1	Tolerance:
	FRANKEL and NORDIN [1980]		Extension	35	0.3–1.8	$M_{flex} = 145–185$ Nm
			Lateral flexion	±20	2.0	
		Static load	Axial rotation	±5	0.9	
Thoracic	KAPANDJI [1974a]	Physiological motion	Flexion	45	1.0–2.1	Tolerance:
	FRANKEL and NORDIN [1980]		Extension	25	0.3–1.8	$M_{flex} = 616$ Nm
			Lateral flexion	±20	2.0	$M_{ext} = 240$ Nm
		Static load	Axial rotation	±35	0.9	
Cervical	KAPANDJI [1974a]	Physiological motion	Flexion	40	1.4	
			Extension	75	2.5	
	WISMANS and SPENNY [1983]	Dynamic load				
		lateral bending	Lateral flexion	±45	0.4–2.2	
		torsion	Axial rotation	±50	0.0–0.5	
	MCELHANEY, DOHERTY, PAVER, MYERS and GRAY [1988]	Dynamic			10.5–67.5 [Nm/rad]	
	MERTZ and PATRIC [1971]	flexion–extension				Tolerance: $M_{flex} = 56.7$ Nm $M_{ext} = 189$ Nm
	BOWMAN, SCHNEIDER, LUSTAK, ANDERSON and THOMAS [1984]	Torsion			2.71–3.74	

TABLE B.18
Articular cartilage and meniscus mechanical properties

Tissue	Authors	Experimental configuration	Characteristics			
			Density [kg/m^3]	Young's modulus [MPa]	Poisson's ratio ν	Other material coefficients
Articular Cartilage	MOW and HAYES [1991]	Tensile Solid matrix		0.3–1.0	0.1–0.4	
	YAMADA [1970]	Compressive		> 1.0		
	VIANO [1986]	Instantaneous		$E = 12$	0.42	$G = 4.1$ MPa $K = 2.5$ MPa
		Asymptotic		$E = 7.1$	0.37	$G = 2.6$ MPa $K = 9.1$ Mpa
	ARMSTRONG, LAI and MOW [1984]		1000	$E = 35$	0.45	
Meniscus	MOW and HAYES [1991]	Tensile		0.1–0.6		

B.6. Articular cartilage and meniscus mechanical properties

Cartilage is known to behave as a biphasic material, where a fluid seeps through a solid porous matrix, which can lead to slow deformation under compressive loads. For the short term behavior in impact studies these time and load dependent properties are not dominant, and only the classical stiffness terms are required. To describe the complete flow and deformational behavior of cartilage and meniscus the biphasic theory was developed by Mow and coworkers. In this theory the solid matrix is linearly elastic and isotropic, the solid matrix and interstitial fluid are intrinsically incompressible and viscous dissipation is due to interstitial fluid flow relative to the solid matrix. (See Table B.18.)

B.7. Inter-vertebral disc mechanical properties

The inter-vertebral discs assure the elastic coherence of the spinal column and they provide a shock absorbing effect. Each disc is composed of the nucleus pulposus, the annulus fibrosus, and a cartilageous end-plate. Between 70 to 90% of the nucleus pulposus by weight is water, and it takes up as much as 40 to 60% of the disc area. The annulus fibrosus is a laminated and hence an anisotropic structure composed of several concentric layers with fibers alternating at plus and minus 30 degree angles of inclination from the horizontal plane. The inner boundary of the annulus fibrosus is attached to the cartilageous end-plate, and the outer surface is directly connected to vertebra body. The discs play dominant role in sustaining the body against compressive load. Under compression, the nucleus pulposus acts like a fluid in a cylinder made of the annulus fibrosus. The discs show greater stiffness for the front/rearward inter-vertebral shearing motion than for the side/side motion, with the annulus fibrosus rather the nucleus pulposus making a major contribution.

The mechanical properties of the inter-vertebral discs are summarized in Table B.19. Table B.20 summarizes age and region dependent tensile properties of the discs and Tables B.21 and B.22 contain the average region dependent compressive and torsional properties (SONODA [1962]). Note that inter-vertebral discs from people in age group 20–39 have the greatest ultimate loads. Discs of females have a breaking moment about five-sixths of that in males. The ultimate torsional strength and angle of twist in whole discs of females are also less than in males.

B.8. Muscle mechanical properties

Skeletal muscles have active and passive properties, Fig. B.3. The active muscle action is usually not of prime importance in car occupant impact simulations, except in low energy collisions, where the activation of the muscles during bracing can modify the injury pattern (example: rear impact/whiplash). In the following tables mostly the passive mechanical behaviour of the skeletal muscles is documented.

A quasi-linear visco-elastic model was proposed to model the passive response of skeletal muscle. Within the physiological muscle length, the passive muscle force is usually much lower than the active force. At high elongation, nearing the physiological limits of joint motion, the passive force increases rapidly and reaches the same level and beyond as the maximum active force, while the active force drops to low values. At high stretch, the axial muscle force is therefore dominated by the passive force. At negative stretch velocities the active muscle force drops to almost zero at a given reference velocity, while at positive stretch velocities this force will increase beyond the activation level at zero stretch velocity. The Hill muscle model is often evoked in simulation of the active and passive kinetics of skeletal muscles. This law is described in Appendix C.

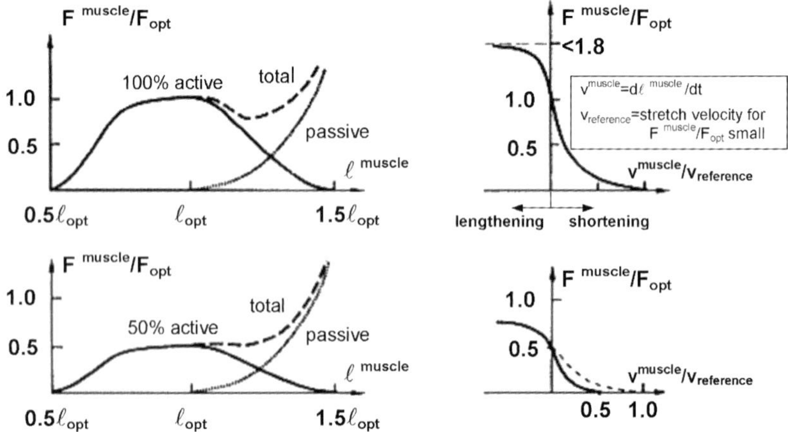

FIG. B.3. Muscle force-length and muscle force-velocity properties (ZAJAC [1989]). (Reproduced by permission of Begell House, Inc.)

TABLE B.19
Inter-vertebral disc mechanical properties

Tissue	Authors	Experimental configuration	Characteristics				
			Density	Young's modulus [MPa]	Poisson ratio ν_{12}	Shear modulus G_{12} [MPa]	Other
Fibers	GALANTE [1967]	Tensile lumbar region		400–500 $E_1 = 500$ $E_2 = 500$	0.3	192	
Ground substance	GOEL, MONROE, GILBERTSON and BRINKMAN [1995]	Compressive lumbar		2–4.2			
	UENO and LIU [1987]	Torsion lumbar		$E_1 = 3$ $E_2 = 3$	0.45	1	
Lamellae	SKAGGS, WEIDENBAUM, IATRIDIS, RATCLIFFE and MOW [1994]	Tensile Posterior of disc Anterior of disc		 $E_1 = 70 \pm 42$ $E_1 = 106 \pm 72$			
	KULAK, BELYTSCHKO, SCHULTZ and GALANTE [1976]			$E_1 = 83$ $E_2 = 2.07$	0.45	1.38	
Annulus fibrosus	EBARA, IATRIDIS, SETTON, FOSTER, MOW and WEIDENBAUM [1996]	Tensile lumbar		$E_{circ} = 5–50$			
Annulus fibrosus	SPILKER, JAKOBS and SCHULTZ [1986]	Modeling		$E_{circ} = 33.4$ $E_z = 0.9$	0.5	0.189	
	LIN, LIU, RAY and NIKRAVESH [1978]			$E_{circ} = 22.4$ $E_z = 11.7$	0.45	3.92	

TABLE B.20
Age and region dependent tensile inter-vertebral disc properties

Region	Characteristics								
	20–39 yr			40–79 yr			Adult average		
	F_u [kg]	σ_u [kg/mm^2]	ε_u [%]	F_u [kg]	σ_u [kg/mm^2]	ε_u [%]	F_u [kg]	σ_u [kg/mm^2]	ε_u [%]
Cervical	105 ± 14.5	0.33 ± 0.02	89 ± 4.2	80 ± 8.6	0.29 ± 0.03	71 ± 3.6	88	0.30	77
Upper thoracic	142 ± 16.3	0.24 ± 0.01	55 ± 3.8	106 ± 9.4	0.20 ± 0.02	41 ± 2.1	118	0.21	46
Lower thoracic	291 ± 21.5	0.26 ± 0.02	57 ± 6.3	220 ± 12.8	0.22 ± 0.01	40 ± 2.4	244	0.23	46
Lumbar	394 ± 24.6	0.30 ± 0.01	68 ± 7.1	290 ± 19.5	0.24 ± 0.01	52 ± 6.2	325	0.26	59
Average	233	0.28	67	174	0.24	51	194	0.25	57
Ratio	1	1	1	0.75	0.85	0.76	0.83	0.89	0.85

F_u = breaking load in [kg]
σ_u = ultimate strength in [kg/mm^2]
ε_u = ultimate elongation in [%]

TABLE B.21
Region dependent compressive inter-vertebral disc properties

Region	Characteristics (40 to 59 years of age)		
	Breaking Load [kg]	Ultimate Strength [kg/mm^2]	Ultimate Contraction [%]
Cervical	320	1.08	35.2
Upper thoracic	450	1.02	28.6
Lower thoracic	1150	1.08	31.4
Lumbar	1500	1.12	35.5
Average	(855)	1.08	32.6

TABLE B.22
Region dependent torsional inter-vertebral disc properties

Region	Characteristics (40 to 59 years of age)		
	Breaking Moment [kg cm]	Ultimate Strength [kg/mm^2]	Ultimate Angle of Twist [deg]
Cervical	51	0.48	34
Upper thoracic	84	0.41	26
Middle thoracic	167	0.44	22
Lower thoracic	265	0.45	17
Lumbar	440	0.48	14
Average	201	0.45	23

TABLE B.23
Ultimate tensile strength of skeletal muscles (KATAKE [1961])

Body segment	Muscles	Ultimate tensile strength [MPa]
Trunk	Sternocleidomastoideus	0.19
	Trapezius	0.16
	Pectoralis major	0.13
	Rectus abdominis	0.14
Upper extremity	Biceps brachii	0.17
	Triceps brachii	0.21
	Flexor carpi radialis	0.15
	Brachioradialis	0.18
Lower extremity	Psoas major	0.12
	Sartorius	0.30
	Gracilis	0.20
	Rectus femoris	0.10
	Vastus medialis	0.15
	Adductor longus	0.13
	Semimembranous	0.13
	Gastrocnemius	0.10
	Tibialis anterior	0.22

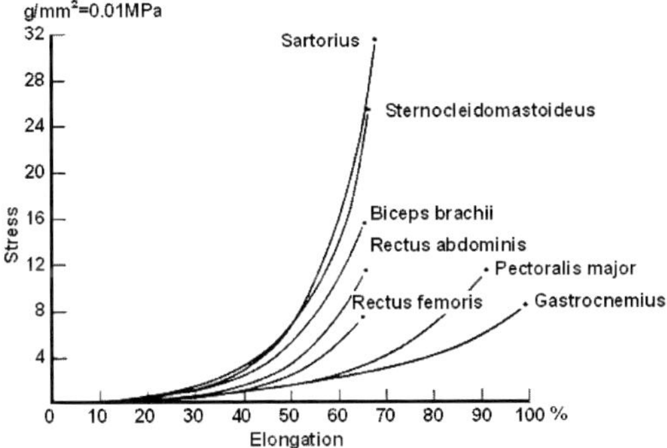

FIG. B.4. Passive stress-strain curves in tension of skeletal muscles for 29 year old persons (YAMADA [1970]). (Reproduced by permission of Lippincott, Williams and Wilkins.)

Table B.23 contains the ultimate tensile strength of a selection of skeletal muscles. The corresponding ultimate passive muscle forces will be obtained by multiplying the strength by the physiological cross section areas of each muscle. Fig. B.4 shows the non-linear passive stress-strain curves of the muscles. Table B.24 contains age differences in passive tensile properties of the rectus abdominis muscle. There is no significant sexual difference in the ultimate strength of skeletal muscles. (See Table B.25.)

B.9. Tendon mechanical properties

Tendons connect muscles to the skeletal bones. There is almost no age effect on the ultimate elongation, δ_u, a moderate age effect on the elastic modulus, E, and a marked

TABLE B.24
Age differences in passive tensile properties of rectus abdominis muscle (KATAKE [1961])

	Age group							Adult
	10–19 yr	20–29 yr	30–39 yr	40–49 yr	50–59 yr	60–69 yr	70–79 yr	average
	Ultimate tensile strength [g/mm^2 = 0.01 Mpa]							
	19 ± 1.2	15 ± 0.6	13 ± 1.0	11 ± 0.6	10 ± 0.5	9 ± 0.3	9 ± 0.3	11
ratio	1.00	0.79	0.68	0.58	0.53	0.47	0.47	
	Ultimate percentage elongation							
	65 ± 1.2	64 ± 1.1	62 ± 0.7	61 ± 0.9	61 ± 1.5	58 ± 1.8	58 ± 1.8	61
ratio	1.00	0.98	0.95	0.94	0.94	0.89	0.89	

TABLE B.25
Summary of studies on mechanical properties of muscles

Authors	Experimental configuration	Characteristics	
		σ_u [MPa]	Other
YAMADA [1970]	Experiments on various animal and human specimens	0.1–0.32	
WINTERS and STARK [1985], WINTERS and STARK [1988]		0.5–1.0	
SCHNECK [1992]	C = damping coefficient K = stiffness of whole muscle	0.2–1.0	C = 10–1000 Ns/m K = 32.5–250 kN/m

age effect on the ultimate tensile strength, σ_u. The ultimate tensile strength can exceed the insertion strength of a tendon. There is a significant strain rate effect on the elastic modulus. The longitudinal tendon strips of the supraspinatus muscle are not of equal strength. (See Table B.26.)

B.10. Skin mechanical properties

Skin is a non-linear elastic material and its response is orthotropic. In human models, skin should be modeled as the enveloping membrane of flesh and fatty tissue. Skin assures the stability of the female breasts, which consists of fatty tissue, enveloped by resistant skin. The behavior of skin is best modeled as a material with perpendicular layers of fibers, each described with non-linear stress-elongation curves (see Figs. B.5–B.7). Some mechanical properties are listed in Table B.27.

B.11. Internal organ mechanical properties

Rather little information is available on the mechanical properties of the internal organs. Table B.28 contains some preliminary data on the lungs, liver and spleen. The data on

TABLE B.26
Mechanical properties of tendons

Type of tissue	Authors	Experimental configuration type and conditions	E [GPa]	ν Poissons ratio	σ_u [MPa] Ultimate strength	δ_u Ultimate elongation
	MOW and HAYES [1991]		1.2–1.8	Incomp.		
Achilles tendon	LEWIS and SHAW [1997a], LEWIS and SHAW [1997b]	Tension (fiber direction) $0.1 \sec^{-1}$ and $1 \sec^{-1}$ embalmed, age 36–100	2.00 ± 0.99	0.4		
Patellar tendon	JOHNSON, TRAMAGLINI, LEVINE, ONO, CHOI and WOO [1994]	Tension (fiber direction) fresh frozen, non-irradiated				
		(i) age 29–50	0.66 ± 0.266		64.7 ± 15	0.14 ± 0.06
		(ii) age 64–93	0.504 ± 0.222		53.6 ± 10	0.15 ± 0.05
Human tendon	VOIGT, BOJSEN-MOLLER, SIMONSEN and DYHRE-POULSEN [1995]	Tension (fiber direction)	1.2		50	0.06 (yield point)
Patellar tendon	WOO, JOHNSON and SMITH [1993]	Tension (fiber direction) Stress–Relaxation	0.58			
Finger flexor	PRING, AMIS and COOMBS [1985]	Tension (fiber direction)				0.13
Supraspinatus	ITOI, BERGLUND, GRABOWSKI, SCHULTZ, GROWNEY, MORREY and AN [1995]	Tension (fiber direction) fresh				
		(i) anterior			16.5 ± 7.1	
		(ii) middle			6.0 ± 2.6	
		(iii) posterior			4.1 ± 1.3	

the liver and spleen are estimations used in a project on virtual abdominal surgery, DAN and MILCENT [2002].

Fig. B.8 shows the non-linear force-displacement curve of an entire human liver under compression between two parallel plates under quasi-static loading, DAN [1995]. The tested liver was pressurized in the sense that the natural in- and outflow of body

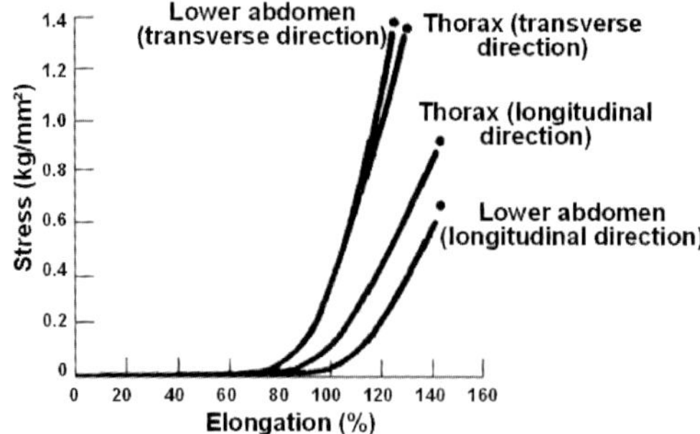

FIG. B.5. Stress-strain curves in tension of the skin of persons 20 to 29 years of age (YAMADA [1970]). (Reproduced by permission of Lippincott, Williams and Wilkins.)

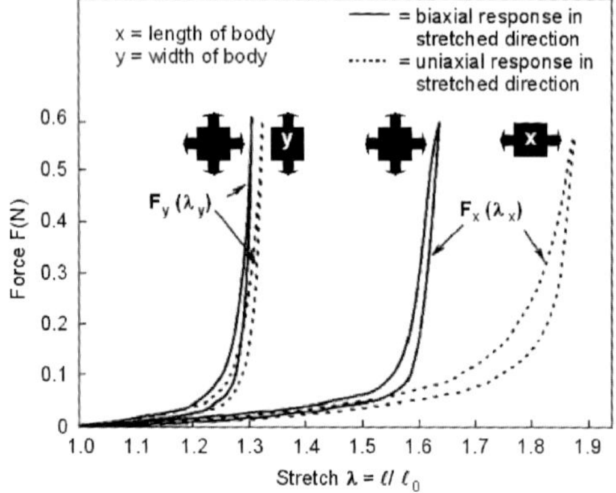

FIG. B.6. Force-stretch relation of rabbit skin at stretch rate 0.2 mm/s (LANIR and FUNG [1974]). (Reproduced by permission of the Journal of Biomechanics.)

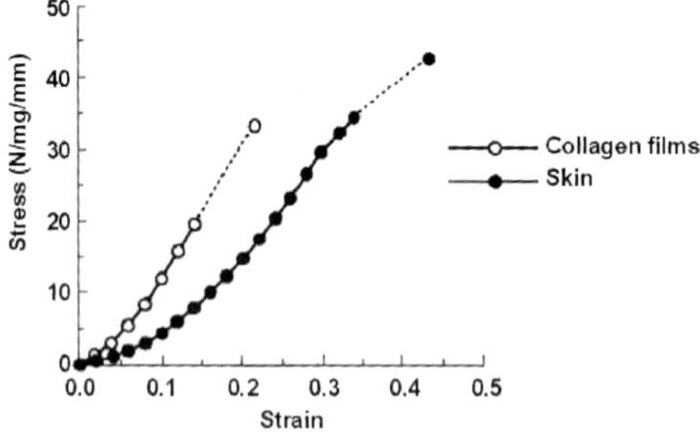

FIG. B.7. Stress-strain relation of rat skin (OXLUND and ANDREASSON [1980]). (Reproduced by permission of Blackwell Publishing Ltd.)

fluids was maintained artificially with a circulating substitute fluid. The liver as a whole body was then loaded. It resists the applied compression load with the combined action of the stored circulating fluid and the resistance of its bulk material.

TABLE B.27
Mechanical properties of skin

Tissue	Authors	Experimental Configuration type and conditions	E [Gpa]	Tensile breaking load per unit width [kg/mm^2]	σ_u [UTS] [kg/mm^2]	δ_{max} Ultimate percentage elongation
Skin of the calf	MANSCHOT and BRAKKEE [1986a], MANSCHOT and BRAKKEE [1986b]	Tension Slow rates in Vivo				
		(i) along tibial axis	Non-linear			
		(ii) across tibial axis	Non-linear			
Skin	YAMADA [1970]	Tension human skin				
		(i) 30–49 years	Non-linear	0.2 to 3.4	0.29 to 1.47	–
		(ii) 10–29 years	Non-linear	–	–	56 to 144
		(iii) Adult average	Non-linear	0.2 to 3.1	0.26 to 1.32	43 to 111

TABLE B.28
Mechanical properties of some organs

Tissue	Authors	Experimental configuration type and conditions	Density ρ [kg/m^3]	Modulus E [Mpa]	Poisson Ratio ν	Shear mod. G [Mpa]
Lungs	HOPPIN, LEE and DAWSON [1975]	Compression (3-D loading)		$E/Pt^* = 4$	0.3	
	HAJJI, WILSON and LAI-FOOK [1979]					$G/07Pt = 1$ to 1.5
Liver	DAN [1999], DAN and MILCENT [2002]	Small test cylinders of parenchyma	1158	0.3E–4 to 5E–4	0.4	
		Venous vessels	1168	0.158	0.49	
		Glisson capsule	1168	1.0	0.49	
Spleen	CARTER [1999]	(Indentation tests for the study of force feedback in virtual surgery)		(much smaller)		

$^*Pt$ = the transpulmonary pressure

Fig. B.9 represents static compression results of small test cylinders of pure liver parenchyma (diameter 12 mm by 15 mm initial height), from which the interstitial fluid can escape, DAN [1999]. It indicates a non linear distribution of the elastic modulus,

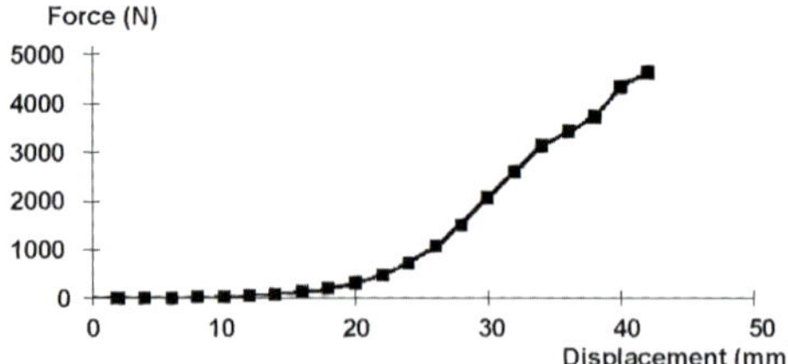

FIG. B.8. Whole human liver force vs. displacement curves (DAN [1995]). (Private communication by the author.)

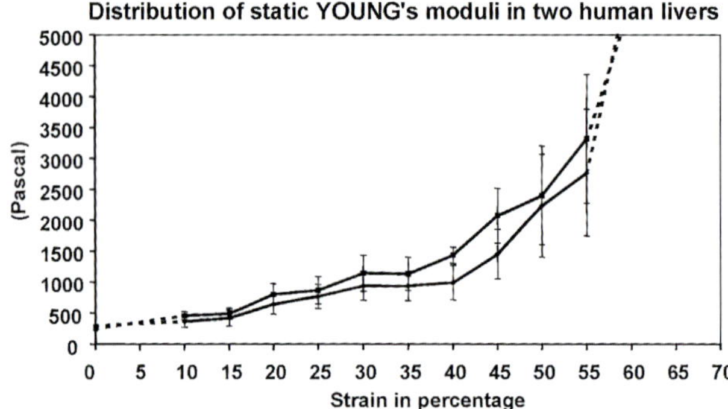

FIG. B.9. Liver parenchyma elastic modulus vs. displacement curves (DAN [1999]). (Private communication by the author.)

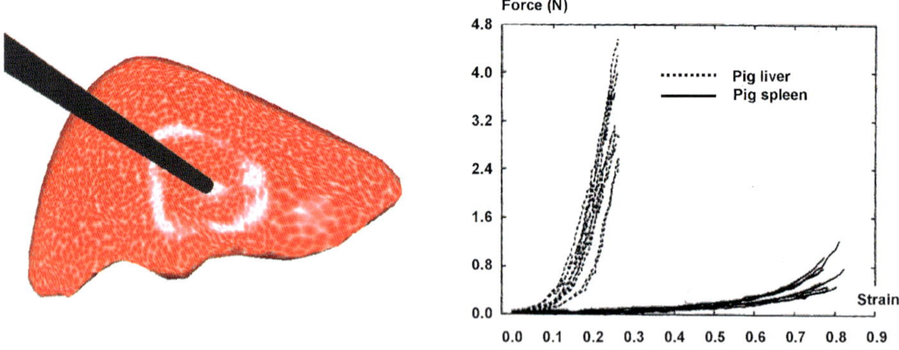

FIG. B.10. Indentation tests on the liver and spleen of pigs (CARTER [1999]). (Reproduced by permission of Fiona Carter of the University of Dundee.)

and the response of the so isolated material is much weaker than the response of the pressurized complete organ.

Fig. B.10 shows results of indentation tests on pig livers and spleens, CARTER [1999], which were performed in projects on the study of haptic force feedback in virtual surgery. The picture shows that the spleen is much less resistant than the liver.

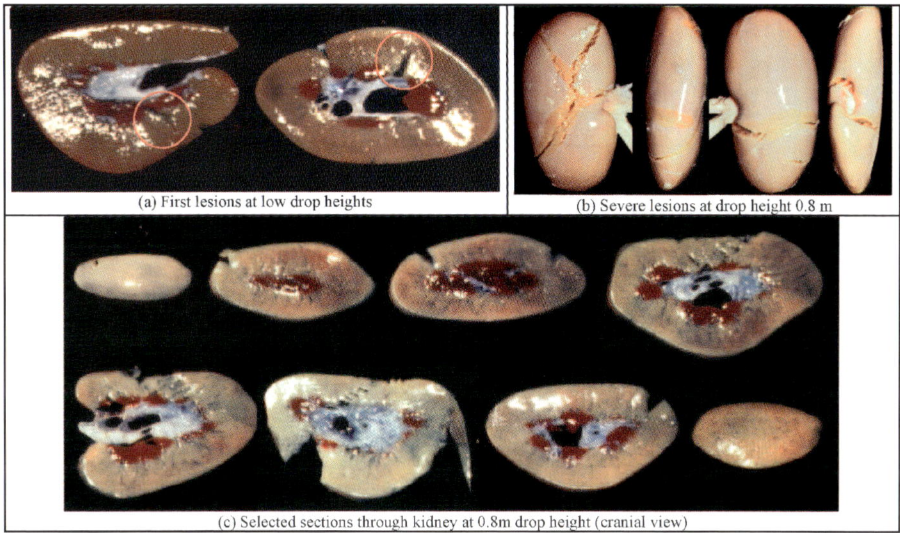

FIG. B.11. Typical drop test lesions in porcine kidneys (BSCHLEIPFER [2002]). (Reproduced by permission of Dr. med. Thomas Bschleipfer.)

FAZEKAS, KOSA, JOBBA and MESZARO [1971a], FAZEKAS, KOSA, JOBBA and MESZARO [1971b], FAZEKAS, KOSA, JOBBA and MESZARO [1972] (in German) published data on the compression resistance of the whole human cadaver liver, kidneys and spleen, respectively. They found that the liver showed superficial ruptures at a compressive stress of 169 kPa and multiple ruptures at 320 kPa. The first superficial ruptures of the spleen occurred at a compressive *stress* of 44 kPa and of the kidneys at a *load* of 60.2 ± 28.2 daN, the latter of which showed multiple ruptures at loads of 109.44 ± 51.4 daN.

BAUDER [1985] (in German) investigated the compressive resistance of the isolated human liver with blunt drop weight impact tests. The tests showed for 3–4 m/s impact velocities mean compressive loads of the organ of 175.6 ± 39.2 daN, which were associated with mean compressive deformations of 29.5 ± 3.5 mm. The observed injuries were contusions, superficial ruptures and crushing of the livers. The thickness of the organs and the portion of connective tissue were important parameters for the severity of the injuries.

BSCHLEIPFER [2002] investigated the lesions inflicted on isolated porcine kidneys under blunt drop test impacts, from heights of 0.1 to 1.0 m, with a cylindrical impactor (∅10 cm) of mass 1.45 kg (1.4 to 14.2 Joule), with and without ligatured urethers. Typical lesions found in the tests are shown in Fig. B.11, for the first lesions at low load, (a), for severe lesions at 0.8 m drop height, views, (b), and sections through the organ, (c).

Fig. B.12, finally, gives an overview on the nonlinear visco-elastic modulus response (Pa) of live porcine livers, based on in vivo semi-infinite elastic body indentation with a vibrating cylindrical indenter, as a function of vibration frequency (Hz) and median

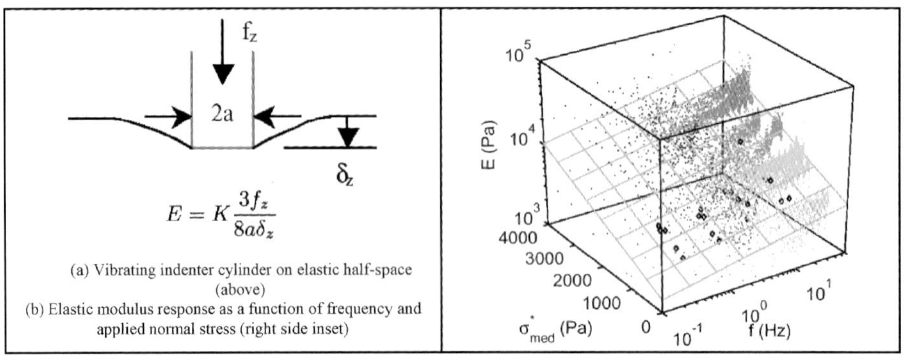

FIG. B.12. In vivo elastic modulus measurements of live porcine livers (OTTENSMEYER and SALISBURY [2001]). (Reproduced by permission of Springer Verlag.)

applied normal stress (Pa) (after OTTENSMEYER and SALISBURY [2001]). This work was performed for calibrating instrument force feedback in virtual surgery.

Summary Table B.29 is taken from YAMADA [1970]. It contains average ultimate strength (stress) and percentage elongation (strain) for a variety of human tissues and organs.

Data about many internal organs of humans are still missing. Table B.30 lists some such data found from animals (YAMADA [1970]).

Further data on liver and kidneys of rhesus monkeys are reported by MELVIN, STALNAKER and ROBERTS [1973]. These data are used by MILLER [2000] in modelling constitutive relationships of the abdominal organs.

B.12. Quasi-linear visco-elastic solids

Many biological tissues exhibit visco-elastic behaviour. A simple such law is described next. The deviatoric response of a linear viscoelastic solid material is governed by Zener's model, which can be considered as a Maxwell spring-dashpot model in parallel with a spring, Fig. B.13.

The "slow motion" response to small strain rates, $\mathrm{d}\varepsilon/\mathrm{d}t$, is governed by the long term shear modulus, $G_\infty(\equiv G_l)$, while the instantaneous response to a step loading, $H(t)$, is according to the long term modulus $G_0(\equiv G_s)$.

The elastic behaviour of this material is described by the deformation rate dependent shear modulus, G, and by the constant bulk modulus, K.

$$\text{Shear modulus} \quad G = \frac{E}{2(1+\nu)} \quad \text{where } E = \text{Young's Modulus}$$

and

$$\text{Bulk modulus} \quad K = \frac{E}{3(1-2\nu)} \quad \nu = \text{Poisson's Ratio.}$$

TABLE B.29

Average adult human mechanical tissue properties (YAMADA [1970])

Tissue	σ_u [kg/mm^2]	δ_{max} [%]	Tissue	σ_u [kg/mm^2]	δ_{max} [%]
Hair	19.7	40	Ureter (L)	0.18	36
Compact bone (femur)	10.9	1.4	Mixed arterial tissue (L)	0.17	87
Chorda tendinea	6.4	33	Venous tissue (L)	0.17	89
Tendinous tissue (calcaneal)	5.4	9	Umbilical cord (mature fetus)	0.15	59
Nail	1.8	14	Mixed arterial tissue (T)	0.14	69
Fascia	1.4	16	Muscular arterial tissue (L)	0.14	102
Nerve (secondary fiber bundle)	1.3	18	Spinal dura mater (T)	0.13	34
Fibrocartilage (annulus fibrosus) (L)	1.3	14	Spongy bone (vertebra)	0.12	0.6
Skin (thorax, neck)	1.3	90	Coronary artery (L)	0.11	64
Spinal dura mater (L)	1.1	21	Renal calyx (L)	0.11	35
Skin (abdomen, back, foot, arm)	0.97	90	Elastic arterial tissue (T)	0.10	82
Skin (leg, hand)	0.74	90	Muscular arterial tissue (T)	0.10	75
Sclera (E)	0.69	17	Cardiac valve (R)	0.094	17
Fibrocartilage (annulus fibrosus) (T)	0.53	12	Elastic arterial tissue (L)	0.08	80
Sclera (M)	0.48	17	Large intestine (L)	0.069	117
Skin (face, head, genitals)	0.38	69	Esophagus (L)	0.06	73
Vertebra	0.35	0.8	Stomach (L)	0.056	93
Cornea	0.35	15	Small intestine (L)	0.056	43
Auricle	0.34	26	Small intestine (T)	0.053	89
Elastic cartilage (auricle)	0.31	26	Renal calyx (T)	0.048	48
Thyroid cartilage (L)	0.30	15	Large intestine (T)	0.045	137
Venous tissue (T)	0.30	66	Ureter (T)	0.045	89
Hyaline cartilage (costal)	0.29	18	Stomach (T)	0.044	127
Intervertebral disc	0.28	57	Tracheal membranous wall (T)	0.036	81
Cardiac valve (C)	0.25	13	Urinary bladder	0.024	126
Tracheal cartilage	0.24	18	Papillary muscle tissue	0.023	30
Amnion (normal labor)	0.24	42	Esophagus (T)	0.018	124
Renal fibrous capsule	0.23	29	Skeletal muscle tissue (rectus abdominis)	0.011	61
Tracheal membrane wall (L)	0.22	61	Cardiac muscle tissue	0.011	64
Tracheal intercartlagnious membrane	0.19	138	Renal parynchyma	0.005	52

(C) = circumferentially, (E) = equatorially, (L) = longitudinally, (M) = meridionally, (R) = radially, (T) = transversely.

TABLE B.30
Mechanical properties of some internal organs of animals (YAMADA [1970])

Tissue	Animal	σ_u [kg/mm^2]	δ_{max} [%]
Liver parenchyma	Rabbits	0.0024	46
Gall bladder	Rabbits	0.21	53
Uterus	Rabbits	0.018	150
Cerebral dura mater	Rabbits	0.038	

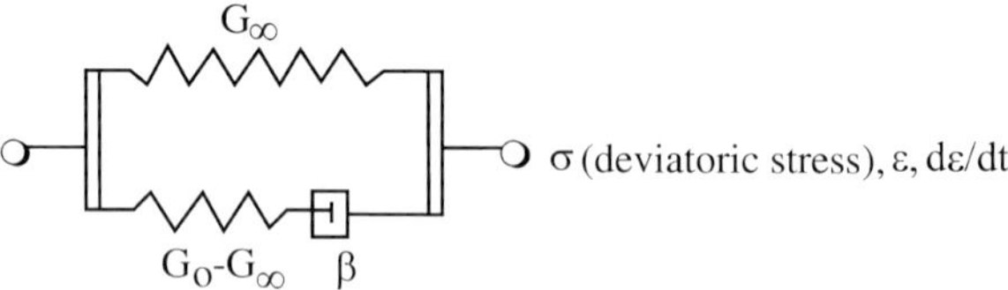

FIG. B.13. Zener type model.

The shear relaxation behaviour is given by the response to a step function, and is described by the shear relaxation modulus

$$G(t) = G_\infty + (G_0 - G_\infty)e^{-\beta t}\left(\equiv G_l + (G_s - G_l)e^{-\beta t}\right).$$

Time t is the current simulation time, but relaxation starts only when the material experiences a deviatoric strain. The decay constant, β, has the unit of (time)$^{-1}$, which must be consistent with the chosen time units.

The deviatoric stress rate, s_{ij}, depends on the shear relaxation modulus as follows

$$s_{ij} = 2\int_{\tau=0}^{t} G(t-\tau)D'_{ij}(\tau)\,\mathrm{d}\tau,$$

where D'_{ij} is the deviatoric velocity strain tensor. The above convolution expression for the deviatoric stress rate may be understood as follows: the deviatoric strain is approximated by a piecewise constant function. The material responds to each step function $H(\tau)$ following the relaxation law. This material model does not describe volumetric viscous effects, as might be present in the compression of foams.

B.13. Further references on biomaterials

The literature on biomaterials and related subjects is abundant. Some references are indicated in this appendix. While many of the indicated references deal with the experimental evaluation of biomaterial properties, others deal with the aspects of their modeling, the use of these materials in biomechanical models and the characterization of trauma and injury.

Many further references for biomaterials could be cited by separate topics on bones, ligaments, brain, joints, spine and inter-vertebral discs, muscles, tendons, skin and organs. These references are not mentioned explicitly in this appendix, but constitute further valuable sources of historical and actual information.

APPENDIX C

The Hill Muscle Model

Skeletal muscles. The Hill muscle model is one of the simplest phenomenological engineering models of the active and passive biomechanical behaviour of skeletal muscles (HILL [1970]). Its simplest implementation is with bar finite elements. More involved implementations can be in 2D and 3D composite finite elements, where the composite fibers are assigned the properties of Hill-type muscle models.

Fig. C.1 gives an overview on the anatomical detail of skeletal muscles.

Active voluntary muscle contraction can be considered a material behaviour that has no parallel in conventional engineering material models. Whereas the passive mechanical impact behaviour of biological tissues can often be approximated using standard engineering material models, active muscle behaviour clearly distinguishes living and non-living materials. For this reason it is interesting to briefly outline the standard Hill muscle model.

Standard Hill muscle model. This model, its implementation into a crash code and its application is described by WITTEK and KAJZER [1995], WITTEK and KAJZER [1997]; WITTEK, HAUG and KAJZER [1999]; WITTEK, KAJZER and HAUG [1999]; WITTEK, ONO and KAJZER [1999]; WITTEK, ONO, KAJZER, ÖRTENGREN and INAMI [2001]. Authors KAJZER, ZHOU, KHALIL and KING [1996] describe the application of modeling of ligaments and muscles under transient loads.

Fig. C.2 summarizes the Hill muscle bar model.

Inset (a) of Fig. C.2 is an overview of the types of skeletal muscles (WIRHED [1985]). Inset (b) shows the schematics of the Hill model for a fusiform tendon-muscle-tendon assembly. Inset (c) depicts the (normalized) active muscle component force versus length diagrams, F_{CE}/F_{max}. Inset (d) contains the active component force versus (normalized) stretch velocity diagram, $F_V(V/V_{max})$, and inset (e) shows the active component activation versus time function, $N_a(t)$.

Inset (b) of Fig. C.2 shows a simple mechanical model of a fusiform muscle with the contractile sub-element, (CE), the parallel elastic sub-element, (PE) and the parallel dashpot sub-element, (DE), of its active "muscle" element, and the nonlinear spring sub-element, (SE), and dashpot sub-element, (DSE), of its two "tendon" elements, switched in series. The tendons are not discussed here because as passive materials their mechanical response can be approximated with standard engineering visco-elastic-damaging material models. Their action can be modeled by arranging serial bars together with the central muscle bar element.

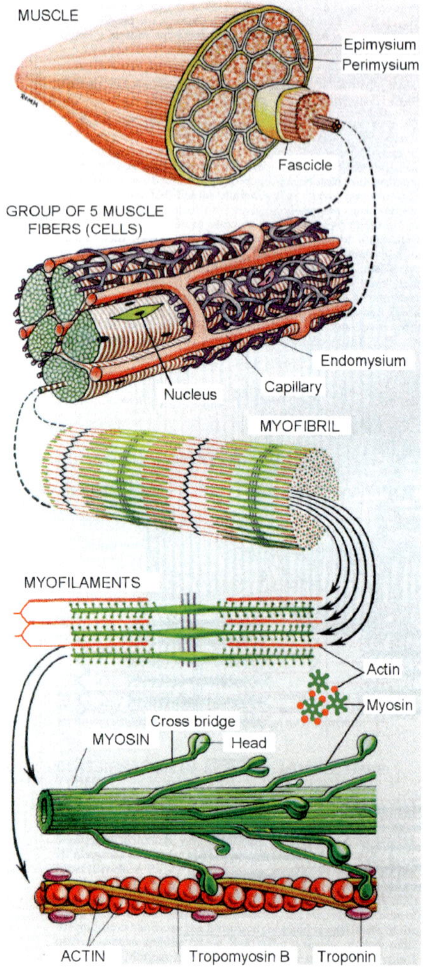

FIG. C.1. Skeletal muscle structure. (Reproduced by permission of the Longman Group UK Ltd.)

For the central "muscle" bar element, the total muscle force consists in an active and a passive component,

$$F_{\text{muscle}} = F_{\text{active muscle}} + F_{\text{passive muscle}} = F_{\text{CE}} + (F_{\text{PE}} + F_{\text{DE}}). \tag{C.1}$$

The normalized "active branch" of the muscle force, $F_{\text{CE}}/F_{\max}$ acting in the contractile sub-element, CE, is modeled by

$$F_{\text{CE}}/F_{\max} = N_{\text{a}}(t) F_V(V/V_{\max}) F_L(L/L_{\text{opt}}) = F_{\text{active muscle}}/F_{\max}. \tag{C.2}$$

In this expression $F_{\max} = \sigma A_{\text{phys}}$ is the maximum muscle force at 100% voluntary muscle activation, with the maximum active muscle stress $\sigma \cong 0.001$ Gpa, which is fairly intrinsic to all skeletal muscles, and A_{phys} = physiological cross section area

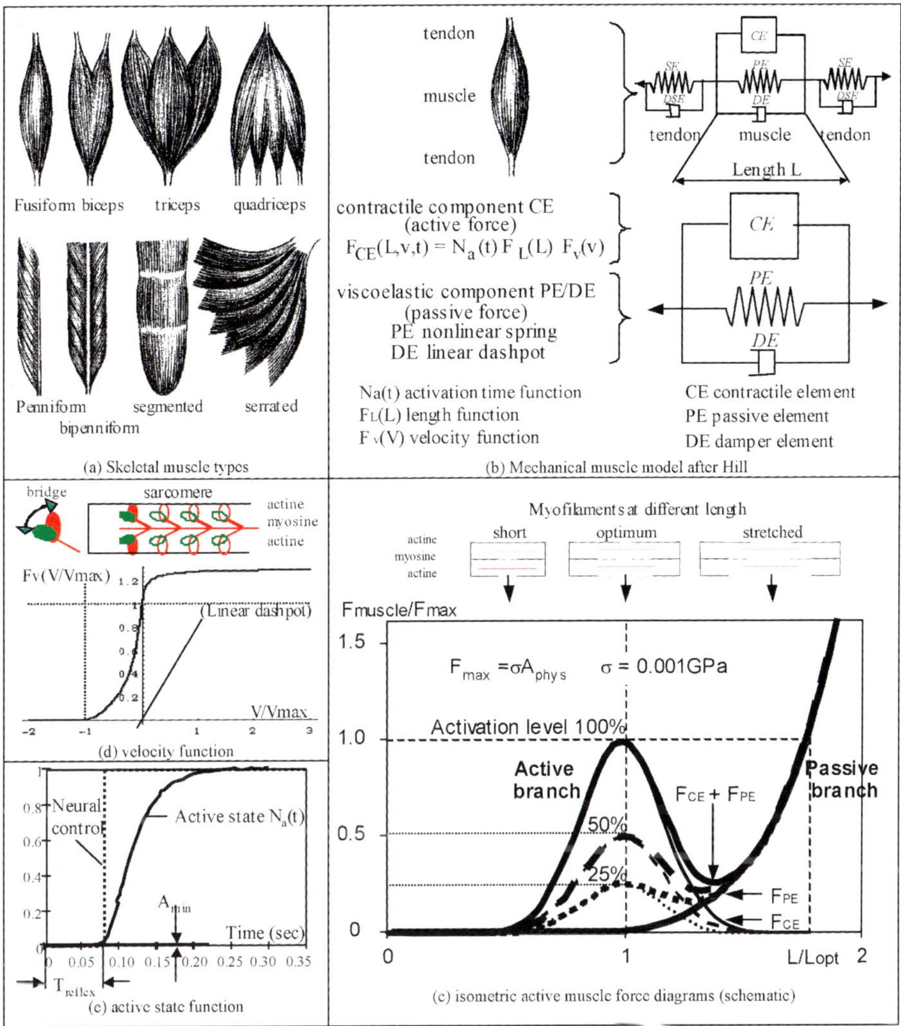

FIG. C.2. Hill's model of skeletal muscles. (Insets (a), (b) anatomical drawings: Reproduced by permission of Rolf Wirhed, WIRHED [1985].)

of the muscle; $N_a(t)$ is the neurological muscle activation state versus time function (voluntary and reflexes); $F_V(V/V_{max})$ is a muscle stretch velocity dependent function, where V_{max} is a reference muscle stretch velocity; $V = \mathrm{d}L/\mathrm{d}t$ is the muscle length rate of change or stretch velocity; $F_L(L/L_{opt})$ is a muscle length dependent shape function, where L is the current length and L_{opt} is the optimal length of the skeletal muscle "at rest", at which the voluntary muscle force can reach its peaks. The optimal muscle length is sometimes attributed to the freely floating position of a dormant astronaut.

Inset (c) of Fig. C.2 shows for the contractile sub-element (CE) the stationary ($F_v = 1$ at $V = 0$) muscle force-elongation curves, $F_{\mathrm{CE}}/F_{\max}$, activated at $N_{\mathrm{a}} = 25$, 50 and 100%, over a normalized length range of L/L_{opt} between about 0.5 and 1.5 with the length dependency curve, $F_L(L/L_{\mathrm{opt}})$ (thin lines). Other length shape functions are possible, depending on a shape factor, C_{sh}, according to

$$F_L(L/L_{\mathrm{opt}}) = \exp\bigl(-(L/L_{\mathrm{opt}} - 1)/C_{\mathrm{sh}}\bigr). \tag{C.3}$$

Inset (c) also shows the normalized "passive branch" contributed as the parallel elastic sub-element (PE) force-length response, which is due to the cohesive material resistance when the active fibers are not activated. The normalized passive forces, $F_{\mathrm{PE}}/F_{\max}$, can be calculated from

$$F_{\mathrm{PE}}/F_{\max} = \bigl(1/\bigl(\exp(C_{\mathrm{PE}}) - 1\bigr)\bigr)\exp\bigl((C_{\mathrm{PE}}/\mathrm{PE}_{\max})\bigl((L/L_{\mathrm{fib}}) - 1\bigr) - 1\bigr). \tag{C.4}$$

In this expression C_{PE} is a shape parameter of the passive force-length curve, $\mathrm{PE}_{\max} = L/L_{\mathrm{fib}}$ at is the muscle stretch when the passive force F_{PE} reaches the value of $F_{\max}$ and L_{fib} is a characteristic fiber length, often set to $L_{\mathrm{fib}} = L_{\mathrm{opt}}$.

If the muscle elongates at a stretch velocity of $V = \mathrm{d}L/\mathrm{d}t$, then the parallel passive dashpot element (DE) responds with the force

$$F_{\mathrm{DE}} = C_{\mathrm{DE}} V, \tag{C.5}$$

where C_{DE} is the damping coefficient of the assumed linear parallel dashpot element, DE.

The curves drawn with thick lines in inset (c) schematize the superimposed active and passive muscle forces under isometric conditions, i.e., when the shortening or lengthening stretch velocities of the muscle are small or zero, $V \approx 0$,

$$F_{\mathrm{muscle}} = F_{\mathrm{CE}} + F_{\mathrm{PE}} + (F_{\mathrm{DE}} = 0) = F_{\mathrm{active\ muscle}} + F_{\mathrm{passive\ muscle}}. \tag{C.6}$$

Inset (d) of Fig. C.2 presents the velocity dependent function, $F_V(V/V_{\max})$, of the active muscle force F_{CE}. This function can be interpreted in classical engineering terms as a nonlinear dashpot, as opposed to the familiar linear dashpot force-velocity curve shown for comparison in the diagram. The curves can be constructed from three branches as follows.

$$F_V(V/V_{\max}) = \begin{cases} 0 & \text{for } v = V/V_{\max} \leqslant -1, \\ (1+v)/(1-v/C_{\mathrm{short}}) & \text{for } -1 < v \leqslant 0, \\ (1+vC_{\mathrm{mvl}}/C_{\mathrm{leng}})/(1+v/C_{\mathrm{leng}}) & \text{for } v > 0, \end{cases} \tag{C.7}$$

where C_{short} is a shape parameter for the non-zero curve segment at shortening stretch velocities, C_{leng} is a shape parameter for the curve segment at lengthening stretch velocities, C_{mvl} is the asymptotic value of the curve for large positive stretch velocities, $v = V/V_{\max} \ll 1$.

The physiological origin of the stretch velocity dependency seems to stem from the actions of the so-called cross-bridges inside the actine-myosine components of the sarcomere cells of the active muscle fibers, see the bottom zooms of Fig. C.1 and insets (c) and (d) of Fig. C.2. At zero stretch velocity, $V = 0$, the muscle can afford an isometric active muscle force, when the cross bridges of the recruited muscle fibers continually connect, flex forward and disconnect the telescoping actine and myosin muscle fiber components. The number of recruited fibers can be modeled with the percentage level of the (normalized) activation function, $0 \leqslant N_a(t) \leqslant 1$. The continuous process of connection, flexion and disconnection creates a forward motion of the myosin fibers into the actine tubes, which counteracts the backward slipping motion due to the constant pull of the section force. This action can be compared roughly to the action of rowers in a boat in still waters, who must keep rowing on the spot in order to create a steady pull on a rope that retains their boat in place. If the retaining rope is cut, the boat will move forward, which corresponds to the shortening of the unconstrained muscle if the external force vanishes. The image of the tied rowers also helps understanding how physiological energy must be spent in order to keep a muscle at the same length under active tension, i.e., when the muscle does no external mechanical work. Although the tied down boat does not move, the rowers will fatigue and eventually stop rowing. Furthermore, it is easy to understand why in inset (c) the isometric active force-length curves, $F_L(L/L_{opt})$, are not constant with the muscle length. If the muscle is longer than optimal, $L > L_{opt}$, then the overlap of the myosin and actine components in a sarcomere decreases in length, and less connecting cross bridges are available to create the active muscle force. On the other hand, if the muscle has shortened, $L < L_{opt}$, the efficiency of the cross bridge action decreases because of the hindrance created by the shortening.

The shape of the velocity dependent function, $F_V(V/V_{max})$, in inset (d) of Fig. C.2 is discussed next. If to a muscle at a given instantaneous length, L, and under a given active force, F, a positive stretch velocity is imposed, $V/V_{max} > 0$, the connected bridges tend to be pulled in fiber direction and the force output at the same voluntary activation level increases by the factor $F_V(V/V_{max}) > 1$. At negative stretch velocities the connecting bridges do not re-connect fast enough to make up for the negative length rate of change of the muscle fibers, and the force output falls drastically and reaches the value of zero at negative stretch velocity $V = -V_{max}$. This again might be compared to rowers in a boat who are more efficient when rowing downstream than upstream. When the face stream velocity becomes equal to the rowing velocity, the action of the rowers will no longer produce any force on the retaining rope.

Inset (e) of Fig. C.2 shows the active muscle state function $N_a(t)$. In a simplified approach, this function depends on a muscle neurological reflex time, T_{reflex}, which is the time that elapses between, say, the onset of an impact event where a defensive muscle action should ideally start and the time when the activation actually starts. After the reflex time has elapsed, a neuro-control flag, $u(t)$, is set equal to one, and the muscle activation process sets in,

$$u(t) = \begin{cases} 0 & \text{for } t \leqslant T_{reflex}, \\ 1 & \text{for } t > T_{reflex}. \end{cases} \tag{C.8}$$

Reflex times for skeletal muscles are known to range from about 25 to 100 milliseconds (ms) and the reflex time is set to 80 ms in inset (e). After the reflex time has passed, the muscle force must be activated. This physiological process takes a certain time, and the maximum muscle force occurs at about 250 ms in inset (e). More details about skeletal muscle excitation and activation can be found in the literature cited in the given references.

The muscle activation function of the Hill model can be calculated from the differential equations

$$\begin{aligned}
\big(\mathrm{d}N_\mathrm{e}(t)/\mathrm{d}t\big) &= \big(u(t) - N_\mathrm{e}(t)\big)/T_\mathrm{ne},\\
\big(\mathrm{d}N_\mathrm{a}(t)/\mathrm{d}t\big) &= \big(N_\mathrm{e}(t) - N_\mathrm{a}(t)\big)/T_\mathrm{a},
\end{aligned} \tag{C.9}$$

where $N_\mathrm{e}(t)$ is the neuro-muscular excitation function, $N_\mathrm{a}(t)$ is the muscle force excitation function, $0.02 \leqslant T_\mathrm{ne} \leqslant 0.05$ s and $0.005 \leqslant T_\mathrm{a} \leqslant 0.02$ s are time constants. For the assumed binary form of the neural control flag $u(t)$, there exists an analytical solution of the form

$$Na(t) = \begin{cases} A_\mathrm{init} & \text{for } t \leqslant T_\mathrm{reflex},\\ 1 + A_\mathrm{a}/B_\mathrm{a} + A_\mathrm{ne}/B_\mathrm{ne} & \text{for } t > T_\mathrm{reflex}, \end{cases} \tag{C.10}$$

where

$$\begin{aligned}
A_\mathrm{a} &= (A_\mathrm{init} - 1)(T_\mathrm{a} - T_\mathrm{ne}) - T_\mathrm{ne},\\
B_\mathrm{a} &= (T_\mathrm{a} - T_\mathrm{ne})\exp\big((t - T_\mathrm{reflex})/T_\mathrm{a}\big),\\
A_\mathrm{ne} &= T_\mathrm{ne},\\
B_\mathrm{ne} &= (T_\mathrm{a} - T_\mathrm{ne})\exp\big((t - T_\mathrm{reflex})/T_\mathrm{ne}\big),\\
A_\mathrm{init} &= A_\mathrm{min} = 0.005,\\
T_\mathrm{ne} &= C_1 + C_2 m C_\mathrm{slow},\\
T_\mathrm{a} &= B_1 + B_2 m (C_\mathrm{slow})^2,\\
C_1 &= 0.025\ \mathrm{s},\\
C_2 &= 0.01\ \mathrm{s},\\
B_1 &= 0.005\ \mathrm{s},\\
B_2 &= 0.0005\ \mathrm{s},\\
m &= \text{muscle mass in grams},\\
C_\mathrm{slow} &= \text{fraction of slow muscle fibers}.
\end{aligned}$$

Applications. The mechanical behaviour of the skeletal muscles in the directions of their fibers can be modeled to first order accuracy by Hill-type muscle bars. Each Hill-type muscle bar element is characterized by the physiological cross section area of the muscle, cut perpendicular to the fibers, and by the muscle fiber stretch and stretch velocity dependent active and passive mechanical properties of the Hill muscle model, described above. The bars cannot, in general, transmit compressive forces.

Fig. C.3 shows a couple of postures and the maximum sustainable limit loads as calculated from the muscled skeleton model, compared to values found in the literature (BOUISSET and MATON [1995], p. 135). The calculated values follow from the application of the optimization process described in Chapter II, Section 6, where the applied loads were incremented until the process could no longer find a solution for the given

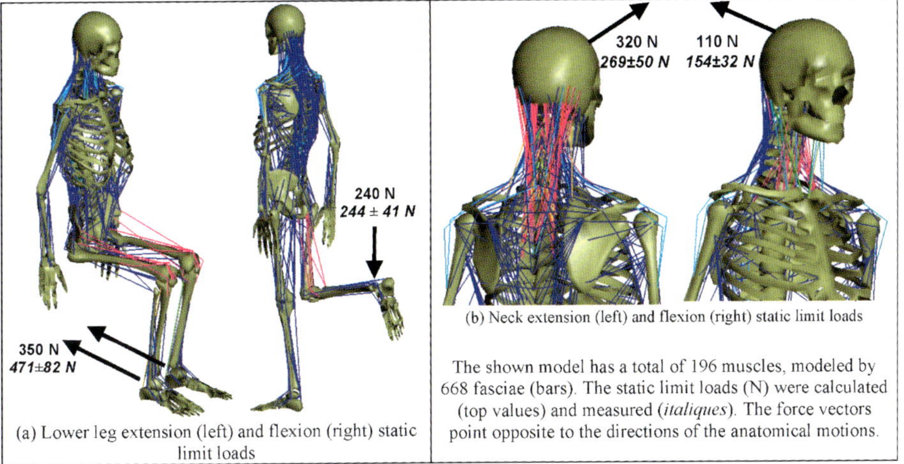

FIG. C.3. Validation of the limit load muscle force calculation (BENES [2002]).

posture, i.e., the respective limit loads were attained. The shown model has a total of 196 anatomical muscle groups, modeled by 668 fasciae (bars). The colors of the bars range between red (100% activation) and blue (no activation). The comparison with the reference values is considered fair in view of the sensitivity of the results to the joint geometry, the lever arms of the muscle bars, the muscle section area and trajectory, the true strategy of activation, the chosen objective function (physiological muscle energy/work) in the optimization process and the uncertainty of the experimental conditions, such as the exact anatomy of the volunteers, the exact posture and the point of load application (BENES [2002]).

References on muscle materials. The following references deal with the structure and the modeling of skeletal muscle: BAHLER, FALES and ZIERLER [1968] on the dynamic properties of skeletal muscle; COLE, BOGERT, HERZOG and GERRITSEN [1996] on modeling of forces in stretched muscles; CRAWFORD AND JAMES [1980] on the design of muscles, HARRY, WARD, HEGLUD, MORGAN and MCMAHON [1990] on cross bridge action; HERZOG [1994] on the biomechanics of the musculo-skeletal system; HAWKINS and BEY [1994], HAWKINS and BEY [1997] on muscle-tendon mechanics and mechanical properties; HILL [1970] on experiments in muscle mechanics; KIRSCH, BOSKOV and RYMER [1994] on stiffness of moving cat muscles; KRYLOW and SANDEROCK [1996] on dynamic force response of muscles under excentric contraction; MA and ZAHALAK [1991] on a distribution-moment model of energetics in skeletal muscle; MORGAN [1990] on the behaviour of muscle under active lengthening; MYERS, VAN EE, CAMACHO, WOOLLEY and BEST [1995] on the structural properties of mammalian skeletal muscle in the neck; RACK and WESTBURY [1969] on the effect of length rate on tension in muscles; SCHNECK [1992] on the mechanics of muscle; VANCE, SOLOMONOV, BARATTA, ZEMBO and D'AMBROSIA [1994] on the comparison of two muscle models; WANK and GUTEWORT [1993] on the simulation of muscular contraction with regard to physiological parameters; WINTERS and STARK [1985], WINTERS and STARK [1988] on muscle modeling and mechanical properties; ZAJAC [1989] on muscle and tendon properties and models; ZUURBIER, EVERARD, VAN DER WEES and HUIJING [1994] on the force-length characteristics in active and passive muscles.

APPENDIX D

Airbag Models

The following paragraphs are based mainly on Pam-Safe documentation of ESI Software.

Airbags. Airbags are considered volumes of ideal gas that are enclosed by a flexible envelope, Fig. D.1. Their physics requires fluid-structure interaction (FSI) simulation. In simple airbag models, pressure and temperature of the gas are assumed to be distributed uniformly throughout the airbag volume. More complex approaches (not described here) model the enclosed gas and the envelope separately, using fluid-solid and multi-physics formulations. The gas can be confined by a bag made of flexible (visco-)elastic industrial fabric or of any other material, such as the walls of hollow organs.

Airbag models can simulate gas inflow and outflow through orifices, and leakage of gas through the fabric of the envelope can be defined. The input data for the gas are atmospheric pressure p, temperature T, constants γ and R, where R is the perfect gas constant from $pV = nRT$, where n is the number of moles in the volume. In the described simple airbag model, the gas is confined in a single chamber and it obeys the thermodynamic equation of a perfect gas. Solution of that equation at each explicit structural solution time step yields a pressure load to be applied to the inside of the envelope. Solution of the equations of motion of the pressurized envelope yields a volume change to be applied to the enclosed gas. This process is repeated over the duration of the simulation.

Airbag gas model. The volume of enclosed gas is subject to the equation of state $pV^{\gamma} = \text{constant}$, where p is the pressure, V is the volume and $\gamma = c_p/c_v$ is the specific heat ratio for a perfect gas under adiabatic conditions, and where c_p and c_v are the specific heat of the gas at constant pressure and at constant volume, respectively. The gas constant is defined as

$$R = c_p - c_v = R_u/W,$$

where R_u is the universal gas constant related to moles and W is the molecular weight of the gas.

For nitrogen gas, N_2, one has in SI-units (m, kg, s) and with the molecular weight $W(N_2) = 0.028014$ kg/mole

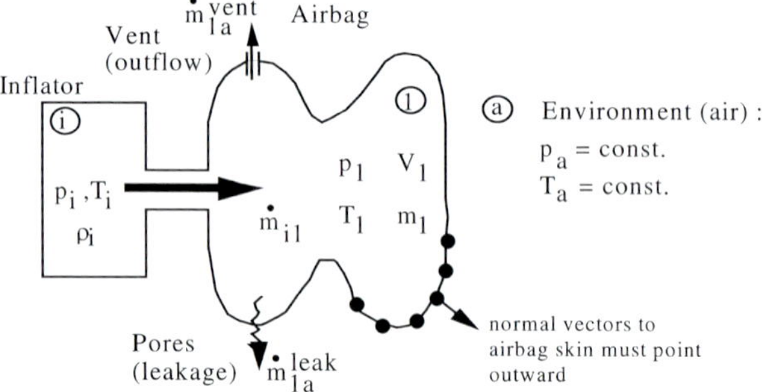

FIG. D.1. Basic airbag model (schematic) (ESI Software).

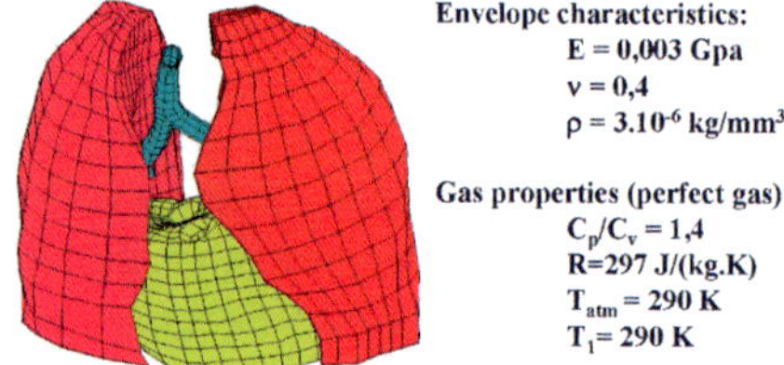

FIG. D.2. Basic airbag model for the lungs (ESI software).

$$c_p = 1038\ \mathrm{J/(kgK)}, \qquad c'_p = 29.08\ \mathrm{J/(mole\ K)},$$
$$c_v = 741\ \mathrm{J/(kgK)}, \qquad c'_v = 20.76\ \mathrm{J/(mole\ K)},$$
$$R = 297\ \mathrm{J/(kgK)}, \qquad R_u = 8.32\ \mathrm{J/(mole\ K)},$$

where J = Joules = Nm (Newton meters), K = °Kelvin and c'_p and c'_v are the molar heat capacities.

The gauge pressure is the differential pressure between the airbag pressure and the atmospheric pressure, Fig. D.1,

$$p_{\mathrm{gauge}} = p_1 - p_{\mathrm{a}}.$$

The specific heat ratio, γ, is defined as

$$\gamma = c_p/c_v = c_p/(c_p - R),$$

which for nitrogen gas (N_2) is equal to 1.4. The meaning of the remaining parameters is illustrated on Fig. D.1.

Bio-bag models. Bio-bag models are derived from airbag models to model hollow organs. Hollow organs have flexible walls and are filled with quasi-incompressible material (blood, body fluids, food), but also (partly or fully) with compressible material (gas in stomach or lungs). Airbag gas models can be used to approximate quasi-incompressible fluid-filled hollow organs by re-setting their input data to model a linear

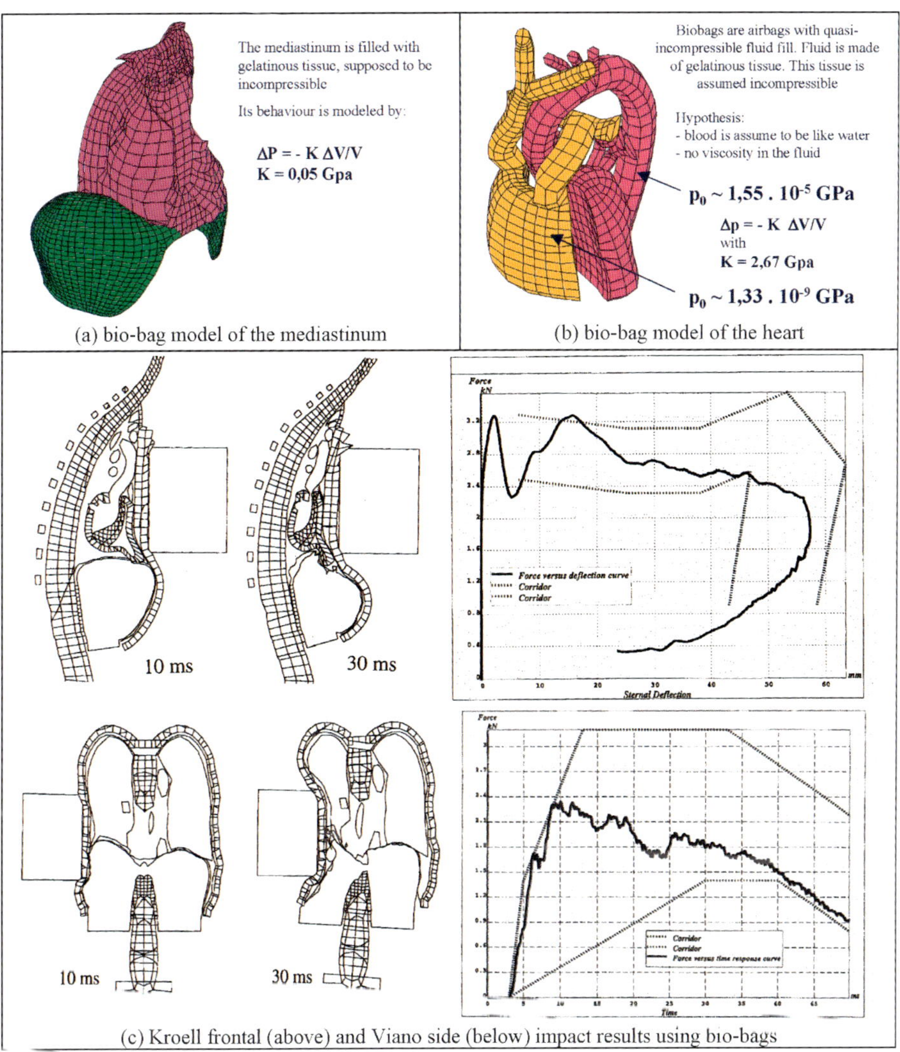

FIG. D.3. Bio-bag models of the mediastinum and heart (ESI software).

pressure-volume equation of the type

$$p = p_0 + K(\rho/\rho_0 - 1),$$

where K is the desired bulk modulus for modeling quasi-incompressibility, p_0 is the initial pressure and ρ and ρ_0 are the mass density in the compressed and uncompressed gas, respectively. This can be achieved approximately by setting c_p to a large value (isothermal conditions) with $\gamma = c_p/c_v$ close to 1.

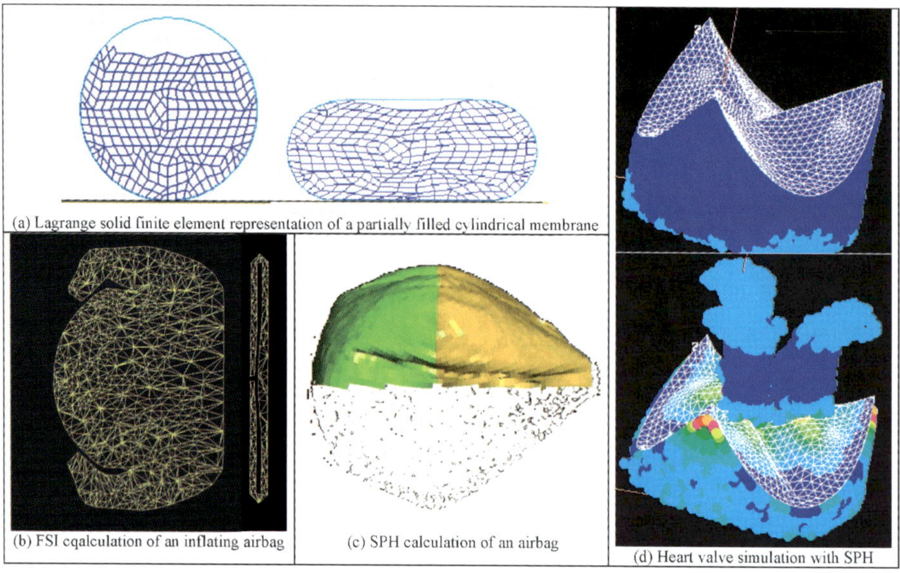

(a) Lagrange solid finite element representation of a partially filled cylindrical membrane

(b) FSI cqalculation of an inflating airbag

(c) SPH calculation of an airbag

(d) Heart valve simulation with SPH

FIG. D.4. Alternative airbag modeling techniques (ESI software).

Examples. The following examples demonstrate the use of "bio-bags" in the modeling of the hollow internal organs.

Fig. D.2 is an approximate mechanical model of the lungs, which uses the basic airbag model with perfect gas properties. A fictitious envelope encloses a space filled with air, and vent holes can be provided to simulate the expulsion of air from a violent compression of the thorax in an impact.

This model is relatively efficient when the lungs must not be simulated in detail, but when only their resistance to compression of the rib cage is of interest.

Fig. D.3 shows bio-bag models of the mediastinum and heart.

The mediastinum is the complex space between the thoracic organs and vessels and it is considered filled with an incompressible gelatinous fluid. It can be modeled by defining a fictitious envelope, Fig. D.3(a), and by assigning the conditions of bio-bags for quasi-incompressibility.

Similarly, the heart can be modeled by a bio-bag with a high degree of incompressibility, Fig. D.3(b). For capturing the effect of expelled blood during a violent chest impact, the modeled heart chambers can be provided with outflow vents. The thorax models can so be calibrated to well represent the results from Kroell frontal and Viano side pendulum impact tests, Fig. D.3(c).

Other fluid-structure interaction modeling techniques. While the described biobag models are efficient but approximate models, the volume enclosed by a hollow organ can be modeled with more precision with *Lagrangian solid finite elements*, which undergo the equation of state given by a fluid, Fig. D.4, inset (a). This model is possible when the fluid transport is small, i.e., the distortion of the Lagrange mesh is limited.

Coupled fluid-structure interaction (FSI) models (LÖHNER [1990]), as shown for an airbag in inset (b), will be the most accurate (and the most expensive) representations of hollow organs. This technique is indicated if the detailed interaction of the moving fluids with the confining wall is of interest (example: aorta rupture). In that case a fluid code and a structure code are coupled, where the fluid code provides the wall pressure loads to the structure code and where the structure code provides the wall positions and velocities to the fluid code at each common solution time step. Finally, the fluid or gas can also be represented by *SPH or FPM "particle" techniques* (MONAGHAN and GINGOLD [1983], MONAGHAN [1988]), insets (c), (d). While in FSI techniques the fluid domain is meshed, inset (b), particle methods do not require a domain mesh. This is particularly convenient when the fluid domain connectivity changes, as in the example of the heart valve, inset (d).

APPENDIX E

Interactions between Parts

Contact simulation. The numerical treatment of contact with various types of contact options was mentioned in Appendix A. Effective treatment of contact is not only of prime importance for modeling impact biomechanics, but certain contact algorithms serve also in assembling the complex geometries of the parts of the human body. Some examples, taken from the HUMOS model (ESI version), explain how different modeling strategies can provide viable solutions to the complex mechanical interactions between organs and parts of the human body.

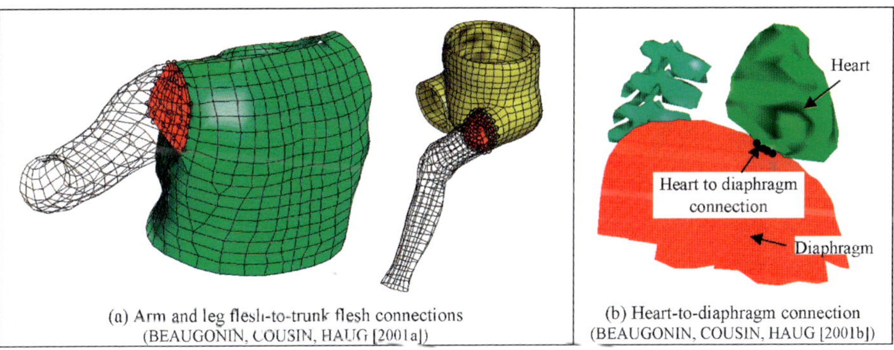

(a) Arm and leg flesh-to-trunk flesh connections (BEAUGONIN, COUSIN, HAUG [2001a])

(b) Heart-to-diaphragm connection (BEAUGONIN, COUSIN, HAUG [2001b])

FIG. E.1. Connections between parts via contact (HUMOS-ESI model).

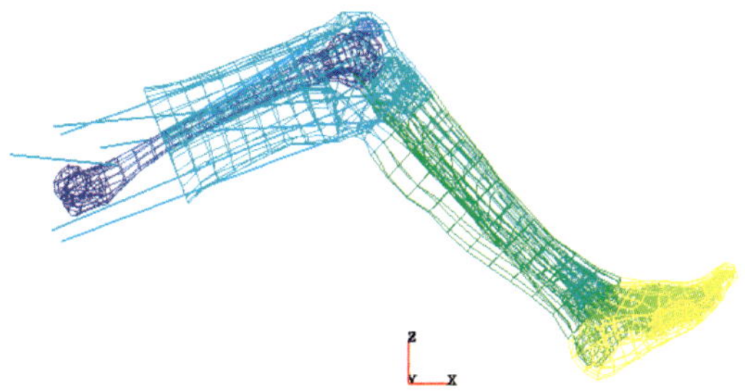

FIG. E.2. Non-matching flesh and bone meshes connected by tied contacts (HUMOS-ESI model).

Mesh merging. Apart from simulating dynamic collisions between moving parts, special contact options are frequently used within models of articulations, to delimit adjacent organs and to conveniently join differently meshed parts, Fig. E.1 (BEAUGONIN, COUSIN and HAUG [2001a], BEAUGONIN, COUSIN and HAUG [2001b]).

In another example, Fig. E.2 shows the independent meshes of the leg bones and the surrounding flesh, BEAUGONIN and HAUG [2001].

This figure demonstrates how different constituents of body parts, such as flesh and bone, are linked with connective membranes and tissue, and are often meshed independently for convenience. Both material constituents are connected by a tied contact option (type 32 in Table A.1, Appendix A).

Example: Lower extremity. To control the interaction between the different components involved in the lower limb segment of the HUMOS model, several types of sliding interfaces have been defined in its PAM-Crash version, Fig. E.3 (BEAUGONIN and HAUG [2001]).

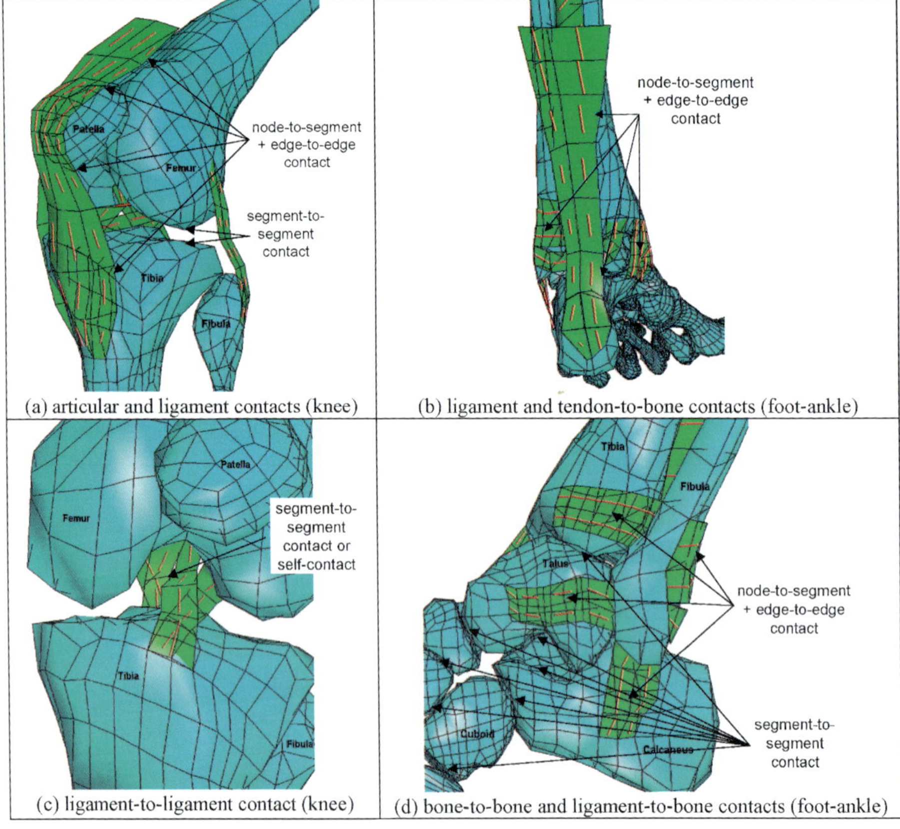

FIG. E.3. Articular, ligamant/tendon-to-bone and self-contacts in the lower extremity (HUMOS-ESI model) (BEAUGONIN and HAUG [2001]).

(i) The sliding interfaces between bones near their articulations are defined with a segment-to-segment contact (type 33 in Table A.1, Appendix A).
(ii) The sliding interfaces between bone and ligament or tendon, and between bone and skin are defined with a segment-to-segment or a node-to-segment contact (type 34 in Table A.1, Appendix A). If necessary, an edge-to-edge contact (type 46 in Table A.1, Appendix A) is added to avoid the penetration between the components.
(iii) The interaction between ligaments is controlled by a segment-to-segment contact or a self-contact (type 36 in Table A.1, Appendix A). The sliding interface between ligament/tendon and skin is defined by a node-to-segment contact.

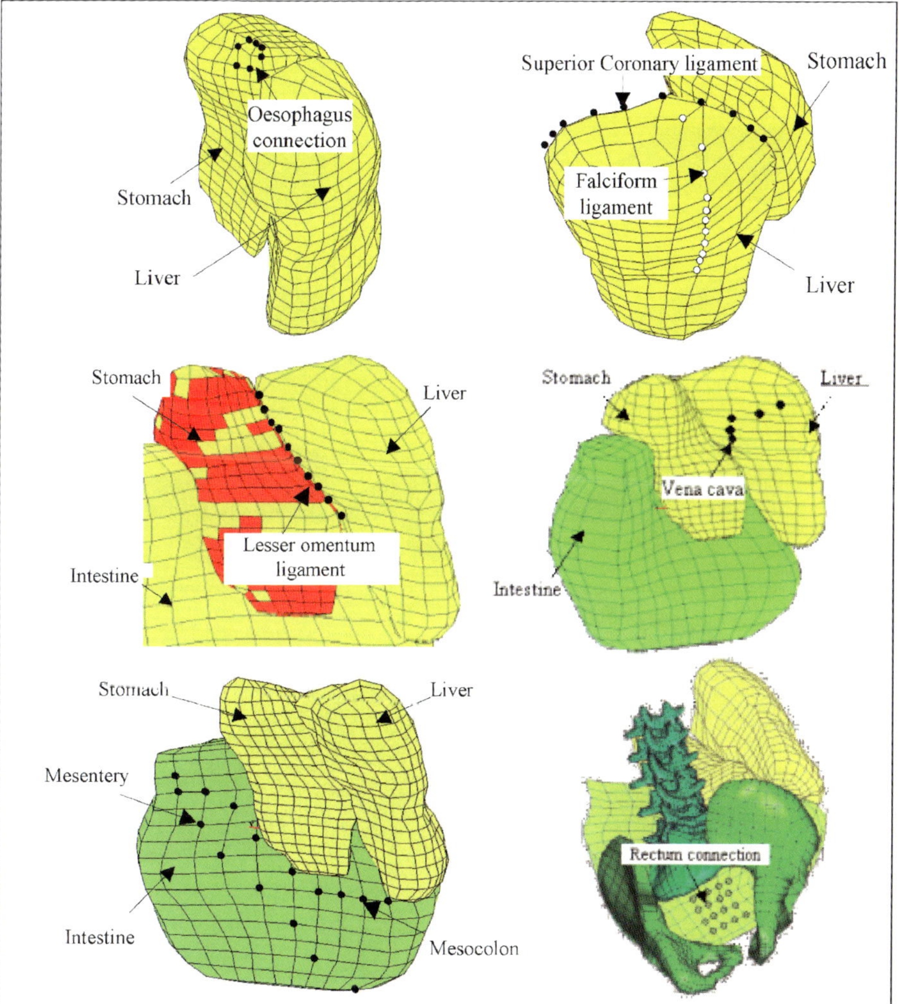

FIG. E.4. Attachments between abdominal organs (HUMOS-ESI model).

Example: Abdominal organs. The abdominal organs are tied together and are loosely tethered to the abdominal walls (peritoneum) by ligaments or folds of the peritoneum. These features can be modeled with tied contact options in the ESI HUMOS model. The tied contact options permit no relative motion between the tied parts, but can be allowed to break when certain contact force limits are exceeded. This option can approximate the rupture of ligaments and other tissue connections. Some of the connections are shown as examples in Fig. E.4 (after BEAUGONIN, ALLAIN and HAUG [2001]).

The abdominal organs, as well as the brain and the thoracic organs, interact with neighbors and cavity walls under considerable relative sliding motions. This can be modeled with the standard sliding contact options.

The organs fill their host cavities with literally no voids or gaps in the sense that the space between the organs is filled with tissue or fluids. This can be modeled approximately with "bio-bags", as shown in Appendix D for the mediastinum, where the

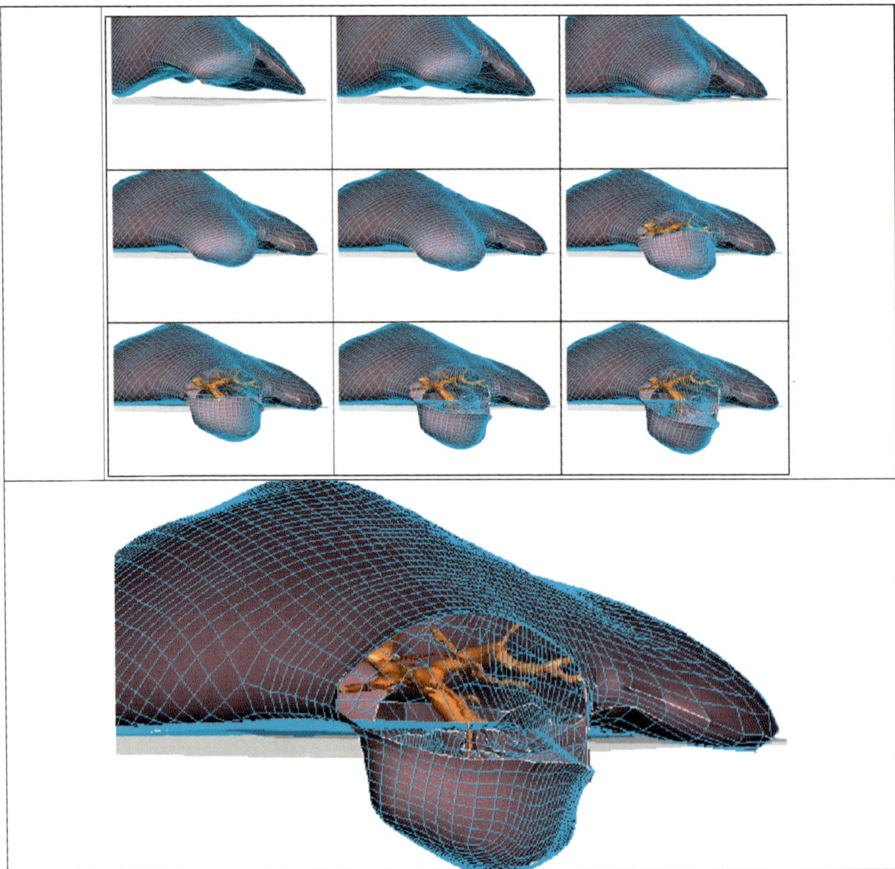

FIG. E.5. Simulation of cut liver with tied meshes between vena porta and parenchyma (CAESARE-ESI model: DAN and MILCENT [2002]).

organ-to-organ contact can be modeled with slide-and-void contacts between the organs, while the incompressibility of the liquid-filled space between them is assured by the bio-bag feature (Appendix D). The same feature was applied in Chapter 3, Section 8 (Fig. 8.8) for the modeling of the Cavanaugh bar impact test on the abdomen.

Example: Liver. For practical reasons (convenience, meshing freedom, mesh size limitations, etc.) vessels inside internal organs are often modeled apart from the bulk matter of the organs. Automatic mesh merging techniques, or tied contact options, can then be applied to tie the non-congruent meshes between the surface of the meshed vessels (often: shells) and the organ bulk matter (solids). Fig. E.5 shows a model of a human liver with the internal arborescence of a systems of vessels (vena porta) exposed through a simulated progressive cut into the parenchyma (from the CAESARE Project DAN and MILCENT [2002]). Glisson's capsule around the liver is modeled with thin membranes. The cut-in-progress was simulated with assigning almost zero resistance to the sectioned elements. In order to expose the incision, the cut portion of the liver is not supported by the horizontal support plate and it deflects through the action of gravity. It is still connected by its uncut portion with the main portion of the organ. The tougher vessels become visible by assigning transparency to the surrounding parenchyma solid elements, which appear only through their exposed surface grids (blue color).

References

ABEL, J.M., GENNARELLI, T.A., SEGAWA, H. (1978). Incidence and severity of cerebral concussion in the rhesus monkey following sagittal plane angular acceleration. In: *Proc. 22nd Stapp Car Crash Conference*, pp. 35–53. Paper No. 780886.

AL-BSHARAT, A.S., HARDY, W.N., YANG, K.H., KHALIL, T.B., TASHMAN, S., KING, A.I. (1999). Brain-skull relative displacement magnitude due to blunt head impact: New experimental data and model. In: *Proc. 43rd Stapp Car Crash Conference*, pp. 101–160. Paper No. 99SC22.

ALLAIN, J.C. (1998). Etude et calibration d'un modèle numérique de thorax, Internal Report, ESI Software S.A., 99, rue des Solets, BP 80112, 94513 Rungis Cedex, France.

ALLEN, B.L., FERGUSON, R.L., LEHMANN, T.R., O'BRIEN, R.P. (1982). A mechanistic classification of closed indirect fractures and dislocations of the lower cervical spine. *Spine* **7** (1), 1–27.

ALLSOP, D.L. (1993). Skull and facial bone trauma: Experimental aspects. In: Nahum, A.M., Melvin, J.W. (eds.), *Accidental Injury – Biomechanics and Prevention* (Springer, Berlin), pp. 247–267 (Chapter 11).

ARMSTRONG, C.G., LAI, W.M., MOW, V.C. (1984). An analysis of the unconfined compression of articular cartilage. *J. Biomech. Engrg.* **106**, 165–173.

ASHMAN, R.B., et al. (1984). A continuous wave technique for the measurement of the elastic properties of cortical bone. *J. Biomech.*, 349.

ASHMAN, R.B., et al. (1986). Ultrasonic technique for the measurement of the structural elastic modulus of cancellous bone. *Trans. Orthopedic Res. Soc.*, 43.

ATTARIAN, D.E., MCCRACKIN, H.J., DEVITO, D.P., MCELHANEY, J.H., GARRETT, W.E. (1985). Biomechanical characteristics of human ankle ligaments. *Foot & Ankle* **6** (2).

BACH, J.M., HULL, M.L., PATTERSON, H.A. (1997). Direct measurement of strain in the postero-lateral bundle of the anterior cruciate ligament. *J. Biomech.* **3** (3), 281–283.

BAHLER, A.S., FALES, J.T., ZIERLER, K.L. (1968). The dynamic properties of mammalian skeletal muscle. *J. Gen. Physiol.* **51**, 369–384.

BANDAK, F.A. (1996). Biomechanics of impact traumatic brain injury. In: *Proceedings of the NATO ASI on Crashworthiness of Transportation Systems, Structural Impact and Occupant Safety, Troia, Portugal*, pp. 213–253.

BANDAK, F.A., EPPINGER, R.H. (1994). A three-dimensional finite element analysis of the human brain under combined translational and rotational acceleration. In: *Proc. 38th Stapp Car Crash Conference*, pp. 145–163. Paper No. 942215.

BANDAK, F.A., TANNOUS, R.E., ZHANG, A.X., DIMASI, F., MASIELLO, P., EPPINGER, R.H. (2001). SIMon: A simulated injury monitor, application to head injury assessment. In: *17th International Technical Conference on the Enhanced Safety of Vehicles, Amsterdam, Holland*.

BANDAK, F.A., TANNOUS, R.E., ZHANG, A.X., TORIDIS, T.G., EPPINGER, R.H. (1996). Use of finite element analysis and dummy test measurements for the assessment of crash injury traumatic brain injury. In: *Advisory Group for Aerospace Research and Development, Mescalero, New Mexico*, pp. 10–10.13.

BARTLEY, M.H., et al. (1966). The relationship of bone strength and bone quantity in health, disease and aging. *J. Gerontol.* **21**, 517.

BATHE, J. (1996). *Finite Element Procedures* (Prentice-Hall, Englewood Cliffs, NY).

BAUDER, B. (1985). The dynamic load tolerance of the human liver (Die dynamisch mechanische Belastbarkeit der menschlichen Leber). PhD Dissertation, Medicine, University of Heidelberg.

BAUDRIT, P., HAMON, J., SONG, E., ROBIN, S., LE COZ, J.-Y. (1999). Comparative studies of dummy and human body models in frontal and lateral impact conditions. In: *Proc. 43rd Stapp Car Crash Conference*, pp. 55–75. Paper No. 99SC05.

BEAUGONIN, M., ALLAIN, J.C., HAUG, E. (2001). Pam–Crash modeling and validation: Pelvis–Abdomen segment. HUMOS Report 5ESI/010131/T3/DA, ESI Software S.A., 99, rue des Solets, BP 80112, 94513 Rungis Cedex, France.

BEAUGONIN, M., COUSIN, G., HAUG, E. (2001a). Pam–Crash: Whole model validation. HUMOS Report 6ESI/010131/T5/DA, ESI Software S.A., 99, rue des Solets, BP 80112, 94513 Rungis Cedex, France.

BEAUGONIN, M., COUSIN, G., HAUG, E. (2001b). Pam–Crash modeling and validation: Thorax-shoulder segment. HUMOS Report 6ESI/010131/T2/DA, ESI Software S.A., 99, rue des Solets, BP 80112, 94513 Rungis Cedex, France.

BEAUGONIN, M., HAUG, E. (2001). Pam–Crash modeling and validation: Lower limb segment. HUMOS Report 5ESI/010131/T5/DA, ESI Software S.A., 99, rue des Solets, BP 80112, 94513 Rungis Cedex, France.

BEAUGONIN, M., HAUG, E., CESARI, D. (1996). A numerical model of the human ankle/foot under impact loading in inversion and eversion. In: *Proc. 40th Stapp Car Crash Conference Proceedings, Albuquerque, NM*, pp. 239–249. Paper No. 962428.

BEAUGONIN, M., HAUG, E., CESARI, D. (1997). Improvements of numerical ankle/foot model: Modeling of deformable bone. In: *Proc. 41st Stapp Car Crash Conference*, pp. 225–237. Paper No. 973331.

BEAUGONIN, M., HAUG, E., HYNCIK, L. (1998). Robby2 – alpha-version. Internal Report, ESI Software S.A., 99, rue des Solets, BP 80112, 94513 Rungis Cedex, France.

BEAUGONIN, M., HAUG, E., MUNCK, G., CESARI, D.A. (1995). Preliminary numerical model of the human ankle under impact loading. In: *Pelvic and Lower Extremity Injuries (PLEI) Conference, Washington, DC, USA*.

BEAUGONIN, M., HAUG, E., MUNCK, G., CESARI, D. (1996). The influence of some critical parameters on the simulation of the dynamic human ankle dorsiflexion response. In: *15th ESV Conference, 96-S10-W-31, Melbourne*.

BEDEWI, P.G., MIYAMOTO, N., DIGGES, K.H., BEDEWI, N.E. (1998). Human femur impact and injury analysis utilizing finite element modeling and real-world case study data. In: *PUCA '98 Proceedings*, Nihon ESI, ESI Software S.A., 99, rue des Solets, BP 80112, 94513 Rungis Cedex, France, vol. 1, pp. 311–320.

BEGEMAN, P., AEKBOTE, K. (1996). Axial load strength and some ligament properties of the ankle joint. In: *Injury Prevention Through Biomechanics Symposium Proceedings, Hutzel Hospital, Wayne State University, Detroit, Michigan, USA*, pp. 125–135.

BEGEMAN, P., BALAKRISHNAN, P., LEVINE, R., KING, A. (1992). Human ankle response in dorsiflexion. In: *Injury Prevention Through Biomechanics Symposium Proceedings, Wayne State University*.

BEGEMAN, P., BALAKRISHNAN, P., LEVINE, R., KING, A. (1993). Dynamic human ankle response to inversion and eversion. In: *Proc. 37th Stapp Car Crash Conference Proceedings*, pp. 83–93. Paper No. 933115.

BEGEMAN, P.C., KING, A.I., PRASAD, P. (1973). Spinal loads resulting from $-gx$ acceleration. In: *Proc. 17th Stapp Car Crash Conference, Warrendale, PA*. Paper No. 730977.

BEGEMAN, P., KOPACZ, J.M. (1991). Biomechanics of human ankle impact in dorsiflexion. In: *Injury Prevention Through Biomechanics Symposium Proceedings, Wayne State University*.

BEHRENS, J.C., et al. (1974). Variation in strength and structure of cancellous bone at the knee. *J. Biomech.* **7**, 201–207.

BEILLAS, P., BEGEMAN, P.C., YANG, K.H., KING, A., ARNOUX, P.-J., KANG, H.-S., KAYVANTASH, K., BRUNET, C., CAVALLERO, C., PRASAD, P. (2001). Lower limb: Advanced FE model and new experimental data. *Stapp Car Crash J.* **45**, 469–494. Paper No. 2001-22-0022.

BEILLAS, P., LAVASTE, F., NICOLOPOULOS, D., KAYVANTASH, K., YANG, K.H., ROBIN, S. (1999). Foot and ankle finite element modeling using CT-scan data. In: *Proc. 43rd Stapp Car Crash Conference*, pp. 1–14. Paper No. 99SC11.

BELYTSCHKO, T., LIN, J.I. (1984). Explicit algorithms for the nonlinear dynamics of shells. *Comput. Methods Appl. Mech. Engrg.* **42**, 225–251.

BELYTSCHKO, T., KULAK, R.F., SCHULTZ, A.B., GALANTE, J.O. (1972). Numerical stress analysis of intervertebral disk. In: *ASME (Biomechanical & Human Factors Division), Winter Annual Meeting, NY*.

BELYTSCHKO, T., KULAK, R.F., SCHULTZ, A.B., GALANTE, J.O. (1974). Finite element stress analysis of an intervertebral disc. *J. Biomech.* **7**, 277–285.

BELYTSCHKO, T., TSAY, C.S. (1983). A stabilization procedure for the quadrilateral plate element with one-point quadrature. *Internat. J. Numer. Methods Engrg.* **19**, 405–419.

BELYTSCHKO, T., WONG, J.S., LIU, W.K., KENNEDY, J.M. (1984). Hourglass control in linear and nonlinear problems. *Comput. Methods Appl. Mech. Engrg.* **43**, 251–276.

BENES, K. (2002). ROBBY 2 Model with Muscles, Internal Report, ESI Software S.A., 99, rue des Solets, BP 80112, 94513 Rungis Cedex, France.

BLACK, J., HASTINGS, G. (eds.) (1998). *Handbook of Biomaterial Properties, Part I* (Chapman & Hall, London).

BOKDUK, N., YOGANANDAN, N. (2001). Biomechanics of the cervical spine, Part 3: Minor injuries. *Clinical Biomech.* **16**, 267–275.

BOUISSET, S., MATON, B. (1995). *Muscles, Posture et Mouvement* (Hermann, Paris).

BOUQUET, R., RAMET, M., BERMOND, F., CESARI, D. (1994). Thoracic and human pelvis response to impact. In: *Proceedings of the 14th International Technical Conference on Enhanced Safety of Vehicles (ESV)*, pp. 100–109.

BOWMAN, B.M., SCHNEIDER, L.W., LUSTAK, L.S., ANDERSON, W.R., THOMAS, D.J. (1984). Simulation analysis of head and neck dynamic response. In: *Proc. 28th Stapp Car Crash Conference*, pp. 173–205. Paper No. 841668.

BROWN, T.D., et al. (1980). Mechanical property distribution in the cancellous bone of the human proximal femur. *Acta Orthop. Scand.* **51**, 429–437.

BSCHLEIPFER, Th. (2002). Das experimentelle stumpfe Nierentrauma: Biomechanik, Traumaverhalten und bildgebende Diagnostik. M.D. Dissertation, Faculty of Medicine, University Ulm, Germany.

BURGHELE, N., SCHULLER, K. (1968). Die Festigkeit der Knochen Kalkaneus und Astragallus. Aus der Klinik fur Orthopädie des Bukarester Unfallkrankenhauses und dem Lehrstuhl fur Festigkeit und Materialprufungswesen der Technischen Hochschule, Bukarest, Rumänien (in German).

BURSTEIN, A.H., et al. (1976). Aging of bone tissue: Mechanical properties. *J. Bone Joint Surgery A* **58**, 82.

CARTER, D.R., HAYES, W.C. (1977). The compressive behavior of bone as a two-phase porous structure. *J. Bone Joint Surgery A* **59**, 954–962.

CARTER, F. (1999). Work at the University of Dundee. Private communication to INRIA, Sophia-Antipolis.

CAVANAUGH, J.M. (1993). The biomechanics of thoracic trauma. In: Nahum, A.M., Melvin, J.W. (eds.), *Accidental Injury – Biomechanics and Prevention* (Springer, Berlin), pp. 362–390 (Chapter 15).

CAVANAUGH, J.M., NYQUIST, G.W., GOLDBERG, S.J., KING, A.I. (1986). Lower abdominal tolerance and response. In: *Proc. 30th Stapp Car Crash Conference*, pp. 41–63. Paper No. 861878.

CAVANAUGH, J.M., WALILKO, T.J., MALHOTRA, A., ZHU, Y., KING, A.I. (1990). Biomechanical response and injury tolerance of the pelvis in twelve sled side impact tests. In: *Proc. 34th Stapp Car Crash Conference*, pp. 23–38. Paper No. 902307.

CESARI, D., BERMOND, F., BOUQUET, R., RAMET, M. (1994). Virtual predictive testing of biomechanical effects of impacts on the human leg. In: *Proceedings of PAM '94*, ESI Software S.A., 99, rue des Solets, BP 80112, 94513 Rungis Cedex, France, pp. 177–188.

CESARI, D., BOUQUET, R. (1990). Behavior of human surrogates thorax under belt loading. In: *Proc. 34th Stapp Car Crash Conference*, pp. 73–81. Paper No. 902310.

CESARI, D., BOUQUET, R. (1994). Comparison of Hybrid III and human cadaver thorax deformation loaded by a thoracic belt. In: *Proc. 38th Stapp Car Crash Conference*, pp. 65–76. Paper No. 942209.

CHANCE, G.O. (1948). Note on a type of flexion fracture of the spine. *Br. J. Radiol.* **21**, 452–453.

CHANDLER, R.F. (1993). Development of crash injury protection in civil aviation. In: Nahum, A.M., Melvin, J.W. (eds.), *Accidental Injury – Biomechanics and Prevention* (Springer, Berlin), pp. 151–185 (Chapter 6).

CHAPON, A., VERRIEST, J.P., DEDOYAN, J., TRAUCHESSEC, R., ARTRU, R. (1983). Research on brain vulnerability from real accidents, ISO Document No. ISO/TC22SC12/GT6/N139.

CHAZAL, J., TANGUY, A., BOURGES, M. (1985). Biomechanical properties of spinal ligaments and a historical study of the supraspinal ligament in traction. *J. Biomech.* **18** (3), 167–176.

CHOI, H.-Y. (2001). Numerical human head model for traumatic injury assessment. *KSME Internat. J.* **15** (7), 995–1001.

CHOI, H.-Y., EOM, H.-W. (1998). Finite element modeling of human cervical spine. *Hongik J. Sci. Technol. (Special volume).*

CHOI, H.-Y., LEE, I.-H. (1999a). Finite element modeling of human thorax for occupant safety simulation. In: *10th Proceedings of PUCA*, Japan; ESI Software S.A., 99, rue des Solets, BP 80112, 94513 Rungis Cedex, France, pp. 469–478.

CHOI, H.-Y., LEE, I.-H. (1999b). Advanced finite element modeling of the human body for occupant safety simulation. In: *EuroPam 1999*, ESI Software S.A., 99, rue des Solets, BP 80112, 94513 Rungis Cedex, France.

CHOI, H.-Y., LEE, I.-H., EOM, H.-W., LEE, T.-H. (1999). A study on the whiplash injury due to the low velocity rear-end collision. In: *3rd World Congress of Biomechanics, Japan*, p. 522.

CHOI, H.-Y., LEE, I.-H, HAUG, E. (2001a). Finite element modeling of human upper extremity for occupant safety simulation. In: *JSAE Spring Conference, No. 34-01*, 20015358.

CHOI, H.-Y., LEE, I.-H., HAUG, E. (2001b). Finite element modeling of human head–neck complex for crashworthiness simulation. In: *First MIT Conference Proceedings*.

CIARELLI, M.J., et al. (1986). Experimental determination of the orthogonal mechanical properties, density, and distribution of human trabecular bone from the major metaphyseal regions utilizing material testing and computed tomography. *Trans. Orthop. Res. Soc.*, 42.

CIARELLI, M.J., et al. (1991). Evaluation of orthogonal mechanical properties and density of human trabecular bone from the major metaphyseal regions with materials testing and computed tomography. *J. Orth. Res.* **9**, 674–682.

CLAESSENS, M. (1997). *Finite Element Modelling of the Human Head under Impact Conditions*. PhD Thesis. Eindhoven University. ISBN 90-386-0369-X.

CLAESSENS, M., SAUREN, F., WISMANS, J. (1997). Modeling the human head under impact conditions: A parametric study. In: *Proc. 41st Stapp Car Crash Conference*, pp. 315–328. Paper No. 973338.

CLEMENTE, C.D. (1981). *Anatomy (A Regional Atlas of the Human Body)*, second ed. (Urban & Schwarzenberg, Baltimore–Munich).

COHEN, D.S. (1987). The safety problem for passengers in frontal impacts: Analysis of accidents, laboratory and model simulation data. In: *11th ESV International Technical Conference on Experimental Safety Vehicles, Washington, DC*.

COLE, G.K., VAN DEN BOGERT, A.J., HERZOG, W., GERRITSEN, K.G.M. (1996). Modelling of force prediction in skeletal muscle undergoing stretch. *J. Biomech.* **29** (8), 1091–1104.

COMPTON, C.P. (1993). The use of public crash data in biomechanical research. In: Nahum, A.M., Melvin, J.W. (eds.), *Accidental Injury – Biomechanics and Prevention* (Springer, Berlin), pp. 49–65 (Chapter 3).

COOPER, P.R. (1982a). Post-traumatic intracranial mass lesions. In: Cooper, P.R. (ed.), *Head Injury* (Williams and Wilkins, Baltimore/London), pp. 185–232.

COOPER, P.R. (1982b). Skull fracture and traumatic cerebrospinal fluid fistulas. In: Cooper, P.R. (ed.), *Head Injury* (Williams and Wilkins, Baltimore/London), pp. 65–82.

COOPER, G.J., PEARCE, B.P., STAINER, M.C., MAYNARD, R.L. (1982). The biomechanical response of the thorax to non-penetrating trauma with particular reference to cardiac injuries. *J. Trauma* **22** (12), 994–1008.

COSTA-PAZ, M., RANALLETTA, M., MAKINO, A., AYERZA, M., MUSCOLO, L. (2002). Displaced patella fracture after cruciate ligament reconstruction with patellar ligament graft, *SICOT Online Report E006 February 2002*, http://www.sicot.org/.

CRANDALL, J.R., DUMA, S.M., BASS, C.R., PILKEY, W.D., KUPPA, S.M., KHAEWPONG, N., EPPINGER, R. (1999). Thoracic response and trauma in airbag deployment tests with out-of-position small female surrogates. *J. Crash Prevention Injury Control* **1** (2), 101–102.

CRAWFORD, G.N.C., JAMES, N.T. (1980). The design of muscles. In: Owen, R., Goodfellow, J., Bullough, P. (eds.), *Scientific Foundations of Orthopaedics and Traumatology* (William Heinemann, London), pp. 67–74.

CURREY, J.D. (1975). The effects of strain rate, reconstruction and mineral content on some mechanical properties of bovine bone. *J. Biomech.*, 81.

CUSICK, J.F., YOGANANDAN, N. (2002). Biomechanics of the cervical spine 4: Major injuries. *Clinical Biomech.* **17**, 1–20.

DALSTRA, M., HUISKES, R. (1995). Load transfer across the pelvic bone. *J. Biomech.* **6** (2), 715–724.

DALSTRA, M., et al. (1993). Mechanical and textural properties of pelvic trabecular bone. *J. Biomech.* **26** (4–5), 523–535.

DAN, D. (1995). Elaboration d'un capteur de pression. Application au foie humain. Projet de fin d'étude D.E.S.S., Collaboration entre L'Université Paris 7 et le LAB Laboratoire d'Accidentologie, de Biomécanique et d'Etude du Comportement Humain de PSA Peugeot Citroën–Renault.

DAN, D. (1999). Caractérisation mécanique du foie humain en situation de choc. PhD Thesis, University Paris 7 – Denis Diderot.

DAN, D., MILCENT, G. (2002). Caesare – chirurgie abdominale et simulation a retour d'effort, French Ministry of Education, Research and Technology Project, Final Technical Report RE/02.1600/A, Contract FSP9E2045, ESI France, 99 rue des Solets, 94513 Rungis Cedex, France.

DAVIDSSON, J., FLOGARD, A., LÖVSUND, P., SVENSSON, M.Y. (1999). BioRID P3—Design and performance compared to Hybrid III and Volunteers in Rear Impacts of $\Delta V = 7$ km/h. In: *Proc. 43rd Stapp Car Crash Conference*, pp. 253–265. Paper No. 99SC16.

DENG, Y.C. (1985). Human head/neck/upper-torso model response to dynamic loading, PhD Thesis, University of California.

DENG, Y.C., GOLDSMITH, W. (1987). Response of a human head/neck/upper-torso replica to dynamic loading – II. Analytical/numerical model. *J. Biomech.* **20**, 487–497.

DIGIMATION/VIEWPOINT CATALOG (2002). External and Skeletal Anatomy, 2002 Edition.

DIMASI, F., MARCUS, J., EPPINGER, R. (1991). 3-D anatomic brain model for relating cortical strains to automobile crash loading. In: *13th ESV International Technical Conference on Experimental Safety Vehicles*. Paper No. 91-S8-O-11.

DOHERTY, B.J., ESSES, S.L., HEGGENESS, M.H. (1992). A biomechanical study of odontoid fractures and fracture fixation. In: *Cervical Spine Research Society*.

DONNELLY, B.R., MEDIGE, J. (1997). Shear properties of human brain tissue. *J. Biomech. Engrg.* **119**, 423–432.

DOSTAL, W.F. (1981). A three-dimensional biomechanical model of hip musculature. *J. Biomech.* **14** (11), 803–812.

DUCHEYNE, P., et al. (1977). The mechanical behavior of intracondylar cancellous bone of the femur at different loading rates. *J. Biomech.* **10**, 747–762.

DVORAK, J., HAYEK, J., ZEHNUDER, R. (1987). CT-functional diagnostics of rotary instability of upper cervical spine. Part II: An evaluation on healthy adults and patients with suspected instability. *J. Spine* **12** (8), 726.

DVORAK, J., PANJABI, M.M., FROEHLICH, D., et al. (1988). Functional radiographic diagnosis of the cervical spine: Flexion/extension. *J. Spine* **13** (7), 748.

EBARA, S., IATRIDIS, J.C., SETTON, L.A., FOSTER, R.J., MOW, V.C., WEIDENBAUM, M. (1996). Tensile properties of nondegenerate human lumbar anulus fibrosus. *J. Spine* **21** (4), 452–461.

ENGIN, A.E. (1979). Passive resistance torques about long bone axes of major human joints. *Aviation, Space and Environmental Medicine* **50** (10), 1052–1057.

ENGIN, A.E. (1980). On the biomechanics of the shoulder complex. *J. Biomech.* **13-7**, 575–590.

ENGIN, A.E. (1983). Dynamic modelling of human articulating joints. *Math. Modelling* **4**, 117–141.

ENGIN, A.E. (1984). On the damping properties of the shoulder complex. *J. Biomech. Engrg.* **106**, 360–363.

ENGIN, A.E., CHEN, S.M. (1988a). On the biomechanics of the human hip complex in vivo – I. Kinematics for determination of the maximal voluntary hip complex sinus. *J. Biomech.* **21** (10), 785–796.

ENGIN, A.E., CHEN, S.M. (1988b). On the biomechanics of the human hip complex in vivo – II. Passive resistive properties beyond the hip complex sinus. *J. Biomech.* **21** (10), 797–806.

ENGIN, A.E., CHEN, S.M. (1989). A statistical investigation of the in vivo biomechanical properties of the human shoulder complex. *Math. Comput. Modelling* **12** (12), 1569–1582.

ENGIN, A.E., PEINDL, R.D. (1987). On the biomechanics of human shoulder complex – I. Kinematics for determination of the shoulder complex sinus. *J. Biomech.* **20** (2), 103–117.

ENGIN, A.E., T'MER, S.T. (1989). Three-dimensional kinematic modelling of the human shoulder complex – Part I: Physical model and determination of joint sinus cones. *J. Biomech. Engrg.* **111**, 107–112.

EPPINGER, R.H. (1976). Prediction of thoracic injury using measurable experimental parameters. In: *6th ESV International Technical Conference on Experimental Safety Vehicles*, pp. 770–779.

EPPINGER, R.H. (1993). Occupant restraint systems. In: Nahum, A.M., Melvin, J.W. (eds.), *Accidental Injury – Biomechanics and Prevention* (Springer, Berlin), pp. 186–197 (Chapter 8).

EPPINGER, R.H., MARCUS, J.H., MORGAN, R.M. (1984). Development of dummy and injury index for NHTSA's thoracic side impact protection research program. In: *Government/Industry Meeting and Exposition, Washington, DC*. SAE 840885.

ESTES, M.S., MCELHANEY, J.H. (1970). Response of brain tissue to compressive loading. ASME Paper No. 70-BHF-13.

EVANS, F.G., et al. (1961). Regional differences in some physical properties of human spongy bone. In: Evans, F.G. (ed.), *Biomechanical Studies of the Musculo–Skeletal System* (CC Thomas, Springfield, IL), pp. 49–67.

EWING, C., THOMAS, D., LUSTICK, L., MUZZY, W. III, WILLEMS, G., MAJEWSKI, P. (1976). The effect of duration, rate of onset, and peak sled acceleration on the dynamic response of the human head and neck. In: *Proc. 20th Stapp Car Crash Conference*, pp. 3–41. Paper No. 760800.

EWING, C., THOMAS, D., LUSTICK, L., MUZZY, W. III, WILLEMS, G., MAJEWSKI, P. (1978). Effect of initial position on the human head and neck response to $+Y$ impact acceleration. In: *Proc. 22nd Stapp Car Crash Conference*, pp. 103–138. Paper No. 780888.

FALLENSTEIN, G.T., HULCE, V.D., MELVIN, J.W. (1970). Dynamic mechanical properties of human brain tissue. *J. Biomech.* **2**, 217–226.

FAZEKAS, I.G., KOSA, F., JOBBA, G., MESZARO, E. (1971a). Die Druckfestigkeit der Menschlichen Leber mit besonderer Hinsicht auf Verkehrsunfälle. *Z. Rechtsmedizin* **68**, 207–224.

FAZEKAS, I.G., KOSA, F., JOBBA, G., MESZARO, E. (1971b). Experimentelle Untersuchungen über die Druckfestigkeit der Menschlichen Niere. *Zacchia* **46**, 294–301.

FAZEKAS, I.G., KOSA, F., JOBBA, G., MESZARO, E. (1972). Beiträge zur Druckfestigkeit dser Menschlichen Milz bei stumpfen Gewalteinwirkungen. *Arch. Kriminol.* **149**, 158–174.

FIROOZBAKSHK (1975). A model of brain shear under impulsive torsional loads. *J. Biomech.* **8**, 65–73.

FOGRASCHER, K. (1998). Development of a simulation model for the head protection system ITS and integration of the component model into the full structural vehicle model, *EuroPam 1998*, ESI Software S.A., 99, rue des Solets, BP 80112, 94513 Rungis Cedex, France.

FOREMAN, S.M., CROFT, A.C. (1995). *Whiplash Injuries: The cervical Acceleration/Deceleration Syndrome*, second ed. (Williams & Wilkins).

FRANK, C.B., SHRIVE, N.G. (1994). Biomechanics of the musculo–skeletal System – 2.3: Ligament. In: Nigg, B.M., Herzog, W. (eds.), *Biomechanics of the Musculo–Skeletal System* (University of Calgary, Alberta, Canada), pp. 106–132.

FRANKEL, V.H., NORDIN, M. (1980). *Basic Biomechanics of the Skeletal System* (Lea & Febiger, Philadelphia, PA).

FUKUBAYASHI, T., KUROSAWA, H. (1980). The contact area and pressure distribution pattern of the knee. *J. Acta Orthop. Scand.* **51**, 871–879.

FUNG, Y.C. (1993a). *Biomechanics – Mechanical Properties of Living Tissues* (Springer, Berlin).

FUNG, Y.C. (1993b). The application of biomechanics to the understanding of injury and healing. In: Nahum, A.M., Melvin, J.W. (eds.), *Accidental Injury – Biomechanics and Prevention* (Springer, Berlin), pp. 1–11 (Chapter 1).

FUNG, Y.C., YEN, M.R. (1984). Experimental investigation of lung injury mechanisms, Topical Report, US Army Medical Research and Development Command. Contract No. DAMD 17-82-C-2062.

FURUKAWA, K., FURUSU, K., MIKI, K. (2002). A development of child FEM model – Part I: Skeletal model of six-year-old child. In: *Proc. 2002 JSME Annual Congress, No. 02-1*, pp. 89–90 (in Japanese).

GADD, C.W. (1961). Criteria for injury potential. In: *Impact Acceleration Stress Symposium*, National Research Council Publication, No. 977 (National Academy of Sciences, Washington, DC), pp. 141–144.

GADD, C.W. (1966). Use of a weighted impulse criterion for estimating injury hazard. In: *Proc. 10th Stapp Car Crash Conference*, pp. 164–174. Paper No. 660793.

GALANTE, J.O. (1967). Tensile properties of human lumbar anulus fibrosus. *Acta Orthop Scand (Suppl.)* **100**.

GALANTE , J.O., et al. (1970). Physical properties of trabecular bone. *Calcif. Tissue Res.* **5**, 236–246.

GALFORD, J.E., MCELHANEY, J.H. (1970). A visco-elastic study of scalp, brain and dura. *J. Biomech.* **3**, 211–221.

GENNARELLI, T.A. (1980). Analysis of head injury severity by AIS-80. In: *24th Annual Conference of the American Association of Automotive Medicine* (AAAM, Morton Grove, IL), pp. 147–155.

GENNARELLI, T.A., THIBAULT, L.E. (1982). Biomechanics of acute subdural hematoma. *J. Trauma* **22** (8), 680–686.

GENNARELLI, T.A., THIBAULT, L.E., ADAMS, J.H., GRAHAM, D.I., THOMPSON, C.J., MARCINCIN, R.P. (1982). Diffuse axonal injury and traumatic coma in the primate. *Ann. Neuron.* **12**, 564–574.

GENNARELLI, T.A., THIBAULT, L.E., TOMEI, G., WISER, R., GRAHAM, D., ADAMS, J. (1987). Directional dependence of axonal brain injury due to centroidal and non-centroidal acceleration. In: *Proc. 31st Stapp Car Crash Conference*, pp. 49–53. Paper No. 872197.

GOEL, V.K., GOYAL, S., CLARK, C., NISHIYAMA, K., NYE, T. (1985). Kinematics of the whole lumbar spine effect of dissectomy. *J. Spine* **10** (6).

GOEL, V.K., MONROE, B.T., GILBERTSON, L.G., BRINKMAN, P. (1995). Interlaminar shear stresses and laminae separation in a disc: finite element analysis of the L3–L4 motion segment subjected to axial compressive loads. *Spine* **20** (6), 689–698.

GOLDSMITH (1972). Biomechanics of Head Injury. In: Fung, Y.C., Perrone, N., Anliker, M. (eds.), *Biomechanics: It's Foundation and Objectives* (Prentice-Hall, Englewood Cliffs, NJ).

GOLDSTEIN, S., FRANKENBURG, E., KUHN, J. (1993). Biomechanics of bone. In: Nahum, A.M., Melvin, J.W. (eds.), *Accidental Injury – Biomechanics and Prevention* (Springer, Berlin), pp. 198–223 (Chapter 9).

GOLDSTEIN, S.A., et al. (1983). The mechanical properties of human tibia trabecular bone as a function of metaphyseal location. *J. Biomech.* **16**, 965–969.

GRANIK, G., STEIN, I. (1973). Human ribs: static testing as a promising medical application. *J. Biomech.* **6**, 237–240.

GRAY'S ANATOMY (1989). Williams, P.L., Warwick, R., Dyson, M., Bannister, L. (eds.), Thirty-seventh ed., Churchill Livington, ISBN 0 443 02588 6, after Figure 5.8A, page 554.

GURDJIAN, E.S., LISSNER, H.R. (1944). Mechanism of head injury as studied by the cathode ray oscilloscope, Preliminary report. *J. Neurosurgery* **1**, 393–399.

GURDJIAN, E.S., ROBERTS, V.L., THOMAS, L.M. (1966). Tolerance curves of acceleration and intracranial pressure and protective index in experimental head injury. *J. Trauma* **6**, 600–604.

GURDJIAN, E.S., WEBSTER, J.E., LISSNER, H.R. (1955). Observations of the mechanism of brain concussion, contusion, and laceration. *Surgery Gynecol. Obstet.* **101**, 680–690.

HAJJI, M.A., WILSON, T.A., LAI-FOOK, S.J. (1979). Improved measurements of shear modulus and pleural membrane tension of the lung. *J. Appl. Physiol.* **47**, 175–181.

HALLQUIST, J.O., GOUDREAU, G.L., BENSON, D.J. (1985). Sliding surfaces with contact–impact in large-scale Lagrangian computations problems. *Comput. Methods Appl. Mech. Engrg.* **51**, 107–137.

HARDY, W.N. (1993). Instrumentation in experimental design. In: Nahum, A.M., Melvin, J.W. (eds.), *Accidental Injury – Biomechanics and Prevention* (Springer, Berlin), pp. 12–48 (Chapter 2).

HARDY, W.N., FOSTER, C.D., MASON, M.J., YANG, K.H., KING, A.I., TASHMAN, S. (2001). Investigation of head injury mechanisms using neutral density technology and high-speed biplanar X-ray. *Stapp Car Crash Conference J.* **45**, 337–368. Paper No. 2001-22-0016.

HARRY, J.D., WARD, A.W., HEGLUD, N.C., MORGAN, D.L., MCMAHON, T.A. (1990). Cross-bridge cycling theories cannot explain high-speed lengthening behavior in frog muscle. *Biophys. J.* **57**, 201–208.

HAUG, E. (1995). Biomechanical models in vehicle accident simulation. In: *PAM – User's Conference in Asia, PUCA '95*, Shin-Yokohama, ESI Software S.A., 99, rue des Solets, BP 80112, 94513 Rungis Cedex, France, pp. 233–256.

HAUG, E., BEAUGONIN, M., TRAMECON, A., HYNCIK, L. (1999). Current status of articulated and deformable human models for impact and occupant safety simulation at ESI group. In: *European Conference on Computational Mechanics, EEVC'99*, August 31 – September 3, 1999, Munic.

HAUG, E., CLINCKEMAILLIE, J.C., ABERLENC, F. (1989a). Computational mechanics in crashworthiness analysis. In: *Post Symposium Short Course of the 2nd Symposium of Plasticity, Nagoya, Japan.*

HAUG, E., CLINCKEMAILLIE, J.C., ABERLENC, F. (1989b). Contact–impact problems for crash. In: *Post Symposium Short Course of the 2nd Symposium of Plasticity, Nagoya, Japan.*

HAUG, E., LASRY, D., GROENENBOOM, P., MUNCK, G., ROGER, J., SCHLOSSER, J., RÜCKERT, J. (1993). Finite element models of dummies and biomechanical applications using PAM-CRASH™. In: *PAM –*

User's Conference in Asia, PUCA '93, Shin-Yokohama, ESI Software S.A., 99, rue des Solets, BP 80112, 94513 Rungis Cedex, France, pp. 123–158.

HAUG, E., TRAMECON, A., ALLAIN, J.C., CHOI, H.-Y. (2001). Modeling of ergonomics and muscular comfort. *KSME Internat. J.* **15** (7), 982–994.

HAUG, E., ULRICH, D. (1989). The PAM–CRASH code as an efficient tool for crashworthiness simulation and design. In: *Second European Cars/Trucks Simulation Symposium, Schliersee (Munich), AZIMUTH* (Springer, Berlin).

HAUT, R.C. (1993). Biomechanics of soft tissues. In: Nahum, A.M., Melvin, J.W. (eds.), *Accidental Injury – Biomechanics and Prevention* (Springer, Berlin), pp. 224–246 (Chapter 10).

HAWKINS, D., BEY, M. (1994). A comprehensive approach for studying muscle-tendon mechanics. *J. Biomech. Engrg.* **116**, 51.

HAWKINS, D., BEY, M. (1997). Muscle and tendon force-length properties and their interactions in vivo. *J. Biomech.* **30** (1), 63–70.

HAYASHI, S., CHOI, H.-Y., LEVINE, R.S., YANG, K.H., KING, A.I. (1996). Experimental and analytical study of a frontal knee impact. In: *Proc. 40th Stapp Car Crash Conference*, pp. 161–173. Paper No. 962423.

HERZOG, W. (1994). Muscle. In: Nigg, B.M., Herzog, W. (eds.), *Biomechanics of the Musculo–Skeletal System* (Wiley, New York), pp. 154–187.

HILL, A.V. (1970). *First and Last Experiments in Muscle Mechanics* (Cambridge).

HIRSCH, C. (1955). The reaction of intervertebral discs to compression force. *J. Bone Joint Surgery A* **37** (6).

HOLBOURN, A.H.S. (1943). Mechanics of head injury. *Lancet* **2**, 438–441.

HOPPIN, F.G., LEE, G.C., DAWSON, S.V. (1975). Properties for lung parenchyma in distortion. *J. Appl. Physiol.* **39**, 742–751.

HORST, M.J. VAN DER, THUNNISSEN, J.G.M., HAPPEE, R., HAASTER VAN, R.M.H.P., WISMANS, J.S.H.M. (1997). The influence of muscle activity on head–neck response during impact. In: *Proc. 41st Stapp Car Crash Conference*, pp. 487–507. Paper No. 973346.

HUANG, Y., KING, A.I., CAVANAUGH, J.M. (1994a). A MADYMO model of near-side human occupants in side impacts. *J. Biomech. Engrg.* **116**, 228–235.

HUANG, Y., KING, A.I., CAVANAUGH, J.M. (1994b). Finite element modelling of gross motion of human cadavers in side impact. In: *Proc. 38th Stapp Car Crash Conference*, pp. 35–53. Paper No. 942207.

HUELKE, D.F., NUSHOLTZ, G.S., KAIKER, P.S. (1986). Use of quadruped models in thoraco-abdominal biomechanics research. *J. Biomech.* **19** (12), 969–977.

HUGHES, T.J.R., PISTER, J.S., TAYLOR, R.L. (1979). Implicit–explicit finite elements in non-linear transient analysis. *Comput. Methods Appl. Mech. Engrg.* **17/18**, 159–182.

HUGHES, T.J.R., TAYLOR, R.L., SACKMAN, J.L., CURNIER, A., KANOKNUKULCHAI, W. (1976). A finite element method for a class of contact–impact problems. *Comput. Methods Appl. Mech. Engrg.* **8**, 249–276.

HVID, I., et al. (1985). Trabecular bone strength patterns at the proximal tibial epiphysis. *J. Orthop. Res.* **3**, 464–472.

HYNCIK, L. (1997). Human articulated rigid body model (ROBBY1). ESI Group Internal Report, ESI Software S.A., 99, rue des Solets, BP 80112, 94513 Rungis Cedex, France.

HYNCIK, L. (1999a). Human shoulder model. In: *15th Conference Computational Mechanics*, pp. 117–124. ISBN 80-7082-542-1 (in Czech).

HYNCIK, L. (1999b). Human inner organs model. In: *15th Conference Computational Mechanics*, pp. 125–132. ISBN 80-7082-542-1.

HYNCIK, L. (2000). Biomechanical human model. In: *8th Conference Biomechanics of Man*, pp. 52–55. ISBN 80-244-0193-2.

HYNCIK, L. (2001a). Rigid body based human model for crash test purposes. *Engrg. Mech.* **5**, 337–342.

HYNCIK, L. (2001b). Biomechanical model of human inner organs and tissues. In: *17th Conference Computational Mechanics*, pp. 113–120. ISBN 80-7082-780-7.

HYNCIK, L. (2002a). Multi-body human model. In: *Preprint of seminar Virtual Nonlinear Multibody Systems*. In: Preprint of NATO Advanced Study Institute **1**, pp. 89–94.

HYNCIK, L. (2002b). Deformable human abdomen model for crash test purposes. In: *4th Conference Applied Mechanics*, pp. 159–164. ISBN 80-248-0079-9.

HYNCIK, L. (2002c). Biomechanical model of abdominal inner organs and tissues for crash test purposes. PhD Thesis, Department of Mechanics of the Faculty of Applied Sciences of the University of West Bohemia in Pilsen.

IRWIN, A.L. (1994), Analysis and CAL3D model of the shoulder and thorax response of seven cadavers subjected to lateral impacts. PhD Thesis, Wayne State University, 1994.

IRWIN, A., MERTZ, H.J. (1997). Biomechanical basis for the CRABI and Hybrid III child dummies. In: *Proc. 41st Stapp Car Crash Conference*, pp. 261–272. Paper No. 973317.

ITOI, E., BERGLUND, L.J., GRABOWSKI, J.J., SCHULTZ, F.M., GROWNEY, E.S., MORREY, B.F., AN, K.N. (1995). Tensile properties of the supraspinatus tendon. *J. Orthop. Res.* **13** (4), 578–584.

IWAMOTO, M., KISANUKI, Y., WATANABE, I., FURUSU, K., MIKI, K., HASEGAWA, J. (2002). Development of a finite element model of the total human model for safety (THUMS) and application to injury reconstruction. In: *2002 International IRCOBI Conference*, pp. 31–42.

JAGER, M. DE (1996). Mathematical head–neck models for acceleration impact. Ph.D. Thesis, Eindhoven University.

JAGER, M. DE, SAUREN, A., THUNNISSEN, J., WISMANS, J. (1994). A three-dimensional neck model: Validation for frontal and lateral impact. In: *38th Stapp Car Crash Conference*, pp. 93–109. Paper No. 942211.

JOHNSON, G.A., TRAMAGLINI, D.M., LEVINE, R.E., ONO, K., CHOI, N.Y., WOO, S.L. (1994). Tensile and viscoelastic properties of human patellar tendon. *J. Orthop. Res.* **12** (6), 796–803.

KAJZER, J. (1991). Impact biomechanics of knee injuries. Doctoral Thesis, Dept. of Injury Prevention, Chalmers University of Technology.

KAJZER, J., SCHROEDER, G., ISHIKAWA, H., MATSUI, Y., BOSCH, U. (1997). Shearing and bending effects at the knee joint at high speed lateral loading. In: *Proc. 41st Stapp Car Crash Conference*, pp. 151–165. Paper No. 973326.

KAJZER, J., ZHOU, C., KHALIL, T.B., KING, A.I. (1996). Modelling of ligaments and muscles under transient loads: application of PAM–CRASH material models. In: *PAM User's Conference in Asia PUCA '96*, Nihon ESI, ESI Software S.A., 99, rue des Solets BP 80112, 94513 Rungis Cedex, France, pp. 223–231.

KALLIERIS, D., SCHMIDT, G. (1990). Neck response and injury assessment using cadavers and the US-SID for far-side lateral impacts of rear seat occupants with inboard-anchored shoulder belts. In: *Proc. 34th Stapp Car Crash Conference*, pp. 93–99. Paper No. 902313.

KANG, H.S., WILLINGER, R., DIAW, B.M., CHINN, B. (1997). Validation of a 3D anatomic human head model and replication of head impact in motorcycle accident by finite element modeling. In: *Proc. 41st Stapp Car Crash Conference*, pp. 329–338. Paper No. 973339.

KAPANDJI, I.A. (1974a). *The Physiology of Joints, vol. 1: Upper Limb* (Churchill Livingstone).

KAPANDJI, I.A. (1974b). *The Physiology of Joints, vol. 2: Lower Limb* (Churchill Livingstone).

KAPANDJI, I.A. (1974c). *The Physiology of Joints, vol. 3: The Trunk and the Vertebral Column* (Churchill Livingstone).

KATAKE, K. (1961). The strength for tension and bursting of human fasciae. *J. Kyoto Pref. Med. Univ.* **69**, 484–488.

KAZARIAN, L.E. (1982). Injuries to the human spinal column: Biomechanics and injury classification. *Exerc. Sport Sci. Rev.* **9**, 297–352.

KAZARIAN, L.E., BEERS, K., HERNANDEZ, J. (1979). Spinal injuries in the F/FB-111 crew escape system. *Aviat. Space Environ. Med.* **50**, 948–957.

KEITHEL, L.M. (1972). Deformation of the thoracolumbar intervertebral joints in response to external loads. *J. Bone Joint Surgery A* **54** (3).

KIMPARA, H., IWAMOTO, M., MIKI, K. (2002). Development of a small female FEM model. In: *Proc. JSAE Spring Congress*, No. 59-02, Paper No. 20025242, pp. 1–4 (in Japanese).

KING, A.I. (1993). Injury to the thoraco-lumbar spine and pelvis. In: Nahum, A.M., Melvin, J.W. (eds.), *Accidental Injury – Biomechanics and Prevention* (Springer, Berlin), pp. 429–459 (Chapter 17).

KIRSCH, R.F., BOSKOV, D., RYMER, W.Z. (1994). Muscle stiffness during transient and continuous movements of cat muscle: Perturbation characteristics and physiological relevance. *IEEE Trans. Biomedical Engrg.* **41** (8), 758–770.

KISIELEWICZ, T.K., ANDOH, K. (1994). Critical issues in biomechanical tests and simulations of impact injury. In: *PAM – User's Conference in Asia, PUCA '94*, Shin-Yokohama, ESI Software S.A., 99, rue des

Solets, BP 80112, 94513 Rungis Cedex, France, pp. 209–222, and: *World Congress on Computational Mechanics, WCCM '94*, Makuhari, Japan, August 1994.

KLEINBERGER, M., SUN, E., EPPINGER, R.H., KUPPA, S., SAUL, R. (1998). Development of improved injury criteria for the assessment of advanced automotive restraint systems. NHTSA report.

KRESS, T.A., SNIDER, J.N., PORTA, D.J., FULLER, P.M., WASSERMAN, J.F., TUCKER, G.V. (1993). Human femur response to impact loading. In: *Int. IRCOBI Conf. on the Biomechanics of Trauma*, Eindhoven, the Netherlands. IRCOBI Secretariat, Bron, France, pp. 93–104.

KROELL, C.K. (1971). Thoracic response to blunt frontal loading. In: Backaitis, S.H. (ed.), *Biomechanics of Impact Injury and Injury Tolerances of the Thorax-Shoulder Complex* (Society of Automotive Engineers), pp. 51–80. Paper No. PT-45.

KROELL, C.K., ALLEN, S.D., WARNER, C.Y., PERL, T.R. (1986). Interrelationship of velocity and chest compression in blunt thoracic impact to swine II. In: *Proc. 30th Stapp Car Crash Conference*, pp. 99–121. Paper No. 861881.

KROELL, C.K., SCHNEIDER, D.C., NAHUM, A.M. (1971). Impact tolerance and response of the human thorax. In: *Proc. 15th Stapp Car Crash Conference*, pp. 84–134. Paper No. 710851.

KROELL, C.K., SCHNEIDER, D.C., NAHUM, A.M. (1974). Impact tolerance and response of the human thorax II. In: *Proc. 18th Stapp Car Crash Conference*, pp. 383–457. Paper No. 741187.

KROONENBERG, A. VAN DEN, THUNNISSEN, J., WISMANS, J. (1997). A human model for low severity rear-impacts. In: *IRCOBI Conference*.

KRYLOW, A.M., SANDEROCK, T.G. (1996). Dynamic force responses of muscle involving eccentric contraction. *J. Biomech.* **30** (1), 27–33.

KULAK, R.F., BELYTSCHKO, T.B., SCHULTZ, A.B., GALANTE, J.O. (1976). Nonlinear behavior of the human intervertebral disc under axial load. *J. Biomech.* **9**.

LANIR, Y., FUNG, Y.C. (1974). Two-dimensional mechanical properties of rabbit skin – II. Experiment results. *J. Biomech.* **7**, 171–182.

LASKY, I.I., SIEGEL, A.W., NAHUM, A.M. (1968). Automotive cardio-thoracic injuries: A medical-engineering analysis. In: *Automotive Engineering Congress, Detroit, MI*. Paper No. 680052.

LEE, S-H., CHOI, H-Y. (2000). Finite element modeling of human head–neck complex for crashworthiness simulation. *Hongik J. Sci. Technol.* **4**, 1–15.

LEE, M.C., HAUT, R.C. (1989). Insensitivity of tensile failure properties of human bridging veins to strain rate: implications in biomechanics of subdural hematoma. *J. Biomech.* **22**, 537–542.

LEE, M.C., MELVIN, J.W., UENO, K. (1987). Finite element analysis of traumatic subdural hematoma. In: *Proc. 31st Stapp Car Crash Conference*, pp. 67–77. Paper No. 872201.

LEE, J.B., YANG, K.H. (2001). Development of a finite element model of the human abdomen. In: *Proc. 45th Stapp Car Crash Conference*, pp. 1–22. Paper No. 2001-22-0004.

LESTINA, D.C., KUHLMANN, T.P., KEATS, T.E., MAXWELL ALLEY, R. (1992). Mechanism of fracture in ankle and foot injuries to drivers in motor vehicles. In: *Proc. 36th Stapp Car Crash Conference*, pp. 59–68. Paper No. 922515.

LEVINE, R. (1993). Injury to the extremities. In: Nahum, A.M., Melvin, J.W. (eds.), *Accidental Injury – Biomechanics and Prevention* (Springer, Berlin), pp. 460–491 (Chapter 18).

LEWIS, G., SHAW, K.M. (1997a). Modeling the tensile behavior of human achilles tendon. *Biomed. Mater. Engrg.* **7** (4), 231–244.

LEWIS, G., SHAW, K.M. (1997b). Tensile properties of human tendo achillis: Effect of donor age and strain rate. *J. Foot Ankle Surgery* **36** (6), 435–445.

LIN, H.S., LIU, Y.K., RAY, G., NIKRAVESH, P. (1978). Mechanical response of the lumbar intervertebral joint under physiological (complex) loading. *J. Bone Joint Surgery A* **60** (1), 41–55.

LINDAHL, O. (1976). Mechanical properties of dried defatted spongy bone. *Acta Orthop. Scand.* **47**, 11–19.

LINDE, F., et al. (1989). Energy absorptive properties of human trabecular bone specimens during axial compression. *J. Orth. Res.* **7**, 432–439.

LISSNER, H.R., LEBOW, M., EVANS, F.G. (1960). Experimental studies on the relation between acceleration and intracranial pressure changes in man. *Surgery Gynecol. Obstet.* **111**, 329–338.

LIZEE, E., ROBIN, S., SONG, E., BERTHOLON, N., LECOZ, J.Y., BESNAULT, B., LAVASTE, F. (1998). Development of a 3D finite element model of the human body. In: *Proc. 42nd Stapp Car Crash Conference*, pp. 1–23. Paper No. 983152.

LIZEE, E., SONG, E., et al. (1998). Finite element model of the human thorax validated in frontal, oblique and lateral impacts: A tool to evaluate new restraint systems. In: *Proceedings of the 1998 International IRCOBI Conference*.

LÖHNER, R. (1990). Three-dimensional fluid-structure Interaction using a finite element solver and adaptive re-meshing. *Comput. Syst. Engrg.* **1** (2–4), 257–272.

LOWENHIELM, P. (1974). Dynamic properties of parasagittal bridging veins. *Z. Rechtsmedizin* **74**, 55–62.

LUNDBERG, A., GOLDIE, I., KALIN, B., SELVIK, G. (1989). Kinematics of the ankle/foot complex: Plantarflexion and dorsiflexion. *Foot & Ankle* **9** (4), 194–200.

LUNDBERG, A., SVENSSON, O., BYLUND, C., GOLDIE, I., SELVIK, G. (1989). Kinematics of the ankle/foot complex Part 2: Pronation and supination. *Foot & Ankle* **9** (5), 248–253.

MA, S., ZAHALAK, G.I. (1991). A distribution-moment model of energetics in skeletal muscle. *J. Biomech.* **24** (1), 21–35.

MA, D., OBERGEFELL, L.A., RIZER, L.A. (1995). Development of human articulating joint model parameters for crash dynamics simulations. In: *Proc. 39th Stapp Car Crash Conference*, pp. 239–250. Paper No. 952726.

MAENO, T., HASEGAWA, J. (2001). Development of a finite element model of the total human model for safety (THUMS) and application to car-pedestrian impacts. In: *17th International ESV Conference*. Paper No. 494.

MAKHSOUS, M., HÖGFORS, C., SIEMIEN'SKI, A., PETERSON, B. (1999). Total shoulder and relative muscle strength in the scapular plane. *J. Biomech.* **32**, 1213–1220.

MANSCHOT, J.F., BRAKKEE, A.J. (1986a). The measurement and modelling of the mechanical properties of human skin in vivo – I. The measurement. *J. Biomech.* **19** (7), 511–515.

MANSCHOT, J.F., BRAKKEE, A.J. (1986b). The measurement and modelling of the mechanical properties of human skin in vivo – II. The model. *J. Biomech.* **19** (7), 517–521.

MARGULIES, S.S., THIBAULT, L.E. (1989). An analytic model of traumatic diffuse brain injury. *J. Biomech. Engrg.* **111**, 241–249.

MARGULIES, S.S., THIBAULT, L.E. (1992). A proposed tolerance criterion for diffuse axonal injury in man. *J. Biomech.* **25** (8), 917–923.

MARGULIES, S.S., THIBAULT, L.E., GENNARELLI, T.A. (1990). Physical model simulations of brain injury in the primate. *J. Biomech.* **23**, 823–836.

MARKOLF, K.L. (1972). Deformation of the thoracolumbar intervertebral joints in response to external loads. *J. Bone Joint Surgery A* **54** (3).

MARKOLF, K.L., MORRIS, J.M. (1974). The structural components of the intervertebral disc. *J. Bone Joint Surgery A* **56** (4).

MARTENS, M., et al. (1983). The mechanical characteristics of cancellous bone at the upper femoral region. *J. Biomech.* **16**, 971–983.

MARTIN, J., THOMPSON, G. (1986). Achilles tendon rupture. *CORR* **210**, 216–218.

MARTINI, F.H., TIMMONS, M.J., TALLITSCH, B. (2003). Human Anatomy, fourth ed. (Pearson Education, Inc., publishing as Benjamin Cummings, Upper Saddle River, NJ), ISBN 0-13 061569 2.

MASSON, C., CESARI, D., BASILE, F., BEAUGONIN, M., TRAMECON, A., ALLAIN, J.C., HAUG, E. (1999). Quasi static ankle/foot complex behavior: Experimental tests and numerical simulations. In: *IRCOBI Conference, Spain*.

MAUREL, W. (1998). 3D modelling of the human upper limb including the biomechanics of joints, muscles and soft tissues, Thèse N° 1906, Ecole Polytechnique Federale de Lausanne.

MCCALDEN, R.W., et al. (1993). Age-related in the tensile properties of cortical bone. *J. Bone Joint Surgery A* **75**, 1193–1205.

MCCLURE, P., SIEGLER, S., NOBILINI, R. (1998). Three-dimensional flexibility characteristics of the human cervical spine in vivo. *Spine* **23** (2), 216–223.

MCELHANEY, J.H. (1966). Dynamic response of bone and muscle tissue. *J. Appl. Physiol.*, 1231.

MCELHANEY, J.H., DOHERTY, B.J., PAVER, J.G., MYERS, B.S., GRAY, L. (1988). Combined bending and axial loading responses of the human cervical spine. In: *Proc. 32nd Stapp Car Crash Conference*, pp. 21–28. Paper No. 881709.

McElhaney, J.H., Melvin, J.W., Roberts, V.L., Portnoy, H.D. (1973). Dynamic characteristics of the tissues of the head. In: Kennedi, R.M. (ed.), *Perspectives in Biomedical Engineering* (MacMillan, London), pp. 215–222.

McElhaney, J.H., Myers, B.S. (1993). Biomechanical aspects of cervical trauma. In: Nahum, A.M., Melvin, J.W. (eds.), *Accidental Injury – Biomechanics and Prevention* (Springer, Berlin), pp. 311–361 (Chapter 14).

McElhaney, J.H., Paver, J.G., McCrackin, H.J., Maxwell, G.M. (1983). Cervical spine compression responses. In: *Proc. 27th Stapp Car Crash Conference*, pp. 163–177. Paper No. 831615.

McElhaney, D.A., et al. (1970). Mechanical properties of cranial bone. *J. Biomech.* **3**, 495–511.

Meany, D.F., Smith, D., Ross, T., Gennarelli, T.A. (1993). Diffuse axonal injury in the miniature pig: Biomechanical development and injury threshold. *ASME Crashworthiness and Occupant Protection Systems* **25**, 169–175.

Melvin, J.W., Lighthall, J.W., Ueno, K. (1993). Brain injury biomechanics. In: Nahum, A.M., Melvin, J.W. (eds.), *Accidental Injury – Biomechanics and Prevention* (Springer, Berlin), pp. 268–291 (Chapter 12).

Melvin, J.W., McElhaney, J.H., Roberts, V.L. (1970). Development of mechanical model of human head – determination of tissue properties and synthetic substitute materials. In: *Proc. 14th Stapp Car Crash Conference*, pp. 221–240. Paper No. 700903.

Melvin, J.W., Stalnaker, R.L., Roberts, V.L. (1973). Impact injury mechanisms in abdominal organs. In: *Proc. 17th Stapp Car Crash Conference Proceedings*, pp. 115–126. Paper No. 730968.

Mertz, H.J., Patric, L.M. (1971). Strength and response of human neck. In: *Proc. 15th Stapp Car Crash Conference*, pp. 207–255. Paper No. 710855.

Mertz, H.J. (1984). A procedure for normalizing impact response data. SAE Paper 840884, Washington, DC.

Mertz, H.J. (1993). Anthropomorphic test devices. In: Nahum, A.M., Melvin, J.W. (eds.), *Accidental Injury – Biomechanics and Prevention* (Springer, Berlin), pp. 66–84 (Chapter 4).

Miller, K. (2000). Constitutive modeling of abdominal organs. *J. Biomech.* **33**, 367–373.

Miller, K., Chinzei, K. (1997). Constitutive modelling of brain tissue: Experiment and theory. *J. Biomech.* **30**, 1115–1121.

Miller, J.A.A., Schultz, A.B., Warwick, D.N., Spencer, D.L. (1986). Mechanical properties for lumbar spine motion segments under large loads. *J. Biomech.* **19** (1), 79–84.

Moffatt, C.A., Advani, S.H., Lin, C.J. (1971). Analytical end experimental investigations of human spine flexure. *Amer. Soc. Mech. Engrg.* **71-WA/BHF-7**.

Momersteeg, T.J.A., Blaskevoort, L., Huiskes, R., Kooloos, J.G., Kauer, J.M.G. (1996). Characterization of the mechanical behavior of human knee ligaments: A numerical–experimental approach. *J. Biomech.* **29** (2), 151–160.

Monaghan, J.J. (1988). An introduction into SPH. *Comput. Phys. Comm.* **48**, 89–96.

Monaghan, J.J., Gingold, R.A. (1983). Shock simulation by the particle method SPH. *Comput. Phys.* **52**, 374–398.

Morgan, D.L. (1990). New insights into the behavior of muscle during active lengthening. *Biophys. J.* **57**, 209–221.

Morgan, R.M., Eppinger, R.H., Hennessey, B. (1991). Ankle joint injury mechanism for adults in frontal automotive impact. In: *Proc. 35th Stapp Car Crash Conference Proceedings*, pp. 189–198. Paper No. 912902.

Moroney, S.P., Schultz, A.B., Miller, J.A.A. (1988). Analysis and measurement of neck load. *J. Orthopaedic Res.* **6**, 713–720.

Moroney, S.P., Schultz, A.B., Miller, J.A.A., Andersson, G.B.J. (1988). Load-displacement properties of lower cervical spine motion segments. *J. Biomech.* **21** (9), 769–779.

Mosekilde, L., et al. (1986). Normal vertebral body size and compressive strength: Relations to age and to vertebral and iliac trabecular bone compressive strength. *Bone* **7**, 207–212.

Mow, V.C., Hayes, W.C. (1991). *Basic Orthopaedic Biomechanics* (Raven Press, New York).

Myers, B.S., McElhaney, J.H., Doherty, B.J., Paver, J.G., Gray, L. (1991). The role of torsion in cervical spinal injury. *Spine* **16** (8), 870–874.

MYERS, B.S., MCELHANEY, J.H., RICHARDSON, W.J., NIGHTINGALE, R., DOHERTY, B.J. (1991). The influence of end condition on human cervical spine injury mechanisms. In: *Proc. 35th Stapp Car Crash Conference*, pp. 391–400. Paper No. 912915.

MYERS, B.S., VAN EE, C.A., CAMACHO, D.L.A., WOOLLEY, C.T., BEST, T.M. (1995). On the structural properties of mammalian skeletal muscle and its relevance to human cervical impact dynamics. In: *Proc. 39th Stapp Car Crash Conference*, pp. 203–214. Paper No. 952723.

MYKLEBUST, J.B., PINTAR, F. (1988). Tensile strength of spinal ligaments. *J. Spine* **13** (5).

NAGASAKA, K., IWAMOTO, M., MIZUNO, K., MIKI, K., HASEGAWA, J. (2002). Pedestrian injury analysis using the human FE model (Part I, development and validation of the lower extremity model and application to damage evaluation of knee ligaments). In: *Proc. JSME 14th Bioengineering Conference*, pp. 141–142 (in Japanese).

NAHUM, M., GADD, C.W., SCHNEIDER, D.C., KROELL, C.K. (1970). Deflection of the human thorax under sternal impact. In: *Proceedings of the International Automobile Safety Conference, Detroit*. SAE 700400.

NAHUM, A.M., MELVIN, J.W. (eds.) (1993). *Accidental Injury: Biomechanics and Prevention* (Springer, New York).

NAHUM, M., SCHNEIDER, D.C., KROELL, C.K. (1975). Cadaver skeletal response to blunt thoracic impact. In: *Proc. 19th Stapp Car Crash Conference*, pp. 259–293. Paper No. 751150.

NAHUM, A.M., SIEGEL, A.W., HIGHT, P.V., BROOKS, S.H. (1968). Lower extremity injuries of front seat occupants. In: *Proc. 11th Stapp Car Crash Conference Proceedings*. Paper No. 680483.

NAHUM, A.M., SMITH, R., WARD, C.C. (1977). Intracranial pressure dynamics during head impact. In: *Proc. 21st Stapp Car Crash Conference*, pp. 339–366. Paper No. 770922.

NEUMANN, P., KELLER, T.S., EKSTROM, L., PERRY, L., HANSSON, T.H., SPENGLER, D.M. (1992). Mechanical properties of the human lumbar anterior longitudinal ligament. *J. Biomech.* **25** (10), 1185–1194.

NEWMAN, J.A. (1993). Biomechanics of head trauma: Head protection. In: Nahum, A.M., Melvin, J.W. (eds.), *Accidental Injury – Biomechanics and Prevention* (Springer, Berlin), pp. 292–310 (Chapter 13).

NHTSA/CIREN (1997). Crash test results from William Lehman Injury Research Center at the Ryder Trauma Center. In: *First Annual CIREN Conference, October 20, 1997*, http://www-nrd.nhtsa.dot.gov/departments/nrd-50/ciren/ciren1.html.

NICOLL, E.A. (1949). Fractures of the dorso-lumbar spine. *J. Bone Joint Surgery B* **31**, 376–393.

NIGG, B.M., SKARVAN, G., FRANK, C.B. (1990). Elongation and forces of ankle ligaments in a physiological range of motion. *Foot & Ankle* **11** (1), 30–40.

NIGHTINGALE, R.W., WINKELSTEIN, B.A., KNAUB, K.E., RICHARDSON, W.J., LUCK, J.F., MYERS, B.S. (2002). Comparative strengths and structural properties of the upper and lower cervical spine in flexion and extension. *J. Biomech.* **35**, 725–732.

NITSCHE, S., HAUG, E., KISIELEWICZ, L.T. (1996). Validation of a finite element model of the human neck. In: *PUCA '96*, Nihon ESI, Japan, ESI Software S.A., 99, rue des Solets, BP 80112, 94513 Rungis Cedex, France, pp. 203–220.

NORDIN, M., FRANKEL, V.H. (1989). *Basic Biomechanics of the Musculoskeletal System*, second ed. (Lea & Febiger, Philadelphia, PA).

NOVOTNY, J.E. (1993). Spinal biomechanics. *J. Biomech. Engrg.* **115**.

NOYES, F.R., GROOD, E.S. (1978). The strength of the anterior cruciate ligaments in humans and rhesus monkey. *J. Bone Joint Surgery A* **58** (68), 1074–1082.

NUSHOLTZ, G.S., WILEY, B., GLASCOE, L.G. (1995). Cavitation/boundary effects in a simple head impact model. *Aviation Space and Environmental Medicine* **66** (7), 661–667.

NYQUIST, G.W., CHENG, R., EL-BOHY, A.A.R., KING, A.I. (1985). Tibia bending: Strength and response. In: *Proc. 29th STAPP Car Crash Conference*, pp. 99–112. Paper No. 851728.

OMMAYA, A.K. (1967). Mechanical properties of tissues of the nervous system. *J. Biomech.* **1**, 127–138.

OMMAYA, A.K., GENNARELLI, T.A. (1974). Cerebral concussion and traumatic unconsciousness: Correlation of experimental and clinical observations on blunt head injuries. *Brain* **97**, 633–654.

OMMAYA, A.K., HIRSCH, A.E. (1971). Tolerances for cerebral concussion from head impact and whiplash in primates. *J. Biomech.* **4**, 13–31.

OMMAYA, A.K., HIRSCH, A.E., FLAMM, E.S., MAHONE, R.H. (1966). Cerebral concussion in the monkey: An experimental model. *Science* **153**, 211–212.

ONO, K., KANEOKA, K., SUN, E.A., TAKHOUNTS, E.G., EPPINGER, R.H. (2001). Biomechanical response of human cervical spine to direct loading of the head. In: *IRCOBI Conference*.

ONO, K., KANEOKA, K., WITTEK, A., KAJZER, J. (1997). Cervical injury mechanism based on the analysis of human cervical vertebral motion and head-neck-torso kinematics during low speed rear impacts. In: *Proc. 41st Stapp Car Crash Conference*, pp. 339–356. Paper No. 973340.

ONO, K., KIKUCHI, A., NAKAMURA, M., KOBAYASHI, H., NAKAMURA, N. (1980). Human head tolerance to sagittal impact reliable estimation deduced from experimental head injury using subhuman primates and human cadaver skulls. In: *Proc. 24th Stapp Car Crash Conference*, pp. 101–160. Paper No. 801303.

ONO, K., et al. (1999). Relationship between localized spine deformation and cervical vertebral motions for low speed rear impacts using human volunteers. In: *IRCOBI Conference, Spain*, pp. 149–164.

OSHITA, F., OMORI, K., NAKAHIRA, Y., MIKI, K. (2002). Development of a finite element model of the human body. In: *Proc. 7th International LS-DYNA Users Conference, Detroit*, pp. 3-37–3-48.

OTTE, D., VON RHEINBABEN, H., ZWIPP, H. (1992). Biomechanics of injuries to the foot and ankle joint of car drivers and improvements for an optimal car floor development. In: *Proc. 36th Stapp Car Crash Conference Proceedings*, pp. 43–58. Paper No. 922514.

OTTENSMEYER, M.P., SALISBURY, J. (2001). In vivo data acquisition instrument for solid organ mechanical property measurement. In: Niessen, W., Viergever, M. (eds.), *MICCAI 2001*. In: Lecture Notes in Comput. Sci. **2208** (Springer, Berlin), pp. 975–982.

OXLUND, H., ANDREASSON, T.T. (1980). The role of hyaluronic acid, collagen and elastin in the mechanical properties of connective tissues. *J. Anat.* **131**, 611–620.

PALANIAPPAN, P. JR., WIPASURAMONTON, P., BEGEMAN, P., TANAVDE, A.S., ZHU, F.A. (1999). Three dimensional finite element model of the human arm. In: *Proc. 43rd Stapp Car Crash Conference*, pp. 351–363. Paper No. 99SC25.

PANJABI, M.M., BRAND, R.A. (1976). Mechanical properties of the human thoracic spine as shown by three-dimensional load displacement curves. *J. Bone Joint Surgery A* **58** (5).

PANJABI, M.M., CRISCO, J.J., VASAVADA, A., ODA, T., CHOLEWICKI, J., NIBU, K., SHIN, E. (2001). Mechanical properties of the human cervical spine as shown by three-dimensional load-displacement curves. *Spine* **26** (24), 2692–2700.

PANJABI, M.M., DVORAK, J., DURANCEAU, J., et al. (1988). Three-dimensional movements of the upper cervical spine. *J. Spine* **13** (7), 726.

PANJABI, M., JORNEUS, L., GREENSTEIN, G. (1984). Physical properties of lumbar spine ligaments. *Trans. Orthop. Res. Soc.* **9**, 112.

PARENTEAU, C.S., VIANO, D.C. (1996). Kinematics study of the ankle-subtalar joints. *J. Biomech. Engrg.*, Ph.D. Thesis work.

PARENTEAU, C.S., VIANO, D.C., PETIT, P.Y. (1996). Biomechanical properties of ankle-subtalar joints in quasi-static loading to failure. *J. Biomech. Engrg.*

PATHRIA, M.N., RESNIK, D. (1993). Radiologic analysis of trauma. In: Nahum, A.M., Melvin, J.W (eds.), *Accidental Injury – Biomechanics and Prevention* (Springer, Berlin), pp. 85–101 (Chapter 5).

PATTIMORE, D., WARD, E., THOMAS, P., BRADFORD, M. (1991). The nature and cause of lower limb injuries in car crashes. In: *Proc. 35th Stapp Car Crash Conference*, pp. 177–188. Paper No. 912901.

PEINDL, R.D., ENGIN, A.E. (1987). On the biomechanics of human shoulder complex – II. Passive resistive properties beyond the shoulder complex sinus. *J. Biomech.* **20** (2), 118–134.

PENN, R.D., CLASEN, R.A. (1982). Traumatic brain swelling and edema. In: Cooper, P.R. (ed.), *Head Injury* (Williams and Wilkins, Baltimore), pp. 233–256.

PENNING, L. (1979). Normal movements of the cervical spine. *Amer. J. Roentgenol.*, 130–317.

PENNING, L., WILMARK, J.T. (1987). Rotation of the cervical spine. *J. Spine* **12** (8), 732.

PIKE, J.A. (1990). Automotive Safety (Society of Automotive Engineers). ISBN 1-56091-007-0.

PINTAR, F.A., YOGANANDAN, N., EPPINGER, R.H. (1998). Response and tolerance of the human forearm to impact loading. In: *Proc. 42nd Stapp Car Crash Conference*, pp. 1–8. Paper No. 983149.

PLANK, G.R., KLEINBERGER, M., EPPINGER, R.H. (1994). Finite element modeling and analysis of thorax/restraint system interaction. In: *The 14th ESV International Technical Conference on the Enhanced Safety of Vehicles* (Munich, Germany).

POPE, M.E., KROELL, C.K., VIANO, D.C., WARNER, C.Y., ALLEN, S.D. (1979). Postural influences on thoracic Impact. In: *Proc. 23rd Stapp Car Crash Conference*, pp. 765–795. Paper No. 791028.

PORTIER, L., TROSSEILLE, X., LE COZ, J.-Y., LAVASTE, F., COLTAT, J.-C. (1993). Lower leg injuries in real-world frontal accidents. In: *28th International IRCOBI Conference on the Biomechanics of Trauma*, pp. 57–78.

PRADAS, M.M., CALLEJA, R.D. (1990). Nonlinear viscoelastic behaviour of the flexor tendon of the human hand. *J. Biomech.* **23** (8), 773–781.

PRASAD, P. (1990). Comparative evaluation of the dynamic responses of the Hybrid II and the Hybrid III dummies. In: *Proc. 34th Stapp Car Crash Conference*, pp. 175–183. Paper No. 902318.

PRASAD, P., CHOU, C.C. (1993). A review of mathematical occupant simulation models. In: Nahum, A.M., Melvin, J.W. (eds.), *Accidental Injury – Biomechanics and Prevention* (Springer, Berlin), pp. 102–150 (Chapter 6).

PRASAD, P., KING, A.I. (1994). An experimentally validated dynamic model of the spine. *J. Appl. Mech.*, 546–550.

PRING, D.J., AMIS, A.A., COOMBS, R.R. (1985). The mechanical properties of human flexor tendons in relation to artificial tendons. *J. Hand Surgery [Br]* **10** (3), 331–336.

PUTZ, R., PABST, R. (2000). *Sobotta – Atlas der Anatomie des Menschen*, twenty-first ed. (Urban&Fischer, München, Jena).

QUHAN, et al. (1989). Comparison of trabecular and cortical tissue moduli from human iliac crests. *J. Orthop. Res.* **7**, 876–884.

RACK, P.M.H., WESTBURY, D.R. (1969). The effect of length and stimulus rate on tension in the isometric cat soleus muscle. *J. Physiol.* **204**, 443–460.

REILLY, D.T., BURSTEIN, A.H., FRANKEL, V.H. (1974). The elastic modulus for bone. *J. Biomech.* **7**, 271–275.

REILLY, D.T., et al. (1975). The elastic and ultimate properties of compact bone tissue. *J. Biomech.* **8**, 395–405.

RENAUDIN, F., GUILLEMOT, H., PÉCHEUX, C., LESAGE, F., LAVASTE, F., SKALLI, W. (1993). A 3D finite element model of the pelvis in side impact. In: *Proc. 37th Stapp Car Crash Conference*, pp. 241–252. Paper No. 933130.

RIETBERGEN, B. VAN (1996). Mechanical behavior and adaptation of trabecular bone in relation to bone morphology. PhD Thesis Nijmegen University. ISBN 90-9010006-7.

RIETBERGEN, B. VAN, MÜLLER, R., ULRICH, D., RÜEGSEGGER, P., HUISKES, R. (1998). Tissue stresses and strain in trabeculae of a canine proximal femur can be quantified from computer reconstructions. *J. Biomech.*

ROAF, R. (1960). A study of the mechanics of spinal injury. *J. Bone Joint Surgery B* **42**, 810–823.

ROBBINS, D.H. (1983). *Anthropometry of Motor Vehicle Occupants, vol. 2: Mid-sized Male, vol. 3: Small Female and Large Male* (UMTRI-83-53-2, University of Michigan).

ROBBINS, D.H., SCHNEIDER, L.W., SNYDER, R.G., PFLUG, M., HAFFNER, M. (1983). Seated posture of vehicle occupants. In: *Proc. 32nd Stapp Car Crash Conference*, pp. 199–224. Paper No. 831617.

ROBBY1 (1997). *User's Guides*, ESI Software S.A., 99, rue des Solets, BP 80112, 94513 Rungis Cedex, France.

ROBBY2 (1998). *User's Guides*, ESI Software S.A., 99, rue des Solets, BP 80112, 94513 Rungis Cedex, France.

ROBIN, S. (1999). Validation data base, Report PSA/990331/T0/DA, LAB PSA Peugeot Citroën – Renault, 132 Rue des Suisses, F-92000 Nanterre.

ROBIN, S. (2001). HUMOS: Human model for safety – A joint effort towards the development of refined human-like car occupant models. In: *17th ESV Conference*. Paper Number 297.

ROBINA (1998). Internal draft report, ESI Software S.A., 99, rue des Solets, BP 80112, 94513 Rungis Cedex, France.

ROCKOFF, S.D., et al. (1969). The relative contribution of trabecular and cortical bone to the strength of human lumbar vertebrae. *Calc. Tiss. Res.* **3**, 163.

ROHEN, J.W., YOKOCHI, C. (1983). In: *Color Atlas of Anatomy* (IGAKU–SHOIN Medical Publishers), p. 349.

ROSE, J.L., GORDON, S.L., MOSKOWITZ, G. (1974). Dynamic photoelastic model analysis of impact to the human skull. *J. Biomech.* **7**, 193–199.

ROUHANA, S.W. (1993). Biomechanics of abdominal trauma. In: Nahum, A.M., Melvin, J.W. (eds.), *Accidental Injury – Biomechanics and Prevention* (Springer, Berlin), pp. 391–428 (Chapter 16).

RUAN, J.S., KHALIL, T.B., KING, A.I. (1991). Human head dynamic response to side impact by finite element modeling. *J. Biomech. Engrg.* **113**, 267–283. PhD Dissertation, Wayne State University.

RUAN, J.S., KHALIL, T.B., KING, A.I. (1993). Finite element modeling of direct head impact. In: *Proc. 37th Stapp Car Crash Conference*, pp. 69–81. Paper No. 933114.

RUAN, J.S., KHALIL, T.B., KING, A.I. (1994). Dynamic response of the human head to impact by three-dimensional finite element analysis. *ASME J. Biomech. Engrg.* **116**, 44–50.

RUAN, J.S., PRASAD, P. (1994). Head injury assessment in frontal impacts by mathematical modeling. In: *Proc. 38th Stapp Car Crash Conference*, pp. 111–121. Paper No. 942212.

RUDOLF, C., FELLHAUER, A., SCHAUB, S., MARCA, C., BEAUGONIN, M. (2002). OOP Simulation with a 5th Percentile Deformable Human Model. In: *EuroPam 2002*, ESI Software S.A., 99, rue des Solets, BP 80112, 94513 Rungis Cedex, France.

SAE ENGINEERING AID 23 (1986). User's Manual for the 50th Percentile HYBRID III Test Dummy.

SANCES JR, A., et al. (1982). Head and spine injuries. In: *AGARD Conference on Injury Mechanisms, Prevention and Cost, Köln, Germany*, pp. 13-1–13-33.

SACRESTE, J., BRUN-CASSAN, F., FAYON, A., TARRIERE, C., GOT, C., PATEL, A. (1982). Proposal for a thorax tolerance level in side impacts based on 62 tests performed with cadavers having known bone condition. In: *Proc. 26th Stapp Car Crash Conference*, pp. 155–171. Paper No. 821157.

SCHMIDT, G., KALLIERIS, D., BARZ, J., MATTERN, R., SCHULZ, F., SCHÜLER, F. (1978). Biomechanics – Determination of the mechanical loadability limits of the occupants of a motor vehicle. Final report at the end of project (31-12-1978) – Research project, No. 3906, Institute of Forensic Medicine, Heidelberg.

SCHNECK, D.J. (1992). *Mechanics of Muscle* (New York University Press, New York).

SCHNEIDER, L.W., KING, A.I., BEEBE, M.S. (1990). Design requirements and specifications, Thorax abdomen development task. Interim report: Trauma assessment device development program. Report No. DOT-HS-807-511.

SCHOENFELD, C.M., et al. (1974). Mechanical properties of human cancellous bone in the femoral head. *Med. Biol. Engrg.* **12**, 313–317.

SCOTT, W.E. (1981). Epidemiology of head and neck trauma in victims of motor vehicle accidents. Head and Neck Criteria. In: Ommaya, A.K. (ed.), *A Consensus Workshop* (US Department of Transportation, National Highway Traffic Safety Administration, Washington, DC), pp. 3–6.

SEDLIN, E.D., et al. (1965). A rheological model for cortical bone. In: *A Study of the Physical Properties of Human Femoral Samples*. In: Acta Orthop Scand. Supplementum **83**.

SEDLIN, E.D., et al. (1966). Factors affecting the determination of the physical properties of femoral cortical bone. *Acta Orthop. Scand.* **37**, 29–48.

SEIREG, A., ARVIKAR, R. (1989). *Biomechanical Analysis of the Musculoskeletal Structure for Medicine and Sports* (Hemisphere, Washington, DC).

SHAH, C.S., YANG, K.H., HARDY, W.N., WANG, H.K., KING, A.I. (2001). Development of a computer model to predict aortic rupture due to impact loading. *Stapp Car Crash J.* **45**, 161–182.

SHELTON, F.E., BUTLER, D.L., FEDER, S.M. (1993). Shear stress transmission in the patellar tendon and mediacollateral ligament. *Adv. Bioengrg. BED ASME* **26**, 271.

SHUCK, L.Z., ADVANI, S.H. (1972). Rheological response of human brain tissue in shear. *ASME J. Basic Engrg.*, 905–911.

SHUCK, L.Z., HAYNES, R.R., FOGLE, J.L. (1970). Determination of viscoelastic properties of human brain tissue, ASME Paper N°70-BHF-12.

SHUGAR, T.A. (1975). Transient structural response of the linear skull-brain system. In: *Proc. 19th Stapp Car Crash Conference*, pp. 581–614. Paper No. 751161.

SKAGGS, D.L., WEIDENBAUM, M., IATRIDIS, J.C., RATCLIFFE, A., MOW, V.C. (1994). Regional variations in tensile properties and biomechanical composition of the human lumbar annulus fibrosus. *Spine* **19**, 1310–1319.

SONNERUP, L. (1972). A semi-experimental stress analysis of the human intervertebral disk in compression. *Experimental Mech.*

SONODA, T. (1962). Studies on the strength for compression, tension and torsion of the human vertebral column. *J. Kyoto Pref. Med. Univ.* **71**, 659–702.

SPILKER, R.L., JAKOBS, D.M., SCHULTZ, A.B. (1986). Material constants for a finite element model of the intervertebral disc with a fiber composite annulus. *J. Biomech. Engrg.* **108**, 1–11.

SPITZER, V.M., WHITLOCK, D.G. (1998). *Atlas of the Visible Human Male* (Jones & Bartlett Publishers), pp. 6, 7.

STALNAKER, R.L., MCELHANEY, J.H., ROBERTS, V.L., TROLLOPE, L.L. (1973). Human torso response to blunt trauma. In: *Proceedings of the Symposium: Human Impact Response – Measurement and Simulation, General Motors Research laboratories, New York* (Plenum Press, London), pp. 181–199.

STALNAKER, R.L., MOHAN, D. (1974). Human chest impact protection criteria. In: *Proc. 3rd International Conference on Occupant Protection* (Society of Automotive Engineers, New York), pp. 384–393.

STATES, J.D. (1986). Adult occupant injuries of the lower limb. In: *Biomechanics and Medical Aspects of Lower Limb Injuries*. SAE 861927.

STOCKIER, R.M., EPSTEIN, J.A., EPSTEIN, B.S. (1969). Seat belt trauma to the lumbar spine: An unusual manifestation of the seat belt syndrome. *J. Trauma* **9**, 508–513.

STRUHL, S., et al. (1987). The distribution of mechanical properties of trabecular bone within vertebral bodies and iliac crest: Correlation with computed tomography density. *Trans. Orthop. Res. Soc.*, 262.

STÜRTZ, G. (1980). Biomechanical data of children. In: *Proc. 24th Stapp Car Crash Conference*, pp. 513–559. Paper No. 801313.

SVENSSON, M.Y., LÖVSUND, P. (1992). A dummy for rear-end collisions – development and validation of a new dummy-neck. In: *Proceedings of IRCOBI Conference, Verona, Italy*.

TADA, Y., NAGASHIMA, T. (1994). Modeling and simulation of brain lesions by the finite element method. *IEEE Engineering in Medicine and Biology* **13** (4).

TARRIERE, C. (1981). Investigation of brain injuries using the CT scanner. In: Ommaya, A.K. (ed.), *Head and Neck Injury Criteria: A Consensus Work-Shop* (US Department of Transportation, National Highway Traffic Safety Administration, Washington, DC), pp. 39–49.

TENNYSON, S.A., KING, A.I. (1976a). A biodynamic model of the spinal column. In: *Mathematical Modeling Biodynamic Response to Impact, SAE Special Publication*. SAE 760711.

TENNYSON, S.A., KING, A.I. (1976b). Electromyographic signals of the spinal musculature during +Gz accelerations. In: *1976 Meeting of the International Society for the Study of the Lumbar Spine, Bermuda*.

THIBAULT, K.L., MARGULIES, S.S. (1996). Material properties of the developing porcine brain. In: *1996 International IRCOBI Conference on the Biomechanics of Impact, Dublin, Ireland*.

THUNNISSEN, J., WISMANS, J., EWING, C.L., THOMAS, D.J. (1995). Human volunteer head–neck response in frontal flexion: A new analysis. In: *Proc. 39th Stapp Car Crash Conference*, pp. 439–460. Paper No. 952721.

T'MER, S.T., ENGIN, A.E. (1989). Three-dimensional kinematic modelling of the human shoulder complex – Part II: Mathematical modelling and solution via optimization. *J. Biomech. Engrg.* **111**, 113–121.

TORG, J.S. (ed.) (1982). *Athletic Injuries to the Head, Neck and Face* (Lea & Febiger, Philadelphia, PA).

TORG, J.S., PAVLOV, H. (1991). Axial load "teardrop" fracture. In: *Athletic Injuries to the Head, Neck and Face*, second ed. (Lea & Febiger, Philadelphia, PA).

TROSSEILLE, X., TARRIERE, C., LAVASTE, F., GUILLON, F., DOMONT, A. (1992). Development of an F.E.M. of the human head according to a specific test protocol. In: *Proc. 36th Stapp Car Crash Conference*, pp. 235–253. Paper No. 922527.

TURQUIER, F., KANG, H.S., TROSSEILLE, X., WILLINGER, R., LAVASTE, F., TARRIERE, C., DÖMONT, A. (1996). Validation study of a 3D finite element head model against experimental data. In: *Proc. 40th Stapp Car Crash Conference*, pp. 283–293. Paper No. 962431.

UENO, K., LIU, Y.K. (1987). A three-dimensional nonlinear finite element model of lumbar intervertebral joint in torsion. *J. Biomech. Engrg.* **109**, 200–209.

UENO, K., MELVIN, J.W., LI, L., LIGHTHALL, J.W. (1995). Development of tissue level brain injury criteria by finite element analysis. *J. Neurotrauma* **12** (4).

UENO, K., MELVIN, J.W., LUNDQUIST, E., LEE, M.C. (1989). Two-dimensional finite element analysis of human brain impact responses: Application of a scaling law. In: *Crashworthiness and Occupant Protection in Transportation Systems*. In: AMD **106** (The American Society of Mechanical Engineers, New York), pp. 123–124.

ULRICH, D. (1998). Evaluation of the mechanical properties of bone with consideration of its microarchitecture. Dissertation for the degree of doctor of the technical sciences of the Swiss Federal Institute of Technology, Zürich, Switzerland.

VANCE, T.L., SOLOMONOV, M., BARATTA, R., ZEMBO, M., D'AMBROSIA, R.D. (1994). Comparison of isometric and load moving length-tension models of two bicompartmental muscles. *IEEE Trans. Biomed. Engrg.* **41** (8), 771–781.

VERRIEST, J.-P., CHAPON, A. (1994). Validity of thoracic injury criteria based on the number of rib fractures. In: Backaitis, S.H. (ed.), *Biomechanics of Impact Injury and Injury Tolerances of the Thorax-Shoulder Complex* (Society of Automotive Engineers), pp. 719–727. SAE PT-45.

VIANO, D.C. (1986). Biomechanics of bone and tissue: A review of material properties and failure characteristics. In: *Symposium on Biomechanics and Medical Aspects of Lower Limb Injuries* (San Diego, California), pp. 33–63. SAE 861923.

VIANO, D.C. (1989). Biomechanical responses and injuries in blunt lateral impact. In: *Proc. 33rd Stapp Car Crash Conference*, pp. 113–142. Paper No. 892432.

VIANO, D.C., LAU, V.K. (1983). Role of impact velocity and chest compression in thoracic injury. *Aviat. Space Environ. Med.* **54** (1), 16–21.

VIIDIK, A. (1987). Properties of tendons and ligaments. In: Skalak, R., Chien, S. (eds.), *Handbook of Bioengineering* (McGraw-Hill, New York), pp. 6.1–6.19.

VIRGIN, W.J., LUDHIANA, P. (1951). Experimental investigations into the physical properties of the intervertebral disc. *J. Bone Joint Surgery B* **33** (4).

VISIBLE HUMAN PROJECT (1994) (public release of male dataset on CD Rom), National Library of Medicine. Visible Human Database. 8600 Rockville Pike, Bethseda, MD 20894.

VOIGT, M., BOJSEN-MOLLER, F., SIMONSEN, E.B., DYHRE-POULSEN, P. (1995). The influence of tendon Youngs modulus, dimensions and instantaneous moment arms on the efficiency of human movement. *J. Biomech.* **28** (3), 281–291.

VOO, L., KUMARESAN, S., PINTAR, F.A., YOGANANDAN, N., SANCES JR, A. (1996). Finite element models of the human head. *Medical Biological Engrg. Comput.* **34**, 375–381.

WAINWRIGHT, S.A., BIGGS, W.D., CURREY, J.D. (1979). *Mechanical Design in Organisms – Pliant Materials* (Edward Arnold, London), pp. 110–143.

WALKE, A.E., KOLLROS, J.J., CASE, T.J. (1944). The physiological basis of concussion. *J. Neurosurg.* **1**, 103–116.

WANK, V., GUTEWORT, W. (1993). Modelling and simulation of muscular contractions with regard to physiological parameters. In: *Proc. of the XIV International Society of Biomechanics Congress, Paris* (International Society of Biomechanics), pp. 1450–1451.

WARD, C.C. (1982). Finite element models of the head and their use in brain injury research. In: *Proc. 26th Stapp Car Crash Conference*, pp. 71–85. Paper No. 821154.

WARD, C.C., THOMSON, R.B. (1975). The development of a detailed finite element brain model. In: *Proc. 19th Stapp Car Crash Conference*, pp. 641–674. Paper No. 751163.

WATANABE, I., ISHIHARA, T., FURUSU, K., KATO, C., MIKI, K. (2001). Basic research of spleen FEM model for impact analysis. In: *Proc. of 2001 JSME Annual Congress*, pp. 115–116 (in Japanese).

WEAVER, J.K., et al. (1966). Cancellous bone: Its strength and changes with aging and an evaluation of some methods for measuring its mineral content. *J. Bone Joint Surgery A* **48**, 289–299.

WIRHED, R. (1985). *Anatomie et Science du Geste Sportif* (Edition VIGOT). ISBN 2-7114-0944-9.

WHITE, A.A., PANJABI, M.M. (1978). *Biomechanics of the Spine* (Lippincott, Philadelphia, PA).

WHITE, A.A., PANJABI, M.M. (1990). Clinical biomechanics of the spine. *J. Biomech.*, Lippincott Company, second ed.

WILLIAMS, J.L., et al. (1982). Properties and an anisotropic model of cancellous bone from the proximal tibia epiphysis. *J. Biomech. Engrg.* **104**, 50–56.

WILLINGER, R., KANG, H.S., DIAW, B. (1999). Three-dimensional human finite-element model validation against two experimental impacts. *Ann. Biomedical Engrg.* **27**, 403–410.

WILSON-MACDONALD, J., WILLIAMSON, D.M. (1988). Severe ligamentous injury of the ankle with ruptured tendo achillis and fractured neck talus. *J. Trauma* **28**, 872–874.

WINTERS, J.M., STARK, L. (1985). Analysis of fundamental human movement patterns through the use of in-depth antagonistic muscle models. *IEEE Trans. Biomedical Engrg.* **12**, 826–839.

WINTERS, J.M., STARK, L. (1988). Estimated mechanical properties of synergistic muscles involved in movements of a variety of human joints. *J. Biomech.* **21**, 1027–1041.

WISMANS, J.S., SPENNY, C. (1983). Performance requirements of mechanical necks in lateral flexion. In: *Proc. 27th Stapp Car Crash Conference*, pp. 56–65. Paper No. 831613.

WISMANS, J., VAN OORSCHOT, H., WOLTRING, H.J. (1986). Omni-directional human head–neck response. In: *Proc. 30th Stapp Car Crash Conference*, pp. 313–331. Paper No. 861893.

WISMANS, J.S.H.M., et al. (1994). *Injury Biomechanics* (Eindhoven University of Technology), p. 49.

WITTEK, A., HAUG, E., KAJZER, J. (1999). Hill-type muscle model for analysis of muscle tension on the human body response in a car collision using an explicit finite element code. *JSME Internat. J. Ser. A: Solid Material Mech.* **43**, 8–18.

WITTEK, A., KAJZER, J. (1995). A review and analysis of mathematical models of muscle for application in the modeling of musculoskeletal system response to dynamic load. In: Högfors, C., Andreasson, G. (eds.), *Proceedings of the 9th Biomechanics Seminar* (Center for Biomechanics Chalmers University of Technology and Göteborg University, Göteborg), pp. 192–216.

WITTEK, A., KAJZER, J. (1997). Modeling of muscle influence on the kinematics of the head–neck complex in impacts. *Mem. School Engrg. Nagoya University* **49**, 155–205.

WITTEK, A., KAJZER, J., HAUG, E. (1999). Finite element modeling of the muscle effects on kinematic responses of the head–neck complex in frontal impact at high speed. *Japan Soc. Automotive Engrg. Rev.*

WITTEK, A., ONO, K., KAJZER, J. (1999). Finite element modeling of kinematics and dynamic response of cervical spine in low-speed rear-end collisions: mechanical aspects of facet joint injury. *J. Crash Prevention Injury Control.*

WITTEK, A., ONO, K., KAJZER, J., ÖRTENGREN, R., INAMI, S. (2001). Analysis and comparison of reflex times and electro-myograms of cervical muscles under impact loading obtained using surface and fine-wire electrodes. *IEEE Trans. Biomedical Engrg.* **48** (2), 143–153.

WOO, S.L.Y., JOHNSON, G.A., SMITH, B.A. (1993). Mathematical modelling of ligaments and tendons. *J. Biomech. Engrg.* **115**, 468–473.

WOO, S.L.Y., PETERSON, R.H., OHLAND, K.J., et al. (1990). The effects of strain rate on the properties of the medial collateral ligament. Skeletally Immature and Mature Rabbits: A Biomechanical and Histological Study. *J. Orthop. Res.* **8**, 712–721.

WYKOWSKI, E., SINNHUBER, R., APPEL, H. (1998). Finite element model of human lower extremities in a frontal impact. In: *IRCOBI Conference*.

YAMADA, H. (1970). In: Evans, F.G. (ed.), *Strength of Biological Materials* (Williams & Wilkins, Baltimore, MD).

YANG, J. (1998). Bibliographic study. Report 3CHA/980529/T1/DB, Chalmers University of Technology, SE-41296 Göteborg, Sweden.

YANG, J., LÖVSUND, P. (1997). Development and validation of a human body mathematical model for simulation of car-pedestrian collisions, in: *IRCOBI Conference*, Hannover.

YANG, K.H., ZHU, F., LUAN, F., ZHAO, L., BEGEMAN, P. (1998). Development of a finite element model of the human neck. In: *Proc. 42nd Stapp Car Crash Conference*, pp. 1–11. Paper No. 983157.

YOGANANDAN, N., HAFFNER, M., MALMAN, D.J., NICHOLS, H., PINTAR, F.A., JENTZEN, J., WEINSHEL, S.S., LARSON, S.J., SANCES JR., A. (1989). Epidemiology and injury biomechanics of motor vehicle related trauma to the human spine. In: *Proc. 33rd Stapp Car Crash Conference*, pp. 22–242. Paper No. 892438.

YOGANANDAN, N., PINTAR, F.A. (1998). Biomechanics of human thoracic ribs. *J. Biomech. Engrg.* **120**, 100–104.

YOGANANDAN, N., SANCES JR., A., PINTAR, F. (1989b). Biomechanical evaluation of the axial compressive responses of the human cadaveric and manikin necks. *J. Biomech. Engrg.* **111**, 250–255.

YOGANANDAN, N., SRIRANGAM, K., PINTAR, F.A. (2001). Biomechanics of the cervical spine Part 2. Cervical spine soft tissue responses and biomechanical modeling. *Clinical Biomech.* **16**, 1–27.

YOO, W.-H., CHOI, H.-Y. (1999). Finite element modeling of human lower extremity. *Hongik J. Sci. Technol.* **3**, 187–212.

ZAJAC, F.E. (1989). Muscle and tendon: Properties, models, scaling and applications to biomechanics and motor control. *Critical Rev. Biomedical Engrg.* **17**, 359–411.

ZEIDLER, F., STÜRTZ, G., BURG, H., RAU, H. (1981). Injury mechanisms in head-on collisions involving glance-off. In: *Proc. 25th Stapp Car Crash Conference*, pp. 825–860. Paper No. 811025.

ZINK, L. (1997). Simulation von Unfällen mit Airbagauslösung für den Out-of-Position Fall. Diploma Thesis prepared at TRW Germany, University of Stuttgart, IVK.

ZHANG, L., BAE, J., HARDY, W.N., MONSON, K.L., MANLEY, G.T., GOLDSMITH, W., YANG, K.H., KING, A. (2002). Computational study of the contribution of the vasculature on the dynamic response of the brain. *Stapp Car Crash J.* **46**, 145–165. Paper No. 2002-22-0008.

ZHANG, L., YANG, K.H., DWARAMPUDI, R., OMORI, K., LI, T., CHANG, K., HARDY, W.N., KHALIL, T.B., KING, A.I. (2001). Recent advances in brain injury research: A new human head model development and validation. In: *Proc. 45th Stapp Car Crash Conference*, pp. 369–394. Paper No. 2001-22-0017.

ZHOU, C., KHALIL, T.B., KING, A.I. (1996). Viscoelastic response of the human brain to sagittal and lateral rotational acceleration by finite element analysis. In: *International IRCOBI Conference on the Biomechanics of Impact, Dublin, Ireland*.

ZHOU, Q., ROUHANA, S.W., MELVIN, J.W. (1996). Age effects on thoracic injury tolerance. In: *Proc. 40th Stapp Car Crash Conference*, pp. 137–148. Paper No. 962421.

ZUURBIER, C.J., EVERARD, P., VAN DER WEES, A.J., HUIJING, P.A. (1994). Length-force characteristics of the aponeurosis in the passive and active muscle condition and in the isolated condition. *J. Biomech.* **27** (4), 445–453.

Soft Tissue Modeling for Surgery Simulation

Hervé Delingette, Nicholas Ayache

INRIA Sophia–Antipolis, 2004, route des Lucioles,
BP 93, 06902 Sophia–Antipolis, France
E-mail addresses: Herve.Delingette@sophia.inria.fr (H. Delingette),
Nicholas.Ayache@inria.fr (N. Ayache)

Foreword

In this chapter, we address a specific issue belonging to the field of biomechanics – modeling living tissue deformation with real-time constraints. This issue was raised by the emergence, in the middle of the 1990s, of a very specific application – the simulation of surgical procedures. This new concept of surgery simulation was in large part advocated by the American Department of Defense (SATAVA [1994]), for which this concept was a key part of their vision of the future of emergency medicine.

Since then, the concept of having surgeons being trained on simulators (just like pilots on flight simulators) has been refined. First, the development of commercial simulators has proved that there was a demand for products that help to optimize the learning curve of surgeons.[1] Second, the emergence of medical robotics and more precisely of minimally invasive surgery robots, has reinforced the need for simulating surgical procedures, since these robots require a very specific hand–eye coordination. Third, there is a large consensus among the medical community that current simulators are not realistic enough to provide advanced gesture training. In particular, the modeling of living tissue, and their ability to deform under the contact of an instrument is one of the important aspect of simulators that should be improved.

[1] This curve represents the number of incidents occurring during the performance of surgery as a function of time. This curve is generally monotonically decreasing under the effect of training and usually reaches an asymptotic value after a certain number N of real interventions. The objective of the simulators is to reduce this number N as much as possible.

Computational Models for the Human Body
Special Volume (N. Ayache, Guest Editor) of
HANDBOOK OF NUMERICAL ANALYSIS, VOL. XII
P.G. Ciarlet (Editor)

ISSN 1570-8659
DOI 10.1016/S1570 8659(03)12005-4

In this chapter, we present different algorithms for modeling soft tissue deformation in the context of surgery simulation. These algorithms make radical simplifications about tissue material property, tissue visco-elasticity and tissue anatomy. The first section of this chapter describes the principles and the components of a surgical simulator. In particular, we insist on the different constraints of soft tissue models in this application, the most important being the real-time computation constraint. In Section 2, we present the process of building a patient-specific hepatic surgery simulator from a set of medical images. The different stages of computation leading to the creation of a volumetric tetrahedral mesh from a medical image are especially emphasized. In Section 3, we detail the five main hypotheses that are made in the proposed soft tissue models. Furthermore, we recall the main equations of isotropic and transversally anisotropic linear elasticity in continuum mechanics. The discretization of these equations are presented in Section 4 based on finite element modeling. Because we rely on the simple linear tetrahedron element, we provide closed form expressions of local and global stiffness matrices. After describing the types of boundary conditions existing in surgery simulation, we derive the static and dynamic equilibrium equations in their matrix form. In Section 5, a first model of soft tissue is proposed. It is based on the off-line inversion of the stiffness matrix and can be computed very efficiently as long as no topology change is required. In such case, in Section 6, a second soft tissue model allows to perform cutting and tearing but with less efficiency as the previous model. A combination of the two previous models, called "hybrid model" is also presented in this section. Finally, in Section 7, we introduce an extension of the second soft tissue model that implements large displacement elasticity.

1. General issues in surgery simulation

1.1. Surgical simulators

1.1.1. Medical impact of surgical simulators

Surgery simulation aims at reproducing the visual and haptic senses experienced by a surgeon during a surgical procedure, through the use of computer and robotics systems. The medical interest of this technology is linked with the development of minimally invasive techniques especially video-surgery (endoscopy, laparoscopy, ...). More precisely, laparoscopy consists in performing surgery by introducing different surgical instruments in the patient abdomen through one centimeter-wide incisions. The surgeon can see the abdominal anatomy with great clarity by watching a high resolution monitor connected to an endoscope introduced inside the patient abdomen. This technique bears several advantages over traditional open surgery. On one hand, it decreases the trauma entailed by the surgical procedure on the patient body. This allows to decrease the patient stay in hospitals and therefore decreases the cost of health care. On the other hand, it reduces the morbidity as demonstrated by the Hunter and Sackier study (BERCI, HUNTER and SACKIER [1994]).

However, if these minimally invasive techniques are clearly beneficial to the patients, they also bring new constraints on the surgical practice. First, they significantly degrade the surgeon access to the patient body. In laparoscopy, for instance, the surgical procedure is made more complex by the limited number of degrees of freedom of each surgical instrument. Indeed, they must go through fixed points where the incisions in the patient's abdomen were done. Furthermore, because the surgeon cannot see his hand on the monitor, this technique requires a specific hand–eye coordination. Therefore, an important training phase is required before a surgeon acquires the skills necessary to adequately perform minimally invasive surgery (corresponding to a plateau in the learning curve).

Currently, surgeons are trained to perform minimally invasive surgery by using mechanical simulators or living animals. The former method is based on "endotrainers" representing an abdominal cavity inside which are placed plastic objects representing human organs. These systems are sufficient for acquiring basic surgical skills but are not realistic enough to represent fully the complexity of the human anatomy and physiology (respiratory motion, bleeding, ...). The latter training method consists in practicing simple or complex surgical procedures on living animals (often pigs for abdominal surgery). This method has two limitations. First, the similarity between the human and animal anatomy is limited and therefore certain procedures cannot be precisely simulated with this technique. Also, the evolution of the ethical code in most countries may forbid the use of animals for this specific training, as it is already the case in several European and North American countries.

Because of the limitations of current training methods, there is a large interest in developing video-surgery simulation software for providing efficient and quantitative gesture training systems (AYACHE and DELINGETTE [2003]). Indeed, such systems should bring a greater flexibility by providing scenarios including different types of pathologies. Furthermore, thanks to the development of medical image reconstruction algorithms, surgery simulation allows surgeons to verify and optimize the surgical strategy of a procedure on a given patient.

1.1.2. Classification of surgical simulators

SATAVA [1996] et al. proposed to classify surgical simulators into three categories (see Fig. 1.1). The first generation simulators are solely based on anatomical information, in particular on the geometry of the anatomical structures included in the simulator. In these simulators, the user can virtually navigate inside the human body but has a limited interaction with the modeled organs. Currently, several first generation surgical simulators are available including commercial products linked to medical imaging systems (CT or MRI scanners) that are focusing on virtual endoscopy (colonoscopy, tracheoscopy, ...). In general, they are used as a complementary examination tools establishing a diagnosis (for instance, when using virtual endoscopy) or as a surgical planning tool before performing surgery.

In addition to geometrical information, second generation simulators describe the physical properties of the human body. For instance, the modeling of soft tissue biomechanical properties enables the simulation of basic surgical gestures such as cutting or suturing. Currently, several prototypes of second generation simulators are being developed including the simulation of cholecystectomy (COVER, EZQUERRA and O'BRIEN [1993], KUHN, KÜHNAPFEL, KRUMM and NEISIUS [1996]), of arthroscopy of the

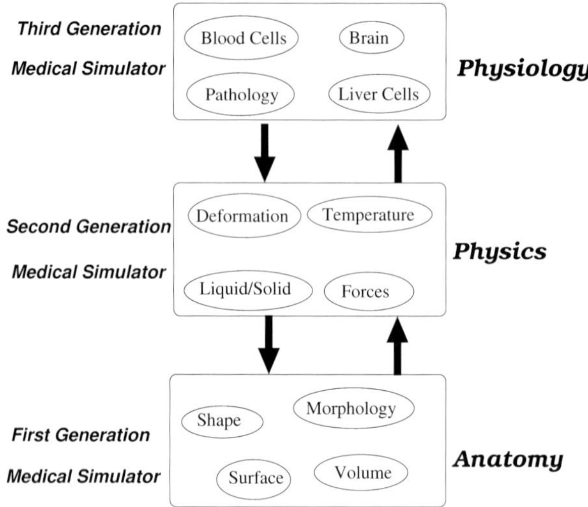

FIG. 1.1. The different generations of medical simulators.

knee (GIBSON, SAMOSKY, MOR, FYOCK, GRIMSON, KANADE, KIKINIS, LAUER and MCKENZIE [1997]) and of gynecological surgery (SZEKELY, BAIJKA and BRECHBUHLER [1999]). Section 2 will shortly describe the hepatic surgery simulator being developed at INRIA.

Third generation of surgical simulators provides an anatomical, physical and physiological description of the human body. There are very few simulators including these three levels of modeling, essentially because of the difficulty to realistically describe the coupling between physiology and physics. A good example of an attempt in this direction is given by the work of KAYE, PRIMIANO and METAXAS [1997] who modeled the mechanical cardiopulmonary interactions. Another important example is the study of the contraction of the right and left ventricles of the heart under the propagation of the action potential which is being carried out by the group of Prof. McCulloch (this work is published in this book) but also by the INRIA ICEMA group (SERMESANT, COUDIÈRE, DELINGETTE and AYACHE [2002], SERMESANT, FARIS, EVANS, MCVEIGH, COUDIÈRE, DELINGETTE and AYACHE [2003]). Finally, it should be noted that a comprehensive effort for creating computational physiological models has been recently launched in the international Physiome Project (BASSINGTHWAIGHTE [2000]).

1.2. *Simulator architecture*

In this section, we present the basic components of simulators for surgical gesture training and especially in the context of minimally invasive therapy. For the acquisition of basic skills, it is necessary to simulate the behavior of "living" tissues and therefore to develop a second generation surgical simulator. However, it raises important technical and scientific issues. The different components of these simulators are summarized in Fig. 1.2.

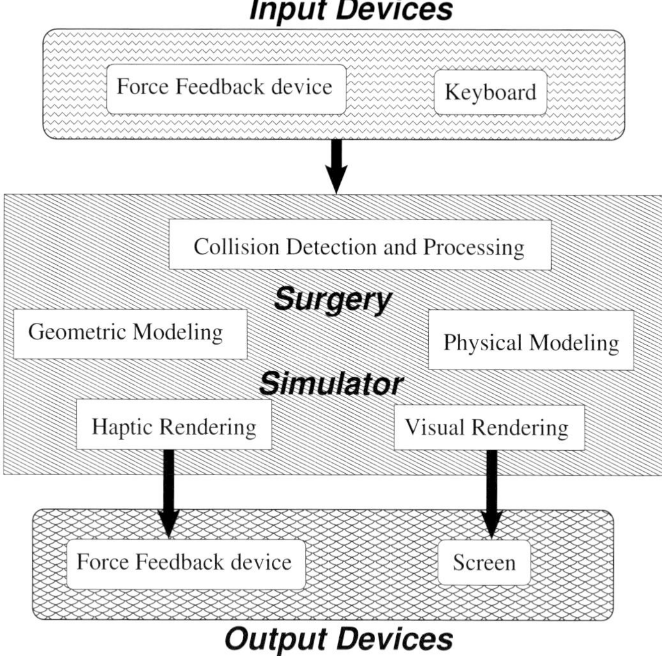

FIG. 1.2. The different components of a second generation surgery simulator.

The input devices in such simulators usually consist of one or several mechanical systems that drive the motion of virtual surgical tools or of virtual endoscopes. In fact, as input devices they do not need to be motorized and they are usually equipped with simple optical encoders or position trackers. A keyboard and electronic mouse are also useful to modify the scenario of the simulation.

The core of a simulator consists of several modules. For instance, a first module provides the enabling tools for the creation of geometric models from medical images (see Section 1.2.1). Another module, detailed in Section 1.3, computes the deformation of soft tissues under the action of virtual instruments. These interactions between virtual instruments and virtual organs, performed in a separate module, mainly consists in detecting collisions followed with modeling contact forces and displacements (see Section 1.2.2).

Finally, a surgical simulator must provide an advanced user interface that includes visual and force feedback (respectively presented in Sections 1.2.3 and 1.2.4). Last but not least, it is necessary to rely on advanced software engineering methodology to make these different modules communicate within the same framework: some of these implementation issues are introduced in Section 1.2.5.

1.2.1. Geometric modeling

In general, the extraction of tridimensional geometric models of anatomical structures is based on medical imagery: CT scanner images, MRI images, cryogenic images, 3D ultrasound images, ... Because medical image resolution and contrast have greatly improved over the past few years, the tridimensional reconstruction of certain structures have become possible by using computerized tools. For instance, the availability in 1995 of the "Visible Human" dataset provided by the National Library of Medicine has allowed the creation of a complete geometric human model (ACKERMAN [1998]). However, the automatic delineation of structures from medical images is still considered an unsolved problem. Therefore, a lot of human interaction is usually required for reconstructing the human anatomy. DUNCAN and AYACHE [2000], AYACHE [2003], provide a survey on the past and current research effort in medical image analysis.

1.2.2. Interaction with a virtual instrument

A key component of a surgery simulation software is the user interface. The hardware interface that drives the virtual instrument essentially consists in one or several force-feedback devices having the same degrees of freedom and appearance as the actual surgical instruments used in minimally invasive therapy (see Fig. 1.3). In general, these systems are force-controlled, sending the instrument's position to the simulation software and receiving reaction force vectors.

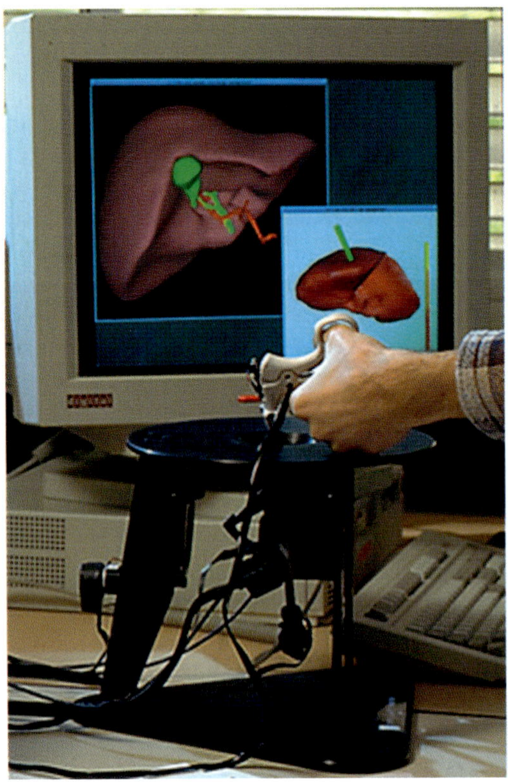

FIG. 1.3. A force feedback system suited for surgery simulation.

Once the position of the virtual instrument is known, it is necessary to detect possible collisions with other instruments or surrounding anatomical structures. In this case, it is particularly difficult to obtain a computationally efficient collision detection algorithm because the geometry of objects may change at each iteration. Therefore, algorithms based on pre-computed data structures (such as the approach proposed in GOTTSCHALK, LIN and MANOCHA [1996]) are not appropriate. LOMBARDO, CANI and NEYRET [1999] proposed an original collision detection method based on the OpenGL graphics library which is especially well-suited for elongated instruments shaped like those used in laparoscopic surgery. Although this technique cannot be used for the detection of self-collisions, several algorithms have been proposed recently (TESCHNER, HEIDELBERGER, MULLER, POMERANETS and GROSS [2003], KNOTT and PAI [2003]) to tackle this complex task.

When a collision is detected, a set of geometrical or physical constraints are applied on soft tissue models. However, modeling the physics of contacts can lead to complex algorithms and therefore purely geometric approaches are often preferred.

1.2.3. Visual feedback

A surgery simulator must provide a realistic visual representation of the surgical procedure. Visual feedback is especially important in video-surgery because it helps the surgeon to acquire a tridimensional perception of his environment. In particular, the effects of shading, shadows and textures are important clues that must be reproduced in a simulator.

The quality of visual feedback is directly related to the availability and performance of graphics accelerators. In the past few years, the market of graphics cards has evolved in three directions: improved price-performance ratio, increased geometric transformation and rasterization performance and the emergence of programmable pixel rendering. Combined with the development of more efficient computer graphics algorithms, we can foresee that realistic visual feedback for surgery simulation could be achieved in the next few years if this additional graphics rendering is focused on the three-dimensional clues used by surgeons to understand the surgical field.

1.2.4. Haptic feedback

Haptic display serves at least two purposes in a surgical simulator: kinesthetic and cognitive. First, it provides the sensation of movement to the user and therefore it significantly enhances surgical performance. Second, it helps to distinguish between tissues by testing their mechanical properties.

However, the addition of a haptic display in a simulation system increases by a large factor its complexity and the required computational power (MARK, RANDOLPH, FINCH, VAN VERTH and TAYLOR II [1996]): it leads to an increase by a factor 10 of the required bandwidth, synchronisation between visual and haptic displays, force computation, ... Only a few papers have assessed the importance of haptic feedback in surgery (MARCUS [1996]). In general, it is accepted that the combination of visual and haptic displays is optimal for surgery training or pre-planning.

In video-surgery, the surgical instruments slide inside a trocard and are constrained to enter the abdomen through a fixed point. This entails substantial friction, specifically for laparoscopy where airtightness must be enforced. The friction of the instruments inside trocards perturbes the forces sensed by the end-user. Despite those perturbations, it appears that it is still necessary to provide force-feedback for realistic user immersion.

1.2.5. Implementation of a simulator

Most of the difficulties encountered when implementing a surgical simulator originate from the trade-off that must be found between real-time interaction and the necessary surgical realism of a simulator.

The first constraint indicates that a minimum bandwidth between the computer and the interface devices must be available in order to provide a satisfactory visual and haptic feedback. If this bandwidth is too small, the user cannot properly interact with the simulator and it becomes useless for surgery gesture training. However, the "real-time" constraint can be interpreted in different ways. Most of the time, it implies that the mean update rate is high enough to allow a suitable interaction. However, it is possible that during the simulation, some events (such as the collision with a new structure) may increase the computational load of the simulation engine. This may result in a lack of synchronicity between the user gesture and the feedback the user gets from the simulator. When the computation time is too irregular, the user may even not be able to use the simulator properly. In order to guarantee a good user interaction, it is necessary to use a dedicated "real-time" software that supervises all tasks executed on the simulator.

The second constraint is related to the targeted application of a simulator: training surgeons to new gestures or procedures. To reach this goal, the user must "believe" that the simulator environment corresponds to a real procedure. The level of realism of a simulator is therefore very dependent on the type of surgical procedures and is also connected with physio-psychological parameters. In any case, increasing the realism of a simulator requires an increase of computational time which is contradictory with the constraint of real-time interaction.

The main difficulty in implementing a simulator is to optimize its credibility, given an amount of graphics and computational resources. Therefore, an analysis of the training scenario should be performed to find the most important elements that contribute to the realism of the simulation.

1.3. Constraints of soft tissue models

In the scope of a surgical simulator, it is not possible to model the biomechanical complexity of living soft tissue. Instead, authors have resorted to simplified models to decrease the implementation complexity and to optimize computational efficiency. A survey on soft tissue modeling can be found in DELINGETTE [1998].

Before presenting the main features of our approach (available in Section 3.1), we list three constraints that should be taken into account when specifying a soft tissue model for surgery simulation.

1.3.1. Visualization constraints

To obtain high quality visual rendering, two techniques are traditionally used: surface and volume rendering. A comparison between these two rendering techniques for surgery simulation is described in SOFERMAN, BLYTHE and JOHN [1998]. Surface rendering is by far the most commonly used technique, and uses basic polygonal elements (triangles, quads, ...) to achieve the rendering of an entire scene. A rule of thumb in surface rendering states that the quality of rendering is proportional to the number of polygonal elements. Unfortunately, the screen refresh rate of a graphics display is inversely proportional to the number of elements.

Therefore, an important concern arises when specifying a soft tissue model: is it compatible with high quality visual rendering? For some models, it is clearly not the case. For instance, the chain–mail algorithm (GIBSON, SAMOSKY, MOR, FYOCK, GRIMSON, KANADE, KIKINIS, LAUER and MCKENZIE [1997]) represents soft tissue with the help of cubic lattices that are allowed to move slightly with respect to their neighbors. For this representation, as well as for particle-based representations (FRANCE, ANGELIDIS, MESEURE, CANI, LENOIR, FAURE and CHAILLOU [2002], DESBRUN and GASCUEL [1995]) and multigrid representations (DEBUNNE, DESBRUN, CANI and BARR [2001]), authors use a two-layer strategy: a volumetric soft tissue model is combined with a surface model dedicated to visual rendering. These two models are often coupled with a linear relationship based on barycentric coordinates: once the shape of a soft tissue model is modified, the surface model is updated in an efficient manner. Similarly, the collision detection is performed on the surface model, but contact forces and displacements are imposed on the volumetric model. However, this approach has two limitations. First, the modeling of contact between a virtual tool and a soft tissue model is usually not satisfactory because the mapping between surface and volumetric model is complex (though mapping from volumetric to surface models is often trivial). Second, this approach makes the modeling of tissue cutting very complex where the surface and volumetric topology is altered.

For soft tissue models based on tetrahedral or hexahedral meshes, the problem of high quality visual rendering is posed in a different manner since the shell of these meshes (made of triangular or quadrangular elements) can be used directly for rendering. However, in general, coarse volumetric meshes are used in order to achieve real-time performances (see next section). Therefore, it is often required to compensate the poor geometrical quality by using specific computer graphics algorithms such as subdivision surfaces (ZORIN, SCHROEDER and SWELDENS [1996]), using avatars (DECORET, SCHAUFLER, SILLION and DORSEY [1999]) or by replacing elements with texture (SILLION, DRETTAKIS and BODELET [1997]). In the case of the hepatic surgery simulator, we have used the PN triangles algorithms (VLACHOS, PETERS, BOYD and MITCHELL [2001]) in order to provide a smooth visual rendering of the liver. The idea behind PN triangles is to subdivide each triangle and its normal vector into subtriangles in order to produce a smoother looking surface (see Fig. 1.4 for an example).

1.3.2. Real time deformation constraints

A surgical simulator is an example of a virtual reality system. To succeed in training surgeons, a simulator must provide an advanced user interface that leads to the immersion of surgeons into the virtual surgical field. To reach this level of interaction, three basic rules must be formulated:

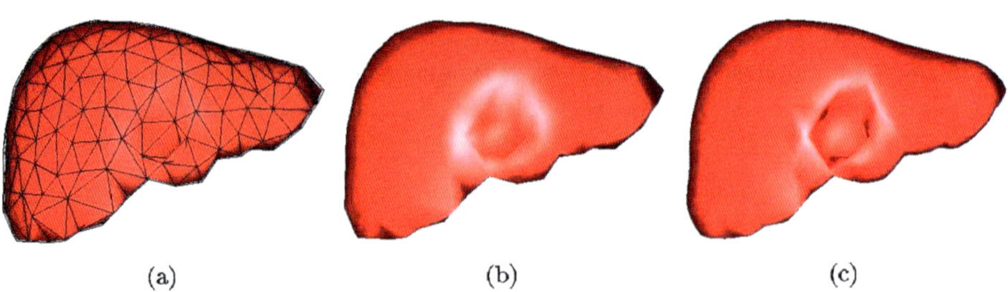

FIG. 1.4. Display of a liver being resected: (a) display of the triangles corresponding to the shell of the liver tetrahedral mesh; (b) surface rendering based on Gouraud shading without PN triangles; (c) surface rendering based on PN triangles with two levels of subdivisions.

Rule 1. *Minimum bandwidth for visual and haptic feedback.* An acceptable bandwidth for visual display is in the range of 20–60 Hz while the acceptable bandwidth for haptic display is on the range of 300–1000 Hz (300 Hz is the free-hand gesture frequency). In fact, this notion of minimal bandwidth depends on the nature of the scene to be displayed: for objects moving slowly on the screen, an update rate of 20 Hz is sufficient. Similarly, a frequency of 300 Hz may be enough to render the contact with very soft objects.

Rule 2. *Low latency.* Latency measures the time between measurements performed by the sensor (for instance, the position of the surgical instrument) and action (visual or haptic display). Latency is critical for user immersion. The hardware configuration of the system can greatly influence latency since communication between elements may be responsible for additional delays. In Fig. 1.5, the architecture of the simulation system used at INRIA (COTIN, DELINGETTE, CLEMENT, TASSETTI, MARESCAUX and AYACHE [1996]) in 1996 is presented. It is composed of one haptic display, a personal computer and a graphics workstation. There are several causes contributing to latency: communication delays between the haptic display and the PC, communication between the PC and the graphics workstation, the delay caused by the graphics display, the computation time for collision detection, force feedback and deformation. Since some of the communication links between elements are asynchronous, the total latency is not the sum of those delays but it is important to reduce them to their minimum values. The latency depends greatly on hardware, specifically on computation and graphics performance.

Rule 3. *Realistic motion of soft tissue.* It is important that the dynamic behavior of a deformable tissue is correctly simulated. To assess the visco-elastic behavior of a material, one can measure the speed at which an object returns to its rest position after it is perturbed. Soft tissue undergoes a damped motion whereas stiff objects react almost instantaneously to any perturbation. At the limit, very stiff objects can be considered to have a quasi-static motion, implying that static equilibrium is reached at each time-step (see Section 5 for a discussion about quasi-static motion).

In terms of soft tissue modeling, two parameters are important for real-time deformation constraints. The first parameter is the *update frequency* f_{u} which controls the rate at which the shape of a soft tissue model is modified. If we write X_t as the position of

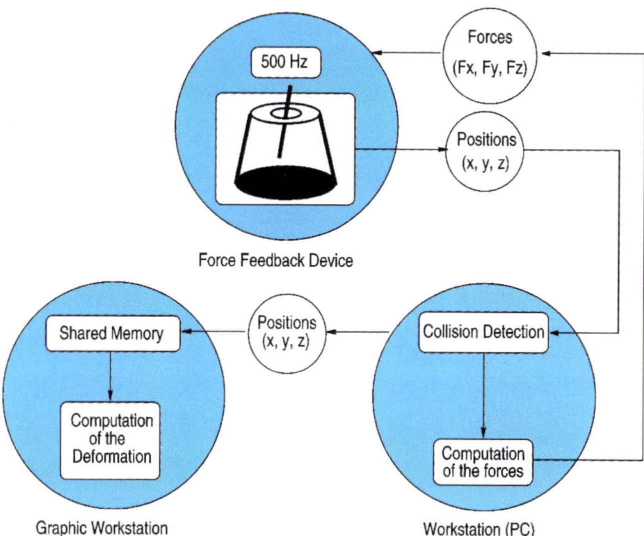

FIG. 1.5. Architecture of a simulator composed of a personal computer driving an haptic device and a graphics workstation.

the tissue model at iteration t, the *computation time* $T_c = 1/f_u$ is the time needed to compute the new position X_{t+1}. The second parameter is the *relaxation time* $T_{\text{relaxation}}$ which is the time needed for a material to return to its rest position once it has been perturbed.

To reach the required bandwidth for haptic and visual rendering (Rule 1) it is necessary that the computation time T_c is bounded by a constant $T_{\text{interaction}}$ that depends on the architecture of the simulator. For instance, in Fig. 1.6 we display three different software architectures for handling soft tissue deformation, visual and haptic feedback.

In a first architecture (Fig. 1.6(a)), all three tasks are performed sequentially, one after the other. The advantage of this approach lies in its simplicity of implementation. However, it has two drawbacks. The main problem is that the computation time T_c must be short enough to follow the minimum frequency for haptic feedback: 300 to 1000 Hz. This implies that $T_{\text{interaction}} \approx 2$ ms which is a very high requirement for a soft tissue model of reasonable size. In fact, to the best of our knowledge, only methods based on pre-computation of the static solution such as the one proposed in Section 5 can comply with this constraint. The second problem with this approach is that a delay in any of the three tasks perturbs the other tasks. For instance, when the user performs tissue cutting, an additional task is needed to update the mesh topology which translates into a delay in the visual and haptic rendering.

The second architecture shown in Fig. 1.6(b) is the most commonly used in today's surgical simulators: the haptic rendering is performed in a different process or different thread than the visual rendering and soft tissue modeling tasks. Its purpose is to sharply decrease the real-time constraint on the soft tissue computation from haptic frequency (≈ 500 Hz, $T_{\text{interaction}} \approx 2$ ms) to visual frequency (≈ 25 Hz, $T_{\text{interaction}} \approx 40$ ms). In

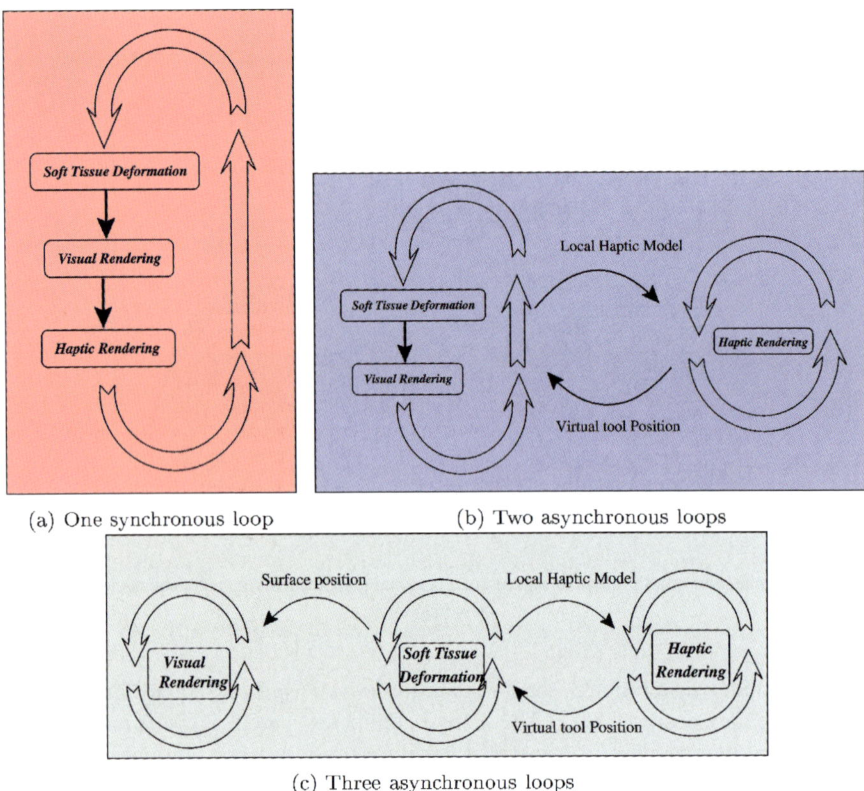

(a) One synchronous loop

(b) Two asynchronous loops

(c) Three asynchronous loops

FIG. 1.6. Different software architecture for handling visual rendering, haptic rendering and soft tissue modeling.

order to keep a satisfactory force feedback, a separate thread or process, running at haptic frequency, computes the force intensity for the haptic device based on a simplified local model. This local model, that may consist of a sphere (SERRANO and LAUGIER [2001]) or a plane (FOREST, DELINGETTE and AYACHE [2002a]) is updated by the soft tissue modeling loop while the position of the virtual surgical tool, necessary to compute its contact with soft tissue, is updated by the haptic rendering process and sent to the process. This asynchronous communication between haptic and visual rendering gives satisfactory results when some temporal smoothing is performed during the computation of force intensity. The main drawback of this approach is that it increases software complexity compared to the previous architecture. However, since only little information must be shared between the two processes, it has been adopted in several simulators, including the current version of the INRIA hepatic surgery simulator.

In the third architecture described in Fig. 1.6(c), the visual and haptic rendering tasks are performed in separate processes or threads in order to remove the latency caused by graphics hardware. Furthermore, this architecture makes the computation of soft tissue deformation more efficient (decrease of T_c) when compared to previous solutions. However, it has little effect on the maximum computation time per iteration $T_{\text{interaction}} \approx 40$ ms since the geometric model still requires to be updated at 25 Hz for visual rendering. This approach is more difficult to implement because the amount of information to transmit to the visual rendering task is quite large. Furthermore, a change in mesh topology during simulation requires to modify the data structure of the computer graphics algorithm. An example of such architecture is provided in BIELSER and GROSS [2000].

To summarize, we can state that a soft tissue model in a surgical simulator must essentially meet two constraints: computation time T_c per iteration less than a constant $T_{\text{interaction}}$, and relaxation time $T_{\text{relaxation}}$ defined by the visco-elastic behavior of the material.

1.3.3. *Tissue cutting and suturing*

The ability to cut and suture tissue is of primary importance for designing a surgery simulation system. The impact of such operations in terms of tissue modeling is considerable since it implies changing tissue topology over time. The cost of such a topological change depends largely on the chosen geometric representation but also on the numerical method that is adopted to compute tissue deformation (see discussion in previous section).

In addition, the tissue behavior must be adapted at locations where cutting or suturing occurs. Little is known about the stress/strain relationship occurring during and after cutting. The basic assumption that is made is that the physical properties of tissue are only modified locally. However, in practice, cutting can modify the boundary conditions significantly between tissue and the surrounding organs which implies considerable change with respect to their ability to deform.

Finally, when cutting volumetric or surface models, it is very likely that the new geometric and physical representation of tissue leads to self-intersections. The detection of self-intersections is very computationally expensive and, therefore repulsive force between neighboring vertices are sometimes added to prevent self-intersections.

1.4. *Computational methods for soft tissue modeling*

Several computational methods can be employed for modeling the deformation of soft tissue. We simplify the taxonomy of these methods by proposing the three classes of algorithms most commonly used:

- *Direct methods*. This category contains all methods that solve the static or dynamic equilibrium equation at each iteration (quasi-static motion). To reach such performance, some kind of pre-computation is performed. The algorithm presented in Section 5 is a direct method as well as the algorithm described in DEBUNNE, DESBRUN, CANI and BARR [2001], RADETZKY [1998].
- *Explicit iterative methods*. With iterative methods, the deformation is computed as the limit (in finite time) of a converging series that have been initialized. The closer the initial value is from the solution the faster the convergence. Iterative methods can be implemented based on implicit or explicit schemes. With

TABLE 1.1
Comparison between the three soft tissue models: direct methods (pre-computed quasi-static model), explicit iterative schemes (tensor–mass and spring–mass models) and implicit iterative schemes (Houbolt or Newmark methods)

	Direct methods	Explicit iterative scheme	Implicit iterative scheme
Computation time	low	low	high
Relaxation time	low	high	low
Cutting simulation	very difficult	possible	difficult

explicit schemes, the next position of the tissue model X_{t+1} is obtained from the application of internal forces estimated at iteration t. These methods encompass the most common algorithms found in the literature for modeling soft tissue deformation, including spring–mass model (KUHNAPFEL, AKMAK and MAA [2000]), tensor–mass models (COTIN, DELINGETTE and AYACHE [2000]) (presented in Section 6), the "chain–mail" algorithm (GIBSON, SAMOSKY, MOR, FYOCK, GRIMSON, KANADE, KIKINIS, LAUER and MCKENZIE [1997]) and others (BRO-NIELSEN [1998]).

- *Semi-implicit iterative methods*. With implicit or semi-implicit schemes, the next position of the tissue model X_{t+1} is obtained from the application of internal forces estimated at iteration $t+1$. Therefore, a linear system of equations needs to be solved entirely or partially (BARAFF and WITKIN [1998]).

In Table 1.1, we present the general features of these three types of numerical methods with respect to the constraints enumerated in Section 1.3. More precisely, the time of computation, the relaxation time (inversely proportional to the speed of convergence towards the rest position) and the ability to support any change of mesh topology during the simulation of cutting or suturing is estimated qualitatively for each method.

Direct methods can support high frequency update f_u and may have a low relaxation time to model stiff material, but they cannot simulate tissue cutting since they rely on the precomputation of some parameters.

On the other hand, explicit iterative methods are well-suited for the simulation of cutting, but these method often suffer from a high relaxation time, which makes their dynamic behavior somewhat unrealistic (jelly-like behavior). This high relaxation time originates from a lack of synchronicity, where the time step Δt used in the discretization of the explicit scheme, is much smaller than the computation time T_c. To obtain satisfactory results, it is often required to use a mesh with a small number of nodes (typically less than 1000 vertices on a standard PC).

Finally, with implicit iterative methods, the time step Δt can be increased by an order of magnitude compared to the explicit case. This allows to obtain much better dynamical behavior, but, on the other hand, they suffer from a higher computation time than explicit methods since a (sparse) linear system of equations needs to be solved at each iteration. Again, to achieve real-time performance, these methods are limited to meshes with a small number of vertices.

2. The INRIA hepatic surgery simulator

2.1. Objectives

In the sequel we use the hepatic surgery simulator developed at INRIA in the Epidaure project[2] as a case study to illustrate the different algorithms and the practical issues involved when building soft tissue models.

The INRIA hepatic surgery simulator was initiated in 1995 as a part of the European project MASTER in collaboration with the IRCAD research center[3] which hosts the European Institute of Tele-Surgery (EITS). The motivations that led us to propose the development of an hepatic surgery simulator were twofold.

First, hepatic pathologies are among the major causes of death worldwide. For instance, hepatocellular carcinoma (HCC) is a primary liver cell cancer and it accounts for most of cancer tumors. It causes the death of 1 250 000 people mainly in Asia and Africa. Furthermore, hepatic metastases (secondary tumorous cells) are mainly caused by colorectal cancers (in 30 to 50% of cases) and patients have little chance to survive hepatic carcinoma without any therapy (0 to 3% of survival for a 5 year period with an average survival period of 10 months).

The second motivation is related to the nature of hepatic resection surgery. Indeed, this surgical procedure involves many generic surgical gestures (large displacement motion, grasping, cutting, suturing) that can be useful in the simulation of different procedures. Also, because of the presence of hepatic parenchyma, the tissue models must be of volumetric nature which departs significantly from previously developed simulators simulating hollow organs like the gall bladder. Finally, tissue being a soft material allows to employ low-end force feedback systems for simulating contact forces between surgical tools and hepatic tissue.

This work has greatly benefited from the INRIA incentive action AISIM[4] which gathered different INRIA teams working in the fields of medical image analysis (Epidaure), robotics (Sharp) (BOUX DE CASSON and LAUGIER [1999]), computer graphics (Imagis) (DEBUNNE, DESBRUN, CANI and BARR [2001]) and numerical analysis (Sinus, Macs) (VIDRASCU, DELINGETTE and AYACHE [2001]).

2.2. Liver anatomy

The liver is the largest gland (average length of about 28 cm, average height of about 16 cm and average greatest thickness of about 8 cm) in the human body. It has numerous physiological functions: to filter, metabolize, recycle, detoxify, produce, store and destroy. It is located in the right hypochondriac and epigastric regions (see Fig. 2.1). The liver has a fibrous coat, the so-called Glisson's capsule. Its rheological behavior is quite different from the glandular parenchyma. Five vessel types run through the liver parenchyma: biliary and lymphatic ducts on one hand, blood vessels (internal portal

[2]Description of the Epidaure project is provided at http://www.inria.fr/epidaure/.

[3]Institut de Recherche Centre le Cancer de l'Appareil Digestif, 1, place de l'Hôpital, 67091 Strasbourg cedex, France, http://www.ircad.com/, funded by Prof. J. Marescaux.

[4]http://www-sop.inria.fr/epidaure/AISIM/.

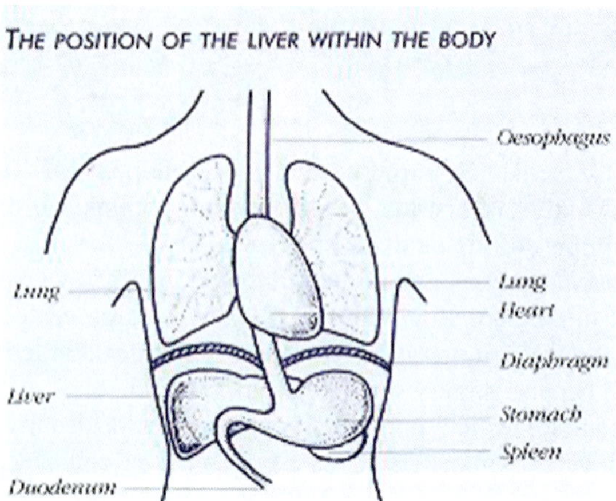

FIG. 2.1. Description of the liver anatomy with its neighboring structures (source Children's Liver Disease Foundation).

supply, hepatic arterial tree and collecting venous network) on the other hand. The portal vein, which conveys blood from the digestive tract to be detoxified and metabolized, is deep to the proper hepatic artery and common bile duct. This hepatic triad runs to the liver; it enters the liver via the hilum. This region is thus supposed to be wholly stable.

2.3. *Creation of an anatomical model of the liver*

In order to produce a model of the liver with anatomical details, the Visible Human dataset (ACKERMAN [1998]) provided by the *National Library of Medicine* was used. This dataset consists of axial MRI images of the head and neck and longitudinal sections of the rest of the body. The CT data consists of axial scans of the entire body taken at 1 mm intervals. The axial anatomical images are scanned pictures of cryogenic slices of the body. They are 24-bit color images whose size is 2048×1216 pixels. These anatomical slices are also at 1 mm interval and are registered with the CT axial images. There are 1878 cross-sections for each modality.

To extract the shape of the liver from this dataset, we used the anatomical slices (cf. Fig. 2.2), which give a better contrast between the liver and the surrounding organs. The dataset corresponding to the liver is reduced to about 180 slices. After contrast enhancement, we apply an edge detection algorithm to extract the contours of the image, and then using a simple thresholding technique, we retain only those with higher-strength contours are considered for further processing. Next, we use semi-automatic deformable contour (KASS, WITKIN and TERZOPOULOS [1988], DELINGETTE and MONTAGNAT [2001]) to extract a smooth two-dimensional boundary of each liver slice. These contours are finally transformed into a set of two-dimensional binary images (cf. Fig. 2.2). The slices generated are then stacked to form a tridimensional binary image (MONTAGNAT and DELINGETTE [1998]) (cf. Fig. 2.3).

In order to capture the shape of the external surface of the liver, one could use a subvoxel triangulation provided by the marching-cubes algorithm (LORENSEN and CLINE

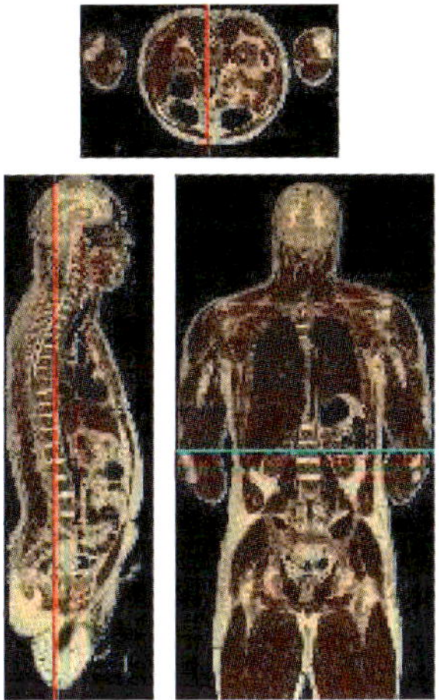

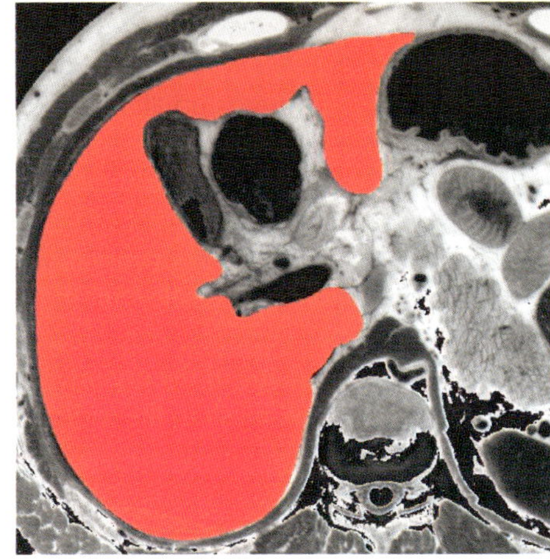

FIG. 2.2. Slice-by-slice segmentation of the liver. The initial data (left) is a high resolution photography of an anatomical slice of the abdomen. The binary image (right) corresponds to the segmented liver cross-section.

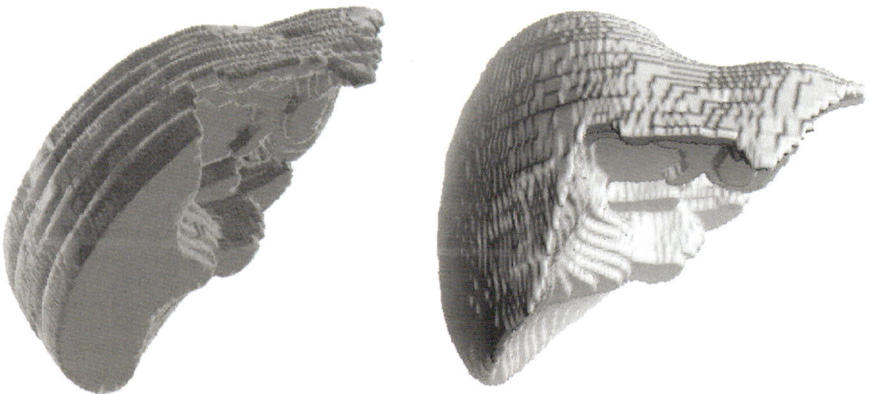

FIG. 2.3. After segmentation, the binary images are stacked (left) to give a 3D binary image. We see the step-effect on the shape of the liver (right) when extracted using the marching-cubes algorithm.

[1987]), however the number of triangles generated is too large for further processing. Moreover, a smoothing of the surface is necessary to avoid staircase effects (see Fig. 2.3). A possible solution consists in decimating an iso-surface model by using a mesh simplification tool (SCHROEDER, ZARGE and LORENSEN [1992]). However, for more flexibility, in both the segmentation and simplification processes, liver reconstruction was performed using *simplex meshes*.

Simplex meshes are an original representation of tridimensional objects developed by DELINGETTE [1999]. A simplex mesh is a deformable discrete surface mesh that is well-suited for generating geometric models from volumetric data. A simplex mesh can be deformed under the action of regularizing and external forces. Additional properties like a constant connectivity between vertices and a duality with triangulations have been defined. Moreover, simplex meshes are adaptive, for example, by concentrating vertices in areas of high curvature (thereby achieving an optimal shape description for a given number of vertices). The mesh may be refined or decimated depending on the distance of the vertices from the dataset. The decimation can also be interactively controlled. Fig. 2.4 shows the effect of the mesh adaptation and where the vertices are nicely concentrated at highly high curvature locations of the liver.

By integrating simplex meshes in the segmentation process, we have obtained smoothed triangulated surfaces, very close to an iso-surface extraction, but with fewer faces to represent the shape of the organs. In our example, the liver model has been created by fitting a simplex mesh to the tridimensional binary image previously described.

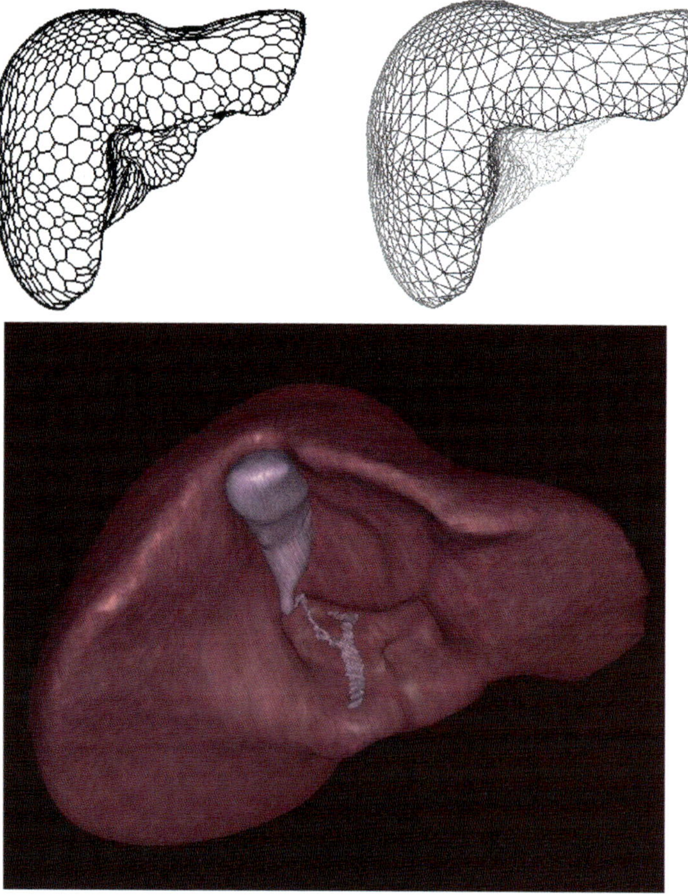

FIG. 2.4. Different representations of the geometric liver model. The simplex mesh (MONTAGNAT and DELINGETTE [1998]) fitting the data (top left) with a concentration of vertices in areas of high curvature, the triangulated dual surface (top right) and a texture-mapped model with anatomical details (gall bladder and ducts) from an endoscopic viewpoint (bottom).

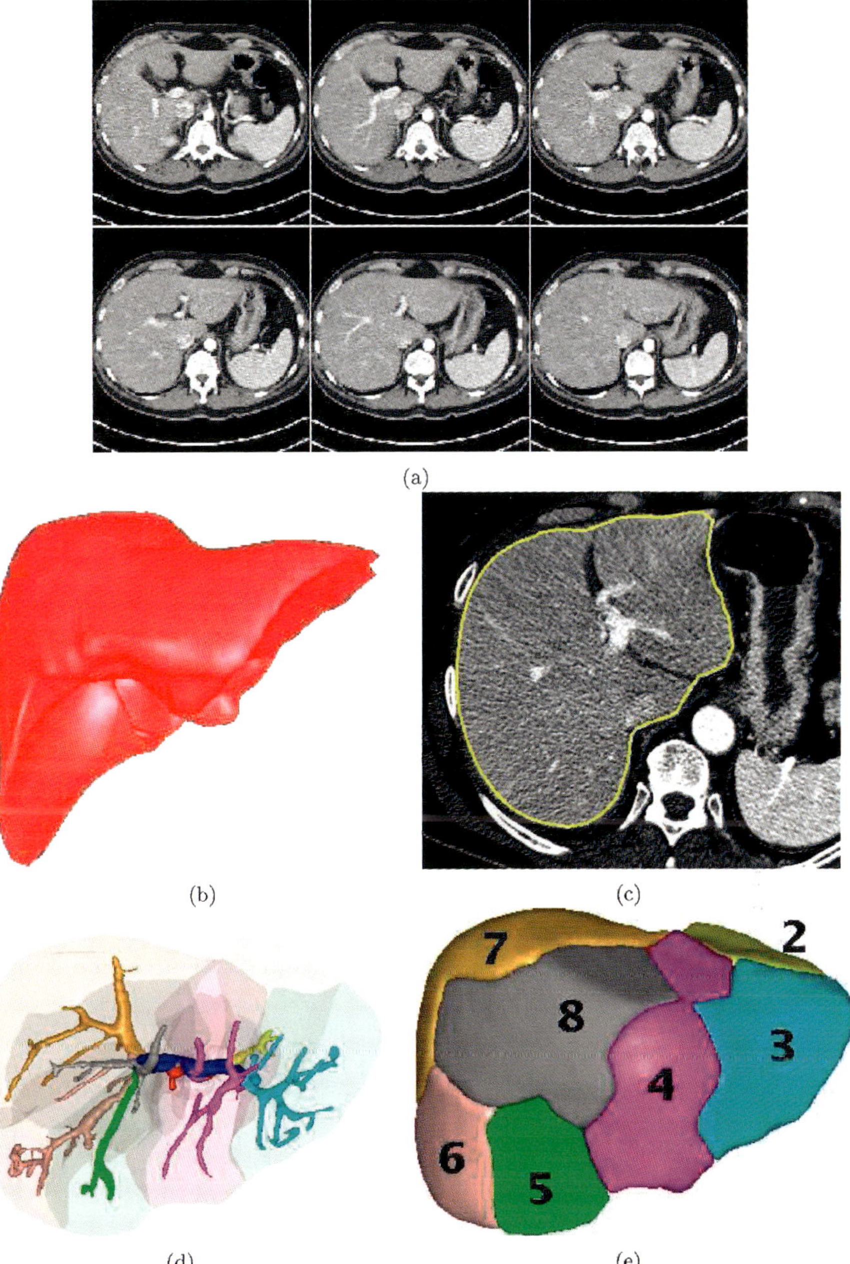

FIG. 2.5. (a) Original CT-scan images of the liver; (b) reconstructed liver model; (c) outline of the liver surface model in a CT-scan image; (d) segmentation of the portal vein (SOLER, DELINGETTE, MALANDAIN, MONTAGNAT, AYACHE, CLÉMENT, KOEHL, DOURTHE, MUTTER and MARESCAUX [2000]); (e) reconstruction of the eight anatomical segments (Couinaud segmentation).

Thanks to the adaptation and decimation properties of the simplex meshes, this model is composed of only 14 000 triangles, whereas the marching-cubes algorithm produced 94 000 triangles (cf. Figs. 2.3 and 2.4).

Although this approach is very useful for building a "generic" liver model, it is essential to integrate "patient-based" models in the simulator. In the framework of this research project, MONTAGNAT and DELINGETTE [1998] have developed a method for extracting liver models from CT scan images. The principle of this algorithm is to deform a generic simplex mesh (for instance, the one extracted from the Visible Human dataset) such that its surface coincides with the liver boundary in the image. The work of SOLER, MALANDAIN and DELINGETTE [1998], SOLER, DELINGETTE, MALANDAIN, MONTAGNAT, AYACHE, CLÉMENT, KOEHL, DOURTHE, MUTTER and MARESCAUX [2000] has extended this work by additionally extracting the main bifurcations of the portal and hepatic veins but also the hepatic lesions and gall-bladder (see Fig. 2.5).

2.4. *Liver boundary conditions*

In the scope of the AISIM project, a reference liver model was created by VIDRASCU, DELINGETTE and AYACHE [2001]. They define the liver environment (VIDRASCU, DELINGETTE and AYACHE [2001]) in order to set up the boundary conditions associated to computational models. The right liver extremity is thick and rounded while the left one is thin and flattened. Both extremities are not submitted to specific loads. The anterior border is thin, sharp and free. The posterior border is connected to the diaphragm by the coronary ligament. The upper surface, covered by the peritoneum, is divided into 2 parts by the suspensory ligament. However, this ligament does not affect the biomechanical behavior of the liver. The lower surface is connected with the gall-blader (GB) within the GB fossa, the stomach, the duodenum, the right kidney and the right part of the transverse colon. These organs are in contact with the liver surface, but they do not interact strongly with the liver; they cannot be considered as being supporting organs. The inferior vena cava (IVC) travels along the posterior surface, very often in a groove. The connection implies another strong fitting condition (clamp).

2.5. *Material characteristics*

The literature on the mechanical property of the liver is relatively poor, but during the past four years, there has been a renewed attention on soft tissue characterization due to the development of new robotics tools and new imaging modalities. The published materials concerning liver biomechanical properties usually include two distinct stages. In a first stage, experimental curves relating strain and stress are obtained from specific experimental setups and in a second stage, parameters of a known constitutive law are fitted to these curves. Concerning the first stage, there are three different sources of rheological data:

- *ex-vivo testing* where a sample of a liver is positioned inside a testing rig,
- *in-vivo testing* where a specific force and position sensing device is introduced inside the abdomen to perform indentation,

TABLE 2.1
List of published articles providing some quantitative data about the biomechanical properties of the liver

Authors	Experimental technique	Liver origin	Young modulus (kPa)
YAMASHITA and KUBOTA [1994]	image-based	human	not available
BROWN, ROSEN, KIM, CHANG, SINANAN and HANNAFORD [2003]	*in-vivo*	porcine liver	≈ 80
CARTER [1998]	*in-vivo*	human liver	≈ 170
DAN [1999]	*ex-vivo*	porcine liver	≈ 10
LIU and BILSTON [2002], LIU and BILSTON [2000]	*ex-vivo*	bovine liver	not available
NAVA, MAZZA, KLEINERMANN, AVIS and MCCLURE [2003]	*in-vivo*	porcine liver	≈ 90
MILLER [2000]	*in-vivo*	porcine liver	not available
SAKUMA, NISHIMURA, KONG CHUI, KOBAYASHI, INADA, CHEN and HISADA [2003]	*ex-vivo*	bovine liver	not available

- *image-based elastometry* where an imaging modality like ultrasound (YAMASHITA and KUBOTA [1994]), Magnetic Resonance Elastometry (MANDUCA, MUTHUPILLAI, ROSSMAN, GREENLEAF and EHMAN [1996]) or CT-scan (O'MAHONY, WILLIAMS and KATZ [1999], HODGSKINSON and CURREY [1992]) provides relevant information to assess the Young modulus of living materials.

A non-comprehensive list of articles describing the liver material characteristics is provided in Table 2.1. From this wide variety of studies, it is difficult to pick one particular constitutive model since each of experimental setup has its advantages and drawbacks. For instance, the rich perfusion of the liver affects deeply its rheology (the liver receives one fifth of the total blood flow at any time) and therefore it is still an open question whether *ex-vivo* experiments can assess the property of living liver tissue, even when specific care is taken to prevent the swelling or drying of the tissue. Conversely, data obtained from *in-vivo* experiments should also be considered with caution because the response may be location-dependent (linked to specific boundary conditions or non-homogeneity of the material) and the influence of the loading tool caliper on the deformation may not be well-understood. Furthermore, both the respiratory and circulatory motions may affect *in-vivo* data.

Furthermore, little is known about the variability of liver characteristics between species (does a porcine liver behave like a human liver?) but also between patients. For instance, studies from NAVA, MAZZA, KLEINERMANN, AVIS and MCCLURE [2003] suggest that a 20% difference in stiffness between normal and diseased livers whereas BROWN, ROSEN, KIM, CHANG, SINANAN and HANNAFORD [2003] show significant differences between *in-vivo* pig livers and *ex-vivo* cow livers.

Another important source of uncertainty in those measurements is the strain state of the liver during indentation. Indeed, as pointed out by BROWN, ROSEN, KIM, CHANG, SINANAN and HANNAFORD [2003], most researchers precondition their liver samples by applying several cycles of indentation in order to have more consistent estimates of stiffness and hysteresis. However, during surgery, (rightfully) surgeons do not precondition living tissues which may imply that only measurements obtained *in-vivo* and *in-situ* through modified surgical instruments (like those developed in CARTER [1998], BROWN, ROSEN, KIM, CHANG, SINANAN and HANNAFORD [2003], NAVA, MAZZA, KLEINERMANN, AVIS and MCCLURE [2003]) are relevant for modeling soft tissue in a surgical simulator.

Finally, the rheology of the liver is not only influenced by its perfusion, but also by the Glisson's capsule. For instance, CARTER [1998] et al. have showed that the stiffness of cylindrical samples of liver parenchyma with part of Glisson's capsule is twice the one without Glisson's capsule, using similar rheological tests (CARTER [1998]).

To conclude, more experimental studies are needed to reach a good understanding of the liver biomechanical properties. Methods based on *in-vivo* and *in-situ* indentations seem to be the most promising ones for building realistic soft tissue models in surgery simulation. All studies demonstrate that the liver is a strongly visco-elastic material, while LIU and BILSTON [2002] suggest that the liver can be considered as linear elastic for strains smaller than 0.2%.

Fortunately, in many surgical simulators, the boundary conditions governing the deformation of soft tissues, consist of imposed displacements only. In such case, the computation of soft tissue deformation requires to solve a homogeneous system of equations $\mathbf{FU} = \mathbf{0}$ which is not sensitive to the absolute value of stiffness materials but to the relative stiffness between materials (GLADILIN [2002]). Hopefully, we can expect that the relative stiffness between the liver and its neighboring organs is less variable and easier to assess, for instance, through medical imagery.

3. Linear elastic models for surgery simulation

3.1. Main features of our approach

In the next sections, we propose three different soft tissue models that are well-suited for the simulation of surgery and which are compatible with the constraints described in Section 1.3. These models bear many common features that are listed below:

- volumetric structures;
- continuum mechanics based deformation;
- finite element modeling;
- linear tetrahedron finite element;
- strong approximation in dynamical modeling.

We explain the motivations of such characteristics in the next sections.

3.1.1. Using volumetric models

We can classify the geometry of anatomical structures depending on their "idealized" dimensionality, even though they all consist of an assembly of tridimensional cells. For instance, at a coarse scale, a blood vessel can be thought as a one-dimensional structure (QUARTERONI, TUVERI and VENEZIANI [2000]) whereas the gall-blader can be represented as a two-dimensional structure (KUHNAPFEL, AKMAK and MAA [2000]) (a closed surface filled with bile). Similarly, the behavior of most parenchymatous organs such as the brain, lungs, liver or kidneys are intrinsically volumetric. But one should notice that at a fine enough scale, all anatomical structures can be considered as volumetric.

In surgical simulators, it is frequent to rely on such dimensionality simplification in order to speed-up computation: tubular surfaces, such as the colon, are modeled as a deformable spline (FRANCE, LENOIR, MESEURE and CHAILLOU [2002]) whereas deformable volumetric structures, such as the liver, are represented with their surrounding surface envelope (KUHNAPFEL, AKMAK and MAA [2000]).

However, such artifices cannot be used in a hepatic resection simulator when the removal of hepatic parenchyma is performed.

3.1.2. Using continuum mechanics

We have chosen to rely on the theory of continuum mechanics to govern the deformation of our volumetric soft tissue models. Other alternative representations exist such as spring–mass models (KUHNAPFEL, AKMAK and MAA [2000]), chain–mail (GIBSON, SAMOSKY, MOR, FYOCK, GRIMSON, KANADE, KIKINIS, LAUER and MCKENZIE [1997]) or long element models (COSTA and BALANIUK [2001]). Spring–mass models correspond to small deformation one-dimensional elastic elements (see Section 6.1.7 for an extended comparison) but are no longer valid for two- or three-dimensional elasticity. These models are especially popular in computer graphics since they are easy to implement and are based on straightforward point mechanics. The chain–mail (GIBSON, SAMOSKY, MOR, FYOCK, GRIMSON, KANADE, KIKINIS, LAUER and MCKENZIE [1997]) is an original quasi-static deformable model based on a hexahedral mesh which is well-suited for stiff material but does not allow any topological change. Long element models (COSTA and BALANIUK [2001]) correspond to valid tridimensional cylindrical elastic models but are used to approximate the deformation of general volumetric shapes.

We chose to base our soft tissue models on continuum mechanics since it offers a well-studied and validated framework for modeling the deformation of volumetric objects unlike the methods cited above. Furthermore, it offers the following advantages:

- *Scalability*: when modifying the mesh topology (refinement or cutting for instance), the behavior of the mesh is guaranteed to evolve continuously.
- *Physical parameter identification*: the elastic parameters of a biomaterial (Young modulus, for instance) can be estimated from various methods (incremental rheological experiments, elastography or solving inverse problems). Parameter identification for spring–mass models is known to be more difficult and requires stochastic optimization (genetic algorithms (LOUCHET, PROVOT and CROCHEMORE [1995]) or simulated annealing (DEUSSEN, KOBBELT and TUCKE [1995])).

3.1.3. Using finite element modeling

Finite Element Modeling (FEM) is certainly the most popular technique for the computation of structure deformation based on the elasticity theory. Furthermore, it is well-formalized and understood and there exists many software implementations although none of them deals with real-time deformation. Nonetheless, there exists alternative approaches such as Boundary Element Modeling (BEM) or the Finite Difference Method (FDM).

The BEM is well-suited for the simulation of linear elastic isotropic and homogeneous materials (for which there exists a Green function) and is indeed a good alternative to FEM when the mesh topology is not modified. In fact, BEM has the important advantage over FEM of not requiring the construction of a volumetric mesh. A more thorough discussion is provided in Section 5.3.2 but this approach is not well-suited when cutting is simulated.

The FDM is well-suited when the domain is discretized over a structured grid in which case partial derivatives can be easily discretized. They often lead to the same equation as FEM when specific finite elements (based on linear interpolation) are employed (BATHE [1982]). On unstructured meshes such as tetrahedral meshes, some extensions of the finite difference method have been proposed (DEBUNNE, DESBRUN, CANI and BARR [2001]) also leading to a similar equation as FEM (see discussion in Section 4.4). With non-linear elasticity however, FEM (PICINBONO, DELINGETTE and AYACHE [2003]) and FDM (DEBUNNE, DESBRUN, CANI and BARR [2001]) differ significantly and no formal proof has been given that the FDM converges towards the right solution as the mesh resolution increases.

3.1.4. *Using linear tetrahedron finite element*

For all finite element models described in the remainder, a simple finite element is used: a 4-node tetrahedron with linear interpolation ($P1$). The Linear Tetrahedron (LT) is known to be a poor element (in terms of convergence) compared to the Linear Hexahedron (LH) for static linear and non-linear elastic analysis (BENZLEY, PERRY, CLARK, MERKLEY and SJAARDEMA [1995]). Also this paper shows that LH performs better than the Quadratic Tetrahedron (10 nodes) even in a static linear elastic analysis.

The motivation for using tetrahedra rather than hexahedra clearly comes from a geometrical point of view. Indeed, meshing most anatomical structures with hexahedra is known to be a difficult task especially for structures having highly curved or circumvoluted parts such as the liver or the brain parenchyma (Fig. 3.1). To obtain a smooth surface envelope, it is then necessary to employ many hexahedra where a smaller number of tetrahedra would suffice. Furthermore, there exist several efficient commercial and academic software (SIMAIL, OWEN [2000]) to fill automatically a closed triangulated surface with tetrahedra of high shape quality (PARTHASARATHY, GRAICHEN and HATHAWAY [1993]). A second motivation for using tetrahedra rather than hexahedra is related to the simulation of cutting soft tissue that involves removing and remeshing of local elements. With hexahedral meshes, it is not possible to simulate general surface of cut without resorting to add new element types (such as prismatic elements). Such multi-element models (BATHE [1982]) would make the matrix assembly and local remeshing algorithms more complex to manage.

Regarding the choice of the interpolation function (linear versus quadratic), our choice has been mainly governed by computational issues. Given that a minimum number of tetrahedra is necessary to get a realistic visual rendering of a structure, the QT element involves one additional node per edge compared to the LT element which on

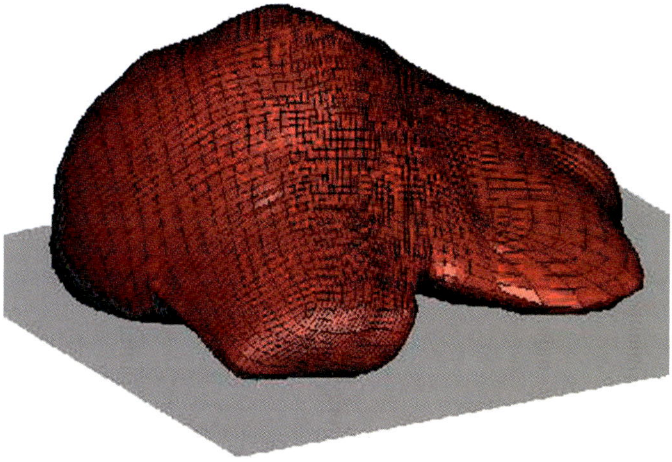

FIG. 3.1. Example of liver meshed with hexahedra (courtesy of ESI SA).

a typical tetrahedral mesh implies at least a sixfold increase of the number of nodes. Furthermore, we believe that the loss of accuracy in the deformation computation entailed by the use of LT elements remains small compared to the large uncertainty on the physical parameter values (Young modulus, ...) existing for most soft tissues.

Finally, by using linear elements, the computation of local stiffness matrices can be done explicitly (analytically) even for non-linear elasticity. Also, the gradient of the displacement field which is constant inside each element (constant strain) has a simple geometric interpretation using area vectors (see Section 4.2). A significant speed-up is therefore obtained when compared to higher order elements that require numerical integration methods such as Gauss quadrature for estimating stiffness matrices.

3.1.5. Using large approximations of dynamic behavior

Despite the development of new *in vivo* rheological equipment (KAUER, VUSKOVIC, DUAL, SZÉKELY and BAJKA [2001]), the dynamical behavior is only known quantitatively for a few anatomical structures: skin, muscle, myocardium, ... The viscoelastic properties of liver tissue have been studied by LIU and BILSTON [2000] but for most organs, constitutive laws of dynamics and their parameters must be hypothesized and validated qualitatively.

In a surgical simulator, the boundary conditions caused by the contact with surgical instruments can change between two iterations. Given that surgeons typically move their instruments at low speed (typically a few millimiters per second) and making the hypothesis that the mass of these instruments is the same or smaller than the mass of anatomical structures, we chose to neglect the dynamics of soft tissue models in two different ways. For a first class of models described in Section 5, we solve the static problem $\mathbf{F} = \mathbf{KU}$ (where $\mathbf{F}$ is the force vector, $\mathbf{K}$ the stiffness matrix and $\mathbf{U}$ the displacement vector) at each iteration thus leading to a quasi-static approximation.

For a second class of models, described in Sections 6.1 and 7, we solve the Newtonian equation of motion $\mathbf{M}\ddot{\mathbf{U}} + \mathbf{C}\dot{\mathbf{U}} = -\mathbf{K}\mathbf{U}$ with the following hypotheses: the mass matrix $\mathbf{M}$ is proportional to the identity matrix while the damping matrix $\mathbf{C}$ is diagonal. Furthermore, in some cases, the computational time T_c is longer than the time step Δt which creates a lack of synchronicity in the simulation.

3.2. *Tridimensional linear elasticity*

The fast computation of soft tissue deformation in a surgical simulator requires that some hypotheses are made about the nature of the tissue material. A first hypothesis, which leads to the two soft tissue models described in Sections 5 and 6, assumes that soft tissue can be considered as linear elastic. The rationale behind this hypothesis is clear: the linear relation between applied forces and node displacements leads to very computationally efficient algorithms. But, linear elasticity is not only a convenient mathematical model for deformable structures: it is also a quite realistic hypothesis. Indeed, all hyperelastic materials can be approximated by linear elastic materials when small displacements (and therefore small deformations) are applied (FUNG [1993], MAUREL, WU, MAGNENAT THALMANN and THALMANN [1998]). It is often admitted as reasonable to consider that a material is linear elastic when observed displacements are less than 5% of the typical object size. In the case of hepatic tissue, a recent publication (LIU and BILSTON [2000]) indicates that the linear domain is only valid for strain less than 0.2%.

Whether this constraint on the amount of displacement is valid or not in a surgical simulator depends both on the anatomical structure and the type of surgery. For instance, when simulating the removal of the gall bladder (cholelysectomy), the liver undertakes small displacements but it is not the case when simulating hepatic resection or any other surgical procedure that requires a large motion of the left lobe.

When large displacements are applied to a linear elastic material, the approximation of hyperelasticity is no longer valid and large errors in the computation of deformation and reaction forces can be perceived both visually and haptically. Section 7.1 describes the shortcomings of linear elasticity in such cases.

To summarize the general equations of linear elastic materials, we proceed in four steps. In Section 3.2.1, we provide some general definitions whereas Sections 3.2.3 and 3.2.4 give the main equations of isotropic and transversally anisotropic material. Finally, the principle of virtual work is formulated in Section 3.2.5.

3.2.1. *Definition of infinitesimal strain*

We consider a three-dimensional body defined in a tridimensional Euclidian space $\mathbb{R}^3$. We describe the geometry of this body in its rest position $\mathcal{M}_{\text{rest}}$ by using material coordinates $\mathbf{X} = (x, y, z)^{\mathrm{T}}$ defined over the volume of space Ω occupied by $\mathcal{M}_{\text{rest}}$.

This body is deformed under the application of boundary conditions: these may be either geometric boundary conditions (also called essential boundary conditions (BATHE [1982])) or natural boundary conditions, i.e., prescribed boundary forces.

We note $\mathcal{M}_{\text{def}}$ the body in its deformed state and $\boldsymbol{\Phi}(x, y, z)$, the deformation function that associates to each material point $\mathbf{X}$ located in the body at its rest position, its new

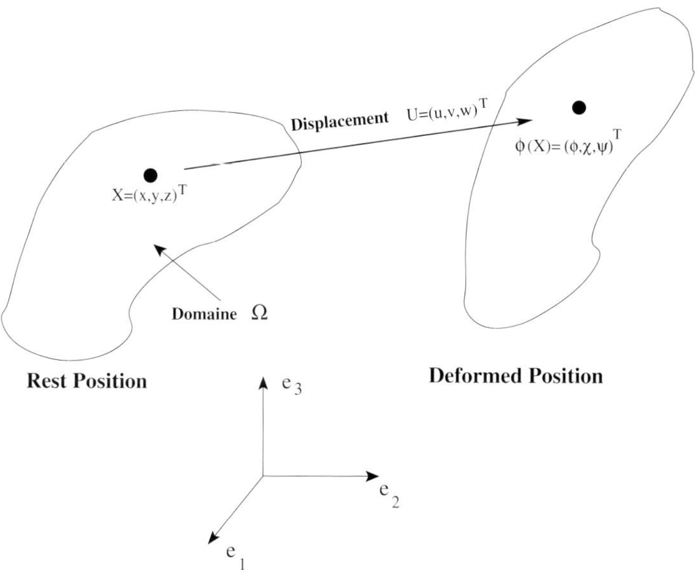

FIG. 3.2. Definition of deformation and displacement between rest and deformed positions.

position $\boldsymbol{\Phi}(\mathbf{X})$ after the body has been deformed

$$\boldsymbol{\Phi} : \Omega \subset \mathbb{R}^3 \mapsto \boldsymbol{\Phi}(\Omega), \quad \mathbf{X} \to \boldsymbol{\Phi}(\mathbf{X}) = \begin{cases} \phi(x, y, z), \\ \chi(x, y, z), \\ \psi(x, y, z). \end{cases}$$

The displacement vector field $\mathbf{U}$ is defined as the variation between the deformed position and the rest position (see Fig. 3.2):

$$\mathbf{U}(\mathbf{X}) : \Omega \mapsto \mathbb{R}^3, \quad \mathbf{X} \to \mathbf{U} = \begin{cases} u(x, y, z), \\ v(x, y, z), \\ w(x, y, z). \end{cases}$$

The observed deformation can be characterized and quantified through the analysis of the spatial derivatives of the deformation function $\boldsymbol{\Phi}(\mathbf{X})$. More precisely, the right Cauchy–Green strain tensor $\mathbf{C}(\mathbf{X})$ which is a symmetric 3×3 matrix (therefore, has 3 real eigenvalues) is simply computed from the deformation gradient

$$\mathbf{C}(\mathbf{X}) = \nabla \boldsymbol{\Phi}^{\mathrm{T}} \nabla \boldsymbol{\Phi}. \tag{3.1}$$

The Green–Lagrange strain tensor $\mathbf{E}(\mathbf{X})$, derived from the right Cauchy–Green strain tensor, allows to analyze the deformation after rigid body motion has been removed:

$$\mathbf{E}(\mathbf{X}) = \frac{1}{2}(\mathbf{C} - \mathbf{I}_3) = \frac{1}{2}\left(\nabla \mathbf{U} + \nabla \mathbf{U}^{\mathrm{T}} + \nabla \mathbf{U}^{\mathrm{T}} \nabla \mathbf{U}\right), \tag{3.2}$$

where $\mathbf{I}_3$ is the 3×3 identity matrix.

In the linear elasticity framework, applied displacements are considered as infinitesimal and the Green–Lagrange strain tensor $\mathbf{E}(\mathbf{X})$ is linearized into the infinitesimal strain tensor $\mathbf{E}_L(\mathbf{X})$. This symmetric 3×3 tensor is simply written as

$$\mathbf{E}_L(\mathbf{X}) = [e_{ij}] = \frac{1}{2}(\nabla\mathbf{U} + \nabla\mathbf{U}^T) = \begin{bmatrix} e_{xx} & e_{xy} & e_{xz} \\ e_{xy} & e_{yy} & e_{yz} \\ e_{xz} & e_{yz} & e_{zz} \end{bmatrix}. \tag{3.3}$$

The diagonal elements e_{ii} of the symmetric matrix describe the relative stretch in the direction of the reference frame, whereas off-diagonal elements e_{ij} describe shearing quantities.

3.2.2. *Definition of infinitesimal stress*

The deformation of a tridimensional body is caused by applying external forces: these forces may be either body forces $\mathbf{F}^B$ (such as gravity forces) or surface forces $\mathbf{F}^S$ (applied pressure) or concentrated forces $\mathbf{F}^P$. As a reaction to external forces, internal forces are created inside the elastic body material.

Through Cauchy theorem (CIARLET [1987]), it is demonstrated that for each volume element inside the deformed body, the force per unit area $\mathbf{t}(\mathbf{X}, \mathbf{n})$ at a point $\mathbf{X}$ and along the normal direction $\mathbf{n}$ is written as

$$\mathbf{t}(\mathbf{X}, \mathbf{n}) = \mathbf{T}(\mathbf{X})\mathbf{n},$$

where $\mathbf{T}(\mathbf{X})$ is the Cauchy stress tensor. The Cauchy stress tensor is a 3×3 symmetric tensor and can be written as

$$\boldsymbol{\Sigma}(\mathbf{X}) = [\sigma_{ij}] = \begin{bmatrix} \sigma_{xx} & \sigma_{xy} & \sigma_{xz} \\ \sigma_{xy} & \sigma_{yy} & \sigma_{yz} \\ \sigma_{xz} & \sigma_{yz} & \sigma_{zz} \end{bmatrix}.$$

The Cauchy stress $\boldsymbol{\Sigma}$ and infinitesimal strain $\mathbf{E}_L$ are conjugated variables (BATHE [1982]) which implies the following relations:

$$\sigma_{ij} = \frac{\partial W}{\partial e_{ij}}, \qquad e_{ij} = \frac{\partial W}{\partial \sigma_{ij}}, \tag{3.4}$$

where $W(\mathbf{X})$ is the amount of elastic energy per unit volume.

3.2.3. *Isotropic linear elastic materials*

For an isotropic linear elastic material, the elastic energy $W(\mathbf{X})$ is a quadratic function of the first two invariants of the infinitesimal strain tensor (CIARLET [1987]):

$$W(\mathbf{X}) = \frac{\lambda}{2}(\operatorname{tr}\mathbf{E}_L)^2 + \mu \operatorname{tr}\mathbf{E}_L^2, \tag{3.5}$$

where λ and μ are the two Lamé coefficients characterizing the material stiffness. These two parameters are simple functions of Young modulus E and Poisson coefficients ν, which belong to the material's physical properties:

$$\lambda = \frac{E\nu}{(1+\nu)(1-2\nu)}, \qquad \mu = \frac{E}{2(1+\nu)},$$
$$E = \frac{\mu(3\lambda + 2\mu)}{\lambda + \mu}, \qquad \nu = \frac{\lambda}{2(\lambda + \mu)}.$$

Through Eq. (3.4), we can derive the linear relationship, known as Hooke's law, between the stress and the infinitesimal strain tensors,

$$\boldsymbol{\Sigma} = \lambda(\operatorname{tr} \mathbf{E}_{\mathrm{L}})\mathbf{I}_3 + 2\mu \mathbf{E}_{\mathrm{L}}. \tag{3.6}$$

Note that the elastic energy can be written simply as a function of the linearized strain and stress tensors,

$$W = \frac{1}{2} \operatorname{tr}(\mathbf{E}_{\mathrm{L}} \boldsymbol{\Sigma}).$$

3.2.4. Transversally anisotropic linear elastic materials

Most anatomical structures like muscles, ligaments or blood vessels are strongly anisotropic material. This anisotropy is caused by the presence of different fibers (collagen, muscle, ...) that are wrapped together within the same tissue. For instance, anisotropic materials have been successfully used to model the deformation of the eye (KAISS and LE TALLEC [1996]), of the heart (HUMPHREY and YIN [1987], HUMPHREY, STRUMPF and YIN [1990], PAPADEMETRIS, SHI, DIONE, SINUSAS, CONSTABLE and DUNCAN [1999]) or the knee ligaments (WEISS, GARDINER and QUAPP [1995], PUSO and WEISS [1998]). In the scope of our hepatic surgery simulator, we have added an anisotropic behavior where the first branches of the portal vein are located inside the hepatic parenchyma.

We have chosen to focus only transversally anisotropic material only where there is one direction $\mathbf{a}_0$ along which the material stiffness differs from the stiffness in the orthogonal plane. Indeed, one major obstacle when modeling anisotropic material is to get reliable data from rheological experiments regarding the directions of anisotropy and the Young modulus in all directions. With transversal anisotropy, it is sufficient to provide a single direction $\mathbf{a}_0$ and an additional pair of Lamé coefficients $\lambda^{\mathbf{a}_0}$ and $\mu^{\mathbf{a}_0}$ (see Fig. 3.3).

The theoretical description of elastic energy of transversally anisotropic material is largely based on the work of SPENCER [1972], SPENCER [1984] and FUNG [1993]. For the sake of clarity, we introduce the notion of direction invariant and the concept of anisotropic stretching and shearing.

We decompose the elastic energy of a transversally anisotropic material as the sum of the isotropic energy, provided by Eq. (3.5) and by a corrective term ΔW_{Ani} which only depends on the variation of Lamé coefficients:

$$\Delta\lambda = \lambda^{\mathbf{a}_0} - \lambda, \qquad \Delta\mu = \mu^{\mathbf{a}_0} - \mu,$$
$$W_{\mathrm{Transv.Ani}}(\mathbf{X}) = W(\mathbf{X}) + \Delta W_{\mathrm{Ani}}(\mathbf{X}, \Delta\lambda, \Delta\mu).$$

Without loss of generality, we can assume that $\mathbf{a}_0$ coincides with the z direction of the reference frame. The isotropic elastic energy can then be written as a function of the stretch e_{zz} and shear (e_{xz}, e_{yz}) in the direction $\mathbf{a}_0$:

$$W(\mathbf{X}) = \left(\frac{\lambda}{2} + \mu\right)\left(e_{xx}^2 + e_{yy}^2 + e_{zz}^2\right) + \lambda(e_{xx}e_{yy} + e_{xx}e_{zz} + e_{yy}e_{zz})$$
$$+ 2\mu\left(e_{xy}^2 + e_{yz}^2 + e_{xz}^2\right).$$

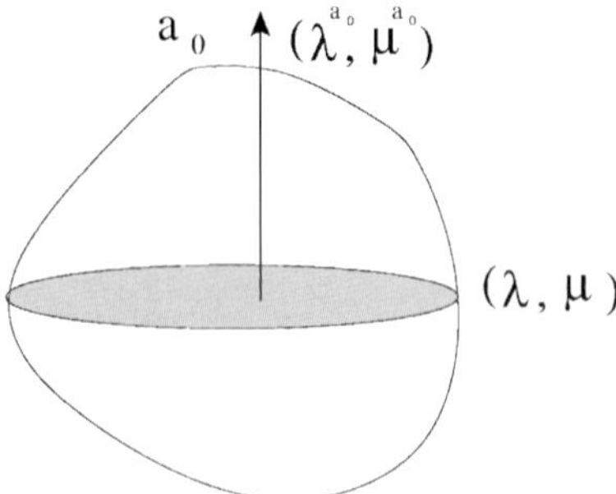

FIG. 3.3. Definition of Lamé coefficients along the direction $\mathbf{a}_0$ are $\lambda^{\mathbf{a}_0}$ and $\mu^{\mathbf{a}_0}$.

The purpose of the corrective term ΔW_{Ani} is to modify the isotropic Lamé coefficients in the direction of anisotropy:

$$\Delta W_{\mathrm{Ani}}(\mathbf{X}) = \left(-\frac{\Delta\lambda}{2} + \Delta\mu\right)e_{zz}^2 + \Delta\lambda e_{zz}(\operatorname{tr}\mathbf{E}_{\mathrm{L}}) + 2\Delta\mu\left(e_{yz}^2 + e_{xz}^2\right).$$

The equation above can be written using the two parameters I_4 and I_5 which characterize the strain tensor in the direction $\mathbf{a}_0$ (PICINBONO, LOMBARDO, DELINGETTE and AYACHE [2002]):

$$I_4 = \mathbf{a}_0^{\mathrm{T}}\mathbf{E}_{\mathrm{L}}\mathbf{a}_0, \tag{3.7}$$

$$I_5 = \mathbf{a}_0^{\mathrm{T}}\mathbf{E}_{\mathrm{L}}^2\mathbf{a}_0. \tag{3.8}$$

The first parameter I_4 is simply the amount of stretch in the direction $\mathbf{a}_0$ whereas the total amount of shearing $e_{xz}^2 + e_{yz}^2$ is given by $I_5 - I_4^2$. With these notations, the corrective term can be written as

$$\Delta W_{\mathrm{Ani}}(\mathbf{X}) = \Delta\lambda I_4 \operatorname{tr}\mathbf{E}_{\mathrm{L}} + 2\Delta\mu I_5 - \left(\frac{\Delta\lambda}{2} + \Delta\mu\right)I_4^2. \tag{3.9}$$

PICINBONO, LOMBARDO, DELINGETTE and AYACHE [2002] proposed to decompose the anisotropic term $\Delta W_{\mathrm{Ani}}(\mathbf{X})$ into a stretching and shearing part:

$$\Delta W_{\mathrm{Ani}}(\mathbf{X}) = \Delta W_{\mathrm{Str.Ani}} + \Delta W_{\mathrm{Sh.Ani}},$$

$$\Delta W_{\mathrm{St.Ani}} = \left(-\frac{\Delta\lambda}{2} + \Delta\mu\right)I_4^2 + \Delta\lambda I_4 \operatorname{tr}\mathbf{E}_{\mathrm{L}},$$

$$\Delta W_{\mathrm{Sh.Ani}} = 2\Delta\mu\left(I_5 - I_4^2\right).$$

In Fig. 3.4, the distinction between stretching and shearing effects of a transversally anisotropic material is pictured by applying a force F_1 and F_2 on a cylinder respectively along and orthogonal to the direction.

3.2.5. *Principle of virtual work*

The equilibrium equation of a deformed body is derived through the *principle of virtual displacements*. This principle states that for any compatible virtual displacement $\hat{\mathbf{u}}(\mathbf{X})$

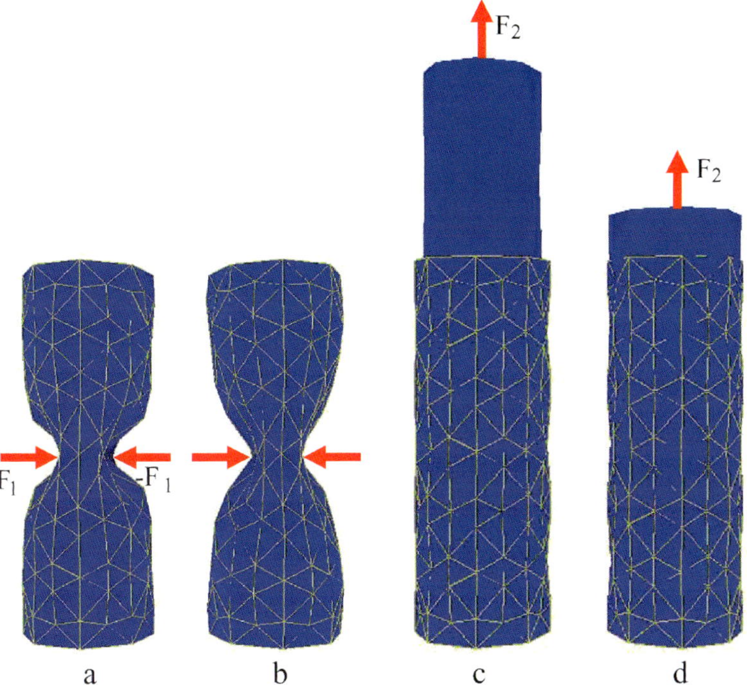

FIG. 3.4. Comparison between isotropic (a and c) and anisotropic (b and d) cylinders (PICINBONO, DELINGETTE and AYACHE [2003]). The same horizontal (respectively vertical) loads F_1 (respectively F_2) are applied in the two leftmost (respectively rightmost) figures.

applied on a body $\mathcal{M}_{\text{def}}$, the total internal virtual work is equal to the total external work. The total internal work is given by the integral of elastic energy over the body volume whereas the external work is created by the application of body and surface forces:

$$\int_{\Omega} \widehat{W}(\mathbf{X})\,\mathrm{d}V = \int_{\Omega} \hat{\mathbf{u}}^{\mathrm{T}}\mathbf{f}^{\mathrm{B}}\,\mathrm{d}V + \int_{\partial\Omega} \hat{\mathbf{u}}^{\mathrm{T}}\mathbf{f}^{\mathrm{S}}\,\mathrm{d}S \tag{3.10}$$

where $\mathbf{f}^{\mathrm{B}}$ and $\mathbf{f}^{\mathrm{S}}$ are the applied body and surface forces. Note that in Eq. (3.10), the virtual displacement field $\hat{\mathbf{u}}(\mathbf{X})$ is supposed to be compatible with the geometric boundary constraints (imposed displacements). Furthermore, this relation is only valid for small virtual displacements such that the linearized strain hypothesis still holds.

4. Finite element modeling

4.1. Linear tetrahedron element

As justified in Section 3.1.4, the computation of soft tissue deformation is based on the finite element method. Anatomical structures of interest are spatially discretized into a

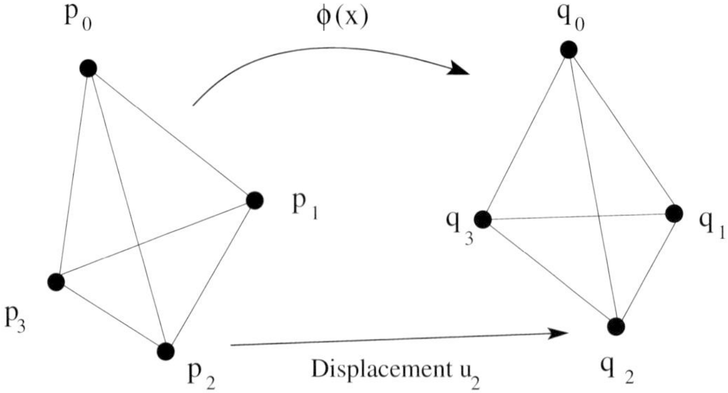

FIG. 4.1. Notations for the position and displacement vectors of a tetrahedron.

conformal tetrahedral mesh. Conformity implies that the intersection of two tetrahedra of that mesh is either empty or consists of a vertex or an edge or a triangle.

Let $\mathcal{M}_{\text{rest}}$ be a conformal tetrahedral mesh at its rest position. The initial position of each vertex is written as $\mathbf{p}_i$ while its position in the deformed position is written as $\mathbf{q}_i$ (see Fig. 4.1). The displacement at each node is then defined as

$$\mathbf{u}_i = \mathbf{q}_i - \mathbf{p}_i.$$

We use a linear tetrahedron finite element, denoted in the literature as P_1. This amounts to assuming a C^0 continuity of the displacement vector across the domain and equivalently assuming constant strain inside each tetrahedron (since the gradient matrix is constant inside each tetrahedron).

More precisely, if $\mathcal{T}$ is a tetrahedron defined by its four vertices $\mathbf{p}_j$, $j = 0, \ldots, 3$, in their *rest position*, then the displacement vector at a given point $\mathbf{X} = (x, y, z) \subset \mathcal{T}$ is defined as

$$\mathbf{u}(\mathbf{X}) = \sum_{j=0}^{3} h_j(\mathbf{X})\mathbf{u}_j,$$

where $h_j(\mathbf{X})$, $j = 0, \ldots, 3$, are the shape functions that correspond to the linear interpolation inside tetrahedron $\mathcal{T}$. These shape functions $h_j(\mathbf{X})$ correspond to the barycentric coordinates of point $\mathbf{X}$ with respect to vertices $\mathbf{p}_i$. The analytical expression of these shape functions is obtained from the linear relation

$$\begin{bmatrix} x \\ y \\ z \\ 1 \end{bmatrix} = \begin{bmatrix} \mathbf{p}_0^x & \mathbf{p}_1^x & \mathbf{p}_2^x & \mathbf{p}_3^x \\ \mathbf{p}_0^y & \mathbf{p}_1^y & \mathbf{p}_2^y & \mathbf{p}_3^y \\ \mathbf{p}_0^z & \mathbf{p}_1^z & \mathbf{p}_2^z & \mathbf{p}_3^z \\ 1 & 1 & 1 & 1 \end{bmatrix} \begin{bmatrix} h_0 \\ h_1 \\ h_2 \\ h_3 \end{bmatrix} = \mathbf{PH},$$

where $\mathbf{p}_i = (\mathbf{p}_i^x, \mathbf{p}_i^y, \mathbf{p}_i^z)^{\mathrm{T}}$ are the coordinates of each tetrahedron vertex. The matrix $\mathbf{P}$ completely encapsulates the shape of the tetrahedron $\mathcal{T}$ at its rest position. Since its determinant $|\mathbf{P}| = 6V(\mathcal{T})$ is the volume of $\mathcal{T}$, for non-degenerate tetrahedra $\mathbf{P}$ can be

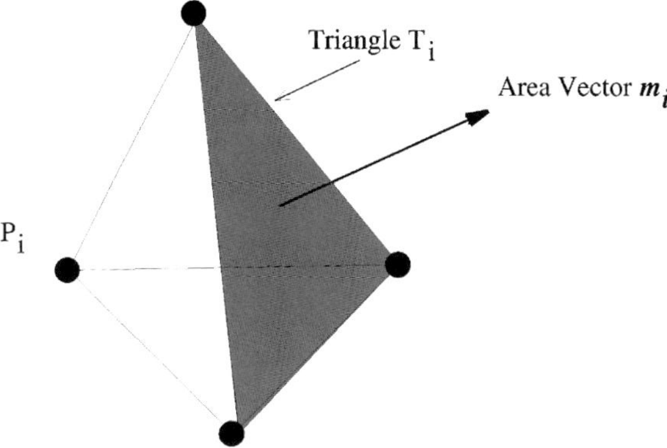

FIG. 4.2. Definition of area vector $\mathbf{m}_i$ on the triangle T_i opposite to vertex $\mathbf{p}_i$ in tetrahedron $\mathcal{T}$.

inverted,

$$[\mathbf{P}^{-1}] = \frac{-1}{6V(\mathcal{T})} \begin{bmatrix} \mathbf{m}_0^x & \mathbf{m}_0^y & \mathbf{m}_0^z & -V_0 \\ \mathbf{m}_1^x & \mathbf{m}_1^y & \mathbf{m}_1^z & -V_1 \\ \mathbf{m}_2^x & \mathbf{m}_2^y & \mathbf{m}_2^z & -V_2 \\ \mathbf{m}_3^x & \mathbf{m}_3^y & \mathbf{m}_3^z & -V_3 \end{bmatrix},$$

where:

- $\mathbf{m}_i = (m_i^x, m_i^y, m_i^z)^{\mathrm{T}}$ is the ith area vector opposite to vertex $\mathbf{p}_i$ (see description below),
- $V_i = (-1)^{i+1}|\mathbf{p}_{i+1}, \mathbf{p}_{i+2}, \mathbf{p}_{i+3}|$ can be interpreted[5] geometrically as 6 times the volume of the tetrahedron made by the origin $\mathbf{o}$ and vertices $\mathbf{p}_{i+1}$, $\mathbf{p}_{i+2}$ and $\mathbf{p}_{i+3}$. To simplify notations, the index $i+k$ should be understood as $(i+k) \bmod 4$.

Area vectors $\mathbf{m}_i$ have a very simple interpretation: they are directed along the outer normal direction of the triangle T_i opposite to $\mathbf{p}_i$ and their norm is equal to twice the area of that triangle (see Fig. 4.2). More precisely, they can be computed as

$$\mathbf{m}_i = (-1)^{i+1}(\mathbf{p}_{i+1} \times \mathbf{p}_{i+2} + \mathbf{p}_{i+2} \times \mathbf{p}_{i+3} + \mathbf{p}_{i+3} \times \mathbf{p}_{i+1}), \tag{4.1}$$

where $\mathbf{p}_{i+1} \times \mathbf{p}_{i+2}$ is the cross product between the two vectors $\mathbf{p}_{i+1}$ and $\mathbf{p}_{i+2}$.

Because they are computed from the inverse of matrix $\mathbf{P}$, these area vectors also capture the shape of $\mathcal{T}$ completely, and thus play a key role when computing the stiffness matrix $\mathbf{K}$. Further properties of area vectors are described in Section 4.2.

The shape functions $h_i(\mathbf{X})$ can then be written as

$$h_i(\mathbf{X}) = -\frac{\mathbf{m}_i \cdot \mathbf{X} - V_i}{6V(\mathcal{T})}, \tag{4.2}$$

where $\mathbf{m}_i \cdot \mathbf{X}$ is the dot product between the two vectors $\mathbf{m}_i$ and $\mathbf{X}$.

If we note that elementary volumes V_i can be expressed as

$$V_i = \mathbf{m}_i \cdot \mathbf{p}_{i+1},$$

[5] $|\mathbf{a}, \mathbf{b}, \mathbf{c}|$ is the triple product of vectors $\mathbf{a}$, $\mathbf{b}$ and $\mathbf{c}$.

then the interpolation of displacement vectors can be written as

$$\mathbf{u}(\mathbf{X}) = -\sum_{i=0}^{3} \frac{\mathbf{m}_i \cdot (\mathbf{X} - \mathbf{p}_{i+1})}{6V(\mathcal{T})} \mathbf{u}_i. \tag{4.3}$$

Finally, the interpolation matrix $\mathbf{H}(\mathbf{X})$ widely used in the finite element literature is defined as

$$\mathbf{u}(\mathbf{X}) = \mathbf{H}(\mathbf{X}) \begin{bmatrix} \mathbf{u}_0 \\ \mathbf{u}_1 \\ \mathbf{u}_2 \\ \mathbf{u}_3 \end{bmatrix},$$

$$\mathbf{H}(\mathbf{X}) = \begin{bmatrix} h_0 & 0 & 0 & h_1 & 0 & 0 & h_2 & 0 & 0 & h_3 & 0 & 0 \\ 0 & h_0 & 0 & 0 & h_1 & 0 & 0 & h_2 & 0 & 0 & h_3 & 0 \\ 0 & 0 & h_0 & 0 & 0 & h_1 & 0 & 0 & h_2 & 0 & 0 & h_3 \end{bmatrix}.$$

4.2. *Properties of area vectors*

Area vectors have a major significance with respect to the geometry of a tetrahedron for instance through the law of cosine. To write essential geometric relations, we need to introduce the following quantities:

- *normal vector* $\mathbf{n}_i$ of triangle T_i defined as the normalized area vector, $\mathbf{n}_i = \mathbf{m}_i/\|\mathbf{m}_i\|$. The normal vector is pointing outward if the tetrahedron $\mathcal{T}$ is positively oriented, i.e., if its volume $V(\mathcal{T})$ is positive;
- *dihedral angle* $\theta_{i,j}$ existing between triangle T_i and T_j and therefore between their normal vectors $\mathbf{n}_i$ and $\mathbf{n}_j$;
- *triangle area* A_i, area of triangle T_i;
- *edge length* $l_{i,j}$ is the length between vertex $\mathbf{p}_i$ and $\mathbf{p}_j$ (see Fig. 4.3);
- *foot height* f_i is the height of vertex $\mathbf{p}_i$ above triangle T_i (see Fig. 4.3).

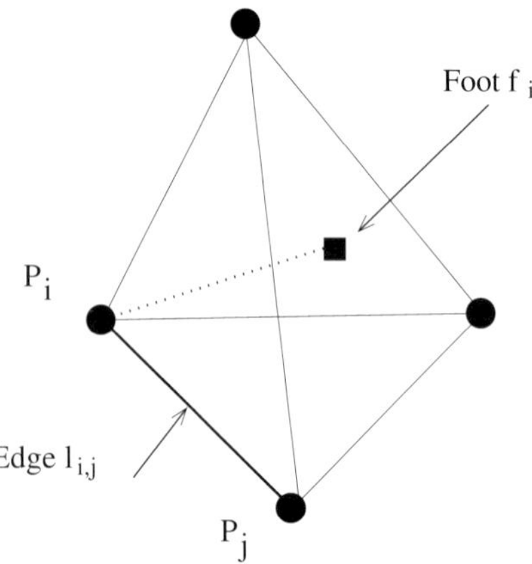

FIG. 4.3. Definition of foot height f_i and edge length $l_{i,j}$ in tetrahedron $\mathcal{T}$.

The definition of area vectors gives the relation

$$\mathbf{m}_i = 2A_i\mathbf{n}_i.$$

Noting that the tetrahedron volume is simply related to the foot height and area, we get

$$\mathbf{m}_i = \frac{2V(\mathcal{T})}{3f_i}\mathbf{n}_i.$$

From the relations $[\mathbf{P}^{-1}][\mathbf{P}] = \mathbf{I}_3$ and $[\mathbf{P}][\mathbf{P}^{-1}] = \mathbf{I}_3$, the following relations are obtained:

$$\sum_i \mathbf{m}_i = 0, \tag{4.4}$$

$$\sum_i \mathbf{m}_i \cdot \mathbf{p}_i = -18V(\mathcal{T}), \tag{4.5}$$

$$\sum_i \mathbf{m}_i \cdot \mathbf{p}_{i+1} = 6V(\mathcal{T}), \qquad (\mathbf{p}_{i+1} - \mathbf{p}_i) \cdot \mathbf{m}_i = 6V(\mathcal{T}),$$

$$\sum_{i \neq j,\ i<j} \mathbf{m}_i \cdot \mathbf{m}_j l_{i,j}^2 = 108V(\mathcal{T})^2, \qquad |\mathbf{m}_i, \mathbf{m}_{i+1}, \mathbf{m}_{i+2}| = (-1)^{i+1}36V(\mathcal{T})^2.$$

The most important result is that all area vectors sum to zero. A result of this property is the *law of cosine*:

$$A_0^2 = A_1^2 + A_2^2 + A_3^2 - 2A_1A_2\cos\theta_{1,2} - 2A_1A_3\cos\theta_{1,3} - 2A_2A_3\cos\theta_{2,3}. \tag{4.6}$$

In fact, area vectors sum to zero for any closed triangulated surface. Indeed, through Green's formulae (BRONSHTEIN and SEMENDYAYEV [1985]), the sum of area vectors can be interpreted as the total flow of a constant field across a closed surface.

4.3. *Computation of stiffness matrix: isotropic case*

We use a displacement based finite element method which is equivalent to the classical Ritz analysis (BATHE [1982]). On a single tetrahedron, the (linear) isotropic elastic energy is equal to

$$W(\mathcal{T}) = \int_{\mathcal{T}} \left(\frac{\lambda}{2}(\operatorname{tr}\mathbf{E}_{\mathrm{L}})^2 + \mu \operatorname{tr}\mathbf{E}_{\mathrm{L}}^2\right) \mathrm{d}V.$$

The gradient of the displacement $\nabla\mathbf{u}(\mathbf{X})$ is constant inside $\mathcal{T}$,

$$\nabla\mathbf{u}(\mathbf{X}) = -\sum_{i=0}^{3} \nabla\frac{\mathbf{m}_i \cdot (\mathbf{X} - \mathbf{p}_{i+1})}{6V(\mathcal{T})}\mathbf{u}_i = -\sum_{i=0}^{3} \frac{1}{6V(\mathcal{T})}\mathbf{m}_i \otimes \mathbf{u}_i,$$

where $\mathbf{m}_i \otimes \mathbf{u}_i = \mathbf{m}_i\mathbf{u}_i^{\mathrm{T}}$ is the tensor product of the two vectors,

$$\mathbf{m}_i \otimes \mathbf{u}_i = \begin{bmatrix} \mathbf{m}_i^x\mathbf{u}_i^x & \mathbf{m}_i^x\mathbf{u}_i^y & \mathbf{m}_i^x\mathbf{u}_i^z \\ \mathbf{m}_i^y\mathbf{u}_i^x & \mathbf{m}_i^y\mathbf{u}_i^y & \mathbf{m}_i^y\mathbf{u}_i^z \\ \mathbf{m}_i^z\mathbf{u}_i^x & \mathbf{m}_i^z\mathbf{u}_i^y & \mathbf{m}_i^z\mathbf{u}_i^z \end{bmatrix}.$$

The infinitesimal strain tensor $\mathbf{E}_{\mathrm{L}}$ is also constant inside $\mathcal{T}$:

$$\mathbf{E}_{\mathrm{L}} = \frac{1}{2}\left(\nabla \mathbf{u} + \nabla \mathbf{u}^{\mathrm{T}}\right) = \frac{-1}{12V(\mathcal{T})} \sum_{i=0}^{3} (\mathbf{m}_i \otimes \mathbf{u}_i + \mathbf{u}_i \otimes \mathbf{m}_i).$$

The first invariant $(\operatorname{tr} \mathbf{E}_{\mathrm{L}})^2$ is simply

$$(\operatorname{tr} \mathbf{E}_{\mathrm{L}})^2 = \frac{1}{144V(\mathcal{T})^2} \left(\sum_{i}^{3} \mathbf{m}_i \cdot \mathbf{u}_i \right)^2 = \frac{1}{144V(\mathcal{T})^2} \sum_{i,j=0}^{3} \mathbf{u}_i^{\mathrm{T}} [\mathbf{m}_i \otimes \mathbf{m}_j] \mathbf{u}_j.$$

The second invariant is slightly more complex to obtain:

$$\begin{aligned} \operatorname{tr} \mathbf{E}_{\mathrm{L}}^2 = \frac{1}{144V(\mathcal{T})^2} \operatorname{tr} \Bigg(\sum_{i,j=0}^{3} &(\mathbf{m}_i \otimes \mathbf{u}_j)(\mathbf{u}_i \cdot \mathbf{m}_j) + (\mathbf{u}_i \otimes \mathbf{m}_j)(\mathbf{m}_i \cdot \mathbf{u}_j) \\ &+ (\mathbf{m}_i \otimes \mathbf{m}_j)(\mathbf{u}_i \cdot \mathbf{u}_j) + (\mathbf{u}_i \otimes \mathbf{u}_j)(\mathbf{m}_i \cdot \mathbf{m}_j) \Bigg) \\ = \frac{1}{72V(\mathcal{T})^2} \sum_{i,j=0}^{3} &\mathbf{u}_i^{\mathrm{T}} \big[(\mathbf{m}_j \otimes \mathbf{m}_i) + (\mathbf{m}_j \cdot \mathbf{m}_i) \mathbf{I}_3 \big] \mathbf{u}_j. \end{aligned}$$

Finally, the linear elastic energy is a quadratic function of the displacement and is written as

$$W(\mathcal{T}) = \frac{1}{72V(\mathcal{T})} \sum_{i,j=0}^{3} \mathbf{u}_i^{\mathrm{T}} \big[\lambda(\mathbf{m}_i \otimes \mathbf{m}_j) + \mu(\mathbf{m}_j \otimes \mathbf{m}_i) + \mu(\mathbf{m}_i \cdot \mathbf{m}_j) \mathbf{I}_3 \big] \mathbf{u}_j,$$

$$W(\mathcal{T}) = \frac{1}{2} \sum_{i,j=0}^{3} \mathbf{u}_i^{\mathrm{T}} \big[\mathbf{B}_{ij}^{\mathcal{T}} \big] \mathbf{u}_j, \tag{4.7}$$

where $[\mathbf{B}_{ij}^{\mathcal{T}}]$ is the 3×3 stiffness matrix of tetrahedron $\mathcal{T}$ between vertices i and j. Noting that $[\mathbf{m}_i \otimes \mathbf{m}_j]\mathbf{a} = \mathbf{m}_i(\mathbf{m}_j \cdot \mathbf{a})$, we can write the local elastic energy in terms of dot products,

$$\begin{aligned} W(\mathcal{T}) = \frac{1}{72V(\mathcal{T})} \sum_{i,j=0}^{3} \big(&\lambda(\mathbf{u}_i \cdot \mathbf{m}_i)(\mathbf{m}_j \cdot \mathbf{u}_j) + \mu(\mathbf{u}_i \cdot \mathbf{m}_j)(\mathbf{m}_i \cdot \mathbf{u}_j) \\ &+ \mu(\mathbf{m}_i \cdot \mathbf{m}_j)(\mathbf{u}_i \cdot \mathbf{u}_j) \big). \end{aligned}$$

Since $\mathbf{m}_i \otimes \mathbf{m}_j = (\mathbf{m}_j \otimes \mathbf{m}_i)^{\mathrm{T}}$, it is clear that local tensors are symmetric matrices: $[\mathbf{B}_{ij}^{\mathcal{T}}] = [\mathbf{B}_{ji}^{\mathcal{T}}]^{\mathrm{T}}$. Therefore, there are only 10 distinct local stiffness matrices with *four vertex matrices* $[\mathbf{B}_{ii}^{\mathcal{T}}]$ and six *edge matrices* $[\mathbf{B}_{ij}^{\mathcal{T}}]$, $i \neq j$.

4.3.1. Local vertex stiffness matrix

Vertex stiffness matrices take the simple form with normal vector $\mathbf{n}_i$:

$$\big[\mathbf{B}_{ii}^{\mathcal{T}}\big] = \frac{A_i^2}{9V(\mathcal{T})} \big((\lambda + \mu)(\mathbf{n}_i \otimes \mathbf{n}_i) + \mu \mathbf{I}_3 \big). \tag{4.8}$$

These matrices have eigenvalues $(\lambda + 2\mu, \mu, \mu)$, $\mathbf{n}_i$ being the first eigenvector, and two directions orthogonal to $\mathbf{n}_i$ being the two other eigenvectors.

4.3.2. *Local edge stiffness matrix*

The stiffness matrix between vertex i and j is

$$\left[\mathbf{B}_{ij}^{\mathcal{T}}\right] = \frac{1}{36V(\mathcal{T})}\big(\lambda(\mathbf{m}_i \otimes \mathbf{m}_j) + \mu(\mathbf{m}_j \otimes \mathbf{m}_i) + \mu(\mathbf{m}_i \cdot \mathbf{m}_j)\mathbf{I}_3\big). \tag{4.9}$$

This matrix has the edge direction $(\mathbf{p}_j - \mathbf{p}_i)/\|\mathbf{p}_j - \mathbf{p}_i\|$ as first eigenvector associated with eigenvalue $(\mu(\mathbf{m}_i \cdot \mathbf{m}_j))/(36V(\mathcal{T}))$. The existence of the other two eigenvectors depends on the sign of the following matrix determinant:

$$\begin{vmatrix} (\lambda+\mu)(\mathbf{m}_i \cdot \mathbf{m}_j) & \mu\|\mathbf{m}_i\|^2 \\ \lambda\|\mathbf{m}_j\|^2 & 2\mu(\mathbf{m}_i \cdot \mathbf{m}_j) \end{vmatrix} = \lambda\mu A_i^2 A_v^2\left(2\left(1+\frac{\lambda}{\mu}\right)\cos^2\theta_{i,j} - 1\right).$$

4.3.3. *Global stiffness matrix*

The elastic energy of the whole deformed body is then computed by summing Eq. (4.7) over all tetrahedra. This total energy $W(\mathcal{M}_{\mathrm{def}})$ may be written with the displacement vector $\mathbf{U}$, gathering all displacement vectors $\mathbf{u}_i$, and a global stiffness matrix $\mathbf{K}$:

$$W(\mathcal{M}_{\mathrm{def}}) = \frac{1}{2}\mathbf{U}^{\mathrm{T}}\mathbf{K}\mathbf{U}. \tag{4.10}$$

This stiffness matrix $\mathbf{K}$ is built by assembling local stiffness matrices $[\mathbf{B}_{ij}^{\mathcal{T}}]$. Because these local matrices are symmetric with respect to the swap of indices $[\mathbf{B}_{ij}^{\mathcal{T}}] = [\mathbf{B}_{ji}^{\mathcal{T}}]$, the global stiffness matrix $\mathbf{K}$ is symmetric.

4.3.4. *Global vertex stiffness matrix*

The 3×3 submatrix $[\mathbf{K}_{i,j}]$ associated with vertices i (row index) and j (column index) is computed as the sum of local stiffness matrices for all tetrahedra containing both vertices. The set of tetrahedra adjacent to a given vertex (respectively edge) is called the *shell* $\mathcal{S}(i)$ of this vertex (respectively edge). In particular, for diagonal submatrices, we get

$$[\mathbf{K}_{i,i}] = \sum_{\mathcal{T}\in\mathcal{S}(i)} \frac{1}{36V(\mathcal{T})}\big((\lambda_{\mathcal{T}} + \mu_{\mathcal{T}})(\mathbf{m}_i \otimes \mathbf{m}_i) + \mu_{\mathcal{T}} A_i^2 \mathbf{I}_3\big).$$

In fact, we can provide a rather simple interpretation of this matrix expression. Its first term can be seen as the inertial matrix (second order moment) of area vectors $\mathbf{m}_i$ weighted by $(\lambda_{\mathcal{T}} + \mu_{\mathcal{T}})/(72V(\mathcal{T}))$ (see Fig. 4.4). Indeed, on a manifold tetrahedral mesh (but not on all conformal tetrahedral meshes), the shell of an interior vertex is homeomorphic to a sphere and it can be easily proved that the sum of its area vectors is null (Minkowsky's sum). Therefore $\mathbf{m}_i \otimes \mathbf{m}_i$ represents the local contribution to an inertia matrix. If vertex $\mathbf{p}_i$ is surrounded by semi-regular tetrahedra, then the matrix of inertia is proportional to identity. Note also that because it is weighted by the inverse of the tetrahedron's volume, it is very sensitive to the disparity in tetrahedra size. The second part is simply the sum of the second Lamé coefficient weighted with the inverse of the tetrahedron volume.

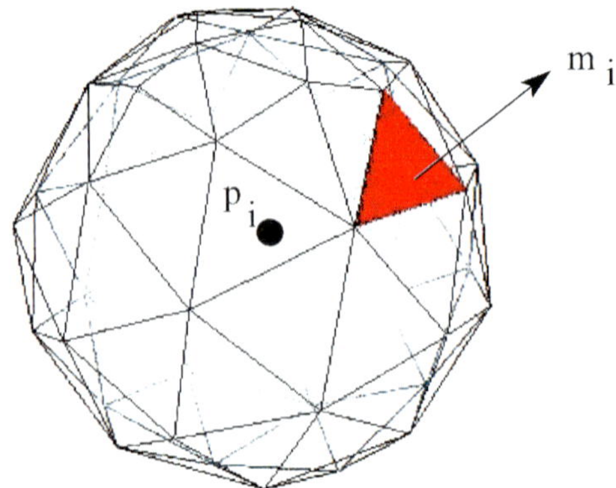

FIG. 4.4. Shell of a vertex $\mathbf{p}_i$ in a tetrahedral mesh: only the opposite triangle is drawn. For each triangle the area vector $\mathbf{m}_i$ is pointing outward. The sum of these area vectors is null and their weighted matrix of inertia determines the global stiffness matrix at $\mathbf{p}_i$.

4.3.5. Global edge stiffness matrix

The off-diagonal terms $[\mathbf{K}_{i,j}]$ of the stiffness matrix correspond to edge stiffness matrices that are the sum of local edge stiffness matrices. The edge direction $(\mathbf{p}_j - \mathbf{p}_i)/\|\mathbf{p}_j - \mathbf{p}_i\|$ is an eigenvector of $[\mathbf{K}_{i,j}]$ associated with the eigenvalue $k_{i,j}$:

$$k_{i,j} = \sum_{\mathcal{T} \in \mathcal{S}(i,j)} \frac{\mu_{\mathcal{T}}(\mathbf{m}_i \cdot \mathbf{m}_j)}{36 V(\mathcal{T})},$$

where $\mathcal{S}(i, j)$ is the set of tetrahedra adjacent to that edge (its shell). The tetrahedron volume $V(\mathcal{T})$ can be written as a function of triangle areas,

$$V(\mathcal{T}) = \frac{2}{3 l_{i,j}^{\text{opp}}} A_i A_j \sin\theta_{i,j},$$

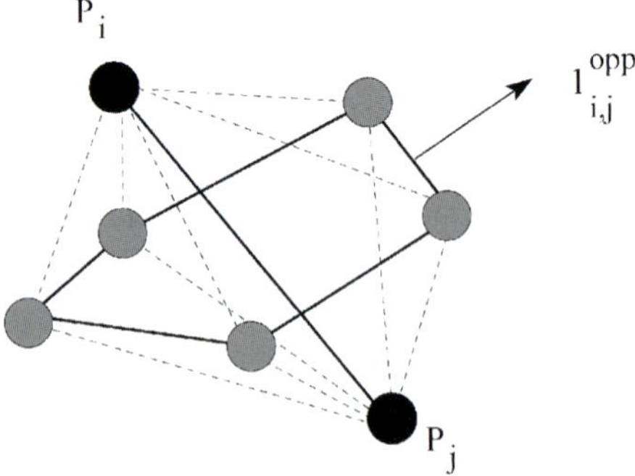

FIG. 4.5. Shell of an edge linking vertices $\mathbf{p}_i$ and $\mathbf{p}_j$ in a tetrahedral mesh: the adjacent tetrahedra are drawn with dashed lines whereas opposite edges to that edge are drawn with solid lines. One of the eigenvalues of edge stiffness matrix depends on the weighted sum of the lengths $l_{i,j}^{\text{opp}}$.

where $l_{i,j}^{\text{opp}}$ is the length of the opposite edge in tetrahedron $\mathcal{T}$ (see Fig. 4.5).

$$k_{i,j} = \frac{1}{6} \sum_{\mathcal{T} \in \mathcal{S}(i,j)} \mu_{\mathcal{T}} l_{i,j}^{\text{opp}} \cot \theta_{i,j}.$$

4.4. *Physical interpretation of isotropic stiffness matrix*

Eq. (4.7) describes the local stiffness matrices and it can be interpreted as the sum of discrete differential operators. Indeed, the isotropic elastic energy related to the first invariant of the infinitesimal strain tensors, can be written as a quadratic functional of the displacement vector:

$$W(\mathbf{X}) = \frac{\lambda}{2}(\operatorname{tr} \mathbf{E}_{\mathrm{L}})^2 + \mu \operatorname{tr} \mathbf{E}_{\mathrm{L}}^2 = \frac{\lambda}{2}(\operatorname{div} \mathbf{u})^2 + \mu \operatorname{tr}(\nabla \mathbf{u}^{\mathrm{T}} \nabla \mathbf{u}) - \frac{\mu}{2} \| \mathbf{curl}\, \mathbf{u} \|^2.$$

The first variation of the elastic force $-\delta W$ can be interpreted as the density of linear elastic force per unit volume, and is given by the Lamé equation:

$$-\delta W = (\lambda + \mu) \nabla (\operatorname{div} \mathbf{u}) + \mu \Delta \mathbf{u}.$$

It is therefore natural to compare the Lamé equation with the expression of the discrete elastic force $\mathbf{F}_i(\mathcal{T})$ acting on a vertex i of tetrahedron $\mathcal{T}$:

$$\mathbf{F}_i(\mathcal{T}) = -\sum_{j=0}^{3} [\mathbf{B}_{ij}^{\mathcal{T}}] \mathbf{u}_j = \frac{-1}{36 V(\mathcal{T})} \sum_{j=0}^{3} \big[\lambda(\mathbf{m}_i \otimes \mathbf{m}_j) + \mu(\mathbf{m}_j \otimes \mathbf{m}_i) + \mu(\mathbf{m}_i \cdot \mathbf{m}_j)\mathbf{I}_3\big] \mathbf{u}_j.$$

The three terms of the local rigidity matrix may be interpreted as follows:

$$\mathbf{F}_i(\mathcal{T}) = \frac{-1}{36 V(\mathcal{T})} \sum_{j-0}^{3} \big[\underbrace{\lambda(\mathbf{m}_i \otimes \mathbf{m}_j)}_{\substack{\mathrm{T}_1 \\ \nabla(\operatorname{div} \mathbf{u}) \\ \text{operator}}} + \underbrace{\mu(\mathbf{m}_j \otimes \mathbf{m}_i)}_{\substack{\mathrm{T}_2 \\ \nabla(\operatorname{div} \mathbf{u}) \\ \text{pseudo-operator}}} + \underbrace{\mu(\mathbf{m}_i \cdot \mathbf{m}_j)\mathbf{I}_3}_{\substack{\mathrm{T}_3 \\ \Delta \mathbf{u} \\ \text{operator}}}\big] \mathbf{u}_j.$$

The first term of the local rigidity matrix corresponds to the integral of the operator $\nabla(\operatorname{div} \mathbf{u})$ over a subdomain of tetrahedron $\mathcal{T}$. Indeed, through Green's second formulae (BRONSHTEIN and SEMENDYAYEV [1985]), the integral over a domain D of that operator can be evaluated along its boundary ∂D:

$$\int_D \nabla(\operatorname{div} \mathbf{u}) = \int_{\partial D} (\operatorname{div} \mathbf{u}) \mathbf{n} \, \mathrm{d}S.$$

However, the divergence operator is actually constant over $\mathcal{T}$ and is equal to

$$\frac{-1}{6 V(\mathcal{T})} \sum_{j=0}^{3} \mathbf{m}_j \cdot \mathbf{u}_j.$$

Furthermore, since the integral $\int \mathbf{n}\,\mathrm{d}S$ over each triangle of $\mathcal{T}$ is equal to $\frac{1}{2}\mathbf{m}_i$, T_1 is equal to one third of the flux of $(\operatorname{div}\mathbf{un})$ through the face of $\mathcal{T}$ opposite to vertex $\mathbf{p}_i$:

$$T_1 = \frac{-1}{36V(\mathcal{T})}\sum_{j=0}^{3}[(\mathbf{m}_i \otimes \mathbf{m}_j)]\mathbf{u}_j = \frac{1}{3}\,\frac{1}{6V(\mathcal{T})}\left(\sum_{j=0}^{3}\mathbf{m}_j \cdot \mathbf{u}_j\right)\frac{\mathbf{m}_i}{2}.$$

Thus, we can provide a straightforward interpretation on the first term of the local rigidity tensor: it corresponds to the integration of the $\nabla(\operatorname{div}\mathbf{u})$ operator over a subdomain $\mathcal{D}_i$ for which

$$\int_{\mathcal{T}\mathcal{D}_i}\mathbf{n}\,\mathrm{d}S = \frac{\mathbf{m}_i}{6}.$$

A natural choice for this subdomain is to consider the shell $\mathcal{S}_i$ of vertex $\mathbf{p}_i$ homothetically scaled down by a ratio of $1/\sqrt{3}$. This subdomain is sketched in Fig. 4.6(a) and (b): the vertices of the subdomain are located at distance of $1/\sqrt{3}$ times the original edge length from vertex $\mathbf{p}_i$. Unfortunately, since $1/\sqrt{3} > 0.5$, the subdomain of two neighboring vertices overlap.

To obtain a non-overlapping subdomain $\mathcal{D}_i$, one should consider the subdomain defined by the middle of each edge, the barycenters of each triangles and the barycenter of the tetrahedron, as proposed by PUTTI and CORDES [1998]. More precisely, as

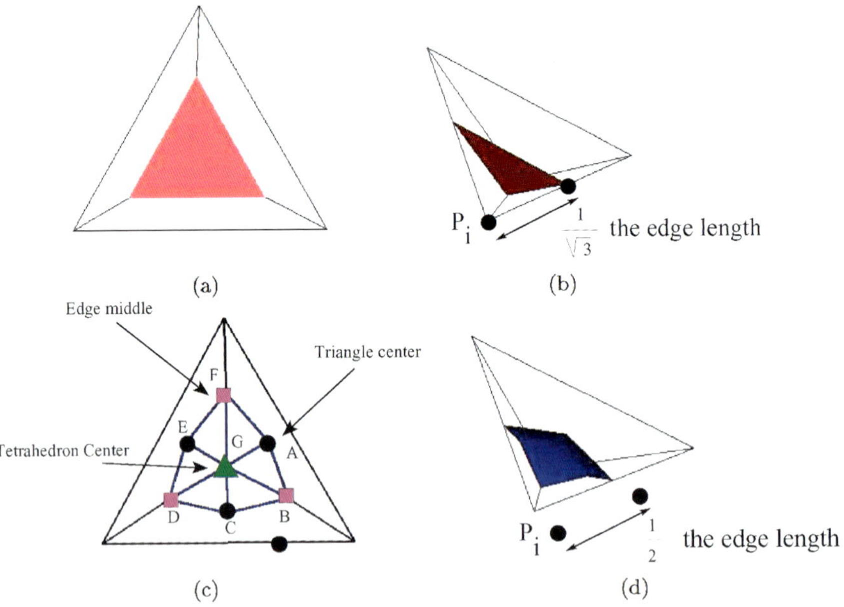

FIG. 4.6. Definitions of two subdomains for which $\int \mathbf{n}\,\mathrm{d}S$ is equal to one third the value through triangle $\triangle_i$, opposite of vertex $\mathbf{p}_i$ in tetrahedron $\mathcal{T}$; (a) and (b): front and side view of the first subdomain consisting of a single triangle corresponding to the homothety $\triangle_i$ with a ratio of $1/\sqrt{3}$; (c) and (d): front and side view of the second non-overlapping subdomain consisting of 6 triangles linking the edge middles, triangle centers and tetrahedron center.

shown in Fig. 4.6(c), the subdomain consists in the six triangles $(FAG, GAB, BGC, CGD, DGE, EGF)$ where A, C, E are the centers of the three triangles adjacent to $\mathbf{p}_i$, B, D, F are the centers of the three adjacent edges and G is the tetrahedron barycenter. This subdomain is called *the barycentric dual cell* in COSMI [2001].

Indeed, the flux over the six triangles may be written as a sum of cross products,

$$\int_{\mathcal{D}_i} \mathbf{n}\, \mathrm{d}S = A \times B + B \times C + C \times D + D \times E + E \times F.$$

Since A, B, C, D, E, F, G are simple barycentric coordinates of the four tetrahedron vertices $\mathbf{p}_i, \mathbf{p}_j, \mathbf{p}_k, \mathbf{p}_l$, it can be simply evaluated as a function of these vertices,

$$\int_{\mathcal{D}_i} \mathbf{n}\, \mathrm{d}S = \frac{1}{6}(\mathbf{p}_j \times \mathbf{p}_k + \mathbf{p}_k \times \mathbf{p}_l + \mathbf{p}_l \times \mathbf{p}_j) = \frac{\mathbf{m}_i}{6}.$$

Thus, to summarize, we have proved so far that term T_1 is the integral of the $\nabla(\operatorname{div} \mathbf{u})$ operator over a non-overlapping subdomain centered on $\mathbf{p}_i$.

The second term T_2 of the local rigidity matrix is the transposed matrix of the first term T_1 but cannot be interpreted in terms of a linear differential operator. In fact, if we write T_2 as $\sum(\mathbf{m}_i \cdot \mathbf{u}_j)\mathbf{m}_j$ we can state that T_2 corresponds to the flux of a scalar field equal to $\frac{1}{12V(\mathcal{T})}(\mathbf{m}_i \cdot \mathbf{u}_j)$ over each face of the subdomain $\mathcal{D}_i$. It should be noticed that T_2 has no equivalent in the continuous formulation (the Lamé equation) and is produced by the evaluation of $\|\operatorname{\mathbf{curl}} \mathbf{u}\|^2$.

The third term T_3 corresponds to the discrete Laplacian operator and its expression originates from the evaluation of $\frac{1}{2}\operatorname{tr}(\nabla\mathbf{u}^{\mathrm{T}}\nabla\mathbf{u})$. The same approach as for the $\nabla(\operatorname{div}\mathbf{u})$ can be applied. First, the integral of the Laplacian operator is integrated over a domain D using the integral Gauss theorem. For the x component u^x of the displacement field, it gives

$$\int_D \Delta u^x \,\mathrm{d}V = \int_D \nabla \cdot (\nabla u^x)\,\mathrm{d}V = \int_{\partial D} (\nabla u^x) \cdot \mathbf{n}\,\mathrm{d}S.$$

If the domain D is included inside a tetrahedron, then the gradient of the displacement field is a constant vector,

$$\nabla u^x = \frac{-1}{6V(\mathcal{T})} \sum_{j=0}^{3} \mathbf{m}_j u_j^x.$$

If we suppose that the domain boundary coincides with triangle $\triangle_i$, opposite to $\mathbf{p}_i$ in tetrahedron $\mathcal{T}$, then we get

$$\int_D \Delta u^x \,\mathrm{d}V = \frac{-1}{6V(\mathcal{T})} \sum_{j=0}^{3} \mathbf{m}_j \frac{1}{6V(\mathcal{T})} \sum_{j=0}^{3} \mathbf{m}_j u_j^x \cdot \frac{\mathbf{m}_i}{2}.$$

Therefore, T_3 corresponds to the integral of the Laplacian operator over a domain $\mathcal{D}$ for which

$$\int_{\partial\mathcal{D}\mathcal{T}} \mathbf{n}\,\mathrm{d}S = \frac{\mathbf{m}_i}{2},$$

for instance the subdomain defined in Fig. 4.6(c), corresponding to the barycentric dual cell of vertex $\mathbf{p}_i$ in tetrahedron $\mathcal{T}$. The Finite Element approximation of the Laplacian operator on tetrahedra was previously studied by PUTTI and CORDES [1998], DEBUNNE, DESBRUN, CANI and BARR [2001] and COSMI [2001].

To summarize, we have proved that the *variational formulation* of linear elasticity over tetrahedral meshes is not completely equivalent to the Finite Difference and Finite Volume methods. Indeed, the latter methods are equivalent to the *differential formulation* of Finite Element method which leads to the following equation of the elastic force:

$$\mathbf{F}_i(\mathcal{T}) = \frac{-1}{36V(\mathcal{T})} \sum_{j=0}^{3} \big[(\lambda+\mu)(\mathbf{m}_i \otimes \mathbf{m}_j) + \mu(\mathbf{m}_i \cdot \mathbf{m}_j)\mathbf{I}_3\big]\mathbf{u}_j. \tag{4.11}$$

The variational formulation of the FEM creates the stiffness matrix from the expression of the elastic energy whereas the differential formulation of the FEM is based on the Lamé differential equation.

4.5. *Computation of stiffness matrix: transversally anisotropic case*

From Section 3.2.4, the density of elastic energy for a transversally anisotropic material can be derived from the isotropic case by adding a corrective term:

$$W(\mathbf{X})_{\text{Transv.Ani}} = W(\mathbf{X}) + \Delta W_{\text{Ani}}(\mathbf{X}),$$

$$W(\mathbf{X})_{\text{Transv.Ani}} = W(\mathbf{X}) + \Delta\lambda I_1 I_4 + 2\Delta\mu\, I_5 - \left(\frac{\Delta\lambda}{2} + \Delta\mu\right) I_4^2,$$

where $\Delta\lambda$ and $\Delta\mu$ are the variation of Lamé coefficient in the direction of anisotropy $\mathbf{a}_0$ and where I_4 and I_5 are the constants defined in Eqs. (3.7) and (3.8). The evaluation of I_4 and I_5 with the linear tetrahedron finite element gives:

$$I_4 = \frac{-1}{6V(\mathcal{T})} \sum_{i=0}^{3} (\mathbf{a}_0 \cdot \mathbf{m}_i)(\mathbf{a}_0 \cdot \mathbf{u}_i),$$

$$(\operatorname{tr} \mathbf{E}_{\mathrm{L}}) I_4 = \frac{1}{72V(\mathcal{T})^2} \mathbf{u}_i^{\mathrm{T}} \big[(\mathbf{a}_0 \cdot \mathbf{m}_j)(\mathbf{m}_i \otimes \mathbf{a}_0)\big]\mathbf{u}_j,$$

$$I_4^2 = \frac{1}{36V(\mathcal{T})^2} \sum_{i,j=0}^{3} \mathbf{u}_i^{\mathrm{T}} \big[(\mathbf{a}_0 \cdot \mathbf{m}_i)(\mathbf{a}_0 \cdot \mathbf{m}_j)(\mathbf{a}_0 \otimes \mathbf{a}_0)\big]\mathbf{u}_j.$$

Similarly for I_5:

$$I_5 = \frac{1}{144V(\mathcal{T})^2} \sum_{i,j=0}^{3} \mathbf{u}_i^{\mathrm{T}} \big[(\mathbf{a}_0 \cdot \mathbf{m}_j)(\mathbf{a}_0 \otimes \mathbf{m}_i) + (\mathbf{a}_0 \cdot \mathbf{m}_i)(\mathbf{m}_j \otimes \mathbf{a}_0) + (\mathbf{m}_i \cdot \mathbf{m}_j)(\mathbf{a}_0 \otimes \mathbf{a}_0) + (\mathbf{a}_0 \cdot \mathbf{m}_i)(\mathbf{a}_0 \cdot \mathbf{m}_j)\mathbf{I}_3\big]\mathbf{u}_j.$$

Thus the additional elastic energy term $\Delta W(\mathcal{T})_{\text{Ani}}$ due to transversal anisotropy can also be written as a bilinear function of vertex displacements,

$$\Delta W(\mathcal{T})_{\text{Ani}} = \frac{1}{2}\sum_{i,j=0}^{3} \mathbf{u}_i^{\text{T}}\left[\mathbf{A}_{ij}^{\mathcal{T}}\right]\mathbf{u}_j \tag{4.12}$$

with the local 3×3 matrix $[\mathbf{A}_{ij}^{\mathcal{T}}]$ being defined as

$$\begin{aligned}\left[\mathbf{A}_{ij}^{\mathcal{T}}\right] = \frac{1}{144V(\mathcal{T})}\Big(&\Delta\lambda(\mathbf{a}_0 \cdot \mathbf{m}_j)(\mathbf{m}_i \otimes \mathbf{a}_0)\\ &- (\Delta\lambda + 2\Delta\mu)(\mathbf{a}_0 \cdot \mathbf{m}_i)(\mathbf{a}_0 \cdot \mathbf{m}_j)(\mathbf{a}_0 \otimes \mathbf{a}_0)\\ &+ \Delta\mu(\mathbf{a}_0 \cdot \mathbf{m}_j)(\mathbf{a}_0 \otimes \mathbf{m}_i) + \Delta\mu(\mathbf{a}_0 \cdot \mathbf{m}_i)(\mathbf{m}_j \otimes \mathbf{a}_0)\\ &+ \Delta\mu(\mathbf{m}_i \cdot \mathbf{m}_j)(\mathbf{a}_0 \otimes \mathbf{a}_0) + \Delta\mu(\mathbf{a}_0 \cdot \mathbf{m}_i)(\mathbf{a}_0 \cdot \mathbf{m}_j)\mathbf{I}_3\Big).\end{aligned}$$

4.5.1. *Local vertex stiffness matrix*

When $i = j$, the vertex stiffness matrix is written as

$$\begin{aligned}\left[\mathbf{A}_{ii}^{\mathcal{T}}\right] = \frac{1}{144V(\mathcal{T})}\Big[&(\Delta\lambda + \Delta\mu)(\mathbf{a}_0 \cdot \mathbf{m}_i)(\mathbf{m}_i \otimes \mathbf{a}_0)\\ &- (\Delta\lambda + 2\Delta\mu)(\mathbf{a}_0 \cdot \mathbf{m}_i)^2(\mathbf{a}_0 \otimes \mathbf{a}_0)\\ &+ \Delta\mu(\mathbf{a}_0 \cdot \mathbf{m}_i)(\mathbf{a}_0 \otimes \mathbf{m}_i) + \Delta\mu\|\mathbf{m}_i\|^2(\mathbf{a}_0 \otimes \mathbf{a}_0)\\ &+ \Delta\mu(\mathbf{a}_0 \cdot \mathbf{m}_i)^2\mathbf{I}_3\Big].\end{aligned}$$

This matrix has $\mathbf{c}_0$, the unit vector orthogonal to both $\mathbf{a}_0$ and $\mathbf{m}_i$ as first eigenvector with eigenvalue $\Delta\mu(\mathbf{a}_0 \cdot \mathbf{m}_i)^2$. The existence of the other two eigenvectors, in the plane defined by $\mathbf{a}_0$ and $\mathbf{m}_i$, depends on the sign of $(2\Delta\mu + \Delta\lambda)(\mathbf{a}_0 \cdot \mathbf{m}_i)^2 - \Delta\lambda\|\mathbf{m}_i\|^2$.

4.5.2. *Global stiffness matrix*

For a transversally anisotropic material, the global stiffness matrix $\mathbf{K}$ is assembled as the sum of local isotropic and anisotropic stiffness matrices:

$$[\mathbf{K}_{i,j}] = \sum_{\mathcal{T}\in\mathcal{S}(i,j)} [\mathbf{B}_{i,j}] + [\mathbf{A}_{i,j}]. \tag{4.13}$$

One should note that the global matrix $[\mathbf{K}_{i,j}]$ contains non-null values only if vertices i and j are linked by an edge of the tetrahedral mesh.

4.6. *Work of gravity forces*

We calculate the potential energy of gravity forces when a displacement field $\mathbf{u}(\mathbf{X})$ is applied on the body $\mathcal{M}_{\text{def}}$. If we write $\mathbf{g}$ the gravity vector ($\|\mathbf{g}\| = 9.8\ \text{m/s}^2$) and ρ the density of the material (assumed constant for the whole body), then the potential energy of a tetrahedron $\mathcal{T}$ is a simple function of the center of mass $\mathcal{T}$:

$$W_g(\mathcal{T}) = \int_{\mathcal{T}} \rho\mathbf{X} \cdot \mathbf{g}\,\mathrm{d}V = \rho\int_{\mathcal{T}} \mathbf{X}\,\mathrm{d}V \cdot \mathbf{g} = \rho V(\mathcal{T})\frac{\mathbf{q}_0 + \mathbf{q}_1 + \mathbf{q}_2 + \mathbf{q}_3}{4} \cdot \mathbf{g}.$$

If we drop the constant part of this energy, which is equivalent to consider the work of gravity forces when a displacement field $\mathbf{u}(\mathbf{X})$ is applied, then we get

$$W_g(\mathcal{T}) = \rho V(\mathcal{T}) \frac{\mathbf{u}_0 + \mathbf{u}_1 + \mathbf{u}_2 + \mathbf{u}_3}{4} \cdot \mathbf{g}$$
$$= \frac{\rho V(\mathcal{T})\mathbf{g}}{4} [\,\mathbf{u}_0 \quad \mathbf{u}_1 \quad \mathbf{u}_2 \quad \mathbf{u}_3\,] \begin{bmatrix} 1 \\ 1 \\ 1 \\ 1 \end{bmatrix}.$$

The potential energy of the whole model $\mathcal{M}_{\text{def}}$ is the dot product of the following two vectors:

$$W_g = \sum_{\mathcal{T}} W_g(\mathcal{T}) = \mathbf{U}^{\mathrm{T}} \mathbf{R}^g = \mathbf{U}^{\mathrm{T}} \begin{bmatrix} \cdots \\ \mathbf{r}_i^g \\ \cdots \end{bmatrix},$$

where $\mathbf{R}^g$ is a vector of size $3N$. More precisely, the sub-vector $\mathbf{r}_i^g$ of $\mathbf{R}^g$ associated with vertex i is proportional to the gravity vector, the coefficient being the volume of its neighboring tetrahedra:

$$\mathbf{r}_i^g = \rho \left(\sum_{\mathcal{T} \in \mathcal{S}(i)} \frac{V(\mathcal{T})}{4} \right) \mathbf{g}. \tag{4.14}$$

4.7. *Work of external surface pressure*

Among external forces acting on deformable soft tissue models, we include a pressure force $\mathbf{f}_p$ which is applied on a part of its surface. We consider that such pressure force has a constant intensity $\|\mathbf{f}_p\| = p$ but its direction may be either constant (contact with a stream of gas) or directed along the surface normal (contact with a solid, fluid or gas at low speed). In the latter case, the force applied on a triangle T is

$$\mathbf{f}_p(T) = p\, \mathbf{n}(T).$$

For a tetrahedral mesh, we consider that such constant pressure $\mathbf{f}_p$ is applied on a set $\mathcal{C}$ of surface triangles. If we consider a triangle $T \in \mathcal{C}$ consisting of vertices $(\mathbf{p}_i, \mathbf{p}_j, \mathbf{p}_k)$, the work of $\mathbf{f}_p$ on this triangle is

$$W_p(T) = \int_T \mathbf{f}_p \cdot \mathbf{u}(\mathbf{X})\, \mathrm{d}A = A(T)\, \mathbf{f}_p \cdot \left(\frac{\mathbf{u}_i + \mathbf{u}_j + \mathbf{u}_k}{3} \right).$$

The work of external surface pressure on the whole model $\mathcal{M}_{\text{def}}$ is then

$$W_g = \mathbf{U}^{\mathrm{T}} \mathbf{R}^p = \mathbf{U}^{\mathrm{T}} \begin{bmatrix} \cdots \\ \mathbf{r}_i^p \\ \cdots \end{bmatrix} \tag{4.15}$$

where $\mathbf{r}_i^p$ is null if vertex $\mathbf{p}_i$ is not adjacent to any triangles in $\mathcal{C}$ and is proportional the sum of triangles area otherwise:

$$\mathbf{r}_i^p = \sum_{\mathbf{p}_i \in T,\, T \in \mathcal{C}} \frac{A(T)\mathbf{f}_p(T)}{3}.$$

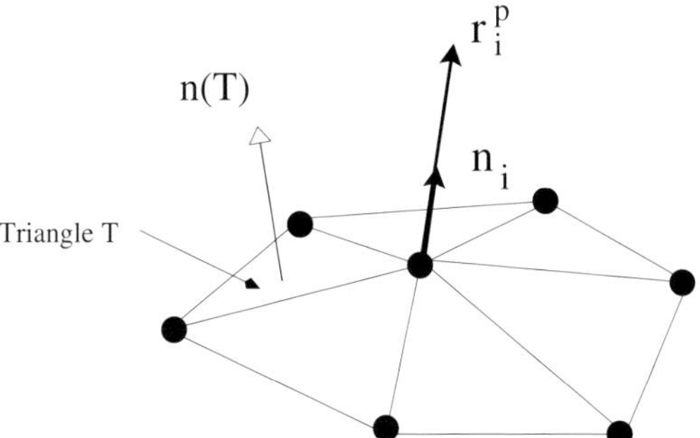

FIG. 4.7. The pressure applied on neighboring triangles results in a force directed along the surface normal at a vertex and proportional to the sum of neighboring triangle area. The vertex surface normal $\mathbf{n}_i$ is computed as the weighted average of triangle normals.

If the pressure force is applied along the surface normal, then vector $\mathbf{r}_i^p$ has an intuitive formulation. The nodal force, resulting from the pressure applied on neighboring triangle, is proportional to the area sum of surrounding triangles and is directed along the surface normal $\mathbf{n}_i$ at vertex $\mathbf{p}_i$ (see Fig. 4.7):

$$\mathbf{r}_i^p = \frac{p}{3}\left(\sum_{\mathbf{p}_i \in T,\, T \in \mathcal{C}} A(T)\right)\mathbf{n}_i,$$

where $\mathbf{n}_i$ is computed as the average of surrounding triangle normals $\mathbf{n}(T)$ weighted by their area,

$$\mathbf{n}_i = \frac{\sum_{T\in\mathcal{C}} \mathbf{n}(T)A(T)}{\sum_{T\in\mathcal{C}} A(T)}.$$

4.8. Mass matrix

The mass matrix is derived from the evaluation of the kinetic energy $\mathcal{E}(\mathcal{M}_{\text{def}})$ on the whole body $\mathcal{M}_{\text{def}}$. The density of kinetic energy $w(\mathbf{X}) = \rho(\dot{\mathbf{u}}(\mathbf{X}))^2$ where $\dot{\mathbf{u}}(\mathbf{X}) = \mathrm{d}\mathbf{u}/\mathrm{d}t$ is the speed of the material point $\mathbf{X}$. It follows that the kinetic energy of tetrahedron $\mathcal{T}$ is a bilinear form of the speed of nodal vertices $\dot{\mathbf{u}}_i$:

$$\mathcal{E}(\mathcal{T}) = \begin{bmatrix} \dot{\mathbf{U}}_0 \\ \dot{\mathbf{U}}_1 \\ \dot{\mathbf{U}}_2 \\ \dot{\mathbf{U}}_3 \end{bmatrix}^{\mathrm{T}} \begin{bmatrix} \mathbf{M}_{0,0}^{\mathcal{T}} & \mathbf{M}_{0,1}^{\mathcal{T}} & \mathbf{M}_{0,2}^{\mathcal{T}} & \mathbf{M}_{0,3}^{\mathcal{T}} \\ \mathbf{M}_{1,0}^{\mathcal{T}} & \mathbf{M}_{1,1}^{\mathcal{T}} & \mathbf{M}_{1,2}^{\mathcal{T}} & \mathbf{M}_{1,3}^{\mathcal{T}} \\ \mathbf{M}_{2,0}^{\mathcal{T}} & \mathbf{M}_{2,1}^{\mathcal{T}} & \mathbf{M}_{2,2}^{\mathcal{T}} & \mathbf{M}_{2,3}^{\mathcal{T}} \\ \mathbf{M}_{3,0}^{\mathcal{T}} & \mathbf{M}_{3,1}^{\mathcal{T}} & \mathbf{M}_{3,2}^{\mathcal{T}} & \mathbf{M}_{3,3}^{\mathcal{T}} \end{bmatrix} \begin{bmatrix} \dot{\mathbf{U}}_0 \\ \dot{\mathbf{U}}_1 \\ \dot{\mathbf{U}}_2 \\ \dot{\mathbf{U}}_3 \end{bmatrix}.$$

This tetrahedron mass matrix has size 12×12 and is composed of 4×4 local mass matrix between vertex i and j, $\mathbf{M}_{i,j}^{\mathcal{T}}$ that are 3×3 diagonal matrices,

$$\mathbf{M}_{i,j}^{\mathcal{T}} = \rho\left(\int_{\mathcal{T}} h_i(\mathbf{X})h_j(\mathbf{X})\,\mathrm{d}V\right)\mathbf{I}_3.$$

To evaluate the integral, we use the 3 barycentric coordinates (h_0, h_1, h_2) as integration variables. Based on Eqs. (4.3) and (4.4), the determinant of the Jacobian matrix is equal to the inverse of $6V(\mathcal{T})$,

$$\left|\frac{\partial h_0}{\partial \mathbf{X}}\,\frac{\partial h_1}{\partial \mathbf{X}}\,\frac{\partial h_2}{\partial \mathbf{X}}\right| = \frac{1}{216V(\mathcal{T})^3}|\mathbf{m}_0\ \mathbf{m}_1\ \mathbf{m}_2| = \frac{1}{6V(\mathcal{T})}.$$

Thus the integral can be computed explicitly using the expression below:

$$\begin{aligned}\int_{\mathcal{T}} h_i(\mathbf{X})h_j(\mathbf{X})\,\mathrm{d}V &= 6V(\mathcal{T})\int_0^1\int_0^{1-h_0}\int_0^{1-h_0-h_1} h_i\,h_j\,\mathrm{d}h_0\,\mathrm{d}h_1\,\mathrm{d}h_2\\ &= \frac{V}{10}\quad \text{if } i=j\\ &= \frac{V}{20}\quad \text{if } i\neq j.\end{aligned}$$

Thus the local mass matrix $\mathbf{M}_{i,j}^{\mathcal{T}}$ is equal to $\frac{\rho V(\mathcal{T})}{10}\mathbf{I}_3$ if $i=j$ and to $\frac{\rho V(\mathcal{T})}{20}\mathbf{I}_3$, otherwise. If we perform *mass lumping* by considering only diagonal elements equal to the sum of row values, then we naturally get $\frac{\rho V(\mathcal{T})}{4}\mathbf{I}_3$, as if the tetrahedron mass is evenly spread over its four vertices.

The kinetic energy of the whole body can be written as a function of the global mass matrix built by assembling the local matrices $\mathbf{M}_{i,j}^{\mathcal{T}}$,

$$\mathcal{E}(\mathcal{M}_{\text{def}}) = \frac{1}{2}\dot{\mathbf{U}}^{\mathrm{T}}\mathbf{M}\dot{\mathbf{U}} = \frac{1}{2}\dot{\mathbf{U}}^{\mathrm{T}}[\mathbf{M}_{i,j}]\dot{\mathbf{U}},$$

where $\mathbf{M}_{i,j}$, the global 3×3 mass matrix between vertex i and j, depends on the volumes of tetrahedra adjacent to vertex i (if $i=j$) or tetrahedra adjacent to edge (i,j) if $i\neq j$:

$$\mathbf{M}_{i,i} = \rho\sum_{\mathcal{T}\in\mathcal{S}(i)}\frac{V(\mathcal{T})}{10}\mathbf{I}_3, \tag{4.16}$$

$$\mathbf{M}_{i,j} = \rho\sum_{\mathcal{T}\in\mathcal{S}(i,j)}\frac{V(\mathcal{T})}{20}\mathbf{I}_3\quad \text{if } i\neq j. \tag{4.17}$$

If we perform *mass lumping* to get a diagonal mass matrix $\mathbf{M}$ (and therefore easily invertible), then the vertex mass is equal to one fourth of the mass of its adjacent tetrahedra:

$$(\mathbf{M}_{i,i})_{\text{lumping}} = \rho\sum_{\mathcal{T}\in\mathcal{S}(i)}\frac{V(\mathcal{T})}{4}\mathbf{I}_3.$$

4.9. Boundary conditions

In a surgical simulator, the boundary conditions of a soft tissue model are related to the existence of contacts with either its neighboring anatomical structures or with surgical tools (Fig. 4.8).

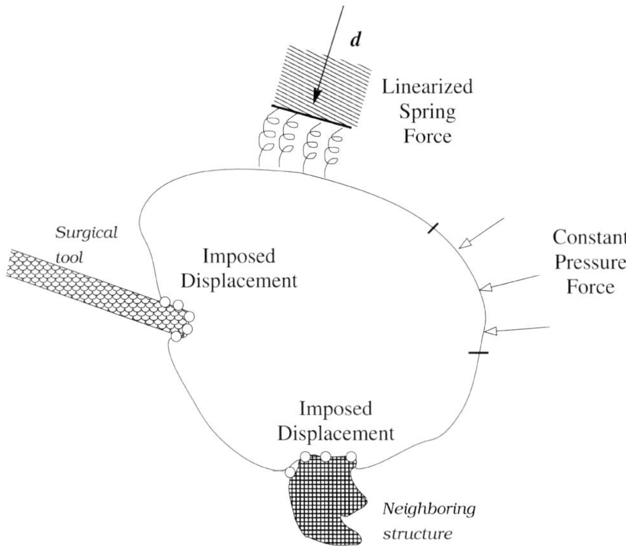

FIG. 4.8. The three different boundary conditions resulting from interaction with neighboring structures or with surgical tools.

We simplify the interaction with other physical material by considering that such an interaction can be represented either in terms of imposed displacements or elastic forces or surface pressure forces. If the material is stiff, or if it is significantly stiffer than the material of interest, we model the contact by imposing given displacements on a set of vertices. For instance, in the case of the liver model, we consider that vertices located near the vena cava (a stiff vessel) are stable (zero displacement).

If neighboring materials are as stiff (or less) than the material of interest, then we model the interaction as a linearized spring force. More precisely, for a boundary vertex $\mathbf{p}_i$, the applied force $\mathbf{r}_i^e$ is directed along a given direction $\mathbf{d}$, with stiffness k_e and rest displacement $\mathbf{u}_i^e$:

$$\mathbf{r}_i^e = -k_e\big(\big(\mathbf{u}_i - \mathbf{u}_i^e\big)\cdot\mathbf{d}\big)\mathbf{d} = -k_e(\mathbf{d}\otimes\mathbf{d})\big(\mathbf{u}_i - \mathbf{u}_i^e\big). \tag{4.18}$$

Using a linearized spring allows to compute the static equilibrium by solving a linear system of equation. Indeed, the stiffness caused by the spring $k_e(\mathbf{d}\otimes\mathbf{d})$ can be added to the global stiffness matrix while the residual force $k_e(\mathbf{d}\otimes\mathbf{d})\mathbf{u}_i^e$ is added to the nodal load at node i. Furthermore, since the stiffness k_e is lower than the Young modulus of the material, the condition number of the updated stiffness matrix is not significantly modified.

In the sequel, we do not consider linearized spring boundary conditions explicitly. Instead, we modify the global stiffness matrix $\mathbf{K}$ into $\mathbf{K}^\star$, and we consider that a nodal force $\mathbf{r}_i^b$ is applied to vertex $\mathbf{p}_i$,

$$\big[\mathbf{K}^\star_{i,i}\big] = \big[\mathbf{K}_{i,i} + k_e(\mathbf{d}\otimes\mathbf{d})\big], \quad \mathbf{r}_i^b = k_e(\mathbf{d}\otimes\mathbf{d})\mathbf{u}_i^e.$$

When a soft tissue model is in contact with some fluids (bile, water, blood, ...) or gas (carbon dioxide, air, ...), we make the hypothesis that a constant pressure is applied along the normal direction of the contact surface. The computation of the nodal forces is detailed in Section 4.7.

Finally, the contact between surgical tools and a soft tissue model may be posed, in theory, either as imposed displacements (geometric method (BATHE [1982])) or as prescribed forces (penalty method (BATHE [1982])). However, in practice, the motion of surgical tools is controlled by the end-user through a force-feedback device. To decrease their cost, these devices are force-controlled and follow a simple open loop: the positions of surgical tools can be sent to a computer while they receive the force level that should be felt by the end-user. In other words, despite the low speed of a surgeon hands the position of a surgical tool varies significantly between two iterations ($dt = 20$ ms) and therefore we found that the penalty method was not suited for deforming a soft tissue model.

Thus, after detecting the collision between soft tissue models and surgical tools, a set of imposed displacements at the collision nodes is computed. This computation is obviously ill-posed since it relies only on geometry (surface–volume intersection) rather than physical principles (Coulomb friction, for instance). Furthermore, a major challenge is to design a stable contact algorithm where a small displacement of the tool entails a small variation of node position. The geometric contact algorithm used in our hepatic surgery simulator, can be found in PICINBONO, LOMBARDO, DELINGETTE and AYACHE [2002].

To summarize, we consider only 2 types of boundary conditions in the remainder:

(1) *Imposed displacement.* We write $\mathcal{V}_d$ the set of vertices $\mathbf{p}_i$ for which the displacement $\mathbf{u}_i^b$ is known. In the scope of surgery simulation, these vertices are always lying on the surface of the mesh.
(2) *Applied nodal forces.* We write $\mathcal{V}_f$ the set of vertices $\mathbf{p}_i$ where an external force r_i^b is applied. Again, we make the hypothesis that applied forces may exist only on surface nodes.

4.10. *Equilibrium equations*

We apply the principle of virtual displacements described in Section 3.2.5 to obtain the finite element formulation of equilibrium equations. In a first stage, we only consider the static equilibrium by neglecting inertial forces. Thus, based on Eq. (3.10), we can state that the virtual elastic energy is equal to the sum of the virtual work of gravity and boundary forces,

$$\frac{1}{2}\widehat{\mathbf{U}}^{\mathrm{T}}\mathbf{K}\widehat{\mathbf{U}} = \widehat{\mathbf{U}}^{\mathrm{T}}\mathbf{R}^{g} + \widehat{\mathbf{U}}^{\mathrm{T}}\mathbf{R}^{b}.$$

Since this equation must hold for any set of compatible displacements, the static equation of equilibrium becomes

$$\mathbf{K}\mathbf{U} = \mathbf{R}^{g} + \mathbf{R}^{b}. \tag{4.19}$$

It is important to note that Eq. (4.19) is written for all nodes including the $\mathcal{V}_d$ nodes where the displacement is imposed. Therefore, in order to compute the unknown displacement vectors (where no displacement is imposed), it is important to write Eq. (4.19) with a distinction between free nodes (subscript f) and constrained nodes (subscript c):

$$\begin{bmatrix} \mathbf{K}_{ff} & \mathbf{K}_{fc} \\ \mathbf{K}_{cf} & \mathbf{K}_{cc} \end{bmatrix} \begin{bmatrix} \mathbf{U}_f \\ \mathbf{U}_c \end{bmatrix} = \begin{bmatrix} \mathbf{R}_f^g + \mathbf{R}_f^b \\ \mathbf{R}_c^g + \mathbf{R}_c^b \end{bmatrix}$$

thus leading to

$$\mathbf{K}_{ff}\mathbf{U}_f = \mathbf{R}_f^g + \mathbf{R}_f^b - \mathbf{K}_{fc}\mathbf{U}_c. \tag{4.20}$$

In the case of a linear tetrahedron finite element, $\mathbf{K}_{fc}\mathbf{U}_c$ is non-zero only for free nodes that are neighbors to fixed nodes. In the remainder, we used simplified notations by dropping the subscript f for the stiffness matrix and displacement vector and by gathering all applied nodes into a single vector:

$$\mathbf{KU} = \mathbf{R}. \tag{4.21}$$

To get the dynamic law of motion, the work of inertial forces $-\frac{1}{2}\dot{\mathbf{U}}^{\mathrm{T}}\mathbf{M}\dot{\mathbf{U}}$ should be added to the work of body forces. By adding the work of damping forces, the following classical equation is obtained:

$$\mathbf{M}\ddot{\mathbf{U}} + \mathbf{C}\dot{\mathbf{U}} + \mathbf{KU} = \mathbf{R}, \tag{4.22}$$

where $\mathbf{C}$ is the damping matrix. In general, we assume that $\mathbf{C}$ follows Rayleigh damping,

$$\mathbf{C} = \gamma_1\mathbf{M} + \gamma_2\mathbf{K}. \tag{4.23}$$

This assumption is important for performing modal analysis but also for ensuring that the damping matrix, as the stiffness matrix, is also sparse.

4.11. Solution of equilibrium equations

The static equilibrium given by Eq. (4.21) is a linear system of equations with a symmetric positive definite stiffness matrix. Since this matrix is sparse, the classical method to solve this equation is to use the conjugated gradient algorithm [SAAD, 1996]

More precisely, when solving the complete system $\mathbf{KU} = \mathbf{R}$, we perform the following steps:

- *Node renumbering* by using the reverse cutting McKee algorithm (SAAD [1996]) in order to decrease the bandwidth of the stiffness matrix.
- *Matrix preconditioning* based on Cholesky factorisation or incomplete LU decomposition (SAAD [1996]).
- *Application of the conjugated gradient algorithm* for solving the linear system of equation. We rely on the *Matrix Template Library* (LUMSDAINE and SIEK [1998]) for an efficient implementation of these algorithms in C++. When the stiffness matrix is poorly conditioned, for instance, for nearly incompressible materials, it is possible that the conjugated gradient algorithm fails. In which case, we resort to using direct methods for solving the system of equation, such as Gauss pivoting (SAAD [1996]).

Despite optimizing the bandwidth and the condition number of the stiffness matrix, the time required for solving the static equation is still too large for real-time computation. For instance, with a liver model composed of a mesh consisting of 1313 vertices, the solution of the linear system of size 3939×3939 requires 9 s on a PC Pentium II (450 MHz) with 140 iterations of the preconditioned conjugated gradient in order to reach an accuracy of 0.001 mm.

Therefore, solving directly the static equation with the conjugated gradient algorithm does not satisfy the real-time constraints mentioned in Section 1.3.2 since $T_c > T_{\text{relaxation}}$. As an alternative, we propose in the next sections, three soft tissue models that satisfy either hard or soft real-time constraints.

5. Quasi-static precomputed linear elastic model

5.1. *Introduction*

Since the complete solution of the static equilibrium equation is too computationally expensive for real-time constraint, a straightforward solution is to perform only few iterations of the conjugated gradient at each time step in order to increase the update rate. This approach, proposed by BARAFF and WITKIN [1998] is well-suited in the context of computer animation but is not applicable for surgery simulation where boundary conditions are constantly changing and are formulated in terms of imposed displacements. Indeed, using a conjugated gradient method would require to modify the stiffness matrix frequently as well as its preconditioning which would considerably reduce its efficiency.

Instead, we propose a *quasi-static precomputed linear elastic model* (COTIN, DELINGETTE and AYACHE [1999a]) that is based on a simple concept which consists in partially inverting the stiffness matrix in a precomputation stage before the simulation.

This model has the following characteristics:

- It is computationally very efficient: the computation complexity during the simulation is proportional to the cube of the number of imposed displacements.
- Only the position of surface nodes is updated during the simulation. In fact, only the data structure of the triangulated surface corresponding to the shell of the tetrahedral mesh is needed online.
- During the simulation the reaction forces at the nodes where the virtual instruments collide are also computed.
- The model is quasi-static, i.e., it computes the static equilibrium position at each iteration.

However, it relies on the following hypotheses:

- The mesh topology is not modified during the simulation. Thus, no simulation of cutting or suturing can be performed on this model.
- The interaction with neighboring tissues or with instruments is translated into modified boundary conditions (displacements or forces) only on surface nodes but not on the boundary conditions of internal nodes.

Therefore, the main limitation of this precomputed model comes from the first hypothesis which states that it is not suited for the simulation of tissue cutting.

5.2. *Overview of the algorithm*

One important feature of the model consists in making a distinction between surface and interior nodes. Thus, for the sake of clarity, we decompose the displacement and load vectors as well as the stiffness matrix according to surface and interior nodes with the s and i subscripts:

$$\begin{bmatrix} \mathbf{K}_{ss} & \mathbf{K}_{si} \\ \mathbf{K}_{is} & \mathbf{K}_{ii} \end{bmatrix} \begin{bmatrix} \mathbf{U}_s \\ \mathbf{U}_i \end{bmatrix} = \begin{bmatrix} \mathbf{R}_s \\ \mathbf{R}_i \end{bmatrix}.$$

It is important to note that only free vertices appear in this matrix as discussed in Section 4.10.

The solution of static equation can be obtained by multiplying the compliance matrix $[\mathbf{G}]$, corresponding to the inverse of the stiffness matrix $[\mathbf{K}]$, with the load vector. This compliance matrix can also be decomposed into surface and interior nodes,

$$\begin{bmatrix} \mathbf{U}_s \\ \mathbf{U}_i \end{bmatrix} = \begin{bmatrix} \mathbf{G}_{ss} & \mathbf{G}_{si} \\ \mathbf{G}_{is} & \mathbf{G}_{ii} \end{bmatrix} \begin{bmatrix} \mathbf{R}_s \\ \mathbf{R}_i \end{bmatrix}. \tag{5.1}$$

The load vector $\mathbf{R}_s$ that applies on free surface nodes can be decomposed into two parts. A first part $\mathbf{R}_s^0$, corresponds to loads that will not evolve during the simulation for instance gravity forces (see Section 4.6), constant pressure forces (see Section 4.7), applied nodal forces (see Section 4.9) or the presence of a non-zero imposed displacement vertex in its neighborhood (see Eq. (4.20)). The second part $\mathbf{R}_s^C$ corresponds to loads that are created by the contact of the soft tissue with surgical tools.

The principle of this soft tissue model is to compute the surface node positions $\mathbf{U}_s$ directly from the contact loads $\mathbf{R}_s^C$ by multiplying this vector with the compliance matrix $\mathbf{G}_{ss}$:

$$\begin{aligned} \mathbf{U}_s &= \mathbf{G}_{ss}\mathbf{R}_s^C + \mathbf{U}_s^0, \\ \mathbf{U}_s^0 &= \mathbf{G}_{ss}\mathbf{R}_s^0 + \mathbf{G}_{si}\mathbf{R}_i. \end{aligned} \tag{5.2}$$

Since the loads on interior nodes $\mathbf{R}_i$ do not evolve during the simulation, $\mathbf{U}_s^0$ is a displacement offset that is computed as the displacement of surface nodes when no contact loads are applied: $\mathbf{R}_s^C = \mathbf{0}$.

The goal of the precomputation stage is to compute the compliance matrix $[\mathbf{G}_{ss}]$.

5.3. *Precomputation stage*

5.3.1. *Description of the algorithm*

In the remainder, we write $[\mathbf{G}_{ss}^{ij}]$ the 3×3 submatrix of $\mathbf{G}_{ss}$ associated to vertex i and j. More precisely, a force $\mathbf{R}_s^j$ applied on vertex j entails an additional displacement of vertex i equal to $[\mathbf{G}_{ss}^{ij}]\mathbf{R}_s^j$.

The algorithm for computing the compliance matrix $\mathbf{G}_{ss}$ is described as Algorithm 1. It consists in solving $3 \times N_s$ times the linear system of equations $\mathbf{KU} = \mathbf{R}$, where N_s is the number of surface vertices. Note that the size of the stiffness matrix $\mathbf{K}$ is $N = N_s + N_i$ whereas the size of the compliance matrix $\mathbf{G}_{ss}$ is $3N_s \times 3N_s$.

```
1:  Set R_i = 0
2:  for all Surface Vertex i do
3:  for all j such that 0 ⩽ j ⩽ 2 do
4:     Set R_s = 0
5:     Set to 1.0 the jth component of the load R_s^i applied to vertex i
6:     Solve the static equilibrium equation KU = R
7:     for all Surface Vertex k do
8:        Store the computed displacement U_k of vertex k into the jth column
          of matrix [G_ss^ki]
9:     end for
10: end for
11: end for
```

ALGORITHM 1. Computation of the compliance matrix $\mathbf{G}_{ss}$.

The solution of equation $\mathbf{KU} = \mathbf{R}$ is performed using the steps described in Section 4.11 including node renumbering and matrix preconditioning. Since the rigidity matrix $\mathbf{K}$ is the same for all $3 \times N_s$ systems of equations, these two steps are performed only once, which significantly speeds-up the computation. Each time a linear system of equation is solved, the displacement of all surface nodes U_s corresponds to a column of matrix $\mathbf{G}_{ss}$. The storage of matrix $\mathbf{G}_{ss}$ requires only $(8 \times 9(N_s)^2)/2$ bytes (each element being stored as a double), since it is a symmetric matrix, as the inverse of a symmetric matrix.

Algorithm 1 can be slightly improved in the following way:

- Applying a unitary force successively along the X, Y and Z directions may cause a loss of accuracy in computing the compliance matrix, because the resulting displacement may be very large or very small depending on the size of the mesh. To obtain meaningful displacements, it is possible to apply a force f_{ref} and then divide the resulting displacement by f_{ref} to compute $\mathbf{G}_{ss}$. A good choice for f_{ref} is $\|[\mathbf{K}_{i,i}]\| * 0.1 * l$, where $[\mathbf{K}_{i,i}]$ is the block diagonal stiffness matrix of vertex i, and l is the estimated diameter of the object. This choice of force scale, produces displacements which are roughly equal to 10% of the diameter.
- It is sometimes necessary to obtain the displacement of some interior nodes during the simulation. This is the case, for instance, when vessels or tumors, located inside an organ, need to be displayed during the simulation. In this case, it is possible in the final loop of the algorithm (lines 7, 8 and 9 of Algorithm 1) to add these inside vertices to the list of surface vertices. Thus, it does not entail the solution of any additional system of equations, but only an additional storage requirement since the compliance matrix becomes a rectangular matrix of size $3N_s \times 3(N_s + N_i^\star)$ where $N_i^\star$ is the number of additional interior nodes.

This precomputation stage is quite computationally expensive and requires between a few minutes up to several hours depending on the number of the mesh vertices and the stiffness of the material. For instance, the liver model presented in Fig. 5.1 is composed of 1394 vertices, 8347 edges and 6342 tetrahedra. Its triangulated surface is composed of 1224 triangles and 614 vertices which is enough to produce a smooth visual ren-

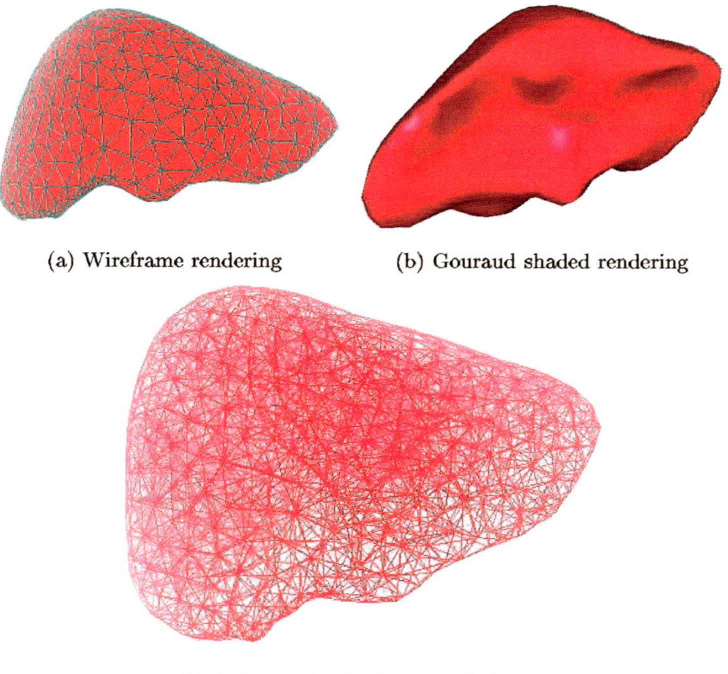

FIG. 5.1. Visualization of a liver model with 1394 vertices and 6342 tetrahedra.

dering. The Poisson ratio of the material is set to 0.45 while its Young modulus is $E = 1000$ kPa. In this case, the precomputation time required nearly 4 h on a Pentium PII 450 MHz, while the compliance matrix is stored in a file of size 13 Mb.

5.3.2. Other methods for computing the compliance matrix

At least two alternative methods have been proposed in the literature to compute the compliance matrix $\mathbf{G}_{ss}$. The first one, proposed by BRO-NIELSEN and COTIN [1996] is based on matrix condensation (MACMILLAN [1955]). More precisely, the compliance matrix $\mathbf{G}_{ss}$ can be directly obtained from the inversion of the stiffness matrix $\mathbf{K}_{ii}$ of interior nodes. From Eq. (5.1), we can derive the following equations:

$$\begin{aligned}
&\mathbf{K}_{ii}\mathbf{U}_i = \mathbf{R}_i - \mathbf{K}_{is}\mathbf{U}_s, \\
&\mathbf{K}_{ss}\mathbf{U}_s + \mathbf{K}_{si}\left(\mathbf{K}_{ii}^{-1}\mathbf{R}_i - \mathbf{K}_{ii}^{-1}\mathbf{K}_{is}\mathbf{U}_s\right) = \mathbf{R}_s, \\
&\left(\mathbf{K}_{ss} - \mathbf{K}_{si}\mathbf{K}_{ii}^{-1}\mathbf{K}_{is}\right)\mathbf{U}_s = \mathbf{R}_s - \mathbf{K}_{si}\mathbf{K}_{ii}^{-1}\mathbf{R}_i.
\end{aligned} \tag{5.3}$$

From Eq. (5.3), we can deduce the expression of the compliance matrix,

$$\mathbf{G}_{ss} = \left(\mathbf{K}_{ss}^{\star}\right)^{-1} = \left(\mathbf{K}_{ss} - \mathbf{K}_{si}\mathbf{K}_{ii}^{-1}\mathbf{K}_{is}\right)^{-1}. \tag{5.4}$$

Therefore, the computation of $\mathbf{G}_{ss}$ requires the inversion of two matrices: the first one of size $3N_i \times 3N_i$ and the second one of size $3N_s \times 3N_s$. This method has the disadvantage of requiring the additional storage of $9(N_i)^2$ numbers in double format, which in general is greater than the size of the compliance matrix: for large meshes, this method may become unpractical. Furthermore, this method is slightly more complex to implement whereas the method proposed in the previous section only requires to solve equation $\mathbf{KU} = \mathbf{R}$ with a sparse matrix $\mathbf{K}$. However, the condensation method is well-suited when the rigidity matrix is very ill-conditioned (Poisson ratio very close to 0.5) in which case the preconditioned conjugated gradient algorithm may fail.

The second algorithm for computing the compliance matrix $\mathbf{G}_{ss}$ is to use the Boundary Element Method (BEM) (CANAS and PARIS [1997]) instead of the Finite Element Method (FEM). The algorithm proposed by JAMES and PAI [1999] creates the stiffness matrix $\mathbf{K}^{\star}_{ss}$ directly from the triangulated surface of the object.

The differences between BEM and FEM are well-understood (HUNTER and PULLAN [1997]). The main advantage of BEM techniques is that they do not require a volumetric tetrahedral mesh but only its triangulated surface. While there exist several free software[6] for automatically creating tetrahedral meshes from triangulated surfaces (SIMAIL, OWEN [2000], JOE [1991]), having a control over the final number of vertices and the quality of tetrahedral elements is still an issue.

On the other hand, BEM techniques have several disadvantages over FEM. First, they make strong hypotheses about the nature of the elastic material: only homogeneous and isotropic linear elastic materials can be modeled. Second, the computation of the compliance matrix, and above all its diagonal elements, is difficult to implement and often numerically unstable because singular integrals must be evaluated over each triangle. The quality of the triangle geometry can influence the stability of this computation. Third, this method cannot compute the displacement of any interior point, which can be a limitation when the displacement of internal structures (vessels, tumors, ...) is needed. Finally, the BEM presented in JAMES and PAI [1999] uses centroid collocation to compute the stiffness matrix. Thus, this matrix allows to compute the displacements of the centroids of all triangles but not the displacements of the triangulation vertices. Therefore, the mesh being deformed is not the original triangulated mesh but its dual mesh which is called a *simplex mesh* (DELINGETTE [1999]). Mapping the displacements of triangle centroids into the displacements of vertices is not trivial since the duality between triangulation and simplex meshes is not a one-to-one mapping from the geometrical standpoint (DELINGETTE [1999]).

To conclude, the algorithm proposed by James et al. is more difficult to implement than our method and it is only suitable for simple material. However, when there is no software program for creating tetrahedral meshes from triangulations, this approach should be used.

5.4. *On-line computation*

5.4.1. *Data structure*

Before starting the simulation, the compliance matrix $\mathbf{G}_{ss}$, previously stored into a file as described in Section 5.3.1, is loaded into a specific data structure. Indeed, this data

[6] A list of available software can be found at the following two URLs: http://www-users.informatik.rwth-aachen.de/~roberts/meshgeneration.html and http://www.andrew.cmu.edu/user/sowen/softsurv.html.

structure only describes the triangulated surface shell of the volumetric tetrahedral mesh with a list of surface vertices and a list of surface triangles. Note that the number of surface vertices is usually greater than N_s because some surface vertices have an imposed displacement. For display purposes, the triangulated data structure may contain additional information such as 2D or 3D texture coordinates as well as parameters describing the rendered material. Finally, the data structure contains a list of imposed displacements and applied nodal forces as a storage of boundary conditions.

For each free vertex of index i, an array of 3×3 matrices $[\mathbf{G}_{ss}^{ji}]$, for all $j \in \{0, \ldots, N_s - 1\}$, is stored inside the vertex data structure. These N_s matrices $[\mathbf{G}_{ss}^{ji}]$ allow to compute the displacement of all surface vertex j, once a force is applied on vertex i.

The data structure optimizes the computation time of deformation but at the cost of being less efficient in terms of memory requirement. Indeed, the compliance matrix $\mathbf{G}_{ss}$ is a symmetric matrix, but it is stored as a non-symmetric matrix in this data structure. To optimize memory at a small additional computational cost, one could alternatively store the symmetric matrix as a double array of 3×3 compliance matrices $[\mathbf{G}_{ss}^{ji}]$ which is filled only if $i < j$.

5.4.2. *Algorithm description and collision processing*

The sketch of the algorithm is given in Algorithm 2 and includes two independent parts. The first part, between lines 1 and 8, consists in detecting and computing the contact between the soft tissue model and each virtual surgical instrument. In Fig. 5.2, we present an example of contact between a liver model and a tool. The collision detection algorithm (LOMBARDO, CANI and NEYRET [1999]) makes the assumption that the handle and the tool extremity can be approximated by a set of cylinders with rectangular section. Its efficiency depends on the availability of graphics cards since it relies on the

1: Reset the list of imposed displacement $l_{\text{displacement}}$ to the empty list
2: Reset the list of applied forces l_{force} to the empty list
3: Reset the position of free surface vertices to their rest position $+ \mathbf{U}_{ss}^{0}$
4: **for all** Surface Tools ST_i **do**
5: **if** collision between the soft tissue model and ST_i **then**
6:     Add imposed displacement to the list $l_{\text{displacement}}$
7: **end if**
8: **end for**
9: **if** $l_{\text{displacement}}$ is not empty **then**
10: Compute the list of applied forces l_{force} from $l_{\text{displacement}}$
11: **for all** Applied forces $\mathbf{F}_j^{\star}$ on vertex j in l_{force} **do**
12:     **for all** Free surface vertex k **do**
13:         Add to current position of vertex k, the displacement $[\mathbf{G}_{ss}^{kj}]\mathbf{F}_j^{\star}$
14:     **end for**
15: **end for**
16: **end if**

ALGORITHM 2. On-line computation of mesh deformation.

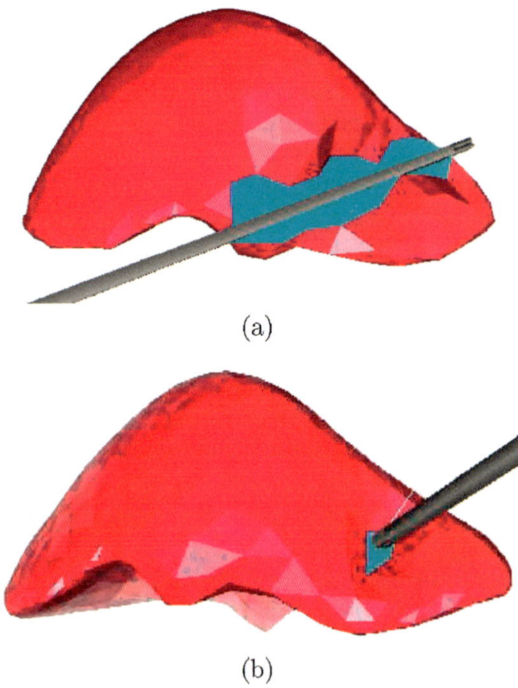

(a)

(b)

FIG. 5.2. Example of collision computation between the handle (a) and the extremity (b) of a surgical tool and a liver soft tissue model (PICINBONO, LOMBARDO, DELINGETTE and AYACHE [2002]). The position of triangles displayed in light gray have been displaced such that the tool is tangent to the liver surface.

OpenGL (WOO, NEIDER and DAVIS [1997]) library. Once a collision has been detected, the collided triangles must be moved such that the tissue model is no longer in contact with the surgical tool. This computation turns out to be quite complex since it not only depends on the tool position but also on its trajectory. The algorithm is described in PICINBONO, LOMBARDO, DELINGETTE and AYACHE [2002]. The outcome of this computation is a list $l_{\text{displacement}}$ of imposed displacements that should apply on each vertex of the collided triangles.

5.4.3. *Imposing displacements*

The second part of Algorithm 2, between lines 9 and 16 computes the position of all surface vertices, given the list of imposed displacements.

The first task corresponding to line 10 consists in computing the set of forces $\{\mathbf{F}_j^\star\}$ that should be applied to each vertex j of $l_{\text{displacement}}$ in order to bring the displacement of these vertices to $\mathbf{U}_j^b$.

To be more didactic, we first consider that only one vertex displacement $\mathbf{U}_j^b$ is imposed on a vertex of index j. Without any collision with a surgical tool, this vertex has a displacement $\mathbf{U}_j^0$ under the application of the *normal* boundary conditions (gravity forces, pressure forces, ... described in Section 4.9). Because the material is linear

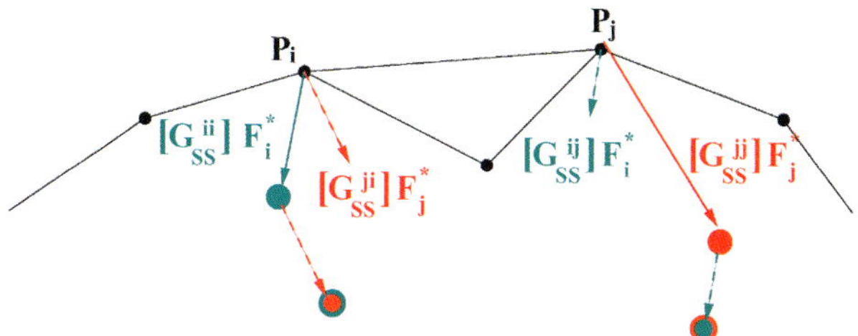

FIG. 5.3. Principle of superposition when applying two forces $\mathbf{F}_i^\star$ and $\mathbf{F}_j^\star$ to the two nodes i and j.

elastic, it follows the superposition principle: the displacements resulting from the application of two sets of nodal forces is the sum of the displacements resulting from the application of each set of forces. Thus, the force $\mathbf{F}_j^\star$ to be computed is the force that should be applied on vertex j in order to create a displacement of that vertex equal to $\mathbf{U}_j^b - \mathbf{U}_j^0$. Because the quantity $[\mathbf{G}_{ss}^{jj}]\mathbf{F}_j^\star$ gives the additional displacement of vertex j resulting from the application of force $\mathbf{F}_j^\star$, the force $\mathbf{F}_j^\star$ is given by

$$\mathbf{F}_j^\star = [\mathbf{G}_{ss}^{jj}]^{-1}(\mathbf{U}_j^b - \mathbf{U}_j^0).$$

When the displacements of two vertices i and j are imposed, the problem is slightly more complex. Indeed, the application of force $\mathbf{F}_i^\star$ on vertex i not only displaces vertex i of the amount $[\mathbf{G}_{ss}^{ii}]\mathbf{F}_i^\star$, but it also moves vertex j by the amount $[\mathbf{G}_{ss}^{ij}]\mathbf{F}_i^\star$ (see Fig. 5.3). Since $\mathbf{F}_j^\star$ also displaces vertex i of $[\mathbf{G}_{ss}^{ij}]\mathbf{F}_j^\star$, to compute the applied force, a 6×6 symmetric linear system of equations needs to be solved,

$$\begin{cases} [\mathbf{G}_{ss}^{ii}]\mathbf{F}_i^\star + [\mathbf{G}_{ss}^{ij}]\mathbf{F}_j^\star = \mathbf{U}_i^b - \mathbf{U}_i^0, \\ [\mathbf{G}_{ss}^{ji}]\mathbf{F}_i^\star + [\mathbf{G}_{ss}^{jj}]\mathbf{F}_j^\star = \mathbf{U}_j^b - \mathbf{U}_j^0. \end{cases}$$

Similarly, when the list of imposed displacements $l_{\text{displacement}}$ contains p elements, then a symmetric linear system of equations of size $3p \times 3p$ needs to be solved to find the set of nodal forces. If we use the set of indices i_j, $j \in [1, \ldots, p]$ to denote the set of vertices where a displacement $\mathbf{U}_{i_j}$ is imposed, then this linear system of equations can be written as

$$\begin{bmatrix} [\mathbf{G}_{ss}^{i_1,i_1}] & [\mathbf{G}_{ss}^{i_1,i_2}] & \cdots & [\mathbf{G}_{ss}^{i_1,i_p}] \\ [\mathbf{G}_{ss}^{i_2,i_1}] & [\mathbf{G}_{ss}^{i_2,i_2}] & \cdots & \vdots \\ \vdots & \vdots & \ddots & \vdots \\ [\mathbf{G}_{ss}^{i_p,i_1}] & \cdots & \cdots & [\mathbf{G}_{ss}^{i_p,i_p}] \end{bmatrix} \begin{bmatrix} \mathbf{F}_{i_1}^\star \\ \vdots \\ \vdots \\ \mathbf{F}_{i_p}^\star \end{bmatrix} = \begin{bmatrix} \mathbf{U}_{i_1}^b - \mathbf{U}_{i_1}^0 \\ \vdots \\ \vdots \\ \mathbf{U}_{i_p}^b - \mathbf{U}_{i_p}^0 \end{bmatrix}. \tag{5.5}$$

In Fig. 5.4, we show an example of a mesh where the same displacement is imposed on three vertices. In this particular case, the direction of computed forces departs strongly from the direction of the prescribed displacement.

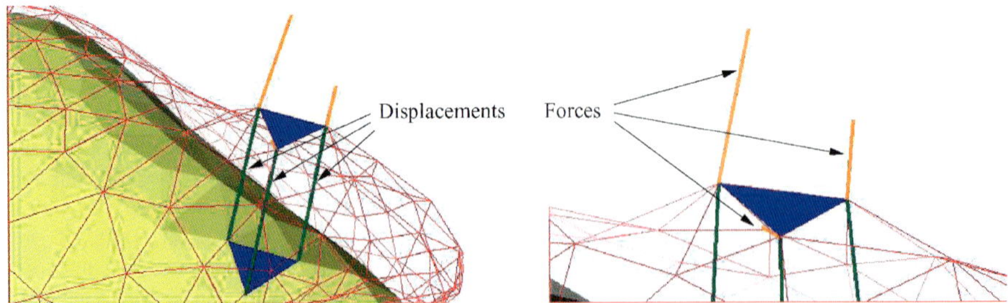

FIG. 5.4. (Right) The same displacement is imposed on the three vertices of a triangle; (left) from Eq. (5.5) we compute, the three forces that should be applied on these three vertices to move them of the given displacement.

5.4.4. Results

Once the set of nodal forces is computed, the additional displacement on all surface (and potentially internal) nodes are computed as described in lines 11 to 15 of Algorithm 2. The number of matrix–vector operations is $p \times N_s$ for p applied forces. In general, p, the number of vertices collided with the surgical tools, is small (from 3 to 20) when compared to N_s (see Fig. 5.5). This is why we chose to store the N_s array of compliance matrix $[\mathbf{G}_{ss}^{ji}]$ at vertex j, in order to optimize the inner loop (lines 12 to 14).

The computational efficiency of this quasi-static precomputed model on the liver mesh shown in Fig. 5.1 is presented in Table 5.1. These performances, measured on three different hardware platforms, correspond to the frequency update that can be achieved when running Algorithm 2 in a loop without any computation for visual and haptic rendering.

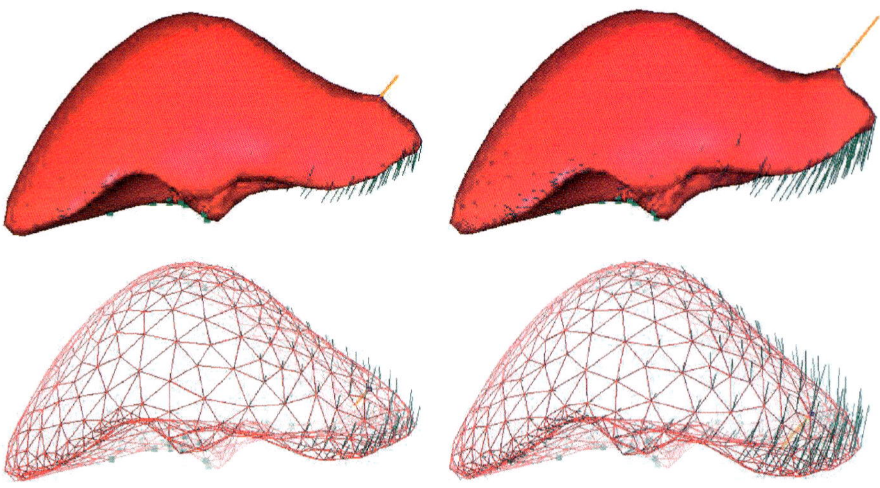

FIG. 5.5. Liver deformation based on a linear elastic pre-computed model (COTIN, DELINGETTE and AYACHE [1999b]). Solid lines indicate the imposed displacements.

TABLE 5.1
Computation efficiency of quasi-static precomputed linear elastic model for different boundary conditions: either when applying nodal forces or when imposing displacements

Simulation frequency (liver model with 614 surface nodes)		Pentium PIII 600 MHz
Force applied on 1 node		3772 Hz
Force applied on 5 nodes		754 Hz
Force applied on 10 nodes		377 Hz
Force applied on 20 nodes		188 Hz
Imposed displacements on	1 node	3759 Hz
	5 nodes	561 Hz
	10 nodes	185 Hz
	20 nodes	40 Hz

When applying one nodal force, corresponding to the execution of lines 12 to 14 in Algorithm 2, the computation time is nearly equal to 0.3 ms. The time required to compute the mesh deformation when applying p forces is strictly proportional to this value: $0.3 \times p$ ms.

When imposing p displacements, which is what occurs in practice in a surgical simulator, the additional computation is the solution of a $3p \times 3p$ linear symmetric system of equations. For $p = 1$, the overhead is very small and hardly perturbs the simulation frequency. However, for larger value of p, the overhead becomes dominant. For 20 vertices, for instance, solving the system of equations of size 60×60 is 3 times more costly than computing the $20 * 614 = 1280$ matrix–vector products and additions.

5.4.5. *Discussion*

As a whole, the proposed method is "very efficient", since it allows real-time visual rendering even for large meshes. When the material is soft enough and when the number of collided vertices remains small (typically less than 15), this model can also be compatible with real-time haptic rendering. In fact, it is one of the few algorithms which are suitable for the first software architecture described in Section 1.3.2 (see also Fig. 1.6(a)) consisting of one synchronous loop including visual and haptic rendering. Furthermore, our approach has one major advantage for haptic rendering computation: it already provides the nodal reaction forces through the algorithm described in Section 5.4.3. Indeed, the set of forces $\mathbf{F}^{\star}_{i_j}$ corresponds to the set of physical forces that have been applied on each node of index i_j in order to deform the soft tissue model: thus, $-\mathbf{F}^{\star}_{i_j}$ corresponds to the nodal reaction force. From this set of forces, one can easily compute the reaction force along the direction of the tool, as well as the torque at the extremity of the tool.

Using the terminology introduced in Section 1.3.2, we can also state that the quasi-static precomputed linear elastic model has a *very low relaxation time* (or equivalently that it has a high speed of convergence). Indeed, each time Algorithm 2 is run, the soft tissue is deformed to its static equilibrium position. Because this algorithm can be run at a high frequency, as seen in Table 5.1, this implies that the relaxation time is very

for all Free surface vertex k **do**
  if $k \notin l_{\text{displacement}}$ **then**
    Let $\mathbf{p}_k$ be the position of vertex k after Algorithm 2
    Let $\mathbf{p}_k^{\text{previous}}$ be the position of vertex k at the previous iteration.
    $\mathbf{p}_k \Leftarrow \gamma \mathbf{p}_k + (1-\gamma)\mathbf{p}_k^{\text{previous}}$
  end if
  $\mathbf{p}_k^{\text{previous}} \Leftarrow \mathbf{p}_k$
end for

ALGORITHM 3. Additional part of Algorithm 2 that adds a visco-elastic behavior controlled by delay parameter γ.

low. In fact, for some soft tissue, this time is too low and degrades the visual realism of the simulation. This is the case, for instance, when the operator grasps and displaces some soft tissue and suddenly ceases the grasping. Because the model has no longer any displacements imposed on its surface, it returns in one iteration to its rest position, while in reality, it takes several milliseconds.

To add some visco-elastic behavior, one can increase the relaxation time artificially by using a delay function. This approach is described in Algorithm 3. For vertices which are not colliding with a surgical tool, the final vertex position is a weighted sum between the position computed by Algorithm 2 and the vertex position at the previous iteration. The weight parameter $0 \leqslant \gamma \leqslant 1$ controls the damping of the material deformation: for $\gamma = 1$, the deformation is not damped (quasi-static motion) while for $\gamma = 0$, the motion is infinitely damped (no motion). Any intermediate value of γ modifies the relaxation time of the material. Note that this damping is not applied to vertices colliding with tools because the collision would otherwise appear visually unrealistic. Algorithm 3 assumes that the model has a damping matrix $\mathbf{C}$ which is proportional to the identity matrix: more sophisticated hypotheses (but often more computationally intensive) could be proposed.

6. Dynamic linear elastic model

In this section, we describe two different soft tissue models that are able to address with the limitation of the previous model: the simulation of tissue cutting. Using the terminology defined in Section 1.3.2, these two methods can be qualified as "Explicit Iterative Methods" sharing the advantage of requiring a small computation time for each iteration but with the drawback of having a low speed of convergence.

The main difference between these two models is that the first can model the visco-elastic behavior of the soft tissue properly whereas the second does not require the evaluation of any time step and is unconditionally stable.

Finally, we propose in Section 6.3 a *hybrid model* which combines any of the two previous models with the precomputed linear elastic model seen in Section 5.

6.1. Tensor–mass model

6.1.1. Introduction

The *tensor–mass model* is based on the dynamic law of motion described in Eq. (4.22):

$$\mathbf{M}\ddot{\mathbf{U}} + \mathbf{C}\dot{\mathbf{U}} + \mathbf{K}\mathbf{U} = \mathbf{R}.$$

This second order differential equation couples the motion of tissue under the influence of inertia $\mathbf{M}\ddot{\mathbf{U}}$, of visco-elasticity $\mathbf{C}\dot{\mathbf{U}}$, elasticity $\mathbf{KU}$ and external loads $\mathbf{R}$.

The most efficient way to solve the equation above is by far to use modal analysis (BATHE [1982]). By making simple assumptions about the damping matrix $\mathbf{C}$, it is possible to simplify the above PDE into a small set of ordinary differential equations with an appropriate change of basis. The proper basis is given by the eigenvectors associated to the largest eigenvalues of the generalized eigenproblem $\mathbf{K}\phi = \omega^2\mathbf{M}\phi$.

However, the eigenproblem must be solved each time the rigidity matrix is modified. Therefore, this approach is not suitable for simulating tissue cutting, since the computation cost to solve the eigenproblem is very high.

Instead, a classical method to solve Eq. (4.22), is to use integration methods: the time dimension is uniformly discretized with a time step Δt, and each term of that equation is supposed to be constant during each time interval. There is an important distinction between *implicit integration schemes* and *explicit integration schemes* depending whether the position of the model at time $t + \Delta t$ requires the solution or not of a global linear systems of equations (see also the discussion in Section 1.3.2).

Implicit schemes are *unconditionally stable* which allows the use of large time steps. In structural analysis, the Houbolt method (HOUBOLT [1950], BATHE [1982]) and the Newmark method (NEWMARK [1959], BATHE [1982]) are the most commonly used. However, these schemes require either to inverse a sparse matrix or to solve at each iteration a linear system of equations. Considering the time required to solve such a linear system (a few seconds for a small-size mesh), these implicit schemes cannot be used for real-time interaction.

Instead, we chose to use *explicit integration schemes* which have several nice properties (ease of implementation, low computational cost) compared to implicit schemes but with the drawback of being *conditionally stable*: the time step must be smaller than a critical time step $\Delta t_{\text{critical}}$. Therefore, smaller time step Δt must be used for explicit schemes which yields a larger relaxation time and a longer time for convergence.

6.1.2. Mass matrix

Regarding the mass matrix, a common choice consists in replacing the symmetric positive definite matrix $\mathbf{M}$ with a diagonal matrix, where each diagonal element is the sum of all row elements in the original matrix: this lumped mass matrix is detailed in Section 4.8.

In order to keep the time step Δt large enough during the simulation, we propose a further simplification of the mass matrix $\mathbf{M}$ by considering that the nodal mass is constant for all nodes, which makes $\mathbf{M}$ proportional to the identity matrix,

$$\mathbf{M} = m_0\mathbf{I}_3,$$

where m_0 is the average mass per node computed as the total mass of the tissue divided by the number of nodes in the initial mesh.

Indeed, the critical time step Δt of the iterative scheme is inversely proportional to the highest eigenvalue of the matrix $\mathbf{M}^{-1}\mathbf{K}$, while the speed of convergence is related to the ratio between the largest to the smallest eigenvalues of the same matrix, also called the *condition number* of that matrix.

From the equation of the nodal stiffness matrix $[\mathbf{K}_{i,i}]$, we can state that the nodal stiffness is proportional to the size (for instance, the largest foot height) of all the tetrahedra surrounding each node:

$$[\mathbf{K}_{i,i}] = \sum_{\mathcal{T} \in \mathcal{S}(i)} \frac{1}{36V(\mathcal{T})} \big((\lambda_{\mathcal{T}} + \mu_{\mathcal{T}})(\mathbf{m}_i \otimes \mathbf{m}_i) + \mu_{\mathcal{T}} A_i^2 \mathbf{I}_3 \big).$$

Thus, the largest eigenvalue of $\mathbf{K}$ is determined by the largest tetrahedra while the condition number is given by the size ratio between the largest and smallest tetrahedra. On the other hand, when performing mass lumping, as in BRO-NIELSEN [1998], the nodal mass of $\mathbf{M}^{-1}$ is inversely proportional to the volume of tetrahedra surrounding each node. Therefore, the power spectrum of $\mathbf{M}^{-1}\mathbf{K}$ largely differs from that of $\mathbf{K}$: the largest eigenvalue of $\mathbf{M}^{-1}\mathbf{K}$ now becomes related to the tetrahedron of smallest size while the condition number is related to the square ratio between the largest and smallest tetrahedra. These properties of $\mathbf{M}^{-1}\mathbf{K}$ have two consequences for the simulation of tissue cutting: both the speed of convergence and the time step Δt decrease as tetrahedra of small size are created.

By choosing a mass matrix proportional to the identity matrix, we keep the spectral properties of the rigidity matrix: the creation of small tetrahedra does not entail any decrease of the time step and limits the decrease of the speed of convergence. However, this choice is a gross approximation of physics since the total mass of the tissue increases as the number of elements increases. As claimed in Section 3.1.5, we prefer to satisfy real-time constraints of the simulation (by keeping a large value of Δt) at the expense of coarse approximations of the tissue dynamic behavior.

6.1.3. Numerical integration

Several explicit iterative schemes can be proposed from Eq. (4.22) depending on the choice of damping matrix and discretization of time derivatives. Below, we propose three explicit schemes that are of interest in the context of surgery simulation. In the remainder, we write ${}^t\mathbf{U}$ the displacement vector at time t.

Euler method. This method uses central finite differences to estimate acceleration but right finite difference to estimate speed. Furthermore, sophisticated damping matrix such as Rayleigh damping can be employed in this scheme:

$$\frac{m_0}{\Delta t^2}\big({}^{t-\Delta t}\mathbf{U} - 2{}^t\mathbf{U} + {}^{t+\Delta t}\mathbf{U}\big) + \frac{1}{\Delta t}(\gamma_1 m_0 \mathbf{I}_3 + \gamma_2 \mathbf{K})\big({}^t\mathbf{U} - {}^{t-\Delta t}\mathbf{U}\big) + \mathbf{K}{}^t\mathbf{U} = {}^t\mathbf{R}.$$

The displacement at time $t + \Delta t$ can be computed through the recurrent equation:

$${}^{t+\Delta t}\mathbf{U} = {}^t\mathbf{U} + (1 - \Delta t \gamma_1)\big({}^t\mathbf{U} - {}^{t-\Delta t}\mathbf{U}\big)$$

$$-\mathbf{K}\left(\frac{\Delta t^2}{m_0}{}^t\mathbf{U}+\frac{\gamma_2\Delta t}{m_0}\left({}^t\mathbf{U}-{}^{t-\Delta t}\mathbf{U}\right)\right)+\frac{\Delta t^2}{m_0}{}^t\mathbf{R}.$$

Euler method with central finite difference. In this case, central finite differences are used to estimate both acceleration and speed, while constant damping is used $\gamma_2 = 0$:

$$\frac{m_0}{\Delta t^2}\left({}^{t-\Delta t}\mathbf{U}-2{}^t\mathbf{U}+{}^{t+\Delta t}\mathbf{U}\right)+\frac{\gamma_1 m_0}{2\Delta t}\left({}^{t+\Delta t}\mathbf{U}-{}^{t-\Delta t}\mathbf{U}\right)+\mathbf{K}{}^t\mathbf{U}={}^t\mathbf{R},$$

which leads to the following update equation:

$${}^{t+\Delta t}\mathbf{U}={}^t\mathbf{U}+\frac{2-\gamma_1\Delta t}{2+\gamma_1\Delta t}\left({}^t\mathbf{U}-{}^{t-\Delta t}\mathbf{U}\right)-\frac{2\Delta t^2}{m_0(2+\gamma_1\Delta t)}\left(\mathbf{K}{}^t\mathbf{U}-{}^t\mathbf{R}\right). \tag{6.1}$$

Runge–Kutta method of order 4. The Runge–Kutta method (PRESS, FLANNERY, TEUKOLSKY and VETTERLING [1991]) is an integration method of fourth order of accuracy, but which requires four evaluations of the Euler recurrent equation. To describe this method, it is necessary to write the original equation as a first order differential equation,

$$\frac{\mathrm{d}}{\mathrm{d}t}\begin{bmatrix}\dot{\mathbf{U}}\\ \mathbf{U}\end{bmatrix}=\begin{bmatrix}\ddot{\mathbf{U}}\\ \dot{\mathbf{U}}\end{bmatrix}=\begin{bmatrix}-\frac{\mathbf{C}}{m_0} & -\frac{\mathbf{K}}{m_0}\\ 1 & 0\end{bmatrix}\begin{bmatrix}\dot{\mathbf{U}}\\ \mathbf{U}\end{bmatrix}+\begin{bmatrix}\frac{\mathbf{R}}{m_0}\\ 0\end{bmatrix}.$$

Now, the state of a soft tissue model at time t is described by two vectors: displacement vector ${}^t\mathbf{U}$ and the velocity vector ${}^t\dot{\mathbf{U}}$. Applying the simple Euler method on this system gives the following relation:

$$\begin{bmatrix}{}^{t+\Delta t}\dot{\mathbf{U}}\\ {}^{t+\Delta t}\mathbf{U}\end{bmatrix}-\begin{bmatrix}{}^t\dot{\mathbf{U}}\\ {}^t\mathbf{U}\end{bmatrix}=\Delta t\begin{bmatrix}\frac{1}{m_0}\left(-\mathbf{C}{}^t\dot{\mathbf{U}}-\mathbf{K}{}^t\mathbf{U}+{}^t\mathbf{R}\right)\\ {}^t\dot{\mathbf{U}}\end{bmatrix}=\begin{bmatrix}\delta v\left({}^t\mathbf{U},{}^t\dot{\mathbf{U}}\right)\\ \delta u\left({}^t\mathbf{U},{}^t\dot{\mathbf{U}}\right)\end{bmatrix}.$$

The fourth order Runge–Kutta method requires to compute the following eight incremental displacement and velocity vectors:

$$\begin{aligned}
&\delta v_1=\delta v\left({}^t\mathbf{U},{}^t\dot{\mathbf{U}}\right), && \delta u_1=\delta u\left({}^t\mathbf{U},{}^t\dot{\mathbf{U}}\right),\\
&\delta v_2=\delta v\left({}^t\mathbf{U}+\frac{\delta u_1}{2},{}^t\dot{\mathbf{U}}+\frac{\delta v_1}{2}\right), && \delta u_2=\delta u\left({}^t\mathbf{U}+\frac{\delta u_1}{2},{}^t\dot{\mathbf{U}}+\frac{\delta v_1}{2}\right),\\
&\delta v_3=\delta v\left({}^t\mathbf{U}+\frac{\delta u_2}{2},{}^t\dot{\mathbf{U}}+\frac{\delta v_2}{2}\right), && \delta u_3=\delta u\left({}^t\mathbf{U}+\frac{\delta u_2}{2},{}^t\dot{\mathbf{U}}+\frac{\delta v_2}{2}\right),\\
&\delta v_4=\delta v\left({}^t\mathbf{U}+\frac{\delta u_3}{2},{}^t\dot{\mathbf{U}}+\frac{\delta v_3}{2}\right), && \delta u_4=\delta u\left({}^t\mathbf{U}+\frac{\delta u_3}{2},{}^t\dot{\mathbf{U}}+\frac{\delta v_3}{2}\right).
\end{aligned}$$

Finally, the velocity and displacement for the next time step are given by the following equation:

$$\begin{bmatrix}{}^{t+\Delta t}\dot{\mathbf{U}}\\ {}^{t+\Delta t}\mathbf{U}\end{bmatrix}=\begin{bmatrix}{}^t\dot{\mathbf{U}}\\ {}^t\mathbf{U}\end{bmatrix}+\frac{1}{6}\begin{bmatrix}\delta v_1\\ \delta u_1\end{bmatrix}+\frac{1}{3}\begin{bmatrix}\delta v_2\\ \delta u_2\end{bmatrix}+\frac{1}{3}\begin{bmatrix}\delta v_3\\ \delta u_3\end{bmatrix}+\frac{1}{6}\begin{bmatrix}\delta v_4\\ \delta u_4\end{bmatrix}.$$

TABLE 6.1
Comparison between three explicit integration methods for soft tissue modeling

	Euler method	Euler central finite differences	Runge–Kutta method
Computation time	low	low	high
Damping	Rayleigh	basic	basic
Time step	small	medium	high

Comparison between the three methods. We summarized in Table 6.1 the properties of the three methods described above. Three qualitative criteria were proposed to outline the advantages and drawbacks of each method. In terms of computation time required to update the position of a model, the first two Euler methods are equivalent while the Runge–Kutta method is at least four times slower. As far as damping is concerned, only the first Euler method allows to use Rayleigh damping while the two other methods can only use diagonal damping matrices. Having a non-diagonal damping matrix helps in keeping a continuous field of velocity throughout the model which improves the visual realism of the simulation. Finally, the Runge–Kutta method is more stable than the Euler method and our experience showed that a tenfold increase of the time step can be observed in the former case. The Euler method with central finite differences allows larger time steps than the Euler method because the velocity computation leaps over position computation by one time step.

6.1.4. *Data structure*

With explicit schemes, the update of the mesh position can be performed locally, at the vertex level, without creating any global matrix. Indeed, for each free vertex of index i, we can take advantage of the sparse nature of the rigidity matrix $\mathbf{K}$, in order to compute the matrix–vector product $\mathbf{KU}$. More precisely, from Eq. (4.13), it is clear that the off-diagonal stiffness matrices $[K_{i,j}]$ are non-null matrices only when there is an edge connecting vertices i and j in the tetrahedral mesh. Therefore, only the set $\mathcal{N}(i)$ of vertices connected to vertex i by an edge is involved when computing the elastic force $\mathbf{F}_i$ applied on vertex i. For instance, the update Eq. (6.1) can be computed for a vertex i as

$$
\begin{aligned}
{}^{t+\Delta t}\mathbf{u}_i &= {}^{t}\mathbf{u}_i + \frac{2-\gamma_1\Delta t}{2+\gamma_1\Delta t}\left({}^{t}\mathbf{u}_i - {}^{t-\Delta t}\mathbf{u}_i\right) \\
&\quad - \frac{2\Delta t^2}{m_0(2+\gamma_1\Delta t)}\left(\sum_{j\in\mathcal{N}(i)}[\mathbf{K}_{i,j}]^t\mathbf{u}_j + [\mathbf{K}_{i,i}]^t\mathbf{u}_i - {}^{t}\mathbf{R}_i\right).
\end{aligned}
$$

The data structure that is suitable for performing this computation follows the data structure required for storing a tetrahedral mesh. The basic structure consists in a double-linked list of vertices, edges and tetrahedra. For each vertex, we store its current position ${}^{t}\mathbf{q}_i$, its rest position $\mathbf{p}_i$ and the symmetric tensor $[\mathbf{K}_{i,i}]$. For each edge, we store its two adjacent vertices (vertex i and vertex j) as well as the tensor $[\mathbf{K}_{i,j}]$, as sketched in Fig. 6.1. We therefore take advantage of the symmetric nature of the stiffness matrix by storing the off-diagonal rigidity matrix only once.

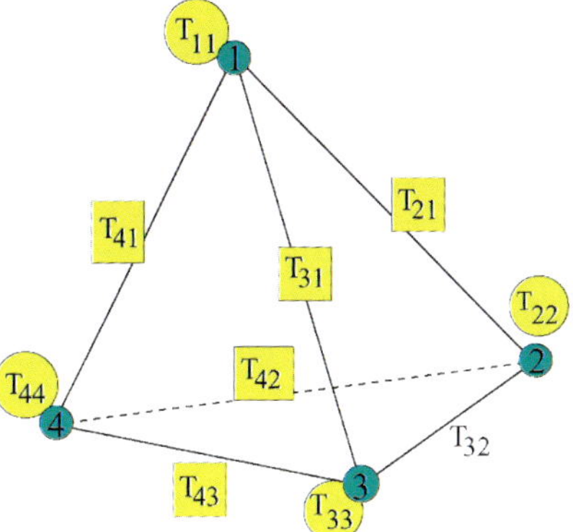

FIG. 6.1. Representation of the data structure of a tensor–mass model. The 3×3 rigidity matrices are stored at each edge and each vertex. The symmetry of the rigidity matrix enables to store only one tensor per edge.

Finally, for each tetrahedron, we store its four vertices and its six edges as well as the Lamé coefficients λ_i, μ_i, the area vectors $\mathbf{m}_i$ and if required the direction of anisotropy $\mathbf{a}_0$.

6.1.5. Cutting and refinement algorithms

One of the basic tasks in surgery simulation consists in cutting and tearing soft tissue. With the dynamic linear elastic model, these tasks can be achieved efficiently.

To perform an hepatectomy (partial resection of the liver), the use of scalpel instruments is not appropriate because of the important vascularization of the liver. Instead, surgeons usually proceed with a set of pliers that smash hepatic cells or with a cavitron device that destroys the hepatic parenchyma with ultrasound energy: in both cases, the resection is performed by removing soft tissue. It is therefore important to simulate the removal of bits of soft tissue located at the vicinity of a surgical tool. To perform this simulation, two basic meshing techniques must be implemented: removal of tetrahedra and local refinement.

At first sight, removing a single tetrahedron from a tetrahedral mesh is straightforward. However, in order to obtain a visually realistic simulation, one should avoid to produce isolated or self-intersecting tetrahedra or even tetrahedra connected through a single vertex. A proper way to keep "visually appealing" meshes is to constrain the mesh to be a *manifold* mesh in addition to being a *conformal* mesh. Indeed, in a manifold mesh, the shell of a vertex located on the mesh surface is homeomorphic a half-sphere (the shell is a sphere for interior vertices) which allows to define unambiguously a surface normal at that vertex. However, by adding this topological constraint, even removing a single tetrahedron is not straightforward as discussed in FOREST, DELINGETTE and AYACHE [2002b]. The detailed description of the topological issues relevant to the operation of tetrahedron removal falls outside the scope of this chapter; instead we present briefly the algorithms related to the computation of soft tissue deformation.

Once a collision between a surgical tool and a set of tetrahedra has been detected, each tetrahedron of the set is removed one after the other. After updating the topological

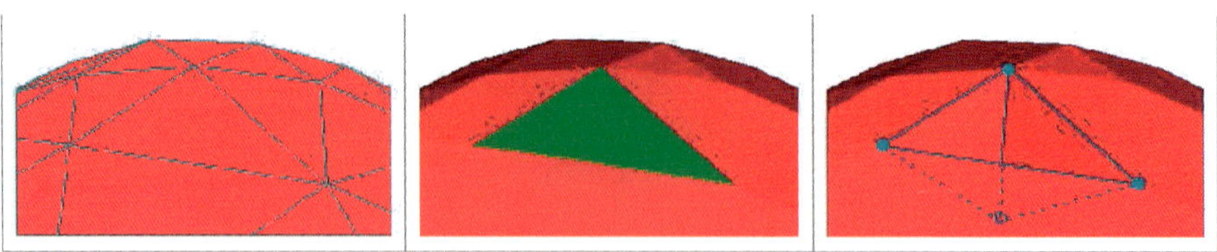

FIG. 6.2. To remove the tetrahedron whose external triangle has been selected (dark gray), it is necessary to update the local rigidity matrices stored at the vertices and edges of that tetrahedron.

structure of the mesh, the local vertex and edge stiffness matrices must also be updated (see Fig. 6.2). When removing tetrahedron $\mathcal{T}$, its 6 edge tensors $[\mathbf{B}_{i,j}^{\mathcal{T}}]$ and 4 vertex tensors $[\mathbf{B}_{i,i}^{\mathcal{T}}]$ are computed based on Eqs. (4.8) and (4.9) and are subtracted from the current edge and vertex local rigidity matrices:

$$[\mathbf{K}_{i,i}] = [\mathbf{K}_{i,i}] - \left[\mathbf{B}_{ii}^{\mathcal{T}}\right], \qquad [\mathbf{K}_{i,j}] = [\mathbf{K}_{i,j}] - \left[\mathbf{B}_{ij}^{\mathcal{T}}\right].$$

These ten local operations are performed efficiently because of the specific data structure associated with a tetrahedron.

The second meshing technique, local refinement, can be used in two cases. First, it can be used offline (before the simulation), to increase the mesh resolution at places of high curvature or near structures of interest (tumors, gall blader, ...). Second, it is often necessary to refine the mesh locally during the removal of soft tissue when the tetrahedra to be removed are too large. In the former case, sophisticated meshing techniques can be employed while in the latter case, real-time constraints allow the application of only basic refinement algorithms. An example of such a basic algorithm consists in adding a vertex at the middle of an edge and then splitting all tetrahedra adjacent to that edge into two tetrahedra (see Fig. 6.3). In this case, the edge and vertex tensors of all tetrahedra adjacent to that edge are first removed and the contributions from all newly created tetrahedra are then added. A more sophisticated refinement algorithm can be found in FOREST, DELINGETTE and AYACHE [2002b].

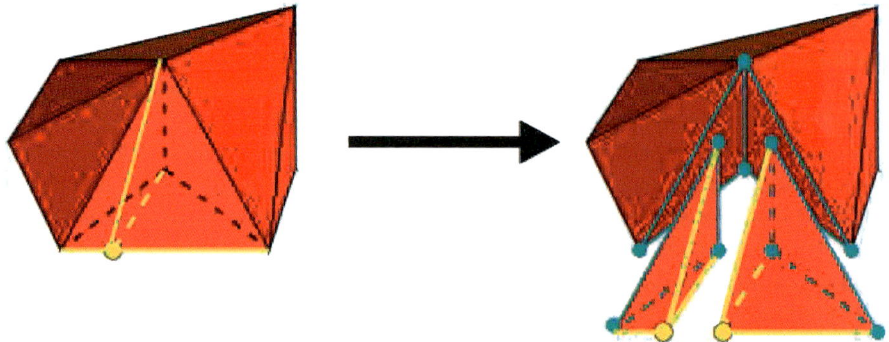

FIG. 6.3. Local refinement of a tetrahedral mesh. An edge is split into two edges by inserting a vertex. The rigidity matrices must be updated for vertices and edges that already existed (drawn in dark grey) while these matrices must be computed for newly created vertices and edges (drawn in light grey).

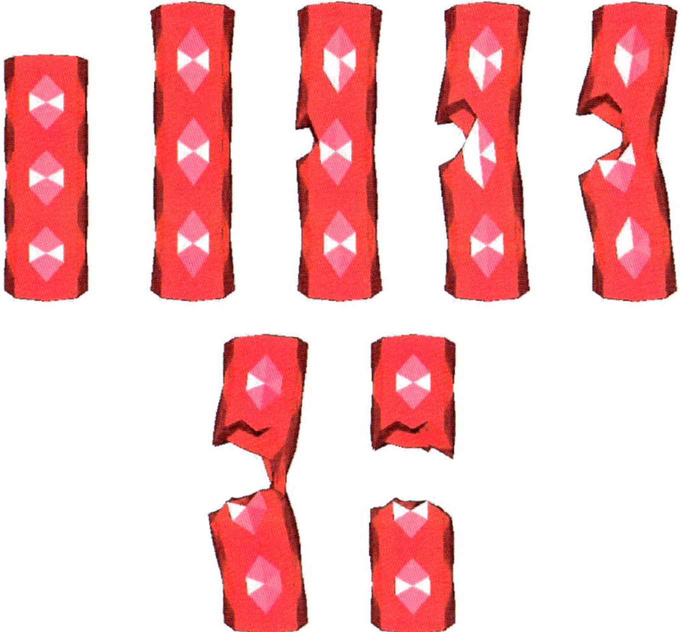

FIG. 6.4. Deformation of a cylinder subject to gravity forces: some tetrahedra are progressively being removed at its center leading to a separation into independent solids.

The proper adjustment of stiffness matrices during the removal of soft tissue reinforces the visual realism of the simulation significantly: this is especially the case when the tissue is cut while being stretched. For instance, in Fig. 6.4, we show the deformation of a cylinder being cut: the cylinder is fixed at its upper part and is under the influence of gravity forces along its main axis.

6.1.6. *Algorithm description*

Before describing the deformation algorithm for a tensor–mass model, we shortly describe the initialization stages in Algorithm 4. Once the vertex and edge stiffness matrices have been assembled, it is necessary to estimate a time step Δt that allow the stability of the iterative schemes described in Section 6.1.3. Finding the critical time step (i.e., the highest possible time step) is actually a difficult task because of the lack of a closed-form expression. However, a practical approach is to estimate the critical time step as a product of an unknown constant with the time step given by the Courant–Friedrich–Levy condition (PRESS, FLANNERY, TEUKOLSKY and VETTERLING [1992]):

$$(\Delta t)_{\mathrm{Courant}} = l_{\max}\sqrt{\frac{\rho}{\lambda + 2\mu}}.$$

Algorithm 5 presents the different loops required to update a tensor–mass model. Unlike the precomputed quasi-static model, it is not necessary to maintain an explicit list of vertices that are displaced by the collision with a surgical tool: it is sufficient

for all Tetrahedron $\mathcal{T}$ **do**
Compute the 4 area vectors $\mathbf{m}_i$
for all Vertex i of $\mathcal{T}$ **do**
  Compute the local rigidity matrix $[\mathbf{B}_{ii}^{\mathcal{T}}]$
  $[\mathbf{K}_{i,i}] \Leftarrow [\mathbf{K}_{i,i}] + [\mathbf{B}_{ii}^{\mathcal{T}}]$
end for
for all Edge between vertices i and j of $\mathcal{T}$ **do**
  Compute the local rigidity matrix $[\mathbf{B}_{ij}^{\mathcal{T}}]$
  $[\mathbf{K}_{i,j}] \Leftarrow [\mathbf{K}_{i,j}] + [\mathbf{B}_{ij}^{\mathcal{T}}]$
end for
end for
Estimate time step Δt.

ALGORITHM 4. Matrix assembly for the tensor–mass model performed before any simulation.

1: **for all** Surface tools ST_i **do**
2: **if** collision between the soft tissue model and ST_i **then**
3:   **if** ST_i represent a cavitron device **then**
4:     Eventually refine locally the mesh near the collision
5:     Remove tetrahedra located near the extremity of ST_i
6:   **end if**
7:   Impose displacements on vertices near the contact zone and raise a flag on these vertices
8: **end if**
9: **end for**
10: **for all** edge e connecting vertex i and j **do**
11: add elastic force $[\mathbf{K}_{i,j}]^t\mathbf{u}_i$ to vertex i
12: elastic force $[\mathbf{K}_{i,j}]^{\mathrm{T}\,t}\mathbf{u}_j$ to vertex j
13: **end for**
14: **for all** vertex i **do**
15: **if** vertex i is free (flag not raised) **then**
16:   compute elastic force $[\mathbf{K}_{i,i}]^t\mathbf{u}_i$
17:   update vertex position ${}^t\mathbf{p}_i$ based on one of the three iterative schemes described in Section 6.1.3
18: **else**
19:   reset flag
20: **end if**
21: **end for**

ALGORITHM 5. On-line computation of tensor–mass model.

(see line 7) to raise a flag stating that these vertices are not free vertices. A second important feature of this algorithm is the existence of a loop on the mesh edges in order to compute the matrix–vector products $\sum_{j\in\mathcal{N}(i)}[\mathbf{K}_{i,j}]^t\mathbf{u}_j$. This approach is more efficient than scanning iteratively the neighbors $\mathcal{N}(i)$ for each vertex i. When using the fourth order Runge–Kutta algorithm, the algorithm from lines 10 to 21 must be modified since it is then necessary to scan four times the edges and vertices of the mesh. For the Euler method, only lines 11 and 12 must be modified in order to compute

$$\mathbf{K}\left(\frac{\Delta t^2}{m_0}{}^t\mathbf{U} + \frac{\gamma_2 \Delta t}{m_0}\left({}^t\mathbf{U} - {}^{t-\Delta t}\mathbf{U}\right)\right)$$

instead of $\mathbf{K}^t\mathbf{U}$.

6.1.7. Comparison between spring–mass and tensor–mass models

We have used the word "tensor–mass model" to designate a finite element model based on Newtonian dynamics and discretized with an explicit scheme. This word has been chosen in order to stress the similarity between a "tensor–mass model" and a "spring–mass model". In particular, it is the purpose of this section to oppose to the widely spread belief stating that "finite element models are slower and more complex to implement than spring–mass models".

A spring–mass model (BARAFF and WITKIN [1998]) consists of a set of masses and a set of springs connecting these masses. The force applied to a point $\mathbf{p}_i$ in a spring–mass system, is given by the relation

$$\mathbf{F}_i = \sum_{j \in \mathcal{N}(i)} k_{ij}\left(\|\mathbf{p}_i\mathbf{p}_j\| - l_{ij}^0\right)\frac{\mathbf{p}_i\mathbf{p}_j}{\|\mathbf{p}_i\mathbf{p}_j\|}, \tag{6.2}$$

where k_{ij} is the stiffness coefficient between vertices i and j, l_{ij}^0 is the length at rest.

Similarly, on a tensor–mass model, the elastic force applied on vertex i is given by

$$\mathbf{F}_i = [\mathbf{K}_{i,i}]\mathbf{u}_i + \sum_{j \in \mathcal{N}(i)} [\mathbf{K}_{i,j}]\mathbf{u}_j. \tag{6.3}$$

By comparing Eqs. (6.2) and (6.3), it is clear that both dynamic models have the same computational complexity which is linear in the number of edges. In practice, we have observed a slight computational advantage for the tensor–mass model, mostly because it does not include any square root evaluation.

However, both approaches differ substantially in terms of biomechanical modeling. Spring–mass systems constitute a discrete representation of an object and their behavior strongly depends on the topology of the spring network. Adding or removing a spring may change the elastic behavior of the whole system drastically. Conversely, a finite element model is a continuous representation of the object and its behavior is independent of the mesh topology (it mostly depends on the mesh resolution). This is an advantage when mesh cutting is performed since it produces continuous and natural deformations.

Because all biomechanical data related to biological soft tissue are formulated as parameters found in continuum mechanics (such as Young's modulus or Poisson coefficients), it is *a priori* difficult to model realistic soft tissue deformations with a spring–mass system. However, several authors (LOUCHET, PROVOT and CROCHEMORE [1995], DEUSSEN, KOBBELT and TUCKE [1995]) have developed genetic or simulated

TABLE 6.2
Comparison between the three soft tissue models: pre-computed quasi-static, tensor–mass and spring–mass models

	Pre-computed	Tensor–mass	Spring–mass
Computational efficiency	+++	+	+
Biomechanical realism	+	+	−
Cutting simulation	−	++	+
Large displacements	−	−	+

annealing algorithms to identify spring parameters (stiffness and damping) from a set of known deformations of an object.

Finally, as previously mentioned, the tensor–mass model is only valid for small displacements. This model is invariant under the application of a global translation, but if a global rotation is applied to the rest shape $\mathcal{M}_{\text{rest}}$, then the forces applied to all vertices will not be null. On the opposite, a spring–mass model under the same displacement would not deform, since the length of the springs are preserved under a rigid transformation. The difference between these three soft tissue models is summarized in Table 6.2.

6.2. *Relaxation-based elastic models*

6.2.1. *Introduction*

In this section, we introduce an alternative algorithm to the tensor–mass model. This algorithm is based on Gauss–Seidel relaxation and has the following properties:

- Its iterative scheme is unconditionally stable. It does not require the estimation of any critical time step.
- The relaxation algorithm is fairly efficient (small computation time required for one iteration) but it is slightly less efficient than a tensor–mass model.
- The algorithm is based on static equilibrium equations whereas tensor–mass models are based on the dynamic law of motion.
- The position of each vertex is updated asynchronously, one vertex after the other.

However, when compared to tensor–mass models, relaxation-based elastic models have two drawbacks. First, their implementation requires the following property for the mesh data structure: each vertex should be able to access efficiently its adjacent edges. This topological "vertex–edge" relationship can be stored in two ways inside a data structure. In a first approach, a list of edges can be stored explicitly at each vertex. After removing or adding tetrahedra, the edge list must be updated for all vertices belonging to these tetrahedra. To achieve this update, each edge must have a list of adjacent tetrahedra which should also be explicitly updated upon the removal or addition of tetrahedra.

In a second approach, the list of edges adjacent to a vertex is recovered through the knowledge of a single tetrahedron adjacent to this vertex. This approach is only applicable if we constrain the tetrahedral mesh to be a manifold mesh (see FOREST, DELINGETTE and AYACHE [2002b] for more details). Indeed, in such case, the neighborhood of a vertex is homeomorphic to a topological sphere or half-sphere. By marching around a vertex from a given tetrahedron, it is possible to obtain all tetrahedra adjacent to a given vertex and consequently the list of all adjacent edges. In this case as in the former case, we do store a list of adjacent edges for each vertex in order to avoid duplicating the search algorithm. However, when a tetrahedron is removed or added, this topological list is reseted and the pointer to the adjacent tetrahedron is eventually updated.

The second drawback of relaxation algorithms is that they require in average 3 times more storage than the tensor–mass model. Indeed, in addition to the symmetric stiffness matrix, a non-symmetric stiffness matrix must be stored.

6.2.2. *Overview of the algorithm*

Following the notations of Eq. (6.3) the static problem $\mathbf{KU} = \mathbf{R}$ can be written at the level of each vertex i as

$$[\mathbf{K}_{i,i}]\mathbf{u}_i + \sum_{j \in \mathcal{N}(i)} [\mathbf{K}_{i,j}]\mathbf{u}_j = \mathbf{R}_i. \tag{6.4}$$

For relaxation algorithms, the displacement of a vertex $\mathbf{u}_i$ is updated independently from other vertices. Therefore, the notation ${}^{t+\Delta t}\mathbf{u}_i$ to describe the position of vertex i at the next time step cannot be used, since formally there is no temporal evolution (and no temporal variable t) in relaxation algorithms. Thus, we note ${}^{+}\mathbf{u}_i$ the next position of vertex i and $\mathbf{u}_i$ its current position.

The principle of relaxation algorithms is quite straightforward: each vertex is moved in order to locally solve Eq. (6.4). Thus, the displacement ${}^{+}\mathbf{u}_i$ is given by

$${}^{+}\mathbf{u}_i = -\sum_{j \in \mathcal{N}(i)} [\mathbf{K}_{i,i}]^{-1}[\mathbf{K}_{i,j}]\mathbf{u}_j + [\mathbf{K}_{i,i}]^{-1}\mathbf{R}_i. \tag{6.5}$$

This is equivalent to minimizing the total mechanical energy by successively optimizing each variable $\mathbf{u}_i$. It is therefore similar to the Iterative Conditional Mode (ICM) algorithm (BESAG [1986]) used in statistical analysis.

If all displacements $\{\mathbf{u}_i\}$ are successively updated according to Eq. (6.4), then this method is equivalent to the Gauss–Seidel relaxation method (SAAD [1996]). More precisely, we can decompose the stiffness matrix $\mathbf{K}$ as the sum of three terms: $\mathbf{K}_D$ a 3×3 block diagonal matrix, $\mathbf{K}_C$ the lower triangle matrix of $\mathbf{K}$ and $\mathbf{K}_C^{\mathrm{T}}$ the upper triangle matrix of $\mathbf{K}$:

$$\mathbf{K} = \underbrace{\begin{bmatrix} [\mathbf{K}_{1,1}] & \mathbf{0} & \cdots & \mathbf{0} \\ \mathbf{0} & [\mathbf{K}_{2,2}] & \ddots & \mathbf{0} \\ \vdots & \ddots & \ddots & \vdots \\ \mathbf{0} & \mathbf{0} & \cdots & [\mathbf{K}_{N,N}] \end{bmatrix}}_{\mathbf{K}_D} + \underbrace{\begin{bmatrix} \mathbf{0} & \mathbf{0} & \cdots & \mathbf{0} \\ [\mathbf{K}_{2,1}] & \mathbf{0} & \ddots & \mathbf{0} \\ \vdots & \ddots & \ddots & \vdots \\ [\mathbf{K}_{N,1}] & [\mathbf{K}_{N,2}] & \cdots & \mathbf{0} \end{bmatrix}}_{\mathbf{K}_C} + \mathbf{K}_C^{\mathrm{T}}.$$

With this notation, the Gauss–Seidel relaxation consists in the application of an iterative equation

$${}^{k+1}\mathbf{U} = (\mathbf{K}_D + \mathbf{K}_C)^{-1}\left(-\mathbf{K}_C^{\mathrm{T}}\, {}^{k}\mathbf{U} + \mathbf{R}\right), \tag{6.6}$$

where ${}^{k}\mathbf{U}$ is the displacement vector at iteration k.

To speed-up convergence, we use over-relaxation (known as the Simultaneous Over-Relaxation algorithm (SAAD [1996])) that consists in anticipating future correction with an overrelaxation parameter ω,

$$ {}^{k+1}\mathbf{U} = (\mathbf{K}_D + \omega \mathbf{K}_C)^{-1}\big(-\omega \mathbf{K}_C^{\mathrm{T}}\,{}^{k}\mathbf{U} + (1-\omega)\mathbf{K}_D\,{}^{k}\mathbf{U} + \omega \mathbf{R}\big). \tag{6.7} $$

This equation translates at the vertex level with the recursion

$$ {}^{+}\mathbf{u}_i = (1-\omega)\mathbf{u}_i - \omega \sum_{j\in\mathcal{N}(i)} [\mathbf{K}_{i,i}]^{-1}[\mathbf{K}_{i,j}]\mathbf{u}_j + \omega[\mathbf{K}_{i,i}]^{-1}\mathbf{R}_i. \tag{6.8} $$

If $\omega = 1$, then the SOR algorithm is equivalent to the Gauss–Seidel relaxation. Convergence is guaranteed for values of ω comprised between 1 and 2, while fastest convergence is obtained for a critical value

$$ \omega_{\mathrm{optimal}} = \frac{2}{1+\sqrt{1-\rho_{GS}}}, $$

where ρ_{GS} is the spectral radius (the modulus of the largest eigenvalue) of the matrix $(\mathbf{K}_D + \omega\mathbf{K}_C)^{-1}\mathbf{K}_C^{\mathrm{T}}$.

The overrelaxation parameter ω controls the dynamics of the soft tissue model. With $\omega \equiv 2$, the model tends to overshoot around the solution whereas with $\omega \equiv 1$, the motion is very damped. In practise, we chose a value of $\omega = 1.2$ as a trade-off between these two behaviors.

6.2.3. *Algorithm description*

The application of the SOR recursive Eq. (6.8) requires the computation of matrices $[\mathbf{K}_{i,i}]^{-1}[\mathbf{K}_{i,j}]$ and $[\mathbf{K}_{i,i}]^{-1}$. For speed-up purposes, these matrices are stored respectively at each vertex and edge. Because the matrix $\mathbf{K}_D^{-1}\mathbf{K}$ is no longer symmetric, at each edge linking vertices i and j, we store the two 3×3 matrices $[\mathbf{K}_{i,i}]^{-1}[\mathbf{K}_{i,j}]$ and $[\mathbf{K}_{j,j}]^{-1}[\mathbf{K}_{i,j}]^{\mathrm{T}}$.

The algorithm of the relaxation-based elastic model is presented as Algorithm 6. A large part is dedicated to the update of these additional matrices each time a topological change of the mesh occurs. A flag is positioned at each vertex and edge in order to indicate whether matrices $[\mathbf{K}_{i,i}]^{-1}[\mathbf{K}_{i,j}]$ and $[\mathbf{K}_{i,i}]^{-1}$ are up-to-date or not. This flag is raised each time a topological change takes place at a vertex or edge level and it is lowered once these matrices are updated.

6.3. *Hybrid models*

6.3.1. *Motivation*

We have previously described two types of *linear elastic models*:

(1) a *quasi-static* pre-computed elastic model which is computationally efficient but that does not allow any change of topology (cutting, tearing) (see Section 5).

(2) two *dynamic* elastic models (tensor–mass and relaxation-based models) that have lower convergence speed but that allow topology changes (see Sections 6.1

for all Surface tools ST_i **do**
if collision between the soft tissue model and ST_i **then**
if ST_i represents a cavitron device **then**
Possibly refine locally the mesh near the collision
Remove tetrahedra located near the extremity of ST_i
end if
Impose displacements on vertices near the contact zone
end if
end for
for all free vertex i **do**
if flag raised at vertex i **then**
compute and store $[\mathbf{K}_{i,i}]^{-1}$
lower flag at vertex i
end if
$\mathbf{u}_i^\star \Leftarrow (1-\omega)\mathbf{u}_i + \omega[\mathbf{K}_{i,i}]^{-1}\mathbf{R}_i$
for all edge e connecting vertex i and j **do**
if flag raised at edge e **then**
if flag raised at vertex j **then**
compute and store $[\mathbf{K}_{j,j}]^{-1}$
lower flag at vertex j
end if
compute and store $[\mathbf{K}_{i,i}]^{-1}[\mathbf{K}_{i,j}]$ and $[\mathbf{K}_{j,j}]^{-1}\ [\mathbf{K}_{i,j}]^{\mathrm{T}}$
lower flag at edge e
end if
$\mathbf{u}_i^\star \Leftarrow \mathbf{u}_i^\star - \omega[\mathbf{K}_{i,i}]^{-1}\ [\mathbf{K}_{i,j}]\mathbf{u}_j$
end for
$\mathbf{u}_i \Leftarrow \mathbf{u}_i^\star$
end for

ALGORITHM 6. On-line computation of the relaxation-based model.

and 6.2). In the remainder, we use tensor–mass models as the method for deforming.

To combine these two approaches, we make a distinction between two types of anatomical structures that usually appear in a surgical simulation:

- Anatomical structures which are the target of the surgical procedure. On these structures, tearing and cutting need to be simulated. In many cases, they correspond to pathological structures and only represent a small subset of the anatomy that needs to be visualized during the simulation.
- Anatomical structures which only need to be visualized or eventually deformed but which are not submitted to any surgical action.

Thus, in a hybrid model, we propose to model the former type of anatomical structures as tensor–mass models whereas the latter type of structures should be modeled as a pre-computed linear model. However, this method is only efficient if the number of tensor–mass elements is kept as low as possible.

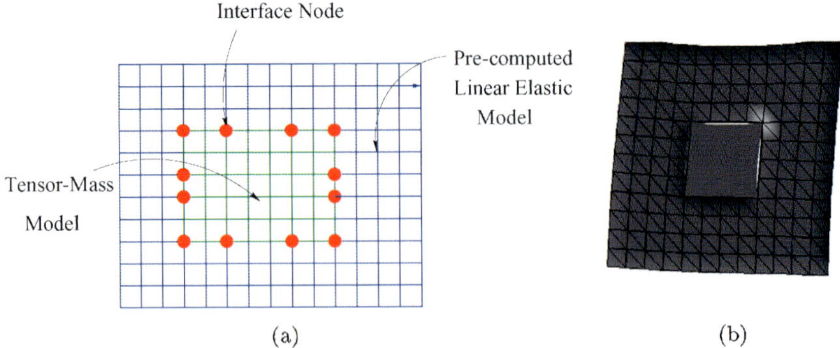

FIG. 6.5. (a) Definition of the interface nodes in a hybrid elastic model; (b) hybrid elastic model with eight interface nodes (COTIN, DELINGETTE and AYACHE [2000]).

6.3.2. *Description*

A hybrid elastic model $\mathcal{M}_{\text{hybrid}}$ is composed of two different types of elements: let $\mathcal{M}_{\text{dynamic}}$ be the set of tensor–mass elements and let $\mathcal{M}_{\text{quasi-static}}$ be the set of pre-computed linear elastic elements. The model $\mathcal{M}_{\text{dynamic}}$ is connected to $\mathcal{M}_{\text{quasi-static}}$ by a set of common vertices called *interface nodes*. These interface nodes define additional boundary conditions for each model. As seen in Fig. 6.5, the two models may not be completely connected along their entire boundaries. In fact, a way to reduce the number of tensor–mass elements, is to associate a fine pre-computed elastic model with a coarse tensor–mass model. As shown in Fig. 6.5(b), this incomplete interface causes some visual artifacts due to the non-continuity between two neighboring parts. However, if the interface zone between the two elastic models is not an important visual cue, a different mesh resolution can be used.

Since both linear elastic models follow the same physical law, their combination should behave exactly as a global linear elastic model. Thus, the additional boundary conditions imposed at the interface nodes must be consistent with responding terms of forces and displacements for both models.

Fig. 6.6 summarizes the computation loop of a hybrid model. Since the pre-computed model $\mathcal{M}_{\text{quasi-static}}$ is more efficient with force boundary conditions than with imposed displacements (see Section 5.4.3), its update is based on forces applied at interface nodes by $\mathcal{M}_{\text{dynamic}}$ but also on imposed displacements resulting from the contact with surgical tools. The applied forces originating from $\mathcal{M}_{\text{dynamic}}$ are computed as reaction forces (opposite of elastic force) at interface nodes. At this stage, the displacement of all surface nodes of $\mathcal{M}_{\text{quasi-static}}$ is computed and the position of interface nodes becomes new displacement constraints for $\mathcal{M}_{\text{dynamic}}$. After $\mathcal{M}_{\text{quasi-static}}$, $\mathcal{M}_{\text{dynamic}}$ is updated based on displacements imposed at the interface nodes by $\mathcal{M}_{\text{quasi-static}}$ and the displacements imposed by the user interaction.

6.3.3. *Examples*

In Fig. 6.7, we present an example of a hybrid cylinder model undergoing deformation caused by gravity forces. The different stages of the deformation process are shown.

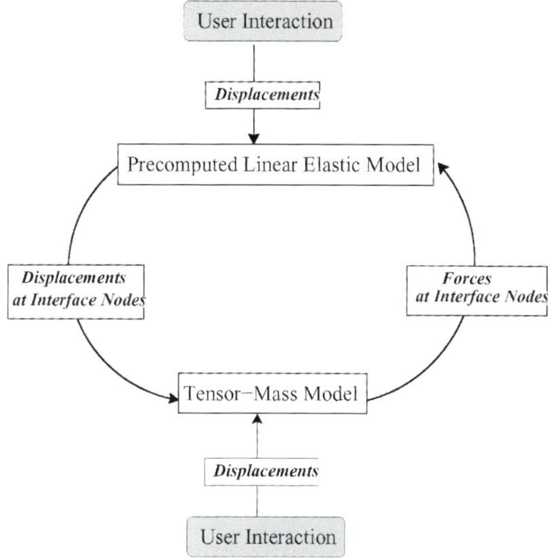

FIG. 6.6. Interaction loop for a hybrid elastic model. Both models are updated alternatively while allowing for user interaction.

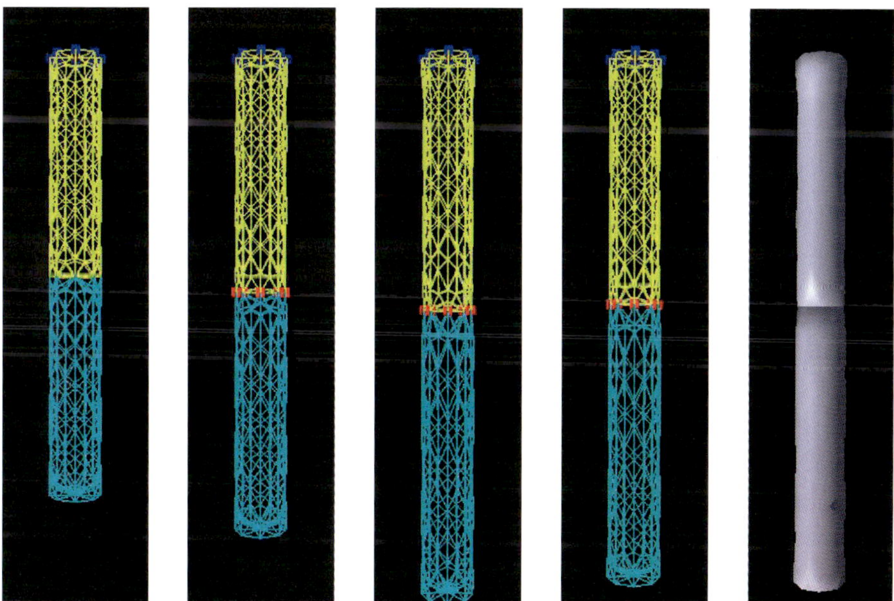

FIG. 6.7. Deformation of a hybrid elastic model under a gravity force: the upper cylinder consists of a pre-computed linear elastic model whereas the lower part is a tensor–mass model. The leftmost figure corresponds to the initial position of the mesh and the rightmost figure to the equilibrium state.

When the equilibrium is reached, as shown in the rightmost figure, forces applied at the interface nodes are null and displacement vectors stabilize to a constant value. In this example, both quasi-static and dynamic models have the same elastic properties and we verified that the equilibrium position is the same as the one that would have been reached by a single quasi-static or dynamic elastic model. Furthermore, this hybrid model converges significantly faster than the corresponding dynamic elastic model.

The second example is related to the simulation of hepatectomy, i.e., the removal of one of the eight anatomical segments – known as Couinaud segments (COUINAUD [1957]) – of a liver. In this example the segment number six has to be removed. A tetrahedral mesh of a liver has been created from a CT scan image. It is composed of 1537 vertices and 7039 tetrahedra – see Fig. 6.8. The tetrahedra of the sixth anatomical segment, which represent 18% (280 vertices and 1260 tetrahedra) of the global mesh, are

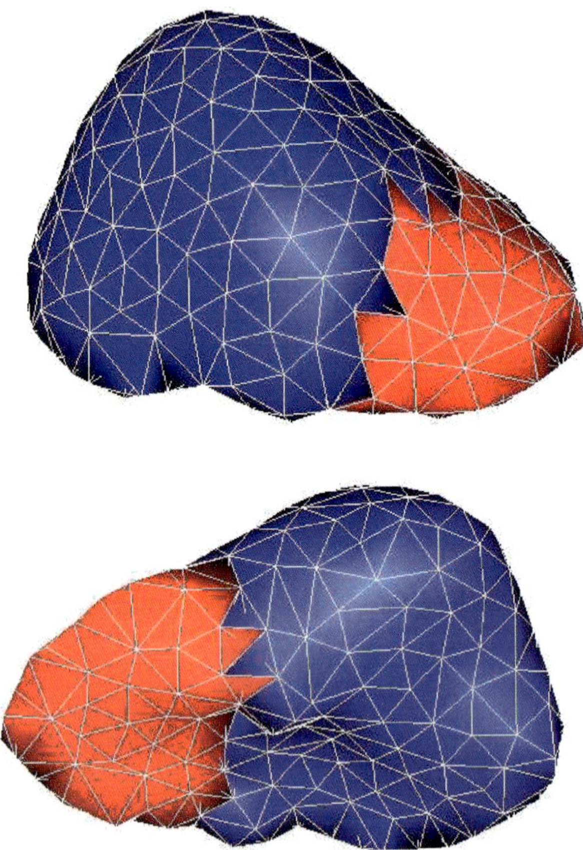

FIG. 6.8. Display of a hybrid liver model. The part displayed in blue corresponds to the pre-computed quasi-static elastic model whereas the red part corresponds to the tensor–mass model. The interface nodes ensure the visual continuity between the two elastic models.

modeled with a tensor–mass model and the remaining tetrahedra with a pre-computed linear elastic model.

In Fig. 6.9, we show different stages of the hepatectomy simulation. The first six pictures show the deformation of the model when the tool collides with the dynamic model. Since both models have the same elastic characteristics, it is not possible to visually distinguish the interface between the two different elastic models.

The last six pictures show the cutting of the liver segment by removing additional tetrahedra. The cutting occurs for the tetrahedron being collided by the tool. One can notice that each part of the hybrid model deforms naturally itself during the resection simulation.

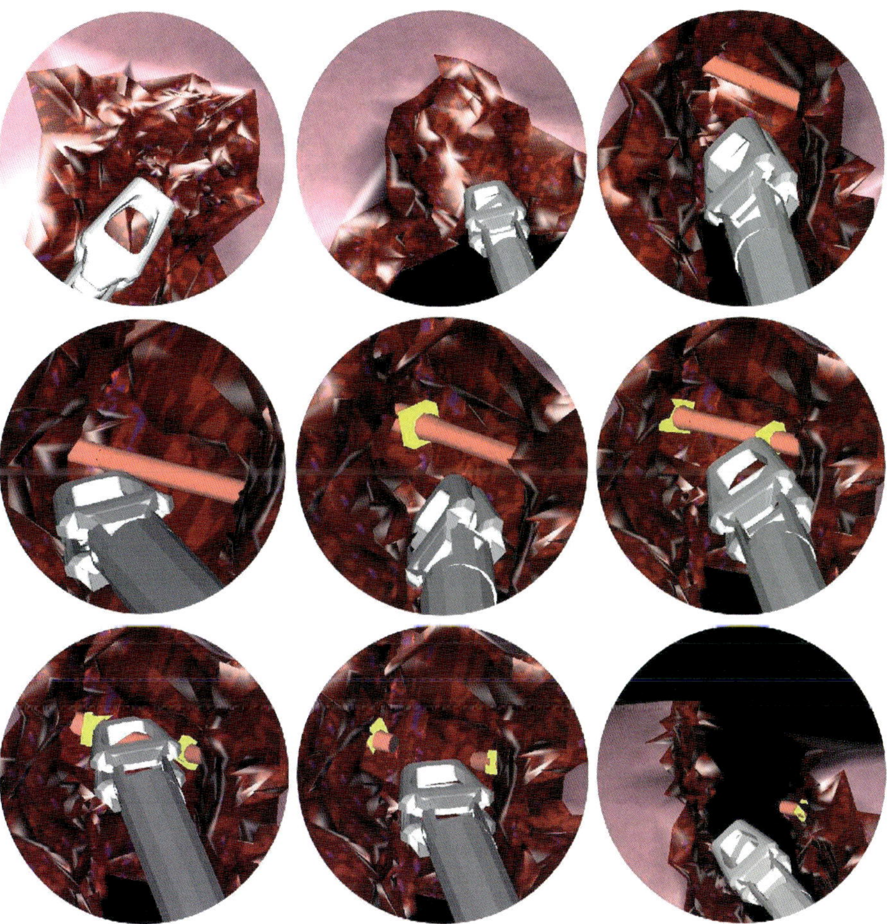

FIG. 6.9. Different stages of the simulation of hepatectomy. In this simulation, we have included lineic models of the main bifurcations of the portal vein (FOREST, DELINGETTE and AYACHE [2002b]). The simulation consists in removing some hepatic parenchyma but also to clamp and cut each vessel.

7. Large displacement non-linear elastic model

7.1. Shortcomings of linear elasticity

The physical behavior of a soft tissue may be considered as linear elastic for small displacements and small deformations (FUNG [1993], MAUREL, WU, MAGNENAT THALMANN and THALMANN [1998]). The hypothesis of small displacements corresponds to displacements that are typically less than 10% of the mesh size.

In the context of surgery simulation, this hypothesis is often violated. For instance, the lobes of the liver are often folded to access underlying structures such as the gall bladder. Also during the resection of a soft tissue, it is common that pieces being cut undergo large rotations either under the action of gravity or under the action of surgical instruments.

In such cases, linear elasticity is not an appropriate physical model because it makes the assumption of infinitesimal strain instead of finite strain. To exhibit the shortcomings of linear elasticity, we produced two examples pictured in Figs. 7.1 and 7.2.

In a first example, we illustrate the action of a global rotation on a linear elastic model. When an object (an icosahedron in Fig. 7.1) undergoes a global rotation, its elastic energy increases, leading to a large variation of volume (as seen in the wireframe mesh of the rightmost figures). Indeed, the infinitesimal strain tensor $\mathbf{E}_\mathrm{L}(\mathbf{X}) = \frac{1}{2}(\nabla\mathbf{U} + \nabla\mathbf{U}^\mathrm{T})$ is not invariant when a global rotation $\mathbf{R}$ is applied since in this case $\nabla\mathbf{U} = \mathbf{R} - \mathbf{I}_3$ and therefore $\mathbf{E}_\mathrm{L}(\mathbf{X}) = \frac{1}{2}(\mathbf{R}+\mathbf{R}^\mathrm{T}) - \mathbf{I}_3 \neq [\mathbf{0}]$. The two invariants $(\operatorname{tr}\mathbf{E}_\mathrm{L})^2$ and $\operatorname{tr}\mathbf{E}_\mathrm{L}^2$ increases under rotation as does the elastic energy.

The second example shows the effect of linear elasticity when only one part of an object undergoes a large rotation (which is the most common case). The cylinder pictured in Fig. 7.2 has its bottom face fixed while a force is being applied at the central top vertex. The arrows correspond to the trajectories of some vertices: because of the linear elastic hypothesis, these trajectories are straight lines. This results in unrealistic distortions of the mesh. Moreover, abnormal deformations are not equivalent in all directions since the object only deforms itself in the rotation plane (Fig. 7.2(c) and (d)).

7.2. St Venant–Kirchhoff elasticity

To overcome the limitations of linear elasticity, we proposed to adopt the St Venant–Kirchhoff elasticity. The St Venant–Kirchhoff model is a generalization of the linear

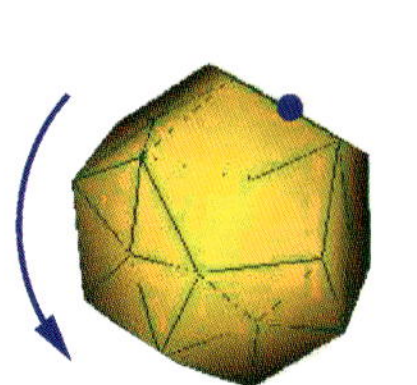
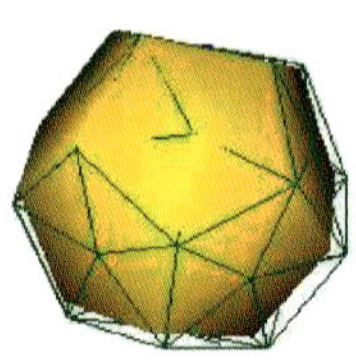
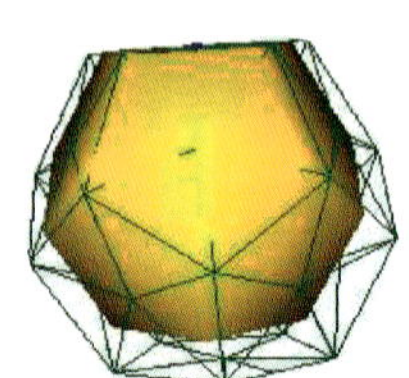
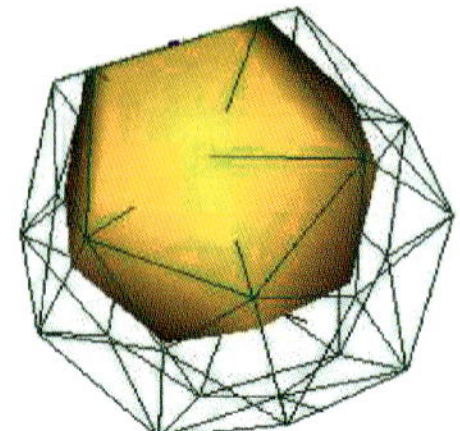

FIG. 7.1. Global rotation of the linear elastic model (wireframe).

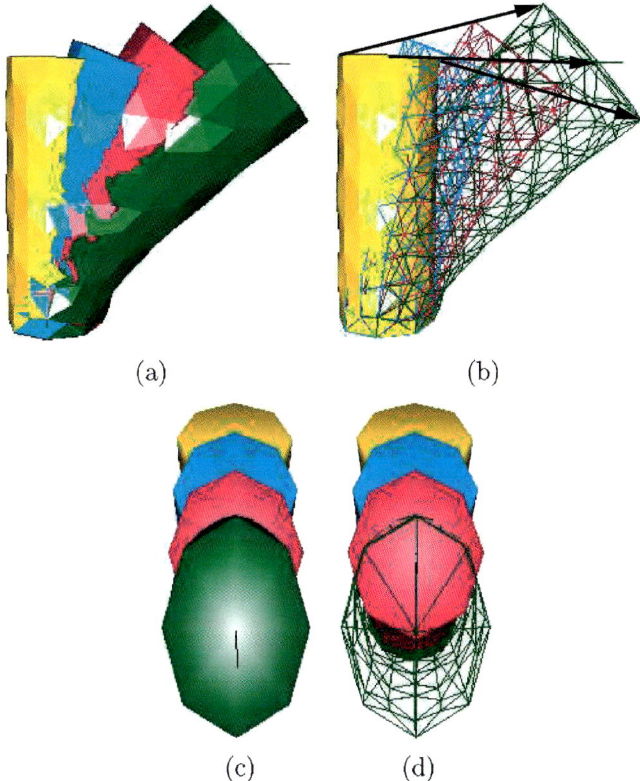

FIG. 7.2. Successive deformations of a linear elastic cylinder (PICINBONO, DELINGETTE and AYACHE [2001]). (a) and (b): side view; (c) and (d): top view.

model for large displacements, and is a particular case of hyperelastic materials. It has been used to model various materials (Table 3.8.4 of CIARLET [1987] provides the constants for materials like steel, glass, lead or rubber) including facial soft tissue (GLADILIN [2002]) and trabecular bone (BAYRAKTAR, ADAMS, GUPTA, PAPADOPOULOS and KEAVENY [2003]). A St Venant–Kirchhoff material relies on the Hooke's law as the definition of elastic energy (see Eq. (3.5) in Section 3.2.3) but the linearized strain tensor $\mathbf{E}_{\mathrm{L}}$ is replaced by the Green–Lagrange strain tensor $\mathbf{E}$:

$$\mathbf{E}(\mathbf{X}) = \frac{1}{2}\left(\nabla\mathbf{U} + \nabla\mathbf{U}^{\mathrm{T}} + \nabla\mathbf{U}^{\mathrm{T}}\nabla\mathbf{U}\right), \tag{7.1}$$

$$W_{\mathrm{NL}}(\mathbf{X}) = \frac{\lambda}{2}(\operatorname{tr}\mathbf{E})^2 + \mu \operatorname{tr}\mathbf{E}^2. \tag{7.2}$$

The Green–Lagrange strain tensor $\mathbf{E}$ is no longer a linear function of the displacement field. A first property is that the elastic energy becomes invariant under the application of rotations. Indeed, when a rigid transformation (with rotation matrix $\mathbf{R}$) is applied to an object, the gradient of the displacement field is $\nabla\mathbf{U} = \mathbf{R} - \mathbf{I}_3$ and therefore the

Green–Lagrange strain tensor remains zero (since $\mathbf{R}\,\mathbf{R}^{\mathrm{T}} = \mathbf{I}_3$),

$$\begin{aligned}
\mathbf{E}(\mathbf{X}) &= \frac{1}{2}\big(\mathbf{R} - \mathbf{I}_3 + \mathbf{R}^{\mathrm{T}} - \mathbf{I}_3 + \big(\mathbf{R}^{\mathrm{T}} - \mathbf{I}_3\big)(\mathbf{R} - \mathbf{I}_3)\big)\\
&= \frac{1}{2}\big(\mathbf{R} + \mathbf{R}^{\mathrm{T}} - 2\mathbf{I}_3 + \mathbf{R}^{\mathrm{T}}\mathbf{R} - \mathbf{R} - \mathbf{R}^{\mathrm{T}} + \mathbf{I}_3\big)\\
&= [\mathbf{0}].
\end{aligned}$$

A second property is that the elastic energy W_{NL} (Section 3.2.3), which was a quadratic function of $\nabla\mathbf{U}$ in the linear case, is now a fourth-order polynomial function with respect to $\mathbf{U}$:

$$\begin{aligned}
W_{\mathrm{NL}} &= \frac{\lambda}{2}(\operatorname{tr}\mathbf{E})^2 + \mu \operatorname{tr}\mathbf{E}^2\\
&= \frac{\lambda}{2}\left[(\operatorname{div}\mathbf{U}) + \frac{1}{2}\|\nabla\mathbf{U}\|^2\right]^2 + \mu\|\nabla\mathbf{U}\|^2 - \frac{\mu}{2}\|\operatorname{rot}\mathbf{U}\|^2\\
&\quad + \mu(\nabla\mathbf{U} : \nabla\mathbf{U}^t\nabla\mathbf{U}) + \frac{\mu}{4}\left\|\nabla\mathbf{U}^t\nabla\mathbf{U}\right\|^2, \qquad (7.3)\\
W_{\mathrm{NL}} &= W_{\mathrm{Linear}} + \frac{\lambda}{2}(\operatorname{div}\mathbf{U})\|\nabla\mathbf{U}\|^2 + \frac{\lambda}{8}\|\nabla\mathbf{U}\|^4\\
&\quad + \mu\big(\nabla\mathbf{U} : \nabla\mathbf{U}^t\nabla\mathbf{U}\big) + \frac{\mu}{4}\left\|\nabla\mathbf{U}^t\nabla\mathbf{U}\right\|^2,
\end{aligned}$$

where W_{Linear} is given by Eq. (3.5) and $A : B = \operatorname{tr}(A^t B) = \sum_{i,j} a_{ij} b_{ij}$ is the dot product of two matrices.

Furthermore, we can extend this isotropic non-linear elastic energy to take into account "transversally isotropic" materials as performed in Section 3.2.4 for the linear elastic model. In fact, Eq. (3.9), which defines the additional anisotropic term, still holds for St Venant–Kirchhoff elasticity. However, for the sake of clarity, we chose to keep only the anisotropic contribution which penalizes the material stretch in the direction given by unit vector $\mathbf{a}_0$:

$$W_{\mathrm{Trans\_iso}} = W_{\mathrm{NL}} + \left(-\frac{\Delta\lambda}{2} + \Delta\mu\right)\big(\mathbf{a}_0^t\mathbf{E}\mathbf{a}_0\big)^2,$$

where $\Delta\lambda$ and Δ are the variations of Lamé coefficients along the direction of anisotropy.

7.3. Finite element modeling

By adopting the same methodology as the one presented in Section 4.3, we provide a closed form expression of the elastic energy of a linear tetrahedron finite element,

$$\begin{aligned}
W_{\mathrm{NL}}(\mathcal{T}) &= \frac{1}{2}\sum_{j,k}\mathbf{U}_j^t\big[\mathcal{B}_{jk}^{\mathcal{T}}\big]\mathbf{U}_k + \frac{1}{2}\sum_{j,k,l}\big(\mathbf{U}_j.\mathcal{C}_{jkl}^{\mathcal{T}}\big)(\mathbf{U}_k.\mathbf{U}_l)\\
&\quad + \frac{1}{2}\sum_{j,k,l,m}\mathcal{D}_{jklm}^{\mathcal{T}}(\mathbf{U}_j.\mathbf{U}_k)(\mathbf{U}_l.\mathbf{U}_m), \qquad (7.4)
\end{aligned}$$

where the terms $\mathcal{B}^{\mathcal{T}}_{jk}$, $\mathcal{C}^{\mathcal{T}}_{jkl}$ and $\mathcal{D}^{\mathcal{T}}_{jklm}$, called "stiffness parameters", are given by

- $\mathcal{B}^{\mathcal{T}}_{jk}$ is a (3×3) symmetric matrix (which corresponds to the linear component of the energy),

$$36V(\mathcal{T})\mathcal{B}^{\mathcal{T}}_{jk} = \lambda(\mathbf{m}_j \otimes \mathbf{m}_k) + \mu\big[(\mathbf{m}_k \otimes \mathbf{m}_j) + (\mathbf{m}_j.\mathbf{m}_k)\mathbf{I}_3\big] + \left(-\frac{\Delta\lambda}{2} + \Delta\mu\right)(\mathbf{a}_0 \otimes \mathbf{a}_0)(\mathbf{m}_j \otimes \mathbf{m}_k)(\mathbf{a}_0 \otimes \mathbf{a}_0),$$

- $\mathcal{C}^{\mathcal{T}}_{jkl}$ is a vector,

$$216\big(V(\mathcal{T})\big)^2\mathcal{C}^{\mathcal{T}}_{jkl} = \frac{\lambda}{2}\mathbf{m}_j(\mathbf{m}_k.\mathbf{m}_l) + \frac{\mu}{2}\big[\mathbf{m}_l(\mathbf{m}_j.\mathbf{m}_k) + \mathbf{m}_k(\mathbf{m}_j.\mathbf{m}_l)\big] + \left(-\frac{\Delta\lambda}{2} + \Delta\mu\right)(\mathbf{a}_0 \otimes \mathbf{a}_0)(\mathbf{m}_j \otimes \mathbf{m}_k)(\mathbf{a}_0 \otimes \mathbf{a}_0)\mathbf{m}_l,$$

- and $\mathcal{D}^{\mathcal{T}}_{jklm}$ is a scalar,

$$1296\big(V(\mathcal{T})\big)^3\mathcal{D}^{\mathcal{T}}_{jklm} = \frac{\lambda}{8}(\mathbf{m}_j.\mathbf{m}_k)(\mathbf{m}_l.\mathbf{m}_m) + \frac{\mu}{4}(\mathbf{m}_j.\mathbf{m}_m)(\mathbf{m}_k.\mathbf{m}_l) + \frac{1}{4}\left(-\frac{\Delta\lambda}{2} + \Delta\mu\right)(\mathbf{a}_0.\mathbf{m}_j)(\mathbf{a}_0.\mathbf{m}_k)(\mathbf{a}_0.\mathbf{m}_l)(\mathbf{a}_0.\mathbf{m}_m).$$

- The last term of each stiffness parameter models the anisotropic behavior of the material.

The elastic force applied at each vertex $\mathbf{p}_i$ of tetrahedron $\mathcal{T}$ is obtained as the derivation of the elastic energy $W_{\mathrm{NL}}(\mathcal{T})$ with respect to the displacement $\mathbf{p}_i$,

$$\mathbf{F}_i(\mathcal{T}) = \underbrace{\sum_j [\mathcal{B}^{\mathcal{T}}_{ij}]\mathbf{U}_j}_{\mathbf{F}^1_i(\mathcal{T})} + \underbrace{\sum_{j,k} (\mathbf{U}_k \otimes \mathbf{U}_j)\mathcal{C}^{\mathcal{T}}_{jki} + \frac{1}{2}(\mathbf{U}_j.\mathbf{U}_k)\mathcal{C}^{\mathcal{T}}_{ijk}}_{\mathbf{F}^2_i(\mathcal{T})} + \underbrace{2\sum_{j,k,l} \mathcal{D}^{\mathcal{T}}_{jkli}\mathbf{U}_l\mathbf{U}^t_k\mathbf{U}_j}_{\mathbf{F}^3_i(\mathcal{T})}. \tag{7.5}$$

The first term of the elastic force ($\mathbf{F}^1_i(\mathcal{T})$) corresponds to the linear elastic case presented in Section 4.4.

7.4. Non-linear tensor–mass model

In this section, we generalize the tensor–mass model introduced in Section 6.1 to the case of large displacement elasticity. The only changes in the tensor–mass algorithm are related to the computation of the elastic force $\mathbf{F}_i$ applied at vertex i.

In the case of linear elasticity, this force was computed by a first scan of all edges to compute the terms $[\mathbf{K}_{ij}]\mathbf{u}_j$ followed by a scan of all vertices to add the terms $[\mathbf{K}_{ii}]\mathbf{u}_i$.

TABLE 7.1
Storage of the stiffness parameters on the mesh

Stiffness parameters distribution	Tensors	Vectors		Scalars		
Vertex p	$\mathcal{B}^{pp}$	$\mathcal{C}^{ppp}$			$\mathcal{D}^{pppp}$	
Edge (p, j)	$\mathcal{B}^{pj}$	$\mathcal{C}^{ppj}$	$\mathcal{C}^{jpp}$	$\mathcal{D}^{jppp}$	$\mathcal{D}^{jjjp}$	$\mathcal{D}^{jpjp}$
		$\mathcal{C}^{jjp}$	$\mathcal{C}^{pjj}$	$\mathcal{D}^{pjjp}$	$\mathcal{D}^{jjpp}$	
Triangle (p, j, k)		$\mathcal{C}^{jkp}$		$\mathcal{D}^{jkpp}$	$\mathcal{D}^{jpkp}$	$\mathcal{D}^{pjkp}$
		$\mathcal{C}^{kjp}$		$\mathcal{D}^{jjkp}$	$\mathcal{D}^{jkjp}$	$\mathcal{D}^{kjjp}$
		$\mathcal{C}^{pjk}$		$\mathcal{D}^{kkjp}$	$\mathcal{D}^{kjkp}$	$\mathcal{D}^{jkkp}$
Tetrahedron (p, j, k, l)				$\mathcal{D}^{jklp}$	$\mathcal{D}^{jlkp}$	$\mathcal{D}^{kjlp}$
				$\mathcal{D}^{kljp}$	$\mathcal{D}^{ljkp}$	$\mathcal{D}^{lkjp}$

We proposed to apply the same principle to the quadratic term ($\mathbf{F}_2^p(\mathcal{T})$ of Eq. (7.5)) and the cubic term ($\mathbf{F}_3^p(\mathcal{T})$). The former requires *stiffness vectors* for vertices, edges and triangles, and the latter requires *stiffness scalars* for vertices, edges, triangles and tetrahedra.

The task of assembling global stiffness parameters is slightly more time consuming than in the linear case, since 31 parameters must be assembled instead of 2; these parameters are presented in Table 7.1.

For vertex, edge and triangle parameters, one needs to add the contributions of all neighboring tetrahedra. For instance, the vertex rigidity vector $\mathcal{C}^{ppp}$ is computed at vertex p as

$$\mathcal{C}^{ppp} = \sum_{\mathcal{T} \in \mathcal{S}(p)} \mathcal{C}^{\mathcal{T}}_{ppp}.$$

For the 6 scalar parameters $\mathcal{D}^{jklp}$ stored at each tetrahedron, no assembly is required since there is no other contribution originating from another tetrahedron.

The computation of the elastic force is performed by successively scanning tetrahedra, triangles, edges and vertices of the mesh. When scanning triangles for instance, the contributions from the three triangles are computed and added to the elastic force of each of its three vertices. The contribution for each element is summarized in Eq. (7.5).

$$\mathbf{F}_i = \mathbf{F}_i^{\text{vertex}} + \mathbf{F}_i^{\text{edge}} + \mathbf{F}_i^{\text{triangle}} + \mathbf{F}_i^{\text{tetrahedron}} \tag{7.6}$$

with

$$\mathbf{F}_i^{\text{vertex}} = \boxed{\begin{array}{l} \text{Vertex contribution} \\ \hline [\mathcal{B}^{pp}]\mathbf{U}_p \\ +\big[(\mathbf{U}_p \otimes \mathbf{U}_p) + \frac{1}{2}(\mathbf{U}_p.\mathbf{U}_p)\mathbf{I}_3\big]\mathcal{C}^{ppp} \\ +2\mathcal{D}^{pppp}\mathbf{U}_p\mathbf{U}_p^t\mathbf{U}_p \end{array}},$$

$$\mathbf{F}_i^{\text{edge}} = \sum_{\text{edges}(p,j)} \left| \begin{array}{l} \text{Edge contribution} \\ [\mathcal{B}^{pj}]\mathbf{U}_j \\ \quad + [(\mathbf{U}_j \otimes \mathbf{U}_p) + (\mathbf{U}_j.\mathbf{U}_p)\mathbf{I}_3]\mathcal{C}^{ppj} + (\mathbf{U}_p \otimes \mathbf{U}_j)\mathcal{C}^{jpp} \\ \quad + (\mathbf{U}_j \otimes \mathbf{U}_j)\mathcal{C}^{jjp} + \frac{1}{2}(\mathbf{U}_j.\mathbf{U}_j)\mathcal{C}^{pjj} \\ \quad + 2[\mathcal{D}^{jppp}(2\mathbf{U}_p\mathbf{U}_p^t\mathbf{U}_j + \mathbf{U}_j\mathbf{U}_p^t\mathbf{U}_p) + \mathcal{D}^{jjpp}\mathbf{U}_p\mathbf{U}_j^t\mathbf{U}_j \\ \qquad + (\mathcal{D}^{jpjp} + \mathcal{D}^{pjjp})\mathbf{U}_j\mathbf{U}_j^t\mathbf{U}_p + \mathcal{D}^{jjjp}\mathbf{U}_j\mathbf{U}_j^t\mathbf{U}_j] \end{array} \right|,$$

$$\mathbf{F}_i^{\text{triangle}} = \sum_{\text{faces}(p,j,k)} \left| \begin{array}{l} \text{Triangle contribution} \\ [(\mathbf{U}_k \otimes \mathbf{U}_j)\mathcal{C}^{jkp} + (\mathbf{U}_j \otimes \mathbf{U}_k)\mathcal{C}^{kjp} + (\mathbf{U}_j.\mathbf{U}_k)\mathcal{C}^{pjk}] \\ \quad + 2[(\mathcal{D}^{pjkp} + \mathcal{D}^{jpkp})(\mathbf{U}_j\mathbf{U}_k^t\mathbf{U}_p + \mathbf{U}_k\mathbf{U}_j^t\mathbf{U}_p) \\ \quad + 2\mathcal{D}^{jkpp}\mathbf{U}_p\mathbf{U}_j^t\mathbf{U}_k \\ \quad + (\mathcal{D}^{kjjp} + \mathcal{D}^{jkjp})\mathbf{U}_j\mathbf{U}_j^t\mathbf{U}_k + \mathcal{D}^{jjkp}\mathbf{U}_k\mathbf{U}_j^t\mathbf{U}_j \\ \quad + (\mathcal{D}^{jkkp} + \mathcal{D}^{kjkp})\mathbf{U}_k\mathbf{U}_k^t\mathbf{U}_j + \mathcal{D}^{kkjp}\mathbf{U}_j\mathbf{U}_k^t\mathbf{U}_k] \end{array} \right|,$$

$$\mathbf{F}_i^{\text{tetrahedron}} = \sum_{\text{tetra}(p,j,k,l)} \left| \begin{array}{l} \text{Tetrahedron contribution} \\ 2[(\mathcal{D}^{jklp} + \mathcal{D}^{kjlp})\mathbf{U}_l\mathbf{U}_j^t\mathbf{U}_k \\ \quad + (\mathcal{D}^{jlkp} + \mathcal{D}^{ljkp})\mathbf{U}_k\mathbf{U}_j^t\mathbf{U}_l \\ \quad + (\mathcal{D}^{kljp} + \mathcal{D}^{lkjp})\mathbf{U}_j\mathbf{U}_k^t\mathbf{U}_l] \end{array} \right|.$$

In terms of data structure, the non-linear tensor–mass model requires the addition of triangles in the mesh topological description. In our case, we chose to store triangles in a hash table which is hashed by the three indices of its vertices in lexicographic order. Furthermore, each tetrahedron owns pointers towards its four triangles and reversely, each triangle owns pointers towards its two neighboring tetrahedra.

During the simulation of resection, tetrahedra are iteratively removed near the extremities of virtual cavitron instruments. When removing a single tetrahedron, 280 floating point numbers are updated to suppress the tetrahedron contributions to the stiffness parameters of the surrounding vertices, edges and triangles:

$$\begin{aligned} &4 * (1 \text{ tensor} + 1 \text{ vector} + 1 \text{ scalar}) \\ &\qquad + 6 * (1 \text{ tensor} + 4 \text{ vectors} + 5 \text{ scalars}) \\ &\qquad + 4 * (3 \text{ vectors} + 9 \text{ scalars}) \\ &\quad = 280 \text{ real numbers.} \end{aligned}$$

By locally updating stiffness parameters, the tissue has exactly the same properties as if the corresponding tetrahedron had been removed at its rest position. Because of the volumetric continuity of finite element modeling, the tissue deformation remains realistic during cutting.

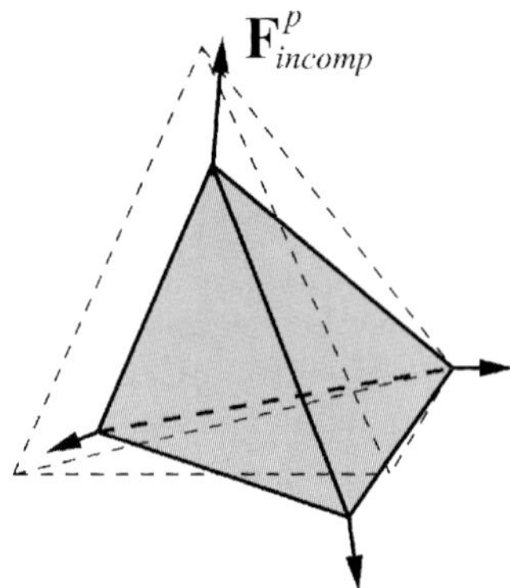

FIG. 7.3. Penalization of volume variation.

7.5. *Incompressibility constraint*

Living tissue, which is made essentially of water is almost incompressible, a property which is difficult to model and which, in most cases, leads to instability problems. This is the case with the St Venant–Kirchhoff model: the material remains incompressible when the Lamé constant λ tends towards infinity. Taking a large value for λ would impose to decrease the time step and therefore to increase the computation time. Another reason to add an external incompressibility constraint to the model is intrinsic to the model itself: the St Venant–Kirchhoff model relies on the Green–Lagrange strain tensor E which is invariant with respect to rotations. But it is also invariant with respect to symmetries, which could lead to the reversal of some tetrahedra under strong constraints.

We chose to penalize volume variation by applying to each vertex of the tetrahedron a force directed along the normal of the opposite face $\mathbf{N}_p$ (see Fig. 7.3), the norm of the force being proportional to the square of the relative volume variation,

$$\mathbf{F}^p_{\text{incomp}} = \text{sign}(V - V_0)\left(\frac{V - V_0}{V_0}\right)^2 \vec{\mathbf{N}}_p. \tag{7.7}$$

Since the volume V is proportional to the height of each vertex facing its opposite triangle, when V is greater than V_0 then the force $\mathbf{F}^p_{\text{incomp}}$ tends to decrease V by moving each vertex along the normal of the triangle facing it. These forces act as an artificial pressure inside each tetrahedron. This method is closely related to Lagrange multipliers, which are often used to solve problem of energy minimization under constraints.

7.6. *Results*

In a first experiment, we wish to highlight the contributions of our new deformable model in the case of partial rotations. Fig. 7.4 shows the same experience as the one presented for linear elasticity (Section 7.1, Fig. 7.2). On the left we can see that the cylinder vertices are now able to follow non-straight trajectories (Fig. 7.4(a)), leading

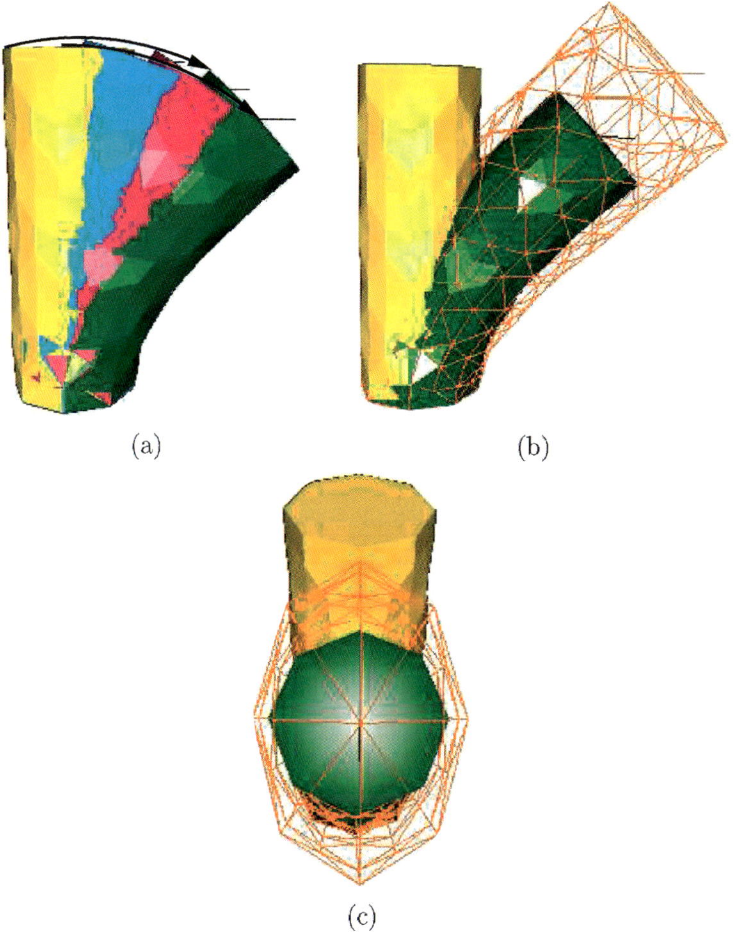

FIG. 7.4. (a) Successive deformations of the non-linear model (PICINBONO, DELINGETTE and AYACHE [2003]). Side (b) and top (c) view of the comparison between linear (wireframe) and non-linear model (solid rendering).

to much more realistic deformations than in the linear (wireframe) case (Figs. 7.4(b) and (c)).

The second example presents the differences between isotropic and anisotropic materials. The three cylinders of Fig. 7.5 have their top and bottom faces fixed, and are submitted to the same forces. While the isotropic model on the left undergoes a "snake-like" deformation, the last two, which are anisotropic along their height, stiffen in order to minimize their stretch in the anisotropic direction. The rightmost model, being twice as stiff as the middle one in the anisotropic direction, starts to squeeze in the plane of isotropy because it cannot stretch anymore.

In the third example (Fig. 7.6), we apply a force to the right lobe of the liver (the liver is fixed in a region near the center of its back side, and Lamé coefficients are:

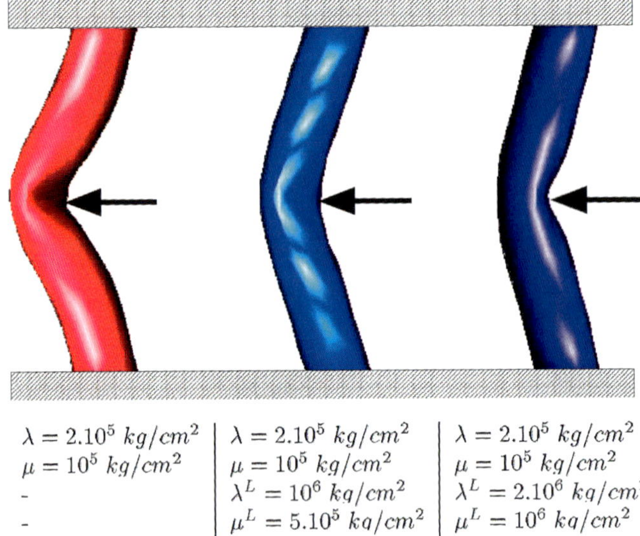

FIG. 7.5. Shearing deformation of tubular structures under the action of the force indicated by the arrow. The leftmost figure corresponds to an isotropic non-linear material while the center and rightmost figures correspond to a non-linear anisotropic material, the direction of anisotropy being the cylinder axis.

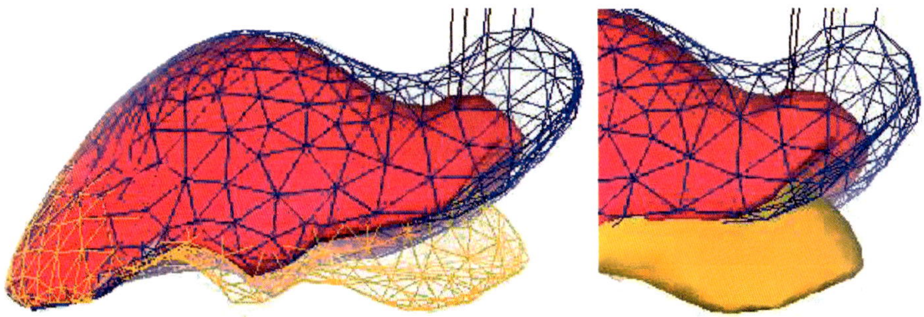

FIG. 7.6. Linear (upper mesh in wireframe), non-linear (Gauraud shaded) liver models and rest shape (lower mesh in wireframe). In both cases, the same forces showed in solid lines are applied to three surface nodes lying on the left lobe (PICINBONO, DELINGETTE and AYACHE [2003]).

$\lambda = 40$ kPa and $\mu = 10$ kPa). Using the linear elastic model, the right part of the liver undergoes a large (and unrealistic) volume increase, whereas with non-linear elasticity, the right lobe is able to rotate partially, while adopting a more realistic deformation.

Adding the incompressibility constraint on the same examples decreases the volume variation even more (see Table 7.2), and also stabilizes the behaviour of the deformable models in highly constrained areas.

The last example is the simulation of a typical laparoscopic surgical gesture on the liver. One tool is pulling the edge of the liver sideways while a bipolar cautery device

TABLE 7.2
Volume variation results. For the cylinder: left, middle and right stand for the different deformations of Figs. 7.5 and 7.6

Volume variation (%)	Linear	Non-linear	Non-linear incomp.
Cylinder (left – middle – right)	7 – 28 – 63	0.3 – 1 – 2	0.2 – 0.5 – 1
Liver	9	1.5	0.7

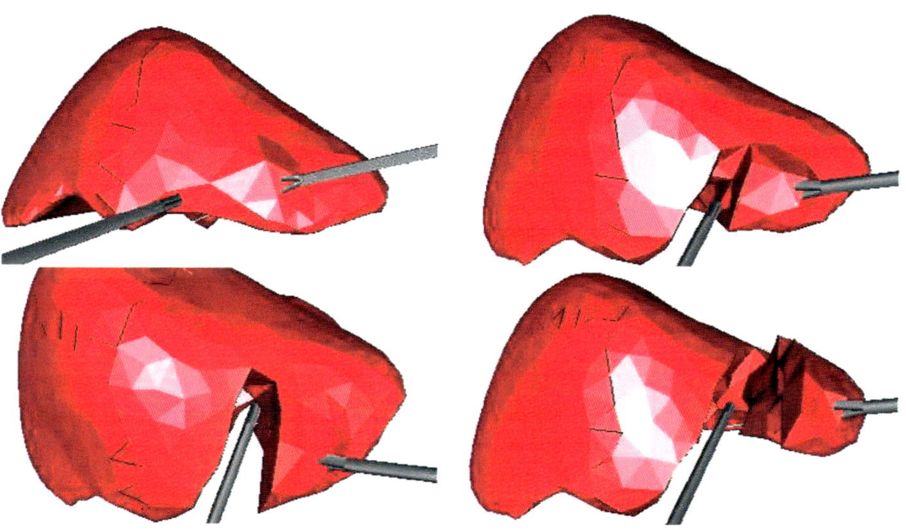

FIG. 7.7. Simulation of laparoscopic liver surgery.

cuts it. During the cutting, the surgeon pulls away the part of the liver he wants to remove. This piece of liver undergoes large displacements and the deformation appears fairly realistic with this new non-linear deformable model (Fig. 7.7).

Obviously, the computation time of this model is larger than for the linear model because the force equation is more complex (Eq. (7.5) in Section 7.3 to be compared with Eq. (6.3) in Section 6.1.7). With our current implementation, the simulation refresh rate is five times slower than with the linear model. Nevertheless, with this non-linear model, we can reach an update cycle of 25 Hz on meshes made of about 2000 tetrahedra (on a PC Pentium PIII 500 MHz). This is enough to achieve real-time visual feedback with quite complex objects, and even to provide a realistic haptic feedback using force extrapolation as described in PICINBONO, LOMBARDO, DELINGETTE and AYACHE [2000].

7.7. Optimization of non-linear deformations

We showed that non-linear elasticity allows to simulate much more realistic deformations than linear elasticity when the model undergoes large displacements. However,

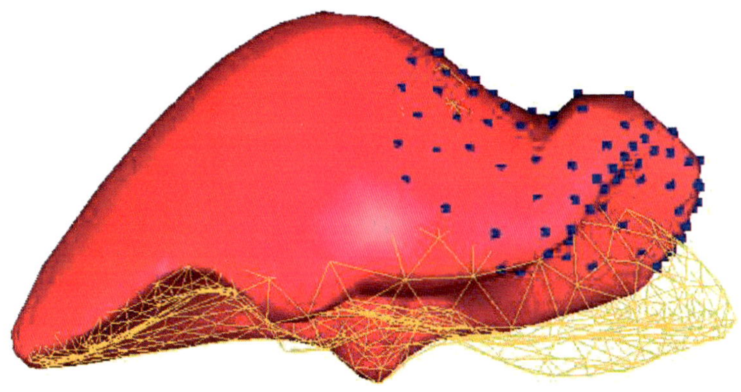

FIG. 7.8. Adaptable non-linear model deformation compared to its rest position (wireframe).

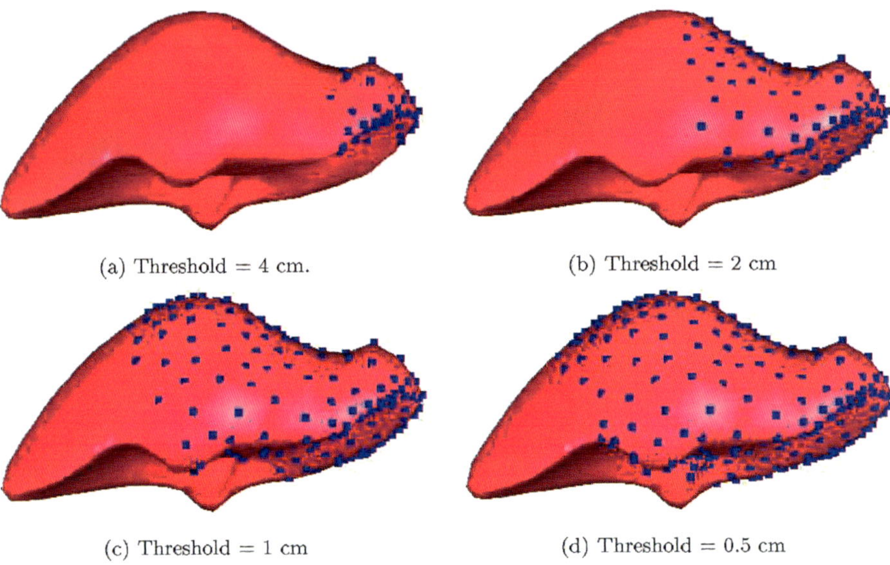

FIG. 7.9. Deformation of the adaptive non-linear model for several values of the threshold.

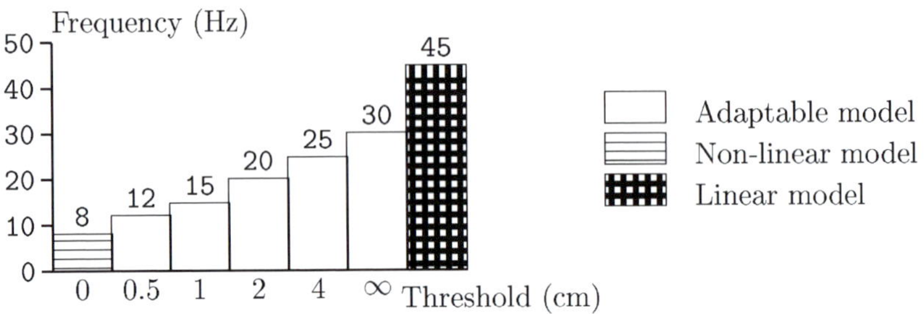

FIG. 7.10. Updating frequencies of the adaptable model for several values of the threshold.

non-linear elasticity is more computationally expensive than linear elasticity. Since non-linear elastic forces tend to linear elastic forces as the maximum vertex displacement decreases to zero, we propose to use non-linear elasticity only at parts of the mesh where displacements are larger than a given threshold, the remaining part using linear elasticity. Thus, we modified the force computation algorithm in the following manner: for each vertex, we first compute the linear part of the force, and we add the non-linear part only if its displacement is larger than a threshold. Fig. 7.8 shows a deformation computed with this optimization (same model as in Fig. 7.6). This liver model is made of 6342 tetrahedra and 1394 vertices. The threshold is set to 2 cm while the mesh is about 30 cm long. The points drawn on the surface identify vertices using non-linear elasticity. With this method, we reach an update frequency of 20 Hz instead of 8 Hz for a fully non-linear model. The same deformation is presented on Fig. 7.9 for different values of the threshold. With this method, we can choose a trade-off between the bio-mechanical realism of the deformation and the update frequency of the simulation. The diagram on Fig. 7.10 shows the update frequencies reached for each value of the threshold, in comparison with the fully linear and the fully non-linear models. Even when this threshold tends towards infinity, the adaptable model is slower than the linear model, because the computation algorithm of the non-linear force is more complex. Indeed, the computation of non-linear forces requires to visit all vertices, edges, triangles and tetrahedra of the mesh, whereas only vertices and edges need to be visited for the

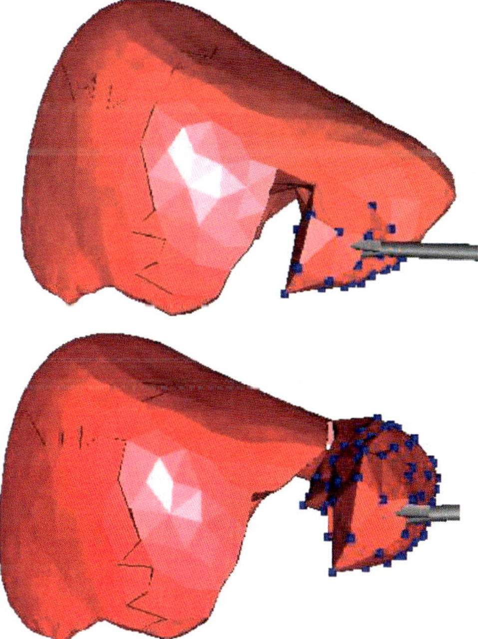

FIG. 7.11. Simulation of hepatectomy based on a non-linear adaptable elastic model. Non-linear elastic force are applied on vertices outlined with a box.

linear model. For the simulation example of Fig. 7.7, this optimization leads to update frequencies varying between 50 and 80 Hz, depending on the number of points modeling non-linear elasticity (Fig. 7.11). The minimal frequency of 50 Hz is reached at the end of the simulation, when all vertices of the resected part of the liver are using large displacement elasticity (on the right of Fig. 7.11).

In general, two strategies can be used to set the value of this threshold. In the first strategy, the threshold is increased until a given update frequency is matched as demonstrated previously. The second strategy is physically-motivated and sets the threshold to 10% of the typical size of the mesh since it corresponds to the extent of displacement for which linear elasticity remains a valid constitutive law.

8. Conclusion

In this chapter, we have presented several algorithms for computing in real-time the deformation of soft tissues in a surgical simulator. We wish to stress two important aspects of these algorithms. First of all, using linear tetrahedra as finite elements helped us to write closed-form expressions of the elastic energy and its derivatives, even in the case of large displacement elasticity. These expressions nicely decouple the physical parameters (Lamé coefficients) from the geometry of each tetrahedron both in its rest position (direction of anisotropy, rest volume, area vectors) and in its deformed state (displacement vectors). Furthermore, it enables to quickly assemble local and global stiffness matrices when the mesh topology has been modified during a cutting simulation.

Second, in the context of surgery simulation, soft tissue deformation algorithms are closely tied with the visualization, collision detection and haptic rendering algorithms. Furthermore, the traditional stages of matrix assembly, matrix preconditioning, system solution and post-processing, cannot be easily decoupled like in classical software packages available in structural mechanics. This implies that the data structure and the flow chart must be carefully designed in order to achieve a reasonable trade-off between these performances. Therefore, building a successful simulator can only be achieved by a multidisciplinary effort covering the fields of biomechanics, numerical analysis, robotics and computer graphics.

An hepatectomy simulator based on the quasi-static precomputed linear elastic model (introduced in Section 5) and the large displacement non-linear elastic models (introduced in Section 7) has been built where the following three basic surgical gestures can be rehearsed: touching soft tissue, gripping soft tissue and cutting parenchyma with a cavitron. Furthermore, we recently added a physical model of the portal vein (FOREST, DELINGETTE and AYACHE [2003]), which allows the user to simulate the clamping and cutting of vessels during the hepatic resection.

However, to increase the training impact and realism of the simulation, it is important to simulate the contact between the liver and neighboring structures such as the gallbladder, the different ligaments, the right kidney, the peritoneum, etc. These additional surface and volumetric models require to extend the soft tissue models introduced in this chapter in two ways.

First, it is necessary to extend the precomputed linear elastic model to include large-displacement non-linear elasticity. Indeed, the linear domain of biological soft tissue is usually rather small, and therefore many surgical gestures can only be simulated by using large-displacement elasticity (like rotating the lobe of the liver or resecting the gall-bladder). The precomputation of non-linear elastic material is not a trivial task since it implies solving a complex third-order algebraic equation in the case of St Venant–Kirchhoff elasticity (see Section 7.2). Instead, it may be possible to find suitable approximations which can be computed efficiently.

Second, it is necessary to extend the concept of hybrid models (introduced in Section 6.3) in order to cope with the deformation of models including several tens of thousands of vertices. Ideally, we would like to provide accurate but computationally expensive soft tissue models in the center of the surgical field where the user performs complex gestures and at the same time to provide less expensive models but potentially less accurate, away from the center of the surgical field. Of course, during surgery, the focus of the surgeon may switch from the gall-bladder to the hepatic parenchyma which implies that those tissue models should evolve dynamically from one level of accuracy to the other. Achieving this level of scalability with the constraint that the topology of these models may change over time, is the main challenge of soft tissue modeling for surgery simulation.

Finally, we would like to stress the importance of validating the different components of a surgical simulator. Concerning soft tissue models, there are at least three levels of validation that need to be achieved. A first validation consists in comparing the soft tissue deformation algorithms that rely on strong hypotheses against well-known finite element packages in order to evaluate the range of approximations that are performed. In the second level of validation, the biomechanical behavior of each anatomical structure must be compared to experimental dataset. Ideally, one would like to validate both boundary conditions and the constitutive law of each biological tissue. However, in practice, this validation is made difficult by the lack of quantitative experimental information. The third level of validation consists in evaluating the dynamic behaviour of each soft tissue during the simulation since some models that appear too soft or too stiff. Finally, and most importantly, it is required to validate the whole simulation system by assessing its ability to succeed in training young residents to perform a given surgical task.

Despite these remaining issues to be solved, we believe that practical surgery simulators will be fully operational and actually part of the surgical studies in the near future.

List of mathematical symbols

f_{u} Update frequency of the soft tissue model
t Discrete or continuous time variable
X_t Position of the model at time t
$T_{\mathrm{relaxation}}$ Relaxation time
T_{c} Computation time
$T_{\mathrm{interaction}}$ Latency caused by the software and hardware architecture
Δt Time step used in the discretization of temporal derivatives
$\mathbf{F}$ Global force vector
$\mathbf{K}$ Global stiffness matrix

$\mathbf{U}$ Global displacement vector
$\mathbf{M}$ Global mass matrix
$\mathbf{C}$ Global damping matrix
$\dot{\mathbf{U}}$ Global speed vector
$\mathcal{M}_{\text{rest}}$ Soft tissue model at its rest position
$\mathcal{M}_{\text{def}}$ Soft tissue model at its deformed position
Ω Region of space for the rest configuration
$\boldsymbol{\Phi}(x, y, z)$ Deformation function that maps point (x, y, z) from the rest configuration to the deformed configuration
$\mathbf{X}$ Point in the rest configuration
$\mathbf{U}(\mathbf{X})$ Displacement function
$\mathbf{C}(\mathbf{X})$ Right Cauchy–Green strain tensor
$\mathbf{E}(\mathbf{X})$ Green–Lagrange strain tensor
$\mathbf{I}_3$ 3×3 identity matrix
$\mathbf{E}_{\text{L}}$ Linearized strain tensor
e_{ij} Element of the linearized strain tensor
$\mathbf{T}(\mathbf{X})$ Cauchy stress tensor
$W(\mathbf{X})$ Density of elastic energy
λ, μ Isotropic Lamé coefficients
E, ν Isotropic Young modulus and Poisson ratio
$\mathbf{a}_0$ Unit vector along the direction of anisotropy for transversally isotropic materials
$\lambda^{\mathbf{a}_0}, \mu^{\mathbf{a}_0}$ Lamé coefficients along the direction of anisotropy
$\Delta\lambda, \Delta\mu$ Difference between the Lamé coefficients along the direction of anisotropy and those in the orthogonal plane
$\Delta W_{\text{Ani}}(\mathbf{X})$ Additional term of the density of elastic energy caused by anisotropy
I_4, I_5 Deformation invariants estimated along the direction of anisotropy
$\mathbf{p}_i$ Point of a tetrahedron in its rest position
$\mathbf{q}_i$ Point of a tetrahedron in its deformed position
$\mathbf{u}_i$ Displacement vector of a vertex of a tetrahedron
$\mathcal{T}$ Tetrahedron as a linear finite element
$h_j(\mathbf{X})$ Shape functions associated with a linear tetrahedron
$\mathbf{P}$ 4×4 matrix describing the shape functions
$V(\mathcal{T})$ Volume of tetrahedron $\mathcal{T}$
$\mathbf{m}_i$ Area vector opposite to vertex i
V_i 6 times the volume of the tetrahedron made by the origin $\mathbf{o}$ and vertices $\mathbf{p}_{i+1}$, $\mathbf{p}_{i+2}$ and $\mathbf{p}_{i+3}$
T_i Triangle opposite to vertex i
$\mathbf{n}_i$ Normal vector at the triangle T_i opposite to vertex i in a tetrahedron
$\theta_{i,j}$ Angle between normal vectors of triangles T_i and T_j
A_i Area of triangle T_i
$l_{i,j}$ Length of the edge connecting vertices i and j
f_i Height of vertex o above triangle T_i
$\mathbf{B}_{i,j}^{\mathcal{T}}$ Element (i, j) of the 3×3 stiffness matrix for a tetrahedron $\mathcal{T}$ made of an isotropic material

$\mathbf{A}^{\mathcal{T}}_{i,j}$ Element (i, j) of the 3×3 stiffness matrix for a tetrahedron $\mathcal{T}$ made of an transversally isotropic material
$\mathbf{K}_{i,j}$ 3×3 global stiffness matrix between vertex i and j
$k_{i,j}$ Eigenvalue along the edge direction of matrix $\mathbf{K}_{i,j}$
$W_g(\mathcal{T})$ Work of gravity forces
$W_p(\mathcal{T})$ Work of external surface pressure
$\mathbf{M}_{i,j}$ 3×3 global mass matrix between vertex i and j
$\mathbf{K}^{\star}_{i,j}$ 3×3 global stiffness matrix between vertex i and j that includes spring boundary conditions
$\mathbf{R}^g$ Global vector of gravity forces
$\mathbf{R}^b$ Global vector of boundary forces

Acknowledgements

We thank Matthias Teschner, Denis Laurendeau and Jean-Marc Schwartz for their priceless comments and for proofreading this article.

The work presented in this paper is a joint work between the authors and mainly two former PhD students: Stéphane Cotin and Guillaume Picinbono. Stéphane Cotin developed the precomputed linear elastic model of Section 5 as well as a first version of the tensor–mass model described in Section 6.1. Guillaume Picinbonno proposed the extension of the tensor–mass model to the case of large displacement elasticity (in Section 7). We also thank Clément Forest and Jean-Christophe Lombardo for their numerous contributions on force-feedback rendering, collision detection as well as mesh data structure. This work was fueled with the stimulating remarks and propositions from our INRIA colleagues who participated in the AISIM and CAESARE joint initiatives: Marie-Paule Cani, Marina Vidrascu, Marc Thiriet, Christian Laugier. Also, we are grateful to Prof. Marescaux, Prof. Leroy and Prof. Luc Soler from the IRCAD research center for their long-term vision and for sharing their expertise of abdominal surgery with us. Finally, we acknowledge the strong support we received from Gilles Khan, INRIA Vice-President for Research, during the different stages of this research work.

References

ACKERMAN, M.J. (1998). The visible human project. *Proc. IEEE: Special Issue on Surgery Simulation* **86** (3), 504–511.

AYACHE, N. (2003). Epidaure: a research project in medical image analysis, simulation and robotics at INRIA. *IEEE Trans. Medical Imaging*, Invited Editorial.

AYACHE, N., DELINGETTE, H. (eds.) (2003). *Int. Symp. on Surgery Simulation and Soft Tissue Modeling*, Juan-Les-Pins, France, June 1998. In: Lecture Notes in Comput. Sci. **2673** (Springer-Verlag, New York).

BARAFF, D., WITKIN, A. (1998). Large steps in cloth simulation. In: *Computer Graphics Proceedings*, SIGGRAPH'98, Orlando, USA, July 1998, pp. 43–54.

BASSINGTHWAIGHTE, J.B. (2000). Strategies for the physiome project. *Ann. Biomed. Engrg.* **28**, 1043–1058.

BATHE, K.-L. (1982). *Finite Element Procedures in Engineering Analysis* (Prentice Hall, New York).

BAYRAKTAR, H., ADAMS, M., GUPTA, A., PAPADOPOULOS, P., KEAVENY, T. (2003). The role of large deformations in trabecular bone mechanical behavior. In: *ASME Bioengineering Conference*, Key Biscayne, FL, USA, June 2003.

BENZLEY, S.E., PERRY, E., CLARK, B., MERKLEY, K., SJAARDEMA, G. (1995). Comparison of all-hexahedral and all-tetrahedral finite element meshes for elastic and elasto-plastic analysis. In: *Proc. 4th Int. Meshing Roundtable*, Sandia National Laboratories, October 1995, pp. 179–191.

BERCI, G., HUNTER, J.G., SACKIER, J.M. (1994). Training in laparoscopic cholecystectomy: Quantifying the learning curve. *J. Endoscopic Surgery* **8**, 28–31.

BESAG, J. (1986). On the statistical analysis of dirty pictures. *J. Roy. Statist. Soc.* **48** (3), 326–338.

BIELSER, D., GROSS, M.H. (2000). Interactive simulation of surgical cuts. In: *Proc. Pacific Graphics 2000*, Hong-Kong, October 2000 (IEEE Computer Society Press), pp. 116–125.

BOUX DE CASSON, F., LAUGIER, C. (1999). Modelling the dynamics of a human liver for a minimally invasive simulator. In: *Proc. Int. Conf. on Medical Image Computer-Assisted Intervention*, Cambridge, UK, September 1999.

BRO-NIELSEN, M. (1998). Finite element modeling in surgery simulation. *Proc. IEEE: Special Issue on Surgery Simulation* **86** (3), 490–503.

BRO-NIELSEN, M., COTIN, S. (1996). Real-time volumetric deformable models for surgery simulation using finite elements and condensation. In: *Eurographics'96, vol. 3*, pp. 57–66.

BRONSHTEIN, I.N., SEMENDYAYEV, K.A. (1985). *Handbook of Mathematics* (Van Nostrand–Reinhold, New York).

BROWN, J.D., ROSEN, J., KIM, Y., CHANG, L., SINANAN, M., HANNAFORD, B. (2003). In-vivo and in-situ compressive properties of porcine abdominal soft tissue. In: *Medicine Meets Virtual Reality*, MMVR'03, Newport Beach, USA, January 2003.

CANAS, J., PARIS, F. (1997). *Boundary Element Method: Fundamentals and Application* (Oxford Univ. Press, London).

CARTER, F.J. (1998). Biomechanical testing of intra-abdominal soft tissue. In: *Int. Workshop on Soft Tissue Deformation and Tissue Palpation*, Cambridge, MA, October 1998.

CIARLET, P.G. (1987). *Mathematical Elasticity, vol. 1: Three-dimensional Elasticity* (North-Holland, Amsterdam). ISBN 0-444-70259-8.

COSMI, F. (2001). Numerical solution of plane elasticity problems with the cell method. *Comput. Methods Engrg. Sci.* **2** (3).

COSTA, I.F., BALANIUK, R. (2001). Lem – an approach for real time physically based soft tissue simulation. In: *Int. Conf. Automation and Robotics*, ICRA'2001, Seoul, May 2001.

COTIN, S., DELINGETTE, H., AYACHE, N. (1999a). Real-time elastic deformations of soft tissues for surgery simulation. *IEEE Trans. Visual. Comput. Graph.* **5** (1), 62–73.

COTIN, S., DELINGETTE, H., AYACHE, N. (1999b). Real-time elastic deformations of soft tissues for surgery simulation. *IEEE Trans. Visual. Comput. Graph.* **5** (1), 62–73.

COTIN, S., DELINGETTE, H., AYACHE, N. (2000). A hybrid elastic model allowing real-time cutting, deformations and force-feedback for surgery training and simulation. *The Visual Computer* **16** (8), 437–452.

COTIN, S., DELINGETTE, H., CLEMENT, J.-M., TASSETTI, V., MARESCAUX, J., AYACHE, N. (1996). Volumetric deformable models for simulation of laparoscopic surgery. In: *Proc. Int. Symp. on Computer and Communication Systems for Image Guided Diagnosis and Therapy, Computer Assisted Radiology*, CAR'96. In: Int. Congr. Ser. **1124** (Elsevier, Amsterdam).

COUINAUD (1957). *Le foie, Études anatomiques et chirurgicales* (Masson, Paris).

COVER, S.A., EZQUERRA, N.F., O'BRIEN, J.F. (1993). Interactively deformable models for surgery simulation. *IEEE Comput. Graph. Appl.* **13**, 68–75.

DAN D. (1999). Caractérisation mécanique du foie humain en situation de choc, PhD thesis, Université Paris 7.

DEBUNNE, G., DESBRUN, M., CANI, M.-P., BARR, A.H. (2001). Dynamic real-time deformations using space and time adaptive sampling. In: *Computer Graphics Proceedings*, SIGGRAPH'01, August 2001.

DECORET, X., SCHAUFLER, G., SILLION, F., DORSEY, J. (1999). Multi-layered impostors for accelerated rendering. *Computer Graphics Forum (Eurographics'99)* **18**, 61–73.

DELINGETTE, H. (1998). Towards realistic soft tissue modeling in medical simulation. *Proc. IEEE: Special Issue on Surgery Simulation* **86**, 512–523.

DELINGETTE, H. (1999). General object reconstruction based on simplex meshes. *Int. J. Comput. Vision* **32** (2), 111–146.

DELINGETTE, H., MONTAGNAT, J. (2001). Shape and topology constraints on parametric active contours. *J. Comput. Vision and Image Understanding* **83**, 140–171.

DESBRUN, M., GASCUEL, M.-P. (1995). Animating soft substances with implicit surfaces. In: *Computer Graphics*, SIGGRAPH'95, Los Angeles.

DEUSSEN, O., KOBBELT, L., TUCKE, P. (1995). Using simulated annealing to obtain a good approximation of deformable bodies. In: *Proc. Eurographics Workshop on Animation and Simulation*, Maastricht, Netherlands, September 1995 (Springer-Verlag, Berlin).

DUNCAN, J., AYACHE, N. (2000). Medical image analysis: Progress over two decades and the challenges ahead. *IEEE Trans. on Pattern Analysis and Machine Intelligence* **22** (1), 85–106.

FOREST, C., DELINGETTE, H., AYACHE, N. (2002a). Cutting simulation of manifold volumetric meshes. In: *Modelling and Simulation for Computer-aided Medicine and Surgery*, MS4CMS'02.

FOREST, C., DELINGETTE, H., AYACHE, N. (2002b). Removing tetrahedra from a manifold mesh. In: *Computer Animation*, CA'02, Geneva, Switzerland, June 2002 (IEEE Computer Society), pp. 225–229.

FOREST C., DELINGETTE H., AYACHE N. (2003). Simulation of surgical cutting in a manifold mesh by removing tetrahedra, *Medical Image Analysis*, submitted for publication.

FRANCE, L., ANGELIDIS, A., MESEURE, P., CANI, M.-P., LENOIR, J., FAURE, F., CHAILLOU, C. (2002). Implicit representations of the human intestines for surgery simulation. In: *Conf. on Modeling and Simulation for Computer-aided Medicine and Surgery*, MS4CMS'02, Rocquencourt, November 2002.

FRANCE, L., LENOIR, J., MESEURE, P., CHAILLOU, C. (2002). Simulation of minimally invasive surgery of intestines. In: Richir, S. (ed.), *Fourth Virtual Reality International Conference*, VRIC'2002, pp. 21–27. ISBN 2-9515730.

FUNG, Y.C. (1993). *Biomechanics – Mechanical Properties of Living Tissues*, second ed. (Springer-Verlag, Berlin).

GIBSON, S., SAMOSKY, J., MOR, A., FYOCK, C., GRIMSON, E., KANADE, T., KIKINIS, R., LAUER, H., MCKENZIE, N. (1997). Simulating arthroscopic knee surgery using volumetric object representations, real-time volume rendering and haptic feedback. In: Troccaz, J., Grimson, E., Mosges, R. (eds.), *Proc. First Joint Conf. CVRMed-MRCAS'97*. In: Lecture Notes in Comput. Sci. **1205**, pp. 369–378.

GLADILIN E. (2002). Biomechanical modeling of soft tissue and facial expressions for craniofacial surgery planning, PhD thesis, Freie Univerisität Berlin, Germany.

GOTTSCHALK, S., LIN, M., MANOCHA, D. (1996). Obb-tree: A hierarchical structure for rapid interference detection. In: *Proc. SIGGRAPH 96*, New Orleans, LA, pp. 171–180. ISBN 0-201-94800-1.

HODGSKINSON, R., CURREY, J.D. (1992). Young modulus, density and material properties in cancellous bone over a large density range. *J. Materials Science: Materials in Medicine* **3**, 377–381.

HOUBOLT, J.C. (1950). A recurrence matrix solution for the dynamic response of elastic aircraft. *J. Aeronautical Sci.* **17**, 540–550.

HUMPHREY, J.D., STRUMPF, R.K., YIN, F.C.P. (1990). Determination of a constitutive relation for passive myocardium: I. A new functional form. *ASME J. Biomech. Engrg.* **112**, 333–339.

HUMPHREY, J.D., YIN, F.C.P. (1987). On constitutive relations and finite deformations of passive cardiac tissue: I. A pseudostrain-energy function. *ASME J. Biomech. Engrg.* **109**, 298–304.

HUNTER, P., PULLAN, A. (1997). *FEM/BEM Notes* (University of Auckland, New-Zeland). Available at http://www1.esc.auckland.ac.nz/Academic/Texts/fembemnotes.pdf.

JAMES, D.L., PAI, D.K. (1999). Artdefo accurate real time deformable objects. In: *Computer Graphics*, SIGGRAPH'99, pp. 65–72.

JOE, B. (1991). Geompack – a software package for the generation of meshes using geometric algorithms. *J. Advanced Eng. Software* **13**, 325–331.

KAISS, M., LE TALLEC, P. (1996). La modélisation numérique du contact œil-trépan. *Revue Européenne des éléments Finis* **5** (3), 375–408.

KASS, M., WITKIN, A., TERZOPOULOS, D. (1988). Snakes: Active contour models. *Int. J. Comput. Vision* **1**, 321–331.

KAUER, M., VUSKOVIC, V., DUAL, J., SZÉKELY, G., BAJKA, M. (2001). Inverse finite element characterization of soft tissues. In: *Proc. 4th Int. Conf. on Medical Image Computing and Computer-Assisted Intervention*, MICCAI'01, Utrecht, October 2001. In: Lecture Notes in Comput. Sci. **2208**, pp. 128–136.

KAYE, J., PRIMIANO, F., METAXAS, D. (1997). A 3d virtual environment for modeling mechanical cardiopulmonary interactions. *Medical Image Analysis (Media)* **2** (2), 1–26.

KNOTT, D., PAI, D. (2003). Collision and interference detection in real-time using graphics hardware. In: *Proc. Graphics Interface*, Halifax, Canada, June 2003.

KUHN, CH., KÜHNAPFEL, U., KRUMM, H.-G., NEISIUS, B. (1996). A 'virtual reality' based training system for minimally invasive surgery. In: *Proc. Computer Assisted Radiology*, CAR'96, Paris, June 1996, pp. 764–769.

KUHNAPFEL, U., AKMAK, H., MAA, H. (2000). Endoscopic surgery training using virtual reality and deformable tissue simulation. *Computers and Graphics* **24**, 671–682.

LIU, Z., BILSTON, L.E. (2000). On the viscoelastic character of liver tissue: experiments and modelling of the linear behaviour. *Biorheology* **37**, 191–201.

LIU, Z., BILSTON, L.E. (2002). Large deformation shear properties of liver tissue. *Biorheology* **39**, 735–742.

LOMBARDO, J.-C., CANI, M.-P., NEYRET, F. (1999). Real-time collision detection for virtual surgery. In: *Computer Animation*, Geneva, Switzerland, May 1999, pp. 82–89.

LORENSEN, W., CLINE, H.E. (1987). Marching cubes: a high resolution 3d surface construction algorithm. *ACM Computer Graphics (SIGGRAPH'87)* **21**, 163–169.

LOUCHET, J., PROVOT, X., CROCHEMORE, D. (1995). Evolutionary identification of cloth animation model. In: *Workshop on Computer Animation and Simulation*, Eurographics'95, pp. 44–54.

LUMSDAINE, A., SIEK, J. (1998). The Matrix Template Library. http://www.lsc.nd.edu/research/mtl/.

MACMILLAN, R.H. (1955). A new method for the numerical evaluation of determinants. *J. Roy. Aeronaut. Soc.* **59** (772).

MANDUCA, A., MUTHUPILLAI, R., ROSSMAN, P., GREENLEAF, J., EHMAN, L. (1996). Visualization of tissue elasticity by magnetic resonance elastography. In: *Proc. of Visualization in Biomedical Imaging*, VBC'96, Hamburg, Germany, pp. 63–68.

MARCUS, B. (1996). Hands on: Haptic feedback in surgical simulation. In: *Proc. of Medicine Meets Virtual Reality IV*, MMVR IV, San Diego, CA, January 1996, pp. 134–139.

MARK, W., RANDOLPH, S., FINCH, M., VAN VERTH, J., TAYLOR II, R.M. (1996). Adding force feedback to graphics systems: Issues and solutions. In: Rushmeier, H. (ed.), *ACM SIGGRAPH Computer Graphics Annual Conference*, SIGGRAPH'96 (Addison–Wesley, Reading, MA), pp. 447–452.

MAUREL, W., WU, Y., MAGNENAT THALMANN, N., THALMANN, D. (1998). *Biomechanical Models for Soft Tissue Simulation*, ESPRIT Basic Research Series (Springer-Verlag, Berlin).

SERRANO, C.M., LAUGIER, C. (2001). Realistic haptic rendering for highly deformable virtual objects. In: *Proc. Int. Conf. on Virtual Reality*, Yokohama, Japan, March 2001.

MILLER, K. (2000). Constitutive modelling of abdominal organs. *J. Biomech.* **33** (3), 367–373.

MONTAGNAT, J., DELINGETTE, H. (1998). Globally constrained deformable models for 3d object reconstruction. *Signal Processing*, 173–186.

NAVA, A., MAZZA, E., KLEINERMANN, F., AVIS, N., MCCLURE, J. (2003). Determination of the mechanical properties of soft human tissues through aspiration experiments. In: *Proc. Conf. on Medical Robotics, Imaging And Computer Assisted Surgery*, MICCAI 2003, Montreal, Canada, November 2003. In: Lecture Notes in Comput. Sci.

NEWMARK, N.M. (1959). A method of computation for structural dynamics. *J. Engrg. Mech. Division* **85**, 67–94.

O'MAHONY, A., WILLIAMS, J., KATZ, J. (1999). Anisotropic elastic properties of cancellous bone from a human edentulous mandible. In: *Proc. ASME Bioengineering'99 Conference*.

OWEN S. (2000). A survey of unstructured mesh generation technology. Technical report, Department of Civil and Environmental Engineering, Carnegie Mellon University.

PAPADEMETRIS, X., SHI, P., DIONE, D.P., SINUSAS, A.J., CONSTABLE, R.T., DUNCAN, J.S. (1999). Recovery of soft tissue object deformation from 3d image sequences using biomechanical models. In: *XVIth Int. Conf. on Information Processing In Medical Imaging*, IPMI'99, Visegrád, Hungary, June 28–July 2, 1999, pp. 352–357.

PARTHASARATHY, V.N., GRAICHEN, C.M., HATHAWAY, A.F. (1993). A comparison of tetrahedron quality measures. *Finite Elements in Analysis and Design* **15**, 255–261.

PICINBONO, G., DELINGETTE, H., AYACHE, N. (2001). Non-linear and anisotropic elastic soft tissue models for medical simulation. In: *IEEE Int. Conf. Robotics and Automation*, ICRA'2001, Seoul, Korea, May 2001. Best conference paper award.

PICINBONO, G., DELINGETTE, H., AYACHE, N. (2003). Non-linear anisotropic elasticity for real-time surgery simulation. *Graphical Models* **65** (5), 305–321.

PICINBONO, G., LOMBARDO, J.-C., DELINGETTE, H., AYACHE, N. (2000). Anisotropic elasticity and forces extrapolation to improve realism of surgery simulation. In: *IEEE Int. Conf. Robotics and Automation*, ICRA'2000, San Francisco, USA, April 2000, pp. 596–602.

PICINBONO, G., LOMBARDO, J.-C., DELINGETTE, H., AYACHE, N. (2002). Improving realism of a surgery simulator: linear anisotropic elasticity, complex interactions and force extrapolation. *J. Visual. Comput. Animation* **13** (3), 147–167.

PRESS, W.H., FLANNERY, B.P., TEUKOLSKY, S.A., VETTERLING, W.T. (1991). *Numerical Recipes in C* (Cambridge Univ. Press, Cambridge, UK).

PRESS, W.H., FLANNERY, B.P., TEUKOLSKY, S.A., VETTERLING, W.T. (1992). *Numerical Recipes in FORTRAN: The Art of Scientific Computing*, second ed. (Cambridge Univ. Press, Cambridge, UK).

PUSO, M.A., WEISS, J.A. (1998). Finite element implementation of anisotropic quasi-linear viscoelasticity using a discrete spectrum approximation. *ASME J. Biomech. Engrg.* **120** (1).

PUTTI, M., CORDES, C. (1998). Finite element approximation of the diffusion operator on tetrahedra. *SIAM J. Scientific Comput.* **19** (4), 1154–1168.

QUARTERONI, A., TUVERI, M., VENEZIANI, A. (2000). Computational vascular fluid dynamics: problems, models and methods. *Computing and Visualization in Science* **2**, 163–197.

RADETZKY, A. (1998). The simulation of elastic tissues in virtual medicine using neuro-fuzzy systems. In: *Medical Imaging'98: Image Display*, San Diego, CA, February 1998.

SAAD, Y. (1996). *Iterative Methods for Sparse Linear Systems* (WPS).

SAKUMA, I., NISHIMURA, Y., KONG CHUI, C., KOBAYASHI, E., INADA, H., CHEN, X., HISADA, T. (2003). In vitro measurement of mechanical properties of liver tissue under compression and elongation using a new test piece holding method with surgical glue. In: *Int. Symp. on Surgery Simulation and Soft Tissue Modeling*, Juan-Les-Pins, France, June 2003. In: Lecture Notes in Comput. Sci. **2673** (Springer-Verlag, Berlin), pp. 284–292.

SATAVA, R. (1994). Medicine 2001: The King Is Dead. In: *Proc. Conf. Virtual Reality in Medicine*.

SATAVA, R. (1996). Medical virtual reality: The current status of the future. In: *Proc. 4th Conf. Medicine Meets Virtual Reality*, MMVR IV, pp. 100–106.

SCHROEDER, W.J., ZARGE, J., LORENSEN, W. (1992). Decimation of triangles meshes. *Computer Graphics (SIGGRAPH'92)* **26**.

SERMESANT, M., COUDIÈRE, Y., DELINGETTE, H., AYACHE, N. (2002). Progress towards an electromechanical model of the heart for cardiac image analysis. In: *IEEE Int. Symp. Biomedical Imaging*, ISBI'02, pp. 10–14.

SERMESANT, M., FARIS, O., EVANS, F., MCVEIGH, E., COUDIÈRE, Y., DELINGETTE, H., AYACHE, N. (2003). Preliminary validation using in vivo measures of a macroscopic electrical model of the heart. In: Ayache, N., Delingette, H. (eds.), *Int. Symp. Surgery Simulation and Soft Tissue Modeling*, IS4TM'03. In: Lecture Notes in Comput. Sci. **2673** (Springer-Verlag, Heidelberg).

SILLION, F.X., DRETTAKIS, G., BODELET, B. (1997). Efficient impostor manipulation for real-time visualization of urban scenery. In: *Proc. Eurographics'97*, Budapest, Hungary, September 1997.

SIMAIL: product of Simulog S.A. – 1, rue James Joule, 78286 Guyancourt cedex, France, http://www.simulog.fr.

SOFERMAN, Z., BLYTHE, D., JOHN, N. (1998). Advanced graphics behind medical virtual reality: Evolution of algorithms, hardware and software interfaces. *Proc. IEEE: Special Issue on Surgery Simulation* **86** (3), 531–554.

SOLER, L., DELINGETTE, H., MALANDAIN, G., MONTAGNAT, J., AYACHE, N., CLÉMENT, J.-M., KOEHL, C., DOURTHE, O., MUTTER, D., MARESCAUX, J. (2000). Fully automatic anatomical, pathological and functional segmentation from ct-scans for hepatic surgery. In: *Medical Imaging 2000*, San Diego, February 2000.

SOLER, L., MALANDAIN, G., DELINGETTE, H. (1998). Segmentation automatique: application aux angioscanners 3d du foie. *Traitement du signal* **15** (5), 411–431 (in French).

SPENCER, A.J.M. (1972). *Deformations of Fibre-Reinforced Materials* (Clarendon Press, Oxford).

SPENCER, A.J.M. (1984). *Continuum Theory of Fiber-Reinforced Composites* (Springer-Verlag, New York).

SZEKELY, G., BAIJKA, M., BRECHBUHLER, C. (1999). Virtual reality based simulation for endoscopic gynaecology. In: *Proc. Medicine Meets Virtual Reality*, MMVR'99, San Francisco, USA, pp. 351–357.

TESCHNER, M., HEIDELBERGER, B., MULLER, M., POMERANETS, D., GROSS, M. (2003). Optimized spatial hashing for collision detection of deformable objects. In: *Proc. Vision, Modeling, Visualization*, VMV'03, Munich, Germany, November 2003.

VIDRASCU, M., DELINGETTE, H., AYACHE, N. (2001). Finite element modeling for surgery simulation. In: *First MIT Conf. on Computational Fluid and Solid Mechanics*.

VLACHOS, A., PETERS, J., BOYD, C., MITCHELL, J.L. (2001). Curved pn triangles. In: *2001 ACM Symp. on Interactive 3D Graphics*.

WEISS, J.A., GARDINER, J.C., QUAPP, K.M. (1995). Material models for the study of tissues mechanics. In: *Proc. Int. Conf. on Pelvic and Lower Extremity Injuries*, Washington, DC, December 1995, pp. 249–261.

WOO, M., NEIDER, J., DAVIS, T. (1997). *OpenGL Programing Guide* (Addison–Wesley, Reading, MA).

YAMASHITA, Y., KUBOTA, M. (1994). Ultrasonic characterization of tissue hardness in the in-vivo human liver. In: *Proc. IEEE Ultrasonics Symposium*, pp. 1449–1453.

ZORIN, D., SCHROEDER, P., SWELDENS, W. (1996). Interpolating subdivision for meshes with arbitrary topology. In: *Proc. 23rd Annual Conf. on Computer Graphics and Interactive Techniques* (ACM Press), pp. 189–192.

Recovering Displacements and Deformations from 3D Medical Images Using Biomechanical Models

Xenophon Papademetris

Departments of Biomedical Engineering and Diagnostic Radiology, Yale University, New Haven, CT, USA
E-mail: xenophon.papademetris@yale.edu

Oskar Škrinjar

Department of Biomedical Engineering, Georgia Institute of Technology, Atlanta, GA, USA
E-mail: oskar.skrinjar@bme.gatech.edu

James S. Duncan

Departments of Electrical Engineering, Biomedical Engineering and Diagnostic Radiology, Yale University, New Haven, CT, USA
E-mail: james.duncan@yale.edu

1. Introduction

The primary emphasis of this chapter is to describe the use of biomechanical models for the estimation of non-rigid displacement fields from sequences of three-dimensional medical images. In both case studies described later in this chapter, namely (i) the estimation of brain shift for neurosurgery and (ii) the estimation of left ventricular deformation, the proper modeling of the underlying tissue is important in order to ensure reliable and robust estimation of the underlying displacement and consequently the deformation. Modeling is needed as the image-derived displacement estimates generated from a number of methods (to be described in Section 2) have the following characteristics:

Computational Models for the Human Body
Special Volume (N. Ayache, Guest Editor) of
HANDBOOK OF NUMERICAL ANALYSIS, VOL. XII
P.G. Ciarlet (Editor)

ISSN 1570-8659
DOI 10.1016/S1570-8659(03)12006-6

- They are *sparse*. Displacements are only available at certain points and not the whole of the material.
- They are *noise-corrupted*. This is an inherent problem in all medical image analysis methods, although the level of noise is very method dependent.
- They may contain only *partial* information. Even where displacements are available, only a certain component of the displacement vector may be known.

The selection of an appropriate model and an appropriate modeling framework are of great importance for the estimation of complete and smooth displacement fields. The rest of this chapter reads as follows.

In Section 2 we present the underlying mathematical framework for the use of continuum mechanical models within this image analysis context. In particular, we present a brief introduction to continuum mechanics (Section 2.1) followed by a description of two frameworks for the integration of image-derived information with a mechanical model. The section continues with a description of a new and unique continuum mechanical model, the active elastic model devised specifically to model actively deforming tissue, which we will later demonstrate in Section 4. In the final part of this section we describe the key numerical technique used in this chapter – the finite element method.

We use two case studies to illustrate the use of this underlying mathematical framework. In Section 3 we describe methodology to compensate for brain shift in image guided neurosurgery and in Section 4 we describe algorithms to estimate the deformation of the left ventricle of the heart. Both sections are structured as follows. First, the background of the problems is given, followed by the overall system design (typically a sequence of image segmentation and mesh generation followed by image-based displacement data extraction). Next, the specifics of the mechanical model are presented, followed by validation results on real and simulated data.

We conclude the chapter with some further thoughts and remarks in Section 5.

2. Mathematical framework

In this section we introduce key concepts from continuum mechanics (Section 2.1) and present two possible frameworks for the integration of such material models with image-derived information (Sections 2.2 and 2.3.) In Section 2.4 we derive a modification of the linear elastic model, *the active elastic model* designed specifically to account for materials which are undergoing active rather than passive deformation.[1] Finally, in Section 2.5 we present an overview of the finite element method – the numerical technique used to solve for the displacements/deformations in the application of this methodology.

[1] We use the term passive deformation to refer to the change in material shape caused by external forces, e.g., gravity. The term active deformation is used to describe the change in shape caused by the object itself, e.g., muscle contraction.

2.1. A brief introduction to continuum mechanics

2.1.1. Deformations

The deformation gradient matrix. In this section we follow the presentations in SPENCER [1980], Chapter 6, and HUNTER, NASH and SANDS [1997]. Consider a body $B(0)$, shown in Fig. 2.1, which after time t moves and deforms to body $B(t)$. A material particle initially located at some position X on $B(0)$ moves to a new position x on $B(t)$. If we further assume that material cannot appear or disappear there will be a one-to-one correspondence between X and x, so we can always write the path of the particle as

$$x = x(X, t). \tag{2.1}$$

We can also define the displacement vector for this particle as

$$u(t) = x(t) - X. \tag{2.2}$$

This relationship is also invertible, given x and t, we can find X. Let us consider two neighboring particles located at X and $X + \mathrm{d}X$ on $B(0)$. In a new configuration $B(t)$ using Eq. (2.1), we can write

$$\mathrm{d}x = \frac{\partial x}{\partial X}\,\mathrm{d}X. \tag{2.3}$$

The Jacobian matrix $F(t) = \partial x(t)/\partial X$ is called *the deformation gradient matrix*. We note that by definition, $F(0) = I$. Using this, we can rewrite Eq. (2.1) more fully as

$$\mathrm{d}x(t) = F(t).\mathrm{d}X, \tag{2.4}$$

$$F_{ij} = \begin{cases} \dfrac{\partial x_i}{\partial X_j}, & F(0) = I, \\[2ex] \dfrac{\partial u_i}{\partial X_j} + \delta_{ij}, & u(0) = 0, \end{cases} \qquad \delta_{ij} = \begin{cases} 1, & i = j, \\ 0, & \text{otherwise.} \end{cases}$$

The mapping defined by Eqs. (2.1)–(2.4) has two components: a rigid motion component and a change in the shape or deformation of the object. For the purposes of capturing the material behavior (to be discussed in Section 2.1), we need to extract from

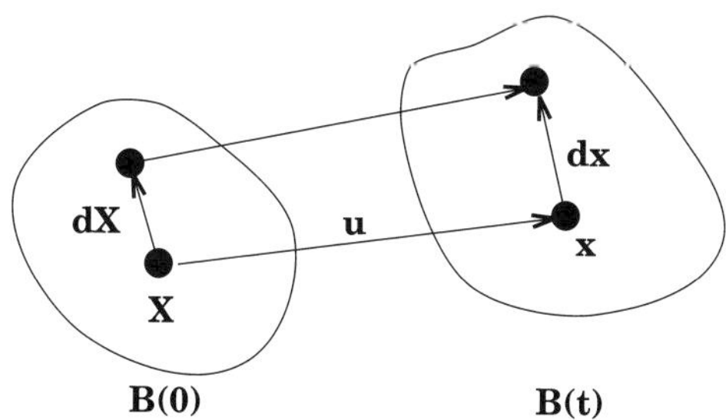

FIG. 2.1. Definition of displacement. Figure reprinted from PAPADEMETRIS, SINUSAS, DIONE and DUNCAN [2001], Estimation of 3D left ventricular deformation from echocardiography, *Medical Image Analysis* 5(1):17–29, ©2001 by permission from Elsevier.

F the component which is a function of the rigid motion and the component which is a function of the deformation.

To extract the deformation component we use the polar decomposition (STRANG [1986]) to write F as

$$F = \underbrace{R}_{\text{Rotation matrix}} \times \underbrace{U}_{\text{Symmetric matrix}}. \tag{2.5}$$

The matrix R is a rotation matrix having the properties $R * R^t = I$, $\det(R) = 1$, and U is a symmetric matrix, i.e., $U^t = U$.

It is also useful to define the *right Cauchy–Green deformation* matrix[2] $G = F^t F$. When we apply the polar decomposition, we get

$$G = F^t F = U^t R^t R U = U^t U. \tag{2.6}$$

This shows that G is independent of the rotation and is purely a function of the deformation. In the case of a pure rotation, i.e., $F = R$, we find that $G = I$. This shows that G in the case of a rotation is equal to identity. We also note that G has three scalar invariants under a rigid coordinate transformation defined as follows:

$$I_1 = \operatorname{trace}(G), \qquad I_2 = \frac{1}{2}\left(\operatorname{trace}(G)^2 - \operatorname{trace}\left(G^2\right)\right), \qquad I_3 = \det(G). \tag{2.7}$$

In particular, in the case of an incompressible material $\det(G) = I_3 = 1$. For completeness we also define the Green strain matrix E as $E = \frac{1}{2}(G - I)$. We next consider the important case of small deformations and rotations.

Small deformations and rotations. If the deformations and the rotations are small (e.g., a maximal length change of the order of < 2–3%, and a maximum rotation of $< 5°$), we use the approximation (SPENCER [1980], Section 6.6)

$$\frac{\partial u}{\partial x} \approx \frac{\partial u}{\partial X}. \tag{2.8}$$

From here we can rewrite $F = RU$ as

$$F = RU = (I + \omega)(I + \varepsilon). \tag{2.9}$$

Here ω is the small rotation matrix and is antisymmetric. ε is the small (infinitesimal) strain matrix and is symmetric. These are defined as

$$\omega = \frac{1}{2}\left(F - F^t\right) = \begin{bmatrix} 0 & \frac{1}{2}\left(\frac{\partial u_1}{\partial x_2} - \frac{\partial u_2}{\partial x_1}\right) & \frac{1}{2}\left(\frac{\partial u_1}{\partial x_3} - \frac{\partial u_3}{\partial x_1}\right) \\ \frac{1}{2}\left(\frac{\partial u_2}{\partial x_1} - \frac{\partial u_1}{\partial x_2}\right) & 0 & \frac{1}{2}\left(\frac{\partial u_2}{\partial x_3} - \frac{\partial u_3}{\partial x_2}\right) \\ \frac{1}{2}\left(\frac{\partial u_3}{\partial x_1} - \frac{\partial u_1}{\partial x_3}\right) & \frac{1}{2}\left(\frac{\partial u_3}{\partial x_2} - \frac{\partial u_2}{\partial x_3}\right) & 0 \end{bmatrix},$$

[2] In continuum mechanics literature this would be defined as the Cauchy–Green deformation tensor. A matrix in that terminology is simply a two-dimensional tensor. In this chapter, we avoid the term tensor and use the term matrix instead to improve general readability.

$$\varepsilon = \frac{1}{2}\left(F + F^t\right) - I = \begin{bmatrix} \frac{\partial u_1}{\partial x_1} & \frac{1}{2}\left(\frac{\partial u_1}{\partial x_2} + \frac{\partial u_2}{\partial x_1}\right) & \frac{1}{2}\left(\frac{\partial u_1}{\partial x_3} + \frac{\partial u_3}{\partial x_1}\right) \\ \frac{1}{2}\left(\frac{\partial u_2}{\partial x_1} + \frac{\partial u_1}{\partial x_2}\right) & \frac{\partial u_2}{\partial x_2} & \frac{1}{2}\left(\frac{\partial u_2}{\partial x_3} + \frac{\partial u_3}{\partial x_2}\right) \\ \frac{1}{2}\left(\frac{\partial u_3}{\partial x_1} + \frac{\partial u_1}{\partial x_3}\right) & \frac{1}{2}\left(\frac{\partial u_3}{\partial x_2} + \frac{\partial u_2}{\partial x_3}\right) & \frac{\partial u_3}{\partial x_3} \end{bmatrix}. \quad (2.10)$$

Often, taking advantage of the symmetries these matrices are written in vector form as

$$e = [\varepsilon_{11}, \varepsilon_{22}, \varepsilon_{33}, \varepsilon_{12}, \varepsilon_{13}, \varepsilon_{23}]^t, \qquad \theta = [0, 0, 0, \omega_{12}, \omega_{13}, \omega_{23}]^t. \quad (2.11)$$

This e is the classical definition for strain in infinitesimal linear elasticity (SPENCER [1980]). Using x, y, z to represent the coordinate axes, e can also be written as

$$e = [\varepsilon_{xx}, \varepsilon_{yy}, \varepsilon_{zz}, \varepsilon_{xy}, \varepsilon_{xz}, \varepsilon_{yz}]^t. \quad (2.12)$$

We note that the objectivity axiom is only approximately satisfied by the small deformation approximation.

2.1.2. *Material models*

So far we have restricted our description to the geometry of the deformation. In this section we extend this to account for what happens when a material deforms and relate the deformation to the change in the internal structure of the material. Before proceeding to give examples of possible material models, we first note that there are some theoretical guidelines which must be observed (ERINGEN [1980]). The most important ones for this work are:

(1) *The axiom of objectivity* – this requires the material model to be invariant with respect to rigid motion or the spatial frame of reference.

(2) *The axiom of material invariance* – this implies certain symmetry conditions dependent on the type of anisotropy of the material, and implicitly reduces the number of free parameters.

The first axiom can be satisfied by postulating an internal or strain energy function W, which depends on the gradient deformation matrix F only through the Green deformation matrix G, the Green strain matrix E, or in small deformation cases the infinitesimal strain matrix ε. The strain energy function serves as the material model. If we postulate an internal energy which is not invariant to a global rotation, we arrive at the following problem. Suppose that work is needed to rotate the object clockwise. From conservation of energy principles, this energy will be returned when the object is turned counter-clockwise. We can keep turning the object counter-clockwise to get more and more energy and in this way we have created a *perpetual motion machine* and not a material model.

Linear elastic energy functions. In this section e will be used to denote the vector form of the infinitesimal strain matrix ε. The simplest useful continuum model in solid mechanics is the linear elastic one. This is defined in terms of an internal energy function W which has the form

$$W = e^t Ce, \quad (2.13)$$

where C is a 6×6 matrix and defines the material properties of the deforming body.[3] The simplest model is the isotropic linear elastic model used widely in the image analysis literature. In this case the matrix C takes the form

$$C^{-1} = \frac{1}{E}\begin{bmatrix} 1 & -\nu & -\nu & 0 & 0 & 0 \\ -\nu & 1 & -\nu & 0 & 0 & 0 \\ -\nu & -\nu & 1 & 0 & 0 & 0 \\ 0 & 0 & 0 & 2(1+\nu) & 0 & 0 \\ 0 & 0 & 0 & 0 & 2(1+\nu) & 0 \\ 0 & 0 & 0 & 0 & 0 & 2(1+\nu) \end{bmatrix}, \tag{2.14}$$

where E is the Young's modulus which is a measure of the stiffness of the material and ν is the Poisson's ratio which is a measure of incompressibility. This is the model that will later be used to model brain deformation in the first case study of this chapter.

A transversely isotropic linear elastic model. For the second case study, involving the left ventricle of the heart, we model the tissue using a transversely elastic material to account for the preferential stiffness in the fiber direction. This is an extension of the isotropic linear elastic model which allows for one of the three material axis to have a different stiffness from the other two. In this case the matrix C takes the form

$$C^{-1} = \begin{bmatrix} \frac{1}{E_p} & \frac{-\nu_p}{E_p} & \frac{-\nu_{pf}}{E_p} & 0 & 0 & 0 \\ \frac{-\nu_p}{E_p} & \frac{1}{E_p} & \frac{-\nu_{pf}}{E_p} & 0 & 0 & 0 \\ \frac{-\nu_{pf}}{E_p} & \frac{-\nu_{pf}}{E_p} & \frac{1}{E_f} & 0 & 0 & 0 \\ 0 & 0 & 0 & \frac{2(1+\nu_p)}{E_p} & 0 & 0 \\ 0 & 0 & 0 & 0 & \frac{1}{G_f} & 0 \\ 0 & 0 & 0 & 0 & 0 & \frac{1}{G_f} \end{bmatrix}, \tag{2.15}$$

where E_f is the fiber stiffness, E_p is cross-fiber stiffness and ν_{pf}, ν_p are the corresponding Poisson's ratios and G_f is the shear modulus across fibers ($G_f \approx E_f/(2(1+\nu_{fp}))$). If $E_f = E_p$ and $\nu_p = \nu_{pf}$ this model reduces to the more common isotropic linear elastic model. The fiber stiffness was set to be 3.5 times greater than the cross-fiber stiffness (GUCCIONE and MCCULLOCH [1991]). The Poisson's ratios were both set to 0.4 to model approximate incompressibility. The fiber orientations used are shown in Fig. 2.2.

2.1.3. Stress and strain

While we have presented a material model formulation in terms of internal energy, an alternative description of the material model is in terms of the stress–strain relationship (SPENCER [1980]). The stress matrix can also be written in vector form (in the same

[3]This class of model is linear as it results in a linear stress–strain relationship, i.e., $\sigma = Ce$. The term elastic refers to the fact that the energy is completely recoverable, i.e., all energy used to compress the material is returned once the compression forces are removed. Further, in this linear elastic formulation thermal effects are ignored, which is equivalent to assuming constant temperature deformation.

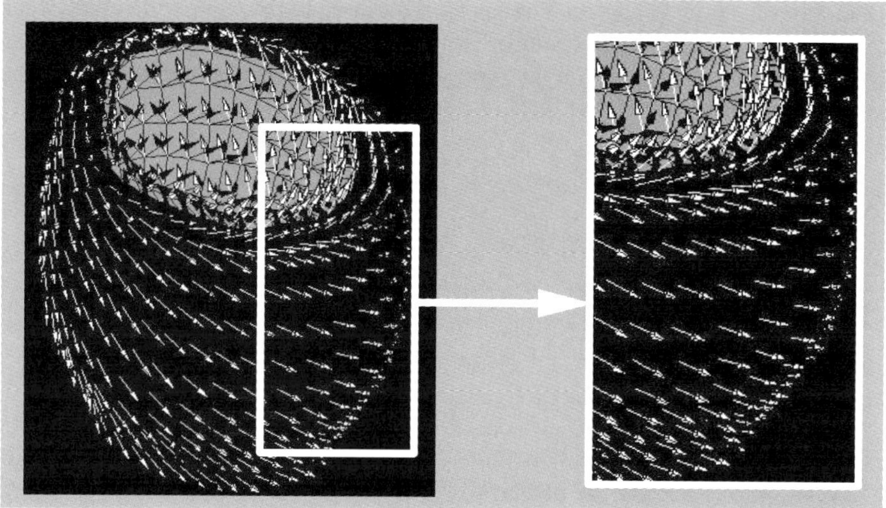

FIG. 2.2. Fiber direction in the left ventricle as defined in GUCCIONE and MCCULLOCH [1991]. Figure reprinted from PAPADEMETRIS, SINUSAS, DIONE and DUNCAN [2001], Estimation of 3D left ventricular deformation from echocardiography, *Medical Image Analysis* 5(1):17–29, ©2001 by permission from Elsevier.

fashion as the strain matrix, see Eq. (2.12)) as

$$\sigma = [\sigma_{xx}, \sigma_{yy}, \sigma_{zz}, \sigma_{xy}, \sigma_{xz}, \sigma_{yz}]^t. \tag{2.16}$$

If one pictures an infinitesimal cube of the material aligned with the coordinate axes x, y, z, the normal stresses σ_{xx}, σ_{yy} and σ_{zz} are equal to the force per unit area acting along the coordinate axes and on the faces of the cube. The shear stresses σ_{xy}, σ_{xz} and σ_{yz} act along the surfaces of the cube.

Given an applied external stress σ, the material deforms resulting in material strain e. If the material can be described using a linear elastic model, the stress–strain relationship has the form

$$\sigma = Ce, \tag{2.17}$$

where C is the 6×6 matrix from equation that models the material properties of the object.

Next we present two possible frameworks for the integration of such material models with image-derived information, the force equilibrium framework (Section 2.2) and the energy minimization framework (Section 2.3). We note that the two are interchangeable in the static case, but that the force equilibrium framework allows for the modeling of material damping such as in visco-elastic materials.

2.2. *The force equilibrium framework*

In the force equilibrium framework the integration of image-derived information with continuum mechanics material models is accomplished by converting the image-derived measurements into boundary conditions for a resulting set of partial differential equations. First, we consider that the underlying material is in a state of static equilibrium, which can be described in terms of the following equilibrium equations (HUEBNER, THORNTON and BYROM [1995]):

$$\begin{aligned}
&\frac{\partial \sigma_{xx}}{\partial x}+\frac{\partial \sigma_{xy}}{\partial y}+\frac{\partial \sigma_{xz}}{\partial z}+F_x=0,\\
&\frac{\partial \sigma_{xy}}{\partial x}+\frac{\partial \sigma_{yy}}{\partial y}+\frac{\partial \sigma_{yz}}{\partial z}+F_y=0,\\
&\frac{\partial \sigma_{xz}}{\partial x}+\frac{\partial \sigma_{yz}}{\partial y}+\frac{\partial \sigma_{zz}}{\partial z}+F_z=0,
\end{aligned} \tag{2.18}$$

where $\boldsymbol{F}=(F_x,F_y,F_z)$ are the externally applied forces.

To obtain the underlying displacement field we use the systems of Eqs. (2.17), (2.18) and (2.10), as well as the definition of C from Eq. (2.14). By eliminating stress (σ) and strain (e) components, one can obtain:

$$\begin{aligned}
&\nabla^2 u_x+\frac{1}{1-2\nu}\frac{\partial}{\partial x}\left(\frac{\partial u_x}{\partial x}+\frac{\partial u_y}{\partial y}+\frac{\partial u_z}{\partial z}\right)+\frac{F_x}{\mu}=0,\\
&\nabla^2 u_y+\frac{1}{1-2\nu}\frac{\partial}{\partial y}\left(\frac{\partial u_x}{\partial x}+\frac{\partial u_y}{\partial y}+\frac{\partial u_z}{\partial z}\right)+\frac{F_y}{\mu}=0,\\
&\nabla^2 u_z+\frac{1}{1-2\nu}\frac{\partial}{\partial z}\left(\frac{\partial u_x}{\partial x}+\frac{\partial u_y}{\partial y}+\frac{\partial u_z}{\partial z}\right)+\frac{F_z}{\mu}=0,
\end{aligned} \tag{2.19}$$

where $\mu=E/(2(1+\nu))$, and E, ν are the Young's modulus and Poisson's ratio, respectively. These three equations are elliptic PDEs in displacements only and are known as Navier equations (VALLIAPPAN [1981]). The image-derived information is used as boundary conditions in the numerical solution of Eq. (2.19). This framework is exploited in case study I and an example solution is presented in Section 3.2.

By proper use of either a finite element or a finite difference discretization scheme, Eq. (2.19) can also be rewritten in matrix form as

$$[K][U]=[F], \tag{2.20}$$

where K is the global stiffness matrix, and U and F are the concatenated displacement and force vectors, respectively. In particular, if the object is discretized to consist of n nodes, the vector U has the form

$$U=[u_{1,x},u_{1,y},u_{1,z},u_{2,x},u_{2,y},u_{2,z},\ldots,u_{n,x},u_{n,y},u_{n,z}]^t, \tag{2.21}$$

where $(u_{p,x},u_{p,y},u_{p,z})$ is the displacement of the pth node. The vector F similarly consists of all the forces that act at each of the nodes. By appropriate manipulation of the matrix K and the vector F, one can also impose displacement boundary conditions as well (HUEBNER, THORNTON and BYROM [1995]). We present an overview of the finite element method in Section 2.5.

2.3. *The energy minimization framework*

In this section we describe a framework in which the goal is to estimate a displacement field u which approximates another displacement field u^m. We will assume that u^m is derived from some image-based algorithm, such as the shape-based tracking algorithm, where the relationships between different displacements are not modeled. We simplify the approximation problem to be a least-squares fit of u to u^m subject to some constraints. This takes the form

$$\hat{u} = \arg\min_{u}\left(\int_V W(\alpha, u, x) + c(x)\left|u^m(x) - u(x)\right|^2 \mathrm{d}v\right), \tag{2.22}$$

where:

- $u(x) = (u_1, u_2, u_3)$ is the vector valued displacement field defined in the region of interest V and x is the position in space.
- $u^m(x) = (u_1^m, u_2^m, u_3^m)$.
- $c(x)$ is the spatially varying confidence in the measurements u^m.
- $W(\alpha, u, x)$ is a positive semi-definite functional which defines the approximation strategy and is solely a function of u, a parameter vector α and the spatial position x. In this work we will use the strain energy function (Eq. (2.13)) to set W.

This is commonly known as the regularization approach and $W(\alpha, u, x)$ is known as the *stabilization functional*. In certain cases the input displacement field u^m is *sparse* and is defined only on a finite number (P) of points p within V. In this case the overall functional takes the form

$$\hat{u} = \arg\min_{u}\left(\int_V W(\alpha, u, x)\,\mathrm{d}v + \sum_{i=1}^{P} c(p_i)\left|u^m(p_i) - u(p_i)\right|^2\right). \tag{2.23}$$

Using principles from the calculus of variations, we can minimize the functionals defined in Eqs. (2.22) and (2.23). In particular, using an appropriate discretization scheme, the derivative form can be re-expressed in the same matrix notation as that of Eq. (2.20), which then allows for the selection of an appropriate numerical solution scheme. We do not discuss the details of the continuous case, instead we present an overview of the finite element method in Section 2.5.

2.3.1. *A probabilistic interpretation of the energy minimization framework*

While we could simply substitute for the functional W in Eq. (2.23) with the appropriate internal energy function as defined in Eq. (2.13) (with C either from Eq. (2.14) or (2.15)), we proceed to describe a probabilistic interpretation of the energy minimization framework. The probabilistic interpretation is useful in indicating how one would integrate noise-corrupted image derived data with a biomechanical model. We exploit this framework in the second case study.

In the probabilistic interpretation of the energy minimization framework we again aim to estimate the output displacements u from a set of measurements u^m. We further assume that we are given the measurement probability density function $p(u^m|u)$, which also corresponds to the noise model for the measurements, and the prior probability density function for u, $p(u)$.[4] We pose this as a Bayesian a-posteriori estimation problem. Within this framework, the solution $\hat{u}$ is the u that maximizes the posterior probability density $p(u|u^m)$. Using Bayes' rule, we can write the posterior probability as

$$\hat{u} = \arg\max_u \left\{ p(u|u^m) = \frac{p(u, u^m)}{p(u^m)} = \frac{p(u^m|u)p(u)}{p(u^m)} \right\}. \tag{2.24}$$

First, we note that $p(u^m)$ is a constant once the measurements have been made and can therefore be ignored in the maximization process. We can rewrite the above expression by taking logarithms to arrive at

$$\hat{u} = \arg\max_u \left(\log p(u) + \log p(u^m|u)\right). \tag{2.25}$$

This expression is now in the same general form as Eq. (2.22). As previously demonstrated by D. GEMAN and S. GEMAN [1984] and applied to medical image analysis problems (e.g., CHRISTENSEN, RABBITT and MILLER [1994], GEE, HAYNOR, BRIQUER and BAJCSY [1997]), there is a correspondence between an internal energy function and a Gibbs probability density function. Given an energy function $W(\alpha, u, x)$ (noting again that this can be expressed using the strain energy function; see Eq. (2.13)), we can write an equivalent prior probability density function $p(u)$ (see Eq. (2.24)) of the Gibbs form (D. GEMAN and S. GEMAN [1984]):

$$\begin{aligned} p(u) &= k_1 \exp\left(-W(\alpha, u, x)\right), \\ \log\left(p(u)\right) &= \log(k_1) - W(\alpha, u, x), \end{aligned} \tag{2.26}$$

where k_1 is a normalization constant.

Next, we define the noise $n = u - u^m$. Then we can model the noise probabilistically, using a multivariate Gaussian distribution, as

$$\begin{aligned} p(n) &= k_2 \exp\left(\frac{-n^t \Sigma^{-1} n}{2}\right), \\ \log p(n) &= \log k_2 - \frac{1}{2} n^t \Sigma^{-1} n, \end{aligned} \tag{2.27}$$

where k_2 is also a normalization constant and Σ is the covariance matrix which in this case can be assumed to be diagonal, with the simplistic assumption that the noise is uncorrelated. The mean of the noise is assumed to be equal to zero. Substituting for n in this expression, we get

$$\log p\left(u^m|u\right) = k_2 - \frac{1}{2}\left(u^m - u\right)^t \Sigma^{-1}\left(u^m - u\right). \tag{2.28}$$

By an appropriate choice of Σ, the second term can be mapped to the data adherence term of Eq. (2.23). In this case Σ^{-1} will be a diagonal matrix with values $c(p_i)$ on the leading diagonal.

[4] We will not define the basic terms of probability here, they can be found in standard textbooks such as PAPOULIS [1991].

Advantages of the probabilistic interpretation. In the soft tissue deformation problem there are usually two types of information: (i) the image derived data which is corrupted by noise and (ii) the material properties of the soft tissue.

The data term is best modeled probabilistically in order to allow for the construction of a proper noise model. Here we can use ideas from the field of Digital Signal Processing (see, for example, OPPENHEIM and SCHAFER [1975]). The material term, however, is best defined in terms of a continuum mechanical model. The ability to generate an equivalent probability density function for an internal energy function, as was done in Eq. (2.26), allows us to take a continuum mechanics model defined in terms of an internal or strain energy function $W(\alpha, u, x)$, and generate a probability density function $p(u)$ which can then be used together with the probabilistic noise model within a Bayesian estimation framework.

2.3.2. *Soft tissue objects as Markov random fields*

In using the Gibbs form (Eq. (2.26)), we have modeled the displacement field of the solid probabilistically as a Markov random field, an example of this is shown in Fig. 2.3. The Markov Random Field (MRF) then can be thought of as the probabilistic analog of the continuum mechanical model. There are two interesting similarities: (i) both can be defined using energy functions and (ii) the energy functions at any given point are functions only of the values of that point and its immediate neighbors. In the case of the MRF point (ii) comes from the fact that the Gibbs probability density function is often defined on first and/or second order cliques which are very local neighborhoods of the point. So if the displacement field is modeled as a MRF, the probability of the displacement of a given point p effectively only depends on the displacement of its neighbors.

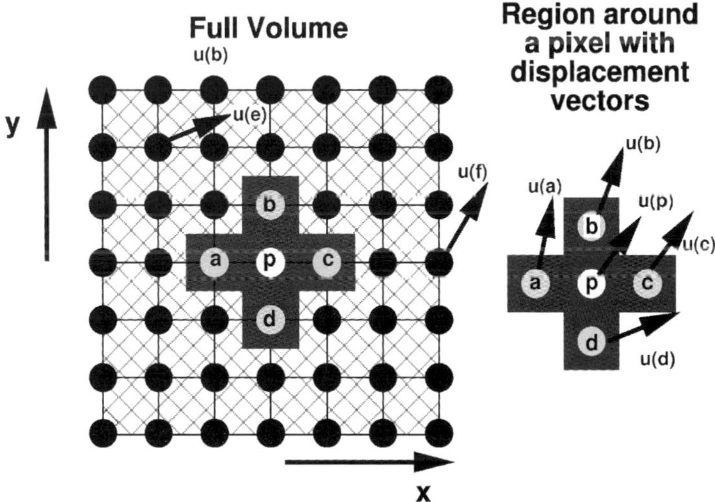

FIG. 2.3. Example of an object discretized by particles shown as black circles. If the displacement field is modeled as a first-order Markov Random Field (MRF) the displacement of a specific particle p depends only on external data and the displacements of its immediate neighbors a, b, c, d.

In the case of the mechanical model described using a strain energy function, the value of the internal energy function, which via exponentiation in Eq. (2.26) becomes the probability density function, at a given point depends only on the local strains. These local strains are only dependent on the displacements of the neighbors of the point and not on the displacements of the whole volume.

2.4. *The active linear elastic model*

The classical linear elastic model described in Eq. (2.13) is a passive model. In the absence of any external force, the material will do nothing. Given no external work, equilibrium is reached at the lowest energy state where the strain vector is identically equal to zero. Such a material model is not accurate in the case of actively deforming objects such as the left ventricle of the heart. In this case, a substantial part of the deformation is actively generated by the muscle and is clearly not a result of external forces. This active deformation does not produce a change in the strain energy of the material and to account for this factor, we need to modify the elastic model appropriately. With this in mind, we propose the active elastic model which takes the form

$$W = \frac{1}{2}\left(e - e^a\right)^t C\left(e - e^a\right), \tag{2.29}$$

where e^a is the active strain component. The active strain component represents the deformation that is not a product of external forces and hence should *not* be penalized by the model. In the absence of external forces, the active elastic model results in a deformation equal to the one actively generated by the object. So in this sense it can deform itself and hence it justifies the label *active*. Given a prior model of the active contraction, the active elastic model can also be used to generate a prediction of the position of the deforming object.

As an aside this model is also appropriate in the case where it is used to regularize an image registration problem where there is no such physical notion of active deformation. Here, the active component e^a can be thought of as the expected magnitude of the deformation.

Using the probabilistic interpretation to understand the active elastic model. By referring back to Section 2.3 and in particular to the expression of the internal energy function as a Gibbs prior (see Eq. (2.26)), we can proceed to understand the rationale for the active elastic model mathematically as follows. We first substitute for the internal energy functions of the active and the passive elastic models into Eq. (2.26). This results in prior probability distributions of the form

$$\text{Passive:} \quad \log p(u) = k_1 + \frac{-e^t C e}{2}, \tag{2.30}$$

$$\text{Active:} \quad \log p(u) = k_2 + \frac{-(e - e^a)^t C (e - e^a)}{2}, \tag{2.31}$$

where k_1 and k_2 are normalization constants.

Note further that the standard multivariate normal distribution (mean $= \mu$, covariance $= \Sigma$) has the form (k_3 is similarly a normalization constant):

$$\log p(u) = k_3 + \frac{-(u-\mu)^t \Sigma^{-1}(u-\mu)}{2}. \tag{2.32}$$

By comparing Eqs. (2.30) and (2.31) to Eq. (2.32), we can see that in both cases the material matrix C plays a similar role to the inverse of the covariance matrix (the stiffer the material is, the greater the coupling between the displacements of neighboring points and hence the smaller the effective component of the covariance matrix), and that in the case of the active model, the active strain e^a acts like the mean of the distribution. In the case of the passive model, the mean is effectively zero. Hence, we can explicitly see that the active elastic model is a generalization of the passive model, by adding the possibility of having a non-zero mean. This is important in describing materials such as the actively contracting tissue of the left ventricle.

2.5. The finite element method

The finite element method is a numerical analysis technique for obtaining approximate solutions to a wide variety of engineering problems (HUEBNER, THORNTON and BYROM [1995]). The key to this method is that the domain of problem is divided into small areas or volumes called *elements*. The problem is then discretized on an element by element basis and the resulting equations *assembled* to form the global solution.

2.5.1. An example problem

In this section we will describe an example problem and outline how it could be solved using the finite element method. We will pose the problem in terms of an energy minimization framework where the goal is to estimate the displacement field $u(x, y, z)$ which is an optimal trade off between an internal energy function[5] $W(C, u)$ and approximating a noisy displacement field $u^m(x, y, z)$ in a weighted least squares sense.

We define the optimal solution displacement field u is the one that minimizes functional $P(u)$. This is defined as

$$P(u) = \int_{\text{vol}} \big(W(C,u) + V\big(u, u^m\big)\big)\,\mathrm{d(vol)},$$

$$W(C,u) = e(u)^t C e(u), \qquad V\big(u, u^m\big) = \alpha\big(u^m - u\big)^2,$$

where $W(C, u)$ is the internal energy function defined by a strain energy function. C is the constitutive law and e is the local strain which is a function of the displacements u. $V(u, u^m)$ is the external energy term. u^m is the original (shape-tracking) displacement estimate and α is the confidence in the match.

[5]Note that although W is defined as function of the strain e, as e is a function of the displacement u, W can also be written as a function of the displacement field u.

2.5.2. Outline of the solution procedure

Step 1. Divide volume into elements (tetrahedra or hexahedra) to provide the basis functions for the discretization. In Fig. 2.4 a myocardium is shown tessellated into hexahedral elements.

Step 2. Discretize the problem by approximating the displacement field in each element as a linear combinations of displacements at the nodes of each element. For a hexahedral element this discretization can be expressed as

$$u \approx \sum_{i=1}^{8} N_i u_i,$$

where N_i is the interpolation shape function for node i and u_i is the displacement at node i of the element. For the isoparametric hexahedral element shown in Fig. 2.5, we define a local coordinate system ξ_i, and in this the shape functions N_i take the form (HUEBNER, THORNTON and BYROM [1995], Section 5.5):

$$N_i(\xi_1, \xi_2, \xi_3) = \frac{1}{8}(1 + \xi_1\xi_{1,i})(1 + \xi_2\xi_{2,i})(1 + \xi_3\xi_{3,i}), \tag{2.33}$$

where $(\xi_{1,i}, \xi_{2,i}, \xi_{3,i})$ are the local coordinates of node i. It is easy to verify that the shape function N_i takes a value of 1 at node i, a value of $\frac{1}{8}$ at the origin and a value of 0 at all other nodes.

Step 3. Write down internal energy equation as the sum of the internal energy for each element:

$$W(u) = \sum_{\text{all elements}} \left[\int_{v_{\text{el}}} e^t C e \, \mathrm{d}(v_{\text{el}}) \right]. \tag{2.34}$$

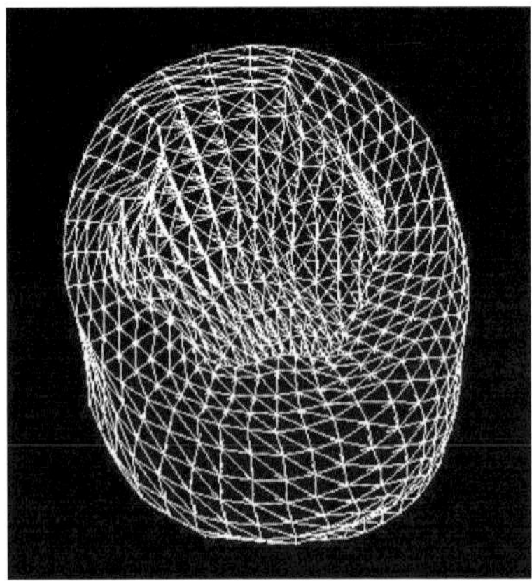

FIG. 2.4. A 3D hexahedral mesh generated by interpolating and filling between the endocardial and epicardial boundaries. Figure reprinted from PAPADEMETRIS, SINUSAS, DIONE and DUNCAN [2001], Estimation of 3D left ventricular deformation from echocardiography, *Medical Image Analysis* 5(1):17–29, ©2001 by permission from Elsevier.

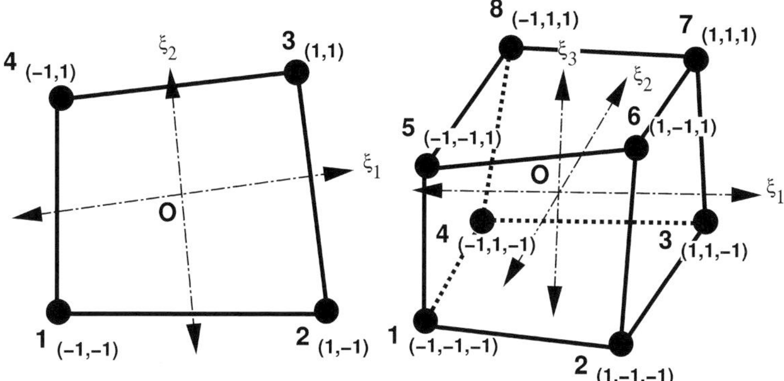

FIG. 2.5. Definition of local element coordinate system ξ_i and node coordinates for the nodes of a 2D 4-node isoparametric element (left) and a 3D 8-node isoparametric element (right). For example, in the 2D case, node 1 has coordinates $(-1, -1)$. The centroid of the element O is the origin of the element specific coordinate system. Note also that the axes are not necessarily orthogonal.

We further note that in an element we can approximate the derivatives of u with respect to components of the global coordinate system x as follows (note that the u_i are constant in this expression):

$$\frac{\partial u}{\partial x_k} = \sum_{i=1}^{8} \frac{\partial (N_i u_i)}{\partial x_k} = \sum_{i=1}^{8} \frac{\partial N_i}{\partial x_k} u_i.$$

However the shape functions N_i are expressed in terms of the local coordinate system ξ. Using the chain rule, we can write

$$\begin{Bmatrix} \dfrac{\partial N}{\partial \xi_1} \\ \dfrac{\partial N}{\partial \xi_2} \\ \dfrac{\partial N}{\partial \xi_3} \end{Bmatrix} = \begin{bmatrix} \dfrac{\partial x_1}{\partial \xi_1} & \dfrac{\partial x_2}{\partial \xi_1} & \dfrac{\partial x_3}{\partial \xi_1} \\ \dfrac{\partial x_1}{\partial \xi_2} & \dfrac{\partial x_2}{\partial \xi_2} & \dfrac{\partial x_3}{\partial \xi_2} \\ \dfrac{\partial x_1}{\partial \xi_3} & \dfrac{\partial x_2}{\partial \xi_3} & \dfrac{\partial x_3}{\partial \xi_3} \end{bmatrix} \times \begin{Bmatrix} \dfrac{\partial N}{\partial x_1} \\ \dfrac{\partial N}{\partial x_2} \\ \dfrac{\partial N}{\partial x_3} \end{Bmatrix} \tag{2.35}$$

or equivalently in matrix notation as $N_\xi = [J] \times N_x$.

Hence we can calculate the desired derivatives N_x from the known derivatives N_ξ by inverting the Jacobian as follows: $N_x = [J]^{-1} N_\xi$. As long as the elements do not have intersecting sides the Jacobian will remain invertible.

Note also that the derivatives of the displacement field u (i.e., $\partial u/\partial x_k$) are a linear function of the nodal displacements u_i. Since the infinitesimal strain tensor consists of only sums and differences of partial derivatives (see Eq. (2.10)) the infinitesimal strain tensor can also be expressed as a linear function of the nodal displacements.[6] This can

[6]The finite strain deformation case is non-linear and does not allow for this simplification. The subsequent expressions are so complicated that it makes the material beyond the scope of this brief overview. The reader is referred to BATHE [1982].

be written in matrix form as $e = Bu$. Substituting this in Eq. (2.34), we get

$$W(u) = \sum_{\text{all elements}} U^{et}\left[\int_{v_{\text{el}}} B^t C B \, \mathrm{d}(v_{\text{el}})\right] U^e = \sum_{\text{all elements}} U^{et}\left[K^e\right] U^e,$$

where K^e[7] is the element stiffness matrix,[8] and U^e is a vector obtained by concatenating all the displacements of the nodes of the element, i.e.,

$$U^e = [u_{1,x}, u_{1,y}, u_{1,z}, \ldots, u_{8,x}, u_{8,y}, u_{8,z}],$$

where $u_i = (u_{i,x}, u_{i,y}, u_{i,z})$ is the displacement of node i.

Step 4. Rewrite the internal energy function in matrix form. First, we define the global displacement vector U as

$$U = [u_{1,x}, u_{1,y}, u_{1,z}, u_{2,x}, u_{2,y}, u_{2,z}, \ldots, u_{n,x}, u_{n,y}, u_{n,z}]^t, \tag{2.36}$$

where n is the total number of nodes for the solid. We also define the global stiffness matrix K as the assembly of all the local element stiffness matrices K^e as

$$K = \sum_{\text{all elements}} \mathcal{I}\left(K^e\right), \tag{2.37}$$

where $\mathcal{I}$ is the re-indexing function. This takes an element K^e_{ij} and adds it to the element K_{kl}, where k and l are the global node numbers of local nodes i and j.[9]

The internal energy can now be written as $W(U) = U^t K U$.

Step 5. Write down the external energy function as a weighted least squares term,

$$V(u) = \sum_{i=1}^{n} \alpha_i \left(u^e_i - u_i\right)^2.$$

If there is no initial displacement estimate for a given node j, set $\alpha_j = 0$.

---

[7] The integration is carried out using Gaussian quadrature (HUEBNER, THORNTON and BYROM [1995]).

[8] Each component of K^e indicates the 'stiffness' between any two nodes. One could in some sense think of K^e_{14} as the stiffness of a spring connecting the x-directions of local nodes 1 and 2. (This '2' is *not* a typo. The first three rows of K^e correspond to the components of the displacement of node 1, the second three to the displacement of node 2, etc. See the definition of U^e.)

[9] Within an element the nodes are always numbered from 1 to 8. However this is a local index (shorthand) to the global node numbers. When the global matrix is assembled the local indices (1 to 8) need to be converted back to the global indices (e.g., 1 to n). K^e has dimensions 24×24 and K has dimensions $3n \times 3n$. K^e_{14}, which is the stiffness between the x-directions of local nodes 1 and 2 would be part of K_{kl} where $k = 3(a-1)+1$ and a is the global index of local node 1 and $l = 3(b-1)+1$, where b is the global index of local node 2. Since nodes appear in more than one element the final value of K_{kl} is likely to be the sum of a number of local K^e_{ij}'s.

Step 6. Rewrite external energy in a matrix form: we define the global initial displacement vector U^m in the same way as U above (see Eq. (2.36)) and the global confidence matrix A to be a diagonal matrix with the confidence values for each displacement (α_i) forming the elements of the leading diagonal as follows:

$$A = \begin{bmatrix} a_1 & & & & & & \\ & a_1 & & & & & \\ & & a_1 & & & & \\ & & & \cdots & & & \\ & & & & a_n & & \\ & & & & & a_n & \\ & & & & & & a_n \end{bmatrix}. \tag{2.38}$$

The external energy can be rewritten as $V(U) = (U^m - U)^t A(U^m - U)$.

Step 7. Form total potential energy equation $P(U) = W(U) - V(U)$.

Step 8. Solve for U. Differentiate $P(U)$ w.r.t. U and set to 0. This results in the final equation

$$KU = A\big(U^m - U\big).$$

This is then solved for U using sparse matrix methods.[10] U represents the values of u at the nodes, and by means of the finite element approximation ($u \approx \sum_{i=1}^{8} N_i u_i$) we can compute the resulting values of the displacement field u anywhere in the volume.

Having described the general common framework for the use of biomechanical models in the estimation of non-rigid displacement fields from medical images, we now proceed to two specific case studies: (i) the estimation of brain shift for image guided neurosurgery and (ii) the estimation of left ventricular deformation.

3. Case study I: brain shift compensation for image guided neurosurgery

3.1. Background

The use of surgical navigation systems has become a standard way to assist the neurosurgeon in navigating within the intraoperative environment, planning and guiding the surgery. One of the most important features of these systems is the ability to relate the position of the surgical instruments to the features in the preoperative images. Ideally, they should provide a 3D display of the neuroanatomical structures of interest and include visualization of surgical instruments within the same frame. In order to be reliably used, the surgical navigation system should be as precise as possible, preferably to within the voxel size of the dataset used (see GRIMSON, ETTINGER, WHITE, GLEASON, LOZANO-PEREZ, WELLS III and KIKINIS [1996]). Most of the

[10] In the case of finite deformations we end up with an expression of the form $K(U) = A(U^m - U)$ which is solved iteratively.

current systems use preoperatively-acquired 3D data and register it to the patient coordinate system (GRIMSON, ETTINGER, WHITE, GLEASON, LOZANO-PEREZ, WELLS III and KIKINIS [1995], GRIMSON, ETTINGER, WHITE, GLEASON, LOZANO-PEREZ, WELLS III and KIKINIS [1996], PETERS, DAVEY, MUNGER, COMEAU, EVANS and OLIVIER [1996], CHABRERIE, OZLEN, NAKAJIMA, LEVENTON, ATSUMI, GRIMSON, KEEVE, HELMERS, RIVIELLO, HOLMES, DUFFY, JOLESZ, KIKINIS and BLACK [1998]). However, they assume that the brain and other intracranial structures are rigid and fixed relative to the skull. The preoperative data is registered to the patient coordinate system at the beginning of the surgery. While this can be done with a precision to within 1 mm at start (GRIMSON, ETTINGER, WHITE, GLEASON, LOZANO-PEREZ, WELLS III and KIKINIS [1996]), the brain deforms within the skull over time and thus the accuracy of the system deteriorates. The median brain shift of points on the brain surface has been estimated to range from 0.3 to 7.4 mm (HILL, MAURER, WANG, MACIUNAS, BARWISE and FITZPATRICK [1997]). It is clear that a system based on the rigid brain assumption cannot achieve a precision better than a few millimeters at the outer structures. Since the deeper brain structures deform less than the outer ones the error is the largest at the cortical surface. Obviously, the brain deforms even more after interventions, e.g., post-resections. The average brain shift for cases in which hematoma or tumors were removed has been reported to be even larger: 9.5 and 7.9 mm, respectively (BUCHOLZ, YEH, TROBAUGH, MCDURMONT, STURM, BAUMANN, HENDERSON, LEVY and KESSMAN [1997]). In our research, we are mainly concerned with (but not limited to) issues surrounding epilepsy surgery where the amount of brain shift of concern is more in line with that cited by HILL, MAURER, WANG, MACIUNAS, BARWISE and FITZPATRICK [1997], although the physical implantation and removal of subdural electrode grids further affect the amount of physical deformation.

Relatively little effort has been put forth to attempt to compensate for the deformation that the brain undergoes during a surgical procedure. One line of investigation incorporates the use of intraoperative MRI (iMRI) to periodically acquire full sets of 3D MR images that can be matched to preoperative anatomical datasets (GERING, NABAVI, KIKINIS, et al. [1999], HATA, NABAVI, WARFIELD, et al. [1999], HILL, MAURER, MARTIN, et al. [1999], FERRANT, NABAVI, MACQ, JOLESZ, KIKINIS and WARFIELD [2001], NABAVI, BLACK, GERING, et al. [2001], MIGA, ROBERTS, KENNEDY, PLATENIKI, HARTOV, LUNN and PAULSEN [2001], WARFIELD, TALOS, TEI, et al. [2002]). The cost-effectiveness and true utility of iMRI remains an open question. Other attempts at this have used intraoperative ultrasound imaging (BUCHOLZ, YEH, TROBAUGH, MCDURMONT, STURM, BAUMANN, HENDERSON, LEVY and KESSMAN [1997]), physical modeling (MIGA, PAULSEN, KENNEDY, HOOPES, HARTOV and ROBERTS [1998], MIGA, ROBERTS, KENNEDY, PLATENIKI, HARTOV, LUNN and PAULSEN [2001]), and includes our own work in using sparse sets of points to update a physical model (SKRINJAR and DUNCAN [1999]), but all of these ideas remain in the very earliest stages of investigation and validation. There is additional work in the field such as AUDETTE, SIDDIQI and PETERS [1999].

3.2. *System description*

Our approach to brain shift compensation employs a 3D biomechanical brain model (SKRINJAR, NABAVI and DUNCAN [2002]) guided by some limited interoperatively acquired data. During the surgery we can use the model output to display preoperative data (deformed according to the current model state). Before the surgery one can acquire anatomical (MRI, CT) and functional (functional MR, SPECT, PET, ...) images, segment them, generate surfaces of the segmented structures of interest, and then deform all of them intraoperatively based on the current model state. If the model deformation prediction is close to the actual brain deformation, then the displayed images and structures of interest (that are deformed according to the current model state) are closer to the current actual brain state than they would be if one did not use the brain shift compensation, making the surgical navigation system more precise and reliable. An example of the effect of the brain shift is shown in Fig. 3.1. We note that the estimation of interior displacements will be based on surface information produced from a 10–12 cm craniotomy that occurs during the first stage of these surgical procedures.

Therefore, we propose a biomechanical-model-based brain shift compensation system composed of the following steps: preoperative image acquisition, segmentation, mesh generation, registration of the model to the intraoperative environment, model setup and guidance, and visualization of model-updated preoperative data.

3.2.1. *Segmentation, visualization and registration*

The first step after the preoperative image acquisition is the segmentation of the brain tissue. For this task we have adopted an approach based on the automated algorithm

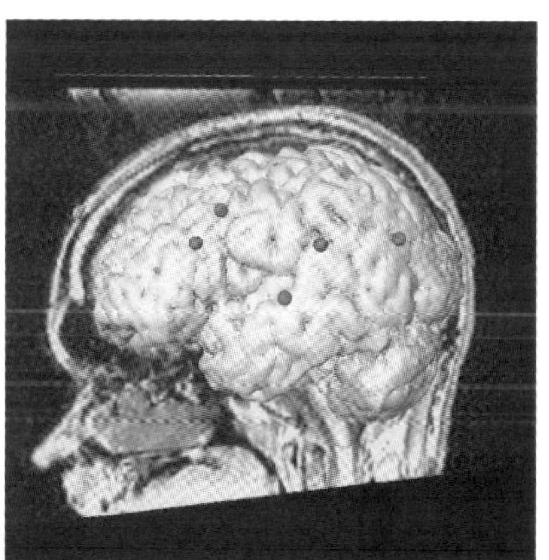
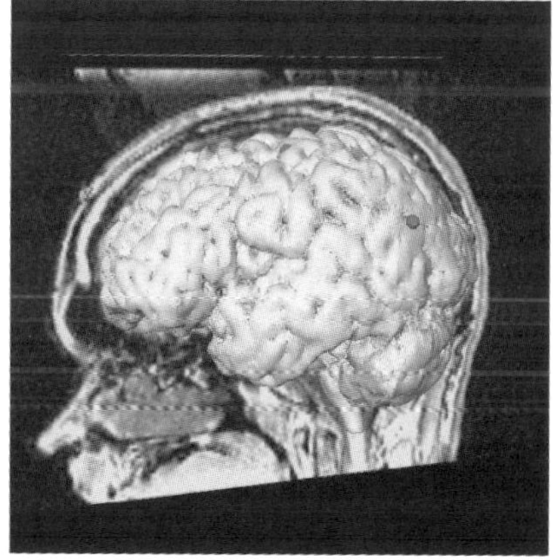

FIG. 3.1. Intraoperatively recorded points on the exposed brain surface at the beginning of the surgery are shown at left, while their positions about 45 minutes later relative to the same pre-deformation brain surface are shown at right. Gravity is perpendicular to the sagittal plane. The points moved in the direction of gravity and they are hidden under the pre-deformation brain surface (only one of the points is still visible in the figure at right). Since the brain deformed (in the direction of the gravity vector), the surface points moved relative to the pre-deformation brain surface. Figure reprinted from SKRINJAR, NABAVI and DUNCAN [2002], Model-driven brain shift compensation, *Medical Image Analysis*, 6(4):361–373, ©2002 by permission from Elsevier.

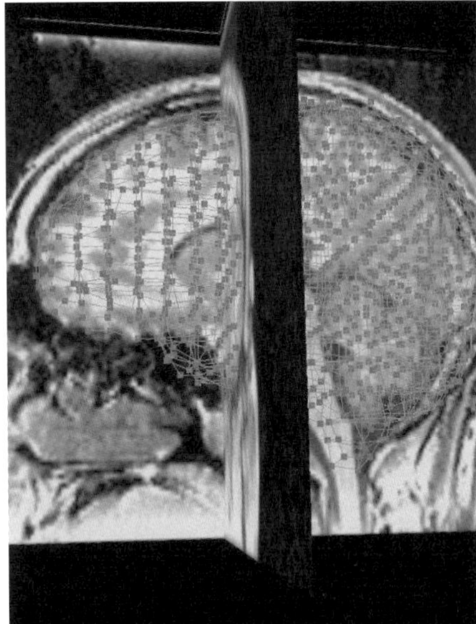
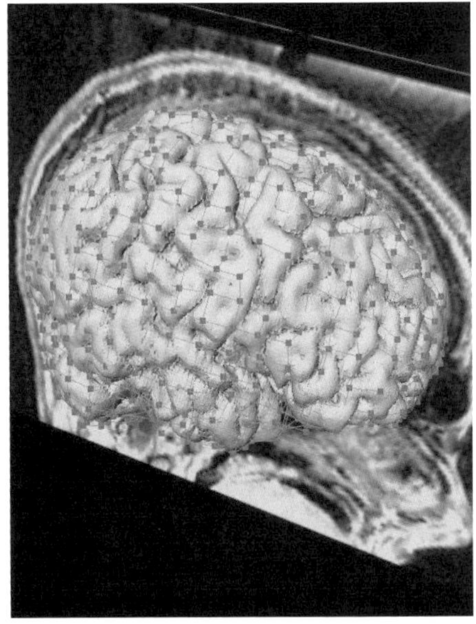

FIG. 3.2. A typical model mesh. The left figure shows the mesh, while the right one shows the mesh and the outer brain surface. The mesh has over 2000 nodes and 1500 elements (bricks). Figure reprinted from SKRINJAR, NABAVI and DUNCAN [2002], Model-driven brain shift compensation, *Medical Image Analysis*, 6(4):361–373, ©2002 by permission from Elsevier.

suggested in STOKKING [1998]. The main idea is to, after thresholding the brain MR image (the threshold selection is the only manual input), a couple (typically three) binary erosions are performed to disconnect the brain tissue from the rest. After that, the largest connected object, which is brain, is selected and then dilated the same number of times as the thresholded image was eroded. The output of the dilations are masked with the thresholded image to insure that the resulting object is within the brain tissue. This simple and fast algorithm produces brain segmentation results of sufficient quality for this project, since the mesh (which generation is based on the brain segmentation) for the biomechanical model does not require the finest geometric details of the brain.

For object surface rendering we have used an improved version of the algorithm presented in GIBSON [1998]. Some of the surfaces produced by this algorithm can be seen in Fig. 3.2.

In order to display and use brain surface data for model guidance, a rigid body transformation between the patient and preoperative image coordinate systems has to be established. For this purpose we used a set of fiducial markers placed on the patient's skin. In the operating room (OR), the marker coordinates were recorded using a mechanical localizer (OMI [1997]). In addition, the markers were manually localized in the preoperative MRI dataset.[11] Then a robust point matching algorithm for resolving the correspondences and finding the optimal rigid body transformation between the two sets of marker locations was applied. The approach relies on the method for computing

[11]Markers have to be visible in both MR and/or CT image data.

the optimal (in the least squares sense) rigid body transformation between two sets of points with known correspondences (ARUN, HUANG and BLOSTEIN [1987]). To establish proper point correspondences, for the three most distant points in the first set, all possible ordered triple of points from the second set are tested (note that there are $N(N-1)(N-2)$ triples, where N is the number of points; this is not computationally demanding, since if $N = 15$, there are less than 3000 triples to check). For an ordered triple of points from the second set, the optimal rigid body transformation is computed from the three selected points in the first set to the triple. This rigid body transformation is used to establish correspondences (based on the closest point criterion) for all the points. Once the correspondences are known for all the points, the optimal rigid body transformation is computed, and the sum of squared distances between corresponding points is stored. This is repeated for all the ordered triples in the second set of points, and the rigid body transformation that yields the smallest sum of squared distances is taken as the final one. Once the rigid body transformation is determined, any point recorded by the localizer can be mapped to the preoperative image coordinate system.

3.2.2. Mesh generation

The next step is to generate the model mesh from the segmented brain tissue. Here we use hexahedral ("brick") elements having 8 nodes at the vertex positions. The segmented object (the brain tissue in this case) is the input to our mesh generator, which generates an unstructured mesh (LISEIKIN [1999]). The algorithm first generates a regular 3D matrix of bricks over the full 3D image. Each brick that has at least a half of its volume inside the segmented object is kept, and others are discarded. The kept bricks will compose the final mesh, while their nodes will be finely readjusted. The nodes are divided into two groups. Each node that has all of its neighboring nodes left is called an interior node, and all other nodes are called surface nodes. Each surface node is moved to the closest point on the surface of the segmented object. Note that surface nodes before moving were not far from the surface of the segmented object. Finally, the interior nodes are smoothed using a Laplacian-type smoother, in order to enhance the regularity of the mesh. A typical output of the mesh generator is shown in Fig. 3.2. The meshes we use do not capture all of the fine details of the segmentation output, but they still achieve a reasonable performance in terms of accuracy and speed.

3.2.3. Image-based displacement estimates

There are different types of intraoperative data available for model guidance: points (e.g., using a localizer and recording brain surface points over time), surface data: obtained by a range system (AUDETTE, SIDDIQI, FERRIE and PETERS [2003]) or by a stereo camera system (SKRINJAR, TAGARE and DUNCAN [2000]), and volumetric data obtained by intraoperative image acquisition systems (intraoperative MRI, CT and ultrasound).

For this model we assume the use of a pair of stereo cameras overlooking the exposed brain surface to acquire intraoperative information about the deforming brain. The idea is to reconstruct and track the exposed brain surface as it deforms during the surgery. If this can be done reliably, one can use the reconstructed brain surface as displacement boundary conditions for the model PDEs. Each time the surgeon moves her or his hands and surgical tools out of the way of the cameras, snapshots from the two cameras are taken, exposed brain surface is reconstructed, the surface is used to guide the model, and once the model is deformed, it can be used to update (properly warp) all preoperative images available.

In these early results we simulate exposed brain surface tracking[12] using intraoperative MR images. In particular, we manually segmented the deformed brain from the intraoperative scan and generated its surface. Since the brain surface did not move significantly, we computed the displacement at each point $\boldsymbol{r}_1$ of the undeformed brain surface S_1 (only at the part of the brain surface that was visible through the craniotomy, i.e., at the exposed brain surface), as $\boldsymbol{\Delta r} = \boldsymbol{r}_2 - \boldsymbol{r}_1$, where $\boldsymbol{r}_2$ is the point on the deformed brain surface S_2 closest to the point $\boldsymbol{r}_1$, i.e., obtained as $\arg_{\mathbf{r}_2 \in S_2} \min \|\mathbf{r}_2 - \mathbf{r}_1\|$. A more advanced version of this approach is given in BESL and MACKAY [1992].

Finally, the computed displacements at the exposed brain surface were used as boundary conditions for the partial differential equations derived using the model to be described in the next section.

3.2.4. *Mechanical model-based integration*

We use a simple linear elastic model of brain deformation based on the following three assumptions: (a) we desire a *relatively simple model*. Due to the complexity of the brain shift phenomenon, not only that it is difficult to model some of the causative factors, but also it is not clear how to set the model parameters (any increase in the model complexity inevitably involves more parameters). Therefore, we base our approach on a simple model, that incorporates the main tissue characteristics (elasticity and near-incompressibility). The complexity of the deformation is made up by intraoperative guidance of the model. (b) We assume a *static model*. Since intraoperative brain deformation is a relatively slow process with negligible dynamic components, we use a static model. (c) We have some *intraoperative input*. The model has to by guided by intraoperative input.

Brain shift is a small deformation relative to the brain size, and it is a good approximation to use a linear stress strain relation and the infinitesimal strain approximation. Although brain tissues are not isotropic, especially white matter due to its fibrous structure, since the fiber directions are not currently available to us, we assume that brain tissues are isotropic materials.[13] Due to the toughness of falx and tentorium, the movement of the two structures is negligible in most cases. For this reason we fix the corresponding parts of the model, i.e., we consider only the brain hemisphere on the side of the craniotomy, and assume that the other brain parts do not deform.[14]

We integrate the material model and the image derived displacements using the force equilibrium framework presented in Section 2.2, resulting in the following partial dif-

[12]In a complete system brain surface tracking would be done by using a pair of stereo cameras.

[13]Note however that the nodes in the model mesh that are located on the walls of the ventricles were set to be free nodes. This allowed for "free" movement (the movement is constrained by the rest of the model) of the walls of the ventricles. For example, this allows for relatively good modeling of the collapse of lateral ventricles, which sometimes happens in brain surgery.

[14]We note that the brain–skull interaction is not modeled directly. Rather, the hemisphere of the brain on the side of the craniotomy is considered, the model nodes corresponding to falx and tentorium are fixed. This indirectly models the effect of the skull on the brain at locations opposite to the craniotomy.

ferential equations:

$$\nabla^2 u_x + \frac{1}{1-2\nu}\frac{\partial}{\partial x}\left(\frac{\partial u_x}{\partial x} + \frac{\partial u_y}{\partial y} + \frac{\partial u_z}{\partial z}\right) + \frac{F_x}{\mu} = 0,$$
$$\nabla^2 u_y + \frac{1}{1-2\nu}\frac{\partial}{\partial y}\left(\frac{\partial u_x}{\partial x} + \frac{\partial u_y}{\partial y} + \frac{\partial u_z}{\partial z}\right) + \frac{F_y}{\mu} = 0,$$
$$\nabla^2 u_z + \frac{1}{1-2\nu}\frac{\partial}{\partial z}\left(\frac{\partial u_x}{\partial x} + \frac{\partial u_y}{\partial y} + \frac{\partial u_z}{\partial z}\right) + \frac{F_z}{\mu} = 0.$$

These equations need to be solved with given displacement boundary conditions. Since they are linear PDEs, and since differentiation is a linear operator, one can separately find the solution $\boldsymbol{u}^1 = (u_x^1, u_y^1, u_z^1)$ for the equations with zero boundary conditions, and the solution $\boldsymbol{u}^2 = (u_x^2, u_y^2, u_z^2)$ for the equations with zero body force, and the total solution will be $\boldsymbol{u} = \boldsymbol{u}^1 + \boldsymbol{u}^2$. However, gravity acts all the time, both before and during the brain deformation, and therefore $\boldsymbol{u}^1$ will be the same in both cases. Since we are interested in the displacement field between the deformed and undeformed state, we do not need to compute $\boldsymbol{u}^1$. Thus, we need to solve only for $\boldsymbol{u}^2$, i.e., solve Eq. (2.19) with the given boundary conditions and zero body force. One should notice that gravity affects $\boldsymbol{u}^2$ through boundary conditions (since the brain deforms partly because of gravity, and a part of the brain surface will be used to define the displacement boundary conditions – there are no explicit force boundary conditions in this case).

Another interesting observation is that Young's modulus does not affect the displacement field ($\boldsymbol{u}^2$), since the body force is zero in this case, and therefore the last terms in Eq. (2.19) containing E (hidden in μ) disappear. Thus, the only model parameter to be set is Poisson's ratio. We have tested a range of values for ν, and the one that yielded the smallest error (a partial validation is presented in Section 3.3) was $\nu = 0.4$, which is a value used by other groups as well (FERRANT, WARFIELD and NABAVI [2000]). We assume that the model is homogeneous since there is no reliable way known to us for setting the model parameter for different brain structures.

3.3. *Experimental results*

In this section we will not present a complete stereo-guided brain deformation compensation system, but rather we will investigate how well a continuum mechanics-based brain model can predict in-volume deformation using only partial (exposed brain) surface data for model guidance. We test our method using intraoperative MR image sequences. We segment the brain and construct a mesh composed of hexahedral ("brick") elements (with 5 mm approximate side lengths) was generated using the segmented data and the in-house mesh generator described earlier. The generated meshes (of the cerebral hemisphere at the side of the craniotomy) had about 6500 nodes and about 5000 "brick" elements. Here we used the anatomical constraints that the falx and tentorium are practically fixed, and we fixed the corresponding model nodes. For this reason it is enough to consider only the half of the brain at the side of the craniotomy, since the other part does not deform. We are aware that, although this assumption holds in most cases, there are exceptions where falx moved during the surgery.

We present here a partial validation of the method using intraoperative MRI for two cases:[15] a sinking brain and a bulging brain. For both cases we generated the model and displacement boundary conditions as explained above. We used ABAQUS (HIBBITT, KARLSSON and SORENSEN [1997]) to compute the model deformation. For a model of about 6500 nodes and about 5000 "brick" elements, it took about 80 seconds to solve the equations on an SGI Octane R12000 computer. This time is almost practically applicable, since it would mean that after about minute and a half after obtaining exposed brain surface data, one would get updated MR images and other preoperative data. In order to validate the computed deformation, we manually selected a set of anatomical landmarks[16] in the preoperative scan of the (undeformed) brain at various locations throughout the volume of the cerebral hemisphere at the side of craniotomy. Then we manually found the corresponding landmarks in the intraoperative scan of the (deformed) brain. Finally, using the displacement field computed by the model, we determined the positions of the "model predicted landmarks" in the deformed brain corresponding to the landmarks in the undeformed brain, and compared them to the corresponding manually set landmarks in the deformed brain.

One can see from Table 3.1 that the maximal true landmark displacement was 3.8 mm (3.6 mm) while the maximal error was 1.4 mm (1.3 mm) for the case of the sinking (bulging) brain. Fig. 3.3 shows an MR image slice of a preoperative brain, the corresponding intraoperative image slice of the deformed brain, and the corresponding model-updated image slice of the deformed brain. The maximal deformation was at the exposed brain surface (about 7 mm for both cases). However, we did not use landmarks close to the exposed brain surface since the exposed brain surface displacement was used as a boundary condition, and the error at such landmarks would be unrealistically small. Rather, we selected landmarks throughout the volume of the cerebral hemisphere at the side of craniotomy away from the exposed brain surface. This is why the maximal landmark displacement was under 4 mm.

TABLE 3.1

Case I (sinking brain) and case II (bulging brain): true landmark displacements ($\boldsymbol{t}$), computed landmark displacements ($\boldsymbol{c}$), and error between true and computed landmark locations ($\boldsymbol{e} = \boldsymbol{c} - \boldsymbol{t}$), for 14 landmarks. All values are in millimeters

		1	2	3	4	5	6	7	8	9	10	11	12	13	14
I	$\|\boldsymbol{t}\|$	0.7	0.9	0.6	0.1	2.3	2.9	2.1	1.0	1.9	2.7	0.8	0.8	2.1	3.8
	$\|\boldsymbol{c}\|$	0.3	0.5	0.7	0.2	1.7	2.4	1.4	0.7	1.3	1.8	0.4	0.5	1.9	3.0
	$\|\boldsymbol{e}\|$	0.8	1.4	0.4	0.2	0.7	1.3	1.4	0.4	1.2	1.3	0.4	0.8	1.0	1.2
II	$\|\boldsymbol{t}\|$	2.7	1.8	0.6	3.6	2.6	0.8	1.3	1.1	1.4	0.7	0.7	0.4	2.4	0.5
	$\|\boldsymbol{c}\|$	2.0	1.6	1.1	2.4	2.6	0.5	0.8	1.2	1.5	0.8	0.5	0.2	2.0	0.3
	$\|\boldsymbol{e}\|$	0.8	1.0	0.6	1.3	0.8	0.4	0.9	0.8	0.9	0.5	0.7	0.5	1.2	0.7

[15] In both cases we used intraoperative MR images after the dura was opened and brain deformed, but before any major resection occurred.

[16] For landmarks, we used points at anatomical structures that can relatively easily be identified in both preoperative and intraoperative images.

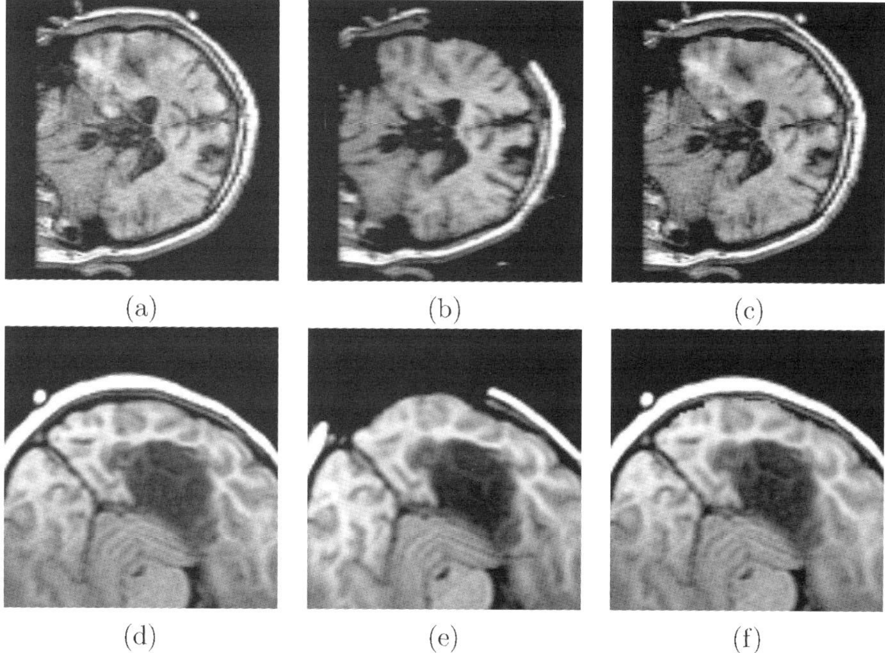

FIG. 3.3. (a) A preoperative coronal slice of a sinking brain, (b) the corresponding intraoperative slice of the deformed brain, (c) the corresponding model-computed slice of the deformed brain. Axial slices (d), (e) and (f) correspond to the bulging brain case (undeformed, deformed and model-computed, respectively). Note that in both cases the exposed brain surfaced in the computed slice moved similarly as the corresponding surface in the intraoperative slice. Figure reprinted from SKRINJAR, NABAVI and DUNCAN [2002], Model-driven brain shift compensation, *Medical Image Analysis*, 6(4):361–373, ©2002 by permission from Elsevier.

4. Case study II: estimation of 3D left ventricular deformation

4.1. Background

Acute coronary artery occlusion results in myocardial injury, which will progress from the endocardium to the epicardium of the heart wall in a wavefront fashion. A primary goal in the treatment of patients presenting with acute myocardial infarction is to reestablish coronary flow, and to interrupt the progression of injury, thereby salvaging myocardium. Unfortunately, there are no universally accepted non-invasive imaging approaches for the accurate determination of the extent of injury. Using conventional measures of regional myocardial function, the extent of myocardial infarction is overestimated. This can be attributed to persistent post-ischemic dysfunction ("stunning"), persistent myocardial hypoperfusion ("hibernation") or mechanical tethering of normal areas by the adjacent injured myocardium. This tethering can be seen at the lateral margins of an infarct, resulting in a viable although dysfunctional border zone. Motion of the viable epicardium can also be constrained by injury of the underlying endocardial myocardial tissue. The location and ultimate transmural extent of the injury has important implications for long term prognosis of patients following myocardial infarction. Those patients with transmural myocardial infarction are likely to dilate their left ventricles over time, a condition termed left ventricular "remodeling". The occurrence of post-infarction remodeling carries a much worse long-term prognosis.

A number of laboratories have shown that a comprehensive quantitative analysis of myocardial strain can more accurately identify ischemic injury than simple analysis of endocardial wall motion or radial thickening (AZHARI, WEISS, ROGERS, SIU and SHAPIRO [1995]). Furthermore, the characterization of segmental strain components has shown great promise for defining the mechanical mechanisms of tethering or remodeling (KRAMER, ROGERS, THEOBALD, POWER, PETRUOLO and REICHEK [1996], MARCUS, GOTTE, ROSSUM, KUIJER, HEETHAAR, AXEL and VISSER [1997]). Experimental animal studies demonstrate that decreased circumferential shortening in myocardial regions adjacent to the infarct zone relative to remote regions is associated with late left ventricular remodeling (KRAMER, et al. [1993]). At present, quantitative non-invasive measurement of 3D strain properties from images has been limited to special forms of magnetic resonance (MR) acquisitions, specifically MR tagging and restricted to mostly research settings.

The MR tagging approach to the measurement of myocardial strain was originally developed, and then vigorously pursued further by two groups, one at the University of Pennsylvania (AXEL [1998]) and the other at Johns Hopkins (MCVEIGH [1998]). In general, there are three different approaches to estimating displacement data from MR tagging. The first approach involves tagging in multiple intersecting planes at the same time, and using the tag intersections as tokens for tracking (e.g., AMINI, CHEN, CURWEN, MANU and SUN [1998], KERWIN and PRINCE [1998], YOUNG, KRAITCHMAN, DOUGHERTY and AXEL [1995]). The second approach involves tagging in multiple intersecting planes, one set of parallel planes at a time. Then, each tagging plane is used separately to estimate the normal direction of motion perpendicular to the plane. This generates a set of partial displacements (i.e., the component parallel to the tag lines is missing) to be combined later (e.g., HABER, METAXAS and AXEL [1998], DENNEY JR. and PRINCE [1995]). The final approach uses a lower resolution modulation technique and attempts to model the tag fading over time using the Bloch equations. The displacements are then extracted using a variable brightness optical flow technique (e.g., PRINCE and MCVEIGH [1992], GUPTA and PRINCE [1995]). The reader is also referred to a recently published book (AMINI and PRINCE [2001]).

As an alternative to MR tagging, several investigators have employed changes in phase due to motion of tissue within a fixed voxel or volume of interest to assist in estimating instantaneous, localized velocities and ultimately cardiac motion and deformation. While the basic ideas were first suggested by VAN DIJK [1984] and NAYLER, FIRMIN and LONGMORE [1986], it was Pelc and his team (PELC, HERFKENS, SHIMAKAWA and ENZMANN [1991], PELC [1991], N.J. PELC, HERFKENS and L. PELC [1992]) that first bridged the technique to conventional cine MR imaging and permitted the tracking of myocardial motion throughout the cardiac cycle. This technique basically relies on the fact that a uniform motion of tissue in the presence of a magnetic field gradient produces a change in the MR signal phase that is proportional to velocity. In general, two approaches have emerged to assemble deformation information from phase contrast images: (i) processing the data directly to estimate strain rate (e.g., WEDEEN [1992], PELC [1991]) and (ii) integrating the velocities over time, via some form of tracking mechanism to estimate displacements (e.g., MEYER, CONSTABLE, SINUSAS and DUNCAN [1996], CONSTABLE, RATH, SINUSAS and GORE [1994], ZHU, DRANGOVA and PELC [1997], HERFKENS, N. PELC, L. PELC and SAYRE [1991]).

The use of computer vision-based techniques to estimate displacement is also possible. One approach to establishing correspondence is to track shape-related features on the LV over time as reported by KAMBHAMETTU and GOLDGOF [1994], COHEN, AYACHE and SULGER [1992]), AMINI and DUNCAN [1992], MCEACHEN, OWEN and DUNCAN [1997] and SHI, SINUSAS, CONSTABLE, RITMAN and DUNCAN [2000]. This is the basis for much of our own work and is expanded later. In general, here preliminary displacements are estimated by matching local curvatures from segmented surfaces from consecutive time frames and then the estimates are smoothed to produce final displacement values. We note that such methods were applied to modalities other than magnetic resonance such as X-ray CT (SHI, SINUSAS, CONSTABLE, RITMAN and DUNCAN [2000], PAPADEMETRIS, SINUSAS, DIONE, CONSTABLE and DUNCAN [2002]) and ultrasound (PAPADEMETRIS, SINUSAS, DIONE and DUNCAN [2001]).

Finally, some investigators have used the intensity of the images directly to track local LV regions. SONG and LEAHY [1991] used the intensity in ultrafast CT images to calculate the displacement fields for a beating heart. In addition, other investigators have used local image intensity or intensity-based image texture from echocardiographic image sequences to track local positions over 2D image sequences (MAILLOUX, BLEAU, BERTRAND and PETITCLERC [1987], MEUNIER [1998]). These efforts, along with some related MR tagging approaches (e.g., GUPTA and PRINCE [1995]) roughly fall into the category of optical flow-based methods. With the exception of methods based on magnetic resonance tagging and to a lesser extent MR phase contrast velocities, none of the other methods is capable of estimating complete three-dimensional deformation maps of the left ventricle.

4.2. System description

Following image acquisition, the images are segmented interactively. From the results of the segmentation we construct a three-dimensional finite element representation of the left ventricle and also estimate initial surface correspondences using the shape-tracking approach. A dense motion field is then estimated using a transversely isotropic, linear-elastic model, which accounts for the muscle fiber directions in the left ventricle. The dense motion field is in turn used to calculate the deformation of the heart wall in terms of strain in cardiac specific directions. We explore each of these steps in more detail next.

4.2.1. Image acquisition

MR imaging was performed on a GE Signa 1.5 Tesla scanner with version 4.8 software using the head coil (26 cm diameter) for transmission and reception. Short axis images, such as those shown in Fig. 4.1, through the left ventricle were obtained with the gradient echo cine technique using the following parameters: TE = 6 msec,

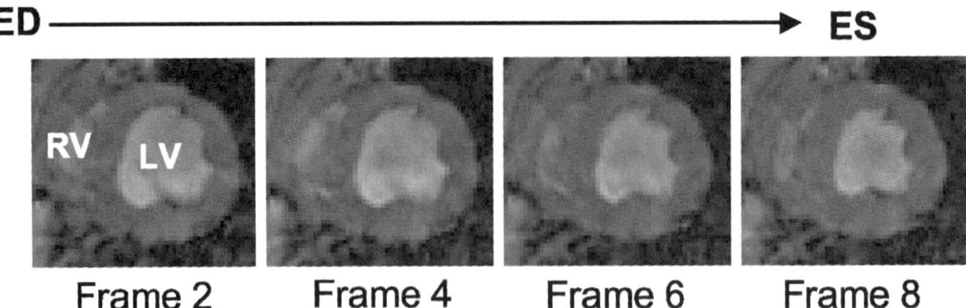

FIG. 4.1. Series of MR short-axis images from End Diastole (ED) to End Systole (ES).

TR = 40 msec, flip angle = 30°, 16 phases collected, 5 mm slices, matrix 256 × 256, 2 averages, FOV = 40 cm. A total of 16 contiguous 5 mm thick slices were collected, by acquiring four sets of staggered short axis slices (4 slices/set) with a separation gap of 20 and 5 mm offset. This sequence provides images with an in-plane resolution of 1.64 × 1.64 mm for a 256 × 256 matrix and a 5 mm resolution perpendicular to the imaging plane. This sequence also provides excellent temporal resolution (16 frames/cardiac cycle, ≈ 40 msec/frame).

4.2.2. *Segmentation and surface reconstruction*

The left ventricle is segmented on a slice by slice basis using a custom designed software platform (PAPADEMETRIS, RAMBO, DIONE, SINUSAS and DUNCAN [1998]). The segmentation algorithm results in a set of planar contours parameterized using b-splines which are subsequently sampled to generate a discrete set of points on each plane.

From these contours we reconstruct the endo- and epicardial surfaces in a two-step procedure as follows: (i) we interpolate between contours to generate in-between contours at the desired sampling distance. This results in an iso-sampled set of points in three dimensions. (ii) We construct a surface mesh by forming triangles between the points. The procedure is illustrated in Fig. 4.2.

4.2.3. *Mesh generation*

We proceed to describe the mesh-generation method used for generating a volumetric model for the left ventricle, in terms of hexahedral elements (PAPADEMETRIS, SINUSAS, DIONE, CONSTABLE and DUNCAN [2002]). The output mesh of this algorithm will be used to describe the geometry of the left ventricle as needed for the estimation of the complete deformation field using finite elements. Here, we describe an algorithm that takes advantage of the ‘cylinder-like’ geometry of the left ventricle to make the problem easier. We first interpolate on a contour-by-contour basis between the endocardial and epicardial surfaces using shape-based interpolation to generate an appropriate number of in-between interpolated surfaces (typically 3 or 4). Because of the greater geometrical complexity of the endocardium, we space the interpolated surfaces to be preferentially closer to the endocardium. We then discretize the contour on the middle slice of the endocardium to the desired number of nodes (typically 35–45). Then we es-

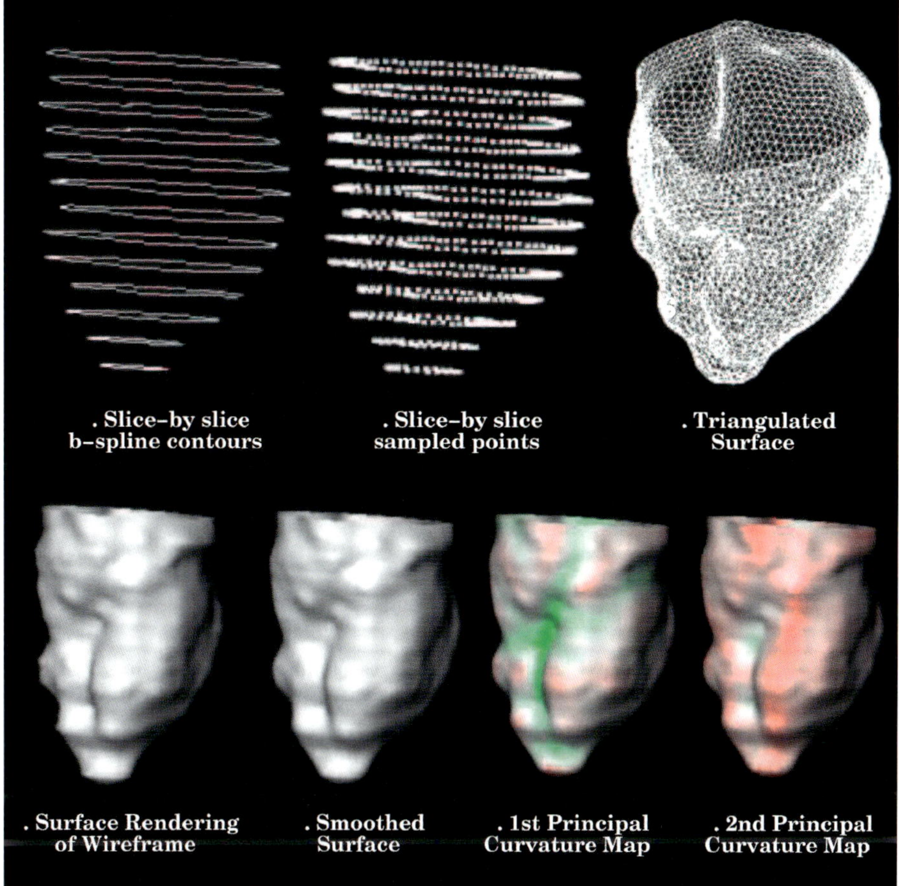

FIG. 4.2. Steps involved in moving from slice by slice contours to full surface representation. (1) Slice-by-slice b-spline parameterized contours as extracted by the segmentation process. (2) Discretized contours as equally-spaced points. (3) Formation of wire-frame by Delaunay triangulation. (4) Surface rendering. (5) Smoothing of surface using non-shrinking smoothing algorithm. (6) + (7) First and second principal curvatures of the surface. Here, green shows negative (i.e., inward) curvature, white shows flat regions and red indicates positive (i.e., outward) curvature. Figure reprinted from PAPADEMETRIS, SINUSAS, DIONE, CONSTABLE and DUNCAN [2002], Estimation of 3D left ventricular deformation from medical images using biomechanical models, *IEEE Transactions on Medical Imaging*, 21(7):786–800, ©2002 by permission from the IEEE.

timate correspondences between the surfaces and connect corresponding points to form hexahedral brick like elements.

4.2.4. *Shape-based tracking*

In this work, the original displacements on the outer surfaces of the myocardium were obtained by using the shape-tracking algorithm whose details were presented in SHI, SINUSAS, CONSTABLE, RITMAN and DUNCAN [2000]. The method tries to track points

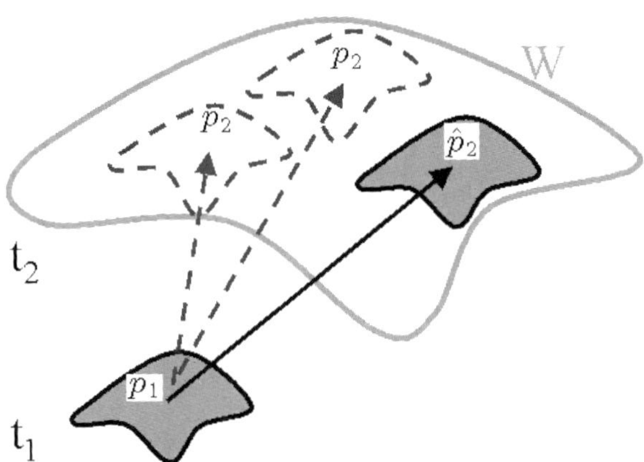

FIG. 4.3. Example of shape-tracking approach. The goal here is to map the original surface to the final surface. For a point p_1 on the original surface a window $\mathcal{W}$ of plausible matching points on the final surface is first generated. Then the point p_2 in W which has the most similar shape-properties to p_1 is selected as the candidate match point. The distance function for shape-similarity is typically based on the curvature(s). Figure reprinted from PAPADEMETRIS, SINUSAS, DIONE and DUNCAN [2001], Estimation of 3D left ventricular deformation from echocardiography, *Medical Image Analysis* 5(1):17–29, ©2001 by permission from Elsevier.

on successive surfaces using a shape similarity metric which tries to minimize the difference in principal curvatures and was validated using implanted markers (SHI, SINUSAS, CONSTABLE, RITMAN and DUNCAN [2000]).

With reference to Fig. 4.3, consider point p_1 on a surface at time t_1 which is to be mapped to a point p_2 on the deformed surface at time t_2. First, a search is performed a physically plausible region $\mathcal{W}$ on the deformed surface, to find the point $\hat{p}_2$ which has the local shape properties closest to those p_1. The shape properties here are captured in terms of the principal curvatures κ_1 and κ_2. The distance measure used is the bending energy required to bend a curved plate or surface patch to a newly deformed state. This is labeled as d_{be} and is defined as

$$d_{be}(p1, p2) = A\left(\frac{(\kappa_1(p_1) - \kappa_1(p_2))^2 + (\kappa_2(p_1) - \kappa_2(p_2))^2}{2}\right). \tag{4.1}$$

The point $\hat{p}_2$ is found by minimizing d_{be} in the region $\mathcal{W}$ which can be expressed as

$$\hat{p}_2 = \arg\min_{p_2 \in \mathcal{W}}\left[d_{be}(p1, p2)\right]. \tag{4.2}$$

Finally, displacement estimate vector for each point p_1, u_1^m is given by

$$u_1^m = \hat{p}_2 - p_1.$$

Confidence measures in the match. The bending energy measures for all the points inside the search region $\mathcal{W}$ are recorded as the basis to measure the *goodness* and *uniqueness* of the matching choices. The value of the minimum bending energy in the search region between the matched points indicates the goodness of the match. Denoting this value as m_g, we have the following measure for matching goodness:

$$m_g(p_1) = d_{be}(p_1, \hat{p}_2). \tag{4.3}$$

On the other hand, it is desirable that the chosen matching point is a unique choice among the candidate points within the search window. Ideally, the bending energy value of the chosen point should be an outlier (much smaller value) compared to the values of the rest of the points. If we denote the mean values of the bending energy measures of all the points inside window W except the chosen point as $\bar{d}_{be}$ and the standard deviation as σ_{be}^d, we define the uniqueness measure as

$$m_u(p_1) = \frac{d_{be}(p_1, \hat{p}_2)}{\bar{d}_{be} - \sigma_{be}^d}. \tag{4.4}$$

This uniqueness measure has a high value if the bending energy of the chosen point is small compared to some smaller value (mean minus standard deviation) of the remaining bending energy measures. Combining these two measures together, we arrive at one *confidence measure* $c^m(p_1)$ for the matched point $\hat{p}_2$ of point p_1:

$$c^m(p_1) = \frac{1}{k_{1,g} + k_{2,g} m_g(p_1)} \times \frac{1}{k_{1,u} + k_{2,u} m_u(p_1)}, \tag{4.5}$$

where $k_{1,g}$, $k_{2,g}$, $k_{1,u}$ and $k_{2,u}$ are scaling constants for normalization purposes. We normalize the confidences to lie in the range 0 to 1.

Modeling the initial displacement estimates. Given a set of displacement vector measurements u^m and confidence measures c^m, we model these estimates probabilistically by assuming that the noise in the individual measurements is normally distributed with zero mean and a variance $\sigma^2 = 1/c^m$. In addition, we assume that the measurements are uncorrelated. Given these assumptions, we can write the measurement probability for each point as

$$p(u^m|u) = \frac{1}{\sqrt{2\pi\sigma^2}} e^{\frac{-(u-u^m)^2}{2\sigma^2}}. \tag{4.6}$$

4.2.5. *Mechanical model-based integration*

We model the left ventricle using the active elastic model described in Section 2.4. The passive properties of the material (captured by the matrix C of Eq. (2.29)) are modeled using the transversely isotropic elastic model described in Eq. (2.15), with preferential stiffness being applied along canonical fiber orientations. Such a fiber model was shown in Fig. 2.2. The active elastic model results in a prior probability density function for the underlying displacement field $p(u)$ which is integrated with the image-derived measurements $p(u^m|u)$ (see Eq. (4.6)) to yield a maximum a-posteriori solution of the form:

$$\hat{u} = \arg\max_u p(u|u^m) = \arg\max_u \left(\frac{p(u^m|u)p(u)}{p(u^m)} \right). \tag{4.7}$$

The prior probability of the measurements $p(u^m)$ is a constant once these measurements have been made and therefore drops out of the minimization process.

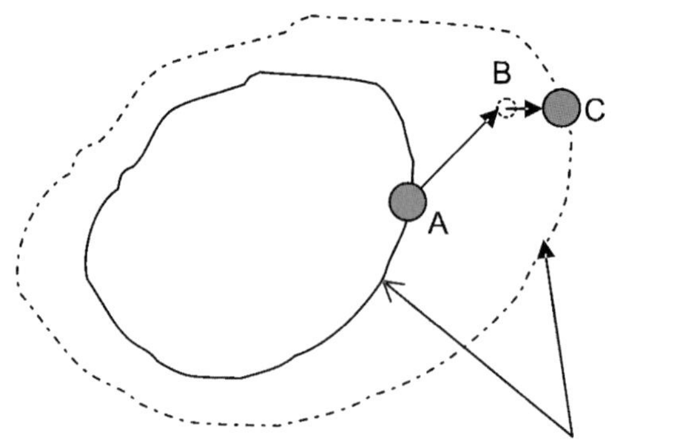

FIG. 4.4. Illustration of the two-step numerical solution technique.

Taking logarithms in Eq. (4.7) and differentiating with respect to the displacement field u results in a system of partial differential equations, which we solve using the finite element method (described in Section 2.5). The first step in the finite element method is the division or tessellation of the body of interest into elements; these are commonly tetrahedral or hexahedral in shape. Once this is done, the partial differential equations are written down in integral form for each element, and then the integral of these equations over all the elements is taken to produce the final set of equations. For more information one is referred to standard textbooks such as BATHE [1982]. The final set of equations is then solved to produce the output set of displacements. In our case the myocardium is divided into approximately 2500 hexahedral elements.

For each frame between end-systole (ES) and end-diastole (ED), a two step problem is posed: (i) solving Eq. (4.7) normally and (ii) adjusting the position of all points on the endo-and epi-cardial surfaces so they lie on the endo- and epi-cardial surfaces at the next frame using a modified nearest-neighbor technique and solving Eq. (4.7) once more using this added constraint. This ensures that there is no bias in the estimation of the radial strain. This is illustrated schematically in Fig. 4.4. Consider the point A on the epicardial contour at time t. (The endo-cardial contours are not shown for the sake of clarity.) After normal solution of Eq. (4.7) it gets mapped to point B which does not lie on the epi-cardial contour at time $t + 1$. The point is then fixed to point C by the modified nearest-neighbor technique and Eq. (4.7) is solved again to adjust the positions of internal points.

4.3. *Experimental results*

In this section we present some preliminary results of the application of this algorithm to left ventricular deformation estimation. We bootstrap the algorithm by using the output produced by our previous work (PAPADEMETRIS, SINUSAS, DIONE, CONSTABLE and DUNCAN [2002]). We label this algorithm as the ‘passive’ algorithm. In the passive algorithm, the images are segmented interactively and then initial correspondence is established using a shape-tracking approach. A dense motion field is then estimated using a passive, transversely linear elastic model, which accounts for the fiber directions in the left ventricle. The dense motion field is in turn used to calculate the deformation of the heart wall in terms of strains.

The output of the 'passive' algorithm consists of a set of vectors $e^p(x_i, t_j)$ representing the strain estimated by the passive algorithm at position x_i and time t_j. Typically we divide the heart into about 800–1000 (i.e., $i \in 1:1000$) elements and use 6–9 time frames ($j \in 1:9$) resulting in a total of approximately 7000 6×1 vectors $e^p = [e^p_{rr}, e^p_{cc}, e^p_{ll}, e^p_{rc}, e^p_{rl}, e^p_{lc}]^t$. The components of e^p are the normal strains in the radial (rr), circumferential (cc) and longitudinal (ll) directions as well as the shears between these direction (e.g., e^p_{rc} is the radial-circumferential shear strain).

These vectors e^p are then used to generate an estimate of the active strain e^a using isovolumic correction and possibly temporal smoothing. In the isovolumic correction procedure at each discrete element position x_i and time t_j we generate an output vector $e^a(x_i, t_j)$ by adjusting the longitudinal strain to create a new set of strain estimates e^a that result in an incompressible deformation. These estimates e^a are used as the mean value for the active elastic model. The variance is determined by the stiffness matrix and is the same as it was for the passive model. We label the results produced by this procedure as *Active*.

Validation. We tested the new algorithm(s) by comparing its output to those obtained using MR tagging (KERWIN and PRINCE [1998]) and implanted markers (PAPADEMETRIS, SINUSAS, DIONE, CONSTABLE and DUNCAN [2002]). In the MR tagging case we used one human image sequence provided to us by Dr. Jerry Prince from John Hopkins University. The images were acquired using 3 orthogonal MR tagging acquisitions and the displacements estimated using an algorithm presented in KERWIN and PRINCE [1998]. From these displacements we estimate the MR tagging derived strains. Images from one of the three acquisitions had the evidence of the tag lines removed using morphological operators, was segmented interactively and the strains were estimated using our previous approach (Passive) (PAPADEMETRIS, SINUSAS, DIONE, CONSTABLE and DUNCAN [2002]). In the case of implanted markers we used 8 canine image sequences with implanted markers as is shown in Fig. 4.5 (see also PAPADEMETRIS, SINUSAS, DIONE, CONSTABLE and DUNCAN [2002]).

We tested two permutations of the active algorithm. For the algorithm labeled *Active* in Fig. 4.7, we used as input the output of the *passive* algorithm after isovolumic correction, without any temporal smoothing. The algorithm labeled as *ActiveT* used the output of the *passive* algorithm with both temporal smoothing and isovolumic correction. Fig. 4.6 illustrates the output of algorithm *ActiveT* at four points in the cardiac cycle as applied to the MR tagging sequence. The output of the tagging method (KERWIN and PRINCE [1998]) at end-systole is presented for comparison.

Fig. 4.7 shows the error between the estimates of our old algorithm labeled *passive* and the two variations of the new active algorithm (*Active* and *ActiveT*), as compared to the output of the tagging algorithm (KERWIN and PRINCE [1998]) and to the estimates obtained using the MR markers. In the case of the tagging algorithm we observe an overall reduction in mean strain error from 9.9% (passive) to 8.1% (active) at end-

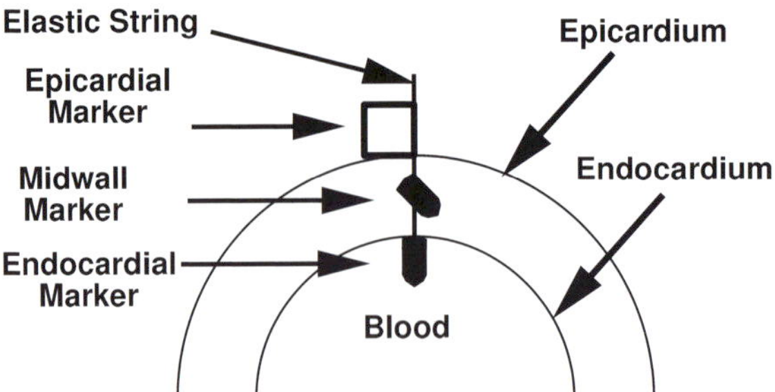

FIG. 4.5. Implantation of Image-Opaque Markers. This figure shows the arrangement of markers on the myocardium. First, a small bullet-shaped copper bead attached to an elastic string was inserted into the blood pool through a needle track. Then the epicardial marker was sutured (stitched) to the myocardium and tied to the elastic string. Finally, the mid-wall marker was inserted obliquely through a second needle track to a position approximately half-way between the other two markers. Figure reprinted from PAPADEMETRIS, SINUSAS, DIONE, CONSTABLE and DUNCAN [2002], Estimation of 3D left ventricular deformation from medical images using biomechanical models, *IEEE Transactions on Medical Imaging*, 21(7):786–800, ©2002 by permission from the IEEE.

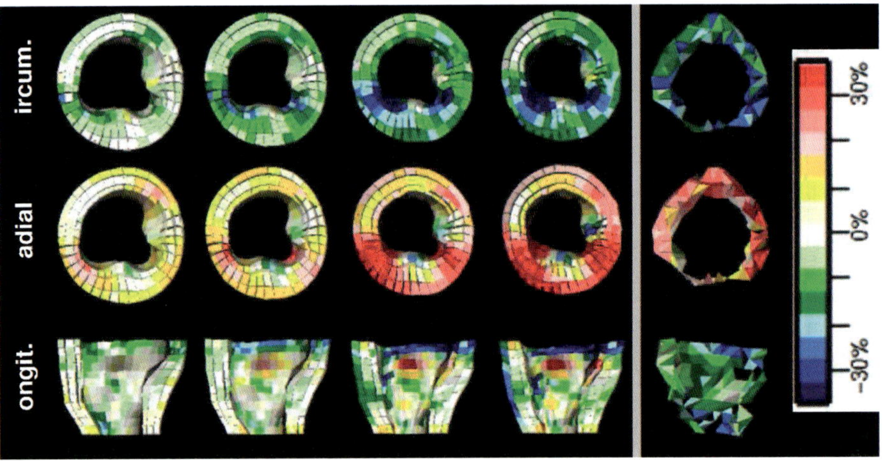

FIG. 4.6. Leftmost four columns: circumferential, radial and longitudinal strain outputs of our active (*Active 2T*) algorithm at four points in the systolic half of the cardiac cycle. Far right column: output of MR tagging based algorithm (KERWIN and PRINCE [1998]) on the same image sequence shown at the last time frame (End-systole). Figure reprinted from PAPADEMETRIS, ONAT, SINUSAS, DIONE, CONSTABLE and DUNCAN [2001], The active elastic model, in: *Information Processing in Medical Imaging*, IPMI'01, Davis, CA, in: Lecture Notes in Computer Science 2082, ©2001 by permission from Springer-Verlag.

systole (frame 10). In the case of the implanted markers we observe a similar reduction from 7.2 to 6.3%.

It is also interesting to note that the MR tagging algorithm (KERWIN and PRINCE [1998]) produces a reduction of myocardial volume of 12% between end-diastole and end-systole, our passive algorithm an increase of approximately 14% and all both ver-

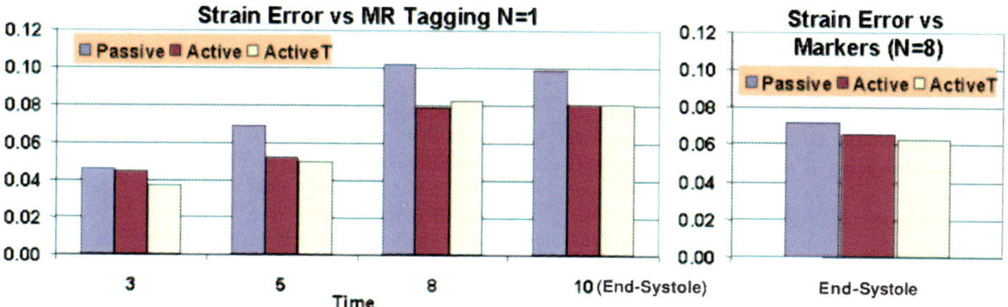

FIG. 4.7. Absolute strain error vs tag data or implanter markers. *Passive* – passive model from PAPADEMETRIS, SINUSAS, DIONE, CONSTABLE and DUNCAN [2002], *Active* and *ActiveT* represent two versions of the active algorithm without and with temporal smoothing. We note that both the active algorithms result in error reduction as compared to the passive algorithm. In the case of the tagging data we plot the absolute error in the cardiac-specific strains whereas in the case of implanted markers we use the principal strains instead (see PAPADEMETRIS, SINUSAS, DIONE, CONSTABLE and DUNCAN [2002]). Figure reprinted from PAPADEMETRIS, ONAT, SINUSAS, DIONE, CONSTABLE and DUNCAN [2001], The active elastic model, in: *Information Processing in Medical Imaging*, IPMI'01, Davis, CA, in: Lecture Notes in Computer Science 2082, ©2001 by permission from Springer-Verlag.

sions of the active algorithm produced small increases ($< 2\%$) showing that the isovolumic correction was effective.

5. Conclusions

The use of biomechanical models to guide the estimation of non-rigid motion and deformation in medical image analysis is now well-established. Additional areas where such models are used are in surgical simulation (COTIN, DELINGETTE and AYACHE [1999]), intra-subject non-rigid brain registration before and after tumor growth (KYRIAKOU and DAVATZIKOS [1998]). Continuum mechanical models have also been used purely for their mathematical properties in the case of the generic non-rigid registration problem (e.g., CHRISTENSEN, RABBITT and MILLER [1996], GEE, HAYNOR, BRIQUER and BAJCSY [1997]).

In this chapter we have particularly focused on the integration of bio-mechanical models with image derived information and have presented two frameworks to achieve this integration: (i) the force equilibrium framework and (ii) the energy minimization framework. The force equilibrium framework results in a set of partial differential equations describing the displacement field and image-derived displacements are used as boundary conditions. The energy minimization framework, which is further interpreted and recast as a Bayesian estimation framework, similarly results in a set of partial differential equations which when discretized using the finite element method yield a system of linear equations. With the use of an appropriate noise model the image-derived measurements can be modeled probabilistically and integrated with the model which in turn is also modeled as a Gibbs prior.

While some of the details of the extraction of image-derived measurements are presented, this is not the major point of the chapter, in this respect our work differs little for other image analysis work utilizing mathematical regularization models such as in PRINCE and MCVEIGH [1992]. The key issue with mathematical regularization models is that the choice of parameters for the regularization functional is fairly arbitrary and *ad-hoc*, the use of biomechanical models offers the advantage that the model parameters are physically meaningful and can be experimentally measured, as is the case with the fiber orientations in the left ventricle.

We expect that the sophistication of the models will continue to improve as researchers begin to tackle more complex problems such as tumor resection. It is also likely that more current sophisticated work in biomechanics such as that in the cardiac modeling area (e.g., HUNTER, MCCULLOCH and NIELSEN [1991]) will become more relevant to state of the art medical image analysis. In this we are encouraged by the appearance of work such as that of SERMESANT, COUDIRE, DELINGETTE and AYACHE [2002] which is beginning the process of integrating not only the biomechanical properties of the left ventricle but also the electrical properties of the tissue within an image analysis problem.

References

ABAQUS/Version 5.7 (1997). Hibbitt, Karlsson & Sorensen, Rhode Island, USA.

AMINI, A.A., CHEN, Y., CURWEN, R.W., MANU, V., SUN, J. (1998). Coupled B-snake grids and constrained thin-plate splines for analysis of 2D tissue deformations from tagged MRI. *IEEE Trans. Medical Imaging* **17** (3), 344–356.

AMINI, A.A., DUNCAN, J.S. (1992). Bending and stretching models for LV wall motion analysis from curves and surfaces. *Image and Vision Computing* **10** (6), 418–430.

AMINI, A.A., PRINCE, J.L. (eds.) (2001). *Measurement of Cardiac Deformations from MRI: Physical and Mathematical Models* (Kluwer Academic, Dordrecht).

ARUN, K.S., HUANG, T.S., BLOSTEIN, S.D. (1987). Least-squares fitting of two 3-d point sets. *IEEE Trans. Pattern Analysis and Machine Intelligence* **9** (5), 698–700.

AUDETTE, M.A., SIDDIQI, K., FERRIE, F.P., PETERS, T.M. (2003). An integrated range-sensing, segmentation and registration framework for the characterization of intra-surgical brain deformations in image-guided surgery. *Computer Vision and Image Understanding* **89** (2–3), 226–251.

AUDETTE, M.A., SIDDIQI, K., PETERS, T.M. (1999). Level-set surface segmentation and fast cortical range image tracking for computing intrasurgical deformations. In: *Medical Image Computing and Computer Assisted Intervention*, pp. 788–797.

AXEL, L. (1998). Physics and technology of cardiovascular MR imaging. *Cardiology Clinics* **16** (2), 125–133.

AZHARI, H., WEISS, J., ROGERS, W., SIU, C., SHAPIRO, E. (1995). A noninvasive comparative study of myocardial strains in ischemic canine hearts using tagged MRI in 3D. *Amer. J. Physiol.* **268**, H1918–H1926.

BATHE, K. (1982). *Finite Element Procedures in Engineering Analysis* (Prentice Hall, Englewood Cliffs, NJ).

BESL, P.J., MACKAY, N.D. (1992). A method for registration of 3-D shapes. *IEEE Trans. Pattern Analysis and Machine Intelligence* **14** (2), 239–256.

BUCHOLZ, R., YEH, D., TROBAUGH, J., MCDURMONT, L., STURM, C., BAUMANN, C., HENDERSON, J., LEVY, A., KESSMAN, P. (1997). The correction of stereotactic inaccuracy caused by brain shift using an intraoperative ultrasound device. In: *CVRMed-MRCAS'97*, Grenoble, France, pp. 459–466.

CHABRERIE, A., OZLEN, F., NAKAJIMA, S., LEVENTON, M., ATSUMI, H., GRIMSON, E., KEEVE, E., HELMERS, S., RIVIELLO, J., HOLMES, G., DUFFY, F., JOLESZ, F., KIKINIS, R., BLACK, P. (1998). Three-dimensional reconstruction and surgical navigation in pediatric epilepsy surgery. In: Wells, W., Colchester, A., Delp, S. (eds.), *MICCAI'98*, pp. 74–83.

CHRISTENSEN, G.E., RABBITT, R.D., MILLER, M.I. (1994). 3D brain mapping using deformable neuroanatomy. *Physics in Medicine and Biology* **39**, 609–618.

CHRISTENSEN, G.E., RABBITT, R.D., MILLER, M.I. (1996). Deformable templates using large deformation kinematics. *IEEE Trans. Image Process.* **5** (10), 1435–1447.

COHEN, I., AYACHE, N., SULGER, P. (1992). Tracking points on deformable objects using curvature information. In: *ECCV'92*. In: Lecture Notes in Comput. Sci. (Springer-Verlag, Berlin), pp. 458–466.

CONSTABLE, T., RATH, K., SINUSAS, A., GORE, J. (1994). Development and evaluation of tracking algorithms for cardiac wall motion analysis using phase velocity MR imaging. *Magn. Reson. Med.* **32**, 33–42.

COTIN, S., DELINGETTE, H., AYACHE, N. (1999). Real-time elastic deformations of soft tissues for surgery simulation. *IEEE Trans. Visualization and Computer Graphics* **5** (1), 62–73.

DENNEY JR., T.S., PRINCE, J.L. (1995). Reconstruction of 3-D left ventricular motion from planar tagged cardiac MR images: An estimation theoretic approach. *IEEE Trans. Medical Imaging* **14** (4), 625–635.

ERINGEN, A.C. (1980). *Mechanics of Continua* (Krieger, New York, NY).

FERRANT, M., NABAVI, A., MACQ, B., JOLESZ, F., KIKINIS, R., WARFIELD, S. (2001). Registration of 3D intraoperative MR images of the brain using a finite-element biomechanical model. *IEEE Trans. Medical Imaging* **20** (12), 1384–1397.

FERRANT, M., WARFIELD, S.K., NABAVI, A.E. (2000). Registration of 3D intraoperative MR images of the brain using a finite element biomechanical model. In: *Medical Image Computing and Computer Assisted Intervention*, pp. 19–28.

GEE, J.C., HAYNOR, D.R., BRIQUER, L.L., BAJCSY, R.K. (1997). Advances in elastic matching theory and its implementation. In: *CVRMed-MRCAS*, Grenoble, France.

GEMAN, D., GEMAN, S. (1984). Stochastic relaxation, Gibbs distribution and Bayesian restoration of images. *IEEE Trans. Pattern Analysis and Machine Intelligence* **6**, 721–741.

GERING, D.T., NABAVI, A., KIKINIS, R., et al. (1999). An integrated visualization system for surgical planning and guidance using image fusion and interventional imaging. In: *Medical Image Computing and Computer Assisted Intervention*, MICCAI'99, Cambridge, UK, pp. 809–819.

GIBSON, S. (1998). Constrained elastic surface nets: Generating smooth surface from binary segmented data. In: *Medical Image Computing and Computer Aided Intervention*, MICCAI'98, pp. 888–898.

GRIMSON, W.E.L., ETTINGER, G.J., WHITE, S.J., GLEASON, P.L., LOZANO-PEREZ, T., WELLS III, W.M., KIKINIS, R. (1995). Evaluating and validating an automated registration system for enhanced reality visualization in surgery. In: *CVRMed*, Nice, France.

GRIMSON, W.E.L., ETTINGER, G.J., WHITE, S.J., GLEASON, P.L., LOZANO-PEREZ, T., WELLS III, W.M., KIKINIS, R. (1996). An automatic registration method for frameless stereotaxy, image guided surgery, and enhanced reality visualization. *IEEE Trans. Medical Imaging* **15** (2), 129–140.

GUCCIONE, J.M., MCCULLOCH, A.D. (1991). Finite element modeling of ventricular mechanics. In: Hunter, P.J., McCulloch, A., Nielsen, P. (eds.), *Theory of Heart* (Springer-Verlag, Berlin), pp. 122–144.

GUPTA, S.N., PRINCE, J.L. (1995). On variable brightness optical flow for tagged MRI. In: *Information Processing in Medical Imaging*.

HABER, E., METAXAS, D.N., AXEL, L. (1998). Motion analysis of the right ventricle from MRI images. In: *Medical Image Computing and Computer Aided Intervention*, MICCAI'98, Cambridge, MA, pp. 177–188.

HATA, N., NABAVI, A., WARFIELD, S., et al. (1999). A volumetric optical flow method for measurement of brain deformation from intraoperative magnetic resonance images. In: *Medical Image Computing and Computer Assisted Intervention*, MICCAI'99, Cambridge, UK, pp. 928–935.

HERFKENS, R., PELC, N., PELC, L., SAYRE, J. (1991). Right ventricular strain measured by phase contrast MRI. In: *Proc. 10th Annual SMRM*, San Francisco, p. 163.

HILL, D., MAURER, C., WANG, M., MACIUNAS, R., BARWISE, J., FITZPATRICK, M. (1997). Estimation of intraoperative brain surface movement. In: *CVRMed-MRCAS'97*, Grenoble, France, pp. 449–458.

HILL, D.L.G., MAURER, C.R., MARTIN, A.J., et al. (1999). Assessment of intraoperative brain deformation using interventional MR imaging. In: *Medical Image Computing and Computer Assisted Intervention*, MICCAI'99, Cambridge, UK, pp. 910–919.

HUEBNER, K.H., THORNTON, E.A., BYROM, T.G. (1995). *The Finite Element Method For Engineers* (Wiley, New York).

HUNTER, P.J., MCCULLOCH, A., NIELSEN, P. (eds.) (1991). *Theory of Heart* (Springer-Verlag, Berlin).

HUNTER, P.J., NASH, M.P., SANDS, G.B. (1997). Computational electromechanics of the heart. In: Panfilov, A.V., Holden, A.V. (eds.), *Computational Biology of the Heart* (Wiley, New York), pp. 346–407.

KAMBHAMETTU, C., GOLDGOF, D. (1994). Curvature-based approach to point correspondence recovery in conformal nonrigid motion. *CVGIP: Image Understanding* **60** (1), 26–43.

KERWIN, W.S., PRINCE, J.L. (1998). Cardiac material markers from tagged MR images. *Medical Image Analysis* **2** (4), 339–353.

KRAMER, C., ROGERS, W., THEOBALD, T., POWER, T., PETRUOLO, S., REICHEK, N. (1996). Remote noninfarcted regional dysfunction soon after first anterior myocardial infarction: A magnetic resonance tagging study. *Circulation* **94**, 660–666.

KRAMER, C., et al. (1993). Regional differences in function within noninfarcted myocardium during left ventricular remodeling. *Circulation* **88**, 1279–1288.

KYRIAKOU, S., DAVATZIKOS, C. (1998). A biomechanical model of soft tissue deformation with applications to non-rigid registration of brain image with tumor pathology. In: *Medical Image Computing and Computer Assisted Intervention*. In: Lecture Notes in Comput. Sci. **1496** (Springer, Berlin), pp. 531–538.

LISEIKIN, V.D. (1999). *Grid Generation Methods* (Springer-Verlag, Berlin).

MAILLOUX, G.E., BLEAU, A., BERTRAND, M., PETITCLERC, R. (1987). Computer analysis of heart motion from two-dimensional echocardiograms. *IEEE Trans. Biomed. Engrg.* **34** (5), 356–364.

MARCUS, J., GOTTE, M., ROSSUM, A.V., KUIJER, J., HEETHAAR, R., AXEL, L., VISSER, C. (1997). Myocardial function in infarcted and remote regions early after infarction in man: Assessment by magnetic resonance tagging and strain analysis. *Magnetic Resonance in Medicine* **38**, 803–810.

MCEACHEN, J., OWEN, R., DUNCAN, J. (1997). Shape-based tracking of left ventricular wall motion. *IEEE Trans. Medical Imaging* **16** (3), 270–283.

MCVEIGH, E.R. (1998). Regional myocardial function. *Cardiology Clinics* **16** (2), 189–206.

MEUNIER, J. (1998). Tissue motion assessment from 3D echographic speckle tracking. *Phys. Med. Biol.* **43**, 1241–1254.

MEYER, F.G., CONSTABLE, R.T., SINUSAS, A.J., DUNCAN, J.S. (1996). Tracking myocardial deformation using phase contrast MR velocity fields: A stochastic approach. *IEEE Trans. Medical Imaging* **15** (4).

MIGA, M., PAULSEN, K., KENNEDY, F., HOOPES, J., HARTOV, A., ROBERTS, D. (1998). Initial in-vivo analysis of 3D heterogeneous brain computations for model-updated image-guided neurosurgery. In: *Medical Image Computing and Computer Assisted Intervention*, pp. 743–752.

MIGA, M., ROBERTS, D., KENNEDY, F., PLATENIKI, L., HARTOV, A., LUNN, K., PAULSEN, K. (2001). Modeling of retraction and resection for intraoperative updating of images. *Neurosurgery* **49** (1), 75–85.

NABAVI, A., BLACK, P., GERING, D.T., et al. (2001). Serial intraoperative MR imaging of brain shift. *Neurosurgery* **48** (4), 787–798.

NAYLER, G., FIRMIN, N., LONGMORE, D. (1986). Blood flow imaging by cine magnetic resonance. *J. Comp. Assist. Tomog.* **10**, 715–722.

OMI (1997). *Operation of the Mayfield® Accis™ Stereotactic Workstation* (OMI® Surgical Products).

OPPENHEIM, A.V., SCHAFER, R.W. (1975). *Digital Signal Processing* (Prentice Hall, Englewood Cliffs, NJ).

PAPADEMETRIS, X., ONAT, E.T., SINUSAS, A.J., DIONE, D.P., CONSTABLE, R.T., DUNCAN, J.S. (2001). The active elastic model. In: *Information Processing in Medical Imaging*, IPMI'01, Davis, CA. In: Lecture Notes in Computer Science **2082** (Springer-Verlag, Berlin).

PAPADEMETRIS, X., RAMBO, J.V., DIONE, D.P., SINUSAS, A.J., DUNCAN, J.S. (1998). Visually interactive cine-3D segmentation of cardiac MR images. *J. Am. Coll. of Cardiology* **31** (2) (Suppl. A).

PAPADEMETRIS, X., SINUSAS, A.J., DIONE, D.P., CONSTABLE, R.T., DUNCAN, J.S. (2002). Estimation of 3D left ventricular deformation from medical images using biomechanical models. *IEEE Trans. Medical Imaging* **21** (7), 786–800.

PAPADEMETRIS, X., SINUSAS, A.J., DIONE, D.P., DUNCAN, J.S. (2001). Estimation of 3D left ventricular deformation from echocardiography. *Medical Image Analysis* **5** (1), 17–29.

PAPOULIS, A. (1991). *Probability, Random Variables and Stochastic Processes*, third ed. (McGraw–Hill, New York).

PELC, N.J. (1991). Myocardial motion analysis with phase contrast cine MRI. In: *Proc. 10th Annual SMRM*, San Francisco, p. 17.

PELC, N.J., HERFKENS, R., PELC, L. (1992). 3D analysis of myocardial motion and deformation with phase contrast cine MRI. In: *Proc. 11th Annual SMRM*, Berlin, p. 18.

PELC, N., HERFKENS, R., SHIMAKAWA, A., ENZMANN, D. (1991). Phase contrast cine magnetic resonance imaging. *Magn. Res. Quart.* **7** (4), 229–254.

PETERS, T., DAVEY, B., MUNGER, P., COMEAU, R., EVANS, A., OLIVIER, A. (1996). Three-dimensional multimodal image-guidance for neurosurgery. *IEEE Trans. Medical Imaging* **15** (2), 121–128.

PRINCE, J.L., MCVEIGH, E.R. (1992). Motion estimation from tagged MR image sequences. *IEEE Trans. Medical Imaging* **11**, 238–249.

SERMESANT, M., COUDIRE, Y., DELINGETTE, H., AYACHE, N. (2002). Progress towards an electromechanical model of the heart for cardiac image analysis. In: *IEEE Int. Symp. Biomedical Imaging*, pp. 10–14.

SHI, P., SINUSAS, A.J., CONSTABLE, R.T., RITMAN, E., DUNCAN, J.S. (2000). Point-tracked quantitative analysis of left ventricular motion from 3D image sequences. *IEEE Trans. Medical Imaging* **19** (1), 36–50.

SKRINJAR, O., DUNCAN, J. (1999). Real time 3D brain shift compensation. In: *Information Processing in Medical Imaging*, IPMI'99, pp. 42–55.

SKRINJAR, O., NABAVI, A., DUNCAN, J.S. (2002). Model-driven brain shift compensation. *Medical Image Analysis* **6** (4), 361–373.

SKRINJAR, O., TAGARE, H., DUNCAN, J. (2000). Surface growing from stereo images. In: *Computer Vision and Pattern Recognition*, CVPR'2000, Hilton Head Island, SC, USA (IEEE Computer Society), pp. 571–576.

SONG, S., LEAHY, R. (1991). Computation of 3D velocity fields from 3D cine CT images. *IEEE Trans. Medical Imaging* **10**, 295–306.

SPENCER, A. (1980). *Continuum Mechanics* (Longman, London).

STOKKING, R. (1998). Integrated visualization of functional and anatomical brain images. PhD thesis, University Utrecht.

STRANG, G. (1986). *Introduction to Applied Mathematics* (Wellesley–Cambridge Press, Wellesley, MA).

VALLIAPPAN, A. (1981). *Continuum Mechanics Fundamentals* (Balkema, Rotterdam).

VAN DIJK, P. (1984). Direct cardiac NMR imaging of heart wall and blood flow velocity. *J. Comp. Assist. Tomog.* **8**, 429–436.

WARFIELD, S.K., TALOS, F., TEI, A., et al. (2002). Real-time registration of volumetric brain MRI by biomechanical simulation of deformation during image guided neurosurgery. *Comput. Visual. Sci.* **5**.

WEDEEN, V. (1992). Magnetic resonance imaging of myocardial kinematics: Technique to detect, localize and quantify the strain rates of active human myocardium. *Magn. Reson. Med.* **27**, 52–67.

YOUNG, A.A., KRAITCHMAN, D.L., DOUGHERTY, L., AXEL, L. (1995). Tracking and finite element analysis of stripe deformation in magnetic resonance tagging. *IEEE Trans. Medical Imaging* **14** (3), 413–421.

ZHU, Y., DRANGOVA, M., PELC, N.J. (1997). Estimation of deformation gradient and strain from cine-PC velocity data. *IEEE Trans. Medical Imaging* **16** (6).

Methods for Modeling and Predicting Mechanical Deformations of the Breast under External Perturbations

Fred S. Azar

Department of Imaging and Visualization, Siemens Corporate Research, 755 College Road East, Princeton, NJ 08540, USA
E-mail: Fred.Azar@scr.siemens.com

Dimitris N. Metaxas

Center for Computational Biomedicine Imaging and Modeling, Division of Computer and Information Sciences, Rutgers The State University of New Jersey, 110 Frelinghuysen Road, Piscataway, NJ 08854-8019, USA
E-mail: dnm@cs.rutgers.edu

Mitchell D. Schnall

Department of Radiology, Hospital of the University of Pennsylvania, MRI Bldg. 1 Founders, 3400 Spruce St., Philadelphia, PA 19104, USA
E-mail: schnall@oasis.rad.upenn.edu

List of symbols

$\mathbf{M}$ Mass matrix
$\mathbf{D}$ Damping matrix
$\mathbf{K}$ Model stiffness matrix
$\mathbf{K}_e$ Elemental stiffness matrix
$\mathbf{q}$ Vector which contains the displacement degrees of freedom
$\mathbf{q}_i = (q_{i,x}, q_{i,y}, q_{i,z})$ Vector-displacement degrees of freedom for node i in the model

Computational Models for the Human Body
Special Volume (N. Ayache, Guest Editor) of
HANDBOOK OF NUMERICAL ANALYSIS, VOL. XII
P.G. Ciarlet (Editor)

ISSN 1570-8659
DOI 10.1016/S1570-8659(03)12007-8

$\mathbf{X}_i = (x_i, y_i, z_i)$ Vector-position of node i in the model
$\mathbf{q}_e$ Element nodal displacement vector
$\mathbf{g}_q$ Inertial forces vector
$\mathbf{f}_q$ Generalized external forces vector
$\mathbf{f}_{i,\text{internal}}$ Internal stiffness force vector on node i
$\mathbf{f}_e$ Element force vector
μ Mass density
V_e Volume of element
$\boldsymbol{\sigma}$ Nodal stress vector
$\boldsymbol{\varepsilon}$ Nodal strain vector
ε_n Strain function for tissue type n
σ_n Stress function for tissue type n
$u = u(x, y, z)$, $v = v(x, y, z)$, $w = w(x, y, z)$ Displacement fields in the x, y and z directions, respectively
E_n Elastic modulus, function of strain for tissue type n
$\mathbf{E}$ Lagrangian strain tensor
$\mathbf{F}$ Deformation gradient tensor

1. Introduction

Breast cancer is the second leading cause of cancer deaths in women today (after lung cancer) and is the most common cancer among women, excluding non-melanoma skin cancers. According to the World Health Organization, more than 1.2 million people were diagnosed with breast cancer in 2001 worldwide. The American Cancer Society estimated approximately 192,000 new diagnosed cases of invasive breast cancer (stages I–IV), and 41,000 deaths from breast cancer among women in the United States in 2001. The incidence rate of breast cancer (number of new breast cancers per 100,000 women) increased by approximately 4% during the 1980s but leveled off to 100.6 cases per 100,000 women in the 1990s (FERLAY, BRAY, et al. [2001]). The death rates from breast cancer also declined significantly between 1992 and 1996, with the largest decreases among younger women. Medical experts attribute the decline in breast cancer deaths to earlier detection and more effective treatments. While breast cancer is less common at a young age (i.e., in their thirties), younger women tend to have more aggressive breast cancers than older women, which may explain why survival rates are lower among younger women. The standard treatment against breast cancer today is to cut out either the tumor or the whole affected area. The only way today to find out for sure if a breast lump or abnormal tissue is cancer, is by having a biopsy. The suspicious tissue, which is removed by a surgeon or radiologist during a biopsy, is then examined under a microscope by a pathologist who makes the diagnosis. A biopsy is done most of the time (except for palpable lesions) with the help of images of the breast obtained using an imaging technique, such as X-rays (PARKER, LOVIN, JOBE, et al. [1990], DRONKERS [1992], DERSHAW [1996]) or Magnetic Resonance Imaging (MRI) (FISCHER, VOSSHENRICH, KEATING, BRUHN, DOLER, OESTMANN and GRABBE [1994], OREL, SCHNALL, NEWMAN, POWELL, TOROSIAN and ROSATO [1994], FISCHER, VOSSHENRICH, DOLER, HAMADEH, OESTMANN and GRABBE [1995]).

Magnetic resonance imaging uses radio waves and magnetic fields to diagnose diseases. Patients are asked to lie on a table during the test, which takes about 30 minutes. They are then advanced into the MRI machine, which contains a strong magnetic field (1.5–4 T). The method consists of injecting a contrast-enhancing dye-like material into the patient's bloodstream and using magnetic resonance imaging to monitor the way in which this material is taken up and cleared out by the tumor tissue. The ability to identify a mass in the breast requires the mass to have a different appearance (or a different contrast) from normal tissue. With MRI, the contrast between soft tissues in the breast is 10 to 100 times greater than that obtained with X-rays (FISCHER, VOSSHENRICH, KEATING, BRUHN, DOLER, OESTMANN and GRABBE [1994], HARMS and FLAMIG [1994], OREL, SCHNALL, NEWMAN, POWELL, TOROSIAN and ROSATO [1994], FISCHER, VOSSHENRICH, DOLER, HAMADEH, OESTMANN and GRABBE [1995]). There are additionally multiple reports of MR imaging-detected breast cancers that are mammographically, ultrasonographically, and clinically occult (HARMS, FLAMIG, et al. [1993], BOETES, BARENTSZ, et al. [1994], GILLES, GUINEBRETIERE, et al. [1994]). Fig. 1.1 shows an MR image of a breast revealing areas of cancer spread (arrows) in addition to the larger tumor. The additional areas of cancer were not visible on the patient's X-ray mammograms.

The main disadvantage of breast MRI is its cost, which today is about 5 times that of X-ray mammography. According to most physicians, mammography is used more widely for breast cancer detection because it is a inexpensive technique, and most health institutions can afford to buy mammographic equipment. Today, however, an area of technical development is in the field of low-cost, dedicated breast MR systems, which could reduce the cost of breast MR imaging dramatically. Sales of magnetic resonance imaging (MRI) scanners reached an all-time high in 1999, topping the $1 billion mark. MRI Industry Report, a quarterly newsletter published by Miller Freeman, reported that

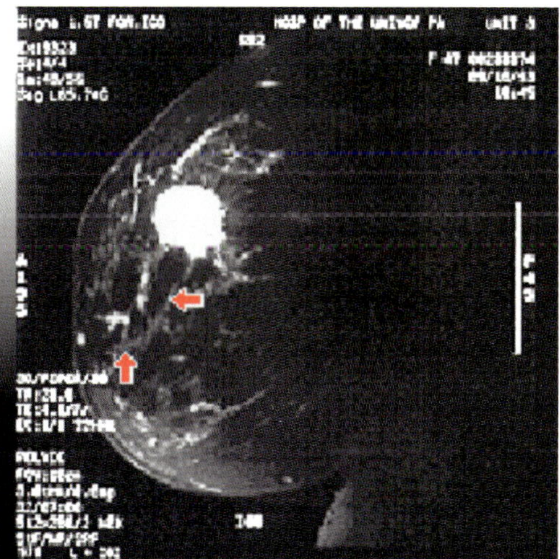

FIG. 1.1. MRI of a breast showing mammographically occult cancer tumors.

928 new MRI systems, valued at $1.07 billion, were installed at US hospitals and clinics in 1999. Sales revenue rose 12.2% due to the strong demand for new MRI systems introduced in 1998 and 1999. Sales appear to be on the rise for more costly MRI systems that feature more and improved capabilities, while sales for less costly systems are tapering off.

In MR breast imaging, the specificity has been reported to vary between 37 and 97% (HARMS, FLAMIG, et al. [1993], BOETES, BARENTSZ, et al. [1994], GILLES, GUINEBRETIERE, et al. [1994]): not all contrast-enhancing lesions prove to be malignant. With such a technique, which is highly sensitive, but not highly specific, an image-guided breast localization and biopsy system is needed to help differentiate between the benign enhancing lesions and carcinomas (OREL, SCHNALL, NEWMAN, POWELL, TOROSIAN and ROSATO [1994]). A whole-body MR system at 1.5 T (Signa; GE Medical Systems, Milwaukee, WI) is used for all needle localizations. The technique requires that the patient lies prone with the breast gently compressed between medial and lateral plates (Fig. 1.2). A multicoil array is used, with two coils placed on the medial plate and one coil on the lateral plate. The lateral plate contains a grid of approximately 1000 18-gauge holes placed at 5 mm intervals, which guide the needle in a plane parallel to the tabletop. The correct hole in the plate is identified and a needle is inserted through that hole into the breast with a pre-calculated depth (Fig. 1.2).

The MR imaging-guided localization technique encounters the following problems:

- The appearance, size and shape of the potential cancer lesion greatly depend on the dynamics of the interaction between the patient's physiology and the contrast-enhancing agent (TOFTS, BRIX, et al. [1999]). The lesion may clearly appear only in the two minutes following the contrast agent injection, then the signal intensity may vary arbitrarily, and it is possible that the apparent boundaries of the lesion may change dramatically.
- The needle is not a very sharp object and cannot be smoothly inserted in the breast. Every time the tip of the needle reaches the interface between two different types of tissue, its further insertion will push the tissue instead of piercing it, causing unwanted deformations until the pressure on the tissue interface is high enough. As soon as the needle pierces the displaced tissue interface, that interface quickly settles back to its original position, or somewhere close to it. The best way to remedy this problem would be to compress the breast as much as possible, which would minimize internal deformations. However, doing that would cause blood to be squeezed out of the breast, and would dramatically alter the appearance and shape of the lesion on the MR image (the perfusion changes to the lesion would disrupt Gd uptake), without mentioning the high level of discomfort for the patient who would be very reluctant to feel the pain for the entire duration of the procedure. The best solution would be to mildly compress the breast and obtain MR images clearly showing the position, shape and extent of the lesion. Then the breast would be highly compressed in order to minimize internal deformations during the needle insertion only. However, the missing link is to predict the displacement of the lesion from the mildly compressed configuration, to the highly compressed configuration.

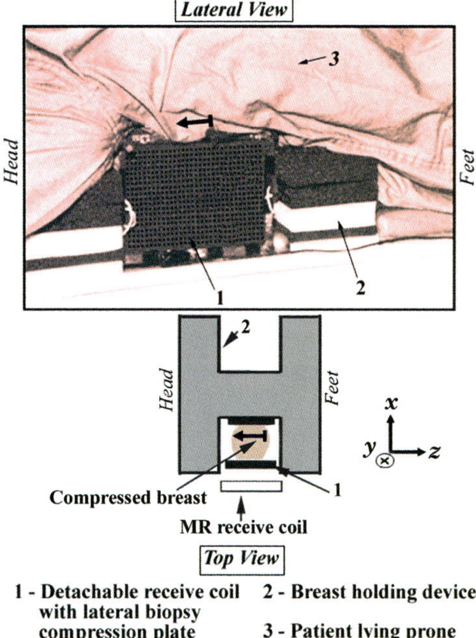

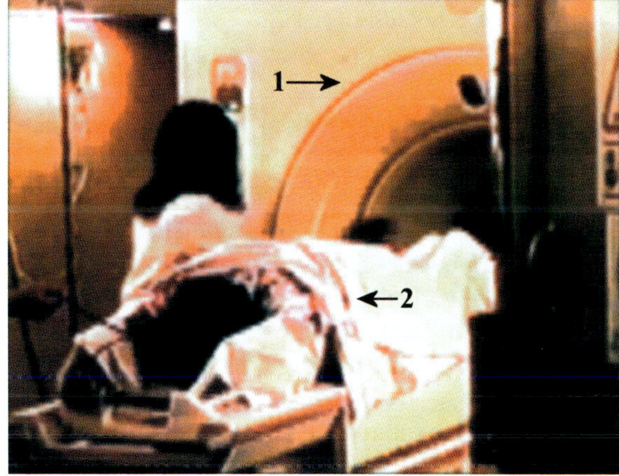

FIG. 1.2. Breast compression setup before a MR-guided needle localization procedure (top). The patient lies prone on the bed with her breast gently compressed between medial and lateral plates, as the bed is being moved into the high-field magnet (bottom).

The above limitations coupled with the deformable structure of the breast makes needle procedures very sensitive to the initial placement of the needle and to the amount of breast compression. It thus becomes relatively uncertain that the tissue specimen removed during the biopsy procedure actually belongs to the lesion of interest, due to the added difficulty of accurately locating the tumor's boundaries inside the breast. It is therefore important to develop techniques, which would solve or bypass the aforementioned problems, increase the level of confidence of a biopsy result (improving the diagnosis), and decrease the cost to society (including health care expenses). The financial savings could be considerable, and the morbidity associated with the biopsy procedure including the lost time from work that occurs as a result of biopsy could be greatly reduced (STAVROS, THICKMAN, RAPP, DENNIS, PARKER and SISNEY [1995]).

We present a virtual deformable breast model of the patient whose geometry is constructed from MR data. The elastic properties of the deformable model are based on the use of finite elements with nonlinear material properties capable of modeling the deformation of the breast under external perturbations. A high-field 1.5 Tesla machine Signa Horizon Echospeed (GEMS, Milwaukee) is used to obtain the 3D breast image sets. The image sets are used to construct the geometry of the finite element model. Contours of the breast are extracted, and each breast slice is segmented to locate the different tissue types, using appropriate custom-written software. The model geometry is then created using a custom-written preprocessor, which allows for a variable mesh size. We also developed a software algorithm (*BreastView*), which models large deformations of the breast model depending on the desired accuracy of the deformation. We hypothesize that the structural complexity of the breast can be simplified to only assign to the model elements, an average value of the mechanical properties of glandular, fatty, and cancerous tissue.

The major novelties in this model include the following:

- Breast plate compression results in a large compression, meaning that the total distance between the two plates decreases by more than 10%. In order to model such large deformations, we divide the large deformation compression into a number of much smaller displacement steps. For every displacement step, we make use of small strain theory. Strain is calculated using Cauchy's infinitesimal strain tensor formula (FUNG [1994]). After every small displacement iteration, the tissues' different material properties are recalculated in all model elements whose maximum principal strain has changed, in order to model the materials' nonlinear behavior. The main advantage of using small strain formulation lies in its simplicity, ease of implementation and fast computation. However, being an incremental formulation, it could lead to an accumulation of discretization errors and in consequence, to a lack of accuracy (SZEKELY, BRECHBUHLER, HUTTER, RHOMBERG and SCHMID [1998]). A way to solve this potential problem would be to use a total Lagrange formulation (as the one we use for estimating the nonlinear material properties of tissue from one displacement iteration to the next), in which every state is related to the initial configuration. However, that would complicate the formulation and slow down the computation of the solution. We show in a silicon phantom study that the incremental errors introduced by small strain formulation can be neglected for the purpose of this model and overall study.
- We present a new breast fatty tissue material model, which takes into account the effect of fat compartmentalization due to Cooper's ligaments in the breast. We show through empirical evidence that fat compartmentalization occurs as the breast is being compressed, and that the new updated fatty tissue material model takes

that phenomenon into account, and performs better than the original fatty tissue model.

- We apply finite element modeling theory to model the deformation of a human female breast in such a way that the entire process takes less than a half-hour (compared to several hours using a commercial finite element modeling package), which according to the clinicians consulted, is a reasonably short time duration.

This model can be used effectively in several different applications:

- *A new method for guiding clinical breast biopsy*. This method involves imaging the patient's breast without any or little compression before a needle procedure, then compressing the breast, and its virtual finite element model (by applying the same pressure to both), and using the displacement of the virtual tumor model to predict the displacement of the real cancer tumor. It is important to note that during the entire procedure (imaging, needle localization, and/or biopsy), the patient remains in the same prone position, and only the equipment 'moves' around the patient. Therefore, perturbations caused by the patient's movements are minor. A model like the one presented here is important to this procedure, in which any improvement in confidence for localizing the cancer tumor could become life-saving.
- *Other applications*. A finite element model of the breast can be a very flexible tool for many applications including registration of different breast MR data sets of the same patient taken under different compression amounts, or registration of different data sets from different imaging modalities. Other possible applications include diagnosis, measurements, surgery planning, simulations of deformation due to inserting a needle, and further away, virtual surgery and tele-surgery. It is important to note that the current model is based on a prone-acquired MRI and virtually all breast surgery is performed in the supine position. Therefore, the rotational invariance of the model would need to be proven for prone-supine deformations.

2. Organization of the chapter

The chapter is organized as follows:

1. In Section 3, the related work is presented and compared to our model. The general flowchart of operation is described.
2. The following six sections present the methods used, going through image acquisition and data extraction (Section 4), 3D mesh domain creation (Section 5), model dynamics (Section 6), internal model forces (Section 7), large deformation models (Section 8), and non-linear material models (Section 9).
3. The following four sections describe phantom and patient studies in detail:
   (a) *A phantom study* (Section 10). A deformable silicon gel phantom was built to study the movement of a stiff inclusion inside a deformable environment under plate compression. The phantom was imaged undeformed, then compressed (14%). The performance of our software algorithm was compared to that of a robust commercial FEM software package. A 3D deformable model of the phantom was built from the resulting MR data using our custom-written software and was virtually compressed using *BreastView*. Another

FEM was built using a commercial pre-processor (PATRAN, MSC, CA) from the phantom's directly measured dimensions, and was virtually compressed using ABAQUS (HKS, Rhode Island). The displacement vectors of the 8 corners of the stiff inclusion and its center were measured both from the MR images and from the two finite element models.

(b) *A patient study* (Section 11). A patient's breast was imaged uncompressed and then compressed 26%. The corresponding deformable model was built and was virtually compressed to match the real compression amount. We tracked the displacement of a small cyst inside the patient's breast, and used the deformable model to predict the cyst's position in the real compressed breast. We also tracked the displacement of two vitamin E pills taped to the surface of the patient's breast. We present a convergence analysis and a material properties sensitivity analysis. The results show that it is possible to create a deformable model of the breast based on the use of finite elements with non-linear material properties, capable of modeling the deformation of the breast in a clinically useful amount of time (less than a half-hour for the entire procedure).

(c) *A clinical breast compression study* (Section 12). Three patient breasts were imaged uncompressed and then compressed. Patients were chosen to have a variety of breast cancers of different shape, size and location in the breast. The displacement of the cancer tumors was recorded in the three patient breasts. A model of each patient's breast was constructed, and then used to predict the deformation and displacement of the cancer tumors after breast compression.

(d) *Registration of breast MR images of the same patient under different compressions* (Section 13). A patient's breast was imaged under two different plate compression amounts. A deformable model of that breast was constructed and compressed using virtual compression plates. The displacement of a small cyst was recorded in the real breast and compared to the displacement of the 'virtual cyst' in the deformable model from one compression state to the other.

4. The following two sections deal with additional issues, including potential sources of error (Section 14) and specific properties of reliability (Section 15).
5. Section 16 summarizes the major novelties in the model, and Section 17 presents the concluding remarks.
6. Finally, Section 18 (Appendix) presents the finite element modeling theory in detail, as well as details of the silicon phantom construction.

3. Related work

Finite element modeling has been used in a very large number of fields. However it is only recently that deformable models have been used to simulate deformations in soft tissue. Physical models are among the first to be used. Among these physical models, elastic (linear and visco-elastic) models have been extensively described in the literature (CHEN and ZELTZER [1992], SPEETER [1992], REDDY and SONG [1995]).

The most widely used representations for deformable volumes are parametric models with B-spline representation (ZIENKIEWICZ [1977]). Other possible models are mass–spring models (MILLER [1988], CHADWICK, HAUMANN and PARENT [1989], LUCIANI, JIMENEZ, FLORENS, CADOZ and RAOULT [1991], NORTON, TURK, BACON, GERTH and SWEENEY [1991], JOUKHADAR [1995]) and implicit surfaces (DESBRUN and GASCUEL [1995]). The mass–spring methods have been used most of the time for surgery simulation due to their simplicity of implementation and their lower computational complexity (KUEHNAPFEL and NEISIUS [1993], BAUMANN and GLAUSER [1996], MESEURE and CHAILLOU [1997]). Other methods have relied on geometry rather than physics to predict breast deformation (BEHRENBRUCH, MARIAS, ARMITAGE, YAM, MOORE, ENGLISH and BRADY [2000]).

Finite element models are less widely used due to the difficulty of their implementation and their larger computing time. There are many powerful commercial FEM packages that allow complex simulations of deformation such as ABAQUS (HKS, Rhode Island); breast tissue is relatively complex, and consists of layers of different tissues interlaced with ligaments and fascias. Very complex models would be needed to model these objects realistically. However, the complexity of the model and the required computational time (which can extend to several days on a SGI workstation) would prohibit these models from being useful clinically. Few models of the breast have actually been implemented using the commercial software packages, and have mostly involved phantom studies (AZAR, METAXAS and SCHNALL [1999], SCIARETTA, BISHOP, SAMANI and PLEWES [1999], WILLIAMS, CLYMER and SCHMALBROCK [1999]). A real-time system has been recently developed for hepatic surgery simulation, and involves deformations of soft tissue (COTIN, DELINGUETTE and AYACHE [1999]). An explicit large displacement model for interactive surgery simulation using fast non-linear finite element models was proposed by PICINBONO, DELINGUETTE and AYACHE [2001], and models for physically realistic simulation of global deformation were proposed by ZHUANG and CANNY [1999]. Finite element models were also proposed for bone and muscle biomechanics research (HOU, LANG, et al. [1998], KABEL, RIETBERGEN, et al. [1999] YUCESOY, KOOPMAN, et al. [2002]).

The desired accuracy of the deformation in the breast model must therefore be balanced against the need for speed. In order to develop a model for deformation of the breast, we must decide on a geometric description of the breast, a mathematical model of the elastic deformation, and a solution algorithm that is both fast and yields a reasonably realistic result. For that reason, what matters most is that the breast deformation results be realistic and available in a clinically useful time (less than thirty minutes for the entire procedure), and that the model be robust and show a consistent and predictable behavior. In our approach, we strive to integrate the requirements for a realistic simulation of deformation, and the reasonably fast time modeling which is a necessity if the model is to be used in a clinical environment: by incorporating the geometric definition of the breast model into the physics-based framework developed in METAXAS [1992], METAXAS and TERZOPOULOS [1993], we create a deformable breast model, capable of reasonably predicting the internal deformations of a real patient breast after plate compression (AZAR, METAXAS and SCHNALL [2001], AZAR, METAXAS and SCHNALL [2002]).

Fig. 3.1 shows the general flowchart of how the FEM of the breast is created (vertical dimension), and the process by which it is virtually compressed (horizontal dimension). We start with the patient's breast MR data which constitutes a 3D image set of parallel slices. Each of these slices is segmented and classified into different tissue types (glandular, fatty and/or cancerous) semi-automatically. Given this data, the 3D mesh of the model is generated automatically, and every element in the mesh is assigned a particular tissue type value. Then the deformation process starts from the given initial and boundary conditions. The large compression plate displacement is translated into a number of much smaller displacement iterations, each of which is applied in turn to the FEM of the breast. For each applied displacement iteration, the equation of motion must be solved and the intermediate model node displacements are calculated. Following that step, the elasticity value of every element in the model is updated given its principal strain value and the non-linear material model of the tissue type it represents. After all the displacement iterations are applied the final compressed model of the breast is ob-

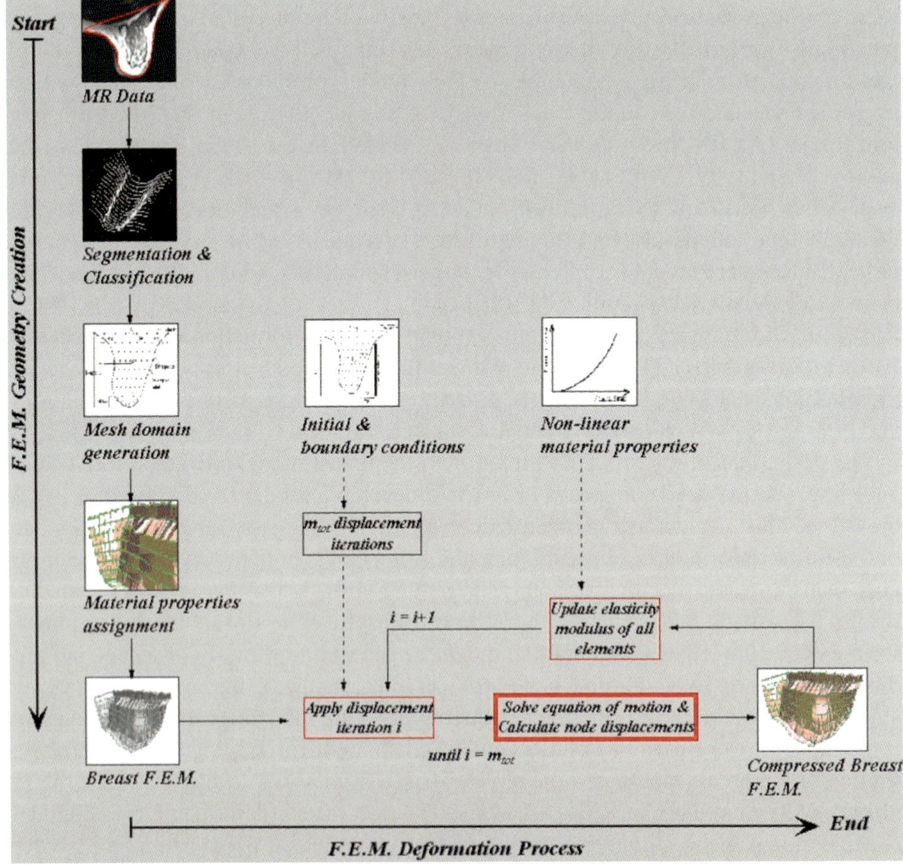

FIG. 3.1. General flowchart of operation.

tained. Because of the high variability of breast shapes and the deformation of the breast when compressed in order to be imaged, the model devised closely follows the contours of the patient breast.

4. Image acquisition and data extraction

The patient data is a set of parallel 2D spoiled gradient echo MR axial slices of the breast. Usually, an axial T1-weighted spin echo sequence is performed with a repetition time of 500 msec, and an echo time of 12 msec, with a 12–16 cm field of view, a 1–3 mm thick section, and a 256×256 matrix. The acquisition ensures a 3D visualization of the patient's breast. First, the MR image 3D set is converted into a set of axial slices (if the original data is not axial) through automatic resampling of the data using a software package such as Scion Image (Scion Corp., Maryland).

The MR images are loaded to a program, which enables segmentation through contours. The contours of the breast boundary are created semi-automatically (using Scion Image and Adobe Photoshop) using a threshold-based technique, and saved to a file for subsequent use. At the same time each MR image is automatically segmented into parenchyma, fat and/or lesion tissue, using a robust segmentation algorithm (each tissue type is assigned a specific gray level value).

The segmentation algorithm used is based on the concept of fuzzy connectedness (CARVALHO, GAU, HERMAN and KONG [1999], SAHA, UDUPA and ODHNER [2000]). This approach is semiautomatic, in the sense that when using the segmentation program (developed by Carvalho et al.) the user must identify seed voxels, which definitely belong to the various objects in the image (fat, glandular, cancerous tissues). The user-selected seed voxels are then used for automatic segmentation of the entire image. This algorithm yields accurate and robust results, and is able to segment a breast full MR 3D volume in less than a minute (on a standard PC workstation), after only spending a few minutes selecting the seed voxels. Although the technique works reasonably well due to the detection of different tissue contrasts, it may not segment lesions well because it is not parametric enough with respect to the pharmacokinetics of the Gadolinium contrast-enhancing agent.

5. 3D mesh domain creation

BreastView, a custom-written program in *C*, takes as input the set of breast contours and the segmentation results. Running on a SGI workstation, it generates the 3D computational domain (mesh) of the breast in a few seconds, allowing to scale the volume elements to any size. The program can also generate a finite element model (FEM) file readable by FEM software ABAQUS. This file contains the definition of the volume elements, as well as the boundary conditions, and the different material properties. Two types of 3D volume elements are used in the model (Fig. 5.1, bottom). Solid 8-node (hexahedral) trilinear isoparametric elements are used to model all breast tissues except for the skin, and are assigned non-linear material properties. 3-node triangular isoparametric elements are used to model the skin, and are assigned non-linear elastic properties in the plane of their triangular surface, and no axial stiffness. The nodes forming the

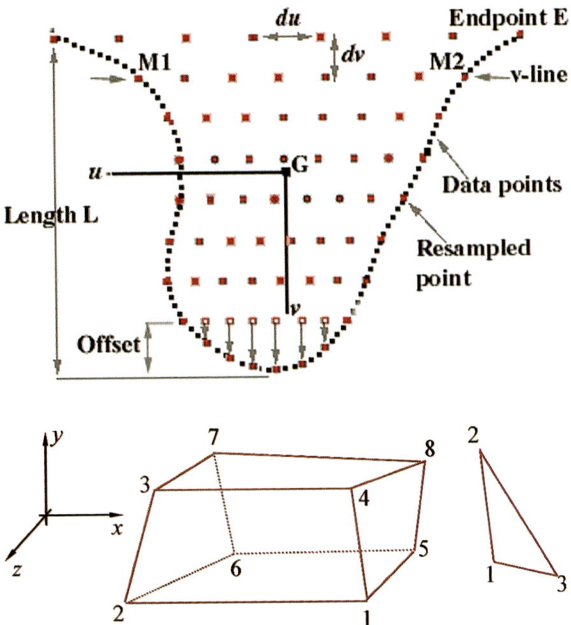

FIG. 5.1. Slice contour with 2D-mesh generation (top), and elements used to build the deformable model (8-node element, 3-node triangle, bottom).

2D mesh representing the skin are the same nodes belonging to the hexahedral elements at the borders of the breast model.

BreastView allows the mesh to be as dense as needed. These volume elements are well-suited for the purpose of the model: if a physician only needs approximate information on how the tumor is moving, the tumor could be included in one volume element (by correctly scaling the mesh density) which would be assigned its own mechanical characteristics. Fig. 5.1 (top) shows a contour and the 2D mesh generated in it.

In order to create the 3D mesh, we first find the principal direction d for the entire contour set (direction of the v-axis, Fig. 5.1, top), from the chest wall to the nipple. Any line with direction d in the plane of a contour, which intersects the contour, will intersect it twice. This direction is that of the line orthogonal to the line, which passes through the two endpoints (the principal direction method using the eigenvectors of the contour set of points did not yield the best direction d). The principal direction d for the entire contour set is obtained by averaging the computed individual slice directions. The center of gravity G of the contour set is calculated. The 3D mesh can now be easily generated following the U (along u) and V (along v) resolution desired, in the orthogonal (u, v) basis centered at G. This algorithm ensures that we have the same number of points on every V-line, and the same number of V-lines on every slice.

In order to avoid that degenerate elements at the edge of the breast may be created during the 3D meshing process, we introduce a predetermined offset for every slice, which is a percentage of the length L of the breast slice, as shown in Fig. 5.1. The row of element nodes generated closest to the edge of the breast slice is at a distance of $L \times$ Offset from the edge of the breast. Finally, every node on this final row is projected along the v-axis onto the edge of the breast slice.

We implemented the meshing in the z-direction (number of slices) in two different ways:

- There can be as many element slices (in the z-direction) as there are segmented data slices.
- We make the meshing in the z-direction variable, increase the mesh density around the point of interest, and decrease it away from the point of interest. This scheme can enable us to decrease the total number of model slices, and hence greatly accelerate the simulation.

Since the point of interest is usually a hard inclusion such as a cyst or a tumor, it is more important for the mesh to be denser in that area than in the slices away from the inclusion, which may have less impact on the displacement of the inclusion. By implementing such an algorithm, we can also decrease the total number of model slices, thereby greatly decreasing the compression simulation time.

Once the 3D mesh is created, each element is assigned a material property corresponding to that of fat, parenchyma or cancerous tissue. In order to determine what type of material property to assign to any one 3D element, an algorithm makes use of the automatically segmented breast MR data. The algorithm calculates what percentage of fat, parenchyma and cancerous tissue lie within the element, by identifying the corresponding number of voxels which lie in the element. The material property that corresponds to the highest percentage is the one assigned to the element. It is clear from this algorithm, that the finer the mesh, and the more precise the assignment of material properties will be.

6. Model dynamics

The governing Lagrange equations of motion are second order differential equations given as follows (TERZOPOULOS, PLATT, et al. [1987], METAXAS [1992], METAXAS and TERZOPOULOS [1993]):

$$\mathbf{M}\frac{\partial^2 \mathbf{q}}{\partial t^2} + \mathbf{D}\frac{\partial \mathbf{q}}{\partial t} + \mathbf{K}\mathbf{q} = \mathbf{g_q} + \mathbf{f_q}, \tag{6.1}$$

where $\mathbf{M}$, $\mathbf{D}$ and $\mathbf{K}$ are the mass, damping and stiffness matrices, respectively. The vector $\mathbf{q}$ contains the displacement degrees of freedom, $\mathbf{g_q}$ are the inertial forces arising from the dynamic coupling between the local and global degrees of freedom, and $\mathbf{f_q}$ are the generalized external forces associated with the model's degrees of freedom.

In our case, given the nature of the problem, and the fact that we consider static deformations only, it makes sense to simplify the equations of motion by setting the mass density to zero, which will still preserve useful first-order dynamics that achieve equilibrium. Setting the mass density μ to zero causes the mass matrix $\mathbf{M}$ and the inertial forces $\mathbf{g_q}$ to vanish (PARK, METAXAS, YOUNG and AXEL [1996a], HABER, METAXAS and AXEL [2000]). This results in the first-order dynamic system

(PARK, METAXAS, YOUNG and AXEL [1996b]),

$$\mathbf{D}\frac{\partial \mathbf{q}}{\partial t} + \mathbf{K}\mathbf{q} = \mathbf{f}_{\mathbf{q}}. \tag{6.2}$$

Because these equations lack an inertial term, the system comes to rest as soon as all the forces equilibrate or vanish. We use **D** as a stabilizing factor only, and do not impose physical damping, which cannot be measured from experimental data. Therefore, we assume that **D** is diagonal and constant over time, and set it equal to the identity matrix ($\mathbf{D} = \mathbf{I}$). Therefore, the breast model is deformed using the following reduced form of Lagrange's equation of motion:

$$\alpha\mathbf{I}\frac{\partial \mathbf{q}}{\partial t} + \mathbf{K}\mathbf{q} = \mathbf{f}_{\mathbf{q}}, \tag{6.3}$$

where the vector $\mathbf{q}$ contains the displacement degrees of freedom, α is a numerical damping constant and the vector $\mathbf{f}_{\mathbf{q}}$ contains the total external forces due to body forces which in our case would be gravity. We are modeling the compression of the breast for a patient in a prone position. The breast is immobile in a state of equilibrium (the skin is mainly responsible for counteracting the effects of the gravitational force), and the plates compressing the breast move in a direction orthogonal to that of the gravitational forces. Therefore the vector $\mathbf{f}_{\mathbf{q}}$ can be set to zero.

We may approximate the equation by using the finite element method described previously. Through this method, all quantities necessary for the Lagrange equations of motion are derived from the same quantities computed independently within each finite element. The various matrices and vectors involved in the Lagrange equations of motion are assembled from matrices computed within each of the elements.

Therefore, in our algorithm we do not assemble the model stiffness matrix **K**, but work with the elemental stiffness matrices $\mathbf{K}_e$, and assemble the forces around the nodes. Compared to the classical theory, this method is equivalent to a relaxation method, i.e., calculating the residuals at the nodes.

Stiffness (represented by the **K** matrix) can also be viewed as an internal resistance which results in a force equal to $-\mathbf{K}\mathbf{q}$. Lagrange's equation of motion can then be written for each and every node i in the model as (HABER, METAXAS and AXEL [2000])

$$\alpha\mathbf{I}\frac{\partial \mathbf{q}_i}{\partial t} = \mathbf{f}_{i,\text{total}} = \mathbf{f}_{i,\text{internal}}, \tag{6.4}$$

where $\mathbf{q}_i$ is the 3D nodal displacement at node i, $\mathbf{f}_{i,\text{internal}} = [-\mathbf{K}\mathbf{q}]_i$ is an internal stiffness force. The nodal equation becomes

$$\alpha\mathbf{I}\frac{\partial \mathbf{q}_i}{\partial t} = \mathbf{f}_{i,\text{internal}}, \tag{6.5}$$

where α is a numerical damping constant, which we set to 1. The time-discretized nodal equation can then be written as

$$\mathbf{q}_{i,t+1} = \mathbf{q}_{i,t} + \Delta t \cdot \mathbf{f}_{i,\text{internal}}. \tag{6.6}$$

At each displacement iteration, we integrate the equation using an Euler method with adaptive step sizing. The step size Δt is inversely proportional to the integration error.

Since the forces multiply the time step, the adaptive step sizing effectively modulates the strength of the forces, thereby making the solution converge much faster than with a simple Euler integration technique (HABER, METAXAS and AXEL [2000]). We compute two estimates of $\mathbf{q}_{i,t+1}$, $\mathbf{q}_{i,a}$ by taking an Euler step of size Δt, and $\mathbf{q}_{i,b}$ by taking two Euler steps of size $\Delta t/2$. Since $\mathbf{q}_{i,a}$ and $\mathbf{q}_{i,b}$ differ from each other by $\mathrm{O}(\Delta t^2)$, a measure of the current error is $e = \sum_{i=1}^{n} |\mathbf{q}_{i,a} - \mathbf{q}_{i,b}|$, where the error is over the sum of the displacement errors of all nodes. If we are willing to have an error of as much as e_A, then the new step size can be written as $\Delta t_{\text{new}} = \sqrt{e_A/e} \cdot \Delta t$. The numerical integration ends when the difference between the sum of all displacements from one time iteration to the next is less than a predetermined threshold.

7. Internal forces due to stiffness

The degrees of freedom of the model are the 3D displacements of the finite element nodes. In our formulation, stiffness is represented as an internal nodal force $\mathbf{f}_{i,\text{internal}}$. The element force vector $\mathbf{f}_\text{e}$ contains the forces on the element nodes: $\mathbf{f}_\text{e} = [\mathbf{f}_{i,\text{internal}}\ \mathbf{f}_{i+1,\text{internal}}\ \cdots\ \mathbf{f}_{i+n-1,\text{internal}}]^\text{T}$ where $\mathbf{f}_{i,\text{internal}}$ is the internal force on node i, and n is the number of nodes in the element. $\mathbf{f}_\text{e}$ is calculated for each element as

$$\mathbf{f}_\text{e} = \mathbf{K}\mathbf{q}_\text{e}, \tag{7.1}$$

where $\mathbf{q}_\text{e}$ contains the element's nodal displacements: $\mathbf{q}_\text{e} = [\mathbf{q}_i\ \mathbf{q}_{i+1}\ \cdots\ \mathbf{q}_{i+n-1}]^\text{T}$. The elemental stiffness matrix, $\mathbf{K}_\text{e}$, incorporates the geometry, material properties of the element and is a triple integral over the volume V_e of the element. It is computed from

$$\mathbf{K}_\text{e} = \iiint_{V_\text{e}} \mathbf{B}^\text{T}\mathbf{D}\mathbf{B}\,\mathrm{d}V. \tag{7.2}$$

The matrix $\mathbf{D}$ contains the material property information (which includes the value of the material's elastic modulus $E_k(\varepsilon_k)$) and relates nodal stresses $\boldsymbol{\sigma}$ to nodal strains $\boldsymbol{\varepsilon}$:

$$\boldsymbol{\sigma} = \mathbf{D}\boldsymbol{\varepsilon}.$$

The matrix $\mathbf{B}$ relates nodal strains to displacements and incorporates Cauchy's infinitesimal strain tensor formulation (see Eq. (7.1)):

$$\boldsymbol{\varepsilon} = \mathbf{B}\mathbf{q}_\text{e}. \tag{7.3}$$

Once the element forces are calculated, each element contributes to the total internal nodal force $\mathbf{f}_{i,\text{internal}}$ for each of its nodes. A detailed derivation of the finite element formulation for calculating the stiffness matrix, $\mathbf{K}$, is presented in ZIENKIEWICZ and TAYLOR [1989].

8. Modeling large deformations

Compressing a breast using compression plates results in a large compression of the breast, meaning that the total compression distance between the two plates decreases by more than 10%. In order to model such large deformations, we divide the large deformation compression into a number of much smaller displacement steps. For every displacement step, we make use of small strain theory. Strain is calculated using Cauchy's infinitesimal strain tensor formula (given in the unabridged notation (FUNG [1994])):

$$\begin{aligned}
&\varepsilon_{xx} = \frac{\partial u}{\partial x}, \qquad \varepsilon_{yy} = \frac{\partial v}{\partial y}, \qquad \varepsilon_{zz} = \frac{\partial w}{\partial z},\\
&\varepsilon_{xy,yx} = \frac{1}{2}\left(\frac{\partial u}{\partial y} + \frac{\partial v}{\partial x}\right), \qquad \varepsilon_{xz,zx} = \frac{1}{2}\left(\frac{\partial u}{\partial z} + \frac{\partial w}{\partial x}\right),\\
&\varepsilon_{yz,zy} = \frac{1}{2}\left(\frac{\partial v}{\partial z} + \frac{\partial w}{\partial y}\right).
\end{aligned} \tag{8.1}$$

$u = u(x, y, z)$, $v = v(x, y, z)$, $w = w(x, y, z)$ are the displacement fields in the x, y and z directions, respectively, from one small displacement iteration to another.

The main advantage of using small strain formulation lies in its simplicity, ease of implementation and fast computation. However, being an incremental formulation, it could lead to an accumulation of discretization errors and in consequence, to a lack of accuracy (SZEKELY, BRECHBUHLER, HUTTER, RHOMBERG and SCHMID [1998]). A way to solve this potential problem would be to use a total Lagrange formulation (as the one we use for estimating the non-linear material properties of tissue from one displacement iteration to the next), in which every state is related to the initial configuration. However that would complicate the formulation and slow down the computation of the solution. We show in the phantom study that the incremental errors introduced by small strain formulation can be neglected for the purpose of this model and overall study.

9. Modeling non-linear material properties

Most biological tissues display both a viscous (velocity dependent) and elastic response, however since we are only interested in slow displacements, the great majority of the forces developed can be attributed purely to the elastic response. All tissues involved in the breast can be considered:

- isotropic (HAYES, KEER, HERMANN and MOCKROS [1972], KROUSKOP, WHEELER, KALLEL, GARRA and HALL [1998]),
- homogeneous (SARVAZYAN, SKOVORODA, EMELIANOV, FOWLKES, PIPI, ADLER, BUXTON and CARSON [1995], SKOVORODA, KLISHKO, GUSAKYAN, MAYEVSKII, YERMILOVA, ORANSKAYA and SARVAZYAN [1995]),
- incompressible (FUNG [1993]),
- to have non-linear elastic properties (FUNG [1981], ZHANG, ZHENG and MAK [1997]).

With these assumptions, it is possible to define the mechanical behavior of breast tissue using a single elastic modulus E_n, which is a function of strain ε_n for tissue type n (σ_n is the stress):

$$E_n = \frac{\partial \sigma_n}{\partial \varepsilon_n} = f(\varepsilon_n). \tag{9.1}$$

This non-linear relationship is calculated for every tissue type from uniaxial stress–strain experiments using tissue samples. The experimental curves are fit to a material model, which can be characterized using a small number of parameters. In order to model the non-linear mechanical behavior in every element, the following steps are followed, after every small deformation increment:

(1) We calculate the Lagrangian strain tensor, $\mathbf{E}$, at the center of the element. $\mathbf{E}$ is a measure of the deformation of a point in the model with respect to its initial position (a detailed description can be found in SPENCER [1980]):

$$\mathbf{E} = \frac{1}{2}\left(\mathbf{F}^{\mathrm{T}}\mathbf{F} - \mathbf{I}\right). \tag{9.2}$$

$\mathbf{F}$ is the (3×3)-deformation gradient tensor, and its components can be written as

$$F_{pq} = \frac{\partial x_p}{\partial X_q}, \tag{9.3}$$

where x_p is one of the three components of $\mathbf{x}$, the final position vector, and X_q is one of the three components of $\mathbf{X}$, the initial position vector. In our case, the final position of an element center point is after a given deformation increment, and the initial center point is before the first deformation increment. $\mathbf{I}$ is the identity matrix.

(2) We calculate the maximum principal strain component, $e_{\max}$, which is given by the largest eigenvalue of the strain tensor $\mathbf{E}$.

(3) Using the experimental stress–strain curve for the particular tissue type, the element is assigned a Young's modulus value corresponding to the slope of the curve at the strain measure $e_{\max}$: $E_n = f(e_{\max})$.

Therefore, after every deformation increment, the element's stiffness value is updated to model the material's non-linear behavior. By doing so, we model a continuously differentiable stress–strain curve (in the limits of our deformations) as a continuous step-wise linear curve (Fig. 9.1).

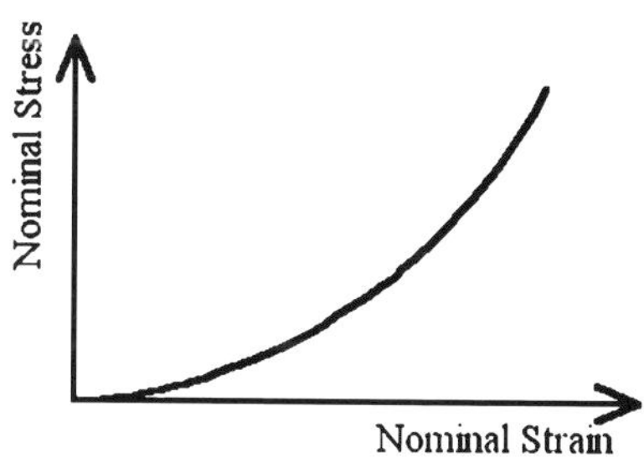

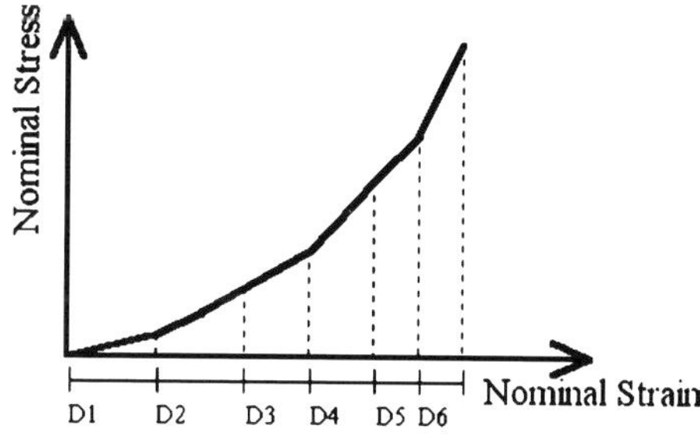

FIG. 9.1. Modeling a non-linear stress–strain curve (left) with a piecewise linear curve (right). D_i is the interval used during the ith deformation increment.

10. A phantom study

10.1. Experimental design and methods

A deformable silicon gel phantom was built to study the movement of a stiff inclusion inside a deformable environment (as a tumor inside the breast) under plate compression (AZAR, METAXAS, MILLER and SCHNALL [2000]). The phantom was imaged undeformed, then compressed. A 3D deformable model of the phantom was built from the resulting MR data, and was virtually compressed using custom-written software (*BreastView*). Another FEM of the phantom was built using a commercial pre-processor (PATRAN) from the phantom's directly measured dimensions, and was compressed virtually using a powerful commercial FEM software package (ABAQUS) which uses implicit integration schemes.

The displacement vectors of the corners of the stiff inclusion and its center were measured both from the MR images and from the two finite element models. This study also serves to validate our mathematical model of large deformations by comparing our results to that of a commercial FEM software package.

The explicit Finite Element formulation, which we implemented, using small strain theory, may not always be as accurate as the implicit one, and time discretization errors can accumulate. The main reason is that our strain formulation is of an incremental form; this could lead to a lack of accuracy (KOJIC and BATHE [1987]). The results of this phantom study show that the compressed model allows us to track the position and motion of the stiff inclusion in the real compressed deformable environment. Furthermore, after comparing the compressed *BreastView* and ABAQUS models, the results also show that using a small strain approximation in our finite element formulation does not introduce a significant error.

10.1.1. Phantom construction

The phantom was designed to have magnetic properties (T1 and T2) similar to those of human breast tissue, to withstand large deformations (20% or greater), and to enable controlled deformations. The gel phantom was build using Sylgard Dielectric Gel 527 (Dow Corning, Midland, MI). A similar silicon gel (model Q7-2218, Dow Corning) has been suggested for use in MR imaging (GOLDSTEIN, KUNDEL, DAUBE-WHITERSPOON, THIBAULT and GOLDSTEIN [1987]), and the same silicon gel was used to validate tagging with MR imaging to estimate material deformation (YOUNG, AXEL, DOUGHERTY, BOGEN and PARENTEAU [1993]). The geometry of the deformable phantom consists of a rectangular box ($84 \times 82 \times 70$ mm) containing a rectangular inclusion ($20 \times 23 \times 20$ mm), which is 4.3 times stiffer than the surrounding silicon (Fig. 10.1). For more details about the silicon phantom construction, see Appendix.

10.1.2. MR imaging of silicon phantom

The full silicon gel phantom was placed in a custom-built pressure device, where a pressure plate could compress the gel phantom in a similar way as with a real breast with the desired amount of deformation. The whole setup was secured firmly and imaged with a

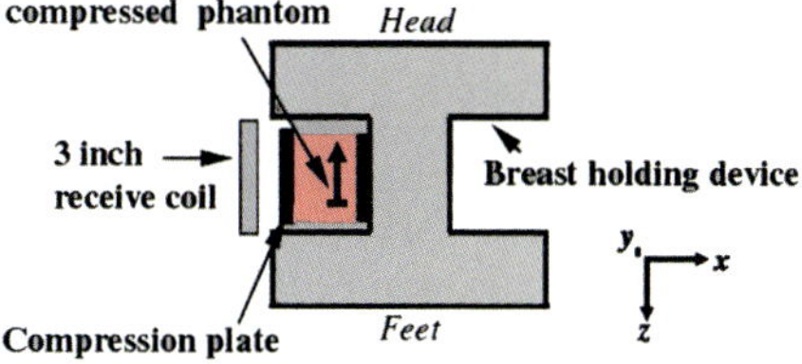

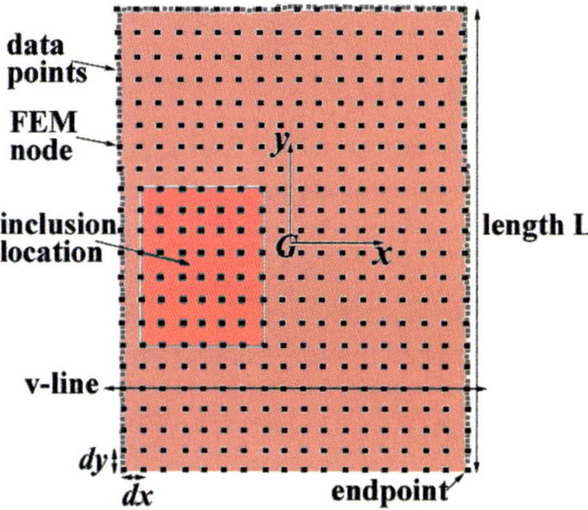

FIG. 10.1. Top view of the setup for imaging the compressed gel phantom (top), and construction of phantom model: 2D axial slice shown (bottom).

whole body 1.5 T superconducting magnet (GEMS, Milwaukee). The silicon gel phantom was first imaged undeformed. The compression plate then applied a deformation width of 14% (9.8 mm) in the x-direction, and the phantom was imaged again. An axial and a coronal T1-weighted fast multi planar gradient echo (FMPGR) sequences were performed in the uncompressed and compressed case.

10.1.3. Material properties of silicon phantom

The elastic properties of the phantom materials were evaluated on an Instron Model 1331 (Cambridge, MA) mechanical testing machine containing a semi-hydraulic computer driven system for very accurate tensile strength measurements. We used a load cell rated to 10 Newtons. This load cell is certified to an accuracy of 0.01 N. Flat cylindrical samples of the silicone gel and the stiffer inclusion underwent uniaxial stress tests. Static load-deformation (stress–strain) curves were obtained.

The silicon gel can be assumed to be an isotropic hyperelastic material, and be described by a "strain energy potential" which defines the strain energy stored in the material per unit of reference volume. The experimental data was fit to a law of rubber-like material known as the Mooney–Rivlin strain energy function (GREEN and ZERNA [1968], SPENCER [1980]). The form of the Mooney–Rivlin strain energy potential is

$$U = C_{10}(\bar{I}_1 - 3) + C_{01}(\bar{I}_2 - 3) + \frac{1}{D_1}\left(J^{\mathrm{el}} - 1\right)^2, \tag{10.1}$$

where U is the strain energy per unit of reference volume; C_{10}, C_{01} and D_1 are temperature-dependent material parameters; $\overline{I}_1$ and $\overline{I}_2$ are the first and second deviatoric strain invariants defined as $\overline{I}_1 = \overline{\lambda}_1^{2} + \overline{\lambda}_2^{2} + \overline{\lambda}_3^{2}$ and $\overline{I}_2 = \overline{\lambda}_1^{(-2)} + \overline{\lambda}_2^{(-2)} + \overline{\lambda}_3^{(-2)}$, where the deviatoric stretches $\overline{\lambda}_i = J^{-1/3}\lambda_i$; J is the total volume ratio; J^{el} is the elastic volume ratio, and λ_i are the principal stretches (defined as the ratios of current length to length in the original configuration in the principal directions of a material fiber). We assume that the silicon gel is incompressible and temperature independent (in the conditions of the experiment). Therefore, the strain energy potential expression can be simplified to

$$U = C_{10}(\overline{I}_1 - 3) + C_{01}(\overline{I}_2 - 3). \tag{10.2}$$

C_{10} and C_{01} are the material parameters to be determined experimentally.

The stress–strain relationship is developed using derivatives of the strain energy function with respect to the strain invariants. We define this relation in terms of the nominal stress T_U (the force divided by the original, undeformed area) and the nominal, or engineering strain ε_U (ratio of deformation length to length in the original configuration). The deformation gradient, expressed in the principal directions of stretch is

$$\mathbf{F} = \begin{bmatrix} \lambda_1 & 0 & 0 \\ 0 & \lambda_2 & 0 \\ 0 & 0 & \lambda_3 \end{bmatrix}, \tag{10.3}$$

where λ_1, λ_2 and λ_3 are the principal stretches. Because we assume incompressibility and isothermal response, $J = \det(\mathbf{F}) = 1$ and, hence, $\lambda_1\lambda_2\lambda_3 = 1$. The deviatoric strain invariants in terms of the principal stretches are then $I_1 = \lambda_1^2 + \lambda_2^2 + \lambda_3^2$ and $I_2 = \lambda_1^{-2} + \lambda_2^{-2} + \lambda_3^{-2}$. The uniaxial deformation mode is characterized in terms of the principal stretches, λ_i, as

$$\lambda_1 = \lambda_U, \qquad \lambda_2 = \lambda_3 = 1/\sqrt{\lambda_U}, \tag{10.4}$$

where λ_U is the stretch in the loading direction. The strain energy potential expression can therefore be expressed solely in terms of λ_U:

$$U = C_{10}\left(\lambda_U^2 + 2\lambda_U^{-1} - 3\right) + C_{01}\left(\lambda_U^{-2} + 2\lambda_U - 3\right). \tag{10.5}$$

To derive the uniaxial nominal stress T_U, we invoke the principle of virtual work ($\delta U = T_U \delta\lambda_U$) so that

$$T_U = \frac{\partial U}{\partial \lambda_U} = 2\left(1 - \lambda_U^{-3}\right) \cdot \left(\lambda_U C_{10} + C_{01}\right). \tag{10.6}$$

Now since the stretch λ_U is related to the nominal strain ε_U by ($\lambda_U = \varepsilon_U + 1$), the nominal stress–strain relationship can finally be written as

$$T_U = 2\left[1 - (1+\varepsilon_U)^{-3}\right] \cdot \left[(1+\varepsilon_U)C_{10} + C_{01}\right]. \tag{10.7}$$

Eq. (10.7) was fit to the experimental stress–strain curves for the two types of silicon gel, using the least sum of squares method. The results are shown in Fig. 10.2. The average parameter values calculated are $C_{10} = 3740 \pm 64\ \mathrm{N/m^2}$, $C_{01} = 1970 \pm 34\ \mathrm{N/m^2}$ for

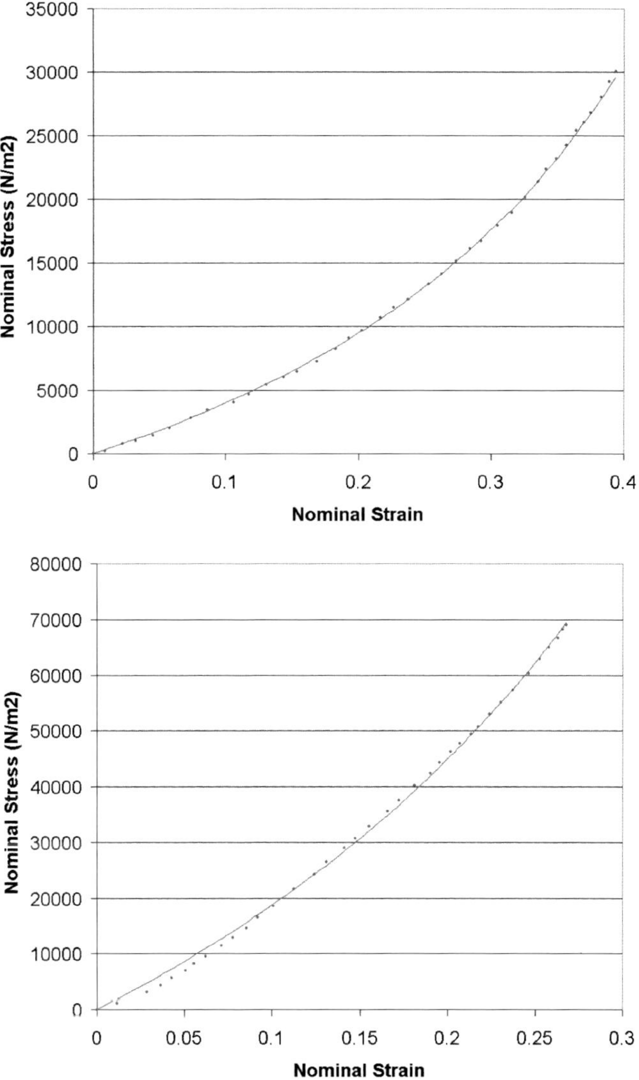

FIG. 10.2. Least-squares fit between the average experimental stress–strain curves (shown as dots), and the Mooney–Rivlin model (shown as a continuous line), for the surrounding silicon gel material (top), and the stiff inclusion gel material (bottom).

the surrounding silicone gel, and $C_{10} = 16\,300 \pm 815\ \mathrm{N/m^2}$, $C_{01} = 10\,490 \pm 524\ \mathrm{N/m^2}$ for the silicone gel inclusion.

10.1.4. Phantom models

The first model was built using the *BreastView* software (see Fig. 10.1, bottom) (AZAR, METAXAS, MILLER and SCHNALL [2000]). The other model of the phantom was built directly from its physical dimensions using a pre-processor program MSC/PATRAN (MSC, CA), which automatically meshed the model. Both FEMs consist of 21 slices (each 4 mm thick), stacked along the z-axis. The number of nodes in the x- and y-directions is 18 and 22, respectively, in order to have square shaped volume elements. The finite element models are made of 7497 elements each. The element material properties were given a homogeneous, isotropic, Mooney–Rivlin hyperelastic model, with the C_{10} and C_{01} constants as measured above. The boundary conditions were applied appropriately, and the 9.8 mm displacement of the pressure plate was modeled in the initial conditions as a 9.8 mm displacement constraint on every node, which belongs to the displaced surface of the phantom.

The finite element modeling simulations were done using:

(1) A robust finite element code ABAQUS/STANDARD V.5.8 [1998], commercially available. Each element was modeled as a hybrid incompressible solid 8-node brick, which allows a fully incompressible constraint at each material calculation point (ABAQUS/STANDARD V.5.8 [1998]).

(2) The *BreastView* software in which 12 displacement iterations were used to compress the model. In each displacement iteration, a compression of 0.8167 mm was applied to the relevant boundary nodes in the model. This represents an average nominal strain of 1.16%, which can be considered small strain. The maximum allowed error of integration e_A in the adaptive algorithm, was chosen to be $(1/1000)$th of the smallest dimension in the model $d_{\min}$, which is the smallest distance between two consecutive nodes: $d_{\min} = 3.90$ mm, and $e_A = 3.90 \times 10^{-3}$ mm. This ensured full convergence of the solution after each displacement iteration.

10.2. Results and discussion

The axial slice going through the center of the inclusion is shown in Fig. 10.3 (top) in the uncompressed and in the compressed mode. As expected the edges of the phantom have changed shape as well as the edges of the tumor. Because silicon is incompressible, the side deformations of the phantom are quite large.

Because it is important in the real case to track the displacement of a cancer tumor in the breast, we tracked the displacement of the inclusion in the phantom. By using image analysis software, we measured the displacement vectors of the center of the inclusion, as well as its eight corners. We used the axial slices to measure the x and y displacements, and the coronal slices to measure the z displacements (the positions of the various points of interest were calculated from the MR images with respect to the position of the non-moving bottom edge of the silicon phantom).

10.2.1. ABAQUS simulation results

Fig. 10.3 (bottom) shows the displacement vectors of the inclusion corners and center from MR data (bottom left), and from the ABAQUS model (bottom right). The average errors in displacements were 0.34, 0.66 and 0.40 mm in the x, y and z directions, respectively, and are within the maximum imaging error.

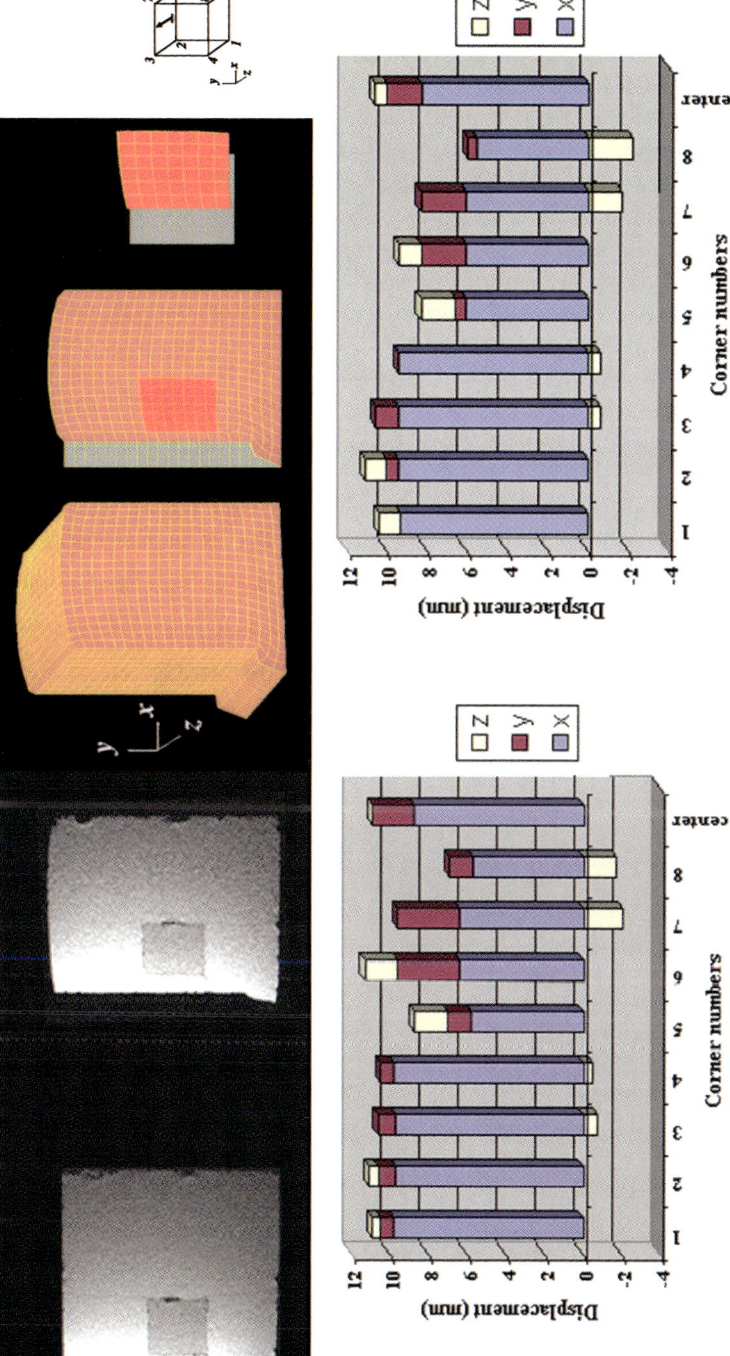

FIG. 10.3. Uncompressed and compressed axial MR slice of phantom (top left), 3D view of model including axial slice through center of inclusion and axial view of inclusion center, before and after compression (top right), displacement vectors of inclusion corners and center from MR data (bottom left), and from the ABAQUS model (bottom right).

10.2.2. BreastView *simulation results*

Fig. 10.4 (top) shows a comparison between the *BreastView* and ABAQUS simulated virtual compressions. The displacement errors between the *BreastView*, ABAQUS simulations and the MR results are shown in Fig. 10.4 (bottom). The results show that the displacement errors between the two simulations and the MR data are very close and within the imaging error magnitude. Furthermore, all of the average displacement errors per node between the two simulations are under 1 mm.

This silicon phantom study shows that:

- the compressed model allows us to track the position and motion of the stiff inclusion in the real compressed deformable environment,
- after comparing the compressed *BreastView* and ABAQUS model results, using a small strain approximation instead of the Lagrangian finite strain expression, in our finite element algorithm does not introduce a significant error in simulations of large deformation.

Although in this phantom study, the compressed model allows us to track the position and motion of the stiff inclusion in the real compressed deformable environment, this result does not necessarily extend flawlessly to an actual tissue model: patient breast compression experiments are still needed for confirmation of our mathematical model applied to human tissue deformation.

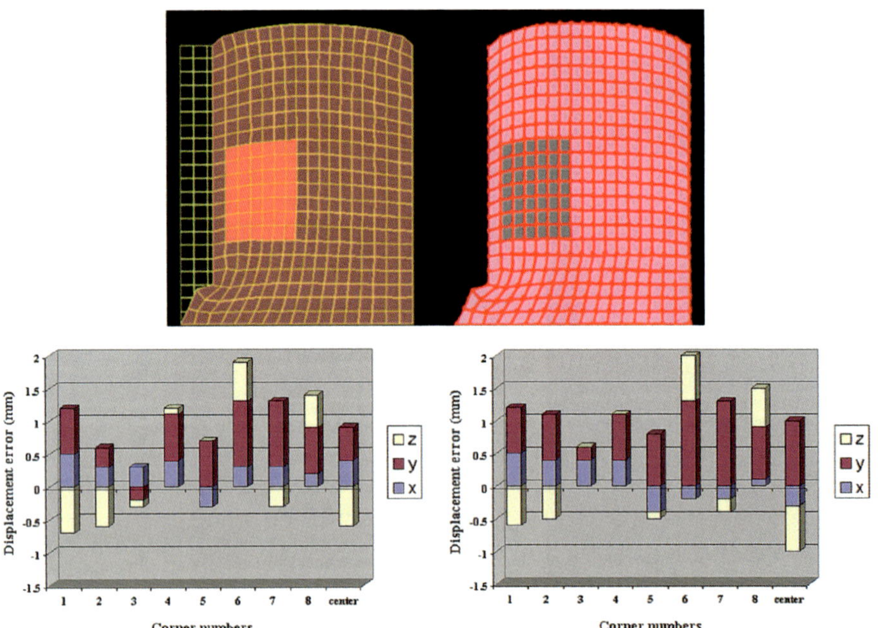

FIG. 10.4. Axial slice going through the inclusion after compression, in the ABAQUS simulation (left model), and in the *BreastView* simulation (right model); displacement errors of inclusion corners and center: ABAQUS–MR (left diagram), *BreastView*–MR (right diagram).

11. A patient study

11.1. Experimental design and methods

A healthy patient's left breast was used to track the displacement of specific landmarks under plate compression. It is usually extremely difficult to accurately and locally track the displacement of tissue in the breast before and after deformation. One may only reliably quantify the displacement of clearly identifiable structures inside the breast, especially if these structures are small and/or rigid enough. We found a small "point-like" cyst inside the breast, which was used as the inner landmark for tracking displacements inside the breast after compression. Additionally, two vitamin E pills, used as landmarks, were taped on the superior and inferior parts of the breast, in order to help track the movement of the breast (vitamin E pills appear as bright spots on the MR images).

11.1.1. Patient breast MR data acquisition
The breast was imaged using a 1.5 T machine Signa Horizon Echospeed (GEMS, Milwaukee). 3D image sets were obtained under plate pressure conditions. The entire breast was imaged medially to laterally, from the visible rib cage to the nipple. The MR acquisition sequence was a 3D fast SPGR (T1 weighted), with a TR of 11.3 ms, a TE of 4.2 ms and a 25 degrees flip angle. We used a phased array multi coil. 124 slices each with a 0.9 mm thickness were acquired sagittally (left to right). Each slice had a field of view of 230 × 230 mm (256 × 256 pixels). This amounts to having cubic voxels with 0.9 mm sides.

11.1.2. Displacements due to plate compression
Usually, two compression plates are used to compress the breast (Table 11.1 shows the plate to plate distances).

We make the following assumptions, which correspond to the actual breast compression setup (see Fig. 10.1, top, and Fig. 11.1):

- the plates are parallel to the y-axis;
- the plates move in the direction of the x-axis, towards the breast, in a way that displacements can be considered static;
- the total compression distance D_{tot} for a plate is from its first point of contact with the breast to its final resting position.

Once the physical parameters are known, such as the plate length L_p, the distance from its tip to the patient's rib cage D_t, and the total compression distance D_{tot}, we

TABLE 11.1
Plate to plate distances before and after compression

	Plate to plate distance (mm)	Compression amount (mm)
Uncompressed	64.8 ± 0.9	0.0
Compressed	47.7 ± 0.9	17.1 ± 1.8

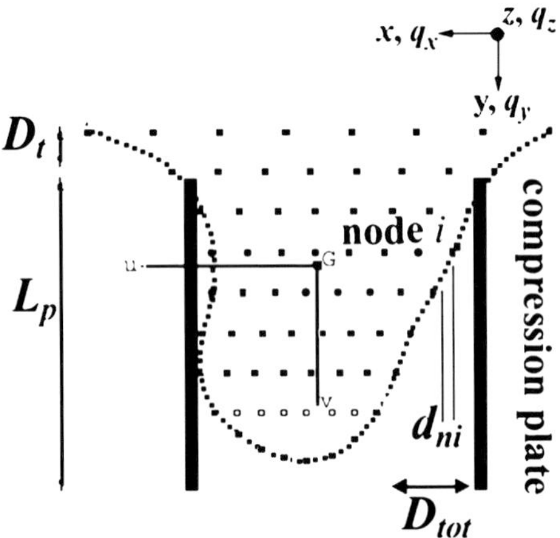

FIG. 11.1. Axial slice of model showing initial displacement parameters.

translate those parameters into prescribed displacements to all individual model surface nodes which come in contact with the plate.

The total compression distance is translated into m_{tot} number of displacement iterations. A simple collision detection algorithm determines which node i comes in contact with the plate at each displacement iteration. Each of these nodes is assigned a prescribed displacement increment d_{ni} repeated in every iteration for which the plate is in contact with the node. Therefore, the number of iterations for every surface node i varies and is denoted m_{ni}.

11.1.3. Boundary conditions

Let $\mathbf{X}_i = (x_i, y_i, z_i)$ be the position of node i in the model ($\mathbf{X}_i$ is also a function of time t), $\mathbf{q}_i = (q_{i,x}, q_{i,y}, q_{i,z})$ represent the displacement degrees of freedom for every node i in the model. The boundary conditions are applied to the displacement degrees of freedom as follows:

- *Base of the breast.* All nodes in the first two planes at the base of the breast model are fixed, and represent the patient rib cage area. This provides the support structure and fixes the breast model in space. Mathematically this is represented by

$$\forall \text{ nodes } i \in \{\text{base of the breast}\}, \quad \begin{cases} q_{i,y} = 0, \\ q_{i,z} = 0. \end{cases}$$

- *Interface between breast and rib cage.* The nodes which are part of the breast tissue in contact with the rib cage, are allowed to slide against the nodes which represent the rib cage. Mathematically this is represented by

$$\forall \text{ nodes } i \in \{\text{breast tissue at the base of the breast}\},$$
$$\{\, y_i \geqslant y_{\text{Rib cage},i} \quad (t = 0).$$

- *Contact with the lateral plate.* Boundary conditions between the virtual plate and the breast model could cause increased displacement errors for the nodes close to the edges of the virtual plate due to possible large rotational effects at the edges. However, since we are concerned with lesion displacement, which is most generally located far enough from the plate edges, we can neglect errors due to rotations in the vicinity of the plate's edges.

Additional experiments were done in order to determine whether the breast skin slides against the compression plates, or whether it sticks to them. Three different patient breasts were compressed and imaged using the same protocol as in interventional procedures. A vitamin E pill was attached to the skin of the patient breast directly in contact with the compression plate. The total average displacement for the vitamin E pills on all breasts (compressed–uncompressed) was 8.8 ± 1.3 mm ($x = 7.8 \pm 0.9$ mm; $y = 4.2 \pm 0.9$ mm). The results show that there is a sliding effect between the skin and the compression plate. We approximate the sliding by allowing the boundary nodes in the model directly in contact with the virtual compression plate, to slide against the plate. Mathematically this is represented by

$$\forall \text{ nodes } i \in \{\text{lateral side of breast}\},$$
$$\textit{if } (\text{node } i \text{ in contact with lateral plate}) \quad \textit{then } \{q_{i,x} = d_{ni},$$

where d_{ni} is the prescribed displacement at every contact iteration for node i.

- *Contact with the medial plate.* In principle, the patient breast is supposed to be pushed against the medial compression plate as much as possible, in order to increase contact with the plate, and decrease motion artifacts when using the lateral plate. In practice, the initial prone lying position of the patient determines how close the breast is to the medial plate; it is quite common that the breast is not completely pushed against the medial plate, and accounts for uncertainties in terms of the contact and boundary conditions between the breast and the medial plate.

  In order to minimize the uncertainty and possible sources of error that may result, we model a real one-plate compression, with a virtual two-plate compression. We suppose that the left plate does all the compression and the right plate remains immobile. Then instead of applying the full compression amount using only the left virtual plate as would be expected, we divide that compression amount evenly, and both left and right virtual plates apply each half of the compression amount (since the breast is compliant, we assume that pressure is equally distributed on each side of the breast). However, after every compression step, the entire breast model is translated back a distance equal to half of the total compression amount, parallel and in the direction of the immobile right plate. In this way, it appears that the model is compressed using only one moving virtual plate. We will compare that method to the one-plate virtual compression, and choose the method, which yields the best results. Mathematically this is represented by

$$\forall \text{ nodes } i \in \{\text{medial side of breast}\},$$
$$\textit{if } (\text{node } i \text{ in contact with medial plate}) \quad \textit{then } \{q_{i,x} = d_{ni},$$

  where d_{ni} is the prescribed displacement at every contact iteration for node i.

The breast was imaged first uncompressed, then imaged under plate compression. The breast compression device, made by GEMS, stabilized the patient's breast well enough to minimize motion artifacts between image sets. The compression plates were able to slide perfectly in the sagittal direction and were locked in position when the desired compression was achieved. The displacements can therefore be considered static.

In order to help track the movement of the breast, we taped to the surface of the breast two vitamin E pills (one towards the superior part, the other one towards the inferior part) which appear as bright spots on the MR images: these serve as landmarks. The pills were taped to the breast in such a way that they do not come into contact with the compression plates, and did not influence the boundary conditions between the compression plates and the breast skin. Gadolinium pills were embedded inside the compression plates. These also appear as bright spots on the MR images, and allowed us to confirm the compression distance between the two plates. The right plate (on the breast's medial side) was kept immobile, while the left plate (on the lateral side) was moved to compress the breast.

11.1.4. Patient breast deformable model

A model of the patient breast was constructed from the given MR data, the boundary conditions, and the applied displacements, using *BreastView*. The different breast tissues inside the breast were modeled. The breast data was segmented semi-automatically, starting at the intersection of the outer side of one compression plate and the breast, all the way to the other plate.

The deformable model of the breast was built using every other slice out of the experimental data set, and discarding the outer slices, which did not contain useful information. A total of 58 slices were used. The breast model consists therefore of 58 slices (each slice in the uv-plane) stacked up along the z-axis. There are 8 nodes in the x-direction and 8 nodes in the y-direction in every slice. The full model contains 3712 nodes, 2793 3D hexahedral elements, 2394 2D triangular elements, and is shown in Fig. 11.2.

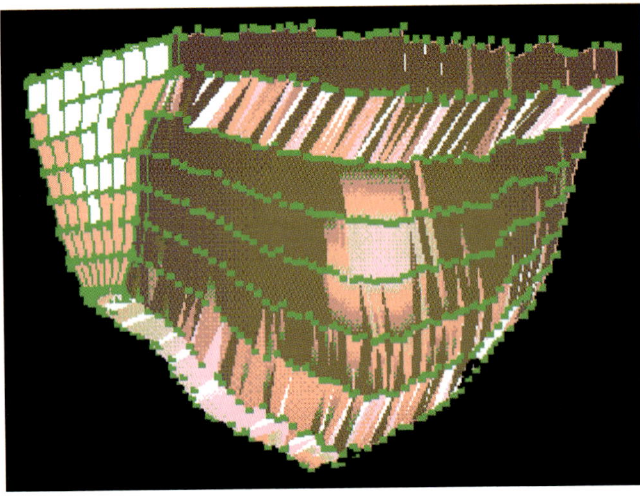

FIG. 11.2. Full breast model.

11.1.5. Mechanical measurements of human breast tissue and skin

Fig. 11.3 (right) shows the major structures of a typical mature pre-menopausal breast. The dimensions and weight of the breast can greatly vary per individual. The breast is an inhomogeneous structure containing many layers of many different kinds of tissue. However, the two predominant types of tissue within the breast are fat and normal glan-

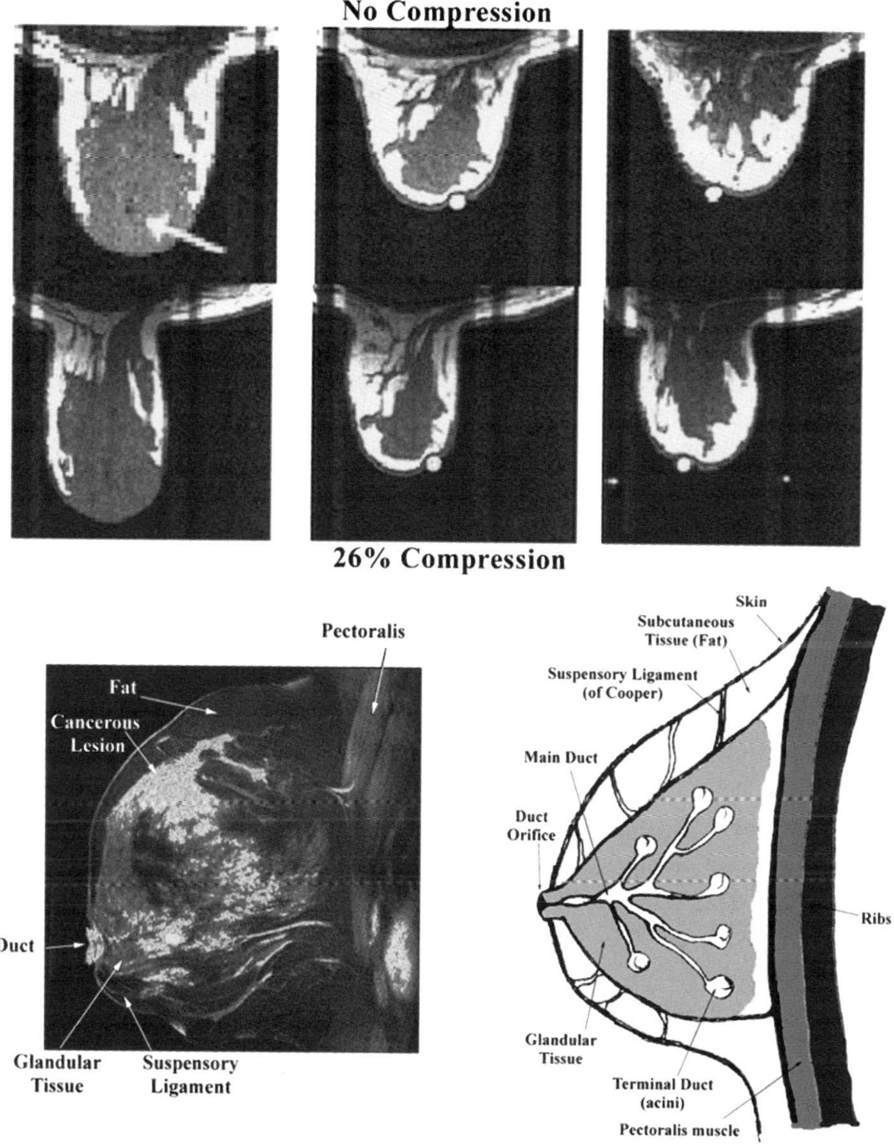

FIG. 11.3. Example of fatty tissue compartmentalization (left), and structure and location of Cooper's ligaments (right).

dular tissue, which supports lactation. The mammary gland forms a cone with its base at the chest wall and its apex at the nipple. Normal skin thickness lies between 0.5 and 1 mm. The superficial layer (fascia) is separated from the skin by 0.5 to 2.5 cm of subcutaneous fat. Tentacle-like prolongations of fibrous tissue extend in all directions from this fascia to the skin; these are called Cooper's ligaments. In the adult mammary gland, there are 15 to 20 irregular lobes, converging to the nipple through ducts 2 to 4.5 mm in diameter (EGAN [1988a]). These ducts are immediately surrounded by dense connective tissue, which acts as a supporting framework. The glandular tissue is supported by estrogen; when a woman reaches menopause the estrogen levels decrease and the glandular tissue atrophies and eventually disappears, leaving only fat and skin (HARRIS, LIPPMAN, MORROW and HELLMAN [1996]).

Carcinomas in affected breasts are usually accompanied by local changes in material properties, due to increased stiffness in the lesion, as well as its shape and size (EGAN [1988b]). There are several types of breast cancer, however the most common are ductal carcinoma (which begins in the lining of the milk ducts of the breast), and lobular carcinoma (which begins in the lobules where breast milk is produced) (NCI [1998]).

It has been observed that mechanical properties of soft tissues are due to their structure rather than to the relative amount of their constituents (FUNG [1987]). Whatever the technique adopted, in most cases, the mechanical properties of soft tissues in vivo can hardly be completely assessed. The constants found in one test only apply to the state of deformation (strain rate, strain range involved (KROUSKOP, WHEELER, KALLEL, GARRA and HALL [1998])), which provides these constants, and no normal response can be uniquely defined for the material. When simple relationships can be written, it is generally only for limited ranges of stresses and strains. Furthermore, the mechanical behavior derived from experiments cannot be readily correlated with in vivo conditions of the tissue where generally the reference state is not completely known (LEE and TSENG [1982]). The assumptions leading to the results of the experiments may therefore no longer be valid in the real conditions of functioning (CRISP [1972]). Experiments are usually limited to a one-dimensional stress field (YAMADA [1970] presents a relatively broad range of data; much of the data are derived from animal experiments and all the information relates to uniaxial tensile properties of soft tissue). It is clear that all stresses and strains in the three dimensions are involved, and should be considered for an accurate characterization of breast tissue. But this would require extensive two- or three-dimensional testing programs, which are difficult to carry out on small specimens that must be maintained in living conditions. It is thus difficult to assert the absolute significance of the material models developed (FUNG [1972]). This is why only average values of experimentally derived material models are used in our finite element analysis.

Few studies have been made on determining the mechanical properties of tissue in the breast, however average values of Young's modulus have been calculated for fat, glandular tissue, and cancer tissue (SKOVORODA, KLISHKO, GUSAKYAN, MAYEVSKII, YERMILOVA, ORANSKAYA and SARVAZYAN [1995], KROUSKOP, WHEELER, KALLEL, GARRA and HALL [1998], LAWRENCE, ROSSMAN, MAHOWALD, MANDUCA, HARTMANN and EHMAN [1999]). Since we need non-linear stress–strain curves describing the mechanical behavior of breast tissue, we will use the mechanical properties of breast tissue determined by WELLMAN and HOWE [1998]. Exponential curves (which have been used in the past for several different tissue types (FUNG [1993])) are used to describe the stress–strain properties of breast tissue, following experimental stress–strain curves obtained from uniaxial loading of breast tissue. The fresh tissue samples were tested in the operating room within 10 minutes of excision, were kept hydrated by periodic application of saline solution, and were tested at room temperature ($21 \pm 2.5\,^{\circ}\mathrm{C}$) (WELLMAN and HOWE [1998], WELLMAN [1999]). The exponential curves describing the elastic modulus E_n for tissue type n, are given by

$$E_n = \frac{\partial \sigma_n}{\partial \varepsilon_n} = b \cdot e^{m\varepsilon_n}, \tag{11.1}$$

where σ_n and ε_n are the nominal stress and strain, respectively, for tissue type n. b and m are the fit parameters determined experimentally for every tissue type:

- $b_{\text{glandular}} = 15\,100\ \mathrm{N/m^2}$;
- $m_{\text{glandular}} = 10.0$ (within 1 standard deviation);
- $b_{\text{fat}} = 4460\ \mathrm{N/m^2}$;
- $m_{\text{fat}} = 7.4$ (within 1 standard deviation).

We used a value of 0.49999 for Poisson's ratio.

The mechanical properties of skin have been studied more thoroughly than those of breast tissue, and several papers have been written on the subject (VERONDA and WESTMANN [1970], AGACHE, MONNEUR, LEVEQUE and DERIGAL [1980], FUNG [1981], SCHNEIDER, DAVIDSON and NAHUM [1984]). Though it is not a homogeneous material, in many cases skin can be simplified to be statistically homogeneous (at higher strains it was found that the slope in the linear region of the stress–strain curve is similar in all directions) (MAUREL, WU, MAGNENAT THALMANN and THALMANN [1998]). The typical experimental stress–strain curve for skin (ELDEN [1977]) is transformed into a piecewise linear stress–strain curve which we will use to describe the mechanical properties of skin in the breast model (as shown in Fig. 11.4).

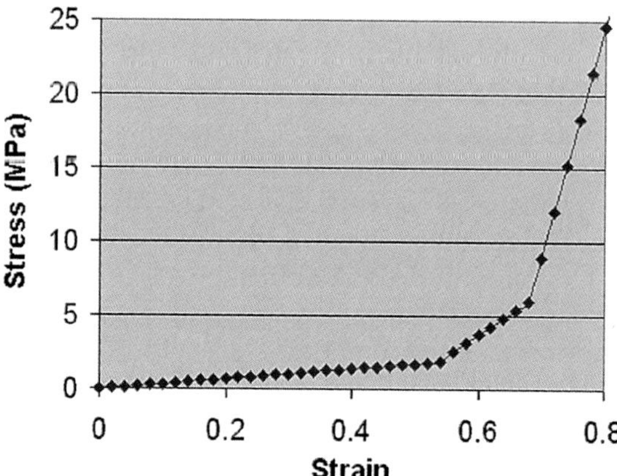

FIG. 11.4. Piecewise linear stress–strain curve for skin (ELDEN [1977]).

TABLE 11.2
Parameters for skin material model (in N/m^2)

Parameter name	Line segment 1	Line segment 2	Line segment 3
a_i (N/m^2)	3.43×10^6	2.89×10^7	1.57×10^8
b_i (N/m^2)	0	-1.36×10^7	-1.01×10^8
Valid strain range	$0 \leqslant \varepsilon \leqslant 0.54$	$0.54 < \varepsilon \leqslant 0.68$	$0.68 < \varepsilon \leqslant 1$

The stress–strain relationship for skin is of the form

$$\sigma_{\text{skin}} = a_i \cdot \varepsilon_{\text{skin}} + b_i, \tag{11.2}$$

where (a_i, b_i) are the parameters for every linear segment of the curve ($i = 1$ to 3). Then the elasticity modulus is given by

$$E_{\text{skin}} = a_i \quad (i = 1 \text{ to } 3). \tag{11.3}$$

The parameter values are shown in Table 11.2.

11.1.6. Modifications of fat tissue model

The rationale underlying the modification of the fatty tissue model is based on the fact that ex-vivo measurements of fatty tissue properties, although accurate, always ignore the supporting structure of fibers in the breast. Also those measurements are almost always made at room temperature, which significantly affects the mechanical properties of fatty tissue, since fatty tissue is almost liquid at body temperature (KROUSKOP, WHEELER, KALLEL, GARRA and HALL [1998]) (on average 10–15 °C higher than room temperature).

None of the experimental measurements of breast tissue encountered in the literature take into account the supporting structure of fibers including Cooper's ligaments to describe the mechanical behavior of fatty tissue in vivo. We also know from the literature, and from testing with other tissue types, how the absolute value of the tissue modulus may be affected by the boundary conditions (confinement) acting on a sample. Confinement can be significantly different from organ to organ depending on its surrounding environment (KROUSKOP, WHEELER, KALLEL, GARRA and HALL [1998]).

We hypothesize that:

- The supporting structure of fibers, including Cooper's ligaments, compartmentalizes fatty tissue, and prevents it from being squeezed out of its location.
- As fatty tissue is being compressed, the local pressure increases and leads to an increase in the apparent stiffness value of fat.

These hypotheses are supported by the experimental evidence of the numerous compressed patient breast images scanned to date using MRI. This data shows fatty tissue clearly not squeezed out but actually remaining in its location as the breast is being compressed, as shown in Fig. 11.3.

We test the hypotheses by updating the material model, such that the stiffness of fatty tissue is made to increase up to an average stiffness value of glandular tissue: as fatty tissue is being compressed and compartmentalized, the local pressure in the compartment increases and leads to an increase in the apparent stiffness value of fat, until the stiffness reaches that of the glandular tissue. Given this hypothesis, we can set the following necessary boundary conditions for E_{fat}:

$$
\begin{aligned}
&\bullet\ \text{If } \varepsilon_{\text{fat}} = 0\text{:} && E_{\text{fat}}(0) = b_{\text{fat}},\\
&\bullet\ \text{If } \varepsilon_{\text{fat}} \leqslant \varepsilon_{\text{limit}}\text{:} && b_{\text{fat}} \leqslant E_{\text{fat}} \leqslant E_{\text{gland}}, \quad \frac{\partial E_{\text{fat}}}{\partial \varepsilon_{\text{fat}}} \geqslant 0,\\
&\bullet\ \text{If } \varepsilon_{\text{fat}} = \varepsilon_{\text{limit}}\text{:} && E_{\text{fat}} = E_{\text{gland}} = b_{\text{gland}} e^{m_{\text{gland}} \cdot \varepsilon_{\text{limit}}} \equiv \alpha,\\
& && \frac{\partial E_{\text{fat}}}{\partial \varepsilon_{\text{fat}}} = \frac{\partial E_{\text{gland}}}{\partial \varepsilon_{\text{gland}}} = b_{\text{gland}} m_{\text{gland}} e^{m_{\text{gland}} \cdot \varepsilon_{\text{limit}}} \equiv \beta,\\
&\bullet\ \text{If } \varepsilon_{\text{fat}} \geqslant \varepsilon_{\text{limit}}\text{:} && E_{\text{fat}} = E_{\text{gland}}.
\end{aligned}
\tag{11.4}
$$

These boundary conditions merely state the fact that E_{fat} is a continuous smooth non-decreasing function.

The simplest equation, which satisfies the above conditions, is a quadratic equation of the form

$$E_{\text{fat}}(\varepsilon_{\text{fat}}) = A \cdot \varepsilon_{\text{fat}}^2 + B \cdot \varepsilon_{\text{fat}} + C. \tag{11.5}$$

Solving for A, B and C given the boundary conditions, yields the following:

$$A = \frac{\beta \cdot \varepsilon_{\text{limit}} + b_{\text{fat}} - \alpha}{\varepsilon_{\text{limit}}^2}, \qquad B = \frac{2\alpha - 2b_{\text{fat}} - \beta \cdot \varepsilon_{\text{limit}}}{\varepsilon_{\text{limit}}}, \qquad C = b_{\text{fat}}. \tag{11.6}$$

There is a condition on $\varepsilon_{\text{limit}}$ which comes from the necessity that $(E_{\text{fat}})' \geqslant 0$; this is equivalent to writing $(E_{\text{fat}})'(0) \geqslant 0$ as long as $A > 0$. This leads to the condition that $\varepsilon_{\text{limit}} \leqslant (2 \cdot (\alpha - b_{\text{fat}}))/\beta$. The curve describing $E_{\text{fat}}(\varepsilon_{\text{fat}})$ is shown in Fig. 11.5 using the experimentally derived values for b_{gland}, m_{gland} and b_{fat}. We solved for the maximum allowed strain limit, and used it to model fatty tissue. We find $\varepsilon_{\text{limit}} = 15.5\%$.

We test the updated fatty tissue model hypothesis, and the value of $\varepsilon_{\text{limit}}$, in the sensitivity analysis, where the effect of varying the material parameters on the model per-

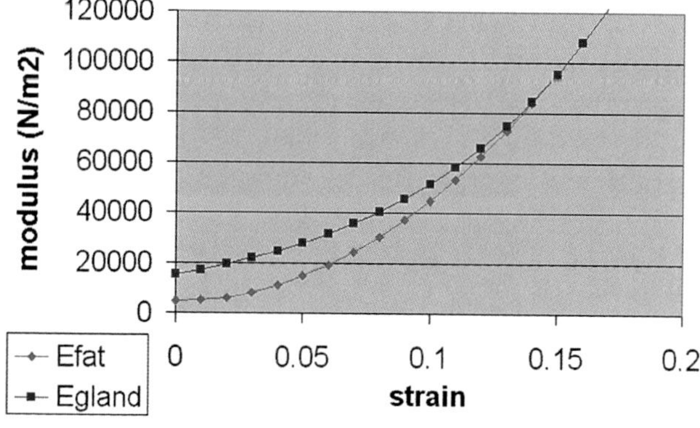

FIG. 11.5. Fat material properties curve.

formance is investigated. Rather than modeling the structure and geometry of Cooper's ligaments, we model their functionality and overall effect.

11.2. Results and discussion

11.2.1. MR results

Table 11.3 shows the displacement vectors of the cyst inside the breast, and the two vitamin E pills after compression. The displacement measurements (in mm) were made using Scion Image software. In order to compute the cyst and vitamin E pills displacements, we calculated the positions of the breast cyst and vitamin E pills with respect to the position of the gadolinium pills embedded in the non-moving compression plate. The edge points of the vitamin E pills closest to the skin were used in the calculations. The images in Fig. 11.6 show the axial cross-sections of the patient's left breast.

11.2.2. Variational study

The effect of varying different parameters is investigated. Several parameters describing the model are varied over a physiologically relevant range, and every time the model simulation is done, the displacement of the landmarks, as well as other relevant performance assessment parameters are recorded.

TABLE 11.3
Displacement vectors (mm) of the landmarks obtained experimentally (compressed–uncompressed)

	Vitamin E pill (superior)			Vitamin E pill (inferior)			Small cyst		
	X	*Y*	*Z*	*X*	*Y*	*Z*	*X*	*Y*	*Z*
Compression	−9.0	3.6	2.7	−7.2	0.9	−3.6	−6.3	1.8	1.8

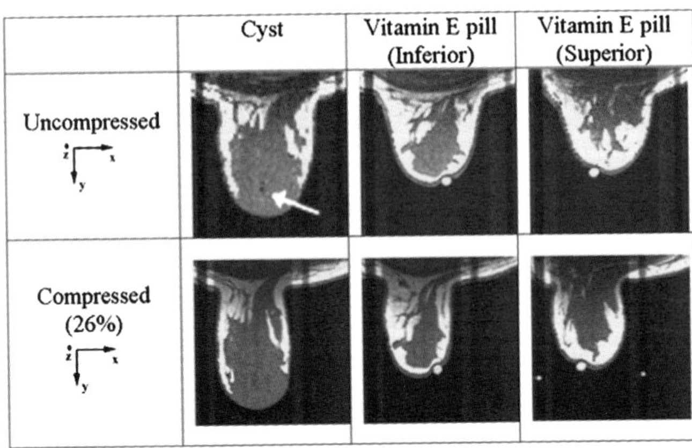

FIG. 11.6. MR slices containing the landmarks, for the uncompressed and compressed cases.

First, we decide which of the one-plate or two-plate virtual compression yields the best results. Then after a convergence analysis we decide whether the slab model (variable z-mesh density) can replace the full model in the virtual compression; this would dramatically decrease the simulation time. Finally, we do a material properties sensitivity analysis to show how sensitive is the model performance to variations in material parameters.

11.2.3. Performance assessment parameters

Two performance assessment parameters are used in the variational study:

- *Displacement difference* ($\mathrm{DISP_{diff}}$): difference (model–real) of displacement, of inclusion center of gravity, where inclusion can be lesion, cyst, or vitamin E pill,

$$\mathrm{DISP_{diff}} = \left(\sqrt{X_{\mathrm{diff}}^2 + Y_{\mathrm{diff}}^2 + Z_{\mathrm{diff}}^2}\right). \tag{11.7}$$

- *%Misclassification* (%MIS): compares the number of misclassified pixels in the model with the compressed MRI,

$$\%\mathrm{MIS} = \mathrm{NMIS}/\mathrm{TNUM}, \tag{11.8}$$

  where NMIS is the number of misclassified pixels in model with respect to compressed MR dataset ((nonfat pixels in fat elements) + (nonglandular pixels in glandular elements) + (nontumor pixels in tumor elements)), and TNUM is the total number of breast pixels in the MR dataset.

The displacement differences are used to assess the performance of the most important result in the model, which is how well the displacement of an inclusion in the real breast, can be predicted using the model. The misclassification percentage is a value, which gives an idea of how well the model globally predicts the displacement of the major structures in the breast.

Therefore, the displacement difference offers a local measure of performance, whereas the misclassification percentage offers a global "macroscopic" measure of the performance of the breast model.

11.2.4. Initial uncompressed model and uncompressed MR images

The uncompressed MR images containing the landmarks and the corresponding uncompressed model slices are shown in Fig. 11.7. The arrows in the figure indicate the location of the landmarks.

All of the simulations in the variational studies were done on a SGI Octane Workstation with 2 195 MHz IP30 processors (MIPS R10000 processors), and 256 Megabytes of memory (RAM). The first step in the variational analysis is to decide what type of model to use, and what type of virtual compression to apply. Given the different parameters outlined above, the virtual compression was done using one then two virtual compression plates, with the full model and the slab model. The displacement differences and %misclassifications are shown in Fig. 11.8.

11.2.5. One-plate vs. two-plate virtual compressions for the full model

The full model consists of 58 slices stacked up along the axial axis. Each slice contains 8×8 nodes, for a total of 3712 nodes, 2793 3D hexahedral elements, and 2394 2D tri-

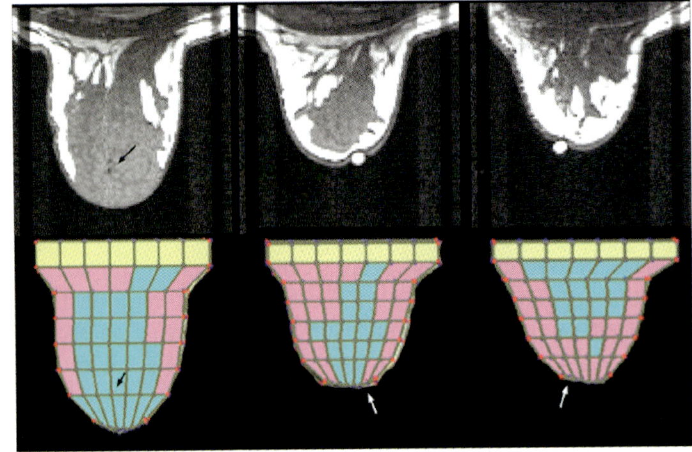

FIG. 11.7. Uncompressed MR images (top), and uncompressed model slices (bottom).

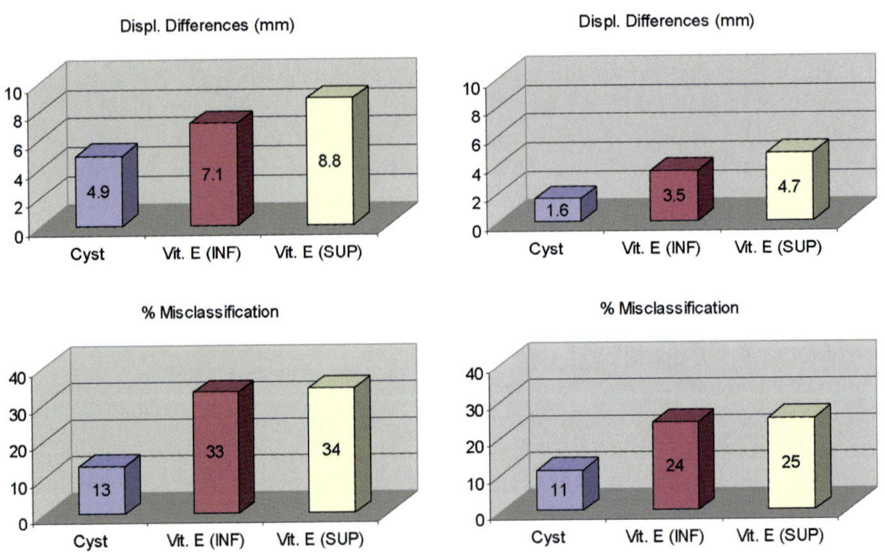

FIG. 11.8. Displacement differences (mm) and %misclassifications in the one-plate virtual compression (left), and in the two-plate virtual compression (right), with the full model.

angular elements. This experiment is aimed at comparing the performance of the virtual one-plate compression simulation to that of the virtual two-plate compression simulation, where the total compression amount is evenly divided between the two virtual compression plates.

Clearly, from the results shown above, the two-plate virtual compression yielded better results than the one-plate virtual compression, using the full model.

The %misclassifications for the slices containing the vitamin E pills are high compared to that of the slice containing the cyst. This higher level of inaccuracy is due to the fact that the former slices belong to the outer edges of the patient's breast; the contact and boundary conditions between these slices and the compression plates is not always clear, and it is very difficult to accurately predict their deformation for that reason.

11.2.6. Convergence analysis

We did a convergence analysis in the (x, y) direction: the (x, y) mesh density was increased, and the model compression was done. We recorded the total displacement of

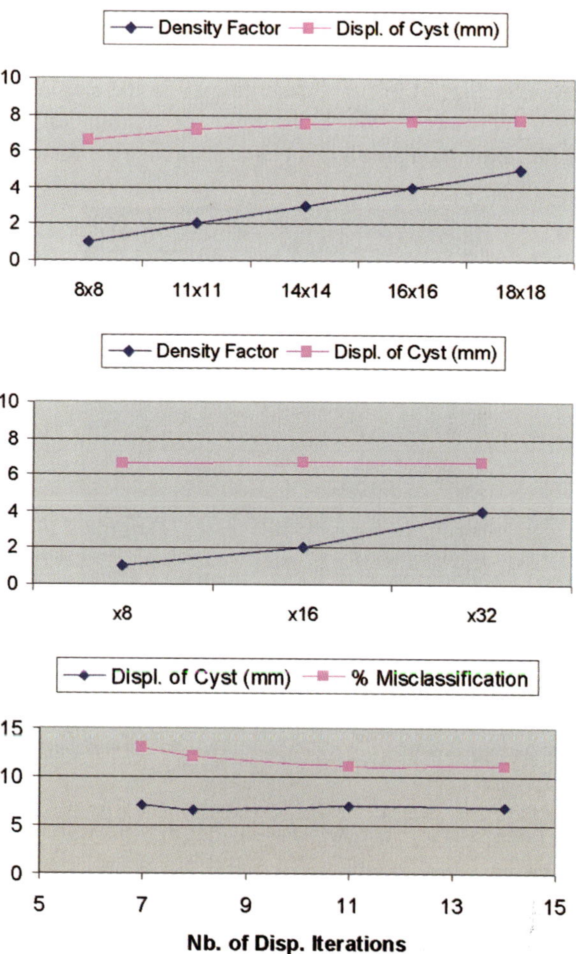

FIG. 11.9. Convergence analysis in the (x, y) direction (top), convergence analysis in the z-direction (middle), and convergence analysis with the number of displacement iterations per simulation step (bottom). The density factor on the top two graphs represents the total number of nodes in the model, normalized by the number of nodes in the model with 8×8 nodes in the (x, y) direction.

the cyst. Another convergence analysis was done in the z direction, using the slab model: the z-direction mesh density was varied using the slab-model algorithm, increasing the number of model slices around the point of interest. We recorded the total displacement of the cyst. The number of displacement iterations was increased (as the displacement per iteration was decreased in order to keep the same total virtual plate compression of 17 mm). The analysis was done on the displacement of the cyst in the model. All the results are shown in Fig. 11.9.

The analysis shows that the simulation indeed converges to the solution, as the model mesh is refined. Also using about 10 displacement iterations per simulation step proves enough for convergence of the solution. Furthermore, using the slab model with only 8 slices instead of 56 does not decrease the performance of the simulation. In order to show the performance of the slab model, we ran the entire simulation using that model. The slab model contains 512 nodes, 343 3D hexahedral elements, and 294 2D triangular elements. As shown in Fig. 11.10, the mesh density in the slab model is the highest around the points of interest (cyst, Fig. 11.10, center or vitamin E pill, Fig. 11.10, right). The results from the simulation are shown below.

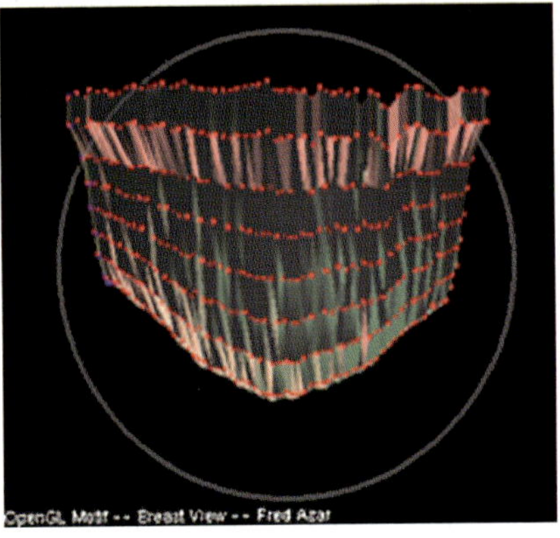

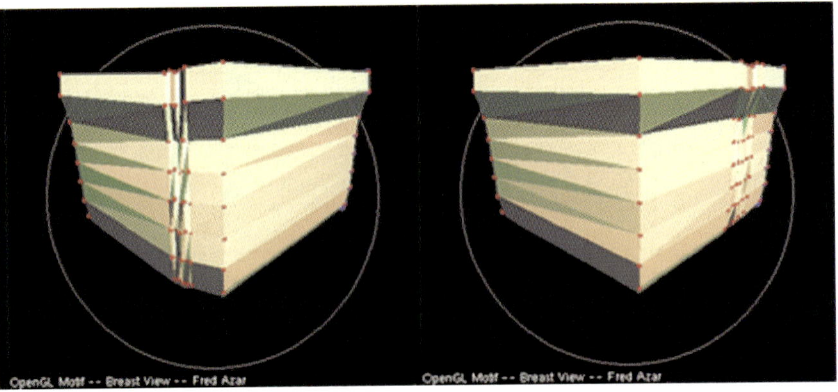

FIG. 11.10. Full breast model mesh (top), and variable meshes around different points of interest (cyst, center and vitamin E pill, bottom).

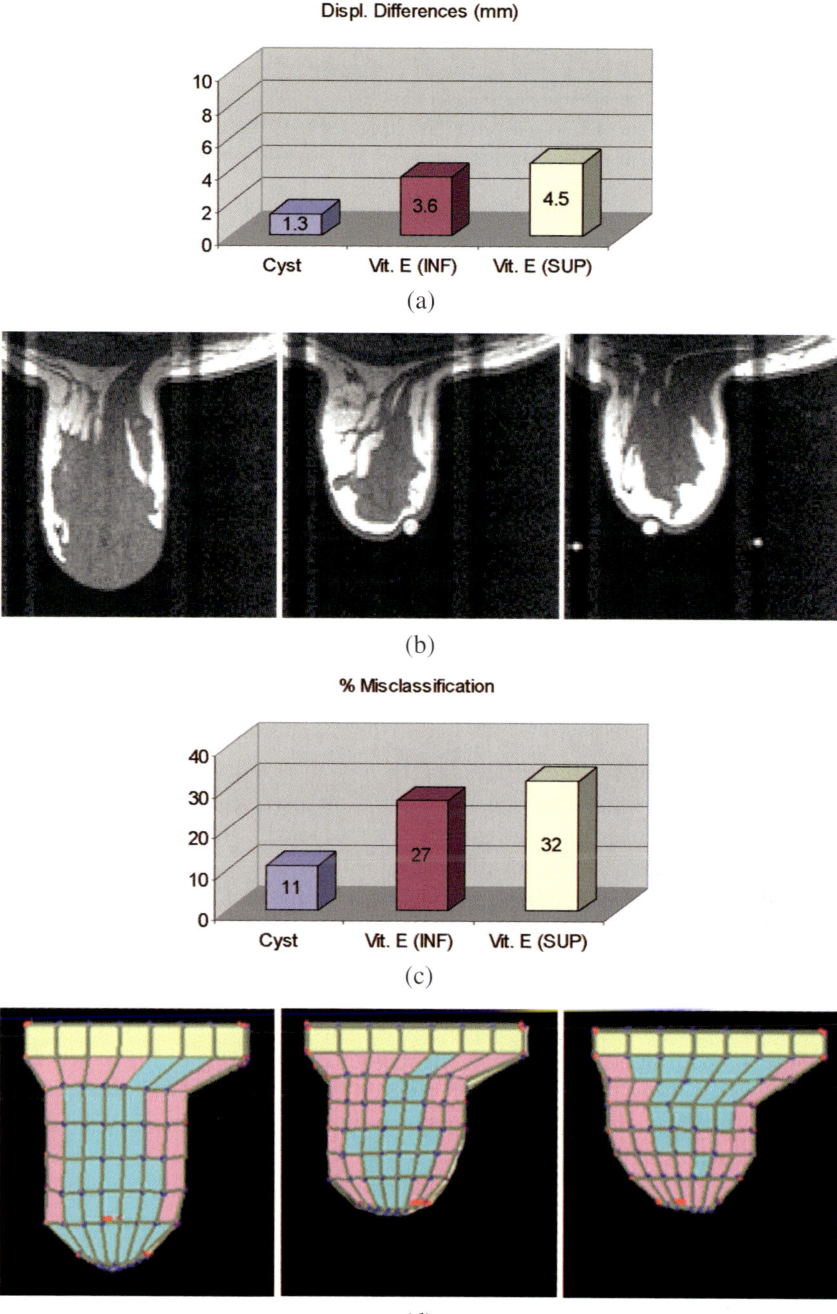

FIG. 11.11. Displacement differences (a), and %misclassification (c) in the two-plate virtual compression with the slab model; compressed MR slices (b), and corresponding virtually compressed model slices (d) after a two-plate virtual compression using the slab model.

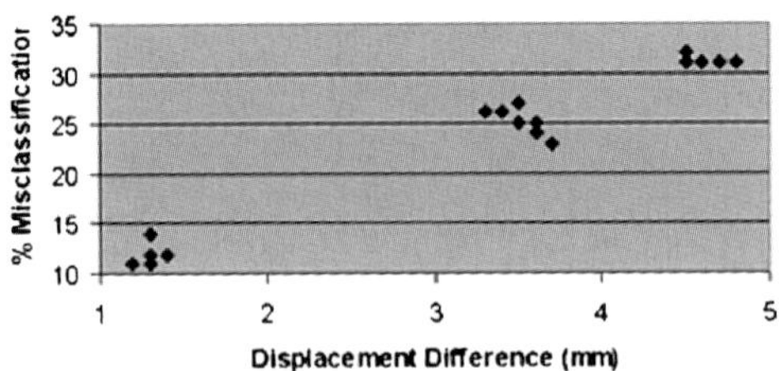

	Cyst	Vit. E pill (INF)	Vit. E pill (SUP)
Disp. difference	1.3 (*0.1*)	3.5 (*0.1*)	4.6 (*0.1*)
%Misclassification	11.8 (*1.0*)	25.1 (*1.2*)	31.4 (*0.5*)

FIG. 11.12. Graph showing sensitivity analysis results (top), and table showing averages and standard deviations (in parentheses) of sensitivity analysis (bottom). Each point in the sensitivity analysis represents a modeling experiment with one of the material parameters varied. We can clearly distinguish three different clusters. The lower left cluster represents results from the cyst displacement, the middle cluster represents results from the vitamin E pill (inferior) displacements, and the upper right cluster represents results from the vitamin E pill (superior) displacements.

When comparing the results in Fig. 11.11 (top) to Fig. 11.8, we see that the performance of the slab model was as good as that of the full model. This enables us to run the simulation dramatically faster since the slab model contains many less elements than the full model.

11.2.7. Material properties sensitivity analysis

The effects of varying the different material properties parameters are investigated. The parameters are varied over a physiologically relevant range, within two standard deviations of their average experimentally derived value. The numerical values of the relevant material properties used in our chosen model were varied within ± 1.2 standard deviations, one at a time. The parameter for the new updated fat model was varied by 30%, then by 60%. The material parameters of interest which were varied are m_{gland}, b_{gland} for the glandular material model, and e_{limit} for the new updated fat model. All of the results are shown in Fig. 11.12, where the displacement differences vs. %misclassifications are graphed. Each point represents a modeling experiment with one of the material parameters varied. We can clearly distinguish three different clusters. The lower left cluster represents results from the cyst displacement, the middle cluster represents results from the vitamin E pill (inferior) displacements, and the upper right cluster represents results from the vitamin E pill (superior) displacements.

12. A clinical breast compression study

12.1. Experimental design and methods

Three patient breasts undergoing breast MRI were imaged uncompressed and then compressed. Patients were chosen to have a variety of breast cancers of different shape, size and location in the breast. Patient ages vary between 49 and 71 years of age. In all cases the breast affected is the left breast. Subsequent biopsy revealed cancer in all cases.

12.1.1. *Initial and boundary conditions*

The breasts were imaged first uncompressed, then imaged under plate compression, as the patients were lying prone. The right compression plate (on the breast's medial side) was kept immobile, while the left plate (on the lateral side) was moved to compress the breast. Table 12.1 shows the compression amounts.

12.1.2. *Patient breast deformable models*

A model of each patient's breast was constructed from the given MR data, the boundary conditions, and the applied displacements, using *BreastView*. The different breast tissues inside the breast were modeled. The breast data was segmented semi-automatically, starting at the intersection of the outer side of one compression plate and the breast, all the way to the other plate. The breast models consist of 8 parallel slices stacked along the z-direction, distributed in such a way as to maximize the number of model slices around the tumor (using the variable meshing algorithm). The number of nodes in the plane of every slice (x, y plane) was chosen as to allow the breast tumor to be fully included inside one 3D element, but at the same time allowing the simulation to run in less than 15 minutes, as shown in Table 12.2.

The displacements of the tumors' center of gravity were recorded both in the real breasts and in the deformable models. All of the simulations were done on a SGI Octane Workstation with 2 195 MHz IP30 processors (MIPS R10000 processors), and 256 Megabytes of memory (RAM).

TABLE 12.1
Breast compression amounts

	Patient 1	Patient 2	Patient 3
Compression amount (mm)	15.3 ± 1.8	14.4 ± 1.8	34.2 ± 1.8
% Compression	24	22	22
Average size of lesion (mm)	7.2	5	13

TABLE 12.2
Breast deformable models

	Patient 1	Patient 2	Patient 3
(x, y) number of nodes	(10, 12)	(10, 10)	(8, 12)
Total number of nodes	960	800	768
Number of 3D elements	693	567	539

12.2. Results and discussion

Fig. 12.1 shows the uncompressed MR slices for the three patients, containing the lesions indicated by an arrow (top), and uncompressed corresponding model slices containing the lesion elements (bottom). Fig. 12.2 shows the compressed MR slices for the three patients, containing the lesions indicated by an arrow (top), and compressed corresponding model slices containing the lesion elements (bottom).

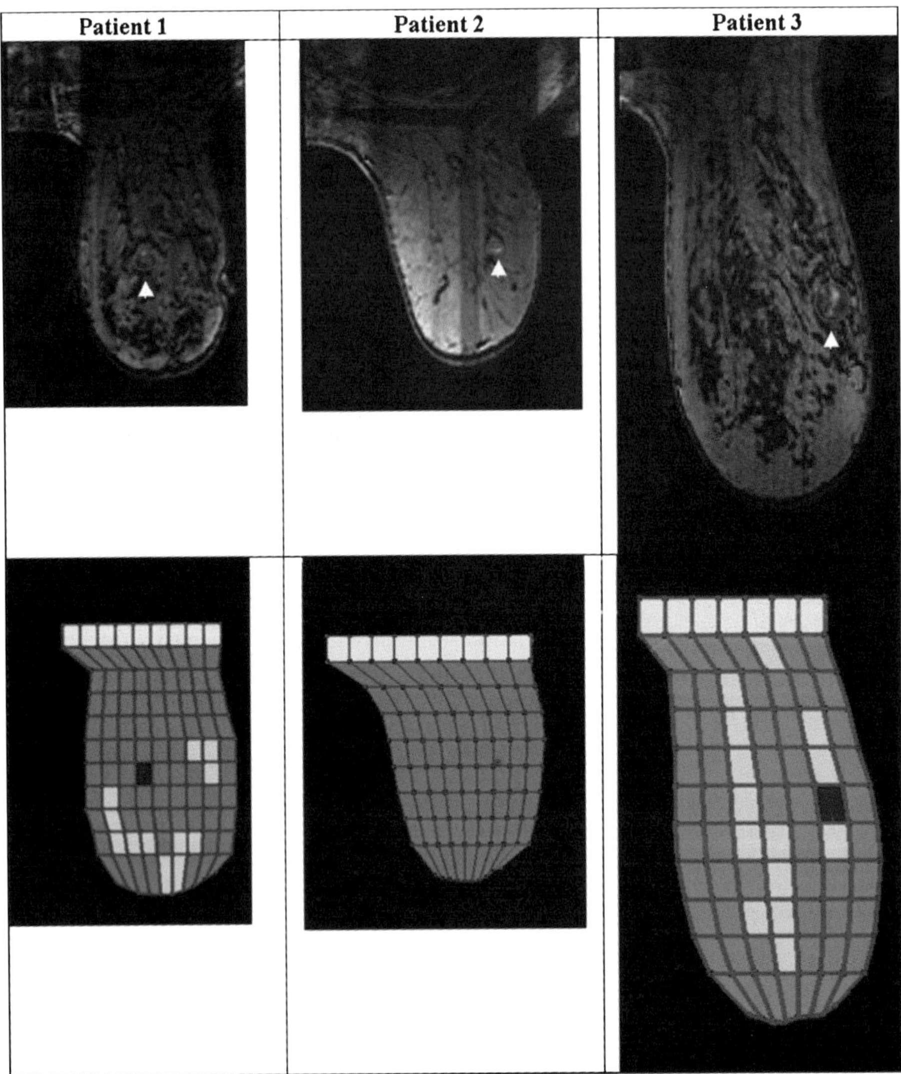

FIG. 12.1. Uncompressed MR slices for the three patients, containing the lesion indicated by an arrow (top), and uncompressed corresponding model slices containing the lesion element (bottom).

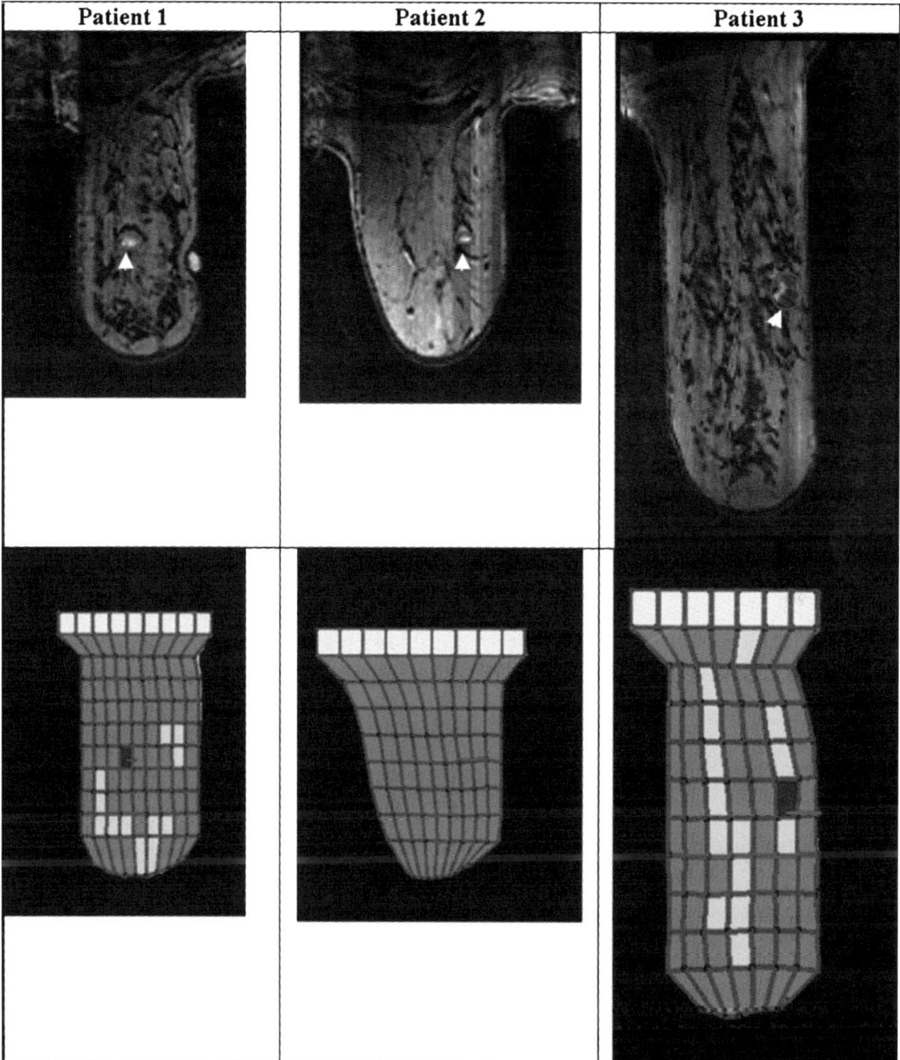

FIG. 12.2. Compressed MR slices for the three patients, containing the lesion indicated by an arrow (top), and compressed corresponding model slices containing the lesion element (bottom).

Table 12.3 shows the experimental lesion displacements, the modeled lesion displacements, and the displacement differences (model–real) $\text{DISP}_{\text{diff}}$ of inclusions center of gravity, $\text{DISP}_{\text{diff}} = (\sqrt{x_{\text{diff}}^2 + y_{\text{diff}}^2 + z_{\text{diff}}^2})$, where $x_{\text{diff}} = x_{\text{model}} - x_{\text{real}}$, $y_{\text{diff}} = y_{\text{model}} - y_{\text{real}}$ and $z_{\text{diff}} = z_{\text{model}} - z_{\text{real}}$. $(x_{\text{model}}, y_{\text{model}}, z_{\text{model}})$ and $(x_{\text{real}}, y_{\text{real}}, z_{\text{real}})$ are the displacement vectors (compressed–uncompressed) of the inclusions center of gravity obtained respectively from the model, and experimentally.

TABLE 12.3
Displacement vectors (mm) of the lesion obtained experimentally, and from the model (compressed–uncompressed), displacement differences $DISP_{diff}$, both in mm and as a percentage of lesion size

	Patient 1			Patient 2			Patient 3		
Lesion displacement (mm)	x	y	z	x	y	z	x	y	z
– experimental	−5.4	1.8	−2.7	−10.8	0.9	0.0	−18.9	5.0	3.6
– model	−4.6	0.3	−0.6	−10.1	0.3	0.5	−14.0	0.1	1.3
$DISP_{diff}$ (mm)		2.7			1.0			7.3	
$DISP_{diff}$ as a % of lesion size		37.5			20			56	

13. Registration of breast MR images of the same patient under different compressions

13.1. Experimental design and methods

A healthy patient's breast was imaged under two different plate compression amounts. A deformable model of that breast was constructed and compressed using virtual compression plates. The displacement of a small cyst was recorded in the real breast and compared to the displacement of the 'virtual cyst' in the deformable model from one compression state to the other.

13.1.1. Initial and boundary conditions

The breast was imaged under two different plate compression amounts: 12 and 26%, resulting in a compression of 9.0 ± 1.8 mm between the two states. The right compression plate (on the breast's medial side) was kept immobile, while the left plate (on the lateral side) was moved to compress the breast.

13.1.2. Patient breast deformable model

A deformable model of the breast was constructed using the least compressed breast MR data, and deformed using virtual compression plates. The breast model consists of 58 slices with 8 nodes in the x-direction, and 8 nodes in the y-direction in every slice. The full model contains 3712 nodes, and 2793 3D elements.

The displacement of a small cyst was recorded both in the real breast and in the deformable model from one compression state (12%) to the other (26%). The simulation was done using the custom-written *BreastView* software. The simulation was done on a SGI Octane Workstation with 2 195 MHz IP30 processors (MIPS R10000 processors), and 256 Megabytes of memory (RAM).

13.2. Results and discussion

13.2.1. Rotational differences in breast positioning

One of the major concerns to this experiment lies in the initial and boundary conditions. In our method, registering two breast image sets of a same patient taken under different pressure plate conditions and at different times, does not account for the possibility of rotational differences in breast positioning. Translational differences are trivial to account for, however, rotational differences may cause significant errors in the modeling results if not taken into account.

In order to assess the amount of rotational differences in breast positioning from one instance to another, four different patients from a previous study were used. The patient breasts were scanned at two different times, weeks apart, and sagittal MR slices from corresponding image sets, and showing the same location in the breast, were compared to each other.

Results are shown below in Fig. 13.1. A qualitative look at the superimposed breast images in Fig. 13.1(c) leads to the surprising conclusion that rotational differences due

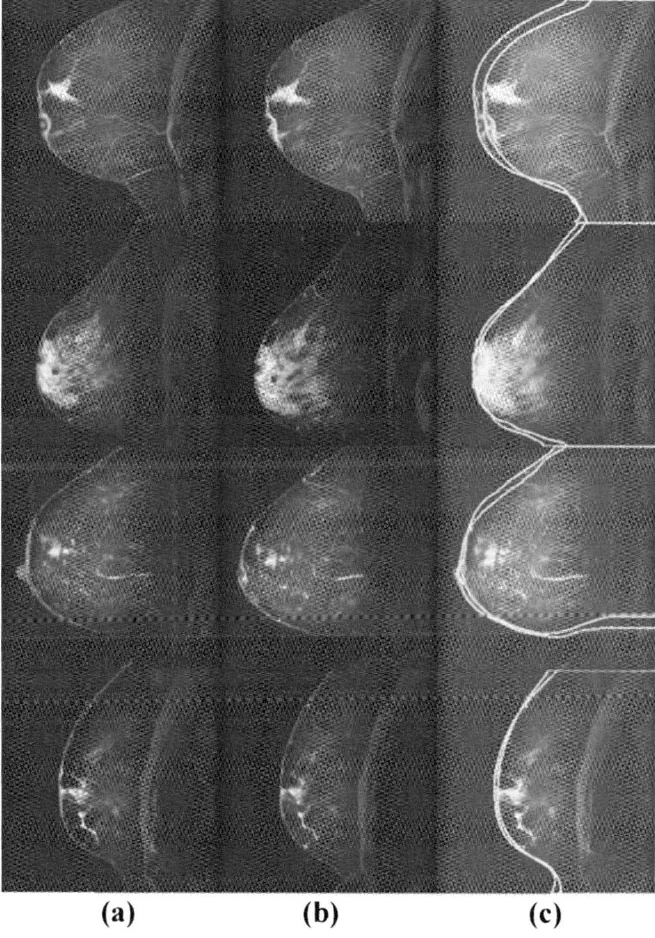

FIG. 13.1. Sagittal cross-sections of four patient breasts taken at two different times ((a) time 1, (b) time 2) weeks apart, and a transparent superposition of the two cross-sections, showing clearly the shift in boundaries from one time to another (c).

to initial positioning of the breast in the compression device are virtually inexistent. There are some translational differences, and differences in shape due to the difference in breast compression amount between the two time points. However, the results clearly show that we can neglect the potential sources of error in the simulation, originating in rotation differences due to breast positioning from one time point to another.

13.2.2. MR and model results

The MR imaging results are shown in Fig. 13.2. Table 13.1 shows the experimental vs. the modeled cyst displacement, and Fig. 13.3 contains the slab of the breast model showing the slice which contains the cyst in the initial 12% compression state (left) and in the final 26% compression state (right).

The three patient breast simulations above show that the displacement differences from experiment to model are smaller than the actual size of the lesions concerned. The simulations therefore show that the model can reasonably predict the displacement of the lesions involved. Table 12.3 indicates also that the displacement differences as a percentage of lesion size are of the order of 50%, which shows that the modeled displacement of the lesion may allow *half of the lesion size* to overlap with the modeled lesion size.

Finally, the model has been shown to predict the displacement of lesions in a patient breast undergoing plate compression, for lesions of about 5 mm or more in size.

The model performance is as reliable as the parameter that is most sensitive to variations in the conditions of the experiment. In our model, it is shown, that the physical representation of the breast is most sensitive in variations of the breast shape. It is also

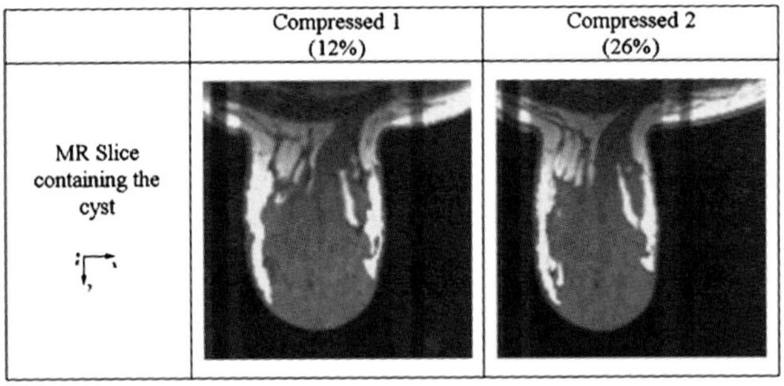

	Compressed 1 (12%)	Compressed 2 (26%)
MR Slice containing the cyst		

FIG. 13.2. MR slice of the patient breast containing the cyst.

TABLE 13.1
Experimental vs. modeled cyst displacement

	Patient breast		
Cyst displacement (mm)	x	y	z
– experimental	−1.8	0.9	1.8
– model	−2.5	1.7	1.4
Displacement difference (mm)		1.4	

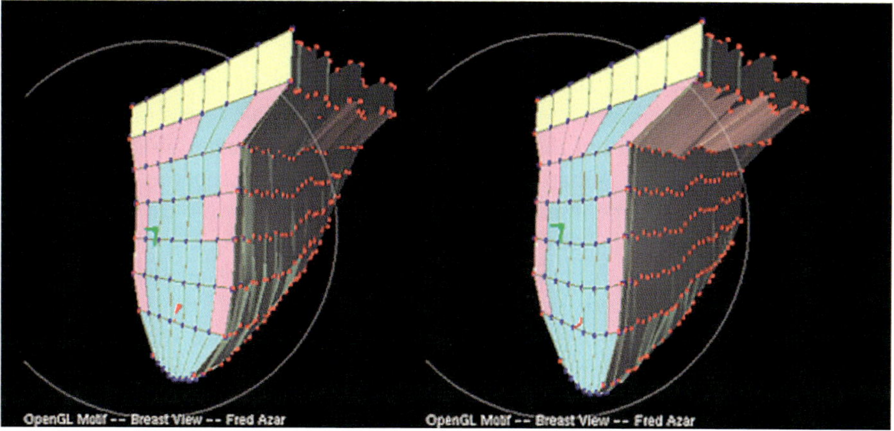

FIG. 13.3. Slab of breast model showing the slice which contains the cyst in the initial 12% compression state (left) and in the final 26% compression state (right).

shown through a material properties sensitivity analysis, that the performance of the model has modest dependence to variations in material properties within 1 to 2 standard deviations. Indeed, the shape and size of a patient breast influence the amount of compression on the breast, and the boundary properties between the breast and the compression plates.

Therefore, additional experiments would be needed in order to assess the reliability of the model by choosing a number of patients with breasts of different sizes, leading to different initial and boundary conditions for the model. We may want to initiate a study involving 12 patients in three categories of breast size: small, medium and large, with four patients in each category. In each category we may want to track lesion displacements for lesions that are located either in the middle of the breast, or towards the outer edges on an axial axis. The results of this study will enable us to determine the relationship between the minimum size of a trackable lesion, the size of the breast, and the location of the lesion in the breast.

It may also appear from the results shown in Table 13.1 that errors in the x, y and z dimensions, respectively, are of the order of 20 to 50%, which seems very high. However, the correct way to look those results is to consider the displacement difference in absolute terms. Indeed, since we are tracking the displacement of a point object, the radius of error is given by the displacement difference, which is 1.4 mm. This error indicates that if the cyst was the center of a lesion with a size of the order of 3 mm (twice the radius of error), it would be possible to track that lesion from the model simulation.

14. Potential sources of error

Three types of discretization errors can occur: first, errors from the non-linear material properties model, then errors from the time domain discretization (solving the dynamic equations), and finally, errors from the finite element method.

14.1. Sources of error from material properties measurements

It is difficult to assert the absolute significance of the material models developed (FUNG [1972]). Furthermore, the actual complexity of breast tissue (anisotropy, inhomogeneity, the number and distribution of Cooper's ligaments) prohibits us from accurately calculating a 3D map of a patient's breast tissue properties. This is why only average values of experimentally derived material models are used in our finite element analysis. The non-linear stress–strain curve describing the mechanical behavior of breast tissue, is discretized into a number of different segments, each corresponding to a displacement iteration. In order to test the effect of material properties model discretization, the number of displacement iterations was increased (as the displacement per iteration was decreased in order to keep the same total virtual plate compression). The analysis was done on the displacement of the cyst in the model.

As shown in the convergence analysis with the number of displacement iterations per simulation step (Fig. 11.9), increasing the number of displacement iterations, which also means increasing the number of linear segments discretizing the non-linear material stress–strain curve, leads to a convergence of the two important result parameters in the model. However, the variation in the parameters do not exceed 15% for the %misclassification, and 14% in the displacement of the cyst. Once again since only average values of experimentally derived material models are used in the analysis, the accuracy involved in the stress–strain curve discretization is not as important as the accuracy involved in approximating the shape of the breast in the model.

The material properties sensitivity analysis (Fig. 11.12) shows that large variations in material properties parameters (± 1.2 standard deviations) do not significantly affect the parameter results. This may be explained by the fact that the breast is under pressure: since glandular and fatty tissue make up the majority of breast tissue, and since the fat material model eventually becomes equivalent to the glandular tissue material model at a certain level of strain, then material properties difference may become of secondary importance in the simulations. If that is the case, then we may be able to simplify the material properties models and thus increase the speed of the breast compression simulation.

14.2. Sources of error from the time domain discretization

The time-discretized nodal equation is

$$\mathbf{q}_{i,t+1} = \mathbf{q}_{i,t} + \Delta t \cdot \mathbf{f}_{i,\text{internal}}. \tag{14.1}$$

At every displacement iteration, we integrate Eq. (14.1) using the adaptive Euler technique (PRESS, TEUKOLSKY, VETTERLING and FLANNERY [1992], DEVRIES [1994]), where the time step Δt, varies according to the amount of integration error.

We can make the integration error sufficiently small to insure convergence of the solution. The numerical integration ends when the difference between the sum of all displacements from one time iteration to the next is less than a threshold (10^{-3}% is

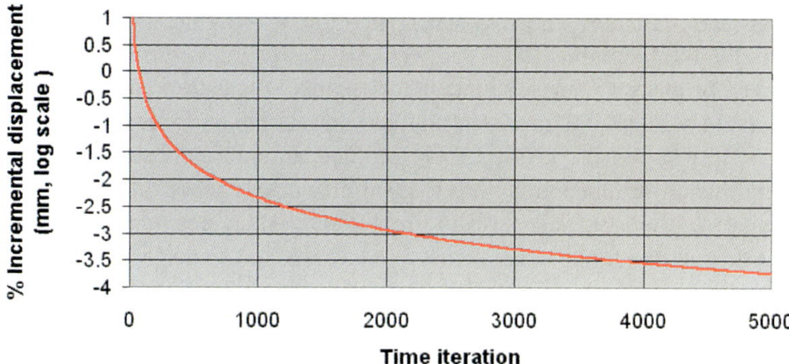

FIG. 14.1. Time convergence curve: % incremental displacement (log scale).

shown to be a good assumption):

$$\textit{difference} = \sum_{i=1}^{n} |\mathbf{q}_{i,t+1}| - \sum_{i=1}^{n} |\mathbf{q}_{i,t}| < \textit{threshold}. \tag{14.2}$$

Fig. 14.1 shows that a difference between two consecutive displacements, of less than $10^{-3}\%$ is enough to claim convergence of the model.

14.2.1. Sources of error from the finite element method

Errors in the finite element method can be divided into two classes:

- Discretization errors resulting from geometric differences between the boundaries of the model and its finite element approximation.
- Modeling errors, due to the difference between the true solution and its shape function representation.

Using smaller elements can reduce discretization errors – the errors tend to zero, as the element size tends to zero. Shape function errors do not decrease as the element size reduces and may thus prevent convergence to the exact solution or even cause divergence. There are two main criteria required of the shape function to guarantee convergence:

- *Completeness.* A complete polynomial of order at least p must be used for the representation of the variable within an element, where p is the order of the highest derivative of the variable appearing in the energy functional (in three dimensions a complete polynomial of order p can be written as

$$f(x, y, z) = \sum_{r=1}^{l} a_r x^i y^j z^k, \quad i + j + k \leqslant p,$$

where the number of terms in the polynomial is $l = (p+1)(p+2)(p+3)/6$.

- *Conformity.* The elements must be conforming, that is, the representations of the variable and its derivatives up to and including order $p - 1$ must be continuous across interelement boundaries, where p is the order of the highest derivative appearing in the functional.

An important property of isoparametric elements is that they provide C^0 continuity, and contain a complete linear polynomial in Cartesian coordinates (ERGATOUDIS, IRONS and ZIENKIEWICZ [1968]). Therefore, isoparametric elements satisfy the two criteria required of the shape functions to guarantee convergence.

In summary, although the breast deformable model has been shown to reasonably predict the displacement and deformation of cancer lesions under plate compression, its performance could be improved by:

- Better characterizing local material properties: presently, only average values of experimentally derived material models are used in our finite element analysis. However, strictly speaking, every little piece of tissue inside a patient's breast has a unique set of material properties, depending on its structure and composition. The only way to better characterize local material properties for every individual patient is through an in vivo technique. This technique should provide live non-invasive quantitative information on the non-linear deformation properties of every location in a patient's breast. Such a technique does not exist yet today, however research towards this goal is underway, through tissue elastography techniques.
- Better defining the boundary and initial conditions: no matter how complex the deformable model of the breast may be, it becomes relatively useless without an accurate quantitative description of the breast's physical interaction with its surrounding environment, i.e., how is the patient positioned, how compressed is the breast, what is the size of the compression plates, how much contact is there between the plates and the breast, what is the friction between the plates and the breast skin, how are the plates moved, etc.
- Using a geometrically more accurate breast model: smaller model elements for example, will decrease discretization errors by allowing a structurally more accurate description of the different breast tissues.

15. Specific properties of reliability

The model performance is as reliable as the parameter that is most sensitive to variations in the conditions of the experiment. In our model, it is shown, that the physical representation of the breast is most sensitive in variations of the breast shape. It is also shown through the material properties sensitivity analysis, that the performance of the model has modest dependence to variations in material properties within 1 to 2 standard deviations.

Indeed, the shape and size of a patient breast influence the amount of compression on the breast, and the boundary properties between the breast and the compression plates. For example, when testing the performance of the two-plate virtual compression in the slab model, the displacement differences in mm were 1.3, 3.6 and 4.5 for the cyst, inferior vitamin E pill, and superior vitamin E pill, respectively. The %misclassifications where 11, 27 and 32%. It is clear that the cyst's displacement was predicted more accurately than the displacement of the vitamin E pills. This difference in accuracy is due to the fact that the vitamin E pills were placed on the outer edges of the patient's breast, where the contact and boundary conditions between the breast and the compression plates are not always clearly defined, and it may become difficult to accurately predict breast deformation at these locations.

Therefore, additional experiments would be needed in order to assess the reliability of the model by choosing a number of patients with breasts of different sizes, leading to different initial and boundary conditions for the model. One would have to measure the boundary conditions through physical measurements of the breast placement configuration.

16. Major novelties in the model

The major novelties in this model include:

- The updated fatty tissue material model, which takes into account the effect of fat compartmentalization due to Cooper's ligaments in the breast. We showed through empirical evidence that fat compartmentalization occurs as the breast is being compressed, and that the new updated fatty tissue material model takes that phenomenon into account, and performs better than the original fatty tissue model.
- The use of small displacement iterations while updating the tissues' different material properties, in order to model the non-linear behavior of tissue material models. The silicon phantom study allowed us to compare predicted displacement of a silicon inclusion from an ABAQUS (large commercially available FEM package) simulation, and from a *BreastView* (our software package) simulation, and showed that our algorithm does yield accurate results.
- The application of finite element modeling theory to model the deformation of a human female breast in such a way that the entire process takes less than a half-hour, which according to the clinicians consulted, is a reasonably short time duration.

17. Concluding remarks

Currently, High Field (1.5 T) Superconducting MR imaging does not allow live guidance during needle breast procedures. The current procedure allows the physician only to calculate approximately the location and extent of a cancerous tumor in the compressed patient breast before inserting the needle. It can then become relatively uncertain that the tissue specimen removed during the biopsy actually belongs to the lesion of interest. A new method for guiding clinical breast biopsy was presented, based on a deformable finite element model of the breast. The geometry of the model is constructed from MR data, and its mechanical properties are modeled using a non-linear material model. This method allows imaging the breast without or with mild compression before the procedure, then compressing the breast and using the finite element model to predict the tumor's position during the procedure.

The final results show that it is possible to create a deformable model of the breast based on the use of finite elements with non-linear material properties capable of modeling and predicting the deformation of the breast. This study also shows that the full procedure can be carried out in less than a half-hour: from start to end, the average times to completion were 12 minutes for segmentation of MR data, 3 minutes for the model mesh creation, and 14 minutes for the model simulation.

The results also suggest that it is possible to use the deformable model of the breast in order to register lesion locations in image sets of the same patient breast taken at different times, and under different pressure plate conditions. We showed qualitatively that rotational differences in initial breast positioning are virtually inexistent, and can therefore be neglected in the simulation.

This deformable model may be used as a new tool to the physician (Azar, Metaxas and Schnall [2001]), who will:

1. image the breast under little or no compression (thus increasing the contrast and visibility of the tumor),
2. build the deformable model of the breast from the data,
3. compress the breast as much as the patient will allow (to minimize deformations caused by the insertion of the needle),
4. 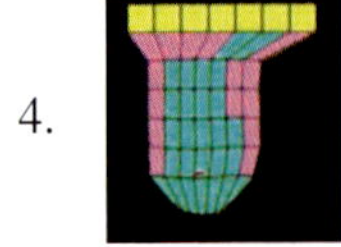 virtually compress the breast model in the virtual environment using the same boundary conditions as in reality,
5. 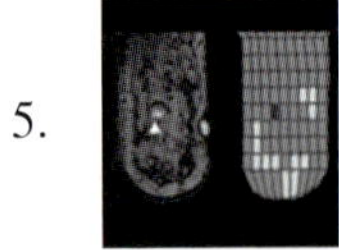 finally use the virtually compressed model to predict position of the real lesion within the real compressed breast during the procedure.

Future development in this project will be to develop a framework for a full-scale real-time finite element simulation of breast tissue deformation in more complex systems involving surgical instruments interacting with the model. The keys for such a development will include scalable parallel solution algorithms (Szekely, Brechbuhler, Hutter, Rhomberg and Schmid [1998]), as well as dedicated parallel hardware. The new system will allow real-time virtual surgical procedures of the breast, allowing the physician or student to fully prepare or train for the real procedure. The real-time capability will then allow the user to connect haptic devices such as a glove with pressure feedback, or a simulated needle with force feedback, and will bring the user one step closer to reality.

18. Appendix

18.1. Modeling 8-node hexahedral solid isoparametric elements

18.1.1. Finite element shape functions

The shape functions are used to interpolate a motion variable (displacement, position) from the nodes to a point in the element. The shape functions for an eight-node brick element (or linear solid element) are:

$$N_i = (1/8)(1 \pm \xi)(1 \pm \eta)(1 \pm \zeta) \tag{18.1}$$

in which $i = 1, 2, \ldots, 8$.

18.1.2. Finite element stiffness matrix

The derivation of the 3D finite element stiffness matrix presented follows the formulation found in COOK, MALKUS and PLESHA [1989]. The finite element stiffness matrix $\mathbf{K}_\mathrm{e}$ is given by

$$\mathbf{K}_\mathrm{e} = \iiint_{V_\mathrm{e}} \mathbf{B}^\mathrm{T}\mathbf{D}\mathbf{B}\,\mathrm{d}V, \tag{18.2}$$

where $\mathbf{D}$ is the stress–strain matrix and $\mathbf{B}$ is the strain–displacement matrix. The stiffness matrix incorporates the material and geometrical stiffness of the element.

18.1.3. Stress–strain matrix

Stress is related to strain through the following relationship, for a linear material:

$$\sigma = \mathbf{D}\boldsymbol{\varepsilon}, \tag{18.3}$$

$$\text{where} \quad \boldsymbol{\varepsilon} = \begin{bmatrix} \varepsilon_{xx} \\ \varepsilon_{yy} \\ \varepsilon_{zz} \\ \varepsilon_{xy} \\ \varepsilon_{zy} \\ \varepsilon_{xz} \end{bmatrix} \tag{18.4}$$

and $\boldsymbol{\sigma}$ is a column vector with the respective strain components. For an isotropic material, $\mathbf{D}$ is defined by

$$\mathbf{D} = \frac{Y(1-\upsilon)}{(1+\upsilon)(1-2\upsilon)} \begin{bmatrix} 1 & \frac{\upsilon}{1-\upsilon} & \frac{\upsilon}{1-\upsilon} & 0 & 0 & 0 \\ & 1 & \frac{\upsilon}{1-\upsilon} & 0 & 0 & 0 \\ & & 1 & 0 & 0 & 0 \\ & & & \frac{1-2\upsilon}{2(1-\upsilon)} & 0 & 0 \\ & \textit{Symmetric} & & & \frac{1-2\upsilon}{2(1-\upsilon)} & 0 \\ & & & & & \frac{1-2\upsilon}{2(1-\upsilon)} \end{bmatrix}, \tag{18.5}$$

where Y is the Young's modulus, and υ is the Poisson ratio.

18.1.4. Strain–displacement matrix

The matrix **B** relates strains to displacements at the nodes

$$\boldsymbol{\varepsilon} = \mathbf{B}\mathbf{q}_{\mathrm{e}}, \tag{18.6}$$

where $\boldsymbol{\varepsilon}$ is given above, and $\mathbf{q}_{\mathrm{e}}$ is the displacements at the element's n nodes. For Cartesian coordinates, u, v and w are displacements in the x, y and z directions, respectively:

$$\mathbf{q}_{\mathrm{e}} = \begin{bmatrix} u_1 \\ v_1 \\ w_1 \\ u_2 \\ v_2 \\ w_2 \\ \vdots \\ u_n \\ v_n \\ w_n \end{bmatrix}. \tag{18.7}$$

The relation between strain and displacements at the nodes involve matrix multiplications as defined below. First, the definition of small strain can be written in matrix form as

$$\boldsymbol{\varepsilon} = \underbrace{\begin{bmatrix} 1 & 0 & 0 & 0 & 0 & 0 & 0 & 0 & 0 \\ 0 & 0 & 0 & 0 & 1 & 0 & 0 & 0 & 0 \\ 0 & 0 & 0 & 0 & 0 & 0 & 0 & 0 & 1 \\ 0 & \frac{1}{2} & 0 & \frac{1}{2} & 0 & 0 & 0 & 0 & 0 \\ 0 & 0 & 0 & 0 & 0 & \frac{1}{2} & 0 & \frac{1}{2} & 0 \\ 0 & 0 & \frac{1}{2} & 0 & 0 & 0 & \frac{1}{2} & 0 & 0 \end{bmatrix}}_{\mathbf{L}:\ 6\times 9} \begin{bmatrix} u,_x \\ u,_y \\ u,_z \\ v,_x \\ v,_y \\ v,_z \\ w,_x \\ w,_y \\ w,_z \end{bmatrix}. \tag{18.8}$$

The derivatives of displacement with respect to global coordinates in this equation can be related to derivatives of displacements with respect to local coordinates by the 9×9 matrix **M**, in the following equation:

$$\begin{bmatrix} u,_x \\ u,_y \\ u,_z \\ v,_x \\ v,_y \\ v,_z \\ w,_x \\ w,_y \\ w,_z \end{bmatrix} = \underbrace{\begin{bmatrix} \boldsymbol{\Gamma}_{11} & \boldsymbol{\Gamma}_{12} & \boldsymbol{\Gamma}_{13} & 0 & 0 & 0 & 0 & 0 & 0 \\ \boldsymbol{\Gamma}_{21} & \boldsymbol{\Gamma}_{22} & \boldsymbol{\Gamma}_{23} & 0 & 0 & 0 & 0 & 0 & 0 \\ \boldsymbol{\Gamma}_{31} & \boldsymbol{\Gamma}_{32} & \boldsymbol{\Gamma}_{33} & 0 & 0 & 0 & 0 & 0 & 0 \\ 0 & 0 & 0 & \boldsymbol{\Gamma}_{11} & \boldsymbol{\Gamma}_{12} & \boldsymbol{\Gamma}_{13} & 0 & 0 & 0 \\ 0 & 0 & 0 & \boldsymbol{\Gamma}_{21} & \boldsymbol{\Gamma}_{22} & \boldsymbol{\Gamma}_{23} & 0 & 0 & 0 \\ 0 & 0 & 0 & \boldsymbol{\Gamma}_{31} & \boldsymbol{\Gamma}_{32} & \boldsymbol{\Gamma}_{33} & 0 & 0 & 0 \\ 0 & 0 & 0 & 0 & 0 & 0 & \boldsymbol{\Gamma}_{11} & \boldsymbol{\Gamma}_{12} & \boldsymbol{\Gamma}_{13} \\ 0 & 0 & 0 & 0 & 0 & 0 & \boldsymbol{\Gamma}_{21} & \boldsymbol{\Gamma}_{22} & \boldsymbol{\Gamma}_{23} \\ 0 & 0 & 0 & 0 & 0 & 0 & \boldsymbol{\Gamma}_{31} & \boldsymbol{\Gamma}_{32} & \boldsymbol{\Gamma}_{33} \end{bmatrix}}_{\mathbf{M}:\ 9\times 9} \begin{bmatrix} u,_\xi \\ u,_\eta \\ u,_\zeta \\ v,_\xi \\ v,_\eta \\ v,_\zeta \\ w,_\xi \\ w,_\eta \\ w,_\zeta \end{bmatrix}. \tag{18.9}$$

The matrix $\mathbf{M}$ is composed of a 3×3 matrix, $\mathbf{\Gamma}$, defined later. The derivatives of displacements with respect to local coordinates can now be related to the nodal displacements (vector $\mathbf{q}_e$) via the shape functions. We know that the interpolation functions define the displacement in the element in terms of nodal displacements. For example, the displacement in the x-direction, u, is

$$u = \sum_{i=1}^{n} N_i u_i = N_1 u_1 + N_2 u_2 + \cdots + N_n u_n \tag{18.10}$$

and the derivative with respect to a local coordinate, ξ, is

$$\frac{\partial u}{\partial \xi} = u_{,\xi} = \sum_{i=1}^{n} N_{i,\xi} u_i = N_{1,\xi} u_1 + N_{2,\xi} u_2 + \cdots + N_{n,\xi} u_n. \tag{18.11}$$

This relation follows for all three displacement derivatives taken with respect to the three local coordinates. It is written in matrix form as

$$\begin{bmatrix} u_{,\xi} \\ u_{,\eta} \\ u_{,\zeta} \\ v_{,\xi} \\ v_{,\eta} \\ v_{,\zeta} \\ w_{,\xi} \\ w_{,\eta} \\ w_{,\zeta} \end{bmatrix} = \underbrace{\begin{bmatrix} N_{1,\xi} & 0 & 0 & N_{2,\xi} & 0 & 0 & \dots & N_{n,\xi} & 0 & 0 \\ N_{1,\eta} & 0 & 0 & N_{2,\eta} & 0 & 0 & \dots & N_{n,\eta} & 0 & 0 \\ N_{1,\zeta} & 0 & 0 & N_{2,\zeta} & 0 & 0 & \dots & N_{n,\xi} & 0 & 0 \\ 0 & N_{1,\xi} & 0 & 0 & N_{2,\xi} & 0 & \dots & 0 & N_{n,\xi} & 0 \\ 0 & N_{1,\eta} & 0 & 0 & N_{2,\eta} & 0 & \dots & 0 & N_{n,\eta} & 0 \\ 0 & N_{1,\zeta} & 0 & 0 & N_{2,\zeta} & 0 & \dots & 0 & N_{n,\xi} & 0 \\ 0 & 0 & N_{1,\xi} & 0 & 0 & N_{2,\xi} & \dots & 0 & 0 & N_{n,\xi} \\ 0 & 0 & N_{1,\eta} & 0 & 0 & N_{2,\eta} & \dots & 0 & 0 & N_{n,\eta} \\ 0 & 0 & N_{1,\zeta} & 0 & 0 & N_{2,\zeta} & \dots & 0 & 0 & N_{n,\xi} \end{bmatrix}}_{\mathbf{Q}:\ 9\times(n^*3)} \mathbf{q}_e. \tag{18.12}$$

Thus, there are three matrix multiplications and the matrix $\mathbf{B}$ is composed of three matrices. Comparing the equations, we get

$$\underbrace{\mathbf{B}}_{6\times(n^*3)} = \underbrace{\mathbf{L}}_{6\times 9}\,\underbrace{\mathbf{M}}_{9\times 9}\,\underbrace{\mathbf{Q}}_{9\times(n^*3)}. \tag{18.13}$$

18.1.5. Derivation of inverse Jacobian

The Jacobian relates derivatives in the global coordinate system (x, y, z) to derivatives in the local coordinate system (ξ, η, ζ) and can be simply derived using the chain-rule. So for a function $\Psi(x, y, z)$:

$$\begin{bmatrix} \psi_{,\xi} \\ \psi_{,\eta} \\ \psi_{,\zeta} \end{bmatrix} = \underbrace{\begin{bmatrix} x_{,\xi} & y_{,\xi} & z_{,\xi} \\ x_{,\eta} & y_{,\eta} & z_{,\eta} \\ x_{,\zeta} & y_{,\zeta} & z_{,\zeta} \end{bmatrix}}_{\mathbf{J}:\ 3\times 3} \begin{bmatrix} \psi_{,x} \\ \psi_{,y} \\ \psi_{,z} \end{bmatrix}. \tag{18.14}$$

In the isoparametric formulation, a point (x, y, z) in the element is interpolated from the nodes with the shape functions:

$$x = \sum_{i=1}^{n} N_i x_i, \qquad y = \sum_{i=1}^{n} N_i y_i, \qquad z = \sum_{i=1}^{n} N_i z_i, \tag{18.15}$$

where the shape functions are functions of the local coordinates. Therefore, the elements in the Jacobian matrix can be evaluated by taking the derivative of the appropriate term in the last equation with respect to the appropriate local coordinates:

$$\mathbf{J} = \begin{bmatrix} \sum_{i=1}^{n} N_{i,\xi} x_i & \sum_{i=1}^{n} N_{i,\xi} y_i & \sum_{i=1}^{n} N_{i,\xi} z_i \\ \sum_{i=1}^{n} N_{i,\eta} x_i & \sum_{i=1}^{n} N_{i,\eta} y_i & \sum_{i=1}^{n} N_{i,\eta} z_i \\ \sum_{i=1}^{n} N_{i,\zeta} x_i & \sum_{i=1}^{n} N_{i,\zeta} y_i & \sum_{i=1}^{n} N_{i,\zeta} z_i \end{bmatrix}. \tag{18.16}$$

This formula for $\mathbf{J}$ can be expanded as follows:

$$\mathbf{J} = \underbrace{\begin{bmatrix} N_{1,\xi} & N_{2,\xi} & N_{3,\xi} & N_{4,\xi} & \dots & N_{n,\xi} \\ N_{1,\eta} & N_{2,\eta} & N_{3,\eta} & N_{4,\eta} & \dots & N_{n,\eta} \\ N_{1,\zeta} & N_{2,\zeta} & N_{3,\zeta} & N_{4,\zeta} & \dots & N_{n,\zeta} \end{bmatrix}}_{\mathbf{D}_n:\ 3\times n} \underbrace{\begin{bmatrix} x_1 & y_1 & z_1 \\ x_2 & y_2 & z_2 \\ x_3 & y_3 & z_3 \\ x_4 & y_4 & z_4 \\ \vdots & \vdots & \vdots \\ x_n & y_n & z_n \end{bmatrix}}_{\mathbf{x}:\ n\times 3}. \tag{18.17}$$

The matrix $\mathbf{x}$ contains the deformed positions of the n nodes in the element. The matrix $\mathbf{\Gamma}$ is finally given by

$$\mathbf{\Gamma} = \mathbf{J}^{-1} = \begin{bmatrix} \mathbf{\Gamma}_{11} & \mathbf{\Gamma}_{12} & \mathbf{\Gamma}_{13} \\ \mathbf{\Gamma}_{21} & \mathbf{\Gamma}_{22} & \mathbf{\Gamma}_{23} \\ \mathbf{\Gamma}_{31} & \mathbf{\Gamma}_{32} & \mathbf{\Gamma}_{33} \end{bmatrix}. \tag{18.18}$$

18.2. Modeling linear triangle isoparametric elements

Fig. 18.1 shows a triangular element $P_1P_2P_3$ in a local orthonormal coordinate system (α, β) with basis unit vectors $\hat{\alpha}, \hat{\beta}$ and its origin at node P_1.

The local node displacement vectors are given by (p_i, q_i) for every node i. The real-world 3D-coordinate system is given by (x, y, z) and the real-world 3D node displacements are given by (u, v, w). The displacement field is given by $p = p(\alpha, \beta)$ and

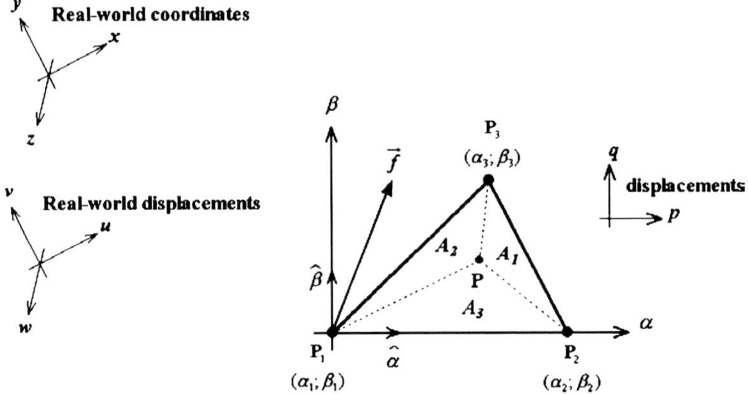

FIG. 18.1. Linear 3-node triangle element.

$q = q(\alpha, \beta)$. They are each interpolated from the *nodal displacement* degrees of freedom p_i and q_j:

$$\begin{bmatrix} p \\ q \end{bmatrix} = \mathbf{N}\mathbf{q}_e, \quad \text{where } \mathbf{N} = \begin{bmatrix} \xi_1 & 0 & \xi_2 & 0 & \xi_3 & 0 \\ 0 & \xi_1 & 0 & \xi_2 & 0 & \xi_3 \end{bmatrix},$$

$$\mathbf{q}_e = \begin{bmatrix} p_1 \\ q_1 \\ p_2 \\ q_2 \\ p_3 \\ q_3 \end{bmatrix}. \tag{18.19}$$

ξ_i are the local area coordinates of any point P inside the triangle element: P divides the triangle element into three sub-areas A_1, A_2 and A_3. Area coordinates are defined as ratios of areas:

$$\xi_1 = \frac{A_1}{A}, \qquad \xi_2 = \frac{A_2}{A}, \qquad \xi_3 = \frac{A_3}{A}, \tag{18.20}$$

where A is the area of the triangle element. Now since $A = A_1 + A_2 + A_3$, the ξ_i are not independent, and they satisfy the constraint equation,

$$\xi_1 + \xi_2 + \xi_3 = 1. \tag{18.21}$$

This constraint equation and the linear relation between Cartesian (α, β) and area (ξ_1, ξ_2, ξ_3) coordinates are expressed by the following equations:

$$\begin{bmatrix} 1 \\ \alpha \\ \beta \end{bmatrix} = \mathbf{A} \begin{bmatrix} \xi_1 \\ \xi_2 \\ \xi_3 \end{bmatrix}, \qquad \begin{bmatrix} \xi_1 \\ \xi_2 \\ \xi_3 \end{bmatrix} = \mathbf{A}^{-1} \begin{bmatrix} 1 \\ \alpha \\ \beta \end{bmatrix}, \tag{18.22}$$

where

$$\mathbf{A} = \begin{bmatrix} 1 & 1 & 1 \\ \alpha_1 & \alpha_2 & \alpha_3 \\ \beta_1 & \beta_2 & \beta_3 \end{bmatrix}, \qquad \mathbf{A}^{-1} = \frac{1}{2A} \begin{bmatrix} (\alpha_2\beta_3 - \alpha_3\beta_2) & \beta_{23} & \alpha_{32} \\ (\alpha_3\beta_1 - \alpha_1\beta_3) & \beta_{31} & \alpha_{13} \\ (\alpha_1\beta_2 - \alpha_2\beta_1) & \beta_{12} & \alpha_{21} \end{bmatrix} \tag{18.23}$$

with $\alpha_{ij} = \alpha_i - \alpha_j$ and $\beta_{ij} \equiv \beta_i - \beta_j$, and $2A = \det(\mathbf{A}) = \alpha_{21}\beta_{31} - \alpha_{31}\beta_{21}$.

Since we know that strains are defined as $\boldsymbol{\varepsilon} = \mathbf{B}\mathbf{q}_e$, and using the chain rule, we obtain the following expression for matrix $\mathbf{B}$, after simplification:

$$\mathbf{B} = \partial\mathbf{N} = \frac{1}{2A} \begin{bmatrix} \beta_{23} & 0 & \beta_{31} & 0 & \beta_{12} & 0 \\ 0 & \alpha_{32} & 0 & \alpha_{13} & 0 & \alpha_{21} \\ \alpha_{32} & \beta_{23} & \alpha_{13} & \beta_{31} & \alpha_{21} & \beta_{12} \end{bmatrix}. \tag{18.24}$$

18.2.1. Material properties matrix E

E is defined as

$$\mathbf{E} = \frac{Y}{(1+\upsilon)(1-2\upsilon)} \begin{bmatrix} (1-\upsilon) & \upsilon & 0 \\ \upsilon & (1-\upsilon) & 0 \\ 0 & 0 & \frac{(1-2\upsilon)}{2} \end{bmatrix}, \tag{18.25}$$

where Y = Young's modulus of elasticity, υ = Poisson ratio.

18.2.2. Element stiffness matrix formulation

Finally, the element stiffness matrix $\mathbf{K}_e$ is given by

$$\mathbf{K}_e = \iiint_{V_e} \mathbf{B}^T\mathbf{D}\mathbf{B}\,dV. \tag{18.26}$$

But $\mathbf{B}$ and $\mathbf{D}$ are constant over the triangle element, and we suppose that the element thickness t (which corresponds to the skin's thickness) is also constant. Then $\mathbf{K}_e$ can be simply written as

$$\underbrace{\mathbf{K}_e}_{6\times6} = At \cdot \underbrace{\mathbf{B}^T}_{6\times3}\underbrace{\mathbf{D}}_{6\times6}\underbrace{\mathbf{B}}_{3\times6}, \tag{18.27}$$

where A is the area of the triangle element.

18.2.3. Forces on element nodes

The forces generated on the element nodes P_1, P_2 and P_3 are then written as

$$\mathbf{F}_e = \mathbf{K}_e\mathbf{q}_e \quad \Longleftrightarrow \quad \begin{bmatrix} \alpha_{F1} \\ \beta_{F1} \\ \alpha_{F2} \\ \beta_{F2} \\ \alpha_{F3} \\ \beta_{F3} \end{bmatrix} = \mathbf{K}_e \begin{bmatrix} p_1 \\ q_1 \\ p_2 \\ q_2 \\ p_3 \\ q_3 \end{bmatrix}. \tag{18.28}$$

18.2.4. Expression for $\mathbf{K}_e$ in the real world coordinates

The node forces generated $(\alpha_{Fi}, \beta_{Fi})$, the node displacement vectors (p_i, q_i) and the node coordinates (α_i, β_i) need to be transformed from and to the real-world 3-dimensional coordinate system (x, y, z). The basis vectors $\hat{\alpha}(x_{\hat{\alpha}}; y_{\hat{\alpha}}; z_{\hat{\alpha}})$ and $\hat{\beta}(x_{\hat{\beta}}; y_{\hat{\beta}}; z_{\hat{\beta}})$ in the (x, y, z) system are given by the following:

$$\begin{aligned} \hat{\alpha} &= \frac{\vec{\alpha}}{\|\vec{\alpha}\|}, \quad \text{where } \vec{\alpha} = \overrightarrow{P_1P_2}, \\ \hat{\beta} &= \frac{\vec{\beta}}{\|\vec{\beta}\|}, \quad \text{where } \vec{\beta} = \overrightarrow{P_1P_3} - \frac{(\overrightarrow{P_1P_2}\cdot\overrightarrow{P_1P_3})}{\|\overrightarrow{P_1P_2}\|^2}\overrightarrow{P_1P_2}. \end{aligned} \tag{18.29}$$

18.2.5. Coordinates of element nodes

The coordinates of the element nodes are given as follows:

$$P_1\begin{cases} \alpha_1 = 0, \\ \beta_1 = 0, \end{cases} \qquad P_2\begin{cases} \alpha_2 = \|\vec{\alpha}\|, \\ \beta_2 = 0, \end{cases} \qquad P_3\begin{cases} \alpha_3 = \dfrac{|\overrightarrow{P_1P_2}\cdot\overrightarrow{P_1P_3}|}{\|\overrightarrow{P_1P_2}\|}, \\ \beta_3 = \|\vec{\beta}\|. \end{cases} \tag{18.30}$$

18.2.6. Coordinates of the nodal displacement field

The coordinates of the nodal displacement field are given as follows:

$$p_i = \begin{bmatrix} u_i \\ v_i \\ w_i \end{bmatrix} [x_{\hat{\alpha}} \quad y_{\hat{\alpha}} \quad z_{\hat{\alpha}}], \qquad q_i = \begin{bmatrix} u_i \\ v_i \\ w_i \end{bmatrix} [x_{\hat{\beta}} \quad y_{\hat{\beta}} \quad z_{\hat{\beta}}]. \tag{18.31}$$

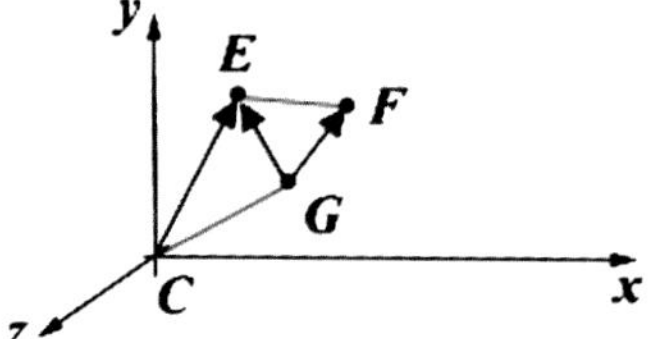

FIG. 18.2. Triangular pyramid.

18.2.7. Real-world coordinates of the resulting nodal forces

The real-world coordinates of the resulting nodal forces are given by

$$\begin{bmatrix} x_{Fi} \\ y_{Fi} \\ z_{Fi} \end{bmatrix} = \begin{bmatrix} x_{\hat{\alpha}} & x_{\hat{\beta}} \\ y_{\hat{\alpha}} & y_{\hat{\beta}} \\ z_{\hat{\alpha}} & z_{\hat{\beta}} \end{bmatrix} \begin{bmatrix} \alpha_{Fi} \\ \beta_{Fi} \end{bmatrix}. \tag{18.32}$$

18.3. Element volume calculations

Calculating the volume of a hexahedral element is used to verify the incompressibility properties of the model, depending on the value of Poisson's ratio: we calculate the volume of every element after every displacement iteration and we verify volume conservation. Volume calculations are also necessary when including body forces in the model, such as gravity.

18.3.1. Volume of a triangular pyramid

In an orthonormal basis, centered at C, the volume of the pyramid $CGEF$ is given by (see Fig. 18.2)

$$V_{EFG} = \frac{1}{6} \left| \overrightarrow{CE} \cdot (\overrightarrow{GE} \times \overrightarrow{GF}) \right|. \tag{18.33}$$

Analytically, we get

$$V_{EFG} = \frac{1}{6} \left| \begin{matrix} (E_x - C_x) \cdot \left[(E_y - G_y) \cdot (F_z - G_z) - (E_z - G_z) \cdot (F_y - G_y) \right] \\ + (E_y - C_y) \cdot \left[(E_z - G_z) \cdot (F_x - G_x) - (E_x - G_x) \cdot (F_z - G_z) \right] \\ + (E_z - C_z) \cdot \left[(E_x - G_x) \cdot (F_y - G_y) - (E_y - G_y) \cdot (F_x - G_x) \right] \end{matrix} \right|. \tag{18.34}$$

18.3.2. Volume of a hexahedral element

First, we find the center point of the element (see Fig. 18.3):

$$\begin{aligned} C_x &= \frac{1}{8}(P_{1x} + P_{2x} + P_{3x} + P_{4x} + P_{5x} + P_{6x} + P_{7x} + P_{8x}), \\ C_y &= \frac{1}{4}(P_{1y} + P_{3y} + P_{5y} + P_{7y}), \\ C_z &= \frac{1}{2}(P_{1z} + P_{5z}), \end{aligned} \tag{18.35}$$

where P_i are the nodes of the element.

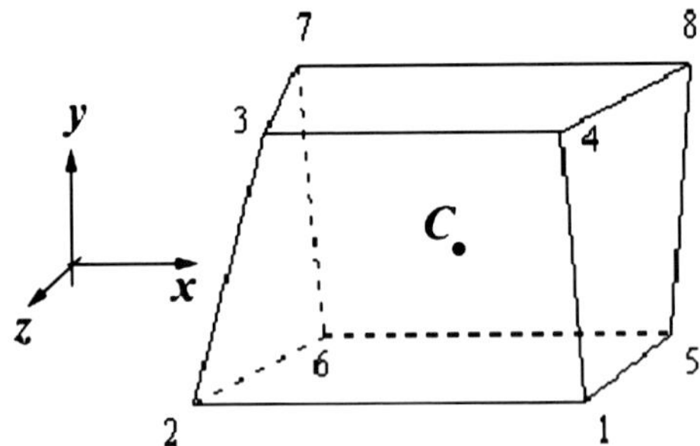

FIG. 18.3. Volume calculation of a hexahedral element.

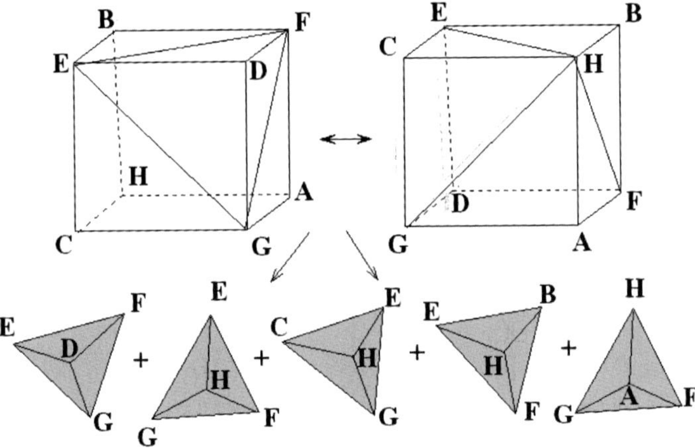

FIG. 18.4. Volume calculation of a hexahedral element using only 5 triangular pyramids.

Finally, the total volume of the element is calculated by dividing the element into several different triangular pyramids, and calculating the volume of each one of them:

$$V_E = (V_{451} + V_{485} + V_{234} + V_{124} + V_{587} + V_{687} + V_{326} + V_{673} + V_{784} + V_{437} + V_{156} + V_{162}). \tag{18.36}$$

We can actually express the volume of a hexahedral element using only 5 triangular pyramids, by dividing the element as shown in Fig. 18.4. Then the total volume of the element is given by

$$V_E = (V_{EFGD} + V_{EFGH} + V_{CEGH} + V_{BFHE} + V_{HAGF}). \tag{18.37}$$

18.4. Is a data point inside or outside an element?

The ability to determine whether a point is inside or outside of an element is very important especially when determining the tissue type of an element: all the data points

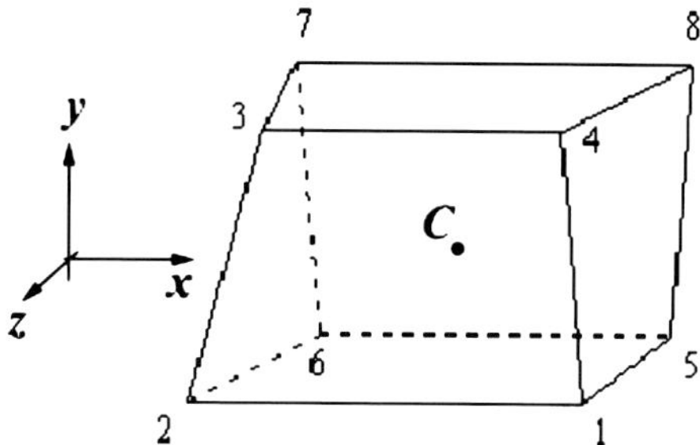

FIG. 18.5. Is C inside the hexahedron?

from the segmented breast image which lie inside the hexahedral element have to be counted, in order to determine what is the highest percentage of tissue type inside the element.

In 2 dimensions.

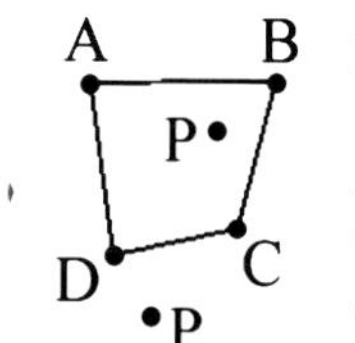

If $PA \times PD$, $PD \times PC$, $PC \times PB$, $PB \times PA$, all have the same direction, then P is inside element $ABCD$; otherwise P is outside $ABCD$.

In 3 dimensions. C is inside the hexahedron, if and only if all of the following quantities are negative (see Fig. 18.5):

$$\begin{array}{llll} \overrightarrow{N_{214}} \cdot \overrightarrow{P_2C}; & \overrightarrow{N_{567}} \cdot \overrightarrow{P_5C}; & \overrightarrow{N_{432}} \cdot \overrightarrow{P_4C}; & \overrightarrow{N_{785}} \cdot \overrightarrow{P_7C} \\ \overrightarrow{N_{158}} \cdot \overrightarrow{P_1C}; & \overrightarrow{N_{237}} \cdot \overrightarrow{P_2C}; & \overrightarrow{N_{841}} \cdot \overrightarrow{P_8C}; & \overrightarrow{N_{762}} \cdot \overrightarrow{P_7C} \\ \overrightarrow{N_{348}} \cdot \overrightarrow{P_3C}; & \overrightarrow{N_{126}} \cdot \overrightarrow{P_1C}; & \overrightarrow{N_{873}} \cdot \overrightarrow{P_8C}; & \overrightarrow{N_{651}} \cdot \overrightarrow{P_6C}, \end{array} \tag{18.38}$$

where P_i are the nodes of the element, and $\overrightarrow{N_{ijk}}$ is the outer unit normal of triangle $P_i P_j P_k$.

18.5. Silicon phantom construction

The geometry of the deformable phantom consists of a rectangular box ($84 \times 82 \times 70$ mm) containing a rectangular inclusion ($20 \times 23 \times 20$ mm), which is 4.3 times stiffer than the surrounding silicon (Fig. 10.1). The gel system is composed of two parts, catalyst (part A) and resin (part B), the ratio of which determines its elastic properties (a decrease in A:B produces stiffer gels). The components, the catalyst (part A) and the resin (part B), both contain silicon copolymers which form cross-links when combined. Since only approximately 2% of the material becomes cross-linked, the material becomes a gel, and its stiffness is directly proportional to the amount of cross-linking (GOLDSTEIN, KUNDEL, DAUBE-WHITERSPOON, THIBAULT and GOLDSTEIN [1987]). Parts

A and B of the gel system were mixed in a ratio 1:1.7. This provided a good combination of flexibility under compression and integrity under gravity, for the surrounding silicon in the phantom. Parts A and B of the gel system were then mixed in a ratio 1:5.7, for the stiff inclusion. The phantom was built as follows:

1. The molds consist of two rectangular boxes, one of size ($84 \times 82 \times 70$ mm) to house the whole silicon phantom, and another smaller mold of size ($20 \times 23 \times 20$ mm) used to make the stiff silicon inclusion. Both molds are made of heat-resistant PVC. The inside walls of the molds were sprayed with Pam oil, and covered with sheets of transparent plastic wrapper, making it much easier to remove the silicon phantoms out of the molds when ready.
2. Parts A and B of the gel system were mixed in a ratio 1:1.7. This provided a good combination of flexibility under compression and integrity under gravity. The mix was stirred for 5 minutes, and then poured into the larger mold so as to fill about half of the mold.
3. Parts A and B of the gel system were then mixed in a ratio 1:5.7, stirred for 5 minutes and then poured into the smaller mold in order to fill it up.
4. Both molds were then heated up at 175 °F for 36 hours, the time needed for the gel systems to cure. The heat acted as a catalyst in the curing process.
5. The stiff silicon gel inclusion was removed from its mold, and then placed inside the larger mold on the top of the already cured silicon gel.
6. Parts A and B of the gel system were again mixed in a ratio 1:1.7, stirred for 5 minutes and then carefully poured into the larger mold in order to fill it up.
7. The large mold was heated up again at 175 °F during 36 hours.
8. The full silicon gel phantom was finally removed from its mold by pulling on the plastic wrap paper, and secured inside a custom-built pressure device (also made of PVC material, which does not cause any extraneous signal when imaged in the MR machine).

Acknowledgements

The authors are thankful to Norm Butler, Allen Bonner, Idith Haber, Reid Miller, Bruno Carvalho and Joe Giammarco for their help in various aspects of this work.

References

ABAQUS/Standard V.5.8 (1998). Hibbitt, Karlsson & Sorensen, vol. II: 14.1.4–1, 14.1.4–17.

AGACHE, P.G., MONNEUR, C., LEVEQUE, J.L., DERIGAL, J. (1980). Mechanical properties and Young's modulus of human skin in vivo. *Arch. Dermatol. Res.* **269**, 221–232.

AZAR, F.S., METAXAS, D.N., MILLER, R.T., SCHNALL, M.D. (2000). Methods for predicting mechanical deformations in the breast during clinical breast biopsy. In: *26th IEEE Annual Northeast Bioengineering Conference*.

AZAR, F.S., METAXAS, D., SCHNALL, M.D. (1999). A finite element model of the breast for predicting mechanical deformations during interventional procedures. In: *Proc. 7th Intl. Soc. Magn. Reson. Med.*, p. 1084.

AZAR, F.S., METAXAS, D.N., SCHNALL, M.D. (2001). A deformable finite element model of the breast for predicting mechanical deformations under external perturbations. *J. Academic Radiology*.

AZAR, F.S., METAXAS, D.N., SCHNALL, M.D. (2002). Methods for modeling and predicting mechanical deformations of the breast under external perturbations. *Medical Image Analysis* **6** (1), 1–27.

BAUMANN, R., GLAUSER, D. (1996). Force feedback for virtual reality based minimally invasive surgery simulator. In: *Medicine Meets Virtual Reality* (IOS Press, Amsterdam).

BEHRENBRUCH, C.P., MARIAS, K., ARMITAGE, P., YAM, M., MOORE, N., ENGLISH, R.E., BRADY, M. (2000). MRI-mammography 2D/3D data fusion for breast pathology assessment. In: *Proc. Medical Image Computing and Computer Assisted Intervention*, MICCAI, pp. 307–316.

BOETES, C., BARENTSZ, J.O., et al. (1994). MR characterization of suspicious breast lesions with a gadolinium-enhanced turboFLASH subtraction technique. *Radiology* **193**, 777–781.

CARVALHO, B.M., GAU, C.J., HERMAN, C.T., KONG, T.Y. (1999). Algorithms for fuzzy segmentation. *Pattern Anal. Appl.* **2**, 73–81.

CHADWICK, J., HAUMANN, D., PARENT, R. (1989). Layered construction of deformable animated characters. *Computer Graphics (SIGGRAPH'89)* **23**, 243–252.

CHEN, D.T., ZELTZER, D. (1992). Pump it up: Computer animation of a biomechanically based model of the muscle using the finite element method. *Computer Graphics (SIGGRAPH'92)* **26**, 89–98.

COOK, R.D., MALKUS, D.S., PLESHA, M.E. (1989). *Concepts and Applications of Finite Elements Analysis* (Wiley, New York).

COTIN, S., DELINGUETTE, H., AYACHE, N. (1999). Real-time elastic deformations of soft tissues for surgery simulation. *IEEE Trans. Visualization and Computer Graphics* **5** (1), 62–73.

CRISP, J.D.C. (1972). Properties of tendon and skin. In: Fung, Y.C. (ed.), *Biomechanics: Its Foundations and Objectives* (Prentice Hall, New York).

DERSHAW (1996). Stereotaxic breast biopsy. *Seminars in Ultrasound, CT, and MRI* **17** (5), 444–459.

DESBRUN, M., GASCUEL, M.P. (1995). Animating soft substances with implicit surfaces. *Computer Graphics (SIGGRAPH'95)*, 287–290.

DEVRIES, P.L. (1994). *A First Course in Computational Physics* (Wiley, New York), pp. 207–225.

DRONKERS (1992). Stereotaxic core biopsy of breast lesions. *Radiology* **183**, 631–634.

EGAN, R.L. (1988a). Breast embryology, anatomy and physiology. In: *Breast Imaging: Diagnosis and Morphology of Breast Diseases*, pp. 30–58, Chapter 4.

EGAN, R.L. (1988b). Malignant breast lesions. In: *Breast Imaging: Diagnosis and Morphology of Breast Diseases*, pp. 227–231, Chapter 14.

ELDEN, H.R. (1977). *Biophysical Properties of Skin* (Wiley–Interscience, New York).

ERGATOUDIS, I., IRONS, B.M., ZIENKIEWICZ, O.C. (1968). Curved isoparametric, 'quadrilateral' elements for finite element analysis. *Int. J. Solids Structures* **4** (1), 31–42.

FERLAY, J., BRAY, F., et al. (2001). *Globocan 2000: Cancer Incidence, Mortality and Prevalence Worldwide, Version 1.0* (IARCPress, Lyon).

FISCHER, U., VOSSHENRICH, R., DOLER, W., HAMADEH, A., OESTMANN, J.W., GRABBE, E. (1995). MR Imaging-guided breast intervention: experience with two systems. *Radiology* **195**, 533–538.

FISCHER, U., VOSSHENRICH, R., KEATING, D., BRUHN, H., DOLER, W., OESTMANN, J.W., GRABBE, E. (1994). MR-guided biopsy of suspect breast lesions with a simple stereotaxic add-on device for surface coils. *Radiology* **192**, 272–273.

FUNG, Y.C. (1972). Stress–strain history relations of soft tissues in simple elongation. In: Fung, Y.C., Perrone, N., Anliker, M. (eds.), *Biomechanics: Its Foundations and Objectives* (Prentice Hall, Englewood Cliffs, NJ).

FUNG, Y.C. (1981). *Biomechanics: Mechanical Properties of Living Tissues* (Springer-Verlag, New York), pp. 203–212.

FUNG, Y.C. (1987). Mechanics of soft tissues. In: Skalak, R., Chien, S. (eds.), *Handbook of Bioengineering* (McGraw–Hill, New York).

FUNG, Y.C. (1993). *Biomechanics: Mechanical Properties of Living Tissues*, second ed. (Springer-Verlag, New York).

FUNG, Y.C. (1994). *A First Course in Continuum Mechanics* (Prentice Hall, Englewood Cliffs, NJ).

GILLES, R., GUINEBRETIERE, J.M., et al. (1994). Non-palpable breast tumors: Diagnosis with contrast-enhanced subtraction dynamic MR imaging. *Radiology* **191**, 625–631.

GOLDSTEIN, D.C., KUNDEL, H.L., DAUBE-WHITERSPOON, M.E., THIBAULT, L.E., GOLDSTEIN, E.J. (1987). A silicone gel phantom suitable for multimodality imaging. *Invest. Radiol.* **22**, 153–157.

GREEN, A.E., ZERNA, W. (1968). *Theoretical Elasticity* (Oxford Univ. Press, London), p. 99.

HABER, I., METAXAS, D., AXEL, L. (2000). Three-dimensional motion reconstruction and analysis of the right ventricle using tagged MRI. *Medical Image Analysis* **4** (4), 335–355.

HARMS, S.E., FLAMIG, D.P. (1994). Staging of breast cancer with MR imaging. *Magn. Reson. Imaging Clin. N. Am.* **2**, 573–584.

HARMS, S.E., FLAMIG, D.P., et al. (1993). MR imaging of the breast with rotating delivery of excitation off resonance: Clinical experience with pathologic correlation. *Radiology* **187**, 493–501.

HARRIS, J.R., LIPPMAN, M.E., MORROW, M., HELLMAN, S. (1996). *Diseases of the Breast* (Lippincott–Raven).

HAYES, W.C., KEER, L.M., HERMANN, G., MOCKROS, L.F. (1972). A mathematical analysis for indentation tests of articular cartilage. *J. Biomechanics* **5**, 541–551.

HOU, F.J., LANG, S.M., et al. (1998). Human vertebral body apparent and hard tissue stiffness. *J. Biomechanics* **31**, 1009–1015.

JOUKHADAR, A. (1995). Energy based adaptive time step and inertia-matrix based adaptive discretization for fast converging dynamic simulation. In: *Proc. Intl. Workshop on Visualisation and Mathematics* (Springer-Verlag, Heidelberg).

KABEL, J., VAN RIETBERGEN, B., et al. (1999). The role of an effective isotropic tissue modulus in the elastic properties of cancellous bone. *J. Biomechanics* **32**, 673–680.

KOJIC, M., BATHE, K.J. (1987). Studies of finite element procedures – stress solution of a closed elastic strain path with stretching and shearing using the updated Lagrangian Jaumann formulation. *Computers & Structures* **26** (1/2), 175–179.

KROUSKOP, T.A., WHEELER, T.M., KALLEL, F., GARRA, B.S., HALL, T. (1998). The elastic moduli of breast and prostate tissues under compression. *Ultrasonic Imaging* **20**, 151–159.

KUEHNAPFEL, U.G., NEISIUS, B. (1993). CAD-based graphical computer simulation in endoscopic surgery. *End. Surg.* **1**, 181–184.

LAWRENCE, A.J., ROSSMAN, P.J., MAHOWALD, J.L., MANDUCA, A., HARTMANN, L.C., EHMAN, R.L. (1999). Assessment of breast cancer by magnetic resonance elastography. In: *Proc. 7th Intl. Soc. Magn. Reson. Med.*, p. 525.

LEE, G.C., TSENG, N.T. (1982). Finite element analysis in soft tissue mechanics. In: Gallagher, R.H., Simon, B.R., Johnson, P.C., Gross, J.F. (eds.), *Finite Elements in Biomechanics* (Wiley, Chichester, UK).

LUCIANI, A., JIMENEZ, S., FLORENS, J.L., CADOZ, C., RAOULT, O. (1991). Computational physics: a modeler simulator for animated physical objects. In: *Eurographics Workshop on Animation and Simulation*, pp. 425–437.

MAUREL, W., WU, Y., MAGNENAT THALMANN, N., THALMANN, D. (1998). Biomechanical Models for Soft Tissue Simulation. In: *Basis Research Series, Esprit* (Springer-Verlag, Berlin).

MESEURE, P., CHAILLOU, C. (1997). Deformable body simulation with adaptive subdivision and cuttings. In: *Proc. Fifth Int. Conf. in Central Europe on Computer Graphics and Visualization* (Pergamon Press, Oxford), pp. 361–370.

METAXAS, D. (1992). Physics-based modeling of nonrigid objects for vision and graphics. PhD thesis, Department of Computer Science, University of Toronto.

METAXAS, D., TERZOPOULOS, D. (1993). Shape and nonrigid motion estimation through physics-based synthesis. *IEEE Trans. Pattern Analysis and Machine Intelligence* **15** (6), 569–579.

MILLER, G. (1988). The motion dynamics of snake and worms. *Computer Graphics (SIGGRAPH'88)* **23**, 169–173.

NCI (1998). *Understanding Breast Cancer Treatment* (National Cancer Institute), pp. 6–7, NIH 98-4251.

NORTON, A., TURK, G., BACON, B., GERTH, J., SWEENEY, P. (1991). Animation of fracture by physical modeling. *The Visual Computer* **7**, 210–219.

OREL, S.G., SCHNALL, M.D., NEWMAN, R.W., POWELL, C.M., TOROSIAN, M.H., ROSATO, E.F. (1994). MR imaging-guided localization and biopsy of breast lesions: Initial experience. *Radiology* **193**, 97–102.

PARK, J., METAXAS, D., YOUNG, A.A., AXEL, L. (1996a). Analysis of left ventricular wall motion based on volumetric deformable models and MRI-SPAMM. *Medical Image Analysis* **1** (1), 53–71.

PARK, J., METAXAS, D., YOUNG, A.A., AXEL, L. (1996b). Deformable models with parameter functions for cardiac motion analysis from tagged MRI data. *IEEE Trans. Medical Image Processing*.

PARKER, S.H., LOVIN, J.D., JOBE, W.E., et al. (1990). Stereotactic breast biopsy with a biopsy gun. *Radiology* **176**, 741–747.

PICINBONO, G., DELINGUETTE, H., AYACHE, N. (2001). Non-linear and anisotropic elastic soft tissue models for medical simulation. In: *IEEE Int. Conf. on Robotics and Automation*, ICRA2001, Seoul, Korea.

PRESS, W.H., TEUKOLSKY, S.A., VETTERLING, W.T., FLANNERY, B.P. (1992). *Numerical Recipes in C: The Art of Scientific Computing* (Cambridge Univ. Press, Cambridge), pp. 707–725.

REDDY, N.P., SONG, G.J. (1995). Tissue cutting in virtual environments. In: *Medicine Meets Virtual Reality IV* (IOS Press), pp. 359–364.

SAHA, P.K., UDUPA, J.K., ODHNER, D. (2000). Scale-based fuzzy connected image segmentation: theory algorithms, and validation. *Computer Vision and Image Understanding* **77**, 145–174.

SARVAZYAN, A.P., SKOVORODA, A.R., EMELIANOV, S.Y., FOWLKES, J.B., PIPI, J.G., ADLER, R.S., BUXTON, R.B., CARSON, P.L. (1995). *Biophysical Bases of Elasticity Imaging* (Plenum Press, New York).

SCHNEIDER, D.C., DAVIDSON, T.M., NAHUM, A.M. (1984). In vitro biaxial stress–strain response of human skin. *Arch. Otolaryngol.* **110**, 329–333.

SCIARETTA, J., BISHOP, J., SAMANI, A., PLEWES, D.B. (1999). MR validation of soft tissue deformation as modeled by non-linear finite element analysis. In: *Proc. 7th Intl. Magn. Reson. Med.*, p. 246.

SKOVORODA, A.R., KLISHKO, A.N., GUSAKYAN, D.A., MAYEVSKII, Y.I., YERMILOVA, V.D., ORANSKAYA, G.A., SARVAZYAN, A.P. (1995). Quantitative analysis of the mechanical characteristics of pathologically changed soft biological tissues. *Biophysics* **40** (6), 1359–1364.

SPEETER, T.H. (1992). Three-dimensional finite element analysis of elastic continua for tactile sensing. *Intl. J. Robotics Res.* **11** (1), 1–19.

SPENCER, A.J.M. (1980). *Continuum Mechanics* (Longman, London), pp. 153–163.

STAVROS, A.T., THICKMAN, D., RAPP, C.L., DENNIS, M.A., PARKER, S.H., SISNEY, G.A. (1995). Solid breast modules: use of sonography to distinguish between benign and malignant lesions. *Radiology* **196**, 123–134.

SZEKELY, G., BRECHBUHLER, CH., HUTTER, R., RHOMBERG, A., SCHMID, P. (1998). Modelling of soft tissue deformation for laparoscopic surgery simulation. In: *Medical Image Computing and Computer-Assisted Intervention*, MICCAI, pp. 550–561.

TERZOPOULOS, D., PLATT, J., et al. (1987). Elastically deformable models. *Computer Graphics (SIGGRAPH'87)* **21**, 205–214.

TOFTS, P.S., BRIX, G., et al. (1999). Estimating kinetic parameters from dynamic contrast-enhanced T(1)-weighted MRI of a diffusible tracer: standardized quantities and symbols. *J. Magn. Reson. Imaging* **10**, 223–232.

VERONDA, D.R., WESTMANN, R.A. (1970). Mechanical characterization of skin-finite deformations. *J. Biomech.* **3**, 111–124.

WELLMAN, P.S. (1999). Tactile imaging. Thesis. Harvard University, Cambridge, MA.

WELLMAN, P.S., HOWE, R.D. (1998). Harvard Bio-Robotics Lab. Tech. Report. #98-121.

WILLIAMS, C., CLYMER, B., SCHMALBROCK, P. (1999). Biomechanics of breast tissue: preliminary study of force-deformation relationship. In: *Proc. 7th Intl. Soc. Magn. Reson. Med.*, p. 524.

YAMADA, H. (1970). *Strength of Biological Materials* (Williams & Wilkins, Baltimore, MD).

YOUNG, A.A., AXEL, L., DOUGHERTY, L., BOGEN, D.K., PARENTEAU, C.S. (1993). Validation of tagging with MR imaging to estimate material deformation. *Radiology* **188**, 101–108.

YUCESOY, C.A., KOOPMAN, B., et al. (2002). Three-dimensional finite element modeling of skeletal muscle using a two-domain approach: linked fiber-matrix mesh model. *J. Biomechanics* **35**, 1253–1262.

ZHANG, M., ZHENG, Y.P., MAK, A.F. (1997). Estimating the effective Young's modulus of soft tissues from indentation tests – nonlinear finite element analysis of effects of friction and large deformation. *Med. Eng. Phys.* **19** (6), 512–517.

ZHUANG, Y., CANNY, J. (1999). Real-time and physically realistic simulation of global deformation. In: *SIGGRAPH'99*.

ZIENKIEWICZ, O.C. (1977). *The Finite Element Method*, third ed. (McGraw–Hill, London).

ZIENKIEWICZ, O.C., TAYLOR, R.L. (1989). *The Finite Element Method*, fourth ed. (McGraw–Hill, New York).

Subject Index